Advances in Intelligent Systems and Computing

Volume 417

Series editor

Janusz Kacprzyk, Polish Academy of Sciences, Warsaw, Poland
e-mail: kacprzyk@ibspan.waw.pl

About this Series

The series "Advances in Intelligent Systems and Computing" contains publications on theory, applications, and design methods of Intelligent Systems and Intelligent Computing. Virtually all disciplines such as engineering, natural sciences, computer and information science, ICT, economics, business, e-commerce, environment, healthcare, life science are covered. The list of topics spans all the areas of modern intelligent systems and computing.

The publications within "Advances in Intelligent Systems and Computing" are primarily textbooks and proceedings of important conferences, symposia and congresses. They cover significant recent developments in the field, both of a foundational and applicable character. An important characteristic feature of the series is the short publication time and world-wide distribution. This permits a rapid and broad dissemination of research results.

More information about this series at http://www.springer.com/series/11156

Luís Paulo Reis · António Paulo Moreira
Pedro U. Lima · Luis Montano
Victor Muñoz-Martinez
Editors

Robot 2015: Second Iberian Robotics Conference

Advances in Robotics, Volume 1

 Springer

Editors
Luís Paulo Reis
University of Minho, School of Engineering
Information Systems Department
Guimarães
Portugal

António Paulo Moreira
University of Porto, Faculty of Engineering
INESC-TEC
Porto
Portugal

Pedro U. Lima
University of Lisbon, Instituto Superior
  Técnico
Institute for Systems and Robotics
Lisboa
Portugal

Luis Montano
University of Zaragoza, School of
  Engineering and Architecture
Computer and Systems Engineering
  Department
Zaragoza
Spain

Victor Muñoz-Martinez
University of Malaga, Superior Technical
  School of Industrial Engineers
Automatic Control and Systems Engineering
  Department
Málaga
Spain

ISSN 2194-5357   ISSN 2194-5365 (electronic)
Advances in Intelligent Systems and Computing
ISBN 978-3-319-27145-3   ISBN 978-3-319-27146-0 (eBook)
DOI 10.1007/978-3-319-27146-0

Library of Congress Control Number: 2015955886

Springer Cham Heidelberg New York Dordrecht London

Springer International Publishing AG Switzerland is part of Springer Science+Business Media
(www.springer.com)

Preface

This book contains a selection of papers accepted for presentation and discussion at ROBOT 2015: Second Iberian Robotics Conference, held in Lisbon, Portugal, November 19th–21st, 2015. ROBOT 2015 is part of a series of conferences that are a joint organization of SPR – "Sociedade Portuguesa de Robótica/Portuguese Society for Robotics", SEIDROB – Sociedad Española para la Investigación y Desarrollo de la Robótica/Spanish Society for Research and Development in Robotics and CEA-GTRob – Grupo Temático de Robótica/Robotics Thematic Group. The conference organization had also the collaboration of several universities and research institutes, including: University of Minho, University of Porto, University of Lisbon, Polytechnic Institute of Porto, University of Aveiro, University of Zaragoza, University of Malaga, LIACC, INESC-TEC and LARSyS.

Robot 2015 builds upon several successful events, including three biennal workshops (Zaragoza- 2007, Barcelona – 2009 and Sevilla – 2011) and the first Iberian Robotics Conference held in 2013 at Madrid. The conference is focussed on the Robotics scientific and technological activities in the Iberian Peninsula, although open to research and delegates from other countries.

Robot 2015 featured three plenary talks by:

- Manuela Veloso, Herbert A. Simon University Professor at Carnegie Mellon University, USA, on "Symbiotic Autonomous Mobile Service Robots";
- Bill Smart, director of the Personal Robotics Group at Oregon State University, USA on "How the Law Will Think About Robots (and Why You Should Care)"; and
- Jon Agirre Ibarbia, co-ordinator of R&D projects at TECNALIA Research & Innovation, Spain, on "Applications in Flexible Manufacturing with Humans and Robots".

Robot 2015 featured 19 special sessions, plus a main/general robotics track. The special sessions were about: Agricultural Robotics and Field Automation; Autonomous Driving and Driver Assistance Systems; Communication Aware Robotics; Environmental Robotics; Social Robotics: Intelligent and Adaptable AAL

Systems; Future Industrial Robotics Systems; Legged Locomotion Robots; Rehabilitation and Assistive Robotics; Robotic Applications in Art and Architecture; Surgical Robotics; Urban Robotics; Visual Perception for Autonomous Robots; Machine Learning in Robotics; Simulation and Competitions in Robotics; Educational Robotics; Visual Maps in Robotics; Control and Planning in Aerial Robotics, the XVI edition of the Workshop on Physical Agents and a Special Session on Technological Transfer and Innovation.

In total, after a careful review process with at least three independent reviews for each paper, but in some cases 4 or 5 reviews, a total of 118 high quality papers were selected for publication, with a total number of authors over 400, from 21 countries, including: Brazil, China, Costa Rica, Croatia, Czech Republic, Ecuador, France, Germany, Italy, India, Iran, The Netherlands, Poland, Portugal, Serbia, Singapore, Spain, Switzerland, United Kingdom, USA and Viet Nam.

ROBOT 2015 was co-located with the RoCKIn Competition 2015, which took place in the Parque das Nações, Lisboa, between 19 and 23 November, nearby the conference venue. RoCKIn is a Coordination Action funded by the European Commission FP7, and its main goal is to foster robotics research, education and dissemination through robot competitions. Thirteen teams from seven countries, including two teams from Mexico, were qualified and competed in RoCKIn@Home and RoCKIn@Work Challenges. Participants from both events had the opportunity to join in social events and to visit both venues, taking advantage of an extraordinary opportunity to follow presentations and actual robot systems showing recent results in this exciting field.

We would like to thank all Special Sessions' organizers for their hard work on promoting their special session, inviting the Program Committee, organizing the Special Session review process and helping to promote the ROBOT 2015 Conference. This acknowledgment goes especially to Vitor Santos, Angel Sappa, Miguel Oliveira, Danilo Tardioli, Alejandro Mosteo, Luis Riazuelo, João Valente, Antonio Barrientos, Luís Santos, Jorge Dias, Raul Morais Santos, Filipe Santos, Germano Veiga, José Lima, Guillermo Heredia, Anibal Ollero, Manuel Silva, Cristina Santos, Manuel Armada, Vicente Matellán, Miguel Ángel Cazorla, Rodrigo Ventura, Nicolas Garcia-Aracil, Alicia Casals, Elena García, José Pedro Sousa, Marta Malé-Alemany, Paulo Gonçalves, Jose Maria Sabater, Jorge Martins, Pedro Torres, Tamás Haidegger, Alberto Sanfeliu, Juan Andrade, João Sequeira, Anais Garrell, Andry Maykol Pinto, Aníbal Matos, Nuno Cruz, Brígida Mónica Faria, Luis Merino, Nuno Lau, Artur Pereira, Bernardo Cunha, Armando Sousa, Fernando Ribeiro, Eduardo Gallego and Oscar Reinoso Garcia.

We would also like to take this opportunity to thank the rest of the organization members (Carlos Cardeira, Brígida Mónica Faria, Manuel Fernando Silva, Daniel Castro Silva and Pedro Fonseca) for their hard and fine work on the local arrangements, publicity, publication and financial issues. We also express our gratitude to the members of all the Program Committees and additional reviewers, as they were crucial for ensuring the high scientific quality of the event and to all the authors and delegates whose research work and participation made this event a success. Last, but not the least, we acknowledge and thank our editor, Springer, that was in charge of these proceedings, and in particular to Dr. Thomas Ditzinger.

November 2015

Luís Paulo Reis
António Paulo Moreira
Pedro U. Lima
Luis Montano
Victor Muñoz-Martinez

Organization

General/Program Chairs

Luís Paulo Reis   University of Minho, Portugal
António Paulo Moreira   University of Porto - FEUP/INESCTEC, Portugal
Pedro U. Lima   University of Lisbon – IST, Portugal
Luis Montano   University of Zaragoza, Spain
Victor F. Muñoz   University of Malaga, Spain

Organizing Committee

Luís Paulo Reis   University of Minho, Portugal (General/Program Chair)
António Paulo Moreira   University of Porto - FEUP/INESCTEC, Portugal
                         (General/Program Chair)
Pedro U. Lima   University of Lisbon – IST, Portugal (General/Program
                         Chair)
Luis Montano   University of Zaragoza, Spain (General/Program Chair)
Victor F. Muñoz   University of Malaga, Spain (General/Program Chair)
Carlos Cardeira   University of Lisbon - IST, Portugal
                         (Local Arrangements Chair)
Brígida Mónica Faria   Polytechnic Institute of Porto, Portugal (Publicity Chair)
Manuel Fernando Silva   Polytechnic Institute of Porto, Portugal (Publications
                         Chair)
Daniel Castro Silva   University of Porto - FEUP, Portugal (Publications
                         Chair)
Pedro Fonseca   University of Aveiro, Portugal (Financial/Registration
                         Chair)

General Robotics Session Organizing Committee

Luís Paulo Reis	University of Minho, Portugal
António Paulo Moreira	University of Porto - FEUP/INESCTEC, Portugal
Pedro U. Lima	University of Lisbon – IST, Portugal
Luis Montano	University of Zaragoza, Spain
Victor F. Muñoz	University of Malaga, Spain

General Robotics Session Program Committee

Alberto Sanfeliu	University Politécnica de Cataluña, Spain
Alexandre Bernardino	Universiy of Lisbon, Portugal
Alfonso García-Cerezo	University Málaga, Spain
Alícia Casals	University Politécnica de Cataluña, Spain
Americo Azevedo	FEUP/INESCTEC, Portugal
Aníbal Matos	University of Porto, Portugal
Anibal Ollero	CATEC-Universidad Sevilla, Spain
Antonio Barrientos	CAR CSIC-UPM, Spain
Antonio Fernando Ribeiro	University of Minho, Portugal
António José Neves	University of Aveiro, Portugal
Antonio R Jiménez	CAR-CSIC, Spain
António Pedro Aguiar	University of Porto – FEUP, Portugal
António Valente	University of Tras dos Montes e Alto Douro, Portugal
Armando Jorge Sousa	University of Porto, Portugal
Artur Pereira	University of Aveiro, Portugal
Brígida Mónica Faria	I.P. Porto, Portugal
Bruno Guerreiro	Universiy of Lisbon, Portugal
Carlos Cardeira	Universiy of Lisbon, Portugal
Carlos Cerrada	UNED, Spain
Carlos Rizzo	University of Zaragoza, Spain
Carlos Sagüés	University Zaragoza, Spain
Cristina Santos	University of Minho, Portugal
Eduardo Zalama	University of Valladolid, Spain
Estela Bicho	University of Minho, Portugal
Eugenio Aguirre	University Granada, Spain
Fernando Caballero	University Sevilla, Spain
Fernando Torres	University Alicante, Spain
Filipe Santos	FEUP/INESCTEC, Portugal
Filomena Soares	University of Minho, Portugal
Francisco Melo	University of Lisbon, Portugal
Hugo Costelha	I.P. Leiria, Portugal

Javier Pérez Turiel	CARTIF, Valladolid, Spain
João Calado	I.P. Lisboa, Portugal
Jon Aguirre	Tecnalia, Spain
Jorge Dias	University of Coimbra, Portugal
Jorge Lobo	University of Coimbra, Portugal
José A. Castellanos	University Zaragoza, Spain
José Luís Azevedo	University of Aveiro, Portugal
José Luis Magalhães Lima	I.P. Bragança, Portugal
José L. Villarroel	University of Zaragoza, Spain
José M. Cañas	University Rey Juan Carlos, Spain
José Nuno Pereira	University of Lisbon, Portugal
José Santos Victor	University of Lisbon, Portugal
Josep Amat	University of Politécnica de Cataluña, Spain
Lino Marques	University of Coimbra, Portugal
Luis Almeida	University of Porto, FEUP, Portugal
Luis Basañez	University Politécnica de Cataluña, Spain
Luis Merino	University Pablo Olavide, Sevilla, Spain
Luis Moreno	University Carlos III de Madrid, Spain
Luis Seabra Lopes	University of Aveiro, Portugal
Manuel Armada	CAR CSIC-UPM, Spain
Manuel Bernardo Cunha	University of Aveiro, Portugal
Manuel Fernando Silva	I.P. Porto, Portugal
Manuel Ferre	CAR CSIC-UPM, Spain
Marcelo Petry	Univ. Federal Santa Catarina, Brazil
Maria Isabel Ribeiro	Universiy of Lisbon, Portugal
Miguel A. Cazorla	University de Alicante, Spain
Nicolás García-Aracil	University Miguel Hernández, Spain
Nuno Lau	University of Aveiro, Portugal
Oscar Reinoso	University Miguel Hernández, Spain
Pascual Campoy	CAR CSIC-UPM, Spain
Paulo Costa	University of Porto, Portugal
Paulo Gonçalves	I.P. Castelo Branco, Portugal
Paulo Jorge Oliveira	University of Lisbon, Portugal
Pedro Costa	University of Porto, Portugal
Pedro Fonseca	University of Aveiro, Portugal
Pedro J. Sanz	UJI, Castellón, Spain
Pere Ridao	University of Girona, Spain
Raul Morais	University of Trás dos Montes e Alto Douro, Portugal
Rafael Sanz	University of Vigo, Spain
Rodrigo Ventura	Universiy of Lisbon, Portugal
Rui Rocha	University of Coimbra, Portugal
Urbano Nunes	University of Coimbra, Portugal
Vicente Feliú	University Castilla la Mancha, Spain
Vicente Matellán	University León, Spain
Vitor Santos	University of Aveiro, Portugal

Agricultural Robotics and Field Automation Session Organizing Committee

Raul Morais Santos   UTAD University, Portugal
Filipe Santos   INESC-TEC, Portugal

Agricultural Robotics and Field Automation Session Program Committee

Angela Ribeiro   CAR-CSIC, Spain
Antonio Valente   UTAD University, Portugal
Armando Sousa   University of Porto - FEUP, Portugal
Carrick Detweiler   University of Nebraska, United States
Dimitrios S. Paraforos   University of Hohenheim, Germany
Eduardo Solteiro Pires   UTAD University, Portugal
Filipe Neves dos Santos   INESC TEC, Portugal
Joris Ijsselmuiden   Wageningen UR, Netherlands
Joaquín Ferruz-Melero   University of Seville, Spain
Raul Morais Dos Santos   INESC TEC/CROB/UTAD, Portugal
Manuel Silva   Inst. Superior de Engenharia do Porto, Portugal
Nieves Pavon   University of Huelva, Spain
Paulo Costa   U. of Porto, FEUP, Portugal
Paulo Moura Oliveira   UTAD University, Portugal
Robert Fitch   Australian Centre for Field Robotics,
                                The University of Sydney, Australia
Tiago Nascimento   Federal University of Paraíba - UFPB, Brazil
Timo Oksanen   Aalto University,Finland

Autonomous Driving and Driver Assistance Systems Session Organizing Committee

Vitor Santos   Universidade de Aveiro, Portugal
Angel Sappa   CVC, Barcelona, Spain
Miguel Oliveira   INESC-TEC, Porto, Portugal

Autonomous Driving and Driver Assistance Systems Session Program Committee

Angelos Amanatiadis	Democritus University of Thrace, Greece
Antonio Valente	UTAD, Portugal
Antonio M. López	CVC and UAB, Barcelona, Spain
Arturo De La Escalera	Universidad Carlos III de Madrid, Spain
Bernardo Cunha	Universidade de Aveiro, Portugal
Carlos Cardeira	IST, Lisboa, Portugal
Cristina Peixoto Santos	Universidade do Minho, Portugal
David Vázquez Bermúdez	CVC, Barcelona, Spain
Fadi Dornaika	University of the Basque Country UPV/EHU & IKERBASQUE, Spain
Frederic Lerasle	LAAS-CNRS, France
Jorge Almeida	DEM, Universidade de Aveiro, Portugal
José Azevedo	Universidade de Aveiro, Portugal
José Álvarez	RSCS, ANU College, Australia
Jose A. Castellanos	University Zaragoza, Spain
Luis Almeida	Universidade do Porto, Portugal
Miguel Angel Sotelo	University of Alcala, Spain
Paulo Dias	IEETA - Universidade de Aveiro, Portugal
Procópio Stein	INRIA, France
Rafael Sanz	Universidad de Vigo, Spain
Ricardo Pascoal	Universidade de Aveiro, Portugal
Ricardo Toledo	CVC and UAB, Barcelona, Spain
Susana Sargento	IT, University of Aveiro, Portugal
Urbano Nunes	University of Coimbra, Portugal

Control and Planning in Aerial Robotics Organizing Committee

Guillermo Heredia	Universidad de Sevilla, Spain
Anibal Ollero	Universidad de Sevilla, Spain

Control and Planning in Aerial Robotics Program Committee

Abdelkrim Nemra	Ecole Militaire Polytechnique, Algiers
Alessandro Rucco	University of Porto, Portugal
Begoña Arrue	Universidad de Sevilla, Spain
Bruno, Guerreiro	Instituto Superior Técnico, Univ. Lisboa, Portugal
Elena Lopez Guillen	Universidad de Alcala, Spain

Eugenio Aguirre   University of Granada, Spain
Fernando Caballero   University of Seville, Spain
Luis Merino   Pablo de Olavide University, Spain
Mario Garzon   Universidad Politécnica de Madrid, Spain
Rita Cunha   LARSyS, Instituto Superior Técnico, Univ. Lisboa,
                              Portugal

Communication Aware Robotics Organizing Committee

Danilo Tardioli   Centro Universitario de la Defensa de Zaragoza,
                              Spain
Alejandro Mosteo   Centro Universitario de la Defensa de Zaragoza,
                              Spain
Luis Riazuelo   University of Zaragoza, Spain

Communication Aware Robotics Program Committee

Carlos Rizzo   University of Zaragoza, Spain
Domenico Sicignano   University of Zaragoza, Spain
Eduardo Montijano   Centro Universitario de la Defensa de Zaragoza,
                              Spain
Enrico Natalizio   Universitè de Technologie de Compiègne, France
Jesus Aisa   University of Zaragoza, Spain
Jorge Ortin García   Centro Universitario de la Defensa de Zaragoza,
                              Spain
Luis Merino   Pablo de Olavide University, Spain
Lujia Wang   The Chinese University of Hong Kong, Hong Kong,
                              China
María T. Lázaro   Sapienza University of Rome, Italy
María-Teresa Lorente   University of Zaragoza, Spain
Pablo Urcola   University of Zaragoza, Spain

Educational Robotics Session Organizing Committee

A. Fernando Ribeiro   University of Minho, Portugal
Armando Sousa   University of Porto - FEUP, Portugal
Eduardo Gallego   Complubot, Spain

Educational Robotics Session Program Committee

A. Fernando Ribeiro   University of Minho, Portugal
Armando Sousa   University of Porto - FEUP, Portugal
Eduardo Gallego   Complubot, Spain
Gil Lopes   University of Minho, Portugal
José Goncalves   ESTiG – I.P. Bragança, Portugal
Paulo Costa   University of Porto - FEUP, Portugal
Paulo Trigueiros   I.P. Porto, Portugal

Environmental Robotics Special Session Organizing Committee

João Valente   Universidad Carlos III de Madrid, Spain
Antonio Barrientos   Universidad Politécnica de Madrid, Spain

Environmental Robotics Special Session Program Committee

Achim J. Lilienthal   Örebro University, Sweden
Angela Ribeiro   Centre for Automation and Robotics
                             (CAR) UPM-CSIC, Spain
Carol Martinez   Pontifical Xavierian University, Colombia
David Gomez   Nat. Res. Inst. Science and Technology for
                             Environment and Agriculture, France
Gonzalo Pajares   Complutense University of Madrid, Spain
Jaime Del Cerro   Polytechnic University of Madrid, Spain
Marc Carreras   University of Girona, Spain
Mario Andrei Garzón   Polytechnic University of Madrid, Spain
Mohamed Abderrahim   Carlos III University of Madrid, Spain
Pablo Gonzalez de Santos   Spanish National Research Council (CSIC), Spain
Paloma de la Puente   Vienna University of Technology, Austria
William Coral   Polytechnic University of Madrid, Spain

Future Industrial Robotics Systems Organizing Committee

Germano Veiga   INESC TEC - Robotics and Intelligent Systems,
                             Portugal
José Lima   INESC TEC - Robotics and Intelligent Systems,
                             Portugal

Future Industrial Robotics Systems Program Committee

Andry Pinto	Universidade do Porto, INESC-TEC, Portugal
António Paulo Moreira	Universidade do Porto, INESC-TEC, Portugal
Fabrizio Caccavale	UNIBAS, Italy
Joerg Roewekaemper	University of Freiburg, Germany
José Barbosa	I.P. Bragança, Portugal
José Miguel Almeida	I.P. Porto, Portugal
Klas, Nilsson	Lund University, Sweden
Luis Rocha	Universidade do Porto, Portugal
Manuel Fernando Silva	I.P. Porto, Portugal
Nuno Mendes	Universidade de Coimbra, Portugal
Pedro Neto	Universidade de Coimbra, Portugal
Simon Bogh	AalBorg University, Denmark
Ulrike Thomas	TU Chemnitz, Germany

Legged Locomotion Robots Session Organizing Committee

Manuel Silva	ISEP/IPP - School of Engineering, Polytechnic Institute of Porto and INESC TEC, Portugal
Cristina Santos	Industrial Electronics Department and ALGORITMI Center, University of Minho, Portugal
Manuel Armada	Centre for Automation and Robotics - CAR (CSIC-UPM), Spain

Legged Locomotion Robots Session Program Committee

Carla M.A. Pinto	Instituto Superior de Engenharia do Porto, Portugal
Filipe Silva	University of Aveiro, Portugal
Filomena Soares	University of Minho, Portugal
Gurvinder S. Virk	University of Gävle, Sweden
José Machado	University of Minho, Portugal
Lino Costa	University of Minho, Portugal
Lino Marques	University of Coimbra, Portugal
Paulo Menezes	University of Coimbra, Portugal
Pedro Figueiredo Santana	Escola de Tecnologias e Arquitetura, Portugal
Rui P. Rocha	University of Coimbra, Portugal
Yiannis Gatsoulis	University of Leeds, United Kingdom

Machine Learning in Robotics Session Organizing Committee

Brígida Mónica Faria	Polytechnic Institute of Porto (ESTSP-IPP), Portugal
Luis Merino	Pablo de Olavide University (UPO), Spain

Machine Learning in Robotics Session Program Committee

Ana Lopes	Institute for Systems and Robotics, Portugal
Armando Sousa	University of Porto, Portugal
Daniel Castro Silva	Univesity of Porto - FEUP, Portugal
Fernando Caballero Benítez	University of Seville, Spain
João Fabro	UTFPR - Federal University of Technology Parana, Brazil
João Messias	Institute for Systems and Robotics, Portugal
Marcelo Petry	Federal University of Santa Catarina, Brazil
Noé Pérez-Higueras	Pablo de Olavide University (UPO), Spain
Nuno Lau	University of Aveiro, Portugal
Pedro Henriques Abreu	University of Coimbra, Portugal
Rafael Ramón-Vigo	Pablo de Olavide University (UPO), Spain

Rehabilitation and Assistive Robotics Organizing Committee

Alicia Casals	Institute for Bioengineering of Catalonia, Spain
Nicolás García-Arazil	Universidad Miguel Hernandez, Spain
Elena García	Centre for Automation and Robotics (CSIC-UPM), Spain

Rehabilitation and Assistive Robotics Program Committee

Aikaterini D. Koutsou	CSIC, Spain
Alicia Casals	Institute for Bioengineering of Catalonia, Spain
Arturo Bertomeu-Motos	UMH, Spain
Diana Ruiz Bueno	University of Saragoza, Spain
Elena García	Centre for Automation and Robotics (CSIC-UPM), Spain
Eloy Urendes Jimenez	CSIC, Spain
Iñaki Diaz	CEIC, Spain
Javier P. Turiel	University of Valladolid, Spain

Jose Maria Sabater-Navarro	UMH, Spain
Luis Daniel Lledó Pérez	UMH, Spain
Nicolás García-Arazil	Universidad Miguel Hernandez, Spain

Robotic Applications in Art and Architecture Session Organizing Committee

Manuel Silva	ISEP/IPP - School of Engineering, Polytechnic Institute of Porto and INESC TEC, Portugal
José Pedro Sousa	Digital Fabrication Lab (DFL/CEAU), Fac. Arquitetura, Universidade do Porto, Portugal
Marta Malé-Alemany	Architect, Researcher and Curator + Polytechnic University of Catalonia, Spain

Robotic Applications in Art and Architecture Session Program Committee

Alexandra Paio	ISCTE - Instituto Universitário de Lisboa, Portugal
André Dias	Polytechnic Institute of Porto and INESC TEC, Portugal
Andrew Wit	Ball State University, USA
António Mendes Lopes	Faculty of Engineering, University of Porto, Portugal
Filipe Coutinho Quaresma	ECATI - Universidade Lusófona, Portugal
Germano Veiga	INESC TEC, Portugal
Gonçalo Castro Henriques	Universidade Federal do Rio de Janeiro, Brazil
Leonel Moura	Universidade de Lisboa, Portugal
Mauro Costa	Universidade de Coimbra, Portugal
Paulo Fonseca de Campos	Faculdade de Arquitetura e Urbanismo, Universidade de São Paulo, Brazil
Sancho Oliveira	ISCTE - Instituto Universitário de Lisboa, Portugal
Wassim Jabi	Welsh School of Architecture, Cardiff University, United Kingdom

Simulation and Competitions Session Organizing Committee

Artur Pereira	Universidade de Aveiro/IEETA, Portugal
Nuno Lau	Universidade de Aveiro/IEETA, Portugal
Bernardo Cunha	Universidade de Aveiro/IEETA, Portugal

Simulation and Competitions Session Program Committee

Antonio Morales	Universitat Jaume I, Spain
Armando Sousa	Universidade do Porto, Portugal
Eurico Pedrosa	Universidade de Aveiro, Portugal
Jorge Ferreira	University of Aveiro, Portugal
José Luís Azevedo	Universidade de Aveiro, Portugal
Luis Moreno	Universidad Carlos III de Madrid, Spain
Nicolas Jouandeau	University Paris8, France
Paulo Goncalves	Polytechnic Institute of Castelo Branco, Portugal
Paulo Trigueiros	Instituto Politécnico do Porto, Portugal
Pedro Fonseca	Universidade de Aveiro, Portugal
Rosaldo Rossetti	Universidade do Porto, Portugal

Social Robotics: Intelligent and Adaptable AAL Systems Organizing Committee

Luís Santos	Institute of Systems and Robotics – University of Coimbra, Portugal
Jorge Dias	Khalifa University Robotics Institute and Institute of Systems and Institute of Systems and Robotics, University of Coimbra, Portugal

Social Robotics: Intelligent and Adaptable AAL Systems Program Committee

Filippo Cavallo	The BioRobotics Institute, Italy
Friederike Eyssel	Bielefeld University, Germany
João Sequeira	Instituto Superior Técnico de Lisboa, Portugal
Jorge Lobo	ISR - University of Coimbra, Portugal

Surgical Robotics Organizing Committee

Alicia Casals	Universidad Politecnica de Catalunha, Spain
Paulo Gonçalves	Instituto Politécnico de Castelo Branco, Portugal
Nicolas Garcia	Universidad Miguel Hernandez de Elche, Spain
Jose Maria Sabater	Universidad Miguel Hernandez de Elche, Spain
Jorge Martins	Instituto Superior Técnico, Univ. Lisboa, Portugal

Surgical Robotics Program Committee

Alicia Casals   Universidad Politecnica de Catalunha, Spain
Jorge Martins   Instituto Superior Técnico, Univ. Lisboa, Portugal
Jose Maria Sabater   Universidad Miguel Hernandez de Elche, Spain
Nicolas Garcia   Universidad Miguel Hernandez de Elche, Spain
Paulo Gonçalves   Instituto Politécnico de Castelo Branco, Portugal
Pedro Torres   Instituto Politécnico de Castelo Branco, Portugal
Tamás Haidegger   ABC Center for Intelligent Robotics, Óbuda
                            University

Urban Robotics Organizing Committee

Alberto Sanfeliu   IRI (CSIC-UPC), Spain
Juan Andrade   IRI (CSIC-UPC), Spain
Joao Sequeira   ISR - Univ. Lisboa, Portugal
Anais Garrell   IRI (CSIC-UPC), Spain

Urban Robotics Program Committee

Alberto Sanfeliu   IRI (CSIC-UPC), Spain
Anais Garrell   IRI (CSIC-UPC), Spain
Joao Sequeira   ISR - Univ. Lisboa, Portugal
Juan Andrade   IRI (CSIC-UPC), Spain

Visual Maps in Robotics Session Organizing Committee

Oscar Reinoso Garcia   Miguel Hernandez University, Spain

Visual Maps in Robotics Session Program Committee

Arturo Gil Aparicio   UMH, Spain
Carlos Sagües   University of Zaragoza, Spain
Fernando Torres Medina   University of Alicante, Spain
Javier González Jiménez   University of Malaga, Spain
Jose Mª Sebastian Zuñiga UPM, Spain

Jose María Martínez Montiel	University of Zaragoza, Spain
Luis Payá Castelló	UMH, Spain
Pablo Gil	University of Alicante, Spain

Visual Perception for Autonomous Robots Session Organizing Committee

Andry Maykol Pinto	INESC-TEC and Faculty of Engineering of the University of Porto, Portugal
Pedro Neto	University of Coimbra, Portugal
Luis Rocha	INESC-TEC, Portugal

Visual Perception for Autonomous Robots Session Program Committee

Aníbal Matos	INESC-TEC and Faculty of Engineering of the University of Porto, Portugal
António Neves	University of Aveiro, Portugal
Armando Pinho	University of Aveiro, Portugal
Bernado Cunha	University of Aveiro, Portugal
Brígida Mónica Faria	Polytechnic Institute of Porto – ESTSP/IPP, Portugal
Hélder Oliveira	INESC-TEC and Faculty of Engineering of the University of Porto, Portugal
Marcelo Petry	Federal University of Santa Catarina and INESC P&D Brasil, Brasil
Nuno Cruz	INESC-TEC and Faculty of Engineering of the University of Porto, Portugal

WAF – XVI Workshop on Physical Agents Organizing Committee

Vicente Matellán	Universidad de León, Spain
Miguel Ángel Cazorla	Universidad de Alicante, Spain
Rodrigo Ventura	Instituto Superior Técnico, Universidade de Lisboa, Portugal

WAF – XVI Workshop on Physical Agents Program Committee

Domenec Puig	Universitat Rovira i Virgili, Spain
Eugenio Aguirre	University of Granada, Spain
Francisco Javier Rodríguez Lera	Universidad de León, Spain
Ismael García-Varea	Univ. de Castilla-La Mancha, Spain
J. Francisco Blanes Noguera	Universidad Politécnica de Valencia, Spain
Joaquin Lopez	University of Vigo, Spain
Jose Manuel Lopez	Basque Country University, Spain
José María Armingol	Universidad Carlos III de Madrid, Spain
Josemaria Cañas Plaza	Universidad Rey Juan Carlos, Spain
Lluís Ribas-Xirgo	Universitat Autònoma de Barcelona, Spain
Pablo Bustos	Universidad de Extremadura, Spain
Rafael Muñoz-Salinas	University of Cordoba, Spain
Roberto Iglesias Rodriguez	University of Santiago de Compostela, Spain

Additional Reviewers

Abdelkrim Nemra
Albert Palomer Vila
Alberto Vale
Ali Marjovi
Alireza Asvadi
Ammar Assoum
Ana Maria Maqueda
André Farinha
André Mateus
Arpad Takacs
Begoña Arrúe
Carlos Martinho
Clauirton Siebra
Cristiano Premebida
Daniel Silvestre
David Fornas
Diana Beltran
Diego Faria
Dolores Blanco
Eduard Bergés
Enric Cervera
Fernando Martín
Friederike Eyssel
Giovanni Saponaro
Guillem Alenya

Guillem Vallicrosa
Iñaki Maurtua
Isabel García-Morales
Iván Villaverde
Jan Veneman
João Quintas
Johannes Kropf
José Antonio Cobano
José Barbosa
José María Martínez Montiel
Juan Carlos García Sánchez
Lorenzo Jamone
Manel Frigola
Miguel Garcia-Silvente
Narcís Palomeras
Pablo Lanillos
Pedro Casau
Pedro Lourenço
Raul Marin
Santiago Garrido
Sedat Dogru
Sten Hanke
Vijaykumar Rajasekaran
Xavier Giralt

Contents

Part I
General Robotics

Lidar-Based Relative Position Estimation and Tracking for Multi-robot Systems

**Alicja Wąsik, Rodrigo Ventura, José N. Pereira,
Pedro U. Lima and Alcherio Martinoli**

Abstract Relative positioning systems play a vital role in current multi-robot systems. We present a self-contained detection and tracking approach, where a robot estimates a distance (range) and an angle (bearing) to another robot using measurements extracted from the raw data provided by two laser range finders. We propose a method based on the detection of circular features with least-squares fitting and filtering out outliers using a map-based selection. We improve the estimate of the relative robot position and reduce its uncertainty by feeding measurements into a Kalman filter, resulting in an accurate tracking system. We evaluate the performance of the algorithm in a realistic indoor environment to demonstrate its robustness and reliability.

Keywords Laser relative positioning system · Estimation · Tracking

1 Introduction

Several applications of cooperative distributed robotic systems (e.g., formation control, coverage) require that each team member is aware not only of its absolute location in a global world frame, but also of the relative locations of its teammates. When wireless communications are reliable each robot can simply regularly share its most recent location estimate with its teammates. Unfortunately, wireless communications are often unreliable, and the amount of communicated data grows quadratically with

A. Wąsik(✉) · J.N. Pereira · A. Martinoli
Distributed Intelligent Systems and Algorithms Laboratory,
École Polytechnique Fédérale de Lausanne, Lausanne, Switzerland
e-mail: {alicja.wasik,jose.pereira,alcherio.martinoli}@epfl.ch

A. Wąsik · R. Ventura · P.U. Lima
Institute for Systems and Robotics, Instituto Superior Técnico, Universidade de Lisboa,
Lisbon, Portugal
e-mail: {rodrigo.ventura,pal}@isr.tecnico.ulisboa.pt

© Springer International Publishing Switzerland 2016
L.P. Reis et al. (eds.), *Robot 2015: Second Iberian Robotics Conference*,
Advances in Intelligent Systems and Computing 417,
DOI: 10.1007/978-3-319-27146-0_1

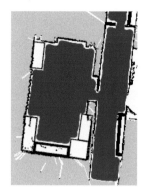

Fig. 1 Left: The Mbot robots. Right: Map of the environment with area allowed for navigation marked in red.

the number of robots in the team. Thus, a solution not based on communications but relying uniquely on data locally acquired by each robot using its own sensors is desirable.

Position estimation and tracking of the robots recently began to attract much attention in the robotics community. Teixid et al. estimated position of circles from a laser scan, stating that attaching a circular marker to a mobile platform facilitates its detection [1]. Yet, their sensor was static and external to the robot, and the experiments limited to a case when the robot has been moved manually, not reflecting its real, dynamic motion. Within the framework of distributed robotic systems, Huang et al. studied multi-robot cooperative localization with a extended Kalman filter (EKF) and two new observability constrained OC-EKF estimators [2]. In their experiments they synthetically produced the relative range and bearing measurements using the differences in the positions of the robots, which were recorded by the overhead camera. He and Du tracked dynamic objects to anticipate possible collisions and choose obstacle avoidance policy, but did not distinguish between a robot or other moving obstacle, such as human [4].

Examples of relative positioning systems operating without wireless communication are also present in the literature. For instance, Fredslund and Mataric performed detection of a neighboring robot using combination of a laser and a camera. The range was acquired using a laser, while the bearing was obtained by reading the tilting of a panning camera and keeping in its center the image of a color-coded marker attached to the other robot. In the work by Soares et al. robots operated in an underwater medium, which makes it problematic to both communicate and localize, yet the local acoustic ranges provided enough information to serve as a basis to perform a formation maneuver [5]. Pugh et al. presented an onboard relative positioning module for miniature robots, which operated using modulated infrared signals [6]. The module enabled accurate multi-robot formations, but was tailored to a specific robotic platform.

In this paper we present a method that enables one robot navigating in an environment shared by other robots to obtain relative positions of the other formation

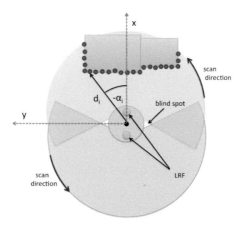

Fig. 2 Illustration of the sensor measurements. The x axis is aligned with the front of the robot (a yellow circle). The two LRF are located (27.5, 0) cm and (-27.5, 0) cm from the origin.

members. Our work relies exclusively on information obtained by a robot from two on-board laser range finders without any assumptions neither about the motion of the robot itself or the other robots or the environment. Since it does not rely on wireless communication, it relaxes the dependence on network reliability to deliver positioning data. For example, if the robots can communicate their self-positioning data, but the communication network is unreliable or has restricted bandwidth, the robots may exchange the data only sporadically and complement the information with the relative positioning system. Combination of both systems would be highly advantageous in a structured indoor environment, resulting in more robust solution than any of the two used separately. Our method is applicable to a general case, such as navigating in an environment with dynamic obstacles that are not part of the multi-robot system but nonetheless may impair the performance. While our detection technique provides an accurate instantaneous estimation, stochastic dynamic filtering helps to deal with random noise, occlusions, and false positives. Problems associated with relative-positioning systems, such as occlusions and limited field-of-view restrain their scalability, however the methods presented in this work could serve as additional back-up positioning system in case of communication issues or even as a fundamental information source, complemented by the data from a more reliable but computationally expensive or network-heavy nodes.

This paper is organized as follows. In Section 2 we present an overview of our robotic platform and its sensing capabilities. In Section 3, we describe the robot detection method using two-dimensional laser measurements and cover in detail the robot tracking algorithm. Experiments with real robots are presented in Section 4. We draw conclusions in Section 5.

6

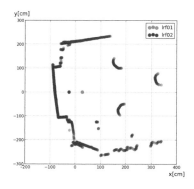 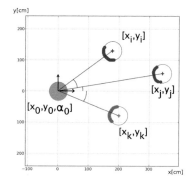

Fig. 3 Left: Raw laser readings in the robots coordinate system. The blue and the green dots indicate positions of the two laser range finders. Right: Result of the circle fitting algorithm. The orange circle is the detecting robot.

2 System Architecture

Our method has been devised and tested on the MBot robots (Fig. 1) designed within the frame of the ongoing FP7 European project MOnarCH (Multi-Robot Cognitive Systems Operating in Hospitals) with the goal of introducing social robots in real human environments and studying the establishment of relationships between them[1].

The robot is equipped with navigation, perception, interaction and low-level safety sensors. For navigation and particularly for mapping, localization and obstacle avoidance, the robot fuses measurements provided by laser range finders, odometry encoders and IMU sensors. The methods presented in this paper are based on the readings of two laser range finders (LRF) URG-04LX-UG01 manufactured by Hokuyo. Each of the two-dimensional LRFs measures 683 distance points in a range from -120° to 120°, where 0° corresponds to the front of the sensor [7]. The sensors are mounted inside the robot at a height of approximately 13 cm above the ground, one in front of the robot heading towards the front and one on the back heading backwards (see Fig. 2). Altogether, both LRFs provide 4 m sensing distance and 360° field-of-view. After translation to the robot's central coordinate system, each individual distance point of the LRF reading is represented in the polar coordinates $S = (d_i, \alpha_i)$, where d_i is the distance of the data point from the origin of the robots coordinate system and α_i is the relative angle. Example of a raw laser scan can be seen in Fig. 3 (left).

The navigation of the MBot robots is based on a standard occupancy grid map, serving for both motion planing [8] and self-localization [9], the latter obtained by combining odometry with AMCL[2]. The robot moves using mecanum wheels, an omnidirectional locomotion system with a maximum speed of 2.5 m/s and maximum acceleration of 1 m/s^2. A complete description of the MBot robot can be found in [9].

[1] MOnarCH, FP7, FP7-ICT-2011-9-601033 (http://monarch-fp7.eu)
[2] AMCL, (http://wiki.ros.org/amcl, retrieved 16 June 2015).

3 Estimation and Tracking Method

We propose a methodology to detect a circular marker on a two-dimensional plane using two laser range finders mounted on a mobile platform. We use one MBot robot to locate all the others in the range of the LRF, as shown in Fig. 3. Although in our case we validated our approach using up to three robots, there are no intrinsic limitations of the method in terms of number of robots, as long as the other platforms are clearly distinguishable and not occluded. The shape of the robot base (approximated by a circle), known a priori, serves as a model for the detection algorithm. Note however that this is not a limitation of the algorithm in terms of generalization as such geometric assumptions can be easily customized for any other robotic platform. In this section, we describe the steps required to process the raw laser scan data to estimate the position of the observed robot. We focus on a scenario involving only two robots and indicate where the algorithm branches for generalization to a multi-robot case. An outline of the estimation technique is presented in Algorithm 1. We denote S the set of Lidar points, CL the set of point-cloud clusters of a size n_{CL} and n_c number of circles fitted to the data.

Algorithm 1. estimation_and_tracking(d,α)

$S = (\mathbf{d}, \boldsymbol{\alpha})$
$S \leftarrow delete\_outliers(S)$
$n_{CL} = 0$
$CL[n_{CL}] \leftarrow (d_0, \alpha_0)$
for $k = 2$ to $len(\mathbf{d})$:   # Point cloud clustering
    if $dist(s_{k-2}, s_{k-1}) > T_{sec}$ and $dist(s_{k-1}, s_k) > T_{sec}$
        $CL[n++] \leftarrow S_k$
    else
        $CL[n] \leftarrow S_k$
for $k = 0$ to n_{CL}:   # Cluster selection
    if $dist(CL[k], CL[len(CL)]) > T_{sel}^{max}$
        or $dist(CL[k], CL[len(CL)]) < T_{sel}^{min}$
        $CL \leftarrow delete(CL[k]), n_{CL} --$
$n_c = 0, [\mathbf{x}_c, \mathbf{y}_c] = \emptyset$
for $k = 0$ to n_{CL}:   # Circle fitting
    $[\mathbf{x}_c, \mathbf{y}_c] \leftarrow fit\_circle(CL[k]), n_c++$
$[\mathbf{x}_c, \mathbf{y}_c] \leftarrow merge\_overlapping\_circles([\mathbf{x}_c, \mathbf{y}_c], n_c)$
$[\mathbf{x}_c, \mathbf{y}_c] \leftarrow is\_in\_area\_allowed([\mathbf{x}_c, \mathbf{y}_c], n_c)$
$[\mathbf{x}_c^g, \mathbf{y}_c^g], n_c \leftarrow global\_to\_local\_coord([\mathbf{x}_c, \mathbf{y}_c])$
$[\mathbf{r}, \boldsymbol{\phi}] \leftarrow euclidean\_to\_polar[\mathbf{x}_c^g, \mathbf{y}_c^g]$

3.1 Data Clustering and Selection

As described in Section 2, the laser scan after initial pre-processing is stored as a tuple (d_i, α_i) for each data point i. Each scan sequence produces a map of the immediate neighborhood with distinguishable shapes, such as walls or objects cluttered within the sensing range (see Fig. 3). After deletion of isolated points caused by a mixed pixel phenomenon [10], which generates a measured range resulting from a combination of the foreground and the background objects, we cluster the scan points using simple nearest neighbor classification by moving a sliding window of a 3-point size and discerning the separated objects using a minimal Euclidean distance threshold. The thresholds are highly implementation dependent, in our case are $T_{sec} = 20$, $T_{sel}^{max} = 200$ and $T_{sel}^{min} = 20$. The acquired segments allow the algorithm to clearly distinguish among different objects, which might be directly adjacent to the walls. Being only attentive to the circular candidates of a certain size, the large objects characterized by excessive segment dimensions as well as object without curvature (the walls) are discarded.

3.2 Circle Fitting

The pre-selected clustered point clouds, having approximately the size of the object to be detected, are each in turn supplied into a least squares optimization algorithm extended with a modification of the Levenberg-Marquardt algorithm [11]. We minimize an objective function F over a space of only two parameters, namely the coordinates (x_c, y_c) of the center of the circle, expressed as a relative polar coordinate in respect to the center of the detecting robot:

$$F(\mathbf{d}, \boldsymbol{\alpha}) = \sum_{i=1}^{N^K} \left(\sqrt{(\varepsilon_x d_i cos(\alpha_i) - x_c)^2 + (\varepsilon_y d_i sin(\alpha_i) - y_c)^2} - R \right)^2 \quad (1)$$

where N^K is the number of data points present in the current cluster K, R is the radius of the circular object (in this case the robot radius) and ε_x and ε_y are the signs reflecting the scan angle convention of the frontal or the rear laser. We use the size of the robot base $R = 60\ cm$. The result of the circle fitting algorithm shown in Fig. 3 illustrates the situation from the Fig. 1 (right), where the robot on the back detects the three robots present in the experimental area.

3.3 Candidate Target Validation

A structured indoor environment, such as the area where we conduct our experiments (see Fig. 1), may generate false detections. Building features as support columns or trash bins unnecessarily increase the number of false positives, therefore we take advantage of the fact that the environment is known a priori. The robots of the MBot size can physically access only limited section of available space; for instance, they cannot move under the structural elements or inside the narrow spaces. These features

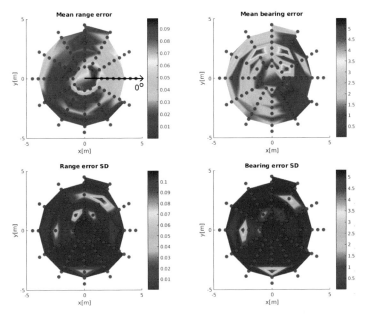

Fig. 4 Error map showing the accuracy (top) and the precision (bottom) of the relative positioning method. The scale for the range error is in meter; the scale for the bearing error is in degree. The heading of the robot is indicated with a zero degree angle.

generate most of the false detections, so we reduce the search space to areas accessible by the MBot robots (Fig. 1 (right)).

Fig. 1 (right) shows a fragment of a standard occupancy grid map available for the robot for the purpose of self-localization, based on which at any point in time, it can estimate its own pose. Having the circle fitting algorithm return a set of coordinates (x_c, y_c) of the detected robots in the local coordinate frame of the detecting robot, we can find a precise location of the robots on the map:

$$\begin{bmatrix} x_c^g \\ y_c^g \end{bmatrix} = \begin{bmatrix} cos(\alpha_0) & -sin(\alpha_0)) \\ sin(\alpha_0) & cos(\alpha_0) \end{bmatrix} \begin{bmatrix} x_c \\ y_c \end{bmatrix} + \begin{bmatrix} x_0^g \\ y_0^g \end{bmatrix} \tag{2}$$

where the superscript g refers to the global coordinate frame and the $[x_0, y_0, \alpha_0]$ is the pose of the observing robot in the global frame. Along the presented methodology we developed a simple tool which allows to analyze a previously marked occupancy grid map to determine whether a given $[x, y]$ point on the map is marked as *accessible*. The map can be edited using any of the image manipulation programs (Fig. 1 shows an example of such a map). We assume that if the (x_c, y_c) is located within the marked area, it becomes a valid measurement of the position of the robot R_k, i.e. (x_k, y_k).

3.4 Range and Bearing

The detection, at this point assumed being a valid relative position of another robot, necessitates an estimation of the associated positioning error. In this section, we present such procedure resulting in the evaluation of the error associated with the detection of the circle position, to be further used in a Bayesian estimator for tracking purposes.

The range and bearing measurement error from the robot R_i to R_j is defined as a difference between the actual relative positions and the estimated values:

$$\tilde{r}(\hat{r}_{ij}, p_i, p_j) = \hat{r}_{ij} - r_{ij}(p_i, p_j) \tag{3}$$

$$\tilde{\phi}(\hat{\phi}_{ij}, p_i, p_j) = \hat{\phi}_{ij} - \phi_{ij}(p_i, p_j) \tag{4}$$

where $p = [x, y]$ is the Euclidean representation of the position and the range and bearing are a transformation of the Euclidean representation in polar coordinates:

$$\begin{bmatrix} r_{ij} \\ \phi_{ij} \end{bmatrix} = T_e^p(p_i, p_j) = \begin{bmatrix} r(p_i, p_j) \\ \phi(p_i, p_j) \end{bmatrix} = \begin{bmatrix} \sqrt{(x_i - x_j)^2 + (y_i - y_j)^2} \\ atan2((y_i - y_j), (x_i - x_j)) \end{bmatrix} \tag{5}$$

The relative positioning error has been evaluated with a set of systematic experiments, where one robot to be detected has been placed manually at distances from 1.5 m to 5 m in steps of 0.5 m and at angles from 0° to 360° in steps of 30°. A total of 100 scans has been acquired for each location. A summary of the results is presented in Fig. 4 as two-dimensional linearly interpolated color maps. In general, we note that the range measurement error is smaller on the sides of the observing robot (i.e. for 60°-120°and 240°-300°), where the observed robot is within the sensing range of both laser range finders. The bearing error has a tendency to increase with the distance. It can be noticed that the frontal sensor has a reduced sensing range of 4 m. Composition of its measurements with those of the rear laser extends the sensing range to 4.5 m on the sides, the same as it is for the rear laser only. The color map distribution serves in the Kalman filter tracking for the estimation of the variance associated with the positioning error measurements. Thus, for the range and bearing measurements we assume the observation noise to be represented by a Gaussian probability density function. We sample the mean μ and the variance σ from the polar error (Fig. 4).

3.5 Multi-target Tracking

The tracking, depending on an application, can be performed in the global coordinate frame or locally, relatively to the tracking robot. The latter approach is typically implemented for tasks requiring the robots to know only the relative positions of the teammates, such as formation control [12]. We perform the tracking in the global coordinate frame because of its generality, that allows the robots to fuse the information coming from various sources.

For our Kalman estimator, we assume a constant velocity and randomized Gaussian acceleration motion model. The state equations of the moving robot are:

$$\begin{bmatrix} x \\ y \\ \dot{x} \\ \dot{y} \end{bmatrix} = \begin{bmatrix} 1 & 0 & \Delta t & 0 \\ 0 & 1 & 0 & \Delta t \\ 0 & 0 & 1 & 0 \\ 0 & 0 & 0 & 1 \end{bmatrix} \begin{bmatrix} x \\ y \\ \dot{x} \\ \dot{y} \end{bmatrix} + \varrho(k), \quad \begin{bmatrix} x \\ y \end{bmatrix} = \begin{bmatrix} 1 & 0 \\ 0 & 1 \end{bmatrix} \begin{bmatrix} x \\ y \end{bmatrix} + \epsilon(k) \qquad (6)$$

The state vector of the observed robot is $x = [x, y, \dot{x}, \dot{y}]^T$, the observation model is $z = [x, y]^T$, $\varrho(k) \sim \mathcal{N}(0, Q(k))$ is a process noise assumed constant with $Q = [0.1\ 0.1\ 0.1\ 0.1]^T$ and $\epsilon(k) \sim \mathcal{N}(0, T_p^e(\xi))$ is a measurement noise with zero mean and the variance sampled from the error map in Fig. 4 transformed into the Euclidean space.

4 Experiments

We considered four distinct scenarios to determine the performance of the detection and tracking method presented above. They demonstrate the cases of detection with one moving robot and one static detecting robot (I), one moving robot and one detecting robot following it (II), fusion of measurements of one observed robot from two detecting robots (III) and detection of two robots by one static robot (IV). Experiments have been carried out in a lab setting cluttered with appliances. For each scenario, we performed 12 runs (unless stated otherwise) and present the time-wise average as well as an aggregate of the results. The experiments took place in an area frequently visited by humans, who either traversed the space or walked around the room performing their duties and not being attentive to the motion of the robots. While the presence of the people caused instantaneous false detections, the Kalman filter was able to reduce their impact to a minimum. We obtain our ground-truth from the self-localization system provided by the navigation software (see Section 2 and [9]). The self-positioning capabilities of the robots, characterized by an accuracy of about 10 cm, serves as a ground truth measurement used in the performance evaluation.

The accuracy of the relative positioning system has been evaluated using standard root-mean-square error between the estimates and the actual values:

$$E_{AV}(t) = \sqrt{\frac{1}{n_d} \sum_{i=1}^{n_d} \| p_i(t) - \hat{p}_i(t) \|^2}, \quad E_T = \frac{1}{T_{max}} \sum_{j=1}^{T_{max}} (E_{AV}(j)) \qquad (7)$$

The $E_{AV}(t)$ is the time-wise average error over the number of experiments n_d and the E_T is the total average error over the duration of a single run.

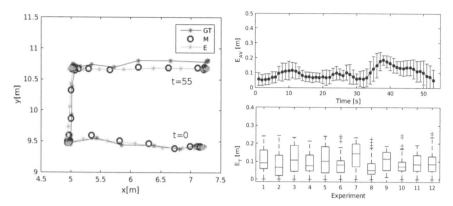

Fig. 5 Single robot detection from a static platform. (Left) Sample trajectory of the observed robot and the position estimate. The observing robot position is [5.9, 7.1]. *GT* is the ground truth, *M* are the measurements and *E* the estimate. This convention remains valid throughout the paper. (Top right) Time-wise average and SD of the error. (Bottom right) The average error aggregated for all the runs.

4.1   Results

I. Single Robot Detection from a Static Platform. In the first scenario we fix the position of the detecting robot and move the observed robot on a simple trajectory as shown in Fig. 5 (left). The sampling frequency is fixed to 1 Hz to reflect the effect of the Kalman filter. The tracking estimates of the observed robot follows very closely the position reported by its self-positioning system, with the error not exceeding 20 cm (see Fig. 5, top right). The Fig. 5 (bottom right) shows the aggregate error of each experiment. The total average error during the period when the robot was detected is around 10 cm, slightly higher than expected given the error map. We hypothesize that the increased error might have been caused by approximation of the robot base with a circle - the actual body of the robot is flattened on the sides.

II. Single Robot Detection from a Mobile Platform. The second case studies the impact of the detecting platform movement on the tracking system. The observed robot follows the trajectory shown in 6 (left) and the detecting robot keeps a constant distance to the estimated position using a simple proportional feedback controller. The sampling frequency is 1 Hz. The detection error is bounded and level off (Fig. 6, bottom right); the average error E_{AV} alternates around the value of 20 cm (Fig. 6, top right). The increase of the detection error in comparison with the static case can be caused by inaccuracies associated with the self-positioning system of the detecting robot. During two runs (outliers of 7 and 9 in Fig. 6 bottom right) we encountered false positive detections that proved necessity of a Kalman filter to reduce their impact on a final performance.

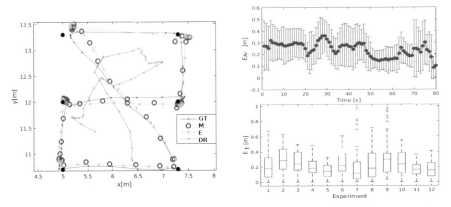

Fig. 6 Single robot detection from a mobile platform. (Left) The trajectory of the observed robot (True), the measurements (Meas.) and the estimate (Estim.), *DR* is the trajectory of the following (detecting) robot.

III. Detection and Fusion from Two Static Platforms. The following set of experiments is conducted as to approximate the effect of sensor fusion from two sensing platforms. To be precise, one observed robot moves around the arena and the two observing robots are static and positioned so that the observed robot for a certain amount of time is outside of the sensing range of at least one of them. The two detecting robots are communicating their Kalman filter estimates and each of them fuses the data for a better approximation of the position of the observed robot. For the two observing robots R_i and R_j, estimating the position of the robot R_k as $[x_i^k, y_i^k]$ and $[x_j^k, y_j^k]$ respectively, the result of the sensor fusion can be calculated by combination of the Gaussians:

$$\frac{1}{\sigma_T^2} = \frac{1}{\sigma_{ik}^2} + \frac{1}{\sigma_{jk}^2}, \quad x_T^k = \sigma_T^2 \left(\frac{x_i^k}{\sigma_{ik}^2} + \frac{x_j^k}{\sigma_{jk}^2} \right) \tag{8}$$

where $\sigma_{mn}^2(r_{mn}, \phi_{mn})$ is the standard deviation associated with the measurement of the robot R_m to the robot R_n obtained from the error map (Fig. 4). The method is fully scalable, up to the capacity of the communication network and can easily accommodate additional nodes. An additional advantage of this method emerges when the observed robot ventures outside of the range of one of the detecting robot. For example, if robot R_i does not obtain an observation, its variance is set to infinity $\lim_{\sigma_{ik}^2 \to \infty} \sigma_T^2 = \sigma_{jk}^2$, the corresponding estimate levels out and the fused estimate in the two-node case attains value of the other measurement as in $\lim_{\sigma_{ik}^2 \to \infty} x_T^k = x_j^k$.

Fig. 7 (left) shows the estimates recorded by the two detecting robots and the fused estimate. The detection range of robot R_1 covers only the upper part of the area (above 10.6 m), below which the fused estimate follows directly that provided exclusively by robot R_2. Otherwise, the fusion significantly improves the overall estimate, lowering the error and reducing its variance (Fig. 7, top right). The difference in comparison to single-platform case is clearly visible in Fig. 7 (bottom right), where the largest error during the runs did not exceed half of that achieved using a single observing robot.

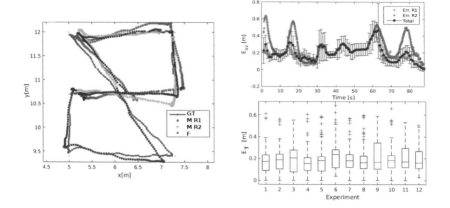

Fig. 7 Detection and fusion from two static platforms. The detecting robots R_1 and R_2 are positioned at $(6.0, 6.7)$ and $(6.4, 13.5)$ respectively. F is the fused estimate. (Top right) shows the average errors of both robots over the runs.

IV. Multi-robot Detection from a Static Platform. The final experiment focuses on a multi-robot detection scenario. We perform 10 runs of an experiment with 2 robots moving on a rectangular trajectory (see Fig. 8, left), with the detecting robot placed in the middle of the arena. Although the tracking performs very well on average (Fig. 8, right), there is a number of false positives (red outliers), caused principally by random objects appearing as having a circular surface from a specific observation point. While these outliers are usually eliminated by the Kalman filter, in this case they might have been caused by temporary occlusions or faulty data association. Occasional false positives do not impair the tracking, but their cause needs to be further investigated.

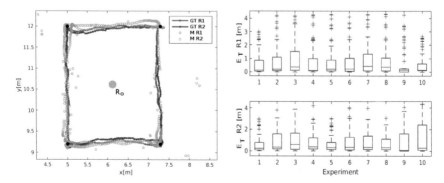

Fig. 8 Multi-robot detection from a static platform. The green mark shows the position of the observing robot at [6.1, 10.7]. (Right) The red outliers are the false detections.

5   Conclusions

In this work we presented a lidar-based relative position estimation and tracking. A distinguishing feature of our work is the fact that each robot only relies on two-dimensional scan provided by a laser range finder for sensory information. By exploiting simple geometric features of the individual robotic platform, we were able to reliably estimate and track the position of the other robots present in the environment. Our method can be easily extended for tracking objects of various sizes and shapes by changing model of the object, allowing for detection and tracking of heterogeneous robots. We evaluated our approach during systematic real-world experiments, where we studied the performance in scenarios involving various combinations of static and mobile robots.

Experimental results show that while the results in terms of accuracy are useful in the targeted range of 20-30 cm, further effort is needed to increase the reliability of the lidar-based relative positioning method, possibly improving on its individual components (classification, tracking etc.). We will further focus on increasing robustness of our method by adding a data association module for tracking multiple robots in cluttered arenas and be able to deal with moving obstacles that can be misclassified as robots. We intend to apply our work as a complementary tool in multi-robot behaviors. In particular, we plan to use the methods presented in this paper as an additional source of information in multi-robot formations behavior [12], significantly reducing its dependence on communication. In contrast to the systems that depend on the local communication, the relative positioning method suffers from the line-of-sight limitation, therefore merging the complementary strengths of both techniques is highly beneficial.

Acknowledgements Supported by ISR/LARSyS Strategic Funds from FCT project FCT[UID/EEA/5009/2013] and FCT/11145/12/12/2014/S and FCT/PD /BD/105784/2014 and by European MOnarCH project FP7-ICT-9-2011-601033.

References

1. Teixid, M., Pallej, T., Font, D., Tresanchez, M., Moreno, J., Palacn, J.: Two-Dimensional Radial Laser Scanning for Circular Marker Detection and External Mobile Robot Tracking. Sensors **12**, 16482–16497 (2012)
2. Huang, G.P., Trawny, N., Mourikis, A.I., Roumeliotis, S.I.: Observability-based consistent EKF estimators for multi-robot cooperative localization. Autonomous Robots **30**(1), 99–122 (2011)
3. Fredslund, J., Mataric, M.J.: A general, local algorithm for robot formations. IEEE Trans. on Robotics and Automation, Special Issue on Advances in Multi-Robot Systems **18**(5), 837–846 (2002)
4. He, F., Du, Z., Liu, X., Ta, Y.: Laser range finder based moving object tracking and avoidance in dynamic environment. In: Proc. of IEEE Int. Conf. on Information and Automation, pp. 2357–2362 (2010)
5. Soares, J.M., Aguiar, A.P., Pascoal, A.M., Martinoli, A.: Joint ASV/AUV range-based formation control: theory and experimental results. In: Proc. of the 2013 IEEE Int. Conf. on Robotics and Automation, pp. 5579–5585 (2013)

6. Pugh, J., Raemy, X., Favre, C., Falconi, R., Martinoli, A.: A Fast Onboard Relative Positioning Module for Multirobot Systems. IEEE/ASME Trans. on Mechatronics **14**(2), 151–162 (2009)
7. Scanning Laser Range Finder URG-04LX-UG01 Specifications. https://www.hokuyo-aut.jp/02sensor/07scanner/download/pdf/URG-04LX_UG01_spec_en.pdf (accessed May 22, 2015)
8. Ventura, R., Ahmad, A.: Towards optimal robot navigation in urban homes. In: Proc. of the 18th RoboCup Int. Symposium (2014)
9. Messias, J., Ventura, R., Lima, P., Sequeira, J., Alvito, P., Marques, C., Carrico P.: A robotic platform for edutainment activities in a pediatric hospital. In: Proc. of the 2014 IEEE Int. Conf. on Auton. Robot Sys. and Competitions, pp. 193–198 (2014)
10. Okubo, Y., Ye, C., Borenstein, J.: Characterization of the Hokuyo URG-04LX laser rangefinder for mobile robot obstacle negotiation. SPIE Def., Sec., and Sens. Int. Soc. for Opt. and Phot. (2009)
11. More, J.J.: The Levenberg-Marquardt algorithm: Implementation and theory. Numerical analysis, pp. 105–116. Springer, Heidelberg (1978)
12. Das, A.K., Fierro, R., Kumar, V., Ostrowski, J.P., Spletzer, J., Taylor, C.J.: A vision-based formation control framework. IEEE Transactions on Robotics and Automation **18**(5), 813–825 (2002)

Vizzy: A Humanoid on Wheels for Assistive Robotics

Plinio Moreno, Ricardo Nunes, Rui Figueiredo, Ricardo Ferreira, Alexandre Bernardino, José Santos-Victor, Ricardo Beira, Luís Vargas, Duarte Aragão and Miguel Aragão

Abstract The development of an assistive robotic platform poses exciting engineering and design challenges due to the diversity of possible applications. This article introduces Vizzy, a wheeled humanoid robot with an anthropomorphic upper torso, that combines easy mobility, grasping ability, human-like visual perception, eye-head movements and arm gestures. The humanoid appearance improves user acceptance and facilitates interaction. The lower body mobile platform is able to navigate both indoors and outdoors. We describe the requirements, design and construction of Vizzy, as well as its current cognitive capabilities and envisaged domains of application.

Keywords Robot design · Mechanical design · Humanoid robots · Mobile robots

1 Introduction

The increasing interest in service robotics for assisting people in daily life tasks is driving research in humanoid-like robots for human-robot interaction. This is a challenging goal since many humanoid technologies are still at its infancy, namely legged locomotion. To address this issue, the trend in assistive robotics has been the use of a mobile base combined with a more human-like upper torso.

Robots like Rollin' Justin [5], Twendy-one [7], the ARMAR III [2] and the iKart[1] mobile platform for the iCub [10] are showcases on the research side, while on

---

P. Moreno(✉) · R. Nunes · R. Figueiredo · R. Ferreira · A. Bernardino · J. Santos-Victor · R. Beira · L. Vargas · D. Aragão · M. Aragão
Institute for Systems and Robotics (ISR/IST), LARSyS, Instituto Superior Técnico, University of Lisbon, Torre Norte Piso 7, Av. Rovisco Pais 1, 1049-001 Lisboa, Portugal
e-mail: {plinio,rnunes,ruifigueiredo,ricardo,alex,jasv}@isr.tecnico.ulisboa.pt,
      ricardo.beira@epfl.ch, luis.vargas@fisherman.pt,
      {daragao,miguelaragao91}@gmail.com
http://vislab.isr.tecnico.ulisboa.pt/

[1] http://wiki.icub.org/wiki/IKart

© Springer International Publishing Switzerland 2016
L.P. Reis et al. (eds.), *Robot 2015: Second Iberian Robotics Conference*,
Advances in Intelligent Systems and Computing 417,
DOI: 10.1007/978-3-319-27146-0_2

17

the comercial side robots like Pepper[2] and REEM[3] are taking this research field to the market. Vizzy belongs to this group of human like upper-body and car-like lower-body robots, which are designed to interact and assist people in their daily tasks (see Figure 1). In comparison to the aforementioned robots, Vizzy has a more friendly and organic appearance, and a mechanical design guided by a modular approach, which facilitates production, storage and assembly. In addition, trajectories generated by the head and arm motions controllers are motivated by those of the humans [11]. In this article we describe the motivations, design and development phases of the Vizzy robot. In Section 2 we describe in detail the design concepts that guided us to the appearance and mechanical design of the robot: friendly and organic approach in the appearance and modularity in the mechanical design. Section 3 describes in detail the mechanical design of the mobile base, upper body and hand. Section 4 explains the sensorimotor capabilities of Vizzy and Section 5 lists the software libraries developed for its cognitive capabilities. Section 6 concludes and summarizes the current developments and future work.

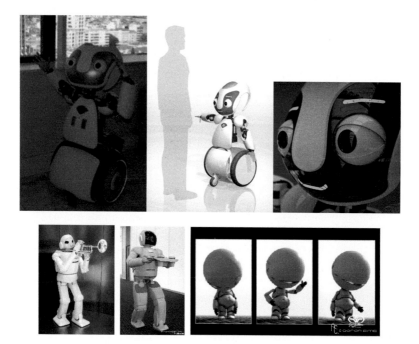

Fig. 1 The top left side image shows Vizzy with open arms, the top-middle a render of the final design compared to a person with 1.75m height, and the top right side a render of the facial expressions. Real vs. fictional robots. On the bottom, from left to right, the Humanoid Toyota partner, Honda Asimo and Marvin from the movie 'The Hitchhiker's guide to the galaxy'[4]. Note the friendlier poses of Marvin compared to Asimo and Toyota partner.

---

[2] https://www.aldebaran.com/en/a-robots/pepper/more-about-pepper

[3] http://www.pal-robotics.com/en/products/reem/

[4] http://www.imdb.com/title/tt0371724/

2 Design: Concepts and Motivation

Aesthetics in humanoid robots can be divided into two main trends: The 'high-tech' appearance of the real robots and the more friendly and organic approach from science fiction. On one hand, the 'high-tech' appearance is driven by technology show-off, functionality and marketing constraints. On the other hand, the science fiction realm is driven by creativity and imagination and less constrained by functionality. Examples of both approaches are shown in Figure 1.

Vizzy's design guidelines combine the friendly and organic approach with the functionality. The main strategic design decision was the adoption of an upper humanoid-like torso and a wheeled platform for locomotion, which is the main constraint in the appearance. The selected mobile platform (Segway RMP 50[5]) with its large wheels inspired the designer team with an upright marsupial (e.g. Kangaroo) that is less anthropomorphic from the waist down. The facial expresions and hand design of Vizzy was influenced by two robots: iCub [10] on the facial expressions and Baltazar [9] on the hand design. The iCub facial expressions are composed by arrays of LEDs in the eyebrows and mouth, litting a subarray of LEDs to generate an expression (see Figure 1). The LEDs are complemented with two plastic shells that act as the eyelids. Both the LEDs and the eyelids were adopted on Vizzy's face expressions. The Baltazar's hand has a planar palm and five underactuated fingers controlled by three motors. The motor pulls one or more strings being attached to the tip of the fingers, moving the three finger limbs with one power source. Vizzy's hands are underactuated too but with several improvements and just four fingers (details in Section 3). The remaining parts of the robot, which include the head, arms and the supporting structures were designed to have a humanoid robot as similar as the humans in terms of degrees of freedom and their corresponding range of motions, as presented in Table 1.

Table 1 Range of human movements vs. Vizzy's range of movements. All the angular values are in degrees

Joint			Standard human	Vizzy
Head		Rotation	-70 to 70	-53 to 53
		Neck flexion	-50 to 60	-18 to 37
		Eye rotation	-40 to 40	-38 to 38
		Eyes flexion	-40 to 40	-38 to 38
Arm	Shoulder	scapula	-45 to 130	-18 to 18
		flexion	-60 to 180	-135 to 75
		abduction	-135 to 90	-75 to 0
		rotation	-90 to 90	-85 to 85
	Elbow	flexion	-150 to 0	-110 to 0
	Forearm	pronation	-90 to 90	-85 to 85
	Wrist	abduction	-20 to 50	-35 to 35
	Wrist	flexion	-70 to 90	-35 to 35

---

[5] http://rmp.segway.com/discontinued-models/

From the engineering point of view, the main concept is the definition of four motor modules, which are replicated over 22 degrees of freedom. Each of these modules are composed by a set of common parts that facilitates the storage and replacing of damaged parts, and reduces the production costs. In addition, the four modules are similar between each other with differences just in size and the presence or absence of a gearbox, which allows to have a general assembly procedure over all types of modules. Then, the assembly procedure for each module just adds specific details. Figure 2 illustrates the similarity and size differences between the modules.

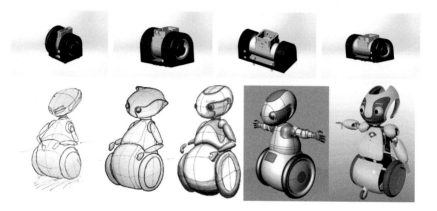

Fig. 2 In the top row, the four different motor modules designed for Vizzy. From left to right, the small, medium, large and extra large modules. For illustrative purposes, the same type of assembly is shown. In the bottom row, the design evolution of Vizzy. The stages shown are: initial sketches, the first concept with closed covers and the final design

Initially, Vizzy had a shell cover that completely wrapped its mechanical and electronic components. This concept later shifted to an open shell strategy that optimised weight and production costs. Figure 2 shows the evolution from the initial drafts until the current version of the robot.

3   Design Specification

This section describes mechanical design contributions for building the mobile base, the torso, arms and the head. Vizzy has a total of 30 degrees of freedom, distributed as follows: 2 dof's for the mobile base, 23 dof's for the torso and the arms and 5 dof's for the head.

3.1 Mobile Base Design

The main guideline for the mobile base design was the autonomy, so the original Segway RMP50 supporting frame was redesigned for safe and robust motion of the torso (~ 30 kg) and for carrying all the electronics and computing resources. The main components are as follows: Segway box, battery, computer and the CAN interface board, which are shown in Figure 3. The first goal was to select a PC and battery set that allow to run experiments continuously during several hours, considering a high performance processor (Intel(R) Core(TM) i7-3930K CPU @ 3.20GHz) and a motherboard (ASRock X79 Extreme 4) with a large number of plugging interfaces. Then, based on the power consumption of the PC and the motors ($\sim 500W$), the battery was chosen to supply power to the PC and the torso motors during 4 hours continuously. The selected battery is the 125255255 (7 cells) from Kokam[6].

Fig. 3 Main components of the mobile base. The supporting frame is coloured with red, and the numbered items correspond to: (1) Segway RMP50 box with wheels, battery and motors (for better visualization one of the wheels was removed), (2) Voltage converter, (3) main battery, (4) PC motherboard, and (5) CAN interface board.

3.2 Upper Body Design

The upper body has a total of 28 degrees of freedom, which are distributed as follows: one for the waist; eight for each arm; tree for each hand; and five for the head. Vizzy is symmetric with respect to the sagittal plane, so the mechanical parts of the left side are replicated on the right side. 22 out of the 28 degrees of freedom are designed as one of the four types of modules. Figure 4 shows the serial assembly of the modules, where the motor names match the listed dof's in Table 2. The remaining 6 dof's correspond to the fingers, which will be explained in the following subsection.

   The common parts over all four type of modules are: the harmonic drive; the DC motor; the encoder; and the two bearings. The additional parts are Faulhaber

---

[6] http://www.kokam.com

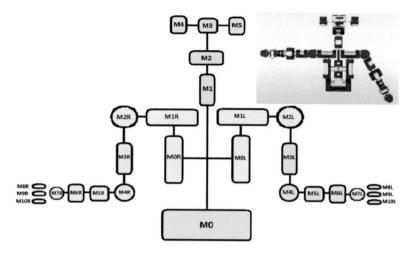

Fig. 4 The motor modules are assembled by following the lines between the modules in the picture. The final assembly of the modules is rendered on the top right of the picture. Each color correspond to the group of modules connected to each CAN bus. All the blue motors are connected to one CAN bus, all the green motors are connected to another CAN bus and so on. There are four CAN buses that control 28 dof.

Table 2 List of the modules and their corresponding body part and joint name. The motor names match the Figure 4

Body part	Motor#	Joint name	Module size
Torso	M0	Waist	Extra Large
Head	M1	Head rotation (pan)	Medium
Head	M2	Head flexion (tilt)	Medium
Head	M3	Eyes flexion (tilt)	Small
Head	M4	Right Eye rotation (pan)	Small
Head	M5	Left Eye rotation (pan)	Small
Arm	M0R / M0L	Shoulder scapula	Large
Arm	M1R / M1L	Shoulder flexion	Large
Arm	M2R / M2L	Shoulder abduction	Large
Arm	M3R / M3L	Shoulder rotation	Medium
Arm	M4R / M4L	Elbow flexion	Medium
Arm	M5R / M5L	Forearm pronation	Small
Arm	M6R / M6L	Wrist abduction	Small
Arm	M7R / M7L	Wrist flexion	Small

gearboxes that scale the torque limit. In the case of the Large and Extra-Large modules, there is one additional Faulhaber gearbox (all small and medium modules do not have any Faulhaber gearbox). Figure 5 shows a planar cut of the both modules, one without the Faulhaber gearbox and the other one with an additional Faulhaber gearbox, and Figure 6 shows a render of all the module types. Notice the difference in assembly options across the same module, which are designed according to the

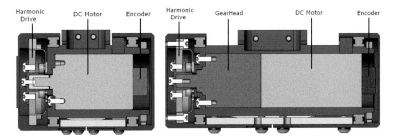

Fig. 5 The drawing on the left side shows the parts of the small and medium modules, while the drawing on the right side shows the parts of the Large and Extra-Large modules. The main difference is the additional Faulhaber gearbox (GearHead). The green colored parts correspond to the cut of the bearings. The blue and red colored parts correspond to the black and gray parts of the renders in Figure 6

Small   Medium   Large   Extra-large

Fig. 6 The four different motor modules designed for Vizzy. From left to right, the small, medium, large and extra large modules. The top row shows the modules that are assembled on the middle top, and the bottom row shows the modules that are assembled on the side.

localization of the module in the upper body. The small, medium and large modules have two assembly options: (i) on the middle top and (ii) on one side of the module.

3.3 Hand Design

Following Baltazar's hand design, Vizzy's hand skills include basic manipulation actions and gesture execution. The manipulation skills include three types of power grasps (cylindrical, spherical and hook) and one type of precision grip (tip-to-tip) [14]. These types of grasps can be executed successfully with four fingers, which are moved with just three motors. Like all the other modules, the four fingers are similar between each other, facilitating the storage and replacement of the damaged parts and reducing the production costs. Figure 7 shows the palmar and dorsal sides of Vizzy's hand, where the motors are on the dorsal side and the contact sensors are visible on the palmar side. Every finger has a string that goes through all the finger limbs, from

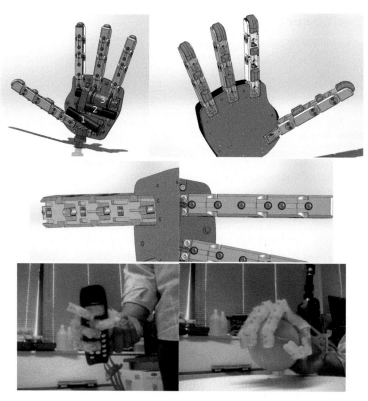

Fig. 7 The top left side image shows the dorsal view, where the numbers correspond to finger motion as follows: (1) thumb, (2) index and (3) the remaining two fingers. The top right side image shows the planar view, where the contact sensors are orange colored. The middle left side image shows the planar hand view, where the cyan colored regions correspond to the string path. The middle right side image shows the dorsal view, where the cyan colored regions show the position of the dental rubber bands. The bottom left side image shows the cylindrical power grasp of a phone. The bottom right side image shows the spherical power grasp of a ball.

the pulley to the finger tip. Figure 7 illustrates the position of the string for the last finger. The motor pulls the string, moving the fingers limbs towards the palm. The final position of the finger limbs can be controlled by changing the velocity of the motor. An improvement added to Vizzy's hand with respect to Baltazar is a group of dental rubber bands on the dorsal side of the fingers, which guarantees that the fingers return to the open hand position. Figure 7 shows the positions of the dental bands. The grasp types already implemented and shown in Figure 7 include the cylindrical and spherical power grasps, which work currently in open loop. The contact sensors will close the control loop.

4 Sensorimotor Description

4.1 Kinematics and Controllers

The Segway RMP 50 has two motors that allow omni-directional motion in the 2D plane. The low-level controller is a velocity controller that accepts the linear + angular target velocities. The Denavit-Hartenberg parameters [6] of Vizzy's torso and their corresponding joint limits are shown in Table 3. Vizzy's upper body joints motors (28 motors) are Faulhaber DC motors that can be controlled in angular position or velocity. The upper body motors are controlled by 10 electronic boards that process the encoder signals and generate the control signals for motor motion, which are amplified with pulse width modulation (PWM) [10]. In addition to the low-level controller interfaces, we implemented mid-level interfaces that control the arms' motion and the gaze: (i) the cartesian controller interfaces [11] for the left and right arm, which are based on the iKin library [12]; and (ii) our gaze controller

Table 3 Vizzy's Denavit-Hartenberg parameters. A and D units are meters and α and θ offset are in degrees

Kin. chain	Joint	A	D	α	θ offset	Limits
	Waist	0	0	90	0	-20 to 20
	Rotation	0	-0.37	90	0	-53 to 53
Left Eye	Neck flexion	0.1362	0	180	201.2	-18 to 37
	Left Eye flexion	0	0.102	90	201.2	-38 to 38
	Left Eye rotation	0	0	90	-90	-38 to 38
	Waist	0	0	90	0	-20 to 20
	Rotation	0	-0.37	90	0	-53 to 53
Right Eye	Neck flexion	0.1362	0	180	201.2	-18 to 37
	Right Eye flexion	0	-0.102	90	201.2	-38 to 38
	Right Eye rotation	0	0	90	-90	-38 to 38
	Waist	0	0.0805	90	0	-20 to 20
	shoulder scapula	0	-0.212	90	0	-18 to 18
	shoulder flexion	0	-0.10256	90	-90	-75 to 135
Left Arm	shoulder abduction	0	0	90	110	0 to 75
	shoulder rotation	0	-0.16296	90	90	-85 to 85
	Elbow flexion	0	0	90	0	0 to 110
	Forearm pronation	0	0.18635	90	0	-85 to 85
	Wrist abduction	0	0	90	-90	-35 to 35
	Wrist flexion	0.1	0	90	180	-35 to 35
	Waist	0	0.0805	90	0	-20 to 20
	shoulder scapula	0	0.212	90	0	-18 to 18
	shoulder flexion	0	0.10256	90	-90	-75 to 135
Right Arm	shoulder abduction	0	0	90	110	0 to 75
	shoulder rotation	0	0.16296	90	90	-85 to 85
	Elbow flexion	0	0	90	0	0 to 110
	Forearm pronation	0	-0.18635	90	0	-85 to 85
	Wrist abduction	0	0	90	-90	-35 to 35
	Wrist flexion	-0.1	0	90	180	-35 to 35

implementation[7], which is largely inspired on the iCub's gaze controller [12]. The arm controller generates trajectories in the cartesian space from the current arm pose to the desired end-effector pose . The gaze controller generates trajectories for the head and eyes motors, moving the head from the current gaze point to a desired gaze point. These controllers were developed as YARP [3] modules.

4.2   Sensors

The sensing capabilities are provided by the following devices:

- One laser scanner Hokuyo URG-04LX, which is located in the front bottom of the mobile base.
- Two PointGrey cameras (Dragonfly 2) that act as the left and right eye of Vizzy, located in the robot's head.
- One ASUS XTION, which is located on the chest.
- One inertial sensor xSens Mti-28A, which is located in the head behind the cameras.
- Twelve contact sensors for each hand that are located on the finger limbs. These sensors are a combination of an Hall effect sensor and a magnet. The magnet is inserted in a silicon shape (shown in Figure 7), which is deformed during contact between the object and the fingers. The deformation will change the Hall sensor readings since the magnet will be positioned closer to the Hall sensor.

5   Perception, Navigation and Manipulation and Human Robot Interaction

The software libraries and modules for perception, navigation and human-robot interaction are available on the official Vizzy github repository [1]. The majority of these skills are based on existing software, and the rest have been implemented at our laboratory. The detailed list as follows:

- Simulation of the Segway RMP 50 in Gazebo[8] (segbot simulator [8]), which was adapted for Vizzy. For the upper-body simulation, our code allows to run fake control boards using ROS controllers[9] and a Gazebo plugin that simulates the actual

---

[7] https://github.com/vislab-tecnico-lisboa/vizzy/tree/master/vizzy_yarp_icub/src/modules/vizzy_iKinCtrlGaze

[8] https://github.com/utexas-bwi/segbot_simulator

[9] http://wiki.ros.org/ros_controllers

control boards[10]. The motion planner interfaces for MoveIt[11] (i.e. configuration files) are available on the fake ROS controllers.

- The adaptive Monte-Carlo localization [4], available as the amcl ROS package[12]
- The Elastic Bands local-planner [13], available as a ROS package[13]
- The monocular ball tracking [16] and grasping YARP[14] modules[15]
- The Aggregated Channel Features (ACF) for pedestrian detection [15][16].

Vizzy's skills are implemented in two differente middlewares: YARP and ROS. This constraint pushed the development of the interoperability between the two middlewares to a working level, but like any interoperability, additional software needs to be written. Recently, we have been developing an automatic code generator that reduces the code to be written in the 'bridging step', which is in the testing phase now[17].

6   Conclusions and Future Work

We present the design guidelines of a humanoid robot on wheels: Vizzy. Its design based on a more organic approach and friendly postures have produced an upper body humanoid with range of motions very similar to the human ranges. The large wheels of the Segway mobile base inspired our team on a marsupial-like shape of the lower body cover. The head covers were designed to match the lower body covers, leading to large head covers that gives Vizzy its unique shape.

The mechanical design follows a modular approach, which facilitates the storage, replacing of damaged parts and production costs. All the modules have similar structure and parts, simplifying the assembly procedure of the robot.

The current skills of the robot include: (i) Indoors navigation using laser, (ii) reaching and grasping for simple shape objects, and (iii) pedestrian detection for human-robot interaction. Current developments include: (i) Complex manipulation skills by considering the contact sensors, and (ii) interoperability of ROS and YARP platforms for upper and lower body simultaneous control. The list of work to do includes: (i) The lights of the facial expressions, (ii) the closed loop for grasping using the touch sensors, (iii) the simultaneous control of the mobile base and upper body for reaching and grasping. Currently Vizzy has been chosen as a test platform for the CMU-Portugal project AHA[18].

---

[10] https://github.com/robotology/gazebo-yarp-plugins

[11] http://moveit.ros.org/

[12] http://wiki.ros.org/amcl

[13] http://wiki.ros.org/eband_local_planner

[14] http://wiki.icub.org/yarpdoc/index.html

[15] https://www.youtube.com/playlist?list=PLU2313o-fhv2grck4iEa5dR8Tpx2bYHZs

[16] https://www.youtube.com/watch?v=Hlfhw6ALJ-c

[17] https://github.com/vislab-tecnico-lisboa/yarp-bottle-generator

[18] http://aha.isr.tecnico.ulisboa.pt/

Acknowledgements This work was supported by FCT [UID/EEA/50009/2013], partially funded by POETICON++ [STREP Project ICT-288382], the FCT Ph.D. programme RBCog and FCT project AHA [CMUP-ERI/HCI/0046/2013].

References

1. Aragão, M., Moreno, P., Figueiredo, R.: Vizzy software repository (2014). https://github.com/vislab-tecnico-lisboa/vizzy
2. Asfour, T., Regenstein, K., Azad, P., Schröder, J., Dillmann, R.: ARMAR-III: a humanoid platform for perception-action integration. In: Proc., International Workshop on Human-Centered Robotic Systems (HCRS), Munich, pp. 51–56 (2006)
3. Fitzpatrick, P., Metta, G., Natale, L.: Towards long-lived robot genes. Robotics and Autonomous Systems **56**(1), 29–45 (2008)
4. Fox, D., Burgard, W., Dellaert, F., Thrun, S.: Monte carlo localization: efficient position estimation for mobile robots. In: AAAI/IAAI, pp. 343–349 (1999)
5. Fuchs, M., Borst, C., Giordano, P.R., Baumann, A., Kraemer, E., Langwald, J., Gruber, R., Seitz, N., Plank, G., Kunze, K., Burger, R., Schmidt, F., Wimboeck, T., Hirzinger, G.: Rollin' justin - design considerations and realization of a mobile platform for a humanoid upper body. In: ICRA Conf., pp. 4131–4137, May 2009
6. Hartenberg, R., Danavit, J.: Kinematic Synthesis of Linkages. McGraw-Hill, New York (1964)
7. Iwata, H., Sugano, S.: Design of human symbiotic robot twendy-one. Robotics and Automation, ICRA 2009, pp. 580–586 (2009)
8. Khandelwal, P., Yang, F., Leonetti, M., Lifschitz, V., Stone, P.: Planning in action language BC while learning action costs for mobile robots. In: ICAPS Conf. (2014)
9. Lopes, M., Beira, R., Praça, M., Santos-Victor, J.: An anthropomorphic robot torso for imitation: design and experiments. In: IROS Conf. IEEE (2004)
10. Parmiggiani, A., Maggiali, M., Natale, L., Nori, F., Schmitz, A., Tsagarakis, N., Santos-Victor, J., Becchi, F., Sandini, G., Metta, G.: The design of the iCub humanoid robot. International Journal of Humanoid Robotics **9**(4), 1250027 (2012)
11. Pattacini, U., Nori, F., Natale, L., Metta, G., Sandini, G.: An experimental evaluation of a novel minimum-jerk cartesian controller for humanoid robots. In: IROS Conference, pp. 1668–1674. IEEE (2010)
12. Pattacini, U.: Modular Cartesian Controllers for Humanoid Robots: Design and Implementation on the iCub. Ph.D. thesis, RBCS, IIT (2011)
13. Quinlan, S., Khatib, O.: Elastic bands: connecting path planning and control. In: ICRA Conference, vol. 2, pp. 802–807, May 1993
14. Smith, L.K., Weiss, E.L., Lehmkuhl, L.D.: Brunnstrom's clinical kinesiology, 5th edn. F.A. Davis, Philadelphia (1996)
15. Taiana, M., Nascimento, J., Bernardino, A.: On the purity of training and testing data for learning: the case of pedestrian detection. Neurocomputing **150**, 214–226 (2015). Part A(0)
16. Taiana, M., Santos, J., Gaspar, J., Nascimento, J., Bernardino, A., Lima, P.: Tracking objects with generic calibrated sensors: An algorithm based on color and 3D shape features. Robotics and Autonomous Systems **58**(6), 784–795 (2010). Omnidirectional Robot Vision

Building Fuzzy Elevation Maps
from a Ground-Based 3D Laser Scan
for Outdoor Mobile Robots

Anthony Mandow, Tomás J. Cantador, Antonio J. Reina, Jorge L. Martínez, Jesús Morales and Alfonso García-Cerezo

Abstract The paper addresses terrain modeling for mobile robots with fuzzy elevation maps by improving computational speed and performance over previous work on fuzzy terrain identification from a three-dimensional (3D) scan. To this end, spherical sub-sampling of the raw scan is proposed to select training data that does not filter out salient obstacles. Besides, rule structure is systematically defined by considering triangular sets with an unevenly distributed standard fuzzy partition and zero order Sugeno-type consequents. This structure, which favors a faster training time and reduces the number of rule parameters, also serves to compute a fuzzy reliability mask for the continuous fuzzy surface. The paper offers a case study using a Hokuyo-based 3D rangefinder to model terrain with and without outstanding obstacles. Performance regarding error and model size are compared favorably with respect to a solution that uses quadric-based surface simplification (QSlim).

Keywords Elevation maps · Mobile robots · 3D scanners · Fuzzy modeling

1 Introduction

Environment mapping is a key issue in mobile robots targeted to unstructured terrains [1, 2]. In this sense, three-dimensional (3D) laser scans provide valuable information for applications such as planetary exploration [3–5] or urban search and rescue [6, 7]. However, as point clouds require coping with a huge amount of spatial data, a simplified and compact representation of navigable terrain is necessary for both motion planning [8] and tele-operation [9].

A. Mandow(✉) · T.J. Cantador · A.J. Reina · J.L. Martínez · J. Morales · A. García-Cerezo
Dpto. Ingeniería de Sistemas y Automática,
Universidad de Málaga, Andalucía Tech, 29071 Málaga, Spain
e-mail: amandow@uma.es
http://www.uma.es/isa

© Springer International Publishing Switzerland 2016
L.P. Reis et al. (eds.), *Robot 2015: Second Iberian Robotics Conference*,
Advances in Intelligent Systems and Computing 417,
DOI: 10.1007/978-3-319-27146-0_3

Elevation maps offer a compact 2.5-dimensional model of terrain surface. In robotics, elevation has been generally represented by regular grids [10–12] or by irregular triangular meshes [4, 13]. Removal of artifacts and mesh simplification algorithms, like mesh decimation [14] and quadric-based polygonal surface simplification (QSlim) vertex clustering [15], can improve the compactness and reliability of these maps [8]. Nevertheless, as tesselated models have limitations in the face of incomplete and uncertain sensor data, some works have proposed using tools that can model elevation in a more robust and natural way. In geographic information systems (GIS), fuzzy logic has been applied to elevation maps for incorporating height uncertainty [16] and assessing visibility between pairs of points [17]. In mobile robotics, elevation grids have been processed with fuzzy rules to assess traversability [18, 19]. Gaussian processes have also been proposed to model terrain from uncertain and incomplete sensor data [20]. Adaptive Network-based Fuzzy Inference Systems (ANFIS) [21] has been employed to recognize objects like trees and buildings from aerial stereo images [22]. Moreover, ANFIS was preliminarily proposed to obtain fuzzy elevation maps of natural terrain from ground-based 3D data [23]. These elevation maps have been already employed for assessing traversability and local path planning [24].

The main contribution of this paper is extending [23] to improve computational speed and performance. To this end, two major modifications have been introduced: *i*) spherical sub-sampling [25] is used to select proper training data from a raw point cloud, and *ii*) rule parameters have been systematically defined by considering triangular sets with standard fuzzy partition [26] as well as zero order Sugeno-type inference. Furthermore, the capability to model a salient obstacle on the terrain surface has been considered in a case study using a 3D Hokuyo-based range-finder. Besides, performance regarding error and computation time are compared with a solution based on the QSlim algorithm.

Following this introduction, section 2 presents the fuzzy terrain modeling method. Then, section 3 describes the experiments including comparisons with QSlim. Finally, Section 4 offers conclusions and future work.

2 Fuzzy Terrain Modeling Method

The proposed fuzzy terrain modeling method is outlined in Fig. 1. The input is a range image from a single 3D scan. It is assumed that the local frame of the 3D rangefinder has its Y and Z axes pointing forwards and upwards, respectively (see Fig. 2). The goal of the method is producing a local elevation map in which ground surface can be represented as a function $z = H(x, y)$, where x and y are the Cartesian coordinates on the XY plane and z is the corresponding elevation. The following subsections develop the major parts of this algorithm.

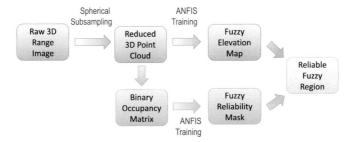

Fig. 1 Fuzzy point cloud terrain modeling algorithm.

2.1 Point Sub-sampling

The purpose of point sub-sampling is to produce a reduced set of representative training points in Cartesian coordinates (x, y, z). The reason for sub-sampling is twofold: improving the speed of fuzzy identification and homogenizing the spatial distribution of training data. In [23], sub-sampling was performed by selecting the highest point within grid cells of a sufficiently high resolution. This approach provides a representative set of points for a smooth terrain but is not so effective to model salient obstacles, such as tree trunks or rocks, which can be filtered by the fuzzy identification method when represented by a small number of samples. Furthermore, finding the maximum value within each cell entails the computational load of processing all scan points in Cartesian coordinates.

Alternatively, spherical sub-sampling [25] performs a fast range-independent data reduction for 3D rangefinders that combine a 2D scan with an additional rotation, as usually found in mobile robots [27]. Moreover, by equalizing the measure-direction density, spherical sub-sampling maintains a proportion of measurements accumu-

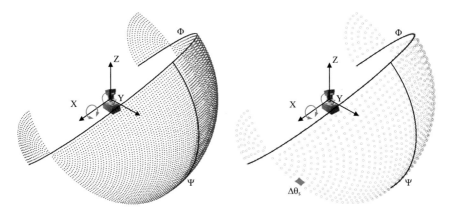

Fig. 2 Measure-direction density. Left: raw scan; right: spherical point sub-sampling.

lated by salient obstacles, so their contribution can be relevant when adjusting a fuzzy surface. Thus, Cartesian coordinates are only computed for sub-sampled points.

The field of view of the 3D sensor is defined by the scope of both scan angles: Φ for the 2D scan and Ψ for the additional rotation, as shown in Fig. 2(left). Then, angular resolution is given by $\Delta\phi$ and $\Delta\psi$ for the first and second rotations, respectively. The amount of data reduction of spherical sub-sampling is established by an equalization factor $0 < p \leq 1$. The equalized angular resolution of the sub-sampled scan (see Fig. 2(right)) is given by $\Delta\theta_s$ as:

$$\Delta\theta_s = \frac{\max(\Delta\psi, \Delta\phi)}{p}. \tag{1}$$

In this way, by setting $p = 1$, the measure-direction density is equalized with the coarsest resolution (either $\Delta\phi$ or $\Delta\psi$) actually provided by the sensor.

2.2 Neuro-Fuzzy Training

The set of sub-sampled Cartesian points is used to adjust a fuzzy model $z = H(x, y)$. Particularly, neuro-fuzzy ANFIS modeling [21] is adopted to identify a fuzzy model with Sugeno inference. ANFIS yields the rule consequents for a given fuzzy structure where membership functions (MFs) have to be specified for each input variable.

The definition of MFs has an impact on the model size; i.e., on the number of parameters. One interesting property of the standard fuzzy partition (SPF) is that triangular or trapezoidal MFs share parameter values with neighbor MFs [26]. In an SPF, the ith MF $\mathbf{F}_i$ defined for a given variable u in the universe U is convex, normal, and it only overlaps with its neighboring fuzzy sets, in such a way that:

$$\forall u \in U, \quad \sum_{\forall i} \mathbf{F}_i(u) = 1. \tag{2}$$

When defining the fuzzy structure, a greater number of MFs improves terrain adjustment at the cost of processing a greater number of rules. This conflict can be coped with by considering that higher detail is more important for the regions that are closest to the robot. Thus, an uneven membership function distribution can provide an appropriate fuzzy structure if the density of MFs for variables x and y is specified depending on the distance to the sensor [23].

Let us define the universe $U_x = [-u_{max}, u_{max}]$ for variable x, and $U_y = [0, u_{max}]$ for y, which corresponds to a $2u_{max} \times u_{max}$ rectangular area in the forward direction of the sensor. Then, uneven SPF MFs can be defined by computing the peak parameter f_i of each triangular MF $\mathbf{F}_i$ (i.e., $F_i(f_i) = 1$) as:

$$f_i = \text{sign}(i) \left(\frac{r^{|i|} - 1}{r^k - 1}\right) u_{max} \tag{3}$$

where $r > 1$ is the expansion ratio and $i = -k, ..., 0, 1, ..., k$ for the x variable and $i = 0, 1, ..., k$ for y. This definition yields $(2k + 1)$ MFs for x and $(k + 1)$ MFs for y. This means $(2k + 1) \times (k + 1)$ rules.

Furthermore, the order of Sugeno inference affects training time as well as the number of rule parameters. In [23], first-order consequents G_i were considered for every rule i as:

$$G_i(x, y) = a_i + b_i x + c_i y, \tag{4}$$

where three parameters (a_i, b_i, c_i) per rule had to be identified. Alternatively, using zero-order Sugeno consequents:

$$G_i(x, y) = a_i \tag{5}$$

requires just one parameter per rule, as $b_i = c_i = 0$. First order Sugeno can be useful for ome applications, like interpolating between piecewise linear controllers, but zero-order consequents can provide good accuracy when approximating nonlinear functions [28]. Furthermore, the output of a zero-order Sugeno model is smooth as long as neighboring MFs in the antecedents are overlapped, as in SPF [29]. Therefore, zero-order Sugeno consequents are proposed as an effective solution for this application.

2.3   Fuzzy Reliability Mask

ANFIS renders a fuzzy surface that can be evaluated for any (x, y) pair in the universe of discourse. This surface filters sensor noise and interpolates missing data from small shadowed areas but can provide erroneous estimations in larger regions with no input data. Since shadowed regions are frequent in ground-based scans of natural terrain proper use of the elevation map requires reliability assessment, which is a fuzzy concept in itself.

A fuzzy reliability mask is proposed as a continuous function $v = V(x, y)$, where $v \in [0, 1]$ for inputs $x \in U_x$ and $y \in U_y$. This function can be trained with ANFIS with the same fuzzy structure as in the elevation map. For this purpose, an occupancy binary matrix is computed from the sub-sampled points. This matrix represents an XY grid with a uniform resolution δ, where ones and zeroes are assigned to cells with or without points, respectively. Then, the training data consists of the set of all matrix values with the corresponding cell center XY coordinates. In the resulting fuzzy model, values of $v = V(x, y)$ close to one indicate high reliability of $H(x, y)$, whereas values close to zero mean unreliable regions.

3 Experimental Results

3.1 *Experimental Setup*

This section discusses the application of fuzzy elevation modeling using 3D range images from a natural terrain —see Fig. 3(left). Four different scans obtained from the same sensor pose are considered: without obstacles, and with a person standing at three different positions with respect to the sensor frame: *close-front*, at a distance of 2.94 m with approximate XY coordinates (0.05m, 2.94m); *close-side*, at 3.94 m with (3.49m, 1.83m); and *far*, at 7.26 m with (−0.14m, 7.26m). Computations have been performed by a QuadCore Intel Core i7 at 2.2 GHz under the Matlab environment, whose Fuzzy Toolbox includes ANFIS.

Range images have been obtained with a 3D laser scanner built by pitching a Hokuyo UTM-30LX 2D rangefinder [30] —see Fig. 3(right). This device has the following specifications: 270° × 135° field of view; measurable ranges between 0.1 m and 30 m; horizontal resolution $\Delta\phi = 0.25°$; and vertical resolution $\Delta\psi$ is adjustable from 0.067° to 4.24°. The scan times at minimum and maximum resolution are 1.54 s and 95.75 s, respectively. In particular, 3D scans have been obtained with $\Delta\psi = 0.278°$, which is similar to $\Delta\phi$, with the sensor standing 1.0 m above the ground. This configuration produces range images with a maximum of 505036 points in 12.43 s.

3.2 *Fuzzy Performance Evaluation*

Performance of the proposed method has been studied for different rule numbers and sub-sampling rates. Two different fuzzy rule structures are considered by using (3) with $r = 1.3$ and a universe of discourse $u_{max} = 10$ m: First, $k = 7$ yields 15×8 rules, and second, $k = 3$ results in 7×4 rules. As for training data, several p values are compared for spherical sub-sampling. Besides, Cartesian points whose coordinates fall out of the 20 m × 10 m rectangular area have been discarded. Fig. 4 illustrates training data for the close-front obstacle case with no sub-sampling and with $p = 0.5$.

Fig. 3 Experimental setup. Left: outdoor scene; right: 3D laser rangefinder.

Table 1 Effect of spherical sub-sampling on ANFIS performance.

Obstacle	p	Training points	No. of rules	RMSE (m²)	Time (s)	Obstacle	p	Training points	No. of rules	RMSE (m²)	Time (s)
None	-	186767	15 × 8	0.0237	176.00	Far	-	185673	15 × 8	0.0369	174.00
			7 × 4	0.0413	116.00				7 × 4	0.0558	117.00
	1	115423	15 × 8	0.0237	83.63		1	114768	15 × 8	0.0369	80.84
			7 × 4	0.0420	50.63				7 × 4	0.0563	47.84
	0.75	65019	15 × 8	0.0238	43.33		0.75	64666	15 × 8	0.0369	41.32
			7 × 4	0.0421	21.33				7 × 4	0.0564	20.32
	0.5	28916	15 × 8	0.0238	22.07		0.5	28770	15 × 8	0.0370	21.12
			7 × 4	0.0421	9.07				7 × 4	0.0564	9.12
	0.1	1180	15 × 8	0.2568	11.90		0.1	1168	15 × 8	0.4484	10.96
			7 × 4	0.0451	5.90				7 × 4	0.0642	5.96
Close-side	-	187600	15 × 8	0.0951	177.00	Close-front	-	189688	15 × 8	0.0871	174.00
			7 × 4	0.1283	115.00				7 × 4	0.1259	119.00
	1	115525	15 × 8	0.0955	80.87		1	117941	15 × 8	0.0879	87.65
			7 × 4	0.1295	48.87				7 × 4	0.1262	50.65
	0.75	65122	15 × 8	0.0955	40.33		0.75	66465	15 × 8	0.0872	44.26
			7 × 4	0.1295	20.33				7 × 4	0.1262	21.26
	0.5	28939	15 × 8	0.0956	21.07		0.5	29554	15 × 8	0.0872	23.05
			7 × 4	0.1295	9.07				7 × 4	0.1262	10.05
	0.1	1174	15 × 8	0.1580	10.89		0.1	1206	15 × 8	0.1568	11.19
			7 × 4	0.1305	4.89				7 × 4	0.1495	5.19

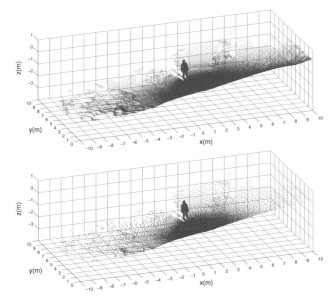

Fig. 4 Training points of the close-front obstacle scene. Top: raw scan points; Bottom: spherical sub-sampling with $p = 0.5$.

The root mean squared error (RMSE) given in Table 1 has been computed from the difference between the Z value of all raw scan points within the universe of discourse and the corresponding fuzzy model elevation. As expected, the RMSE for the 15×8 rulebase is better than the 7×4 model. Besides, fuzzy surfaces improve RMSE for smooth terrain, i.e., with no close salient obstacles. On the other hand, the sub-sampling rate has an important effect on computation time, but it does not generally produce important RMSE differences within the same rule structure and scene. This indicates that spherical sub-sampling gives a representative subset of the complete scan. The only exception to these results is with 15×8 rules and sub-sampling factor 0.1, where the 120-rule fuzzy model is overadjusted to a small data sample (around 1200 points, which is only about 0.6% of the raw data set).

From this analysis, if fast rough terrain model is needed, the 7×4 rule model with 0.1 sub-sampling could offer an appropriate solution. For a compromise between acceptable model accuracy and computation time, good results are given by the 15×8 rule model with 0.5 sub-sampling (see Fig. 5).

3.3 Comparison with QSlim

The proposed method has been compared with a QSlim-based solution. QSlim [15] performs mesh surface simplification by iterative contraction of vertex pairs using plane-based error quadrics. QSlim has been employed for terrain simplification of topographical maps [31, 32]. However, QSlim cannot be directly applied to the Delaunay mesh corresponding to raw ground-based scans because of artifacts due to shadows and salient objects [8]. Therefore, preprocessing is necessary to produce a proper surface to start with. In this paper, a proper Delaunay mesh is built from the XY coordinates of the maximum height points within grid cells of resolution $\delta = 0.1 \, \text{m}$. The resulting mesh has been fed to the QSlim algorithm to produce models with both 100 and 1000 faces, as illustrated by Fig. 6.

Table 2 compares results obtained with the proposed method, the solution proposed in [23], and the QSlim approach. QSlim is an efficient method, so most of the computation time corresponds to preprocessing. As this preprocessing is very similar to sub-sampling in [23], both methods achieve similar computation times. Spherical sub-sampling improves these times, which is particularly noticeable for 7×4 rules and $p = 0.1$. Regarding accuracy, the best results are obtained by the new method with 15×8 rules and $p = 0.5$, where RMSE is greatly improved with respect to [23] for the same number of rules. The accuracy of QSlim with 1000 faces is comparable to the best ANFIS performance with no close obstacles. However, the proposed method prevails when the model includes close obstacles. Table 2 also offers model size as the number of parameters. The proposed model requires one consequent parameter per rule. Triangular MFs are defined by three parameters, but as neighbor MFs share parameters in an SPF, only one parameter per MF is actually needed. These are improvements over [23], which required three consequent parameters and plus two values per Gaussian MF. As for QSlim, it consists of a list of vertices and a list of faces. Vertices are defined by three Cartesian coordinates and faces are specified as three vertex indices. The number of parameters of the proposed model improves [23] and clearly outperforms QSlim.

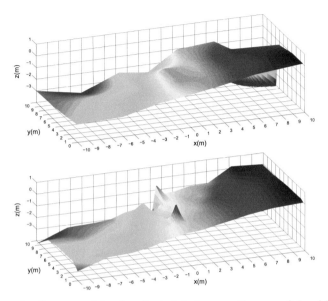

Fig. 5 Fuzzy elevation maps of the close-front obstacle scene. Top: $p = 0.1$ and 7×4 rules; bottom: $p = 0.5$ and 15×8 rules.

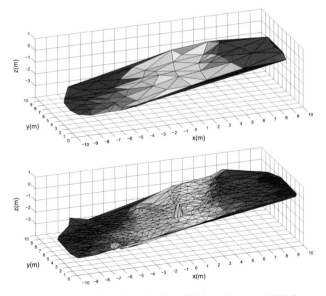

Fig. 6 QSlim maps of the close-front obstacle. Top: 100 faces; bottom: 1000 faces.

Table 2 Comparison between ANFIS and QSlim-Based Approaches.

Scene	Modeling method	Characteristics	RMSE (m^2)	Time (s)	Number of parameters
No obstacle	ANFIS	15×8, $p = 0.5$	0.0238	22.07	143
		7×4, $p = 0.1$	0.0451	5.90	39
		15×8 ([23])	0.0468	31.01	406
	QSlim	1000 faces	0.0250	29.17	4590
		100 faces	0.0357	29.17	483
Far obstacle	ANFIS	15×8, $p = 0.5$	0.0370	21.12	143
		7×4, $p = 0.1$	0.0642	5.96	39
		15×8 [23]	0.0660	31.20	406
	QSlim	1000 faces	0.0383	32.21	4572
		100 faces	0.0783	32.03	471
Close-side obstacle	ANFIS	15×8, $p = 0.5$	0.0956	21.07	143
		7×4, $p = 0.1$	0.1305	4.89	39
		15×8 [23]	0.1256	32.14	406
	QSlim	1000 faces	0.1349	29.14	4578
		100 faces	0.1959	29.14	477
Close-front obstacle	ANFIS	15×8, $p = 0.5$	0.0872	23.05	143
		7×4, $p = 0.1$	0.1495	5.19	39
		15×8 [23]	0.1303	33.07	406
	QSlim	1000 faces	0.1212	33.18	4569
		100 faces	0.2038	33.16	477

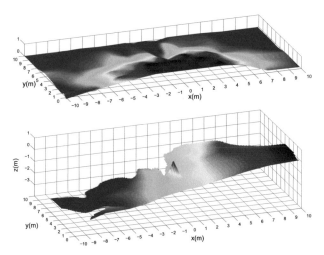

Fig. 7 Top: fuzzy reliability mask of the close-front obstacle scene with $p = 0.5$. Bottom: Application of the mask to the elevation map with 15×8 rules for $v \geq 0.1$.

3.4 Application of the Reliability Mask

ANFIS training of fuzzy elevation from data-less regions may not be reliable and can even create artifacts, like those around $(10\,m, 10\,m)$ in Fig. 5(top) or behind the person like in Fig. 5(bottom). Fig. 7 shows the fuzzy reliability mask computed for the close-front obstacle scene with $p = 0.5$ and $\delta = 0.1\,m$ as well as its to the ANFIS model of 15×8 rules with $v \geq 0.1$. The result is a reliable fuzzy elevation map without modelling artifacts.

4 Conclusions

This paper has addressed terrain modeling with fuzzy elevation maps from a ground-based 3D laser scan. Fuzzy surfaces provide a non-discrete representation of terrain elevation that greatly reduces the number of parameters with respect to raw point clouds. In particular, substantial performance improvements over [23] have been contributed. Proposed enhancements are spherical sub-sampling for training data selection, triangular membership functions with standard fuzzy partition, and zero order Sugeno-type inference. Furthermore, the problem of fuzzy surface artifacts in shadowed regions has been addressed by computing a fuzzy reliability mask from the same set of sub-sampled data.

Scans from a Hokuyo-based 3D rangefinder have been used to model natural terrain with and without salient obstacles. The proposed method outperforms QSlim and the previous ANFIS model in the adjustment to the original scan data, as indicated by the root mean squared error, especially with close obstacles. Besides, the new solution reduces significantly the number of model parameters. Future work will explore fuzzy identification with no predefined structure.

Acknowledgements This work was partially supported by the Spanish CICYT project DPI 2011-22443 and the Andalusian project PE-2010 TEP-6101.

References

1. Serón, J., Martínez, J.L., Mandow, A., Reina, A.J., Morales, J., García-Cerezo, A.: Automation of the arm-aided climbing maneuver for tracked mobile manipulators. IEEE Transactions on Industrial Electronics **61**(7), 3638–3647 (2014)
2. Santamaria-Navarro, A., Teniente, E.H., Morta, M., Andrade-Cetto, J.: Terrain classification in complex three-dimensional outdoor environments. Journal of Field Robotics **32**(1), 42–60 (2015)
3. Tong, C.H., Barfoot, T.D., Dupuis, E.: Three-dimensional SLAM for mapping planetary work site environments. Journal of Field Robotics **29**(3), 381–412 (2012)
4. Rekleitis, I., Bedwani, J.-L., Dupuis, E., Lamarche, T., Allard, P.: Autonomous over-the-horizon navigation using LIDAR data. Autonomous Robots, 1–18 (2012)

5. Ishigami, G., Otsuki, M., Kubota, T.: Path planning and navigation framework for a planetary exploration rover using a laser range finder. Springer Tracts in Advanced Robotics **92**, 431–447 (2014)
6. Nagatani, K., Tokunaga, N., Okada, Y., Yoshida, K., Hada, Y., Yoshida, T., Koyanagi, E.: Teleoperation of all-terrain robot using continuous acquisition of three-dimensional environment under time-delayed narrow bandwidth communication. In: IEEE Int. Workshop on Safety, Security & Rescue Robotics, Denver, USA (2009)
7. Birk, A., Pathak, K., Vaskevicius, N., Pfingsthorn, M., Poppinga, J., Schwertfeger, S.: Surface representations for 3D mapping. Künstliche Intelligenz **24**, 249–254 (2010)
8. Gingras, D., Lamarche, T., Bedwani, J.L., Dupuis, E.: Rough terrain reconstruction for rover motion planning. In: Proc. Canadian Conference on Computer and Robot Vision, Ottawa, Canada, pp. 191–198 (2010)
9. Vaskevicius, N., Birk, A., Pathak, K., Schwertfeger, S.: Efficient representation in three-dimensional environment modeling for planetary robotic exploration. Advanced Robotics **24**(8–9), 1169–1197 (2010)
10. Pfaff, P., Triebel, R., Burgard, W.: An efficient extension to elevation maps for outdoor terrain mapping and loop closing. Int. Journal of Robotics Research **26**(2), 217–230 (2007)
11. Ishigami, G., Nagatani, K., Yoshida, K.: Path planning and evaluation for planetary rovers based on dynamic mobility index. In: IEEE Int. Conf. on Intelligent Robots and Systems, pp. 601–606 (2011)
12. Souza, A., Gonçalves, L.M.G.: Occupancy-elevation grid: an alternative approach for robotic mapping and navigation. Robotica (2015)
13. Gerbaud, T., Polotski, V., Cohen, P.: Simultaneous exploration and 3D mapping of unstructured environments. In: IEEE Int. Conf. on Systems, Man and Cybernetics, vol. 6, pp. 5333–5337 (2004)
14. Ciampalini, A., Cignoni, P., Montani, C., Scopigno, R.: Multiresolution decimation based on global error. The Visual Computer **13**, 228–246 (1997)
15. Garland, M., Heckbert, P.S.: Surface simplification using quadric error metrics. In: ACM SIGGRAPH Conf. on Computer Graphics, pp. 209–216 (1997)
16. Santos, J., Lodwick, W.A., Neumaier, A.: A new approach to incorporate uncertainty in terrain modeling. In: 2nd Int. Conf. on Geographic Information Science, Boulder, USA, pp. 291–299 (2002)
17. Anile, M., Furno, P., Gallo, G., Massolo, A.: A fuzzy approach to visibility maps creation over digital terrains. Fuzzy Sets and Systems **135**, 63–80 (2003)
18. Liu, H., Yang, J., Zhao, C.: A generic approach to rugged terrain analysis based on fuzzy inference. In: 8th Int. Conf. on Control, Automation, Robotics and Vision, vol. 2, pp. 1108–1113 (2004)
19. Gu, J., Cao, Q., Huang, Y.: Rapid traversability assessment in 2.5D grid-based map on rough terrain. Int. Journal of Advanced Robotic Systems **5**(4), 389–394 (2008)
20. Vasudevan, S., Ramos, F., Nettleton, E., Durrant-Whyte, H.: Gaussian process modeling of large scale terrain. Journal of Field Robotics **26**(10), 812–840 (2009)
21. Jang, J.-S.R.: ANFIS: Adaptative-network-based fuzzy inference system. IEEE Transactions on Systems, Man and Cybernetics **23**(3), 665–685 (1993)
22. Samadzadegan, F., Azizi, A., Hahn, M., Lucas, C.: Automatic 3D object recognition and reconstruction based on neuro-fuzzy modelling. ISPRS Journal of Photogrammetry and Remote Sensing **59**(5), 255–277 (2005)
23. Mandow, A., Cantador, T.J., García-Cerezo, A., Reina, A.J., Martínez, J.L., Morales, J.: Fuzzy modeling of natural terrain elevation from a 3D scanner point cloud. In: 7th IEEE Int. Symposium on Intelligent Signal Processing (WISP), Floriana, Malta (2011)
24. Martínez, J.-L., Mandow, A., Reina, A., Cantador, T.-J., Morales, J., García-Cerezo, A.: Navigability analysis of natural terrains with fuzzy elevation maps from ground-based 3D range scans. In: Proc. IEEE/RSJ International Conference on Intelligent Robots and Systems, Tokyo, Japan, pp. 1576–1581 (2013)

25. Mandow, A., Martínez, J.L., Reina, A.J., Morales, J.: Fast range-independent spherical sub-sampling of 3D laser scanner points and data reduction performance evaluation for scene registration. Pattern Recognition Letters **31**(11), 1239–1250 (2010)
26. Xiu, Z.-H., Ren, G.: Stability analysis and systematic design of Takagi-Sugeno fuzzy control systems. Fuzzy Sets and Systems **151**(1), 119–138 (2005)
27. Morales, J., Martínez, J.L., Mandow, A., Reina, A.J., Pequeño-Boter, A., García-Cerezo, A.: Boresight calibration of construction misalignments for 3D scanners built with a 2D laser rangefinder rotating on its optical center. Sensors **14**(11), 20 025–20 040 (2014)
28. Derbel, N.: A discussion on Sugeno fuzzy logic approximations of nonlinear systems. In: 5th Int. Multi-Conf. on Systems, Signals and Devices, Amman, Jordan (2008)
29. Jang, J.-S.R., Sun, C.-T., Mizutani, E.: Neuro-fuzzy and soft computing: a computational approach to learning and machine intelligence. Prentice Hall (1997)
30. Morales, J., Martínez, J.L., Mandow, A., Pequeño-Boter, A., García-Cerezo, A.: Design and development of a fast and precise low-cost 3D laser rangefinder. In: IEEE Int. Conf. on Mechatronics, Istanbul, Turkey, pp. 621–626 (2011)
31. Little, C., Peters, R.: Simulated mobile self-location using 3D range sensing and an a-priori map. In: Proc. IEEE International Conference on Robotics and Automation, Barcelona, Spain, vol. 2005, pp. 1459–1464 (2005)
32. Biniaz, A.: Slope preserving terrain simplification - an experimental study. In: Proc. Canadian Conference on Computational Geometry, Vancouver, Canada, pp. 59–62 (2009)

Physics-Based Motion Planning: Evaluation Criteria and Benchmarking

Muhayyuddin Gillani, Aliakbar Akbari and Jan Rosell

Abstract Motion planning has evolved from coping with simply geometric problems to physics-based ones that incorporate the kinodynamic and the physical constraints imposed by the robot and the physical world. Therefore, the criteria for evaluating physics-based motion planners goes beyond the computational complexity (e.g. in terms of planning time) usually used as a measure for evaluating geometrical planners, in order to consider also the quality of the solution in terms of dynamical parameters. This study proposes an evaluation criteria and analyzes the performance of several kinodynamic planners, which are at the core of physics-based motion planning, using different scenarios with fixed and manipulatable objects. RRT, EST, KPIECE and SyCLoP are used for the benchmarking. The results show that KPIECE computes the time-optimal solution with heighest success rate, whereas, SyCLoP compute the most power-optimal solution among the planners used.

Keywords Physics-based motion planning · Benchmarking · Kinodynamic motion planning

1 Introduction

Robotic manipulation requires precise motion planning and control to execute the tasks, either for industrial robots, mobile manipulators, or humanoid robots. It is necessary to determine the way of safely navigating the robot from the start to the goal state by satisfying the kinodynamic (geometric and differential) constraints, as

M. Gillani · A. Akbari · J. Rosell(✉)
Institute of Industrial and Control Engineering,
Universitat Politècnica de Catalunya, Barcelona, Spain
e-mail: {muhayyuddin.gillani,aliakbar.akbari,jan.rosell}@upc.edu

J. Rosell—This work was partially supported by the Spanish Government through the projects DPI2011-22471, DPI2013-40882-P and DPI2014-57757-R. Muhayyuddin is supported by the Generalitat de Catalunya through the grant FI-DGR 2014.

43
L.P. Reis et al. (eds.), *Robot 2015: Second Iberian Robotics Conference*,
Advances in Intelligent Systems and Computing 417,
DOI: 10.1007/978-3-319-27146-0_4

well as to incorporate the physics-based constraints imposed by possible contacts and by the dynamic properties of the world such as gravity and friction [1, 2]. These issues significantly increase the computational complexity because certain collision-free geometric paths may not be feasible in the presence of these constraints.

Physics-based motion planning has emerged, therefore, as a new class of planning algorithms that considers the physics-based constraints along with the kinodynamical constraints, i.e. it is an extension to the kinodynamic motion planning [3] that also involves the purposeful manipulation of the objects by considering the dynamical interaction between rigid bodies. This interaction is simulated based on the principal of basic Newtonian physics and the results of simulation are used for planning. The performance of a physics-based planner largely depends on the choice of the kinodynamic motion planner, that is implicitly used for sampling the states and the construction of the solution path. The state propagation is performed using dynamic engines, like ODE [4], that incorporates the kinodynamical and physics-based constraints.

Motion planning in its simplest form (i.e. as a geometric problem) is PSPACE-complete [5]. The incorporation of kinodynamic constraints and the physics-based properties make it even more complex and computationally intensive, and for complex systems even the decidability of the physics-based planning problem is questionable [6]. Therefore, to make the physics-based planning computationally tractable, it is crucial to use the most appropriate and computationally efficient kinodynamic planner. In previously proposed physics-based planning approaches, different kinodynamic motion planners and physics engines have been used.

A few studies provided comparative analysis of the performance of some kinodynamical motion planners within different physics-based planning frameworks. For instance, the physics-based planning algorithm proposed in [7], that used nondeterministic tactics and skills to reduce the search space of physics-based planning, was evaluated (in term of planning time and tree length) using two different kinodynamic motion planners, Behavioral Kinodynamic Rapidly-exploring Random Trees (BK-RRT [8]) and Balanced Growth Trees (BGT [7]). The physics engine PhysX [9] was used as state propagator. Another physics-based planning approach [10] integrated the sampling-based motion planing with the discrete search using the workspace decomposition in order to map the planning problem onto a graph searching problem. This work evaluated the performance (in term of planning time) using RRT, Synergistic Combination of Layers of Planning (SyCLoP [11]) and a modified version of the SyCLoP as kinodynamic motion planners. The propagation step was performed using the Bullet [12] physics engine. A third approach proposed a physics-based motion planning framework that used manipulation knowledge coded as an ontology [13]. This approach performed a reasoning process over the knowledge to improve the computational efficiency and has shown a significant improvement in performance (in term of planning time and generated trajectory), as compared to the simple physics-based planning. Two kinodynamic motion planners were used, Kinodynamic Planning by Interior-Exterior Cell Exploration (KPIECE [14]) and RRT. The Open Dynamics Engine (ODE) was used as state propagator.

All the above stated studies basically measured the time complexity of different kinodynamic motion planners. Since physics-based planning simultaneously evaluates the kinodynamical and the physics-based constraints, the evaluation based on just planning time may not be sufficient. New evaluation criteria is required because a number of other dynamical parameters (such as power consumed, action, smoothness) may significantly influence the planning decisions, like in the task planning approaches proposed in [15, 16] that use a physics-based reasoning process to determine the feasibility of a plan by evaluating the cost in terms of power consumed and the action. With this in mind, the present study proposes a new benchmarking criteria for the physics-based planning that incorporates the dynamical properties of the system (to determine the quality of the solution) as well as the computational complexity. It is used to compare different kinodynamic motion planners (RRT, EST, KPIECE and SyCLoP) within the physics-based planning framework presented in [13] based on a reasoning process over ontological manipulation knowledge.

2   Kinodynamic Motion Planning

Motion planning problems deal with computing collision-free trajectories from a given start to a goal state in the configuration space (C), the set of all possible configurations of the robot [17]. The geometrically accessible region of C is called C_{free} and the obstacle region is known as C_{obs}. The sampling-based algorithms such as Probabilistic RoadMaps [18] and the Rapidly-exploring Random Trees [19] have shown significant performance when planning in high-dimensional configuration spaces. These algorithms connect collision-free configurations with either a graph or a tree to capture the connectivity of C_{free} and find a path along these data structures to connect the initial and the goal configurations.

Kinodynamic motion planning refers to the problems in which the motion of the robot must simultaneously satisfy the kinematic constraints (such as joint limits and obstacle avoidance) as well as some dynamic constraints (such as bounds on the applied forces, velocities and accelerations [20]). Tree-based planners are best suited to take into account kynodynamic constraints [1], since the dynamic equations are used to determine the resulting motions used to grow the tree. The general functionality of sampling-based kinodynamic planners is to search a state space S of higher dimensions that records the system's dynamics. The state of a robot for a configuration $q \in C$ is defined as $s = (q, \dot{q})$. To determine a solution, the planning will be performed in state space, in a similar way as in C. This section briefly reviews the existing most commonly used kinodynamic motion planners, that can be categorized into three classes: a) RRT and EST belong to the class of planning algorithms that sample the states; b) KPIECE belongs to the class that samples the motions or path segments; c) SyCLoP is an hybrid planner that splits the planning problem into a discrete and a continuous layer.

Rapidly Exploring Random Trees (RRT): It is a sampling-based kinodynamic motion planning algorithm [21] that has the ability to efficiently explore the high dimensional configuration spaces. The working mechanism of RRT-based algorithms is to randomly grow a tree rooted at the start state ($q_{start} \in C$), until it finds a sample at the goal state ($q_{goal} \in C$). The growth of the tree is based on two steps, *selection* and *propagation*. In the first step a sample is randomly selected (q_{rand}), and its nearest node in the tree is then searched (q_{near}). The second step applies, from q_{near}, random controls (that satisfy the constraints) during a certain amount of time. Among the configurations reached, the one nearest to q_{near} is selected as q_{new} and an edge from q_{near} to q_{new} is added to the tree. Using this procedure, all the paths on the tree will be feasible, i.e. by construction they satisfy all the kinodynamic constraints.

Expansive-Spaces Tree Planner (EST): This approach constructs a tree-shaped road-map T in the *state×time* space [22] [23]. The idea is to select a milestone of T and from there randomly apply sampled controls for a certain amount of time. If the final state is in free-space, it will be added as a milestone in T. The selection of the milestone for expansion is done in a way that the resultant tree should neither be too dense nor sparse. This kinodynamic planner works in three steps: *milestone selection*, *control selection* and *endgame connection*. In the first step, a milestone m in T is selected with probability inversely proportional to the number of neighboring milestones of m. In the second step, controls are randomly sampled and applied from the selected milestone m. Since by moving under kinodynamic constraints it may not be possible to reach exactly the goal state, the *endgame connection* step is the final step that defines a region around the goal in such a way that any milestone within this region is considered as the goal state.

Kinodynamic Motion Planning by Interior-Exterior Cell Exploration (KPIECE): This planner is particularly designed for complex dynamical systems. KPIECE grows a tree of motions by applying randomly sampled controls for a randomly sampled time duration from a tree node selected as follows. The state space is projected onto a lower-dimensional space that is partitioned into cells in order to estimate the coverage. As a result of this projection, each motion will be part of a cell, each cell being classified as an interior or exterior cell depending on whether the neighboring cells are occupied or not. Then, the selection of the cell is performed based on the *importance* parameter that is computed based on: 1) the *coverage* (the cells that are less covered are preferred over the others); 2) the *selection* (the cells that have been selected less number of times are preferred); 3) the *neighbors* (the cells that have less neighbors are preferred); 4) the *selection time* (recently selected cells are preferred); 5) the *expansion* (easily expanded cells are preferred over the cells that expand slowly). The cell that has maximum importance will be selected. The process continues until the tree of motion reaches the goal region.

Synergistic Combination of Layers of Planning (SyCLoP): This is a meta approach that considers motion planning as a search problem in a hybrid space (of a continuous and a discrete layer) for efficiently solving the problem under kinody-namical constraints. The continuous layer is represented by the state space (that is explored by a sampling-based motion planner like RRT or EST) and the discrete layer is determined by the decomposition of the workspace. The decomposition is used to compute a cost parameter called *lead* that guides the motion planner towards the goal. SyCLoP works based on the following steps: *lead computation* and *region selection*. The lead is computed based on the coverage and the frequency of the selection. The former is obtained by the sampling-based motion planner (continuous layer) and the latter is computed by determining how many times a cell has been selected from discrete space. The selection of the region will be performed based on the available free volume of the region (high free volume regions are preferred for the exploration). The process continues until the planner finds a sample in the goal region. SyCLoP will be recalled SyCLoP-RRT or SyCLoP-EST based on the planner used in the continuous layer.

2.1 Ontological Physics-Based Motion Planning

Physics-based motion planning is composed of a new class of planning algorithms that basically go a step further towards physical realism, by also taking into account possible interactions between bodies and possible physics-based constraints (such as gravity and friction that conditions the actions and its results). The search of collision-free trajectories is not the final aim; now collisions with some objects may be allowed, i.e. these algorithms also consider the manipulation actions (such as push action) in order to compute the appropriate trajectory. The incorporation of the dynamic interaction (for manipulation) between rigid bodies and other physics-based constraints increase the dimensionality of the state space and the computational complexity. In some cases, particularly for the systems with complex dynamics, the problem may even be not tractable.

The ontological physics-based motion planning is a recently proposed approach that tries to cope with aforementioned challenges [13]. This approach takes the advantage of Prolog-based reasoning process over the knowledge of objects and manipulation actions (this knowledge is represented in the form of an ontology). The reasoning process is used to improve the computational efficiency and to make the manipulation problem computationally tractable. It applies a hybrid approach consisting of two main layers which are a knowledge-based reasoning layer and the motion planning layer.

The knowledge-based reasoning layer uses the manipulation ontology to derive a knowledge, called abstract knowledge, that contains information of the objects and their properties (such as their manipulatable regions, e.g. the regions from where an object can be pushed), and the initial and goal states of the robot. The abstract knowledge, moreover, categorizes the objects into fixed and manipulatable objects, being the manipulatable objects further divided into freely and constraint-oriented manipulatable ones (e.g. some objects can be pushed from any region while others may only be pushed from some given region and in some predefined directions). Furthermore, abstract knowledge also determines the geometrical positions of the objects to distinguish whether the goal state is occupied or not.

The motion planning layer includes a reasoning process that infers from the abstract knowledge. The inferred knowledge is called instantiated knowledge and is a dynamic knowledge that is updated at each instance of time. Motion planning layer employs a sampling-based kinodynamic motion planner (like KPIECE or RRT) and a physics engine used as state propagator. After the propagation step, the new state is accepted by the planner if all the bodies satisfy the manipulation constraints imposed by the instantiated knowledge (e.g. a car-like object can only be pushed forward or backward and therefore any state that results with a collision with its lateral sides is disallowed). In this way the growing of the tree-like data structure of the planner is more efficient since useless actions are pruned.

3 Benchmarking Parameters

A wide variety of the kinodynamic motion planners is available, the planning strategy of these algorithms is conceptually different from one another, and this can significantly affect the performance of the physics-based planning. A criteria is proposed to evaluate the performance of the kinodynamic motion planners for the physics-based planning. It is suggested that, along with the computational complexity, it is also necessary to evaluate the dynamical parameters that determine the quality of the computed solution path. This is determined by estimating the power consumed by the robot while moving along the computed path, by the total amount of action (i.e. dynamic attribute of trajectory explained later in this section) of the computed trajectory and by the smoothness of the trajectory. The computational complexity is computed based on the planning time and the average success rate of each planner. A planner is said to be most appropriate if it is optimal according to the above said criteria.

The choice of physics engine (such as ODE, Bullet and PhysX) may not affect the simulation results because the design philosophy of all of them is based on the basic physics, and the performance and accuracy may vary a little but the simulation results of all the physics engines are almost the same.

The following performance parameters have been established to evaluate different kinodynamic motion planners within the framework of the ontological physics-based motion planner. The trajectories given by the kinodynamic motion planners are described by a list of forces and their duration that have to be consecutively applied to move the robot (either in a collision-free way or possibly pushing some manipulatable objects):

- **Action:** It is a dynamical property of a physical system, defined in the form of a functional $\mathcal{A}$ that takes a sequence of moves that define a trajectory as input, and returns a scalar number as output:

$$\mathcal{A} = \sum_{i}^{n} |\mathbf{f_i}| \Delta t_i \varepsilon_i, \tag{1}$$

where $\mathbf{f}_i$, Δt_i and ε_i are, respectively, the applied control forces, their duration and the resulting covered distances.

– **Power Consumed:** The total amount of power consumed $\mathcal{P}$ by the robot to move from start to the goal state is computed as:

$$\mathcal{P} = \sum_{i}^{n} \frac{\mathbf{f_i d_i}}{\Delta t_i}, \tag{2}$$

where $\mathbf{f}_i$, $\mathbf{d}_i$ and Δt_i are, respectively, the applied control forces, the resultant displacement vectors and the time duration.

– **Smoothness:** The smoothness $\mathcal{S}$ of a trajectory can be measured as a function of jerk [24], the time derivative of acceleration:

$$\mathcal{J}(t) = \frac{d\,a(t)}{dt}. \tag{3}$$

For a given trajectory τ the smoothness is defined as the sum of squared jerk along τ:

$$\mathcal{S} = \int_{t_i}^{t_f} \mathcal{J}(t)^2\, dt, \tag{4}$$

where t_i and t_f are the initial and final time, respectively.

– **Planning Time:** It is the total time consumed by the ontological physics-based planner to compute a solution trajectory.
– **Success Rate:** It is computed based on the number of successful runs.

4   Results and Discussion

In this section the results of benchmarking of kinodynamical motion planners for ontological physics-based planning is presented. The benchmarking is performed using *The Kautham Project* [25], a C++ based open source platform for motion planning that includes geometric, kinodynamic, and ontological physics-based motion planners. It uses the planning algorithms offered by the Open Motion Planning Library (OMPL) [26], an C++ based open source motion planning library, and ODE as state propagator for the physics-based planning.

4.1 Simulation Setup

The simulation setup consists of a robot (green sphere), free manipulatable bodies (blue cubes), constraint oriented manipulatable bodies (purple cubes), and the fixed bodies (triangular prisms, walls, and floor). The benchmarking is performed with the three different scenes shown in Fig. 1, that are differentiated based on the degree of clutter. The robot is depicted at its initial configuration, being the goal robot configuration painted as a yellow circle. Fig. 1-a describes the simplest scene that consists of a robot, free manipulatable bodies, and fixed bodies. The second scene, represented in Fig. 1-b, consists of a robot, free manipulatable bodies, fixed bodies, and a constraint-oriented body. In this scene the narrow passage is occupied by the constraint-oriented manipulatable body (it can only be pushed vertically, along y-axis) that has to be pushed away by the robot in order to clear the path towards the goal. It is important to note that since there is not any collision-free path available from the start to the goal state, the geometric as well as the kinodynamic planners are not able to compute the path; only physics-based planners has the ability to compute the path by pushing the object away. The final scene is depicted in Fig. 1-c and has the highest degree of clutter. The goal is occupied with a constraint-oriented manipulatable body (it can only be pushed horizontally, along x-axis); in order to reach the goal region the robot needs to free it by pushing the body away. As before, no collision-free path exists. The same planning parameters are used for all the planners: goal bias equal to 0.05, sampling control range between -10N and 10N, and propagation step size equal to 0.07s.

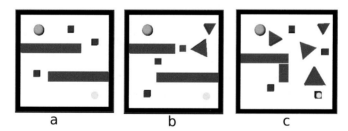

Fig. 1 Simulation setup used for the benchmarking. The robot (green sphere) is depicted at the start configuration while the yellow circle represents the goal.

4.2 Benchmarking Results

The ontological physics-based motion planner is run 10 times for each scene and for each of the kinodynamic motion planners summarized in Section 2. The average values of the benchmarking parameters are presented in the form of histograms. Fig. 2 shows as sequence of snapshots of a sample execution of the ontological physics-based motion planner using KPIECE. In order to estimate the coverage of configuration space, and the solution trajectory Fig. 3 depicts the configuration spaces

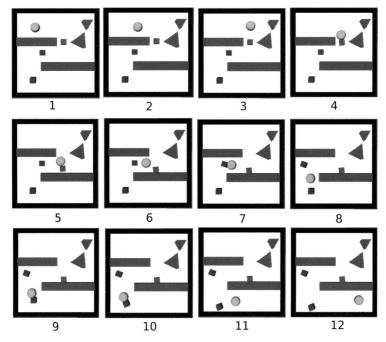

Fig. 2 Sequence of snapshots of the execution using ontological physics-based motion planner.

and solution paths computed by different kinodynamic planners. Fig. 4 shows the average planning time (being the maximum allowed planning time set to 500 s). All the planners have been able to compute the solution within the maximum range of planning time except EST. Among all the planners KPIECE computed the solution most efficiently. The success rate of the planners is described in Fig. 5. It is computed for each scene individually based on the number of successful runs (i.e. how many times the planner computes the solution within the maximum limit of time), then the average success rate of each planner is determined by computing the average successful runs for the three scenes. Results show that the KPIECE has the highest overall success rate. The SyCLoP-RRT also shows an impressive success rate, whereas, EST has zero success rate.

The results of the dynamical parameters, action and power, are shown in Fig. 6 and Fig. 7, respectively. Regarding action, the solution trajectory with minimum amount of action is considered as the most appropriate. Among the planners that computed the solution within the given time, on average, KPIECE has the minimum amount of action and SyCLoP-RRT has the maximum one. Regarding power, it is desirable that the robot should consume a minimum amount of power while moving along the solution. Our analysis shows that SyCLoP-RRT finds the power-optimal trajectory whereas KPIECE is the worst. The results for RRT and SyCLoP-EST are almost the same. Regarding trajectory smoothness, Fig. 8 shows the comparison.

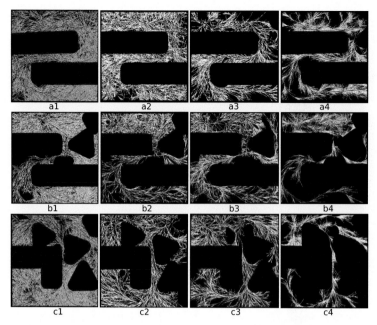

Fig. 3 Configuration space and solution path: each row corresponding to the scene, the columns in each row (left to right) represents the configuration space and solution path using KPIECE, RRT, SyCLoP-RRT, and SyCLoP-EST respectively.

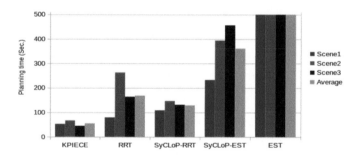

Fig. 4 Average planning time (10 runs) for each scene and the overall averaged planning time (three scenes).

The SyCLoP-RRT computes the most smooth trajectory among all the planners whereas, KPIECE show the worst results in term of smoothness. Since EST was not able to compute the solution within the given time, the action, power and smoothness for EST are set to infinity and not shown in the histograms.

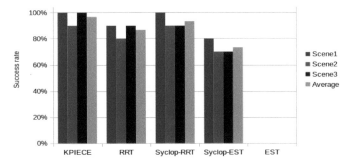

Fig. 5 Success rate of the planners for each scene and the overall averaged success rate (three scenes).

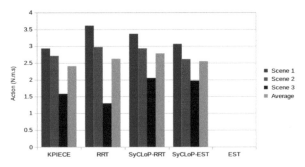

Fig. 6 Average amount of action for each scene and the overall averaged amount of action (three scenes).

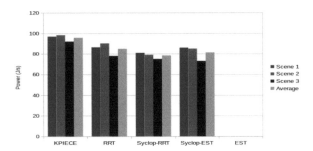

Fig. 7 Average power consumed by the robot while moving along the solution path and the overall average (three scenes).

4.3 Discussion

We proposed a benchmarking criteria for physics-based planning. Based on the proposed criteria, the performance of physics-based planning is evaluated using different kinodynamic motion planners. Our analysis shows that in terms of planning time, success rate and the action value, KPIECE is the most suitable planner. SyCLoP-RRT

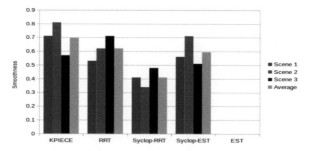

Fig. 8 Smoothness measure for each scene and the overall average (three scenes).

shows significant results in terms of smoothness of the computed trajectory and power consumed by the robot moving along the computed path. The planning time and success rate for SyCLoP-RRT is also impressive. RRT shows average results throughout the evaluation. SyCLoP-EST is good in terms of action value and power consumption but the value of planning time is very high and it has a low success rate. The EST was not able to compute the solution within time and has zero success rate.

5 Conclusion and Future Work

This paper proposed an evaluation criteria for the physics-based motion planners. The proposed benchmarking criteria computes dynamical parameters (such as power consumed by the robot to move along the solution path) for the evaluation of the quality of the computed solution path. Further, based on the proposed benchmarking criteria, the performance of the ontological physics-based motion planner (using different kinodynamic motion planners) is evaluated and the computed properties of the each kinodynamic motion planner are discussed in detail. For now the evaluation criteria was implemented on simple scenes and with the push action as the sole manipulation action; as future work the proposed benchmarking criteria will be implemented for mobile manipulators to also benchmark the grasping and pick and place manipulation actions.

References

1. Tsianos, K.I., Sucan, I.A., Kavraki, L.E.: Sampling-based robot motion planning: Towards realistic applications. Computer Science Review **1**(1), 2–11 (2007)
2. Ladd, A.M., Kavraki, L.E.: Motion planning in the presence of drift, underactuation and discrete system changes. In: Robotics: Science and Systems, pp. 233–240 (2005)
3. Zickler, S., Veloso, M.M.: Variable level-of-detail motion planning in environments with poorly predictable bodies. In: Proc. of the European Conf. on Artificial Intelligence Montpellier, pp. 189–194 (2010)
4. Russell, S.: Open Dynamic Engine (2007). http://www.ode.org/

5. Reif, J.H.: Complexity of the move's problem and generalizations. In: Proc. of the 20th Annual IEEE Conf. on Foundations of Computer Science, pp. 421–427 (1979)
6. Cheng, P., Pappas, G., Kumar, V.: Decidability of motion planning with differential constraints. In: Proc. IEEE Int. Conf. on Robotics and Automation, pp. 1826–1831 (2007)
7. Zickler, S., Veloso, M.: Efficient physics-based planning: sampling search via non-deterministic tactics and skills. In: Proc. of The 8th Int. Conf. on Autonomous Agents and Multiagent Systems, vol. 1, pp. 27–33 (2009)
8. Zickler, S., Veloso, M.: Playing creative soccer: randomized behavioral kinodynamic planning of robot tactics. In: RoboCup 2008: Robot Soccer World Cup XII, pp. 414–425. Springer (2009)
9. NVIDIA: Physx. https://developer.nvidia.com/physx-sdk
10. Plaku, E.: Motion planning with discrete abstractions and physics-based game engines. In: Proc. of the Int. Conf. on Motion in Games, pp. 290–301. Springer (2012)
11. Plaku, E., Kavraki, L., Vardi, M.: Motion planning with dynamics by a synergistic combination of layers of planning. IEEE Tran. on Robotics **26**(3), 469–482 (2010)
12. Erwin, C.: Bullet physics library (2013). http://bulletphysics.org
13. Muhayyudin, Akbari, A., Rosell, J.: Ontological physics-based motion planning for manipulation. In: Proc. of IEEE Int. Conf. on Emerging Technologies and Factory Automation (ETFA) (2015)
14. Sucan, I., Kavraki, L.E.: A sampling-based tree planner for systems with complex dynamics. IEEE Transactions on Robotics **28**(1), 116–131 (2012)
15. Akbari, A., Muhayyudin, Rosell, J.: Task and motion planning using physics-based reasoning. In: Proc. of the IEEE Int. Conf. on Emerging Technologies and Factory Automation (2015)
16. Akbari, A., Muhayyuddin, Rosell, J.: Reasoning-based evaluation of manipulation actions for efficient task planning. In: ROBOT2015: Second Iberian Robotics Conference. Springer (2015)
17. Lozano-Pérez, T.: Spatial Planning: A Configuration Space Approach. IEEE Trans. on Computers **32**(2), 108–120 (1983)
18. Kavraki, L., Svestka, P., Latombe, J.C., Overmars, M.: Probabilistic roadmaps for path planning in high-dimensional configuration spaces. IEEE Trans. on Robotics and Automation **12**(4), 566–580 (1996)
19. Lavalle, S.M., Kuffner, J.J.: Rapidly-exploring random trees: progress and prospects. In: Algorithmic and Computational Robotics: New Directions, pp. 293–308 (2001)
20. Donald, B., Xavier, P., Canny, J., Reif, J.: Kinodynamic motion planning. Journal of the ACM **40**(5), 1048–1066 (1993)
21. LaValle, S.M., Kuffner, J.J.: Randomized kinodynamic planning. The Int. Journal of Robotics Research **20**(5), 378–400 (2001)
22. Hsu, D., Latombe, J.C., Motwani, R.: Path planning in expansive configuration spaces. In: Proc. of the IEEE Int. Conf. on Robotics and Automation, vol. 3, pp. 2719–2726. IEEE (1997)
23. Hsu, D., Kindel, R., Latombe, J.C., Rock, S.: Randomized kinodynamic motion planning with moving obstacles. The Int. Journal of Robotics Research **21**(3), 233–255 (2002)
24. Hogan, N.: Adaptive control of mechanical impedance by coactivation of antagonist muscles. IEEE Trans. on Automatic Control **29**(8), 681–690 (1984)
25. Rosell, J., Pérez, A., Aliakbar, A., Muhayyuddin, Palomo, L., García, N.: The kautham project: a teaching and research tool for robot motion planning. In: Proc. of the IEEE Int. Conf. on Emerging Technologies and Factory Automation (2014)
26. Sucan, I., Moll, M., Kavraki, L.E., et al.: The open motion planning library. IEEE Robotics & Automation Magazine **19**(4), 72–82 (2012)

FuSeOn: A Low-Cost Portable Multi Sensor Fusion Research Testbed for Robotics

Jose Luis Sanchez-Lopez, Changhong Fu and Pascual Campoy

Abstract Nowadays, the utilization of multiple sensors on every robotic platform is a reality due to their low cost, small size and light weight. Multi Sensor Fusion (MSF) algorithms are required to take advance of all the given measurements in a robust and complete manner. These high demanding developed algorithms need to be tested under real and challenging situations and environments. To the knowledge of the authors, none of the available datasets fulfills are fully suitable to be used for MSF applied to Robotics, due to the lack of multiple sensor measurements provided by light weight, small size and low cost sensors; due to their restricted motions (typically planar movements); or to their limited environmental conditions.

The contributions of the paper are twofold. First, a low-cost portable and versatile testbed has been developed for MSF research with various types of sensors. Second, a group of datasets for MSF research for Robotics have been made public as a common framework for algorithm testing after a comparison with the existing databases in the state of the art, highlighting the differences and advantages of the one presented in this paper, that are: low-cost sensors for general use on Robotics, fully 3 dimensional movements (six degrees of freedom), as well as challenging indoor and outdoor small and large environments.

Keywords Multi sensor fusion · Datasets · Robotics · Testbed

1 Introduction

The latest advances on Micro-Electro-Mechanical Systems (MEMS) and the diminishing size, weight and cost of sensor technologies, are motivating the installation

J.L. Sanchez-Lopez(✉) · C. Fu · P. Campoy
Computer Vision Group, Centre for Automation and Robotics,
CSIC-UPM, Calle Jose Gutierrez Abascal 2, 28006 Madrid, Spain
e-mail: jl.sanchez@upm.es
http://www.vision4uav.eu

© Springer International Publishing Switzerland 2016
L.P. Reis et al. (eds.), *Robot 2015: Second Iberian Robotics Conference*,
Advances in Intelligent Systems and Computing 417,
DOI: 10.1007/978-3-319-27146-0_5

of more sensors in all the developed robots. In the particular case of multirotor type Unmanned Aerial Systems (UAS), this is a tangible reality. For example, the old Asctec Autopilot[1] has a single Inertial Measurement Unit (IMU), compared to the new cutting-edge autopilot, the Asctec Trinity[2] with triple redundant IMUs. Another example is the famous Open Hardware PixHawk Autopilot[3], which includes double redundant IMUs.

In the past years, the information given by every sensor was treated individually for very limited purposes (e.g. an optical flow sensor onboard a UAS to measure the ground speed, that reports a wrong measurement if the UAV is hovering over a moving obstacle). Nowadays, the measurements given by all the sensors are analyzed together by means of Multi Sensor Fusion (MSF) algorithms to improve the state estimation (of the robot and / or its environment) and to reduce the effect against the failure of the sensors (e.g. if the onboard IMU measurement is fused with the optical flow measurement, the robot could infer that the obstacle is moving under it).

MSF algorithms have a high number of difficult requirements, which convert the development of these algorithms a challenging task that still continue to be under research. They require to be able to manage a big

Fig. 1 The FuSeOn testbed during an experiment.

amount of information because each sensor outputs its raw measurements that need to be processed. They have to deal with sensors that work at a different rates, with different time delays, and with different measurement noises. The robot state might not be calculated directly using the sensor measurements, having to infer it. MSF algorithms must detect sensor failures, processing this information in a special manner. To take advance of the MSF algorithms, they are required to be fast enough to work in real time. Good precision and performance in the state estimation is also compulsory. Finally, MSF algorithms have to deal with real systems that typically have non-linearities in their erroneous or incomplete models. This implies that an auto-calibration capability of the sensors and the models is a desired feature on these algorithms that allows to improve their precision.

There exist multitude works in the field of Multi Sensor Fusion, with a tangible growth in the last years. Some works are related to the well studied Multi GPS-INS Fusion [10, 18]. In the last years, the computer vision has experienced a new renaissance with new localization and mapping algorithms, specially with monocular and RGB-D cameras. To recover the scale factor with monocular cameras, a lot of works have been published on Visual-INS Fusion [13, 20] with a promising success.

The final objective of most of the works on Multi Sensor Fusion is to apply them to the Robotics [6]. Multi Sensor Fusion is specially useful in UAS, where its 6 DoF

---

[1] http://wiki.asctec.de/display/AR/AscTec+AutoPilot

[2] http://www.asctec.de/en/asctec-trinity/

[3] https://pixhawk.org/modules/pixhawk

movement is hard to be tracked and estimated. Combining UAS with Visual-INS sensors is a current trend [1, 2], but most advanced works include in the sensor fusion algorithm, the information given by many other sensors [7, 19].

To ease the research on Multi Sensor Fusion, and to be able to compare different proposed algorithms, datasets are specially useful. Nevertheless, to the knowledge of the authors, none of them fulfills completely the requirements highlighted in Section 3.

As the authors believe on the Open Science concept [3], this paper introduces two useful contributions for the scientific community:

The first one is the hardware and software design of the FuSeOn testbed (Section 2), a specifically designed low cost testbed for the research on Multi Sensor Fusion for Robotics. Its portability permits to easily be carried everywhere by a person to acquire data or to test his algorithms with any desired movement. Its modularity and versatility ensure the possibility to reconfigure it or add it new features or sensors with a low effort. The hardware design is detailed to serve as a starting point for other researchers interested on working on Multi Sensor Fusion with enough resources to build their own testbed. The software sources has been also made available to the scientific community.

The second contribution of this work is double. First of all, the state of the art available datasets are analyzed in Section 3. Then, in Section 4, a collection of datasets for the research on Multi Sensor Fusion for Robotics is presented, aimed to researchers without enough resources to build their own testbed, as well as, to serve as a reference to compare the developed algorithms by testing them under the same conditions.

2 System Description

The FuSeOn testbed (Figure 2) has been designed taking into account the following requirements:

- It has to be mobile and portable, to be able to move it anywhere to acquire and process the data in order to allow the researchers to work in a very wide number of scenarios. A person has to be able to carry it with a small effort, so that the FuSeOn board can reach the same places than the person.
- It has to be modular and versatile enough to be able to be modified, in order to add new sensors, computers or to reconfigure it in case of a new research need. Additionally, must be mechanically simple to maintain it with the less possible effort.
- Its endurance time has to be enough for most of the MSF tests (e.g. typically, UAS flights last less than one hour).
- The usage of the testbed has to be simple, allowing the researchers to easily control, monitor and supervise not only the system, but also the ongoing test.
- Its computational power has to be enough to run most of MSF algorithms (equivalent to a current average laptop).

– It has to be cost-effective.

The requirements cited above have been satisfied thanks to its design using methacrylate layers (Figure 2), in which each of the parts of the board has been divided, allowing the desired modularity and versatility, as well as easing its maintenance. Its small size (less than $40 \times 25 \times 40 \ cm$) and light weight (less than $4 \ kg$) let a person to carry it everywhere (Figure 1). Its power system makes the FuSeOn board fully portable, as well as, giving it an endurance time higher than one hour with the same battery. Its communication system permits a simple remote usage, control, monitoring and supervision by means of the ground computers, as well as the connection of the testbed with the Internet. The computational power needs are

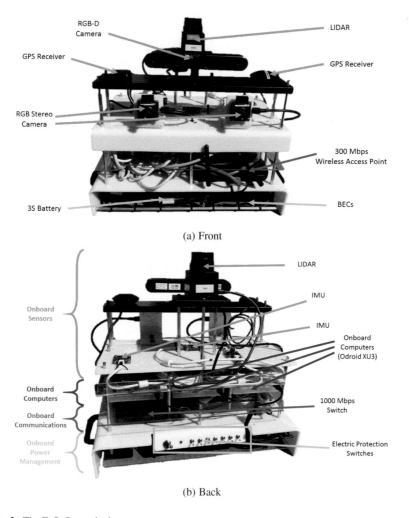

(a) Front

(b) Back

Fig. 2 The FuSeOn testbed.

satisfied with three small onboard computers added to the ground computers, plus the capability of cloud computing. Its software has also been very carefully designed setting an easy process management and inter processes communication. Last but not least, the main part of the cost of the platform is due to the installed sensors, being then cost-effective.

The FuSeOn testbed is divided on three subsystems: The **FuSeOn Board** is the portable and mobile part of the testbed, in charge of the sensing and the real time processing of the sensed data. It includes the onboard sensors (Section 2.1), the onboard computers Section 2.2), and the onboard power management (Section 2.3). The **FuSeOn Ground Stations** are a semi-portable part of the testbed formed by the ground computers (Section 2.4), which mission is the interaction with the user for controlling and supervising the FuSeOn board, as well as the heavy (and low frequency) processing of the sensed data. Finally, the **FuSeOn Communication System** allows the communication within the FuSeOn system and with the external networks (Section 2.5).

The enumeration of the specifications of each device installed on the FuSeOn testboard is out of the scope of this paper and can be consulted on the Internet following the links provided on the footnotes.

2.1   Onboard Sensors

The FuSeOn includes four proprioceptive sensors (gyroscopes and accelerometers of two IMUs) and six exteroceptive sensors (compasses of two IMUs, two RGB Cameras, one RGB-D Camera and one LIDAR), all of them with an Universal Serial Bus (USB) interface, what eases the hardware design:

– **Global Positioning System (GPS) Receivers**: Two different GPS receivers have been installed for MSF challenges on different locations on the FuSeOn: A Phidgets Inc, GPS receiver (P/N: 1040)[4] and a US GlobalSat Inc, GPS receiver (P/N: BU-353S4)[5]. The main difference between these two sensors is the update rate, being the Phidgets 1040 ten times faster than the US GlobalSat BU-353S4.
– **Inertial Measurement Units (IMU)**: Similarly to the GPS receivers, two different IMUs were installed on the FuSeOn: The main IMU sensor, a Phidgets Inc, Spatial 3/3/3 HR (P/N: 1044)[6] was installed in the center of coordinates of the FuSeOn. An additional Phidgets Inc, Spatial 3/3/3 (P/N: 1042)[7] was also installed on away its center of coordinates. The Phidgets 1044 is much more precise than the Phidgets 1042.

---

[4] http://www.phidgets.com/products.php?product_id=1040

[5] http://usglobalsat.com/p-688-bu-353-s4.aspx

[6] http://www.phidgets.com/products.php?product_id=1044

[7] http://www.phidgets.com/products.php?product_id=1042

- **RGB Cameras**: Two uEye cameras, UI-1221LE-C-HQ[8] were installed on the FuSeOn creating a stereo pair. The decision to mount two similar cameras in a stereo pair configuration is to allow the use of both one or two monocular algorithms, as well as a stero algorithm for visual state estimation.
- **RGB-D Camera**: An ASUS XTion Pro Live[9] has been installed on the FuSeOn, aiming to the same scene than the two RGB cameras.
- **LIDAR (laser rangefinder)**: A Hokuyo URG-04LX-UG01 has been installed on the FuSeOn.

2.2   Onboard Computers

Three Odroid XU3 micro computers[10] were installed to receive and process in real-time the information given by the mounted sensors. These computers, combine a low-price (less than 200€), small size and weight (less than 10 x 8 x 2 cm. and 200 g.) and good performance (higher than any smartphone on the market), in addition to five USB ports to connect all the sensors. All computers run Linux LUbuntu 14.04.2 LTS for ARM devices.

2.3   Onboard Power Management

A 3S 5000 mAh 20C LiPo battery provides the system with enough energy to power all the sensors, onboard computers and onboard electronics during more than one hour, allowing the portability and mobility of the FuSeOn board. The full electric circuit is protected by a several switches. The installed Battery Eliminator Circuits (BECs) have the mission of stabilizing the non-constant voltage given by the battery to a constant voltage that the onboard electronics require to work.

2.4   Ground Computers

The monitoring and supervision of the FuSeOn board, as well as the user interaction, is done by means of one or more ground computers mechanically detached to it, connected to the FuSeOn network by means of a WiFi adapter (Section 2.5). In addition, these computers can be used to run low frequency (less than 10 Hz.) heavy computational processes that the onboard computers are not able to. The ground computers run Linux Ubuntu 14.04.2 LTS.

---

[8] https://en.ids-imaging.com/IDS/spec_pdf.php?sku=AB.0010.1.22500.00

[9] https://www.asus.com/Commercial_3D_Sensor/Xtion_PRO_LIVE/specifications/

[10] http://www.hardkernel.com/main/products/prdt_info.php?g_code=G140448267127

2.5   Communication System

The communications system of the FuSeOn has two main objectives: First, it has to make available with the smallest possible delay all the sensor measurements in all the computers (onboard and ground computers) within the FuSeOn network. Secondly, it has the mission to communicate all the computers within the FuSeOn network with other networks to increase its capabilities, allowing then the use of advanced networking features like cloud computing or the Internet of things (IoT).

The onboard computers are connected in a wired network by means of a 1000 Mbps switch, minimizing the delays. An onboard 300 Mbps Wireless Access Point connected to the switch, creates its own WiFi a/b/g/n network, allowing the ground computers to have fast access to the FuSeOn network. Finally, a ground router enables the communication between the FuSeOn network with other networks.

2.6   Software

The software developed for the FuSeOn testbed uses ROS[11] as middle-ware, what not only eases the communications and management of processes, but also is a trend in the scientific community, being a multitude of software packages available ready to be used. A Plug & Play philosophy eases the use of the FuSeOn testbed.

Each sensor has its own ROS package to read the measurements given by the hardware, publishing them into a predefined ROS topic. The connected devices to the onboard computers are managed by udev, the device manager of the Linux kernel, by means of udev rules that launch Linux scripts to run the sensor drivers when attached to the onboard computer. A Devices Manager ROS node, running on a master onboard computer, manages and supervises the status of all the sensors and computers.

3   Available Datasets Comparison

To the knowledge of the authors, there is no complete and specific public dataset for the research on Multi Sensor Fusion applied to small-size or low-cost robotics (including UAS).

Table 1 shows a summary of the most famous state of the art available datasets, highlighting their main features. Some of these datasets, have a reduced number of available sensor measurements (because they are specific for the research in a concrete area); in some others, the movement of the testbed is limited (e.g. typically ground mobile robots only have three degrees of freedom).

---

[11] http://www.ros.org/

Table 1 The well-known datasets for the research on Multi Sensor Fusion for Robotics

Dataset	Description[1]
TUM RGB-D [17]	A variety of **indoor** environments
	Microsoft Kinect: RGB-D images (30Hz, 640×480) and accelerometer data. **GT**: Sensor trajectory by motion capture system
	`http://vision.in.tum.de/data/datasets/rgbd-dataset`
NYU Depth v1 [15] & v2 [14]	A variety of **indoor** environments
	Microsoft Kinect, and a subset of the video data is accompanied by dense multi-class labels.
	`http://cs.nyu.edu/ silberman/datasets/`
CMU Visual Localization [12]	Multi-environment **outdoor** sequences by a **ground robot**
	GPS, Gyroscopes and Magnetometers, Lidars and Omni-directional camera.
	`http://3dvis.ri.cmu.edu/data-sets/localization/`
New College vision and laser [16]	**Outdoor** scenarios by a **ground robot**
	Stereo imagery captured (20Hz); 5-view omni-directional images (5Hz); range and intensity data from two lasers scanners (75Hz).
	`http://www.robots.ox.ac.uk/NewCollegeData/`
The Rawseeds Project	Bicocca (**indoor**) and Bovisa (**outdoor + mixed**) by a **ground robot**
	12 ultrasound transducers; one IMU; multiple Cameras System; and multiple Laser Range Finders. **GT**: drawings and recovered robot trajectories with expensive LRF.
	`http://www.rawseeds.org/`
Victoria Park Sequence [11]	**Outdoor** environment with trees by a **ground robot**.
	Laser; GPS; and steering and speed of a ground vehicle.
	`http://www-personal.acfr.usyd.edu.au/nebot/ victoria_park.htm`
Malaga Dataset 2009 [4] & 2013 [5]	**Outdoor** urban scenario by a **ground robot**.
	One stereo camera, IMU and five laser scanners. **GT**: by a set of three expensive Real Time Kinematics (RTK) GPS receivers.
	`http://www.mrpt.org/Paper:Malaga_Dataset_2009/` `http://www.mrpt.org/MalagaUrbanDataset/`
Ford Campus & North Campus Vision and LIDAR	**Outdoors** by a **ground robot** with loop closures.
	Two IMU, a 3D-lidar scanner, two 2D lidars, and an omnidirectional camera system
	`http://robots.engin.umich.edu/SoftwareData/Ford` `http://robots.engin.umich.edu/SoftwareData/NCLT`
KITTI [9]	**Outdoors** by a **ground robot** in rural areas and on highways
	two Grayscale cameras; two Color cameras; GPS/IMU; one 3D Laserscanner
	`http://www.cvlibs.net/datasets/kitti/`

[1] GT = Ground Truth.

[2] Other available datasets: `http://projects.asl.ethz.ch/datasets/doku.php?id=home`
`http://projects.asl.ethz.ch/datasets/doku.php?id=related_links`
`http://its.acfr.usyd.edu.au/datasets/`

With few exceptions, the environment of the datasets is not challenging enough in terms of MSF (e.g. typically transitions between indoor and outdoor environments are not available). In addition, nearly all of the most complete accessible datasets include measurements given by very expensive and heavy sensors, that would unlikely be mounted on every robot (e.g. a multirotor UAS cannot carry a heavy and expensive 3D Laser Scanner). Finally, some of the available datasets include Ground Truth. However, they are the measurements given by a high-precision sensor (e.g. a Motion Capture System) as well as the building plans, in indoor environments. In outdoor environments, the Ground Truth is provided by high-precision and high cost GPS-INS measurements (like D-GPS), as well as 3D laser scanner measurements.

4 Multi Sensor Fusion Datasets

As the second contribution of the paper, several datasets have been recorded and made public to the scientific community interested on low-cost multi sensor fusion with 6 DOF movements. The datasets are intended to be a reference for all Multi Sensor Fusion works, allowing to test all the developed algorithms under the same conditions to compare them in an objective way. The datasets can be downloaded following this link: http://vision4uav.eu/?q=dataimage. To reach all research interests, the authors promise to do their best to record new more datasets if required by the scientific community.

The datasets have been recorded in two different formats: a ROS bag[12] that eases the reading of the data using ROS; and a custom log plus images that eases the data analysis. In addition to the timestamped sensor measurements, the datasets provide all available sensor parameters and calibrations.

The datasets are classified based on the environment in: **indoor environments datasets** (Section 4.1), **outdoor environments datasets** (Section 4.2), and **datasets with transitions between indoors and outdoors environments** (Section 4.3).

The datasets have been recorded in the environment of the School of Industrial Engineering of the Technical University of Madrid (address: Calle Jose Gutierrez Abascal 6, 28006 Madrid (Spain)), see Figure 3. All the datasets always start with the FuSeOn board located on a planar surface will all the sensors acquiring correct measurements if signal is available. The IMUs have been previously calibrated to zero; the GPS sensors have had enough time to acquire satellites signal (if available); and the cameras and LIDAR have had enough time to be started. In addition, to ease the depth estimation with the RGB cameras, a calibrated ArUco visual marker [8] has been placed in front of the cameras. The FuSeOn board has been carried by a human operator during the tests, who is always away to the LIDAR measurements and the camera images.

Sensor failures have not been recorded, because they can be easily simulated by the datasets users. Only environment dependent sensor malfunctions are included, like RGB-D camera malfunctions in outdoor scenarios, or GPS sensors signal lost in indoor scenarios.

The proposed datasets do not include Ground Truth because the authors consider that high-precision measurements are still measurements that can be used as MSF

---

[12] http://wiki.ros.org/rosbag

algorithms inputs. Under the authors point of view, to compute the "Ground Truth" to test the performance of a developed algorithm, the only correct option is the comparison against other algorithms using a particular set of sensors measurements depending on the research interest of each dataset user.

4.1 Indoor Environments Datasets

In indoor environments, all the sensors are expected to have a correct operation, except the GPS sensors due to the lack of satellite signal.

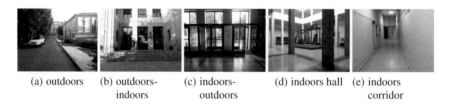

(a) outdoors   (b) outdoors-   (c) indoors-   (d) indoors hall   (e) indoors
                     indoors   outdoors   corridor

Fig. 3 Examples of the environments of the recorded datasets.

Three different datasets have been acquired, with diverse objectives:

- **Short sequence:** The first sequence that must be used in the research of MSF algorithms. The FuSeOn board moves in a small area. Some challenging situations appear occasionally, like image occlusions or fast movements.
- **Long exploratory sequence:** A higher level dataset. It is a sequence with a long exploratory movement through small corridors and large halls. Some other challenging situations appear occasionally, like moving objects, image occlusions or fast movements.
- **Long and closed exploratory sequence:** The most challenging one. It combines a long exploratory movement with the pass through previously visited areas, called loop closure.

4.2 Outdoor Environments Datasets

Unlike indoor environments, outdoor ones, expects a malfunction of the RGB-D sensor due to the interference of the IR solar radiation with its projected pattern. However, GPS sensors are supposed to have enough satellite signal to give a correct measurement.

Similarly to indoor datasets, three different purposes datasets have been acquired: a **Short sequence**; a **Long exploratory sequence**; and a **Long and closed exploratory sequence** with similar challenges.

4.3 Transition between Indoor and Outdoor Environments Datasets

To lead the research on Multi Sensor Fusion to its top level, a combination of indoor and outdoor environment has to be tackled. Transitions between indoors and outdoors are complex because some sensors suddenly stop correctly working, while others start given new correct measurements. Acquired images are very different in terms of lighting. Outdoor environments are larger than indoors, and their obstacles have also bigger sizes and faster movements.

Two datasets have been acquired, depending on the transition: an **Indoors - Outdoors transition**, and an **Outdoors - Indoors transition**.

5 Conclusions and Future Work

In this paper, a low-cost testbed for the research on Multi Sensor Fusion for Robotics, called FuSeOn, has been presented. This testbed has a well studied design that ensures its portability, allowing a person to easily carry it everywhere to acquire data or to test his algorithms. In addition, the testboard design guarantee its modularity and versatility, promising a low maintenance, and the possibility to reconfigure it or to add it new features or sensors.

The second presented contribution is the comparison between the available datasets, demonstrating the need of a new collection of datasets for the research in MSF applied to low-cost and small-size Robotics. As the last contribution, these needed datasets have been acquired using the presented FuSeOn testbed. They include the measurements given by all the sensors mounted on the FuSeOn testbed, together with its calibration parameters. The datasets have been recorded in very different conditions, including indoor and outdoor scenarios, as well as, transitions between them; with an adequate number of light weight, small size and low cost sensor measurements; and with variated 6 DoF movements.

Both, the hardware design, and the software source code running on the testbed, as well as the recorded datasets, have been made public available for the use of the scientific community, as part of the authors belief on the Open Science concept. In addition, the authors take on a commitment of do their best to improve the recorded datasets, as well as the FuSeOn testbed, if requested by the scientific community.

There exist two main possibilities to continue with the work presented on this paper. The first one, is related with future improvements on the FuSeOn testbed or in the acquisition of new datasets. The FuSeOn testbed can be enhanced by adding more sensors like thermal cameras, or optical flow sensors. The size reduction of the FuSeOn board could open a whole world of applications like using it onboard UAVs, at the cost of loosing modularity and versatility. Additionally, improvements on sensors calibration and "Ground Truth" computation for comparison purposes are interesting future works. The second possibility is to apply the presented FuSeOn testbed on the research on Multi Sensor Fusion.

Acknowledgments The authors would like to thank the Consejo Superior de Investigaciones Cientificas (CSIC) of Spain for the JAE-Predoctoral scholarships of one of the authors and their research stays, and the Spanish Ministry of Science MICYT DPI2014-60139-R for project funding. Special thanks to Angel Luis Martinez for manufacturing the FuSeOn board, and to Miriam Uzuriaga for testbed name suggestions.

References

1. Abeywardena, D., Dissanayake, G.: Tightly-coupled model aided visual-inertial fusion for quadrotor micro air vehicles. In: Field and Service Robotics, pp. 153–166. Springer (2015)
2. Abeywardena, D., Wang, Z., Kodagoda, S., Dissanayake, G.: Visual-inertial fusion for quadrotor micro air vehicles with improved scale observability. In: 2013 IEEE International Conference on Robotics and Automation, pp. 3148–3153. IEEE (2013)
3. Bartling, S., Friesike, S., Bartling, S., Friesike, S.: Opening Science. Springer (2014)
4. Blanco, J.-L., Moreno, F.-A., Gonzalez, J.: A collection of outdoor robotic datasets with centimeter-accuracy ground truth. Autonomous Robots 27(4), 327–351 (2009)
5. Blanco, J.-L., Moreno, F.-A., González-Jiménez, J.: The málaga urban dataset: High-rate stereo and lidars in a realistic urban scenario. International Journal of Robotics Research 33(2), 207–214 (2014)
6. Ciftcioglu, O., Sariyildiz, I.S.: Data sensor fusion for autonomous robotics (2012)
7. Fang, Z., Yang, S., Jain, S., Dubey, G., Maeta, S., Roth, S., Scherer, S., Zhang, Y., Nuske, S.: Robust autonomous flight in constrained and visually degraded environments
8. Garrido-Jurado, S., Muñoz-Salinas, R., Madrid-Cuevas, F.J., Marín-Jiménez, M.J.: Automatic generation and detection of highly reliable fiducial markers under occlusion. Pattern Recognition 47(6), 2280–2292 (2014)
9. Geiger, A., Lenz, P., Stiller, C., Urtasun, R.: Vision meets robotics: The kitti dataset. International Journal of Robotics Research (2013)
10. Guerrier, S.: Improving accuracy with multiple sensors: study of redundant MEMS-IMU/GPS configurations. In: Proceedings of the 22nd International Technical Meeting of The Satellite Division of the Institute of Navigation, pp. 3114–3121 (2009)
11. Guivant, J., Nebot, E.: Simultaneous localization and map building: test case for outdoor applications. In: IEEE Int. Conference on Robotics and Automation (2002)
12. Huber, D., Badino, H., Kanade, T.: The CMU visual localization data set
13. Klein, G.S.W., Drummond, T.W.: Tightly integrated sensor fusion for robust visual tracking. Image and Vision Computing 22(10), 769–776 (2004)
14. Kohli, P., Silberman, N., Hoiem, D., Fergus, R.: Indoor segmentation and support inference from RGBD images. In: ECCV (2012)
15. Silberman, N., Fergus, R.: Indoor scene segmentation using a structured light sensor. In: Proceedings of the International Conference on Computer Vision - Workshop on 3D Representation and Recognition (2011)
16. Smith, M., Baldwin, I., Churchill, W., Paul, R., Newman, P.: The new college vision and laser data set. The International Journal of Robotics Research 28(5), 595–599 (2009)
17. Sturm, J., Engelhard, N., Endres, F., Burgard, W., Cremers, D.: A benchmark for the evaluation of RGB-D SLAM systems. In: Proc. of the International Conference on Intelligent Robot Systems, October 2012
18. Waegli, A., Guerrier, S., Skaloud, J.: Redundant MEMS-IMU integrated with GPS for performance assessment in sports. In: 2008 IEEE/ION Position, Location and Navigation Symposium, pp. 1260–1268. IEEE (2008)
19. Weiss, S., Achtelik, M.W., Chli, M., Siegwart, R.: Versatile distributed pose estimation and sensor self-calibration for an autonomous MAV. In: 2012 IEEE International Conference on Robotics and Automation, pp. 31–38. IEEE (2012)
20. Weiss, S., Siegwart, R.: Real-time metric state estimation for modular vision-inertial systems. In: 2011 IEEE International Conference on Robotics and Automation, pp. 4531–4537. IEEE (2011)

Reasoning-Based Evaluation of Manipulation Actions for Efficient Task Planning

Aliakbar Akbari, Muhayyuddin Gillani and Jan Rosell

Abstract To cope with the growing complexity of manipulation tasks, the way to combine and access information from high- and low-planning levels has recently emerged as an interesting challenge in robotics. To tackle this, the present paper first represents the manipulation problem, involving knowledge about the world and the planning phase, in the form of an ontology. It also addresses a high-level and a low-level reasoning processes, this latter related with physics-based issues, aiming to appraise manipulation actions and prune the task planning phase from dispensable actions. In addition, a procedure is contributed to run these two-level reasoning processes simultaneously in order to make task planning more efficient. Eventually, the proposed planning approach is implemented and simulated through an example.

Keywords Task planning · Reasoning process · Manipulation

1 Introduction

Increasing complexity of daily manipulation tasks requires a robot to become more capable, robust, as well as autonomous in order to carry out various manipulation actions, e.g., pushing different sort of objects holding unique characteristics in human-like environments. In solving a given complex manipulation problem, high-level beside low-level planning are required, and their combination plays a crucial role in realizing a solution plan in terms of finding a sequence of actions and a way of execution them. In this scope, many studies apply recent and efficient task

A. Akbari · M. Gillani · J. Rosell(✉)
Institute of Industrial and Control Engineering,
Universitat Politècnica de Catalunya, Barcelona, Spain
e-mail: {aliakbar.akbari,muhayyuddin.gillani,jan.rosell}@upc.edu

J. Rosell—This work was partially supported by the Spanish Government through the projects DPI2011-22471, DPI2013-40882-P and DPI2014-57757-R. Muhayyuddin is supported by the Generalitat de Catalunya through the grant FI-DGR 2014.

69
L.P. Reis et al. (eds.), *Robot 2015: Second Iberian Robotics Conference*,
Advances in Intelligent Systems and Computing 417,
DOI: 10.1007/978-3-319-27146-0_6

planning approaches such as hierarchical-based task planning or GraphPlan in order to combine them with motion planning by different techniques (e.g., [1][2][3]).

Manipulation actions comprise contacts between a robot and manipulatable objects, so motion planning has to be aware of the possible interaction among rigid bodies. To deal with this issue, a physics-based reasoning engine is employed that enables the access to such information aiming to evaluate feasibility or effect of actions. In this line, the work in [4] presented an approach, called *smart motion planning*, that is able to incorporate ontology-based knowledge within a physics-based motion planner with the purpose of allowing contact between the robot and the objects only from where these objects can be manipulated. Also in a similar way, a lightweight reasoning process has been proposed by [5] that evaluates stability, reachability, and also visibility with respect to manipulation actions that conditions the motion planning.

Solving mobile manipulation problems by evaluating several possible alternative plans has been recently deliberated as an interesting research line. For instance, the Task Motion Multigraph (TMM [6]) has been introduced to simultaneously plan the motions of alternative task plans, although without considering physics-based issues. On the contrary, the recent work [7] suggested the integration of task planning with a physics-based reasoning module. The approach first collects all possible plans and then compares them based on their feasibility, in order to realize the best one according to the minimum total action cost (the pruning of unnecessary actions while planning is not considered). Following this latter work, the present study introduces a framework based on a version of GraphPlan that uses a reasoning process (based on an ontological knowledge modelling) and a physics-based motion planner to identify unnecessary actions (inessential, ineffective, or infeasible) and exclude them from the planning graph. This pruning of the task planning search space makes task planning more efficient.

2   Problem Statement and Solution Overview

A mobile manipulator is considered that is able to deal with two actions, to push removable objects and to freely move around. The problem to be solved is to efficiently find a sequence of actions to perform a given task. Then, the first point to consider is to discriminate between feasible actions and those which do not provide useful results. To illustrate this issue, the example represented in Figure 1 is assumed in which there are two rooms separated by a corridor and a mobile manipulator that must traverse to move from the initial to the goal region. To achieve this, the robot needs to push away some objects in order to free its path towards the goal. The following constraints hold for the manipulatable objects: they can only be pushed from their manipulatable regions and, according to them, along the x or y directions. Regarding the possible pushing actions, it can be seen that, for instance, pushing obstacle A from m_{A1} becomes an ineffective action because it does not clear the

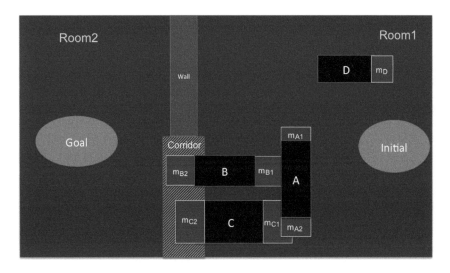

Fig. 1 A manipulation problem where a robot must find a feasible path to reach the goal region avoiding the fixed wall.

access to either manipulatable region m_{B1} or m_{C1}. Thus, such type of actions should be identified and omitted in the task planning search space.

To cope with this type of problem, the use of manipulation knowledge in the form of an ontology is proposed in order to represent all the manipulation problem components (e.g. manipulatable regions or constraints) and to facilitate the reasoning process. Also, the use of a physics-based engine is envisioned to reason about the feasibility of actions and its interleaving with the task planning based on GraphPlan can reduce the planning graph by pruning those actions that are inessential, ineffective, or infeasible, and the corresponding branches. In this way, the high-level planning module will maintain only valid actions and the search of the plan will be easier.

3   Manipulation Modeling Using Ontologies

An ontological knowledge-based management system can be integrated within task planning in order to improve its efficiency. The current proposal envisions the use of ontology-based knowledge in order to hand over to the task planning algorithm sufficient knowledge concerning the manipulation of objects, i.e., knowledge concerning the way an object can be manipulated. This may be determined by the object type and be given by its properties, e.g., *free manipulatable objects* can be pushed in any direction and from any of its sides, while *constraint-oriented manipulatable objects* can only be pushed in some directions and from some given sides (such as object D in Figure 1 that can only be pushed horizontally from the manipulatable region m_D).

3.1 Ontology Concept

Ontology deals with the concern regarding things existence. It has been recognized as an approach to expose explicit knowledge about the world in Artificial Intelligence with respect to characterizing concepts as well as relations. Ontologies are able to be stored in the Web Ontology Language (OWL) [8] being a family of knowledge representation on the basis of description logics. The intention of OWL is to access and share ontological knowledge among several systems through the world wide web. This knowledge is modeled over classes and individuals. Classes (declared by *owl:Class*) are dedicated to comprise those objects sharing analogous knowledge, while individuals (declared by *owl:NamedIndividual*) describe particular elements of classes. Correspondence between classes, between classes and individuals, as well as between individuals are performed with properties (declared by *owl:ObjectProperty*) in the OWL. OWL can be designed by one of the most popular ontology editors called *Protégé* [9] enabling knoweledge development. On the advantages of this editor is to visualize ontology as a form of graph to represent relations between knowledge in the database.

The management of an ontology-based knowledge described in the OWL and its later use for reasoning purposes can be done using the Knowrob tool [10]. This software tool is developed based on SWI Prolog and the Semantic Web library and enables fundamental predicates to access such knowledge using Prolog language, e.g., the query *owl_subclass_of(?SubClass, ?Class)* explores all available subclasses of a class, *owl_individual_of(?Indv, ?Class)* seeks to list all individuals of a class, and *class_properties(?Class, ?Properties, ?Value)* determines the value of a class under particular properties.

3.2 Components of the Manipulation Ontology

A manipulation ontology using OWL has been designed as shown in Figure 2. The classes *ManipulationWorld* and *ManipulationPlanning* with their subclasses represent the knowledge about the world and about high-level planning components, respectively. The OWL files are accessible at the following web adress: `https://sir.upc.edu/projects/ontologies/`.

The subclasses of the *ManipulationWorld* class are:

- *"Regions"*: Class that represents various types of regions. For instance, *ManipulatableRegion* belongs to an object and is the region where the robot should be located in order to apply forces to push it. *CriticalRegion* corresponds with regions which should be free from obstacles, e.g., the corridor depicted in Figure 1.
- *"ObjectsType"*: Class that collects information about the objects type.
- *"Path"*: Class that involves a number of abstract paths connecting two different regions (the actual path can be acquired by querying a motion planner). A property is considered for the individuals of the class that defines whether the regions which

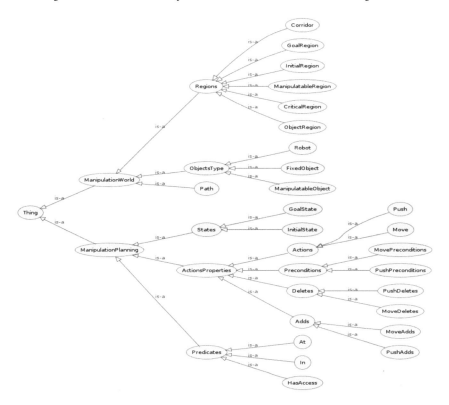

Fig. 2 Taxonomy of knowledge-based manipulation

the path connects are critical or not. This property is updated after each push action.
Related to Figure 1, the following individuals of this class are used in the solution.

- *P1, P2,* and *P3*: Paths that connect the initial region to the regions m_D, m_{A1}, and m_{A2}, respectively.
- *P4* and *P5*: Paths that connect the region m_D to the regions m_{A1} and m_{A2}, respectively.
- *P6* and *P7*: Paths that connect the region m_{A1} to the regions m_{B1} and m_{C1}, respectively.
- *P8* and *P9*: Paths that connect the region m_{A2} to the regions m_{B1} and m_{C1}, respectively.
- *P10*: Path that connect the region m_{B1} to the goal region when the robot is in the room 2.
- *P11*: Path that connect the region m_{C1} to the goal region when the robot is in the room 2.

Likewise, the subclasses of the *ManipulationPlanning* class are:

– *"Predicates"*: Class that expresses a number of parameters with associated names devoted to describe preconditions and side effects of actions beside conditions of initial and goal states. The following predicates have been defined:

- *HasAccess(Robot, Path)*: Returns true if the *Robot* is located at the initial region of the *Path*.
- *At(Robot, Region, Path)*: Says whether the *Robot* has reached the *Region* through the *Path*.
- *In(Object, Region)*: Returns true if the *Object* is at the *Region*.

– *"States"*: Class that contains the conditions about the initial and goal states of the problem.
– *"ActionProperties"*: Class used to bind actions with their needs and side effects. Two actions have been defined to solve manipulation problems as that of Figure 1: the *Move* action used to convey the robot to different regions and the *Push* action whose purpose is to clear critical regions.

- *Move(Robot, Region, ThroughPath):*
  Precondition: *HasAccess(Robot, ThroughPath)*
  Add: *At(Robot, Region, ThroughPath)*
  Delete: _

- *Push(Robot, Obj, ManipRegion, CriticalRegions, ToAccessPath, ThroughPath):*
  Precondition: *At(Robot, ManipRegion, ThroughPath)*
  Add: *HasAccess(Robot, ToAccessPath)*
  Delete: *In(Obj, CriticalRegions)*

4 Reasoning Process about Manipulation Actions

Two types of reasoning processes are proposed to evaluate manipulation actions. First, a high-level inference process determines whether an action is relevant or not. It is called essential reasoning and is performed based on the pre- and post-conditions of actions available through the Planning Graph generated by the task planning algorithm used (GraphPlan [11]). Second, a low-level reasoning based on a physics-based engine determines if an essential action can be executed or not, called feasibility reasoning, and if feasible whether the expected results are accomplished or not, called geometric reasoning. The integration of both layers is discussed in the next Section.

4.1 High-Level Reasoning

Essential reasoning process is performed in the Planning Graph generated by the GraphPlan algorithm. The Planning Graph is a layered graph with two types of levels: state-levels with sets of literals representing conditions, and action-levels with sets of possible actions. Edges connecting a state-level with the following action-level represent the pre-conditions of an action; edges connecting an action-level with the following state-level represent the post-conditions (or effects) of an action. Maintenance action is always included to retain literals for the subsequent level.

There are different constraints between actions and between literals. Two actions are constrained when:

– An effect of one action negates an effect of the other action called **"inconsistent effect"** constraint.
– An effect of one action deletes a precondition of the other action called the **"interference"** constraint.
– They have mutually exclusive preconditions called **"competing needs"** constraint.

Two literals, furthermore, are constrained if:

– One of them is the negation of the other one. It is called the **"inconsistent support"** constraint.

The construction phase of the GraphPlan procedure begins from the initial state and expands consecutively by interleaving state and action levels forming the Planning Graph, until the goal state conditions are met. Then, the search phase of GraphPlan uses a backward search to find the solution sequence. If this search fails because not all the planning constraints are satisfied, then the construction phase resumes.

The high-level reasoner can be applied during the construction phase to prune those actions that are not relevant for the task to be solved, i.e., those that are inessential. This is done by evaluating, in a given action-layer, whether the post-conditions of a non-maintenance action can only lead to actions already found in the action-layer. If this is the case, the corresponding action is marked as inessential and pruned from the Planning Graph. Also, the pre-conditions of the pruned action are deleted, as well as all the maintenance branches leading to them and they do not appear again. In this way, only fruitful actions are kept and any dispensable action branch is removed.

4.2 Low-Level Reasoning

The physics-based engine is embedded inside the low-level planning layer with the purpose of providing robust inference capability over actions. Physics-based motion planning is the more realistic approach for robot motion planning because it is able to evaluate interactions between rigid bodies and, therefore, can both be used for move actions to plan collision-free paths as well as for push actions where collisions with manipulatable objects occur.

Then, in order to determine whether an action is feasible or not, it is planned using a physics-based motion planner and the feasibility of the resulting path can then be evaluated as follows. Based on the control forces, applied duration, and distance covered, the dynamic costs of push and move actions, c_p and c_m respectively, are computed in terms of power consumed and action [7]:

$$c_p = \frac{n\mathbf{f}\mathbf{d}}{\Delta t} \qquad c_m = \sum_{i}^{n} |\mathbf{f_i}| \Delta t_i \varrho_i \qquad (1)$$

where, $\mathbf{f}$ is the applied force, Δt is the time duration, $\mathbf{d}$ represents the displacement covered in the result of $\mathbf{f}$, and n is the number of times the force shall be applied to complete the push action. In case of c_m, ϱ represent the distance. If the planner is not able to find a path or the required power is above the robot specifications, then the action is considered infeasible.

For those actions that are feasible, their effects are analyzed by geometrically evaluating the position of the objects and regions, e.g., it is verified if after a push action a critical region is free from obstacles. Then, an action is called ineffective if its effects are not met. In case of finding infeasible or ineffective action, the same pruning process is performed to remove corresponding actions.

5 Efficient Task and Motion Planning

Task planning is devoted to determining the sequence of actions to perform a given task, while motion planning is devoted to determine the motions to execute a given action. The interleaving of both planning levels may result in a robust and computationally efficient solution. The proposal is sketched in Algorithm 1 that takes the initial and goal states, s_{init} and s_{goal} respectively, and the whole set of possible actions A as input and returns the solution plan π.

The core of the computation is to appraise each selected action along the corresponding effects by the proposed reasoning process in order to prune dispensable actions. This results in a smaller Planning Graph, making both the construction phase and the search phase more efficient. When the reasoning process determines that an action is inessential, infeasible, or ineffective, then it is pruned. Figure 3 illustrates the constructed Planning Graph for the example represented in Figure 1. For the sake of saving space, only the relevant information has been exposed (e.g. maintenance actions are neglected).

The Planning Graph is established with the initial state (s_{init}) [line 1] and the construction phase procedure is launched until a layer is found that satisfies the goal constraints [line 27]. At every iteration i, the set A_i of actions whose pre-conditions appear in the previous state-level S_{i-1} is selected [Line 4] and then evaluated [Lines 5-23]. First, the effects e are obtained [line 6] to be forwarded to the sequence of reasoning process:

Algorithm 1. Efficient task and motion planning

Input: The initial and goal states s_{init}, s_{goal}. Set of possible actions A. The maximum allowed iteration m
Output: The solution plan π

1: $S_0 \leftarrow s_{init}$
2: $A_{valid} = \emptyset$
3: **for** $i \leftarrow 1$ **to** m **do**
4:     $A_i \leftarrow \{a \in A \mid precond(a) \subseteq S_{i-1}\}$
5:     **for each** $a \in A_i$ AND $a \notin A_{valid}$ **do**
6:         $e \leftarrow$ effects(a)
7:         essentialReasoner(a,e)
8:         **if** a is inessential **then**
9:             pruneActions(a)
10:             **continue**
11:         **end if**
12:         feasibilityReasoner(a,e)
13:         **if** a is infeasible **then**
14:             pruneActions(a)
15:             **continue**
16:         **end if**
17:         geometryReasoner(a,e)
18:         **if** e is ineffective **then**
19:             pruneActions(a)
20:             **continue**
21:         **end if**
22:         $A_{valid} \leftarrow a$
23:     **end for**
24:     $A_i \leftarrow$ maintenanceActions(S_{i-1})
25:     $S_i \leftarrow (l \mid l \in$ effects(A_i))
26:     checkMutexes(A_i, S_i, S_{i-1})
27:     **if** goalFound(s_g, S_i) **then**
28:         $S_g = S_i$
29:         backtrack(s_g)
30:         **if** constraintSatisfaction(S_g, S_0) **then**
31:             $\pi \leftarrow$ extractPlan(S_g, S_0)
32:             **return** π
33:         **end if**
34:     **end if**
35:     $i \leftarrow i + 1$
36: **end for**
37: **return** NULL

– *essentialReasoner(a,e)* checks whether an action is essential or not [lines 7-11], e.g., the action *PushD* from m_D becomes inessential regarding the existing actions *PushA* from m_{A1} or from m_{A2} since they can always be done irrespective of performing the *PushD* action.

– *feasibilityReasoner(a,e)* evaluates the feasibility of an action by determining whether the robot is capable or not to perform it [lines 12-16], e.g., the action *PushC* is infeasible with respect to the robot's capability since it is a very big and heavy object.

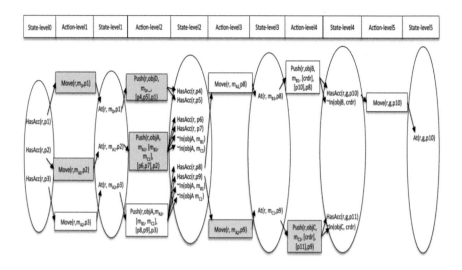

Fig. 3 The constructed Planning Graph. Pruned actions are specified in blue.

- *geometryReasoner(a,e)* checks whether the effects of an action are reached or not [lines 17-21], i.e., the action *PushA* from m_{A1} is ineffective because after pushing the object downward, regions m_{B1} and m_{C1} are not freed because the object collides with the wall and gets stuck over these regions.

The action that is not valid is pruned using function pruneActions(*a*), that deletes the action from the Planning Graph as well as the pre-conditions and all the maintenance branches leading to them. The actions that are valid are stored in the set A_{valid} [line 22], in order not to evaluate them again if the action further appears, and maintenance actions are later added [line 24]. Then the next state-level is created from the set A_i and the constraints between actions and literals is computes by function *checkMutexes(A_i, S_i, S_{i-1})* as done in the standard GraphPlan algorithm.

In the case of goal conditions are met, backtracking is performed. If the initial state-level S_0 is reached in this backtracking process and all the constraints are met the plan is extracted [lines 29-33].

6   Implementation and Simulation Results

The implementation of the proposed algorithm entails four phases: high-level planning, low-level planning, middleware, and executive. In the high-level planning, ontological knowledge concerning the manipulation world is represented using OWL and developed with *Protégé*. Knowrob software is applied to manage such knowledge and use it for the task planning algorithm. In the low-level planning, The Kautham

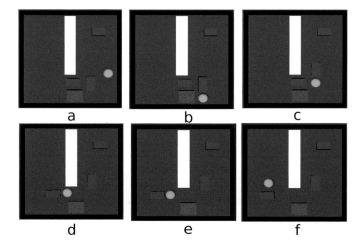

Fig. 4 Simulation results of the final plan

Project [12], an open-source tool for motion planing environment, is employed that uses the *Open Motion Planning Library (OMPL)* [13] as the core set of planning algorithms. OMPL allows planning under geometric constraints as well as under differential constraints, including those that required dynamic simulations (OMPL uses the *Open Dynamic Engine* for the dynamic simulation). Transferring information between both planning levels is performed by the middleware phase using ROS-based communications [14]. The high-level module is encapsulated as a ROS client and the low-level module works as a ROS service in order to evaluate manipulation actions. Finally, the plan is forwarded to an executive module that is the responsible of executing the whole task.

Simulation results corresponding to the scenario represented in Figure 1 are shown in Figure 4 as a sequence of snapshots. The SyCLoP-RRT [15] is used as kinodynamic motion planner for physics-based planning because recent benchmarking study of the kinodynamic motion planners for physics based planning [16] shows that SyCLoP-RRT computes the power optimal and smooth solution. The average planning time is about 35 seconds for the executive plan.

7   Conclusion and Future Work

The current paper presents a simultaneous task and motion planner. Task planning is done with the GraphPlan algorithm, that has been modified to make it computationally more efficient. This is achieved by reasoning over the action and pruning those that are dispensable. This reasoning process is performed at two levels. A high level that allows to find those actions that are not relevant and a low-level using a physics-based motion planner that allows to find those that are either infeasible or ineffective. The proposed algorithm has been implemented and illustrated through a manipulation problem. As a future work, to tackle several feasible plans, heuristics-based task planning is going to be considered that takes into account the costs of actions to find the most feasible actions while planning the whole task.

References

1. Kaelbling, L.P., Lozano-Pérez, T.: Hierarchical task and motion planning in the now. In: International Conference on Robotics and Automation (ICRA), pp. 1470–1477. IEEE (2011)
2. Shafii, N., Passos, L.S., Reis, L.P., Lau, N.: Humanoid soccer robot motion planning using graphplan. In: Proceedings of the 11th International Conference on Mobile Robots and Competitions, pp. 84–89 (2011)
3. Di Marco, D., Levi, P., Janssen, R., van de Molengraft, R., Perzylo, A.: A deliberation layer for instantiating robot execution plans from abstract task descriptions. In: Proceedings of the International Conference on Automated Planning and Scheduling: Workshop on Planning and Robotics (PlanRob) (2013)
4. Muhayyudin, Akbari, A., Rosell, J.: Ontological physics-based motion planning for manipulation. In: Proceedings of the International Conference on Emerging Technologies and Factory Automation (ETFA). IEEE (2015)
5. Mosenlechner, L., Beetz, M.: Fast temporal projection using accurate physics-based geometric reasoning. In: International Conference on Robotics and Automation (ICRA), pp. 1821–1827. IEEE (2013)
6. Şucan, I.A., Kavraki, L.E.: Mobile manipulation: encoding motion planning options using task motion multigraphs. In: International Conference on Robotics and Automation (ICRA), pp. 5492–5498. IEEE (2011)
7. Akbari, A., Muhayyudin, Rosell, J.: Task and motion planning using physics-based reasoning. In: Proceedings of the IEEE International Conference on Emerging Technologies and Factory Automation (ETFA) (2015)
8. McGuinness, D., Van Harmelen, F., et al.: Owl web ontology language overview. W3C recommendation **10**(10) (2004)
9. Stanford 2007: Protégé (2007). http://protege.stanford.edu/
10. Tenorth, M., Beetz, M.: Knowrob knowledge processing for autonomous personal robots. In: IEEE/RSJ International Conference on Intelligent Robots and Systems (IROS), pp. 4261–4266 (2009)
11. Blum, A.L., Furst, M.L.: Fast planning through planning graph analysis. Artificial Intelligence **90**(1), 281–300 (1997)
12. Rosell, J., Pérez, A., Aliakbar, A., Muhayyuddin, Palomo, L., García, N.: The kautham project: a teaching and research tool for robot motion planning. In: Emerging Technology and Factory Automation (ETFA), pp. 1–8. IEEE (2014)
13. Şucan, I.A., Moll, M., Kavraki, L.E.: The Open Motion Planning Library. Robotics & Automation Magazine **19**(4), 72–82 (2012)
14. Quigley, M., Conley, K., Gerkey, B., Faust, J., Foote, T., Leibs, J., Wheeler, R., Ng, A.Y.: Ros: an open-source robot operating system. In: ICRA Workshop on Open Source Software, vol. 3, p. 5 (2009)
15. Plaku, E., Kavraki, L., Vardi, M.: Motion planning with dynamics by a synergistic combination of layers of planning. IEEE Transactions on Robotics **26**(3), 469–482 (2010)
16. Muhayyuddin, Akbari, A., Rosell, J.: Physics-based motion planning: evaluation criteria and benchmarking. In: ROBOT2015: Second Iberian Robotics Conference. Springer (2015)

Control of Robot Fingers with Adaptable Tactile Servoing to Manipulate Deformable Objects

Ángel Delgado, Carlos A. Jara, Fernando Torres and Carlos M. Mateo

Abstract Grasping and manipulating objects with robotic hands depend largely on the features of the object to be used. Especially softness and deformability are crucial features to take into account during the manipulation tasks. Positions of the fingers and forces to be applied when manipulating an object are adapted to the caused deformation. For unknown objects, a previous recognition stage is needed to set features of the object, and manipulation strategies can be adapted depending on that recognition stage. This paper presents an adaptable tactile servoing control scheme that can be used in manipulation tasks of deformable objects. Tactile control is based on maintaining a force value at the contact points which changes according to the object softness, a feature estimated in an initial stage.

Keywords In-hand manipulation · Grasping · Tactile servoing · Deformable object

1 Introduction

In the field of in-hand manipulation tasks with robotic hands, most of the works developed so far are centered on rigid objects. For this kind of objects, studies of how to evaluate conditions of the grasp are centered in concepts like form and

C.A. Jara · F. Torres
Physics, Systems Engineering and Signal Theory Department,
University of Alicante, San Vicente del Raspeig, Spain
e-mail: {carlos.jara,fernando.torres}@ua.es

A. Delgado(✉) · C.M. Mateo
University Institute of Computing Research, University of Alicante,
San Vicente del Raspeig, Spain
e-mail: {angel.delgado,cm.mateo}@ua.es

© Springer International Publishing Switzerland 2016
L.P. Reis et al. (eds.), *Robot 2015: Second Iberian Robotics Conference*,
Advances in Intelligent Systems and Computing 417,
DOI: 10.1007/978-3-319-27146-0_7

force closure, hybrid position-velocity control and impedance control [1]. In this case, the forces are controlled to keep the equilibrium in the system hand-object. Three main models of the finger are used: hard finger, hard finger with friction and soft finger. On the hard finger models, only lineal velocities are considered, while in the soft finger also rotational velocities and moments are employed.

For deformable objects, a multisensory system is needed, mainly to control the state of the object on each step. Some works have been developed to study and classify grasps using tactile data and internal information from the robot [2, 3]. In [4], the authors use a human inspired scheme to control the movements of a gripper after collecting and processing tactile information. When a deformable object is manipulated, it is recommended to have a model of the object, or if the object is unknown, it is necessary to estimate its softness. In [5], a strategy to estimate the stiffness of some objects using a gripper and a robot finger is defined. Other approaches to estimate object features are based on the kind of finger used, because finger determines how a grasp configuration must be modeled. This is the case of the approach presented in [6], where the behavior of artificial fingers is compared with the behavior of human fingers.

When a tactile sensor is used to retrieve information about the forces at the contact points, the specifications of this device define how the data must be treated. A wide number of sensors have been studied and used for research works [7, 8]. From the information given by the sensors, it is possible to retrieve data of the contact such as position, magnitude or direction [9]. This data can also be used to get features of the object grasped or manipulated, such as shape or pose [10]. Tactile information given by the sensor can be used to apply control techniques. In [11], a tactile-driven control is applied to a robotic arm with a tactile array sensor on its end, where the authors use a tactile servoing method to stabilize position, magnitude and orientation of the forces detected by the sensor.

Besides tactile data, vision information can be very helpful to analyze grasps and to control the movements of the robotic hand according to the deformation that is detected by a vision sensor. In [12] the deformation is detected with a fusion of the information given by a tactile sensor and a vision sensor. In this kind of works, vision and tactile information are integrated as a feedback for control mechanisms during manipulation tasks [13].

In the field of robotic grasps and in-hand manipulation, the use of tactile data to obtain information about the softness and deformation of an object and also how to use this information as a reference to control the motion of the fingers is an open topic where there is still research to do. This field of study is inspired on the human grasping and manipulation scheme, which essentially depends on tactile sensation rather than vision or proprioception. Most advanced estimations can also be tested apart from stiffness estimation, such as in [14], where tactile sensors are used to detect movement inside a container, for example, liquid. The paper presents a control strategy only based on tactile information that can be used in manipulation tasks of deformable objects. The controller designed is based on an adaptable tactile servoing control for maintaining a force value at the contact points which changes according to the object softness, a feature estimated in an initial stage.

2   System Description

The robotic system used to test this framework is based on a Shadow robot hand [15], and a Tekscan Grip tactile sensor [16]. In order to develop grasping and in-hand manipulation tasks, two articulated robots Mitsubishi PA-10 are used to add more mobility to the system. One of the robots holds the Shadow hand at its end effector, while the other holds a Kinect range sensor at its end. The Kinect is used to find the object on the real world, so that the hand-robot system can be moved to the initial grasp position.

The Shadow hand has five articulated fingers and twenty degrees of freedom. Four of the fingers (first, middle, ring and little) have coupled joints at the end of them. This design and the position limits for each joint of the hand were chosen to imitate more precisely the movements of a human finger. The Tekscan Grip sensor is attached to the hand to give a map of the forces applied by the hand during the execution of the manipulation tasks. This sensor has the shape of a human hand, which is suitable for the Shadow hand, as long as the size of the robot hand is quite similar to the human one. The sensor is divided in eighteen regions that cover palm and fingers. The sensor is connected by Ethernet to the computer, and the samples are obtained at a high sampling rate (850 Hz). Sensor values can be extracted as a whole map or as a single region map.

3   Tactile Servoing Control Framework

In this section the tactile servoing control scheme is described. The tactile servoing control is applied once the contact points with the object are reached by the fingers of the hand, and an estimation of the softness of that object is given. A grasp planner and a recognition strategy are used with those purposes. These two first stages, which were presented in other papers, are briefly described and referenced, so that the whole framework can be understood. The control stage, that is the basis of the article, is described in detail.

3.1   Grasp Planner

The mission of the grasp planner [17] is to determine the contact points on the object to be grasped and the movements that are needed to reach those points. A three dimensional vision system is used to find the position of the object, and the robot hand kinematics is used to get the desired joint configuration. The grasp planner algorithm is based on the human hand grasp system.

3.2   Tactile-Based Grasp Strategy for Deformability Estimation

Once the robot-hand system is situated at the contact points with the object, and the object is unknown, the estimation stage is executed [18]. The main idea of this stage is to get an estimated value of the object softness. To do that, a motion algorithm for the robot fingers is executed, controlling position and forces of the joints until reaching an initial force value. Then, the final values of positions and forces applied are employed to get a relative value of the deformation which was caused (Fig. 1). This value is set as a force threshold which is used as reference for the tactile servoing. The estimation works for objects with uniformly distributed mass.

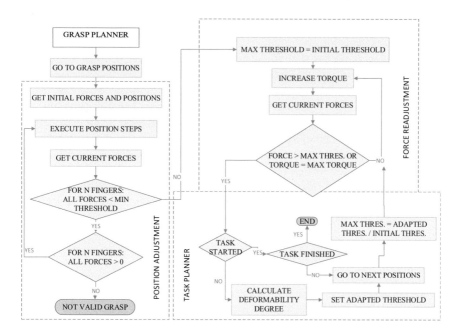

Fig. 1 Diagram of the algorithm used to estimate deformability

3.3   Finger Control with Adaptable Tactile Servoing

The tactile servoing system is applied for each finger used in the task. Each finger is controlled by a tactile-position controller. Reference position and initial force value for each finger are set to the controller by the global task planner, which will be described in this section. Each controller is executed in a parallel way. To prevent a possible loss of contact, the tactile servoing uses the pressure map of each region in contact with the object to control the state of each contact point. The task planner adapts the reference forces of each finger depending on the force configuration of the whole group. The task planner will be explained in detail below.

Feature Extraction from the Tactile Map

The Tekscan Grip sensor attached to the hand (Fig. 2) is divided in different regions. In this approach, only the information of the fingertips' regions is used. For each region, the tactile pressure value is stored in a 4 by 4 matrix. As the size of each region and cell of the region are known, the pressure and force value for each of the cells of the array can be obtained, and then the total force and pressure of the entire region. The position of the applied force is set as the position of the cell with the maximum value.

Fig. 2 Shadow hand with the Tekscan tactile sensor used for the experiments

To obtain a real and reliable value of force, the sensor must be first calibrated. Each cell of the sensor gives a raw value that varies according to the applied force. The response of each cell is different, so the first step is to balance the sensor using a uniform force distributed over the sensor surface. After the sensor is balanced, a balancing coefficient is applied to the raw value. Once the sensor is balanced, it has to be calibrated using a known weight. For each region a calibration curve is obtained, mapping from raw value to pressure value.

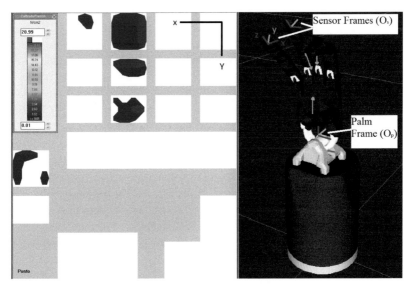

Fig. 3 Visualization of the map pressure (left) and representation of the forces on the virtual model of the hand (right)

The force exerted by the fingertip is computed using the pressure measurements of the tactile sensors. These measurements are multiplied by the area of each cell, so the forces are obtained. The sum of these forces is considered as the force applied by the fingertip. Moreover, the contact points are supposed to be located in the cell of maximum pressure exerted (Fig. 3).

Control Scheme for Tactile Servoing

The control scheme for each of the fingers is shown in the next figure.

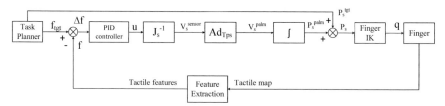

Fig. 4 Scheme of the control loop with tactile information feedback used for each finger

The aim of the controller is to get the sensor motion V_s^{sens} that minimizes the error in the features Δf (Fig. 4). Only lineal velocity is considered, because the fingers of the hand only have four degrees of freedom and two coupled joints on four of them. In this way, the rotational movements are very limited, and not much significant for this control scheme. The obtained velocity values are integrated to get the new positions of the sensor frame and, after using finger kinematics, the computed positions of the finger joints are sent to the position controller of the real hand. Besides, the task planner can send new target positions for the sensor P_s^{tgt} in order to change the distribution of the contact points.

The vector of features $\mathbf{f} = [f_x, f_y, f_m]$ denotes the position and magnitude of the applied force in the sensor. Position features (f_x, f_y) are set as the position of the cell with the maximum value in the sensor region located at each fingertip, and the magnitude is set as the sum of the forces of all the cells of the sensor region S.

$$f_m = \Sigma_{ij \in R} f_{ij} \tag{1}$$

The feature error $\Delta f(t) = [\Delta f_x, \Delta f_y, \Delta f_m]$ is obtained as the difference between the current features vector f, and the target features vector $\mathbf{f_{tgt}} = [f\_x_{tgt}, f\_y_{tgt}, f\_m_{tgt}]$. The target position ($f\_x_{tgt}, f\_y_{tgt}$) of the force is the position of the central cell of the sensor region, in order to maintain the contact point in the center of the sensor and fingertip. The magnitude f_m_{tgt} is an adaptable value given by the task planner at each iteration. The way the task planner computes this value for each finger is explained in the next subsection.

A PID controller is executed to minimize the error vector. Each feature is controlled by a different configuration of the PID controller. The parameters of the PID that controls the error in force magnitude Δf_m are set in order to obtain smooth responses, because an oscillation may cause undesired deformations on the object. The obtained control variable **u** is used in the next step to compute the motion of the fingers.

To map the response of the controller to a sensor velocity that minimizes the error, a task inverted Jacobian J_s^{-1} must be defined. The axes of the plane of the sensor region (x, y) correspond respectively with the axes X_s and Z_s of the sensor frame O_s (see Fig. 3), so errors in force position are mapped in the X_s and Z_s axes. Mapping the force magnitude value is not trivial, different configurations of J_s^{-1} were tested to obtain an appropriate and realistic movement of the fingers. If the error in force is related only with the Z_s axis of O_s, the finger movements are not the desired ones. Each error step related with the force is related with the Z_s and Y_s axes, in order to achieve a trajectory that can be followed by the fingers. In the case of the thumb, the matrix elements are negative, because the thumb is moved in the opposite direction to the rest of the fingers, so that a closing movement is produced. The next equation shows how the error is mapped:

$$V_s^{sens} = J_s^{-1} \cdot \Delta f = \begin{pmatrix} 1 & 0 & 0 \\ 0 & 1 & 0 \\ 0 & 1 & 1 \end{pmatrix} \cdot \begin{pmatrix} \Delta f_x \\ \Delta f_m \\ \Delta f_y \end{pmatrix} \qquad (2)$$

Considering the PID controller, the obtained velocity $V_s^{sens} = [V_x, V_y, V_z]$ in the sensor frame is shown in the next equation:

$$V_s^{sensor} = J_s^{-1} \cdot \left(K_p \cdot \Delta f(t) + K_i \cdot \int \Delta f(t) dt + K_d \cdot (\Delta f(t) - \Delta f(t-1)) \right) \qquad (3)$$

The obtained velocity in the sensor frame O_s is transformed to the velocity V_s^{palm} on the reference frame of each finger, placed on the palm of the hand O_p. To transform the velocity, the adjoint matrix Ad_{Tps} is used, which is related with the current transformation matrix between O_p and O_s:

$$Ad_{Tps} = \begin{pmatrix} R_{ps} & \hat{p}_{ps} R_{ps} \\ 0 & R_{ps} \end{pmatrix} \qquad (4)$$

From the sensor velocity relative to the palm reference, the new position (P_s^{palm}, P_s) for the fingertip is obtained integrating the value for each axis between t and t-1, and the position for each joint (**q**) is obtained with the inverse kinematics of the hand.

The finger's workspace is limited, so the position of the fingertip must be controlled to be inside the workspace, and a low resolution is allowed to get inverse kinematics solutions. An adapted inverse kinematics solver is used for the four fingers with coupled joints.

Task Planner and Adaptable Behavior

The task planner controls both the target position and fingertips' force at each step of a task. Each step of a manipulation task is defined by a configuration of the fingers positions and forces to be applied by each of them. In this paper only is

discussed how the task planner changes the target forces. Each configuration uses
the initial force value obtained after the recognition stage (see also Section 3.2) for
all the fingers. As commented before, each finger is controlled independently in
parallel. The position reference has higher priority, once the position is reached
the force reference is followed and is used with more priority than the position
reference until a new position reference is given by the task planner.

The adaptable behavior of the controller consists in changing the reference
forces for the fingers during the execution of one configuration. The force reference depends on the state of the configuration on the contacts and it is controlled
by the task planner (Fig. 5). A configuration is considered valid when the force
applied by the thumb and another finger is not zero, and at least one of the fingers
has achieved the initial target value. When this condition is satisfied, the task
planner changes the target forces to the current forces. This behavior must be understood as a control of the deformation on the object. With a valid configuration
the object is grasped in a secure way. If any of the fingers had not achieved the
target force, the controller would send new position steps, and this would cause
higher and undesired deformations.

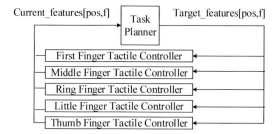

Fig. 5 Task planner diagram

4   Experimental Results

In this section some experiments carried out to test this control scheme are explained. Two different experiments are shown, representing how the control system acts at a concrete step of a task, when a configuration composed by positions
of the fingers and forces to be applied are set.

In the first case a deformable cube must be grasped using three points of contact (Fig. 10), and the initial reference force is set to 1 newton after the recognition
stage was carried out (see also Section 3.2). Figure 6 shows the evolution of the
force values for each finger in movement and the evolution of the error in the
feature related to the force magnitude. It can be seen where the targeted force is
changed by the task planner, after one of the fingers reached its target value. In
this example, middle finger reaches the targeted 1 newton in the interval of time
between second 1 and 2. After that, the task planner changes the target values, and
it can be seen in the error evolution that the errors converge to 0 at the second 2,

after the adjustment. A small interval from -0.1 Newton to 0.1 Newton is accepted as a valid result, because the sensors also have a relative error in response, inherent for its physical features. In the Figure 6, it can be seen that the error values converge to 0 or proximal to 0 value, but it also can be seen that at certain points the error increases. After experimentation, it has been observed that these changes in error are caused by irregular deformations of the object that produce variations in the sensor response.

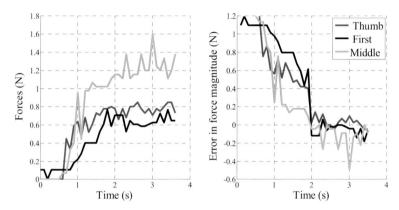

Fig. 6 Evolution of force values and error in the force magnitude feature for each finger

Figure 7 shows the output velocities for each of the fingers. The velocity V_x is close to 0, because the force is centered in the sensor 'x' axis. The velocities V_y and V_z have bigger values that tend to zero. Low scale in the error is used, to get smooth movements. The output velocities have maximum values about 0.01 and -0.01 m/s.

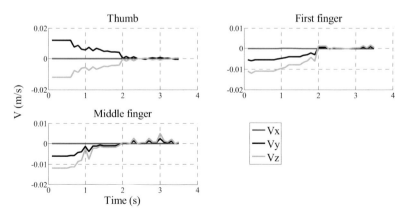

Fig. 7 Output velocities for each finger during the first experiment

In the second experiment shown, a foam is grasped with four fingers and an initial target force value of 1.2 newton. In this case it can be seen that high frequency changes are measured in the forces of ring and middle finger (Fig. 8). This variations are caused by the effect of the group of the applied forces that are deforming the object. In the error graphic it can be seen that the applied forces and the error tend to be stabilized after the second 1, but still some important variations appear, caused by the deformation. In Figure 9 the evolution of the output velocities for all the fingers is shown. As in the first experiment, small values for V_x, and higher values for V_y and V_z are obtained to correct the error.

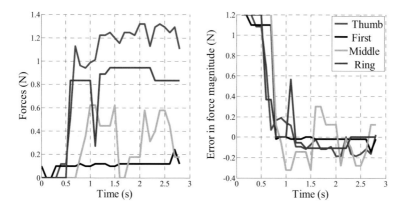

Fig. 8 Evolution of force values and error in the force magnitude feature for each finger

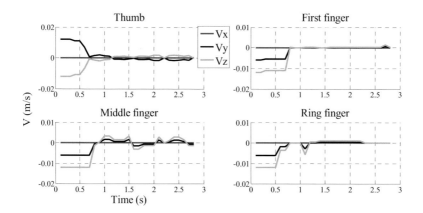

Fig. 9 Output velocities for each finger during the second experiment

Fig. 10 Robot hand used in the experiments, and set of objects for the tests

5   Conclusions

In this paper a control strategy based in tactile information was presented. This control strategy does not depend on using a model of the object, only depends on a good estimation of its features, like size and softness. A good estimation of these features allows the task planner presented and the tactile servoing control scheme to obtain acceptable results, as seen in the experiments, where the stability of the object grasped by the hand is achieved.

The main problem for this strategy is the use of badly estimated features of the object, which could cause undesired results when the object is manipulated. Therefor a good estimation of the features is a key point to research.

Regarding the tactile servoing, it is demonstrated that is a reliable control strategy to get stable configurations with the object, but there are some issues to take into account. One is the problem when the position obtained after a control loop is not accessible by the finger kinematics. The workspace of a single finger is very limited, so is very common to find this problem. To solve this problem, an approximation based in moving joints directly can be added. The second main problem is the uncontrolled variation in the values that are obtained from the sensor, which affects the system stability. Therefor a small scale in the error is used to minimize the effect of high variations.

Acknowledgements Research supported by Spanish Ministry of Economy, European FEDER funds and the Valencia Regional Government, through projects DPI2012-32390 and PROMETEO/2013/085.

References

1. Yoshikawa, T.: Multifingered robot hands: Control for grasping and manipulation. Annual Reviews in Control **34**(2), 199–208 (2010)
2. Dang, H., Weisz, J., Allen, P.K.: Blind grasping: stable robotic grasping using tactile feedback and hand kinematics. In: Proceedings - IEEE International Conference on Robotics and Automation (2011). http://doi.org/10.1109/ICRA.2011.5979679

3. Cutkosky, M.R., Hyde, J.M.: Manipulation control with dynamic tactile sensing. In: International Symposium on Robotics Research (1993)
4. Romano, J.M., Hsiao, K., Niemeyer, G., Chitta, S., Kuchenbecker, K.J.: Human-inspired robotic grasp control with tactile sensing. IEEE Transactions on Robotics (2011). http://doi.org/10.1109/TRO.2011.2162271
5. Pedreño-Molina, J.L., Guerrero-González, A., Calabozo-Moran, J., López-Coronado, J., Gorce, P.: A neural tactile architecture applied to real-time stiffness estimation for a large scale of robotic grasping systems. Journal of Intelligent and Robotic Systems: Theory and Applications (2007). http://doi.org/10.1007/s10846-006-9040-x
6. Han, H.-Y., Kawamura, S.: Analysis of stiffness of human fingertip and comparison with artificial fingers. In: Proceedings of IEEE International Conference on Systems, Man, and Cybernetics, IEEE SMC 1999, vol. 2, pp. 800–805 (1999). doi:10.1109/ICSMC.1999.825364
7. Yousef, H., Boukallel, M., Althoefer, K.: Tactile sensing for dexterous in-hand manipulation in robotics—A review. Sensors and Actuators A: Physical **167**(2), 171–187 (2011). http://doi.org/10.1016/j.sna.2011.02.038
8. Howe, R.D.: Tactile Sensing and Control of Robotic Manipulation. Journal of Advanced Robotics **8**(3), 245–261 (1994)
9. Wettels, N., Loeb, G.E.: Haptic feature extraction from a biomimetic tactile sensor: force, contact location and curvature. In: 2011 IEEE International Conference on Robotics and Biomimetics, ROBIO 2011 (2011). http://doi.org/10.1109/ROBIO.2011.6181676
10. Liu, H., Song, X., Nanayakkara, T., Seneviratne, L.D., Althoefer, K.: A computationally fast algorithm for local contact shape and pose classification using a tactile array sensor. In: Proceedings - IEEE International Conference on Robotics and Automation, pp. 1410–1415 (2012)
11. Li, Q., Schürmann, C., Haschke, R., Ritter, H.: A control framework for tactile servoing. In: Proceedings of Robotics: Science and Systems (2013)
12. Luo, Y., Nelson, B.J.: Fusing force and vision feedback for manipulating deformable objects. Journal of Robotic Systems **18**(3), 103–117 (2001)
13. Li, Q., Elbrechter, C., Haschke, R., Ritter, H.: Integrating vision, haptics and proprioception into a feedback controller for in-hand manipulation of unknown objects. In: Proc. IEEE/RSJ Int. Conf. on Intelligent Robots and Systems (IROS), Tokyo, pp. 2466–2471 (2013)
14. Chitta, S., Sturm, J., Piccoli, M., Burgard, W.: Tactile sensing for mobile manipulation. IEEE Transactions on Robotics **27**(3), 558–568 (2011)
15. Shadow Robot: Dexterous Hand. http://www.shadowrobot.com/products/dexterous-hand/
16. Tactile Sensor Tekscan. http://www.tekscan.com/grip-pressure-measurement
17. Mira, D., Delgado, A., Mateo, C.M., Puente, S.T., Candelas, F.A., Torres, F.: Study of dexterous robotic grasping for deformable objects manipulation. In: 2015 23th Mediterranean Conference on Control and Automation (MED), pp. 262–266, June 16-19, 2015. doi:10.1109/MED.2015.7158760
18. Delgado, A., Jara, C.A., Mira, D., Torres, F.: A tactile-based grasping strategy for deformable objects' manipulation and deformability estimation. In: Proceedings of the 12th International Conference on Informatics in Control, Automation and Robotics, pp. 369–374 (2015). doi:10.5220/0005562103690374

Path Planning for Mars Rovers Using the Fast Marching Method

Santiago Garrido, David Álvarez and Luis Moreno

Abstract This paper presents the application of the Fast Marching Method, with or without an external vectorial field, to the path planning problem of robots in difficult outdoors environments. The resulting trajectory has to take into account the obstacles, the slope of the terrain (gradient of the height), the roughness (spherical variance) and the type of terrain (presence of sand) that can lead to slidings. When the robot is in sandy terrain with a certain slope, there is a landslide (usually small) that can be modelled as a lateral current or vectorial field in the direction of the negative gradient. Besides, the method calculates a weight matrix W that represents difficulty for the robot to move in certain terrain and is built based on the information extracted from the surface characteristics. Then, the Fast Marching Method is applied with matrix W being a velocities map. Finally, the algorithm has been modified to incorporate the effect of an external vectorial field.

Keywords Path planning · Fast marching · Planetry exploration

1 Introduction

Planetary exploration with robots has increased in the last decades. Different robots, such as the the the Mars Exploration Rovers (MER) [1], Spirit and Opportunity [2], have been placed in planets, like Mars, and asteroids, like Ceres and Vesta, in order to obtain valuable samples for the scientific community and enlarge our knowledge of the universe. These activities require safe and accurate movements to be executed

S. Garrido · D. Álvarez(✉) · L. Moreno
RoboticsLab, Carlos III University of Madrid, 28911 Leganés, Spain
e-mail: {sgarrido,dasanche,moreno}@ing.uc3m.es
http://roboticslab.uc3m.es/roboticslab/

This work is funded by the project number DPI2010-17772, by the Spanish Ministry of Science and Innovation, and also by RoboCity2030-II-CM project (S2009/DPI-1559), funded by Programas de Actividades I+D en la Comunidad de Madrid and co-funded by Structural Funds of the EU.

© Springer International Publishing Switzerland 2016   93
L.P. Reis et al. (eds.), *Robot 2015: Second Iberian Robotics Conference*,
Advances in Intelligent Systems and Computing 417,
DOI: 10.1007/978-3-319-27146-0_8

combined with on-board decision making, specially due to the long time the information needs to travel. These robots are equipped with different sensors to evaluate their environment. Stereo-vision system are commonly used [3], although other type of sensors such as active vision LIDAR systems and six-axis Inertial Measurement Units are also being tested [4].

Once the environment is recognised, path planning is essential for exploration and reaching interesting goals. The objective of a path planner for a mobile robot operating in environments with obstacles, is to calculate collision-free trajectories with the best possible characteristics. General desired path characteristics are safety (in terms of obstacle avoidance) and shortness (for energy optimization). Besides, in the case of planetary exploration, it is common to take into account the height [5, 6, 7, 8], the roughness (unevenness) [6, 7, 8] of the terrain, and the possible sliding of the robot due to, i.e. sandy slopes (height gradient) [6].

The path planer presented is based on the Fast Marching Method (FMM), which essentially computes the time of arrival of a wave to the different points in the environment. Then, the gradient descent method is used for path extraction. The intrinsic nature of FMM and a proper definition of the velocities at which the wave (robot) can move, makes the computed path to assure safety, and shortness [9]. In order to define the speed over the environment, the method uses the height information and evaluates the roughness of the surface based on its spherical variance. It also takes into account the gradient of the height of the area, and the distance to obstacles. Then, it combines these data with the robot movement limitations to generate a weighted matrix. This matrix can be seen as a cost function and defines the different velocities at each point of the map.

Finally, robots not always can avoid difficult terrains. This may lead to small slidings while moving. An approach for considering these slippings is presented. These undesired movements are modelled as a vectorial field which depends on the terrain and the heigh gradient of the surface, and its effect is considered using the Fast Marching Method subjected to a vectorial field (FMVF).

Statement of Contributions: This work introduces a method for considering important characteristics of the environment (in planetary exploration) and taking them into account by the path planning method, Fast Marching. Besides, new considerations in the computation of Fast Marching subjected to a Vectorial Field are presented, and its application to the robot sliding modelling is detailed.

The remainder of the paper is organised as follows. Section 2 introduces an explanation about the FMM and how this algorithm can be implemented. Next, section 3 expalins the formulation of the Fast Marching method subjected to a Vectorial Field. Following, section 4 details the velocities matrix and how it is formed. Section 5 presents numerical simulations that show promising results. Finally, the main conclusions of this paper are summarised in section 6.

2   The Eikonal Equation and the Fast Marching Planning Method

In robotics, the path planner of the mobile robot must drive it in a smooth, safe and fast way to the goal point. In nature, there are phenomena that work in the same way, such as electromagnetic waves. If at the goal point there is an antenna that emits an electromagnetic wave, then the robot could drive himself to the destination following the waves to the source. This idea is especially interesting because the potential magnetic field has all the good properties desired for the trajectory, such as smoothness (that is, C^{∞}) and the absence of local minima. In a similar manner, Fermat's principle in optics says that the path of a beam of monochromatic light follows the path of least time. When the refractive index is constant, the wave-fronts are circles, which represent different arrival times, and the paths are straight lines. In the case of two media with different indices of refraction, rays are bent as shown figure 1-a), resulting in the refraction of the path, known as Snell's law effect. When the refractive index varies continuously, the light beam is also continuously bend, as shown in in figure 1-b).

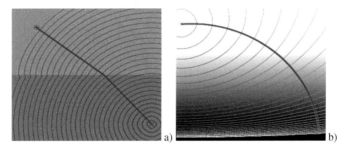

Fig. 1 Paths (or light rays) obtained with the Fast Marching Method when a) there are two media of different refractive indexes (air and water, for example) and b) when the refractive index changes continuously.

One way to characterise the position of a front in expansion is to compute the arrival time T, in which the front reaches each point of the space. It is evident that, for one dimension, we can obtain the equation of the arrival function T in an easy way, simply considering the fact that the distance x is the product of the speed F and the time T.

$$x = F \cdot T \tag{1}$$

Then, the spatial derivative of this function becomes the gradient as:

$$1 = F\frac{dT}{dx} \tag{2}$$

and therefore, the magnitude of the gradient of the arrival function $T(x)$ is inversely proportional to the speed:

$$\frac{1}{F} = |\nabla T| \tag{3}$$

For multiple dimensions the same concept is valid, since the gradient is orthogonal to the level sets of the arrival function $T(x)$. In this way, we can characterise the movement of the front as the solution of a boundary conditions problem. If speed F depends only on the position, then equation 3 can be reformulated as the Eikonal equation:

$$|\nabla T| F = 1. \tag{4}$$

The Fast Marching method (FMM) is a numerical algorithm for solving the Eikonal equation, originally on a rectangular orthogonal mesh, introduced by Sethian in 1996 [10]. The FMM is an $O(n)$ algorithm, as demonstrated in [11], where n is the total number of grid points. The algorithm relies on an upwind finite difference approximation to the gradient as a first order solution of the differential equation.

The FMM is used for problems in which the speed function never changes of sign, and so the wave front is always moving forwards (there are no reflections). This allows us to transform the problem into a stationary formulation, because the wave front crosses each grid point only once. The wave propagation given by the FMM gives us a distance function that corresponds to the Geodesic distance measured with the metric given by the the refraction matrix, which indicates the velocity at which the wave front moves forward.

Since its introduction, the FMM has been applied with success to a wide variety of problems that arise in geometry, computer vision and manufacturing processes (see [12] for details). Several advances have been made to the original technique, including the adaptive narrowband methodology and the FM method for solving the static Eikonal equation [10], and also different implementations have been tried out to make it faster [13].

2.1 Algorithm Implementation on an Orthogonal Mesh

The Fast Marching Method models any phenomena that can be described as a wave front propagating normal to itself with a speed function $F = F(i, j)$. The main idea is to methodically build the solution using only upwind values. Let $T(i, j)$ be the time at which the curve crosses the point (i, j), then it satisfies $|\nabla T| F = 1$, the Eikonal equation.

It is important to point out that the propagation happens from smaller to bigger values of T. While evolving, the algorithm classifies the points of the mesh into three sets: frozen, open and unvisited. Frozen points are those where the arrival time has been computed and is not going to change in the future. Unvisited points are those that have not been processed yet. Finally, open are those points belonging to the propagating wave front, which can be considered as an interface between frozen and unvisited regions of the mesh. Initially, the source X_s is marked as frozen and assigned a time of arrival of 0. All points adjacent to it (von Neumann neighbourhood is considered), are marked as open and their T value is computed using equation 5.

$$\left(\frac{T - T_x}{\Delta x}\right)^2 + \left(\frac{T - T_y}{\Delta y}\right)^2 = \frac{1}{F_{ij}^2} \tag{5}$$

where T_x and T_y are the minimum arrival time of the neighbours in X and Y axes respectively, and F_{ij} is the velocity of propagation in the cell for which the arrival time is being computed.

At each iteration, the open point with the smallest value of $T(x)$ is labelled as frozen and its neighbours are analysed and tagged as open. The process continues until all points are marked as frozen or the goal is reached.

Figure 2a) explains evolution of the method: in the first subfigure the dark blue point is the start point of the wave, and its neighbours are marked in grey. In the 2nd subfigure, their value T is computed and they are coloured in light blue, while their neighbourhoods are labelled in argy. The algorithm continues and the points with different time of arrival are marked with a different colour. If the final solution is plotted using the time of arrival as the third axis, the funnel potential with one global minimum shown in figure 2b) is obtained.

Finally, since the time of arrival function has a funnel-like shape, the robot's path is extracted using the gradient descent method. So far, no kinematic constraints are included in the path planning, however, due to the nature of the wave expansion, no abrupt turns appear in the final path as long as the velocity field changes smoothly.

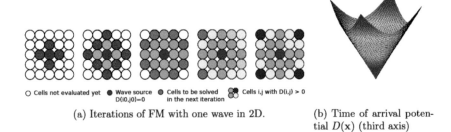

○ Cells not evaluated yet ● Wave source ● Cells to be solved ⊗ Cells i,j with D(i,j) > 0
                              D(i0,j0)=0   in the next iteration

(a) Iterations of FM with one wave in 2D.   (b) Time of arrival potential $D(x)$ (third axis)

Fig. 2 Scheme of Fast Marching propagation with an initial point. Different colours (blue to red) represent different arrival times in increasing order.

3 Fast Marching Method subjected to a Vectorial Field (FMVF)

The Fast Marching method subjected to a Vectorial Field (FMVF) was developed by Vladimirsky [14], though he called it Anisotropic Fast Marching. The problem is that the same name has been used by Peyré [15] for a different algorithm. We also believe that the name does not correspond to its real meaning. Besides, the name Vectorial Fast Marching was also used by Valero [16] for oriented propagation algorithms based on FMM. Therefore, we have considered preferable to use the surname 'subjected to a Vectorial Field', which we think corresponds better to its physical meaning.

From a physical point of view, the velocity of propagation over any medium belongs to $[0, c]$, where c is the light speed, but it can be normalised to the interval $[0, V]$, where V represents the maximum velocity of propagation of our vehicle. Besides, the refractive index $n = c/F$ belongs to the interval $[0, 1]$, with 1 representing the maximum possible speed of the wave propagation (or robot velocity). Therefore, the cost function $f = 1 - n$ belongs to the interval $[0, 1]$, being $f = 0$ the minimum cost and $f = 1$ is the maximum cost of the function. Thus, the optimal route minimizes the time between the starting point x_s and end point x_g:

$$T = V \cdot \min_{C(s) \subset D} \int_{x_s}^{x_g} f(C(s)) ds \qquad (6)$$

where $C(s)$ represents any path in the domain D.

It is interesting to note that the authors (Vladimirsky [14] and Petres [17], [18]) that treated this subject previously, did not continue using this technique. We think this is because they perform a normalisation of the magnitude of the external vectorial field without taking into account the magnitude of the cost function of the map. This makes a vectorial field with an intensity of 1 to have the same effect on the final path than one with an intensity of 10. However, in our opinion, the function that it is necessary to normalise is the total cost function:

$$\tilde{f} = f_{dif} + f_{vect} \qquad (7)$$

where f_{dif} is the cost function due to the difficulty of the terrain and the obstacles, and it is computed as $f_{dif} = 1 - W$, being W a cost matrix that will be explained in section 4. The second part corresponds to the external vectorial field.

This way, the influence of the vectorial field over the velocity of the vehicle depends on the magnitude of the two of them and on the angle between them, i.e., it depends on scalar product, and therefore the f_{vect} can be defined as:

$$f_{vect}(i, j) = 1 - \langle \nabla T_{i,j} \cdot \mathbf{F}_{i,j} \rangle \qquad (8)$$

Physically, this is equivalent to say that a force favours the vehicle when both external vectorial field and vehicle are pointing to the same direction.

It is very important to highlight that the new cost function defined in equation 7 must to be always positive, because in the Fast Marching methods the wave-front cannot move backwards. More details on the algorithm can be consulted in Petres [17, 18].

In figure 3-a) the expansion of a wave (FMM) is shown. Note there is a rectangular obstacle in the middle. Then, the expansion of a wave when it is subjected to a unitary vectorial field which points to the right in the upper part, to the left in the lower part, and with no external field in the middle, is shown in figure 3-b). It can be noted how the wave expands faster in the upper part than in the lower one.

Furthermore, in this paper, we have applied the FMVF method on a matrix of difficulty for trajectories that optimise a cost function while they are subject to a vectorial field, which in the case of paths for a Rover on Mars can be caused by the sideslip due to the sand.

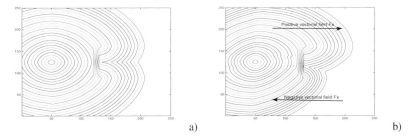

a)   b)

Fig. 3 Fast marching expansion wave with a rectangular obstacle in the middle a) and Fast Marching subjected to a vectorial field (FMVF) expansion wave b) subjected to a unitary vectorial field to the right in the upper part, zero in the middle and a unitary vectorial field to the left in the lower part.

4 Matrix W: The Difficulty Map

As expressed in section 1, the proposed technique is based on the FMM. This algorithm uses a velocity (refraction) index of the wave front to compute the time-of-arrival in the map. If we build this map based on the 3D characteristics of the environment and the robot limitations, desirable path attributes are obtained. The trajectory is not going to be the simple geodesic (computed when we have constant velocity of the wave), but it is going to be modified according to the robot and task needs. The proposed method creates a weight matrix W, which is built based on: the *spherical variance*, the *saturated gradient of height*, and the *height*.

4.1 Spherical Variance

In [19] a method to calculate the roughness degree is presented. The spherical variance finds the roughness of a surface to determine the level of difficulty for the robot to move through it. The spherical variance is obtained from the orientation variation of the normal vector at each point. In a uniform terrain (low roughness), the normal vectors will be approximately parallel and, for this reason, they will present a low dispersion. On the other hand, in an uneven terrain (high roughness) the normal vectors will present great dispersion due to changes in their orientation. The spherical variance is obtained as follows:

1. Given a set of n normal vectors to a surface, defined by their three components $\overrightarrow{N_i} = (x_i, y_i, z_i)$, the module of the sum vector R is calculated by:

$$R = \sqrt{\left(\sum_{i=0}^{n} x_i\right)^2 + \left(\sum_{i=0}^{n} y_i\right)^2 + \left(\sum_{i=0}^{n} z_i\right)^2} \tag{9}$$

2. Next, the mean value is normalized following equation 10, so the value of the result is within $[0, 1]$.

$$\frac{R}{n} \in [0, 1] \tag{10}$$

3. Finally, the spherical variance ω is defined as the complementary of the previous result.

$$\omega = 1 - \frac{R}{n} \tag{11}$$

When $\omega = 1$, there exists a maximum dispersion that can be considered as the maximum roughness degree, and when $\omega = 0$, a full alignment exists and the terrain will be completely flat.

4.2 Saturated Gradient

The gradient of a scalar field f is defined to be a vector field whose components are the partial derivatives of f. That is:

$$\nabla f = (\frac{\partial f}{\partial x_1}, ..., \frac{\partial f}{\partial x_n}) \tag{12}$$

The saturated gradient of the height magnitude consists of giving a limit value to the gradient of each point over the surface. The gradient limit depends on the robot capabilities; the maximum inclination that the robot is able to cross will be the limit value for the saturated gradient.

4.3   Construction of Matrix W

As previously explained, with this matrix the algorithm can modify the path that the robot is going to follow across the 3D surface by having a proper velocity value for each point on the surface. We can add as many characteristics as we need to get different paths including several attributes of the terrain.

In our case, the saturated gradient, the spherical variance, and the height are three matrices G, Sv, and H with the same size as the environment map. To build matrix W, we give a weight factor to each surface characteristic and we can determine which one is the more important depending on the task requirements. The values of the matrix's elements are normalized, and therefore they vary from 0 to 1, so after the combination based on equation 13, the values of matrix W are also within this range. The components of matrix W with a value of 0 will be points in the environment with maximum speed. Hence, these are points which the robot can cross without any problem and at its maximum speed. The components of W with a value of 1 will be points in the *vertex* with a minimum speed. This means that the robot will not be able to pass across them.

$$W = a_1 \cdot G + a_2 \cdot Sv + a_3 \cdot H \qquad (13)$$

where a_1, a_2 and a_3 are weighting factors, and they fulfil $\sum_i a_i = 1$.

After matrix W is generated, the FMM is run over the modified map to calculate the best trajectory. With the FMM we can assure that the path found will be the one of less time in the W metrics.

5   Numerical Simulations

Several tests on different environments have been carried out in order to test the usability of the algorithm and the influence of the weighted matrix and the external vectorial field on the planned paths.

Figure 4 shows the behaviour of the FMM when it is subjected to a external vectorial field using as initial map the room shown in Fig. 4a). In this case, the velocities matrix W only takes into account the distance to obstacles, and it is shown in figure 4b), where yellow and green colours indicate areas with more clearance. Using this difficulty map, the FMVF method computes the propagation of a wave taking into account a vectorial field. In figure 4c), there is no external field and the wave expansion forms circles of different colours. Each colour indicates a different time of arrival of the wave. Figure 4e) shows the wave expansion when there is an external field pointing upwards. It can be seen how the light blue area expands farther in the upper part, meaning it takes less time to arrive to that area. Finally, figures 4d) and f) show the different path extracted in each case. It is easy to observe how these paths are affected by the external field.

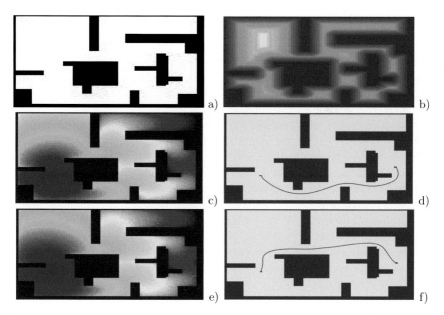

Fig. 4 Path calculated using FMM when there is an external vectorial field. a) Original binary map of the room. b) Velocities map taking into account the distance to obstacles. c) and e) Wave expansion (arrival time) with an external field that goes downwards and upwards respectively. d) and f) Path extracted from the wave expansion.

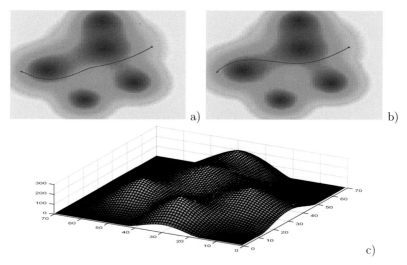

Fig. 5 Path calculated penalising the change in height, when there is an external field a) downside, b) upside and c) original map.

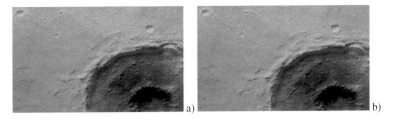

Fig. 6 Different paths in the Mars Gale crater taking into account the lateral landslide proportional to the gradient. a) without landslide, b) with landslide.

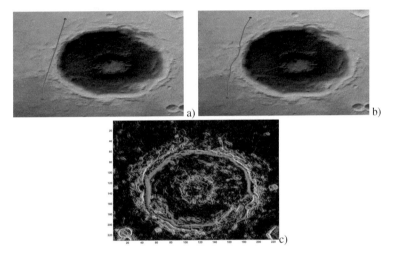

Fig. 7 Path calculated using mountains map penalising the change in height, when there is an external field a) downside, b) upside and c) original map.

Next, the effect of the external vectorial field is combined with an elevation map represented as a 2.5D surface that can be seen in figure 5c). In this case, the height is used for building the W map. The *base* is the height of the initial point $x_s = (i_0, j_0)$ and the change in height is penalised. The difficulty matrix calculated as the absolute value of the elements height matrix $H(i, j)$ minus the height of the initial point: $W(i, j) = |H(i, j) - H(x_s)|$. In Fig. 5a) and b), two different paths using this difficulty matrix are shown. The changes in height along the resulting paths are very small because these changes are penalised by the matrix W. When an external vectorial field is applied, the path is modified. In Fig. 5a) the applied current goes downside, and in Fig. 5b) goes upside.

Then, experiments are done using a height map of the area around the Gale crater in Mars, which can be seen in figure 6. In this case, possible slidings while moving inside the crater are modelled applying a vectorial field proportional to the gradient of the terrain surface. The different resultant paths can be seen in figures 6a) and b).

Finally, both the roughness and the changes of height are considered together in the construction of the difficulty map, which can be seen in figure 7c). Different values a_i in the weighting factor when computing the map W lead to different paths. In figure 7a), the changes in height along the path are more penalised than in figure 7b), in which the roughness of the terrain is more penalised.

6 Conclusions

The algorithm we have presented is a novel methodology for considering the most important aspects of path planning for exploration of planets and asteroids. The main characteristics considered as: height, roughness, magnitude of slope, path length and clearness among obstacles. They are used to build a velocities map that is later used by the Fast Marching Method for computing the time of arrival of a wave expansion. Then, the path is extracted using gradient descent. Also, possible slidings due to slopes and terrain conditions are modelled as an external vectorial field, and taken into account using Fast Marching Method subjected to a Vectorial Field. Numerical simulations show promising results on the application of this methodology. Different results show the influence of the combination of several characteristics of the environment in the resulting path. In all cases, obstacles and difficult terrain are avoided. Besides, possible sliding problems are taken into account in the path planning phase.

References

1. Volpe, R., Estlin, T., Laubach, S., Olson, C.: Enhanced mars rover navigation techniques. In: IEEE International Conference on Robotics and Automation, San Francisco, vol. 1, pp. 926–931 (2000)
2. Maimone, M.W., Leger, C., Biesiadecki, J.: Overview of the mars exploration rovers' autonomous mobility and vision capabilities. In: IEEE International Conference on Robotics and Automation, Space Robotics Workshop, Roma (2007)
3. Carsten, J., Rankin, A., Ferguson, D., Stentz, A.: Global path planning on board the mars exploration rovers. In: IEEE Aerospace Conference, Big Sky, pp. 1–11 (2007)
4. Rekleitis, I., Bedwani, J.L., Dupuis, E., Allard, P.: Path planning for planetary exploration. In: Canadian Conference on Computer and Robot Vision, Windsor, pp. 61–68 (2008)
5. Muñoz, P., Rodríguez, M.D., Castaño, B., Martínez, A.: Fast path-planning algorithm for future mars exploration. In: International Symposium on Artificial Intelligence, Robotics and Automation in Space, Turin (2012)
6. Ishigami, G., Nagatani, K., Yoshida, K.: Path planning for planetary exploration rovers and its evaluation based on wheel slip dynamics of the terrain. In: IEEE Int. Conference on Robotics and Automation, Roma, pp. 2361–2366 (2007)
7. Potiris, S., Tompkins, A., Goktogan, A.: Terrain-based path planning and following for an experimental mars rover. In: Australasian Conference on Robotics and Automation, Melbourne, pp. 1–10 (2014)
8. Gennery, D.B.: Traversability Analysis and Path Planning for a Planetary Rover. Autonomous Robots 6, 131–146 (1999)
9. Garrido, S., Moreno, L., Blanco, D.: Exploration of a cluttered environment using voronoi transform and fast marching method. Robotics and Autonomous Systems 56, 1069–1081 (2008)

10. Sethian, J.A.: Theory, algorithms and aplications of level set methods for propagating interfaces. Acta numerica, Cambridge University Press, pp. 309–395 (1996)
11. Yatziv, L., Bartesaghi, A., Sapiro, G.: A fast o(n) implementation of the fast marching algorithm. J. of Computational Physics **212**, 393–399 (2005)
12. Sethian, J.: Level set methods. Cambridge University Press (1996)
13. Gómez, J.V., Álvarez, D., Garrido, S., Moreno, L.: Fast methods for eikonal equations: an experimental survey (2015). ArXiV abs/1506.03771
14. Vladimirsky, A., Sethian, J.A.: Ordered upwind methods for static Hamilton-Jacobi equation: Theory and algorithms. SIAM J. of Numerical Analysis **41**(1), 325–363 (2003)
15. Peyre, G.: Advanced Signal, Image and Surface Processing. Ceremade, Université Paris-Dauphine (2010)
16. Valero-Gomez, A., Gómez, J.V., Garrido, S., Moreno, L.: The Path to Efficiency: Fast Marching Method for Safer, More Efficient Mobile Robot Trajectories. IEEE Robotics and Automation Magazine **20**(4), 111–120 (2013)
17. Petres, C., Pailhas, Y., Evans, J., Petillot, Y., Lane, D.: Underwater path planning using fast marching algorithms. In: Oceans European Conference, vol. 2, pp. 814–819 (2005)
18. Petres, C., Pailhas, Y., Patron, P., Petillot, Y., Evans, J., Lane, D.: Path Planning for Autonomous Underwater Vehicles. IEEE Transactions on Robotics **23**(2), 331–341 (2007)
19. Castejon, C., Boada, B., Blanco, D., Moreno, L.: Traversable region modeling for outdoor navigation. J. Intelligent and Robotic Systems **43**, 175–216 (2005)

Expressive Lights for Revealing Mobile Service Robot State

Kim Baraka, Ana Paiva and Manuela Veloso

Abstract Autonomous mobile service robots move in our buildings, carrying out different tasks across multiple floors. While moving and performing their tasks, these robots find themselves in a variety of states. Although speech is often used for communicating the robot's state to humans, such communication can often be ineffective. We investigate the use of lights as a persistent visualization of the robot's state in relation to both tasks and environmental factors. Programmable lights offer a large degree of choices in terms of animation pattern, color and speed. We present this space of choices and introduce different animation profiles that we consider to animate a set of programmable lights on the robot. We conduct experiments to query about suitable animations for three representative scenarios of our autonomous symbiotic robot, CoBot. Our work enables CoBot to make its state persistently visible to humans.

Keywords Mobile service robots · Expressive lights · Robot internal state · Non-verbal communication · Human-robot interaction

1 Introduction

Collaborative service robots are meant, through symbiotic autonomy [15], to effectively collaborate with humans in order to successfully perform their tasks.

K. Baraka(✉)
Robotics Institute, Carnegie Mellon University, Pittsburgh, PA, USA
e-mail: kbaraka@andrew.cmu.edu

A. Paiva
INESC-ID, Instituto Superior Tecnico, Porto Salvo, Lisboa, Portugal
e-mail: ana.paiva@inesc-id.pt

M. Veloso
Computer Science Department, Carnegie Mellon University, Pittsburgh, PA, USA
e-mail: mmv@cs.cmu.edu

© Springer International Publishing Switzerland 2016   107
L.P. Reis et al. (eds.), *Robot 2015: Second Iberian Robotics Conference*,
Advances in Intelligent Systems and Computing 417,
DOI: 10.1007/978-3-319-27146-0_9

With symbiotic autonomy, a two-way symmetric relationship holds: the robot servicing the human and the human servicing the robot. While our own collaborative robots, the CoBots [15], move in our buildings, successfully carrying out different tasks and traversing multiple floors, there is a need for revealing their internal states in several situations where they are meant to collaborate with humans. Currently, our CoBots communicate mainly verbally, speaking instructions out to both task solicitors (people who request tasks from the robot) and task helpers (the human actors in the symbiotic autonomy process). However, these mobile robots have many features in their internal state, including map representations, task and sensor information, which are all not visible to humans. One of our important goals is to find a good way of expressing information and internal state of these robots through features visible to humans. For this purpose, verbal communication has its limits. One of them is proximity: on-robot verbal communication is limited to the intimate, personal and social domains [2]. There are some cases where communication in the public domain is required (e.g. robot calling for help), and verbal communication (or even on-screen display) is helpless in this case. Another limitation of verbal communication is its transient nature (the expression lasts the duration of a sentence).

To remedy these problems, we propose to use lights as a medium of communication from robot to humans, as a way to reveal to the latter the internal state of the robot. Unlike however most of the existing robots that use lights for expression of state, CoBot is a mobile robot that interacts with humans in different specific manners: requests help (to activate objects in its tasks), influences change in the user's motion (in relation to its own motion) or provides useful information (task-related or general). The spectrum of entities potentially expressed through these lights is hence greatly diverse and non-simplistic.

The rest of the paper is organized as follows. Section 2 discusses related work. Section 3 describes the design of our light interface for revealing the internal state of the robot. In section 4, we focus on three scenarios in which CoBot finds itself and investigate the *what* and *how* of internal state expression. Section 5 shows experimentation with the goal of selecting appropriate light animations for these scenarios. Finally, section 6 presents conclusions and future research directions.

2   Related Work

Most human-oriented technology generally makes use of some form of light indicators. Lights are used in personal electronic devices ranging from cell phones to toasters, and their expressivity can be greatly exploited [8]. Expressive lights have also been used in wearable technology (on apparel for instance [4]) and interactive art installations [9] [1]. Another important but different use of light is for stage or scene lighting, which still shares common expressive features with indicator lights like color, intensity and time-varying patterns [5]. As robots themselves become more human-oriented, designers and researchers started integrating lights on robots (like has been done with the NAO or the AIBO robots) for non-verbal communication.

More recent works have considered more functional uses of lights specifically on robots, which we describe next.

The purpose of using lights on robots varies in the different works we found, but almost all uses of expressive lights on robots still remain rudimentary. First, some work has been done on making impressions on the user rather than explicitly trying to express something tangible. The type of expressions used in this work are called artificial subtle expressions (ASE's) and are not linked to an internal state of the robot [11]. Second, explicit expression of emotions in robots and artificial agents has recently become an active area of research in the field of affective computing [3]. For instance, lights on a robot have been used to express people's emotions in a cafe-style room [13]. Another example of affective expression through lights is the use of lights as a complement to facial expressions [10]. Third, lights can sometimes be used for strictly functional purposes. Examples include debugging or improving human-robot interaction in a practical way, where for example a blinking LED is used to avoid utterance collisions in verbal human-robot communication [7]. Finally, lights can be used to communicate intent, such as flight intent of an aerial robot [14].

Most of the works presented above mainly focus on the "what" component of expression (what to express). Equally important to that is the "how" component (how to express). An in-depth analysis and study of possible ways of using a single point light source to express a wide array of messages was found [8]. This work is a good starting point for thinking of ways to use a set of lights as a genuinely expressive and functional medium.

3   Light Interface for State Revealing

3.1   *Formalization*

Robot Internal State. We assume that the *full* state of the robot at a particular time can be represented as the tuple:

$$S(t) = \langle \mathcal{F}(t), \mathcal{P}(t) \rangle$$

where: $\mathcal{F}(t) = (f_1(t), ..., f_n(t))$ is a vector of (discrete) state *features* that determines the type of state in which the robot is; $\mathcal{P}(t) = (p_1(t), ..., p_n(t))$ is a vector of (possibly continuous) state *parameters* which modulate the state within the state type defined by $\mathcal{F}$. The reason why we distinguish between features and parameters is the following. We would like to associate a light animation type to a state type (determined solely by feature variables), while also having a way of modulating this animation with possibly varying parameters without having to define micro-states for each parameter value. Both feature and parameter variables are functions of sensor and/or task execution information. Some parameters might be relevant to one or more state types, and irrelevant to others, depending on the value of $\mathcal{F}(t)$.

From $\mathcal{S}(t)$, we are only interested in $\mathcal{S}'(t) \subset \mathcal{S}(t)$, which we call the *relevant* state. It represents the set of variables we wish to make transparent to the outside world, or externalize. We write $\mathcal{S}'(t)$ as: $\mathcal{S}'(t) = \langle \mathcal{F}'(t), \mathcal{P}'(t) \rangle$ where $\mathcal{F}'(t)$ and $\mathcal{P}'(t)$ are the *relevant* state features and parameters, respectively. The optimal choice of $\mathcal{S}'(t)$ from $\mathcal{S}(t)$ is a separate research question that is outside the scope of this paper. To each state in the relevant state space, we would like to associate an animation of the lights. We are hence looking for a mapping $\mathcal{M} : \mathcal{S}' \to \mathcal{A}$, where $\mathcal{A}$ is the set of possible animations that we consider.

Light Animations. An animation A of a set of n lights is defined as a time-varying n-by-3 matrix of light intensities:

$$A(t) = \begin{pmatrix} i_{1R}(t) & i_{1G}(t) & i_{1B}(t) \\ \vdots & \vdots & \vdots \\ i_{nR}(t) & i_{nG}(t) & i_{nB}(t) \end{pmatrix}$$

where the rows represent the indices of the individual lights (which we call pixels from now on) and the columns represent the R, G and B components of each pixel. Similar to [8], we focus on a limited set of possible intensity functions $i_{jk}(t)$. Here we consider basic types of functions that could be later modulated or combined if needed. They are summarized in Table 1.

Table 1 Shapes considered for each $i_{jk}(t)$

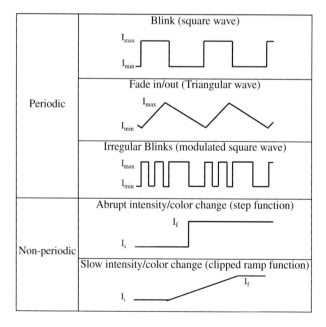

For each animation A, we restrict ourselves to the case where all $i_{jk}(t)$'s in $A(t)$ have the same shape among the ones presented in table 1. We also allow a possible offset between the rows of $A(t)$ if we want to achieve a spatial scan over the lights in space. Note that if the ratios $I_{R,max} : I_{G,max} : I_{B,max}$ and $I_{R,min} : I_{G,min} : I_{B,min}$ are maintained in these animations, it will result in a monochromatic animation. If the ratio is not respected however, we see changes in color throughout the animation.

3.2   Proposed Interface

Fig. 1 shows all parts of the proposed interface. A node running on the robot itself (1) collects state information $\mathcal{S}'(t)$ at every time step. Any change in $\mathcal{S}'(t)$ will trigger a command from (1) to the microcontroller (2), notifying it only of the variables in $\mathcal{S}'(t)$ which changed. The microcontroller keeps track of $\mathcal{S}'(t)$ locally (in synchronization with (1)'s copy of it). Also, although state variables are constantly updated, only data variables which are relevant to the current state are updated. (2) acknowledges that it correctly received the command by responding to (1) with an acknowledgement (ACK in the figure). (2) is programmed with the state-animation mapping $\mathcal{M}$ mentioned in the previous section, and triggers the animation $\mathcal{M}(\mathcal{S}'(.))$ in the programmable lights (3) at each state change. The animation then runs continuously until interrupted by a subsequent state change.

We implemented the interface described above on our collaborative mobile service robot, CoBot, which runs the ROS operating system. The state collection node is a simple ROS node that subscribes to different topics published by other nodes which provide enough information to infer $\mathcal{S}'(t)$ at every time step. An Arduino Uno board was used as the microcontroller, communicating serially with both (1) and (3) in Fig. 1. The program on the microcontroller alternates between a cycle in which it listens

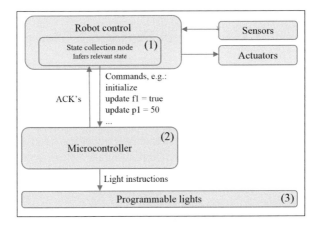

Fig. 1 Control diagram of the proposed programmable lights interface

to possible updates and a cycle in which it refreshes (3) (being a sequential device, it cannot perform both simultaneously). For the programmable lights, the Adafruit NeoPixels strip, a linear LED strip with individually controllable pixels, was chosen. Compared to other options like luminous fabrics or LED panels, a linear strip is both simple in structure and flexible to adopt different mounting alternatives on CoBot. The NeoPixels strip moreover provides high light intensity thanks to its density of 144 LEDs/m (35 Watts/m max). The 63 cm strip was mounted vertically over the body of the CoBot as shown in the hardware diagram of Fig. 2.

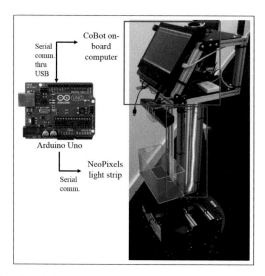

Fig. 2 Hardware interface design

4 Opportunistic Cases for Light Expression: What and How to Express?

There is a wide spectrum of aspects of the robot's internal state that could be expressed through lights (virtually any set of variables in the full state of the robot). In practice, the limited medium we are dealing with (a single strip of lights) gives rise to a trade-off between legibility of expression and diversity of the interface vocabulary. From the robot's perspective, this can be seen as a trade-off between transparency of the robot's internal state (*what* it reveals) and understandability of the elements it externalizes (*how* well it reveals it). As a consequence, choosing *what* to express is an important step before thinking of *how* to express. The choice of states that will require animation is directly coupled to the possible impact the addition of lights could make in these states, in relation to the tasks carried by the robot and their associated needs.

In this paper, we focus on three situations that we believe are representative of the situations in which a mobile service robot like CoBot generally finds itself while performing its diverse tasks.

4.1 Selected Scenarios

Waiting for Human Input (Scenario "Waiting"). CoBot is a symbiotic autonomous robot [15] that proactively asks for help given its limitations. It often finds itself in situations where it is waiting for a human input to carry on its tasks. For example, as it does not have arms, it cannot press the elevator buttons when travelling from floor to floor and hence asks for help verbally when it is in the elevator hall. Such spoken request is not always effective because of the transient nature of the communication and the limited auditory range. The presence of lights, bright enough to be persistently seen from far away, might provide a more effective way of expressing CoBot's need for help.

Blocked by a Human Obstacle (Scenario "Blocked"). CoBot's navigation is often impeded by humans who stand in its way. In these situations, and as CoBot does not deviate more than a predefined threshold from its planned route, it will not be able to move unless the person moves out of its way. CoBot issues a verbal request ("Please excuse me.") with the hope that the human opens a path for it to pass. Again, the verbal command could be complemented with lights as a more effective way to express the robot's interrupted state.

Showing Task Progress to a User (Scenario "Progress"). This is a scenario where there is a need to display the progress of the robot on a task (for example, CoBot escorting a visitor to a room). The presence of a progress indicator has been shown to reduce user anxiety [12]. In the escort example, as the visitor follows the robot and does not know the location of the destination room, he/she is ignorant of how much is left to navigate. We investigate the use of lights to display progress, as a function of the estimated distance from the task goal.

4.2 Relevant State Choice

For the three scenarios described above, the relevant state S' is represented by the following set of state variables (both features and parameters).
The relevant feature variables $\mathcal{F}'$ are:
- path_blocked (abbr. pb): a boolean variable indicating whether CoBot's path is impeded,
- el_waiting (abbr. ew): a boolean variable indicating whether the robot is waiting for human input at an elevator
- task_type (abbr. tt) : a variable indicating the type of task being executed.

We are interested in the value "esc", which indicates that a person is currently being escorted.
The relevant parameters $\mathcal{P}'$ used are esc_tot_time (abbr. et) and esc_rem_
time (abbr. er) which indicate respectively the estimated total and remaining time
for the current escort task.

4.3 Animation Framework

Parametrization of $\mathcal{A}$. To simplify the search for suitable animations for each of
the scenarios presented above, it is useful to focus on a finite set of parameters that
fully define the animation. Finding suitable animations will hence reduce to finding
suitable values for these parameters.

   We used a different parametrization for scenarios "waiting" and "blocked" than we
did in scenario "progress", given that the nature of the expression differs considerably.
The parametrizations used are described next.

– Scenarios "waiting" and "blocked": For these scenarios, we opt for a periodic animation function. Furthermore, all pixels have identical animations, i.e. all rows
  of $A(t)$ are equal functions of time. The parameters we look at are: the animation
  function shape wv (selected from the possible options shown in Table 1), the dynamics parameter D (defined as the percentage of the period where the function is
  high or rising), the period T and the R:G:B ratio or color (R, G, B). The maximum
  intensity $I_{max} = I_{R,max} + I_{G,max} + I_{B,max}$ is set to a constant and I_{min} is set to be
  zero for R,G and B.
– Scenario "progress": For this scenario, we consider non-periodic animation functions and do not require the animation functions to be synchronized across pixels.
  We look at the following parameters: progress display method $disp$ (how is the
  progress towards the goal expressed?), direction of progress displayed u_{disp} (only
  if the progress is displayed spatially - e.g progress bar), the initial color $(R, G, B)_i$
  (strip color at the beginning of the escort) and the final color $(R, G, B)_f$ (strip
  color when the goal is reached).

Animation Algorithms. As discussed in previous sections, there is a direct mapping
between $\mathcal{F}'$ and $\mathcal{A}^*$, where $\mathcal{A}^*$ is our parametrized set of animations. The following
animation methods are triggered by values taken by $\mathcal{F}'$:
   -anim_waiting(wv, D, T, R, G, B)
   -anim_blocked(wv, D, T, R, G, B)
   -anim_progress$(disp, u_{disp}, (R, G, B)_i, (R, G, B)_f, et, er)$

Algorithm 1. Animation control algorithm

```
1: while true do
2:     (F′, P′) = UPDATE_STATE()
3:     if pb == 1 then  anim_blocked(wv,D,T,R,G,B)
4:     else
5:         if ew == 1 then anim_waiting(wv,D,T,R,G,B)
6:         else
7:             if tt == "esc" then anim_progress(...,et,er)
```

Note that et and er are the state parameters $\mathcal{P}'$ which modulate the corresponding method anim_progress, while the rest of the method arguments are all design parameters (to be determined in the next section). The first two methods, linked to a periodic animation as mentioned above, only execute one period of the animation. The last method only performs an update to the light strip in response to change in the parameters. Putting these in a loop structure performs the required animation, as shown in Algorithm 1. Note that scenarios can overlap (e.g. being blocked while escorting), so some prioritization is needed.

5   Study: Animation Selection

Methodology: In order to select suitable parameters for the animations presented above, we conducted a study with a video-based survey. Participants were first given detailed description about the situation of the robot in each scenario and then asked to watch videos showing the robot in each of the scenarios defined above, while answering a survey through the form of a spreadsheet.

Preliminary Study: A preliminary study was conducted with the people who have the most expertise for our purposes, namely the CoBot developers. Eight developers participated in the survey, and submitted their choices. To validate our design choices, we recruited 30 more people to include in the study. The results across both studies were consistent. The extended study is described next.

Participants: A total of 38 participants took part in this study. 61% study or work in a robotics-related field, 18% are in a design-related field and 21% are in an engineering-related field. Ages range from 19 to 50 years with an average of around 25 years. 18% are from North America, 32% are from Europe, 29% are from the Middle East and 21% are from Asia. 68% are male and 32% are female.

Survey Design: Participants were asked to give their input on three aspects of the animation: animation type, speed and color. For each scenario, 3 different types of animations where shown with the same neutral color (light blue). Nuances of 3 different speeds were also shown within each type. The participants were asked to select the one that they thought would fit best the robot's expression purposes in the given scenario. Participants were also shown 6 possible light colors (in the form of a static image of the robot) and were asked to select the most appropriate for each scenario. We make the reasonable assumption that the choice of color for the animation is independent of the actual animation selected, which helps reduce the amount of choices to be shown. Indeed, while animation type (or pattern) and speed both relate to modulations in time and intensity, color seems to be much less intertwined to the other two. Furthermore, according to color theory [16], color on its own plays a strong role in expression. Next, we list and justify the choices of animation types shown to the participants.

- Scenario "waiting": A regular blinking animation (Blink); a siren-like pattern; a rhythmic (non-regular) blinking animation. We believe these to be good candidates for grabbing attention because of the dynamic aspect, the warning connotation and the non-regular pattern respectively.
- Scenario "blocked": A faded animation (that we call "Push") that turns on quickly and dies out slower (giving the impression of successively pushing against an obstacle); an "aggressive" blink (fast blink followed by slow blink); a simple color change at the time the robot gets blocked. We believe these to be good candidates for inciting the human to move away from the path.
- Scenario "progress": A bottom-up progress bar where lights gradually fill from top to bottom proportionally to the distance from the goal; a top-down progress bar where lights fill from the top towards the bottom; a gradual change from an initial color to a final color, again proportionally to the distance from goal.

The parameter values associated with these animations are summarized in Table 2. In addition to the animation summarized in the table, the following colors were shown for each scenario as static images of the lighted robot: Red (R), Orange (O), Green (G), Light Blue (B), Dark Blue (B') and Purple (P).

Table 2 Parameter values for the animation choices shown

Scenario "Waiting"			Scenario "Blocked"		
wv	D	T (s)	wv	D	T (s)
Blink			Push		
	0.5	2/1.6/0.6		0.25	1.5/1/0.5
Siren			Aggressive Blink		
	0.5	2/1.6/0.6		0.5	2/1.6/0.6
Rhythmic Blink			Color change		
	0.5	3/2.5/1.5		1	-

Scenario "Progress"	
$disp$	u_{disp}
prog_bar	bottom_up
prog_bar	top_down
color_change	-

5.1 Results

Table 3 shows the selected best choices, which were consistent between the preliminary and the extended study. Fig. 3 show the distribution of the results in the extended study. In the following discussion, p-values are obtained from a Chi-Square goodness-of-fit test against a uniform distribution.

In Fig. 3, we show the results for the animation type. For the scenario "waiting" (p-value of 0.0137), among the participants who chose the winning animation "Siren", 64% chose the slower speed, 29% the medium speed and 7% the faster speed. For the scenario "blocked" (p-value of 0.0916), among the participants who chose the winning animation "Push", 27% chose the slower speed, 40% the medium speed and 33% the faster speed. For the scenario "progress" (p-value of 1.10^{-6}), the participants chose the bottom-up progress bar animation. All p-values obtained are below 0.10, which indicates a strongly non-uniform distribution of preferences for each scenario, and this can clearly be seen in Fig. 3.

The results for colors similarly show a clear preference for one option in each case. For instance, light blue was selected for the "waiting" scenario. This result supports the statement in [4] that cold colors are better than warm colors at grabbing attention. Also, red was selected as the best color for the "blocked" scenario. This is consistent with the fact that red is often perceived as demanding [16] or stimulating [4], which are both desirable in this scenario.

The results of the study show that some animation design alternatives can be eliminated, while a small set can be considered valid. Although there is generally a clear preference for one of the choices in each scenario, the study was informative of the distribution of preferences, which enables us to possibly generate animations probabilistically instead of only committing to a single one. Also, the scenarios we looked at are quite generic and are commonly encountered in interactions involving a social robot and a human. However, before extrapolating our results to other platforms, we need to ensure that other factors (e.g. strip size or placement, light diffusion mechanism ...) do not influence the perception of the expression. These results can

Table 3 Selected best animations for each scenario

Scenario	Animation and parameters			
"waiting"	Light blue "Siren" with period 2s			
	wv	D	T (s)	Color
		0.5	2	Light Blue
"blocked"	Red "Push" with period 1.5s			
	wv	D	T (s)	Color
		0.25	1.5	Red
"progress"	Bottom-up progress bar			
	$disp$	u_{disp}	In. Color	Fin. Color
	prog_bar	bottom_up	Red	Green

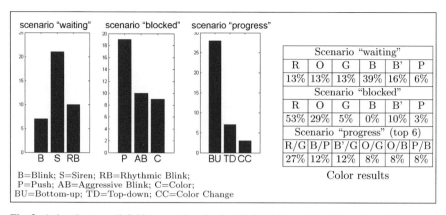

B=Blink; S=Siren; RB=Rhythmic Blink;
P=Push; AB=Aggressive Blink; C=Color;
BU=Bottom-up; TD=Top-down; CC=Color Change

Fig. 3 Animation type (left histograms) and color (right table) results. Animation codes are explained in the figure. Color codes correspond to those used in Section 5

still however serve as a starting point for the design of future social robotic systems which use lights as a means of communication.

6   Conclusion and Future Work

We have proposed a design for an interface between a collaborative mobile robot and programmable lights to be used for expressively revealing the robot's internal state. We have focused on three scenarios in which our collaborative robot, CoBot, finds itself and which could use the help of lights for expressing parts of its internal state. Finally, we presented a study to select appropriate parameters for the light animations in each of the three scenarios we consider.

Our ultimate future goal of using expressive lights on a mobile service robot is threefold. It can be summarized by the three I's: Inform, Influence and Interact. Firstly, Informing consists in having some transparency to the robot's internal state. Secondly, Influencing consists in changing human behavior to the robot's advantage. Thirdly, Interacting possibly includes an affective component of communication. In the current paper, although we have superficially touched at all three points mentioned above, the evaluation was mainly relevant to the first component. It would be interesting as a next step to evaluate the second component, i.e. to what extent our lights can influence or change human behavior.

Acknowledgements This research was partially supported by NSF award number NSF IIS-1012733 and by the FCT INSIDE ERI and UID/CEC/50021/2013 grants. The views and conclusions contained in this document are those of the authors only.

References

1. Betella, A., Inderbitzin, M., Bernardet, U., Verschure, P.F.: Non-anthropomorphic expression of affective states through parametrized abstract motifs. In: 2013 Humaine Association Conference on Affective Computing and Intelligent Interaction (ACII), pp. 435–441. IEEE (2013)
2. Bethel, C.L.: Robots without faces: non-verbal social human-robot interaction (2009)
3. Castellano, G., Leite, I., Pereira, A., Martinho, C., Paiva, A., McOwan, P.W.: Multimodal affect modeling and recognition for empathic robot companions. International Journal of Humanoid Robotics **10**(01), 1350010 (2013)
4. Choi, Y., Kim, J., Pan, P., Jeung, J.: The considerable elements of the emotion expression using lights in apparel types. In: Proceedings of the 4th International Conference on Mobile Technology, Applications, and Systems, pp. 662–666. ACM (2007)
5. De Melo, C., Paiva, A.: Expression of emotions in virtual humans using lights, shadows, composition and filters. In: Affective Computing and Intelligent Interaction, pp. 546–557. Springer (2007)
6. Fong, T., Nourbakhsh, I., Dautenhahn, K.: A survey of socially interactive robots. Robotics and Autonomous Systems **42**(3), 143–166 (2003)
7. Funakoshi, K., Kobayashi, K., Nakano, M., Yamada, S., Kitamura, Y., Tsujino, H.: Smoothing human-robot speech interactions by using a blinking-light as subtle expression. In: Proceedings of the 10th International Conference on Multimodal Interfaces, pp. 293–296. ACM (2008)
8. Harrison, C., Horstman, J., Hsieh, G., Hudson, S.: Unlocking the expressivity of point lights. In: Proceedings of the SIGCHI Conference on Human Factors in Computing Systems, pp. 1683–1692. ACM (2012)
9. Holmes, K.: The mood of the chinese internet lights up the facade of beijing's water cube (2013). http://motherboard.vice.com/blog/video-the-great-mood-building-of-china
10. Kim, M.G., Sung Lee, H., Park, J.W., Hun Jo, S., Jin Chung, M.: Determining color and blinking to support facial expression of a robot for conveying emotional intensity. In: The 17th IEEE International Symposium on Robot and Human Interactive Communication, RO-MAN 2008, pp. 219–224. IEEE (2008)
11. Kobayashi, K., Funakoshi, K., Yamada, S., Nakano, M., Komatsu, T., Saito, Y.: Blinking light patterns as artificial subtle expressions in human-robot speech interaction. In: 2011 IEEE RO-MAN, pp. 181–186. IEEE (2011)
12. Myers, B.A.: The importance of percent-done progress indicators for computer-human interfaces. In: ACM SIGCHI Bulletin, vol. 16, pp. 11–17. ACM (1985)
13. Rea, D.J., Young, J.E., Irani, P.: The roomba mood ring: an ambient-display robot. In: Proceedings of the Seventh Annual ACM/IEEE international Conference on Human-Robot Interaction, pp. 217–218. ACM (2012)
14. Szafir, D., Mutlu, B., Fong, T.: Communicating directionality in flying robots. In: Proceedings of the Tenth Annual ACM/IEEE International Conference on Human-Robot Interaction, pp. 19–26. ACM (2015)
15. Veloso, M., Biswas, J., Coltin, B., Rosenthal, S., Kollar, T., Mericli, C., Samadi, M., Brandao, S., Ventura, R.: Cobots: collaborative robots servicing multi-floor buildings. In: 2012 IEEE/RSJ International Conference on Intelligent Robots and Systems (IROS), pp. 5446–5447. IEEE (2012)
16. Wright, A.: The colour affects system of colour psychology. In: AIC Quadrennial Congress (2009)

Low-Cost Attitude Estimation
for a Ground Vehicle

Javier Rico-Azagra, Montserrat Gil-Martínez, Ramón Rico-Azagra, and Paloma Maisterra

Abstract This paper deals with accurate attitude estimation in unmanned ground vehicles using low-cost inertial measurement units. Using Euler angles representation, direct estimations are firstly performed from a single sensor, accelerometer or gyroscope. Then, the low frequency components of the first one and the high frequency components of the second one are fused through an explicit complementary filter (ECF), which uses the quaternion representation. A feedback control structure implements the ECF whose controller parameters determine the filter cut-off frequencies. Finally, a scheduling of controllers in the ECF structure overcomes the shortcomings of accelerometer direct estimation at low frequencies. It provides reliable attitude, although the vehicle movement is accelerated. Illustrative experiments are driven with a Traxxas Car equipped with an Ardupilot Mega 2.5 board.

Keywords Attitude estimation · Complementary filter · Gain scheduling · Unmanned ground vehicle

1 Introduction

Attitude estimation is required in several fields as aerospace [1], robotics [2], unmanned systems [3], image capture [4] or motion detection [5]. The increasing popularity of Unmanned Vehicles (UV) has favoured new developments on estimation strategies to gain efficiency, accuracy and to reduce computational cost. On the other hand, a rapid development of Micro Electro Mechanical Systems (MEMS) has introduced small size and low-cost measurement devices. This becomes specially attractive in the miniaturization of estimation systems.

J. Rico-Azagra(✉) · M. Gil-Martínez · R. Rico-Azagra · P. Maisterra
Electrical Engineering Departament, University of La Rioja, San José de Calasanz, 31,
26004 Logroño, Spain
e-mail: {javier.rico,montse.gil}@unirioja.es, {ramon.rico,paloma.maisterra}@alum.unirioja.es

© Springer International Publishing Switzerland 2016   121
L.P. Reis et al. (eds.), *Robot 2015: Second Iberian Robotics Conference*,
Advances in Intelligent Systems and Computing 417,
DOI: 10.1007/978-3-319-27146-0_10

Low-cost estimation systems exclusively take the information out of an Inertial Measurement Unit (IMU). Their main drawbacks are a low resolution, high noise levels and dynamic drifts. The sensor pack consists of a 3-axis accelerometer and a 3-axis gyroscope. There will be shown the mathematical foundations of *direct attitude estimation* with the information from a single sensor. Expected theoretical deficiencies will be checked with experimental data, collected with the embedded device *ArduPilot Mega 2.5*. The board incorporates a 8-bit processor *ATmega2560* and the IMU *MPU6000*.

The use of filters will deal with some of low-cost IMU shortcomings [6]. Filter strategies can be shorted out in two main groups: those based on the Extended Kalman Filter (EKF) [7, 8] and those using Complementary Filters (CFs) [9–11]. The former ones yield accurate estimations but demand a high computational cost, which often can not be afforded by low-cost devices. As an example, let us cite the 8-bit microprocessor in this work. On the other hand, CFs require less computational cost, and then, allow a robust implementation in low-cost boards. That is why CFs are catching popularity among software developers for UVs.

Firstly, an accurate estimation of roll and pitch angles will require a CF each one of them. The CF consists of a feedback PI-control structure that handles a low-pass filtering of angle direct estimation from accelerometer and a high-pass filtering of angular velocity direct estimation from gyroscope. Secondly, an Explicit Complementary Filter (ECF) will execute the full attitude estimation in a single CF structure. In this case, a quaternion representation for the body attitude [12] also avoids trigonometric bugs in the estimation algorithms that were based on Euler angles. The cut-off frequency of previous CFs must be selected depending on the sensor characteristics and the type of motion.

Here the *ArduPilot Mega 2.5* board is assembled on a *Traxxas Grinder* car (Fig. 1), which follows an Ackerman configuration. A *GPS LEA 6-a* achieves the global position and a *3DR Radio 433Mhz kit* provides a wireless communication with the ground station for telemetry. Major challenges on true attitude estimation arise during circular motion and changes of speed or direction, since the body experiments normal and tangential accelerations. These spoil accelerometer contribution at low frequencies in ECF. A gain-scheduling of controllers will fix this shortcoming.

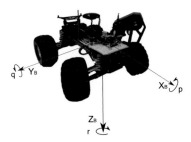

Fig. 1 Unmanned Ground Vehicle and Body Frame of Reference.

2   Direct Attitude Estimation

2.1   Direct Estimation with Gyroscope

Let us denote $\omega_b = [p, q, r]^t$ the angular velocity in the body frame (Fig. 1), which is registered by the 3-axis gyroscope. Euler's angles $\rho = [\phi, \theta, \psi]^t$ denote the body attitude (roll, pitch, yaw) in an inertial frame. Then, ω_b is rotated as

$$
\begin{bmatrix} \dot{\phi} \\ \dot{\theta} \\ \dot{\psi} \end{bmatrix} = \begin{bmatrix} 1 & \sin\phi \tan\theta & \cos\phi \tan\theta \\ 0 & \cos\phi & \cos\phi \tan\theta \\ 0 & \sin\phi \sec\theta & \cos\phi \sec\theta \end{bmatrix} \begin{bmatrix} p \\ q \\ r \end{bmatrix},
\tag{1}
$$

to achieve the rotational speed ω in the inertial frame. The integration of (1) provides the attitude.

However, the gyroscope measures the true variable $\omega_b(t)$ contaminated with high frequency noise $\mu_b(t)$ and a bias b_b whose change over time is being neglected. Rotating each of these components as in (1), it gives $\omega(t)$, $\mu(t)$ and b, respectively. Then, integrating their sum as

$$
\hat{\rho}(t) = \int (\omega(t) + \mu(t) + b)\, dt = \rho(t) + bt,
\tag{2}
$$

a direct estimation of attitude $\hat{\rho}$ is provided from gyro measurements. Supposing an ideal white noise μ_b, its rotated integration becomes zero. ρ is the integration of the rotated true variable ω_b. The rotated bias integration, despite b being small, becomes a drift over the time. This increasing error hampers the use of gyroscope as a single attitude sensor. Apart from the drift, other gyro deficiencies are scaling errors and axis misalignments.

2.2   Direct Estimation with Accelerometer

The rigid solid dynamic in the body frame obeys to

$$
f = m(\dot{v}_b + \omega_b \times v_b),
\tag{3}
$$

where $v_b = [u, v, \omega]^t$ is the linear velocity, $\dot{v}_b$ is the linear acceleration, m is the body mass and f are all forces on the body.

Thus, the accelerometer measures

$$
a_b = \dot{v}_b + \omega_b \times v_b + g_b,
\tag{4}
$$

which reduces to the gravity in the body frame g_b when the body is at rest. That is

$$\begin{bmatrix} a_x \\ a_y \\ a_z \end{bmatrix} = g \begin{bmatrix} -\sin\theta \\ \cos\theta\sin\phi \\ \cos\theta\cos\phi \end{bmatrix}. \tag{5}$$

From its components, roll and pitch angles are estimated as

$$\hat{\phi} = \arctan\left(\frac{a_y}{a_z}\right) \tag{6}$$

$$\hat{\theta} = \arctan\left(\frac{-a_x}{\sqrt{a_y^2 + a_z^2}}\right), \tag{7}$$

considering that the arctangent function is implemented by the function *atan2*.

Therefore, only under near stationary conditions attitude estimations (6) (7) from accelerometer measurements are accurate enough. However, when other accelerations apart from gravity are captured, the estimations (6) (7) fail. An additional high frequency shortcoming is that accelerometers are very sensitive to vibrations generated by moving parts of actuators. Physically attaching anti-vibratory elements to the IMU usually mitigates these negative effects.

2.3 Real-Life Experiments of Direct Attitude Estimation

Experimental data are collected with the *ArduPilot Mega 2.5* board. Departing from the board at rest and zero attitude angles, the first experiment (Exp. #1) consists of two rotations (first in pitch and then in roll). After each rotation the board is driven back to the initial state. The rotation forces are hand-made. Thus, linear body accelerations are small and vehicle actuator vibrations do not affect. Fig. 2 (1st column) shows the results (for the moment, dismiss GS and ECF data, which belong to later tests and explanations). As expected, direct angle estimations from the accelerometer are accurate enough along the whole experiment and fully reliable when the body is at rest. However, direct estimations from gyroscope only perform well during transients. Its bias error is clearly noticeable when the body is back at rest; this error actually responds to a drift error, i.e. it would increase uninterruptedly for an infinite time experiment.

The second experiment (Exp. #2) reproduces an oscillatory movement of the board onto a plane. Results are in Fig. 2 (2nd column). Whereas the true attitude is zero, $\theta = 0$ and $\phi = 0$, the accelerometer senses the gravity combined with linear accelerations, which spoils the estimation. On the other hand, direct estimations from gyroscope match true attitude, since accumulative drift errors only turn up in stationary regime.

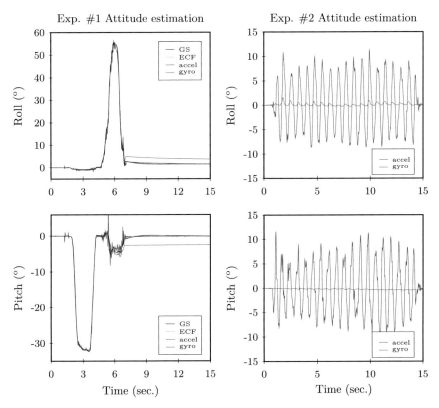

Fig. 2 Attitude estimation under no significant body accelerations.

3 Complementary Filters

3.1 Complementary Filters for Each Axis

According to previous results, the low-pass filtering of accelerometer direct estimation and the high-pass filtering of gyroscope direct estimation would achieve a reliable attitude estimation. This task can be performed by complementary filters (CF). Considering the nature of each sensor measurement, the feedback control structures in Fig. 3 implement CFs, being $\hat{x}_1(s)$ the angle estimation from accelerometer and $\hat{x}_2(s)$ the rotational speed estimation from gyroscope. The coupled version (Fig. 3.a) is much more common, since it has the ability of disturbance rejection such as the drift error. Inside it, selecting a PI controller with the structure

$$c = k_p + \frac{k_i}{s}, \tag{8}$$

the fusion of direct estimations achieves

$$\hat{x}(s) = f_1(s)\,\hat{x}_1(s) + f_2(s)\,\hat{\dot{x}}_2(s) = \frac{(k_p s + k_i)\,\hat{x}_1(s)}{s^2 + k_p s + k_i} + \frac{s\,\hat{x}_2(s)}{s^2 + k_p s + k_i}. \quad (9)$$

The low-pass filter f_1 and the high-pass filter f_2 are 2nd order and complementary, i.e. $f_1(s) + f_2(s) = 0$.

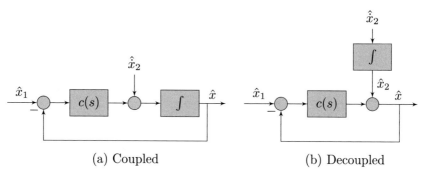

(a) Coupled   (b) Decoupled

Fig. 3 Attitude complementary filters.

The controller parameters determine the filter cut-off frequency, which can be selected in accordance with the IMU and the type of motion. Getting complex poles without resonance peak achieves a good filtering performance. Thus, being $s^2 + 2\xi\,\omega_n\,s + \omega_n^2$ the desired characteristic equation, its damping coefficient must meet $\xi \leq 1/\sqrt{2}$ and the desired cut-off frequency is ω_n. Setting $\xi = 1/\sqrt{2}$, the controller parameters clear up as

$$k_p = \omega_n\sqrt{2}, \quad k_i = \omega_n^2, \quad (10)$$

whenever

$$k_i = 0.5k_p^2. \quad (11)$$

3.2 Explicit Complementary Filter

Euler angles carry on some problems. The *gimbal lock* is the most remarkable, which blocks the axes as the equation is not defined at $\theta = 90°$; see p.e. (1). A minor shortcoming is an accumulative error due to trigonometric function use. All these faults can be fixed using quaternions for attitude representation [12].

In [9] a CF on space SO(3) was proposed. Later it was extended and implemented using quaternions [10], being known as explicit complementary filter (ECF).

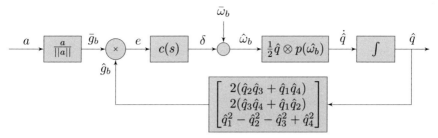

Fig. 4 Explicit complementary filter structure [10].

Another advantage is ECF operates on vectors. Thus, a single control loop is taken on, as Fig. 4 shows. Similarly to the traditional one, the ECF is based on the direct estimation of the gravity direction

$$\bar{g}_b = \frac{a}{|a|}, \tag{12}$$

using the 3-axis accelerometer measurements. The filtered body attitude is represented through the quaternion $\hat{q}$. Then, the filtered gravity direction in the body is

$$\hat{g}_b = \hat{q} \otimes g \otimes \hat{q}^*, \tag{13}$$

which achieves

$$\hat{g}_b = \begin{bmatrix} 2(\hat{q}_2\hat{q}_3 + \hat{q}_1\hat{q}_4) \\ 2(\hat{q}_3\hat{q}_4 + \hat{q}_1\hat{q}_2) \\ \hat{q}_1^2 - \hat{q}_2^2 - \hat{q}_3^2 + \hat{q}_4^2 \end{bmatrix}. \tag{14}$$

The error vector

$$e = \bar{g}_b \times \hat{g}_b \tag{15}$$

represents the relative rotation between $\bar{g}_b$ and $\hat{g}_b$. A PI controller c is used to guarantee zero steady state error and sets the filter cut-off frequency.

Combining the control action δ and the direct estimation of velocity vector $\bar{\omega}_b$ from gyroscope measurements, the angular velocity on the body $\hat{\omega}_b$ is filtered. Then,

$$\dot{\hat{q}} = \frac{1}{2} q \otimes \mathbf{p}(\hat{\omega}_b), \tag{16}$$

where $\mathbf{p}(.)$ is the pure quaternion operator $\mathbf{p}(\omega_b) = [0, \hat{\omega}_b]$. This equation relates velocity to orientation and it is the equivalent to (1) in quaternions. Finally, integrating (16), the body attitude is estimated through the quaternion $\hat{q}$.

3.3 Gain-Scheduling of Controllers

A shortcoming of CFs is the feedback loop set-point must necessarily be the direct attitude estimation from accelerometers (Fig. 3 and Fig. 4), since this signal is more accurate on static conditions than that from gyroscopes. However, the accelerometer estimations are poor when the system is continuously accelerated, which corrupts the final attitude estimation despite the gyroscope being working fine.

One solution is a Gain-Scheduling (GS) of controllers [13], which modifies the CF bandwidth depending on the measured acceleration.

In particular, the acceleration vector from the IMU must have unitary module in g units when the only force on the system is the gravity. Otherwise the accelerometer direct estimation should be disconnected or its influence should be limited. Accordingly, the auxiliary variable for the controller-scheduling is

$$\alpha = |\sqrt{a_x^2 + a_y^2 + a_z^2} - 1|. \tag{17}$$

Depending on the on-line computed value of α, a different controller c is switched inside the feedback control structure of the complementary filter.

4 Experimental Results

4.1 Comparison of Alternatives

The adduced benefits of CFs are here illustrated. In particular, an ECF is implemented (Fig. 4), whose PI controller is designed as

$$c(s) = 1 + \frac{0.5}{s}, \tag{18}$$

which achieves a bandwidth of 0.7 rads/s for the low-pass filtering of accelerometer. This cut-off frequency is selected on purpose larger than the 0.1 rads/s that [13] recommends since our first experiment will present slight linear accelerations. On the other side, a scheduling of controllers is implemented inside a ECF structure, being α the auxiliary variable (17) such that

- If $\alpha \leq 0.02$, the system is not significantly accelerated. The $c(s)$ controller in (18) works and the cut-off frequency that switches between accelerometer and gyroscope is 0.7 rads/s.
- If $\alpha > 0.02$, the accelerometer must be disconnected for the estimation since the system is accelerated. The controller switches to $c(s) = 0$ and the filter cut-off frequency becomes zero.

The real-life experiment was detailed in Sect. 2.3 (Exp. #1) and the comparison results were in Fig. 2 (1st column). The ECF overcomes the estimation error due to gyro drift and the high-frequency distortion due to accelerometer. The experiment is an ideal situation where α does not exceed the threshold of 0.02 many times. Thus, the controller *switching* turns up few times and ECF performs similarly to GS.

4.2   Attitude Estimation of a Ground Vehicle

The equipment of the ground vehicle was detailed in Sect. 1. The motion surface is flat. The suspension system allows maximum inclinations of $\pm15°$ around X_B (roll) and Y_B (pitch) axes in the reference system on Fig. 1. Then, roll angle deviations from zero should answer to changes of direction. Counter Clock Wise (CCW) vehicle turns should yield positive roll angles. On the other hand, pitch angle deviations from zero should answer to changes on the forward speed. Positive accelerations should produce positive pitch angles.

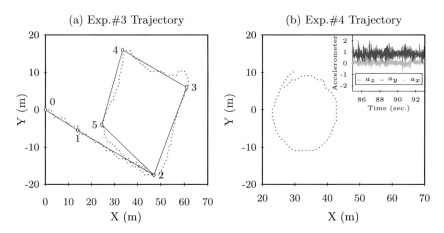

Fig. 5 2D Trajectories for the ground vehicle.

On a first test (Exp. #3) the vehicle describes the trajectory in Fig. 5.a. The waypoints (WP) are numbered and linked with a continuous line. The dotted line corresponds to the path tracking that GPS measures. The motion is mainly along straight lines with some CCW turns. Thus, roll angle is expected to be positive and less than 15°. Sudden braking is avoided. Thus, pitch angle is expected to be always positive due to forward accelerations and near 0°.

Attitude estimations are compared in Fig. 6. Over the time interval $0 - 20$ sec., the vehicle is stopped on WP0, and the direct estimation from the gyroscope accumulates a large drift error. In contrast, the other estimators achieve a good behaviour ($\phi = \theta \approx 0$).

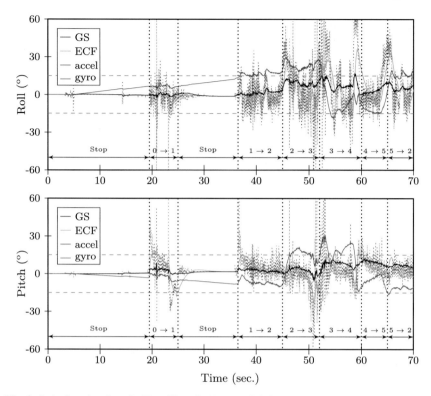

Fig. 6 Attitude estimations for Exp. #3 on the Traxxas Grinder car.

During the interval 19.5 sec. $\leq t \leq 24$ sec. the vehicle moves from WP0 to WP1, where the vehicle stops again. In this stretch the accelerometer direct estimation shows high noise components and is distorted by the linear acceleration. So, despite roll angle being expected to remain zero, it reaches impossible values near to $\pm 60°$. Similarly, despite pitch angle being expected positive and small, it even exceeds the physical limits of $\pm 15°$, which are marked as discontinuous lines. ECF estimations are admissible for the roll angle. However, the 0.7 rad/s cut-off frequency is not enough to filter high-frequency deficiencies from accelerometer, and the pitch angle is wrongly estimated. The GS fixes this fault by reducing the cut-off frequency. Over the stop interval (24 sec.$\leq t \leq 38$ sec.) all estimators recover the true values except the gyro that integrates the bias error. However, during forward motion it did not.

From $t = 38$ sec. on, the vehicle performs the way-point trajectory in Fig. 5.a from WP1. A nearly constant linear velocity was attempted during driving. Direct estimations from accelerometer and gyroscope are dismissed. GS estimation clearly improves ECF performance. ECF estimations of both angles are not reliable as they exceed the physical limits. Negative roll values are inadmissible in CCW rotation. Let us pay especial attention to the time periods when the vehicle is turning. Here the accelerometer measurements deviate highly from the solely gravity component due to centripetal acceleration. Thus, the roll angle is wrongly estimated by ECF. Similarly, changes on linear acceleration during turning spoil ECF pitch angle estimation. GS overcomes these bugs thanks to the accelerometer disconnection when the measured acceleration exceeds the gravity. Its attitude estimations are feasible and roll angles are always positive.

A specific experiment is carried out to confirm the challenges of attitude estimation during circular motion. A CCW trajectory is driven at maximum speed (Exp. #4) . The GPS records are in Fig. 5.b, together with the raw data from accelerometers. Let us note as the measurement $a = [0, 1, 1]^t$ clearly differs from $g = [0, 0, 1]^t$. Thus, direct estimations from accelerometers (accel) and ECF exhibit a large offset error in the roll angle in Fig. 7. The ECF only filters the high-frequency components of the accelerometer direct estimation. The low-frequency offset can only be cancelled with the full removal of accelerometer direct estimation, as GS handles. Low frequency components captured by the gyro turn out in sinusoidal drift errors. Thus, direct estimation from gyroscopes (gyro) are clearly inadmissible, but their contribution is crucial in the sensor fusion. As a conclusion, the attitude estimations from a scheduling of controllers (GS) inside a ECF structure are fully reliable.

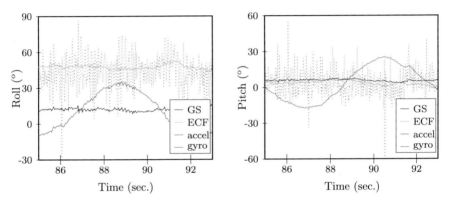

Fig. 7 Attitude estimations for Exp. #4 on the Traxxas Grinder car.

5 Conclusions

This paper proposed several alternatives of complementary filters to achieve reliable attitude estimations from a low-cost inertial measurement unit (IMU) used in autonomous vehicle navigation.

The algorithms were implemented in an *ArduPiltot Mega 2.5* board. It incorporated a 8-bit microprocessor and an IMU of 3-axis accelerometers and 3-axis gyroscopes. *Direct attitude estimations* were firstly computed using a single type of sensor. It turned out high-frequency estimation errors from accelerometers and low-frequency estimation errors from gyroscopes. *Complementary filters* (CFs) were added to take advantage of each sensor optimal bandwidth. CFs developed low-pass and high-pass filtering through feedback control structures. Each attitude angle required a CF. Afterwards an *explicit complementary filter* (ECF) put together the 3-axis attitude estimation in a single feedback control structure. *Quaternions* were used to avoid the *gimbal-lock*. A direct attitude estimation from accelerometer was the set-point in the feedback control loop since most systems come back to zero linear and angular acceleration apart from gravity, and then, estimations converged to the true values. This was not the case in circular motion or during accelerated movements. In these cases, the accelerometer filter bandwidth had to be reduced, even to zero in strictly circular motion. A scheduling of controllers in the ECF structure fixed those shortcomings. The controller parameters determined the desired cut-off frequencies of the filters. The performance of these alternatives for attitude estimation were proved on a *Traxxas Grinder* driving.

Acknowledgements The authors are grateful for the assistance provided by La Rioja Government through Project ADER 2012-I-IDD-00093.

References

1. Gebre-Egziabher, D., Hayward, R.C., Powell, J.D.: Design of multi-sensor attitude determination systems. IEEE Transactions on Aerospace and Electronic Systems **40**(2), 627–649 (2004)
2. Barshan, B., Durrant-Whyte, H.F.: Inertial navigation systems for mobile robots. IEEE Transactions on Robotics and Automation **11**(3), 328–342 (1995)
3. De Marina, H.G., Pereda, F.J., Giron-Sierra, J.M., Espinosa, F.: Uav attitude estimation using unscented kalman filter and triad. IEEE Transactions on Industrial Electronics **59**(11), 4465–4474 (2012)
4. Hurak, Z., Rezac, M.: Image-based pointing and tracking for inertially stabilized airborne camera platform. IEEE Transactions on Control Systems Technology **20**(5), 1146–1159 (2012)
5. Fang, L., Antsaklis, P.J., Montestruque, L.A., McMickell, M.B., Lemmon, M., Sun, Y., Fang, H., Koutroulis, I., Haenggi, M., Xie, M., Xie, X.: Design of a wireless assisted pedestrian dead reckoning system - the navmote experience. IEEE Transactions on Instrumentation and Measurement **54**(6), 2342–2358 (2005)
6. Crassidis, J.L., Markley, F.L., Cheng, Y.: Survey of nonlinear attitude estimation methods. Journal of Guidance, Control, and Dynamics **30**(1), 12–28 (2007)
7. Choukroun, D., Bar-Itzhack, I.Y., Oshman, Y.: Novel quaternion kalman filter. IEEE Transactions on Aerospace and Electronic Systems **42**(1), 174–190 (2006)
8. Li, W., Wang, J.: Effective adaptive kalman filter for mems-imu/magnetometers integrated attitude and heading reference systems. Journal of Navigation **66**(1), 99–113 (2013)
9. Mahony, R., Hamel, T., Pflimlin, J.: Nonlinear complementary filters on the special orthogonal group. IEEE Transactions on Automatic Control **53**(5), 1203–1218 (2008)
10. Euston, M., Coote, P., Mahony, R., Kim, J., Hamel, T.: A complementary filter for attitude estimation of a fixed-wing UAV. In: 2008 IEEE/RSJ International Conference on Intelligent Robots and Systems, IROS, pp. 340–345 (2008)
11. Madgwick, S.O.H., Harrison, A.J.L., Vaidyanathan, R.: Estimation of IMU and MARG orientation using a gradient descent algorithm. In: IEEE International Conference on Rehabilitation Robotics (2011)
12. Kuipers, J.: Quaternions and Rotation Sequences: A Primer with Applications to Orbits, Aerospace, and Virtual Reality. ser. Princeton paperbacks, Princeton University Press (2002)
13. Yoo, T.S., Hong, S.K., Yoon, H.M., Park, S.: Gain-scheduled complementary filter design for a mems based attitude and heading reference system. Sensors **11**(4), 3816–3830 (2011)

Detection of Specular Reflections in Range Measurements for Faultless Robotic SLAM

Rainer Koch, Stefan May, Philipp Koch, Markus Kühn and Andreas Nüchter

Abstract Laser scanners are state-of-the-art devices used for mapping in service, industry, medical and rescue robotics. Although a lot of work has been done in laser-based SLAM, maps still suffer from interferences caused by objects like glass, mirrors and shiny or translucent surfaces. Depending on the surface's reflectivity, a laser beam is deflected such that returned measurements provide wrong distance data. At certain positions phantom-like objects appear. This paper describes a specular reflectance detection approach applicable to the emerging technology of multi-echo laser scanners in order to identify and filter reflective objects. Two filter stages are implemented. The first filter reduces errors in current scans on the fly. A second filter evaluates a set of laser scans, triggered as soon as a reflective surface has been passed. This makes the reflective surface detection more robust and is used to refine the registered map. Experiments demonstrate the detection and elimination of reflection errors. They show improved localization and mapping in environments containing mirrors and large glass fronts is improved.

Keywords SLAM · Error-free mapping · Multi-echo laser scanner · Reflectance filter · Specular reflection · Reflective objects

R. Koch(✉) · S. May · P. Koch · M. Kühn
Technische Hochschule Nürnberg Georg Simon Ohm, Kesslerplatz 12, 90489 Nürnberg, Germany
e-mail: rainer.koch@th-nuernberg.de
http://www.th-nuernberg.de

A. Nüchter
Informatics VII – Robotics and Telematics, Julius-Maximilians University Würzburg,
Am Hubland, 97074 Würzburg, Germany
e-mail: andreas@nuechti.de
http://www.uni-wuerzburg.de

© Springer International Publishing Switzerland 2016
L.P. Reis et al. (eds.), *Robot 2015: Second Iberian Robotics Conference*,
Advances in Intelligent Systems and Computing 417,
DOI: 10.1007/978-3-319-27146-0_11

1   Introduction

Mapping is an essential task in mobile robotics. It is used in service robotics, e.g., in industrial, medical, and rescue applications. Before training or exploration, the environment is partly or completely unknown. SLAM (Simultaneous Localization and Mapping) is one of the most frequently applied approaches to provide an environmental representation to service robots. Nevertheless, customizing most environments is necessary to reduce interferences from objects with specular reflective and transparent surfaces, e.g., glass, mirrors, and shiny metal. This is one reason why robots are still not ad-hoc integrable in most applications.

In addition, glass surfaces are reflective or transparent depending the laser beam's incident angle. Objects behind the glass surface are only occasionally visible. Even worse is the aspect that the glass surface is rated as volatile (disappearing) object or not seen at all. This carries the risk of navigating a robot into it. Hence most service robots are reliant on a second sensor principle, like ultrasonic arrays, to respect these situations. Despite the fusion with other sensor principles, it is difficult

Fig. 1 Robot equipped with laser scanner facing a unframed mirror.

to register a laser scan based map without reflection influences. Therefore the environment is modified. If the laser beams hit the surface in an angle associated to total reflection, returned measurements provide wrong distance data. At certain positions phantom-like objects appear in the map. Figure 2 depicts this effect for three state-of-the-art SLAM approaches using the same dataset: CRSM-SLAM (Critical Rays Scan Match-SLAM), Hector-SLAM and TSD-SLAM (Truncated Signed Distances-SLAM). Phantom-like areas are marked with a red rectangle. The location of the mirror is marked with a blue rectangle and magnified on the top left. Hector-SLAM creates a static map, i.e., points added once to the map remain ad infinitum. The mirror is partly recognizable in the Hector-SLAM map due the fact that at some positions the laser beam was not deflected. In comparison, CRSM- and TSD-SLAM build a dynamic map. Changes in the environment are respected in both approaches, e.g., if objects are moved. Therefore, the mirror disappears if its surface is not measurable at certain perspective views. This is likely the case for passing a mirror.

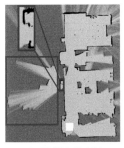 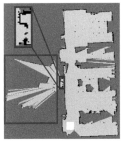 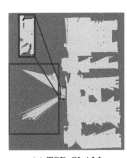

| (a) CRSM-SLAM | (b) Hector-SLAM | (c) TSD-SLAM |

Fig. 2 Maps registered with the same dataset and with different SLAM approaches in environment containing a mirror. The mirror is marked blue and the reflections are marked red.

In the following, we present a reflectance detection approach applicable to multi-echo laser scanners in order to remove above mentioned effects. Section 2 outlines related work. Section 3 describes the two filter stages used for mirror detection. In Section 4 experiments demonstrate the applicability to environments with a large proportion of reflective and transparent surfaces. Finally, Section 5 summarizes results and gives an outlook for future work.

2 Related Work

As far as reflection is concerned, there are two different strands of research - stationary and mobile systems. Often, the environment is adapted to prevent influences when working with stationary systems. Therefore, research in this field has less impact on mapping. Covering objects is unwanted when mapping with mobile systems because it requires a lot of effort to deal with all specular reflective and transparent objects, especially when operating outside.

To avoid the need to cover surfaces for mapping, Yang et al. [1] presented an approach which fuses a laser scanner with an ultrasonic sensor. Two individual grid maps are created. With the assumption that reflective objects are flat and framed, the data from the two sensors are compared w.r.t. consistency. Mirrors are detected and tracked online, while resulting errors are recalculated only offline. In further research Yang et al. [2] extended their algorithm for advanced mirror detection and identification of mirror images. Each gap in the wall is assumed to be a specular reflective object. Therefore, no ultrasonic sensor is required anymore. Once such a mirror candidate is detected, the space behind the gap is analysed for a mirrored image, i.e., the search for similarity between both sides of the opening. Objects with symmetry w.r.t. a line might be identified wrongly.

Another online applicable approach was implemented by Forster et al. [3]. At specific angles reflections can be identified based on the returning intensity of the laser. A subset of these angles is tracked – on occurrence mirrors are assigned in dependency of the laser beam's intensity. An object with diffuse reflectivity causes false identification if it is placed directly behind the transparent object.

Kähhammer et al. [4] presented an approach which recognises framed mirrors with a predefined size in 3D point cloud data. A panorama range image is generated and searched for jumping edges. In case of a positive search, the contour of the mirror frame will be extracted. Finally, objects are verified by considering their size and shape. This only applies to framed squared mirrors with a known size. Glass or other objects are not considered.

Tatogulu et al. [5] use the best fitting illumination model to modulate the surface. Lambertian diffuse reflection models, Blinn-Phong models [6], Gaussian models [7] and Beckmann specular reflection models [8] are fitted to the data set to identify the characteristics of the scanned surface. While this system is quite effective for diffuse surfaces, it does not handle specular reflections.

We present a specular reflectance detection approach applicable to multi-echo laser scanners in order to identify and filter mirrored objects. In contrast to above mentioned approaches, recognition of specular reflections is possible, also for frameless and free-standing objects.

3 Approach

The mirror detector uses a Hokuyo 30LX-EW multi-echo laser scanner. For each data take the Hokuyo records up till three echoes of the returning light wave, including distance and intensity. The first two echoes s_1 and s_2, which are further called scan tuple s, are taken to detect reflective objects:

$$s = \{s_1, s_2\} \tag{1}$$

with

$$s_1 = \{d_{1,i} | i = 1 \cdots N\}, \tag{2}$$

$$s_2 = \{d_{2,i} | i = 1 \cdots N\}, \tag{3}$$

where $d_{1|2,i}$ are distance measurements and N is the number of measurement points.

Differences in scan messages indicate surface reflection properties. While a specular reflective object causes differences in both scan messages, diffuse reflective objects provide consistency. The problem in detecting specular reflections is that it depends on the laser beam's incident angle to the surface and the refractive index. If the angle is too big, the light will be totally reflected according to the reflection law. Hence, the robot will only detect the mirrored object. If the angle is smaller, there are three potential cases of measurements also depending on the material. For a transparent object the robot can receive a point on the surface, a point behind the surface, or a mirrored point. If the object is nontransparent, the last case disappears. Therefore, the robot has to pass the surface to ensure that it is seen at least once from the "right" perspective. Hence, it is not possible to eliminate all reflective errors on the fly. That is why the mirror detector is set up in two filter stages and two mapping stages, cf. Figure 3. The pre-filter runs on the fly and filters current scans. The post-filter is triggered after a reflective object has been passed, e.g. by an passing-algorithm or a loop-closure, to reduce remaining errors.

Figure 3a shows the processing chain of the pre-filter with its mapping stage. The resulting map is without reflection errors, which are detectable in a single data take. Hence, the map include less erroneous data than a map using data directly provided by the Hokuyo. But the map is not completely cleaned of reflective influences. The

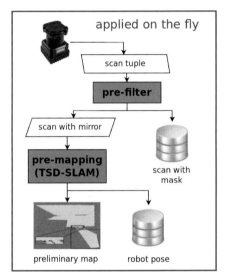

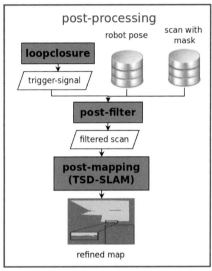

(a) Pre-filter and pre-mapping.   (b) Post-filter and post-mapping.

Fig. 3 Processing chains of the mirror detector: Pre-filtering removes affections on the fly. Post-filtering refines the resulting map after a trigger signal.

mapping stage assigned to the post-filter chain, cf. Figure 3b, considers a set of scan tuple. All measurements originating from specular reflective objects are filtered in retrospect, i.e., specular reflective objects detected in any scan tuple are propagated to the whole history. This supplies a map free of any reflection errors available at new trigger events, e.g., from a loop closure module. All modules are implemented as ROS-nodes and are publicly available as open-source packages at http://www.github.com/autonohm/ohm_mirror_detector.git.

3.1 Pre-filter

The pre-filter receives a scan tuple from the laser scanner. First, it removes sparse points, i.e., isolated points without other points nearby. These points are likely to be artefacts for example from jumping edges. This happens when neighbouring measurements cross an object edge and provide discontinuity in depth. Further, the corresponding points in the scan tuple are subtracted and analysed. A difference between s_1 and s_2 points out that the laser beam was reflected. This happens when it hits, e.g., a reflective or transparent surface. The first echo has to be from the error causing object, since it was hit first by the laser beam. The second echo includes a point more fare away, an affected point. It is identified according to the distance between the scan messages:

$$\Delta d_i = d_{2,i} - d_{1,i}, \tag{4}$$

$$f(\Delta d_i) = \begin{cases} d_{1,i} \rightarrow \text{valid,} & \text{if } \Delta d_i \leqq \text{threshold} \\ d_{1,i} \rightarrow \text{mirror, } d_{2,i} \rightarrow \text{affected,} & \text{if } \Delta d_i > \text{threshold.} \end{cases} \tag{5}$$

Glass fronts and mirrors are assumed to be planar surfaces. They relate to line segments in a scan message. The two distinct end points c_1 and c_2 are determined with a RANSAC-based algorithm, i.e., finding the line parameters fitting best to the set of mirror points in Cartesian space.

With robot position p a sector is spanned up, cf. Figure 4. Finally, the scan is checked again to classify the points into three groups: valid, mirror plane, and affected points. Valid points are located in the green hatched area. They are free of reflection influences. The second group contains points on the mirror, glass, or reflective plane, e.g., blanc metal. They are found in the solid blue area. All other points remain in the red crossed area and shall be assigned to the third group. The pre-filter processing time is less than 4.5ms, which caused mainly by the RANSAC-algorithm.

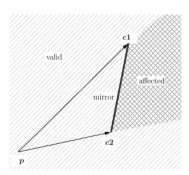

Fig. 4 Classification of points based on the mirror line corners.

3.2 Pre-mapping

The first two point groups, containing valid and mirror points, are forwarded as scan messages to the pre-mapping module. As a result, the preliminary map is generated on the fly. The pre-mapping employs an unadjusted version of TSD-SLAM as described in [9]. Nevertheless other mapping algorithms can be integrated as well. In order to do so, the SLAM module must provide the scanner's pose.

3.3 Loop Closure

A simple loop closure algorithm compares the current robot position with the previous robot positions. Therefore, it records a complete history of the robot pose. If the new position is within a limited range to previous positions, a trigger signal is broadcasted to the post-filter.

3.4 Post-filter

A mirror is only detectable, if the incoming laser beam hits the mirror in a particular angle. Hence, the post-filter builds a history of all pre-filtered scans and poses with dynamic length. If a dataset includes mirror points the post-filter identifies the line associated to the mirror plane. The corners of the line are simply the outer points in the scan because the dataset is ordered according the increase of angle and the scan has been cleaned in the pre-filter. Received mirror corners are added at the first occurrence. Corners nearby existing mirror corners are fused together.

While building the history, the post-filter awaits a trigger signal. It is provided for instance by the loop closure module or any other external trigger. Afterwards the post-filter starts to refine the scans in the history. It uses the same algorithm as in the pre-filter module to span up a sector and classify the points into the three groups. This will be repeated for every set of distinct end points. The processing time of the post-filter differs. Without a running publisher the processing time is less than 30μs. Afterwards it rises up to maximum of 1.5ms. Finally, the refined scans are transferred to the post-mapping module. Thus, the sensor's final pose is determined by the SLAM-module on the basis of the refined scans.

3.5 Post-mapping

Post-mapping also applies an untouched version of the TSD-SLAM approach. It delivers a refined map. The SLAM module can be replaced by any other approach. The map is registered with refined scans from the history and therefore without reflection errors.

4 Experiments and Results

This chapter consists of three sections to qualify the experiments with the mirror detector. The first experiment uses a "sandbox setup" to test the mirror detector on a defined scene. Experiment 2 has been performed in an office-like environment. Experiment 3 applies the approach to a corridor with a large glass front.

4.1 Experiment 1: "Sandbox"

The "sandbox" is used to evaluate the mirror detection approach on a simple scene with a defined mirror location. The setup is shown in Figure 5 and the mirror is marked with a blue square.

As already described, the mirror cannot be seen from every pose, even it is in the

Fig. 5 "Sandbox"-experiment with mirror and simple scene.

field of view of the laser scanner. Only if the angle of the incoming laser beam hits the mirror plane in a particular angle, the scanner will broadcast different values in its echoes. As a result, the mirror is identified. If a mirror plane was detected, cf. Figure 6a, the history is searched for erroneous points. Therefore, the post-filter stores the history of all scans.

We assumed a planar surface of the reflective object. That is why, two boundary points c_1 and c_2, cf. Figure 6b, are enough to describe the subject. These corners will be updated when more points on the mirror plane are determined or added as new ones. When receiving a trigger signal, the post-filter, cf. Figure 6c, masks all points in the history. Subsequently, it broadcasts a refined scan, including valid and mirror plane points (green) and a scan with erroneous points (red). The post-mapping uses the valid scans to build up a map including the mirror plane. The erroneous points are not used yet, but it is our intent to recalculate them to their true position and embed them in mapping. Figure 7a evinces the preliminary map and Figure 7b the refined map to show the difference.

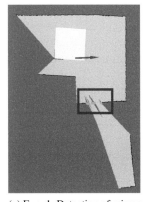 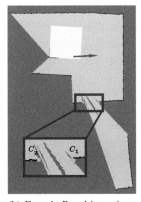 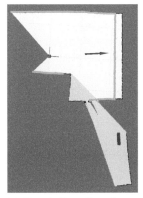

(a) Exp. 1: Detection of mirror plane.

(b) Exp. 1: Resulting mirror corner points.

(c) Exp. 1: Post-filtered scan (green), erroneous points (red).

Fig. 6 Exp. 1: Different steps of the mirror detection.

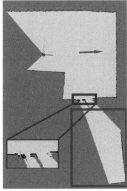 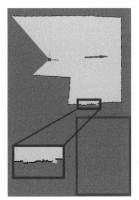

(a) Exp. 1: Preliminary map without the mirror surface and including reflections.

(b) Exp. 1: Refined map including the mirror surface and free of reflections.

Fig. 7 Exp. 1: Comparison between preliminary map and refined map. The mirror is marked blue and reflections are marked red.

4.2   Experiment 2: Office-Like Environment

The map of experiment 2 contains three office rooms with a mirror, cf. blue squared in Figure 8. It is necessary to note that the mirror is not planar to the wall, as visible in the magnified area of the refined map. Comparing the preliminary map in Figure 9a and the refined map in Figure 9b the effect of the mirror detector is visible. There are no remaining reflections in the refined map (red square). In addition the mirror plane is completely mapped (blue square). In this case the robot will not try to navigate through the mirror plane.

Fig. 8   Exp. 2: Office room with mirror.

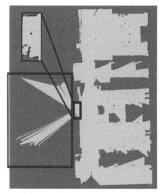

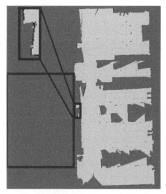

(a) Exp. 2: Preliminary map without the mirror surface and including reflections.

(b) Exp. 2: Refined map including the mirror surface and free of reflections.

Fig. 9   Experiment 2 - office-like environment. The mirror is marked blue and reflections are marked red.

4.3   Experiment 3: Room with Corridor

In this experiment a corridor with a large glass front was mapped, cf. Figure 10. Such glass fronts are a major reason for erroneous measurements. Therefore, they are normally covered by hand.

Fig. 10 Exp. 3: Corridor with glass front.

Figure 11a displays the preliminary map of the mirror detector. The glass front is marked with a blue square. Above the glass front another part of the corridor is mapped. This is correct and also visible in Figure 10. As previously described, this is a possible result if the robot faces a transparent surface. The mirror detector is not able to distinguish between mirror and glass yet. Therefore it erases the points behind the detected surface, cf. Figure 11b. In addition, it marks the points on the surface

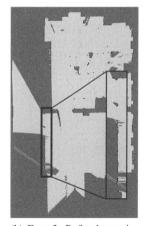

(a) Exp. 3: Preliminary map without the glass surface.

(b) Exp. 3: Refined map including the glass surface.

Fig. 11 Experiment 3 - room with corridor. The glass front location is marked blue.

and therefore the post-mapping includes them. This is wanted, since it will prevent the robot to think there is a free path. Still some area behind the glass remained in the refined map. This is caused by the fact, that the laser beams had not yet hit the surface in the desired angle. In this particular case, the glass behind the open door was previously seen, but when passing, the door blocked the laser beam from reaching the desired angle. That is why, the glass area cannot be identified as such.

This experiment emphasizes that further investigation is required. It is necessary to distinguish between transparent and mirror surfaces. Thereby, a discrimination of the points behind the surface is possible. Further, there is a need to improve the broadcast of the trigger signal. A trigger module is needed to assure that a certain position before the surface has been reached. In that case it is verified, that it is not a reflective area.

5 Conclusions and Future Work

The mirror detector identifies transparent and specular reflective objects like glass, mirrors, or shiny surfaces such as blank metal. Free standing unframed objects of different size are detectable. Two filter stages are implemented. The first filter reduces errors in the current scans on the fly. A subsequently applied SLAM module builds a preliminary map and provides the robot poses for the post-filter. The post-filter records the pre-filtered scans and the robot pose in a history. Besides, it calculates the corner points of the mirror plane. As soon as a trigger signal from the loop closure occurs, the post-filter delivers refined scans to the second mapping stage. The result is a refined map without reflection errors. The TSD-SLAM module allows to map in a dynamic world. However, the influences of moving reflective objects are not tested yet.

Future work is concentrated on advanced loop closure techniques, classification of transparent and reflective materials, as well as detecting shaped reflective objects. Hence, the plane detection algorithm will be replaced. These will help to archive some difficult but interesting tasks, such as mapping modern glass galleries or historic buildings. Castles and palaces are full of reflective and transparent objects, e.g., chandeliers, golden artwork, or mirror cabinets. The aim is to support robust localisation and mapping in such areas.

References

1. Yang, S.W., Wang, C.C.: Dealing with laser scanner failure: mirrors and windows. In: IEEE International Conference on Robotics and Automation ICRA (2008)
2. Yang, S.W., Wang, C.C.: On solving mirror reflection in lidar sensing. IEEE/ASME Transactions on Mechatronics 16(2), 255–265 (2011)
3. Foster, P., Sun, Z., Park, J.J., Kuipers, B.: Visagge: visible angle grid for glass environments. In: 2013 IEEE International Conference on Robotics and Automation (ICRA), pp. 2213–2220 (2013)

4. Käshammer, P.F., Nüchter, A.: Mirror identification and correction of 3d point clouds. ISPRS - International Archives of the Photogrammetry, Remote Sensing and Spatial Information Sciences **XL–5/W4**, 109–114 (2015)
5. Tatoglu, A., Pochiraju, K.: Point cloud segmentation with lidar reflection intensity behavior. In: 2012 IEEE International Conference on Robotics and Automation (ICRA), pp. 786–790 (2012)
6. Blinn, J.F.: Models of light reflection for computer synthesized pictures. In: Proceedings of the 4th Annual Conference on Computer Graphics and Interactive Techniques, SIGGRAPH 1977, New York, NY, USA, pp. 192–198. ACM (1977)
7. Torrance, K.E., Sparrow, E.M.: Theory for off-specular reflection from roughened surfaces. J. Opt. Soc. Am. **57**(9), 1105–1112 (1967)
8. Beckmann, P., Spizzichino, A.: The scattering of electromagnetic waves from rough surfaces. Pergamon Press (1963)
9. Koch, P., May, S., Schmidpeter, M., Kühn, M., Martin, J., Pfitzner, C., Merkl, C., Fees, M., Koch, R., Nüchter, A.: Multi-robot localization and mapping based on signed distance functions. In: 2015 IEEE International Conference on Autonomous Robot Systems and Competitions (ICARSC), March 2015. https://www.waset.org/conference/2015/03/istanbul/ICARSC

Hardware Attacks on Mobile Robots: I2C Clock Attacking

Fernando Gomez-Bravo, R. Jiménez Naharro, Jonathan Medina García, Juan Gómez Galán and M.S. Raya

Abstract This paper presents a study on various types of attacks on the security of a robotic system. The work focuses on hardware attacks, and particularly considers vulnerable features of the I2C protocol. An analysis of the effects of these attacks, when they are applied on a differential driving mobile robot that aims to track a path, is presented. The paper shows how, with such actions, it is possible to modify the behavior of the robot without leaving traces of the attack and also maintaining the system without any damage.

Keywords Hardware attacks · Mobile robot vulnerability · Robot reliability

1 Introduction

Nowadays, robotic control platforms are usually microcontroller- or microprocessor-based systems. The most common way of attacking such machines involves the use of computer virus, so that disruptions in the control software are achieved. The obvious solution against this attack is to use update anti-virus software. However, until now, the most of robot implementations do not include mechanisms to improve its security in front of hackers. This tendency is changing due to the high use of robotic systems in applications with sensitive information [1, 2] (such as banks, military environments or surveillance systems), and then the interest to attack these systems is increasing. Taking into account that the virus appears firstly, and then the anti-virus is designed to protect the system from the new virus, there is a great security fail during the adaptation time of the anti-virus software. Depending on the system to protect, this time of vulnerability must not be tolerated. Therefore, a solution is a system fully implemented in hardware that does not include any software system. Obviously, this solution avoids the efficiency of any software attack. However, the fully hardware systems can also be attacked by hackers with the same objectives than a software attack. The attacks that exploit the hardware vulnerabilities are known as hardware attacks. Though this kind of attacks is targeted at hardware systems, hardware attack can also be used in software systems

F. Gomez-Bravo(✉) · R.J. Naharro · J.M. García · J.G. Galán · M.S. Raya
Department of Electronic Engineering, Computer Systems and Automation,
Huelva University, Huelva, Spain
e-mail: {fernando.gomez,naharro,jonathan.medina,jgalan,msraya}@diesia.uhu.es

© Springer International Publishing Switzerland 2016   147
L.P. Reis et al. (eds.), *Robot 2015: Second Iberian Robotics Conference*,
Advances in Intelligent Systems and Computing 417,
DOI: 10.1007/978-3-319-27146-0_12

because microcontrollers or microprocessors are implemented using hardware techniques. In fact, one of the first hardware attacks reported consisted in obtaining information from microprocessor operations [3]. The main objectives of the attacks can be very different but all of them can be summarized in the same effect: the final behavior of the system is different to the one desired by users. Some examples are:

- Transmitting sensitive information (such as keys or password) to non-authorized person.
- Increasing the vulnerability of encryption algorithms or in general a communication process.
- Obtaining sensitive information inside the system.

Several mechanisms to implement a hardware attack have been reported in literature, such as hardware trojan (including a malicious hardware block in the system in order to alter its global behavior) [4], replay attacks (supplanting the identity of an authorized block by a non-authorized block in a communication process) [5], injection failures attacks (injecting a failure in the system to increase its vulnerability) [6], and reverse engineering (obtaining the behavior from the signal values) [7].

Reliability of mobile robots is an open issue recently discussed by the scientific community. Hardware attacks directly affect robot reliability. An important question regarding this matter is the communication between the control system and the peripherals, monitoring and acting over the environment. In this context, I2C (Inter-Integrated Circuit) protocol is a commonly used mechanism to establish this communication process. In fact, the use of I2C low level controller in mobile robotics has spread in last years.

This paper aims to bring out the possibility of attacking a robotic system by using hardware attack techniques. The interest to perform an attack in a robot is justified due to the use of these systems in a huge variety of application fields. More concretely, the attack will be performed in the communication process using the I2C protocol. Though this work is focused on the I2C protocol, different types of protocols can be objectives of this type of attacks. The application presented in this paper is related with mobile robot navigation along a previously defined path. It represents a typical situation in many of the current robotic applications (industry, agriculture, service, etc.), performed whether in outdoor or indoor scenarios [9, 10]. It is supposed that user's intentions involve visiting certain areas of the scenario by the robot. Thus, it is also supposed that a planning algorithm has provided a path, and that a path-tracking algorithm is applied so that the robot follows this trajectory with precision. Hence, those who want to interfere with user's intentions would try to modify the course of the robot, in a way that user does not realize the interference (apart from the effect on the robot motion). Accordingly, the planned path should remain invariable, and the external interference may be applied to the low level actuator, on a short period of time, so that the disturbance takes place temporarily leaving no traces. Finally, a hard constraint to the attacker could be to maintain the robot in a safe course; even following the planned path, but making robot repeat motion or avoid visiting a specific place.

The paper is organized as follow. Section 2 is devoted to present the control architecture and the basis of the I2C protocol. Section 3 introduces the details of the attack and describes how it is implemented in the present mobile robot application. Section 4 shows details about the experimental platform that validates the authors' hypothesis. In section 5 some experimental results are shown. Finally, some conclusions are drawn in section 6.

2 Mobile Robot Control Architecture and I2C Protocol

2.1 Mobile Robot Control Architecture

According to the traditional architecture of a mobile robotic platform [8,12], (see Fig. 1) high level controllers are the responsible of taking decision about the robot motion, as well as to set the correspondent actions to be executed by the actuators. In the case of having motors as actuators, and depending on the robotic application, the high level controller defines the motors' speed or the angular position of the motors' axe. Then, these values are transferred to the low level controller so that motors make the robot moves in a correct way.

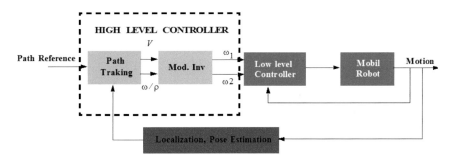

Fig. 1 Traditional mobile robot control architecture

In many cases [9, 10, 13], both functionalities, high and low level control are implemented in different CPUs. Moreover, specific electronics devices can be found in the market for motors with low level control purposes [11]. Then, a communication procedure has to be applied for transferring these data from one to another CPU, and a communication channel will be involved in this process. In this context, protecting this communication and making the channel free of security bugs is a relevant matter that has to be taken into account for developing reliable and safe robots. There are numerous robotic examples where I2C low-level controllers are applied to set the actions of the actuators. In this work, I2C drivers are used for controlling the velocity of the robot wheels. Temporary attacks will be performed against the communication process between the path-tracking algorithm and the I2C controller in order to change the final course of the robot. The implementation of this attack will treat to be as dangerous as possible, and hence, it must pass fully unobserved. Then, no damage must be observed, neither in the robot structure nor in the hardware of control. The user should perceive the attack as a bizarre behavior of the robot-but never as the result of an external interference.

2.2 I2C Protocol

I2C protocol is based on two signals, generally named SDA (data line) and SCL (clock line) signals, and eventually a ground signal (all modules must have the same reference of voltage). All modules are connected to the same bus lines as shown in Fig. 2a, and generally, any module enables the connection of a new module to the bus. There are two kind of modules: a master module that generates SCL signal and controls the transmissions; and one or several slave modules that will be the source (read operations) or destination (write operations) of information. SDA signal is generated by both master and slave modules. The logic values in I2C protocol are ground by low level and high impedance by high level. For the high impedance level the different modules have not to use the same polarization source, and thus the modules which require high voltage can be connected to the same bus than modules which require low voltage.

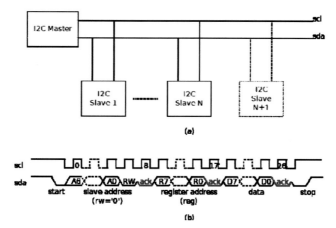

Fig. 2 (a) Architecture of a system based on I2C protocol. (b) Signaling of a write operation in a I2C protocol.

The early characteristics of I2C protocol (an open protocol and the use of high impedance by high level) are a high vulnerability in the system. Firstly, an open protocol enables the connection of any module (authorized or non-authorized) to the bus. Secondly, the high impedance enables to overwrite the information by low levels without any consequence in the signaling of the protocol.

The behavior of a write operation using an I2C protocol is shown in Fig. 2b. Before the beginning of communication, both SCL and SDA signals are high in an idle state. Firstly, the master sends a start condition (SDA signal falls while SCL signal is high) to begin a new communication process (in this case a write operation). In this moment, all slaves monitor SDA signal in order to determine if the master wants to communicate with them. After that, SCL signal has the same behavior than a clock signal whose frequency enables the correct operation of both master and slave modules. Concerning this signal, SDA signal can only change while SCL signal is at low level; otherwise, the change will be considered as a control flow of the protocol (start, restart and stop conditions). This operation involves the transmission of three packets of nine bits (eight bits of data and one bit of acknowledge). The first packet (generated by the master) is composed by the slave address to communicate plus a control bit indicating the type of operation (in our case a low level indicating a write operation). After these eight bits, the control of SDA line is obtained by the slave whose address was the sent one. This slave sends an acknowledge (SDA signal to low level) indicating that it has recognized the communication process. If no slave sends the acknowledge, the master understands that the slave is not accessible. From now, the rest of slaves will wait until that a stop condition indicates the end of the communication. The second packet contains the register address of the slave where the master wants to write data. Again, the master sends this address coded by eight bits and it waits to the acknowledge of the slave. The third packet contains the data that will be stored in the early register. Again, the master sends data coded by eight bits and it waits to the acknowledge. Finally, the master sends the stop condition (SDA signal rises while SCL signal is high) indicating the end of the transmission. After that, both SDA and SCL lines are at high level in the idle state waiting a new communication process.

Though I2C protocol has more complex functionalities, this paper focuses in those basic described above. Among the more complex, clock stretching is a remarkable functionality. This technique allows the slave to stop the SCL signal because it needs a higher time to complete its operation. In this case, the master waits until the SCL signal is free to continue sending the next data packet. Later, it will be described how this procedure is affected by an attack to the SCL signal.

3   Hardware Attacks

In the considered scenario, the hacker must perform several actions in order to guarantee the correct action of the attack. Firstly, hacker must choose the moment in which to initiate the attack. This moment can be obtained using reverse engineering. Secondly, hacker must avoid that the orders from the master module arrive to the slave module. This action involves the use of hardware trojan in order to implement the attack. The attack implemented consists in injecting a failure modifying the frequency of the clock line (SCL signal). Finally, the hardware trojan will send the acknowledges to the master module as it was the slave module, that is, a replay attack.

3.1 Atack on the I2C Protocol

Firstly, the attack cannot be performed directly by the hacker due to the mobility of the system considered. Hence, the attack must be implemented in a module as a hardware trojan. Generally, a hardware trojan is usually included in the fabrication phase by an employee (with authorization because the foundry is not recommendable, or without authorization). However, the open sense of I2C protocol enables to include any hardware trojan after the fabrication phase, that is, during the utilization phase. Once the hardware trojan (attack module from now) is installed in the bus, the hacker can control the attack without accessing to the system. Obviously, this module has not an associated address because it is unknown for the rest of the system.

The attack module has three basic functions. Firstly, the module will monitor the transmissions as a common slave, but the monitoring function is to determine the slave of the communication because the attack will prevent some communications with a certain slave. The correct address and the moment of the attack can be determined using techniques of reverse engineering because the attack module has access to data transmitting. The attack module can send this information to the hacker by a wireless channel to process it. Then the hacker can send to the attack module the necessary information (address and attack moment). In our case, the attack module has already the address of the sensitive slave (motor controller) and the order of attack will arrive in real time. Secondly, in the case of attack, the module must cut the communication between master and slave module; otherwise, the module must do nothing. The communication will be cut avoiding the changes in SCL signal forcing a low level. This is possible due to the high level is high impedance. With this action, the first packet is not completed and no slave will send the acknowledge. This action can be seen as an injection failure attack varying the frequency of clock signal. Thirdly, the attack module must generate the acknowledges in the case of attack. In this situation, the master believes that the communication process is performing and it does not repeat the order. Finally, the attack module must enable the communication of the stop condition in order to all slaves are ready for a new communication process.

Fig. 3 shows the behavior of two different transactions (from the logic analyzer equipment): one of them (red) has been attacked; while the other one (green) has not. In both cases, the slave address is 0xB0, the register address is 0x00, and data is 0x07. The signals in Fig. 3 are the order of attack, SDA line in bus, SCL line generated in the master, SCL line in the bus (altered by the attack module), and the identification of the acknowledge cycle.

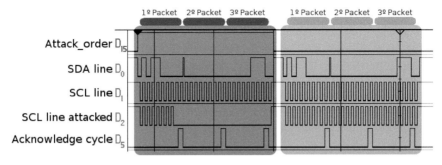

Fig. 3 Behavior of two transactions: an attacked transaction (red); and a safe transaction (green).

In the red transaction, the attack module reads the slave address (0xB0) that must be attacked because the attack order is active. In this cycle (the seventh cycle of the first packet), the attack is performed, and hence SCL line in the bus is down during the rest of the transaction. As no slave detects a complete transaction, no slave sends the acknowledge; thus, the attack module sends the acknowledge in order to the master detects a normal transaction. Finally, when the attack module determines the end of the transaction, it enables the pass of the stop condition from the master. In this case, the clock stretching does not eliminate the vulnerability because the attack module does not release the SCL signal, and hence, the master waits indefinitely for the SCL signal to be free, and the command is not sent (that is the objective of the attack).

In the green transaction, the attack module acts in the same way as in the red transaction, that is, it determines the slave address of the transaction and it detects the end of the communication. The only difference is that the attack is not performed because the attack order is not active. In this case, the slave (motor controller) sends the acknowledges in the different packets because in this case the complete transaction arrives to the slave.

3.2   Hacking the Robot Motion

In this work, the robot has been controlled by using the pure-pursuit algorithm [8], a very well-known path tracking algorithm, whose efficiency has been extensively demonstrated. This algorithm takes into account the current robot's configuration and determines the correspondent low level action that makes the robot follow a desired path. However, any other robot controller could be applied without modifying the conclusions of this work, as the attacks have been performed on the communication hardware and not on the control software.

A differential drive robot has been considered, accomplishing the traditional kinematics model. Therefore two motors are used as actuators, and the speeds of both are determined by the control algorithm. The low level controller (see Fig. 1) will receive these values, being the responsible of generating and sending the velocity commands to the motors. Then, one master and one slave I2C devices are used to support the command transmission to the actuators.

At this stage, selective hardware attacks can be implemented by hacking the clock of the I2C bus. The objective of this research is the study of the effect of implementing selective hardware attacks to avoid communication between the controller and the right motor, the left motor or both of them. As it is shown below, if these attacks are properly executed, the course of robot can be modified conveniently. The main constraint for designing the attacks is that neither the user nor the controller should detect that a hardware attack has been performed, with the exception of the change of the robot course. As a consequence, the following conditions must be taken into account for the attack implementation:

1) During and after the attack, the robot should keep navigating in a safe way.
2) The consequent of the attack is one of these options:
    a) The robot will not reach a specified location.
    b) The robot will repeat the way to visit a certain area.

4   Experimental Platform

A static experimental platform has been built in order to verify the authors' hypothesis. The platform simulates the motion of a differential drive robot in different virtual environments, so that digital and analogic instrumentation can be used in order to measure and characterize the behaviors of the hardware against attacks (see Fig. 5). Thus, an architecture similar to the one presented in Fig. 1 has been implemented, by including a high level controller, a low level controller and two motors (representing the wheels actuators). The scheme and a picture of the experimental platform are shown in Fig. 4 and 5 respectively.

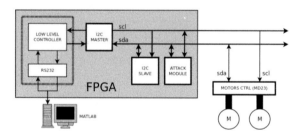

Fig. 4 Scheme of the experimental platform.

It is divided into three different zones (see Fig. 4): a PC that executes Matlab software to implement the path tracking algorithm; a FPGA device to implement hardware modules (such as the attack module), and a standard slave device (more concretely the motor controller MD23 [11]). The FPGA device is aimed to implement the hardware modules to obtain a better control of the signals. These modules are an UART based on RS232 protocol (to communicate with Matlab in

PC), a low level controller (to adapt the orders from Matlab to the rest of modules), an I2C master (to control the communication through I2C protocol), an I2C slave (to test the system inside the FPGA device) and the attack module (to implement the attacks to I2C communications). The platform also includes several measurement elements, highlighted in red in Fig. 5: a logic analyzer that monitors the main signals in the communication process, the same PC that monitors the velocity of the motors, and a current measuring device (between the supply source and the motor controller) that monitors the current consumed by the motors.

Fig. 5 The experimental platform.

The way of working with this platform is the following: Once a scenario is determined, a path is generated, so that the robot has to follow it accurately. This path is given to the high level controller program running in Matlab. This program executes an iterative loop, which takes into account the current robot position and applies the pure pursuit algorithm to determine the desired speeds for the wheels. These speeds are communicated to the FPGA device using a RS232 protocol. The UART implemented in the FPGA receives the speeds and the low level controller decoded to insert them in the I2C bus through the I2C master. This module generates SCL and SDA signals and they go out the FPGA device to the motors driver. This driver reads the orders by I2C protocol and changes the speed of the motors. The MD23 controller allows measuring the angular position of the motor axes. Then, the tracking algorithm asks this angular position. This value is sent from motors' driver to the I2C master. The low level controller codes these positions, which are sent to Matlab in PC via RS232 protocol. In Matlab, these angular positions are translated to Cartesian position by applying dead reckoning techniques, so that the position of the robot is estimated. Hence, new values of motors speeds are obtained by the path-tracking algorithm. This flow is repeated until the robot arrives to the final position of the path. In this application, the path is generated in order to ensure that the robot goes over some places in a certain order. The reverse engineering will be used to obtain this order, and hence, to determine the optimum moment and duration of the attack. At the selected time, when the master tries to communicate with the engine to be attacked, the attack module acts on the clock, preventing such communication. After that, it sends the acknowledgment. Because of this fact, the attack achieves two effects:

- The master module believes that there has been no problem and that the motor follows the references.
- The slave module receives nothing and keeps the previous speed value.

5   Experimental Results

Many different experiments have been performed in several virtual environments. The idea was to find those types of paths and scenarios in which temporary hardware attacks accomplish the conditions defined in section 3. Due to the lack of space, only three experiments, the most significant, will be presented. The most interesting result was found in the scenario of the experiment shown in Fig. 6a). This figure shows with arrows the initial and the final position. In this scenario, the robot evolves describing what can be called as a roundabout-like trajectory, in which the robot moves around of a specific area (marked also in Fig. 6a). In this experiment, navigation was implemented without any attack, the path-tracking algorithm succeeded in controlling the motion of the robot very close to the planned trajectory. The trajectory described by the robot has been calculated by applying a dead-reckoning methodology that uses the signal provided by the motors' encoders and applies a classical Extended Kalman Filter. Fig. 6b) illustrates, in red line, the reference provided by the high level controller and in blue line, the encoder signals (up right motor, down left motor).

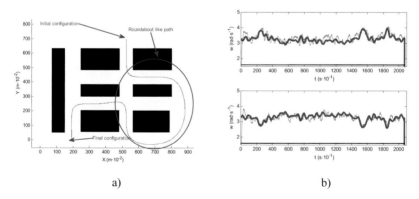

a)   b)

Fig. 6 Experiment without attack: a) planned trajectory; b) signals from the encoders.

In the first experiment with attacks, only one motor was interfered. Surprisingly, the high level controller was capable of compensating the perturbation provoked by the attack, by taking into account the position estimation and controlling only one motor. The controller modifies the velocity of the non-attacked motors so that the system follows the path, with a non-neglected error, but succeeding in moving the robot to the end. A remarkable fact: remember that the controller knows nothing about the existence of the attack. It affects only to the communication process in the I2C bus. Fig. 7a) illustrates in blue the planned trajectory and in red the path of the robot, and the moment in which the attack takes place. Fig.7b) presents, in red, the evolution of the reference for the motors speed, and in blue the encoders signal of the right and left motors (up and down respectively). Below this signals, the state of the attack module is also represented: 'low level' means no attack, 'high level' means that the module is working. Observe that, the right encoder signal does not follow the control reference during the attack, while the control signal of the left motor evolves following the references. The robot trajectory suggests that this control architecture presents a robust behavior against the hardware attack of one motor.

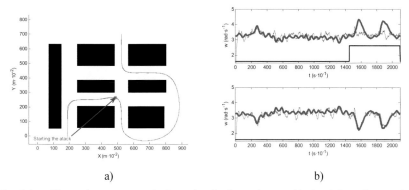

a)   b)

Fig. 7 Attacking only one motor: a) planned and robot trajectory; b) signal from the encoders.

In the following experiments, the attack was implemented on both motors. Two different moments for the attack were chosen. In the second experiment, the clock attack took place before the roundabout of the path, and disappeared when the robot was closer to the end of the roundabout. Then, the path tracking algorithm makes the robot follow the path to the end, preventing the system to navigate along the roundabout. Fig.8a) shows in red the path executed by the robot, and Fig. 8b) represents the velocity references (red) and the encoder signals of both motors (blue). Note that, during the attack, none of the motors follow the high level controller reference. In Fig 8a) it can be appreciated that this attack results in preventing the robot to visit the roundabout.

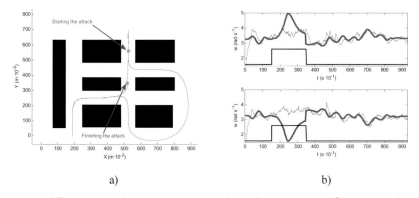

a)   b)

Fig. 8 Avoiding the roundabout: a) planned and robot trajectory; b) signal from the encoders.

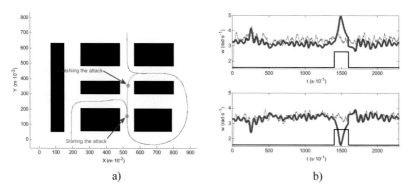

a)   b)

Fig. 9 Visiting the roundabout twice: a) planned and robot trajectory; b) signal from the encoders.

In the third experiment, the attack took place close to the end of the roundabout and disappeared when the robot is closer to beginning of the roundabout. In this way, the path tracking algorithm makes the robot follow the roundabout again, visiting this area twice. Fig.9a) illustrates in red the path of the robot, and Fig. 9b) shows the velocity references (red) and the encoders' signals (blue) respectively. These results confirm that a convenient selection of the attack moments can make the robot change the expected behavior. The resulting motions appear to be "natural" in both of experiments; in fact the robot returns to the desired trajectory; only a bizarre behavior appears that no one could understand without knowing the existence of the attack.

6 Conclusions

This paper describes different ways of performing hardware attacks on robotic control platforms. Special attention has been paid to illustrate an attack on the clock signal in the I2C bus. This type of attack has been applied to an experimental platform that allows studying the consequences of such actions in mobile robots navigation. Some peculiar circumstances have been characterized. When the robot is attacked in these situations, it navigates safely, but it is prevented from visiting a particular area, or forced to visit twice the same area. Thus, the I2C vulnerability opens the door for possible manipulation of the path followed by the robot, with no appearance of any external intervention. The typical countermeasure to this attack is a filter that eliminates the signal when the frequency is different to its nominal value. However, this technique is not useful because the attack and the countermeasure have the same result. Therefore, it is necessary to look for an alternative to defense the system. Currently, authors work on the use of a frequency sensor for detecting clock attacks.

References

1. Sheppard, B., Thompson, T.: Cyber Security for Robots: Scenarios for 2030. Robotic Business Review. http://www.roboticsbusinessreview.com/article/cyber_security_for_robots_scenarios_for_2030
2. Nobile, C.: Robots Vulnerable to Hacking. Robotic Business Review. http://www.roboticsbusinessreview.com/article/robots_vulnerable_to_hacking/
3. Anderson, R., Kuhn, M.: Tamper Resistence a Cautionary Note. In: Proceedings of 2nd USENIX Workshop on Electronic Commerce, vol. 2, pp. 1–11 (1996)
4. Tehranipoor, M., Koushanfaar, F.: A Survey of Hardware Trojan Taxonomy and Detection. IEEE Tran. on Design and Test of Computers 27(1), 10–25 (2010)
5. Bruschi, D., Cavallaro, L., Lanzi, A., Monga, M.: Replay Attack in TCG Specification and Solution. In: Proceedings 21st Annual Computer Security Applications Conf. 11–137 (2005)
6. Karaklajic, D., Schmidt, J.M., Verbauwhede, I.: "Hardware Designer's Guide to Fault Attacks". IEEE Trans. on VLSI Systems 21(12), 2295–2306 (2013)
7. Huang, A.: Hacking the Xbox: an introduction to reverse engineering. N. St. Press (2002)
8. Ollero, A., Mandow, A., Muñoz, V.F., De Gabriel, J.G.: Control architecture for mobile robot operation and navigation. Robotics and computer-integrated manufacturing 11(4), 259–269 (1994)
9. Mandow, A., Gomez-de-Gabriel, J.M., Martinez, J.L., Munoz, V.F., Ollero, A., García-Cerezo, A.: The autonomous mobile robot AURORA for greenhouse operation. IEEE Robotics & Automation Magazine 3(4), 18–28 (1996)
10. Cuesta, F., Gómez-Bravo, F., Ollero, A.: Parking maneuvers of industrial-like electrical vehicles with and without trailer. IEEE Transact. on Ind. Electr. 51(2), 257–269 (2004)
11. Motor Controller MD23. http://www.robot-electronics.co.uk/htm/md23tech.htm
12. Ollero, A.: Robótica: manipuladores y robots móviles. Marcombo (2001)
13. Ollero, A., Arrue, B.C., Ferruz, J., Heredia, G., Cuesta, F., López-Pichaco, F., Nogales, C.: Control and perception components for autonomous vehicle guidance. Appliction to the ROMEO vehicles. Control Engineering Practice 7(10), 1291–1299 (1999)

Integration of 3-D Perception and Autonomous Computation on a Nao Humanoid Robot

David S. Canzobre, Carlos V. Regueiro, Luis Calvo-Varela
and Roberto Iglesias

Abstract The humanoid robot Nao is a great platform for robotics research, in particular it provides an important testbed for computer vision, machine learning and human robot interface. Nevertheless, its limited sensorization and computation power reduces its autonomy severely. To overcome some of these limitations, in this paper we describe the integration of a RGB-D camera together with a mini-PC, into the Nao robot. Our objective is to get a Nao robot being able to carry out autonomously (onboard) all the tasks that involve 3D environment perception. As an example we used two applications: (1) mimic of human movements, (which will be used on learning by demonstration), and (2) RTABmap SLAM algorithm. Finally, we also tested the Nao's walking stability when it was equipped with all the new elements.

Keywords Aldebaran Nao · Humanoid robot · ROS framework · Mini-PC · RGB-D cameras · 3D Perception · Autonomous robot

1 Introduction

Nowadays we have robots that can build a car or explore deep space, but unfortunately we do not have a general purpose robot that does our job by us. To achieve that, we still need to achieve some milestones, like for example the interaction with a human being or the autonomous navigation.

D.S. Canzobre · L. Calvo-Varela · R. Iglesias
CITIUS (Centro Singular de Investigación en Tecnoloxias da Información),
15782 Santiago de Compostela, Spain
e-mail: {david.sanchez.canzobre,luis.calvo.varela,roberto.iglesias}@usc.es

C.V. Regueiro(✉)
Department of Electronics and systems, Universidade da Coruña, 15071 Coruña, Spain
e-mail: cvazquez@udc.es

© Springer International Publishing Switzerland 2016   161
L.P. Reis et al. (eds.), *Robot 2015: Second Iberian Robotics Conference*,
Advances in Intelligent Systems and Computing 417,
DOI: 10.1007/978-3-319-27146-0_13

For the autonomous navigation of a robot we need sensors that perceive the environment and a powerful on board computer to process the sensor data. The idea of implementing 3D perception on a humanoid robot is natural, but very challenging if such robot does not include convenient sensors for such task: stereo vision, RGB-D cameras or 3D range laser. In this work we focused on the integration of a RGB-D camera on a humanoid robot, in our case the Aldebaran's Nao. As the Nao's onboard computer does not have enough power to process the data from the RGB-D camera, we decided to mount on his back a mini-PC. Our aim is to achieve an autonomous platform that allows us to test our algorithms in the robot.

As proof of concept, we focused on two tasks: The first one was a simple human-robot interaction application that allows Nao robot to detect the arms of a person standing in front of it, so that Nao can mimic the movements it observes. This application will be an important part of learning by demostration tasks, that is, an user can teach the robot how to carry out a task. The second one was a simultaneous mapping and localization task (SLAM). The objective was to create a map of the environment, for navigation and localization purposes. To do this task we used a well known SLAM algorithm: RTABmap. On the other hand, we have tested both the stability of the Nao's walking behaviour and the impact of walking movements on the SLAM task.

We have structured the article in the following sections: section 2 describes the related works. Our solution for the computational autonomy is described in section 3. Section 4 is dedicated to the RGB-D cameras that we tested, their software and how they were assembled on the Nao's head. The experimental results are discussed on section 5 and, finally the conclusions and future work are described in the last section.

2 Related Works

We inspired on several works that needed a huge computation power and therefore used an external mini-PC to do the heavy tasks. Tomic et al. [1] used a three Gumstix Overo Tide boards attached to an UAV onboard computer to do the image processing tasks. Dávila Chacón, on his Master Thesis [2], used a Nao with a fit-PC2i mini-PC attached on its chest and the battery which provided the power was located on its back. Nevertheless, this layout of battery and processing unit makes the sonars of the Nao useless and reduce Nao's walking stability. The weight of the battery it is not reflected, but the fit-PC2i weighs 0.370 kg, which is much more than Odroid U3 plus Odroid UPS set that we suggest in this article

Our mimic application is based on the formulas presented by Rodriguez et al. [3] for the imitation of human arms movements by the Nao. Few works have integrated SLAM on a Nao. Mostly, because Nao does not have the best sensors for such task. Wen et al. [4] used a range laser and the Nao's cameras to do a SLAM task. Tjernberg [5] shows how to do SLAM with Nao, employing RGB-D cameras. In this paper, our objective is to suggest a hardware and software platform that allows the Nao robot to be fully autonomous and to perceive its environment in three dimensions. To carry out the SLAM task we will use a well known solution: Rtabmap [6].

Table 1 Features of embedded systems that we tested.

	Odroid XU3	Odroid U3
CPU	Samsung Exynos-5422: Cortex-A15 and Cortex-A7 big.LITTLE	Samsung Exynos4412 Prime Cortex-A9 Quad-core
GPU	Mali-T628 MP6	Mali 400 MP
RAM	2 GBLPDDR3	2GB LPDDR2
Internal Storage	eMMC up to 64GB	eMMC up to 64GB
External Storage	micro SD slot up to 64GB	micro SD slot up to 64GB
Network	10/100 Ethernet	10/100 Ethernet
Ports	USB3 Host x1, USB2 Host x4, USB3 OTG x1, PWM for Cooler Ethernet RJ-45, Headphone Jack, 30Pin: GPIO/IRQ/SPI/ADC	USB2 Host x3, Device x1, Ethernet RJ-45, Headphone Jack GPIO, UART, I2C, SPI (Board Revision 0.5 or higher)
Power supply	5V/4A	5V/2A
Dimensions	9.4 x 7cm	8.3 x 4.8cm

3 Computational Autonomy

The computation power of Nao's embedded system[1] is very limited. Therefore we decided to use a mini-PC attached to the Nao's body. In this section we are going to present both the hardware and the software options to achieve Nao's computation autonomy.

3.1 Embedded Systems

We can split the mini-PCs into a two big families: ARM architecture and x86 architecture. The problem with ARM architecture is the lack of support for some software components that we need (for instance Nao's framework called NaoQi). On the other hand, we could have used a x86 development board, but with this boards we had the trouble of power consumption (for example Gizmosphere gizmo 2 needs a 12 V 2 A power supply). Another important features were the small size and light weight.

For our tests with Nao we chose two ARM boards from Hardkernel [9] (Odroid XU3 and U3). Table 1 shows the features of the boards tested. We chose the Odroid U3 board because the battery that we needed to feed the Odroid XU3 is heavier than the Odroid U3 UPS.

3.2 Power Supply

In order to achieve complete power autonomy we have decided to add a dedicated power bank. The features that we considered to choose a battery for each embedded

---

[1] Intel Atom Z530 1.6GHz CPU, with 1 GB RAM and 2 GB flash memory.

Table 2 Features of the batteries that we tested.

	Anker Astro3	Odroid UPS
Capacity	12800 mAh	2 x 1500 mAh
Input	5V/2A	5V/2A
Output	5V/4A	5V/2A
Autonomy	3 h	11.5 h
Dimensions	11.1 x 8.3 x 2.6 cm	8.3 x 4.8 x 1 cm
Weight	298 gr	64 gr

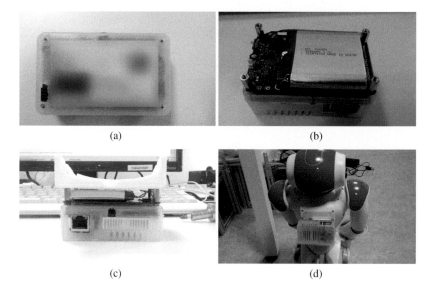

(a)   (b)

(c)   (d)

Fig. 1 Odroid U3 assembled on Nao: a) Perforations on the U3's case, b) U3 plus battery, c) Nao's bag, d) Final montage.

board were: enough power to feed the embedded system, small size, light weight and, finally, the possibility of recharging the battery while the mini-PC powered by it is still running. To feed the Odroid XU3 we need a 5V/4A battery and we chose the Anker Astro3. For the Odroid U3 we need a 5V/2A battery, so we chose the Odroid UPS [9]. Table 2 shows the technical specifications of both batteries.

3.3 Assembly on Nao's Body

We built a backpack to assembly the embedded system plus the battery. To build this backpack we used a 3D printer using the model provided by Konstantinos Chatzilygeroudis [7] to attach Odroid U3 with the UPS to the Nao's back. To attach the backpack to the Nao's back we replaced the battery's cover screws with Metric A-2 17mm long screws.

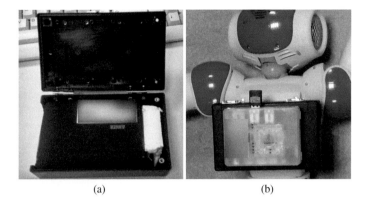

| (a) | (b) |

Fig. 2 Odroid XU3 assembly on Nao: a) Case for the Anker battery, b) Final montage.

Table 3 Hardware for the computation autonomy weights.

XU3 + Anker Battery	Weight	U3 + UPS battery	Weight
Nao's bag	29 g	Nao's bag	29 g
WiFi dongle	3 g	WiFi dongle	3 g
Odroid XU3	71 g	Odroid U3	53 g
Odroid XU3's case	39 g	Odroid U3's case	24 g
Anker battery	298 g	Odroid UPS	64 g
Anker battery's cable	29 g	Odroid UPS's cable	20 g
Anker battery's case	141 g		
Total	**610 g**	**Total**	**188 g**

We would have preferred using the Odroid U3 with its case to protect the mini-PC, but unfortunately the case was not designed to use with the UPS. Hence, we had to modify it by doing several perforations. Figure 1 shows this process step to step.

To assembly the Odroid XU3 with the Anker battery we had to use an additional case to hold the battery (fig. 2(a)). As the weight is critical for the Nao, we weighted every hardware element. Table 3 shows the results.

3.4 Software for Embedded Boards

In both embedded systems we installed Lubuntu 14.04 LTS[2], which was downloaded from the Hardkernel's official web [9]. Regarding the robotic framework we selected ROS, version Indigo [12], so we could access to all its algorithms and code. To interact with the Nao, the natural option was to use the wrapper for NaoQi [13, 14], the Nao's official framework.

---

[2] Lubuntu is a lightweight Ubuntu flavour that uses LDXE desktop instead Unity.

Table 4 Features of tested RGB-D cameras.

	Microsoft Kinect	Asus Xtion Pro Live	Creative Senz3D
Field of view	43° V, 57° H	45° V, 58° H	58° V, 74° H
Depth minimum distance	0.8 m	0.8 m	0.15 m
Depth maximum distance	4.0 m	3.5 m	1.5 m
Frame rate colour stream	30 FPS	30 FPS	30 FPS
Frame rate depth stream	30 FPS	30 FPS	30 FPS
Resolution colour stream	640 x 480	1280 x 1024	1280 x 720
Resolution depth stream	640 x 480	640 x 480	320 x 240
Weight (without base)		204 g	179 g
Helmet weight		54 g	57 g

Initially we have tried to install NaoQi in the Odroid machines but we had the problem that Aldebaran did not release an ARM version of NaoQi. For these reason, we could not run the NaoQi's ROS wrapper on the Odroid systems. We have explored two options to overcome this problem: on one hand we installed ROS Indigo, NaoQi and its wrapper of ROS in a x86 laptop. The benefit of this option is that we can develop and test code very fast, but the main drawback is that Nao is not fully autonomous and the wireless connection is a limitation factor. Regarding the second option we have analysed, it consisted on installing a basic set of ROS and the NaoQi's ROS wrapper on the Nao's embedded system (with an Atom Intel CPU). We followed the "How to" in the ROS official web [11] and applied two known patches: to change the "emerge" default directory to the directory where the Nao's internal SD card is mounted, and to install the tinyxml package and created a symbolic link where the ROS Nao nodes search it. Nevertheless, with the basic installation, the Nao's cameras node did not work. We had to install its dependencies: *ros-image-common-1.11.3* and *camera_info_manager_py*. Finally, the ROS installed on our Nao is fully functional.

4   3D Perception

In this section we are going to show the different RGB-D cameras that we tested, the assembly on the Nao's head, the software that we used to test the RGB-D cameras and finally the test that we did with the RGB-D cameras.

4.1   RGB-D Cameras

For this work we considered three RGB-D cameras: Microsoft Kinect [15], Xtion Pro Live [16] and Creative Senz3D[3] [17]. Table 4 shows the technical features of

---

[3] According to SoftKinetic official web site the Creative Senz3D it is the same camera than Soft-Kinetic DS325, but with different outer casing and Product ID.

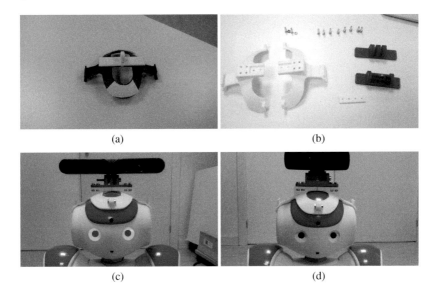

(a)   (b)

(c)   (d)

Fig. 3 Nao's helmet for RGB-D cameras: a) First prototype, b) Helmet parts, c) Nao with Asus Xtion Pro Live, d) Nao with Creative Senz3D.

the three cameras. Finally we discarded the Microsoft Kinect due to its weights. Our intention is to use the Asus Xtion Pro Live for both, mapping and localization, while the Creative Senz3D will be used in manipulation tasks to detects small objects (for example when Nao has to grasp a tactile pointer [8].

4.2 Assembly on Nao's Head

To assemble the cameras on Nao's head we used a helmet. Our helmets are based on the one designed by K. Chatzilygeroudis [7]. Figure 3 shows the selected cameras mounted on Nao's head and the parts of Nao's helmet. As with the mini-PC we weighted all components to mount the cameras on Nao's head. In table 4 we show the results.

In the initial tests, we observed that with the first prototype of Asus Xtion helmet (fig. 3(a)) the camera oscillated during the Nao's walking movement and eventually made the robot lose balance. Because of this we had to re-design the helmet and we had to add two reinforcements (fig. 3(b)).

4.3 Software for RGB-D Cameras

The first piece of software to install was the driver of each RGB-D camera, and its corresponding wrappers to ROS. We selected the Openni2 driver [18] and its wrapper

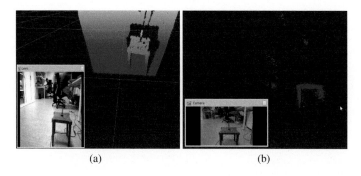

(a)   (b)

Fig. 4 Comparison of RGB-D cameras: a) Asus Xtion Pro Live, b)Creative Senz3D.

for ROS [19] (openni2_launch) to access to the Asus Xtion Pro Live camera. To get access to the Creative Senz3D camera we used the official SoftKinetic DS325 SDK [20]. There is not an "official" wrapper for ROS for this camera, but we selected the stack *ros_depthsense_camera*, developed by Walter Lucetti [21]. There is another ROS wrapper from Fraunhofer IPA[4], but unfortunately it does not work on an ARM architecture.

Additionally, we have installed libraries to make specific tasks and which are supported or have a wrapper for ROS. For the task of real-time human body detecting and tracking, we selected the NITE library (version 1.5.2.23) [22].

To process the big volume of point cloud data generated by the RGB-D cameras, we chose the Point Cloud Library [23]. When we tested the Creative Senz3D camera, we discovered that its ROS wrapper (*ros_depthsense_camera*) it is not compatible with PCL[5].

Finally, for simultaneous mapping and localization tasks, we selected the well known RTABmap algorithm [6]. Nevertheless, when we tried to run on the Odroid U3 the RTABmap ROS wrapper it fails due to a bus error. We could not correct this problem, as it is very recent and it is not well documented. Fortunately, the standalone RTABmap applications work correctly on the ARM embedded boards. As a temporary solution, while this malfunction is fixed, we can employ the RTABmap ROS wrapper on a x86 machine (our development laptop), mainly for visualization purposes.

5   Experimental Results

We tested the stability of the robot with the two sets of mini-PC plus battery (Table 3). With the Odroid XU3 and the Anker battery (610 g) attached to the Nao's back the

---

[4] https://github.com/ipa320/softkinetic

[5] At the moment of write this paper we have yet not found the issue of this malfunction.

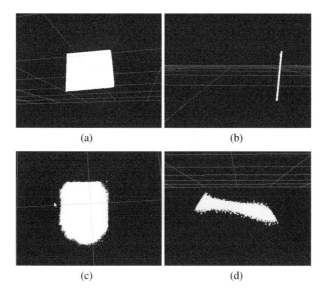

Fig. 5 Comparative between the selected RGB-D cameras: Asus Xtion Pro Live (Upper row) and Creative Senz3D (bottom row). Point clouds of the ground plane viewed from the front (left column) and from the side (right column).

robot oscillates significantly and eventually it falls down. With the Odroid U3 and the Odroid UPS (188 g) the robot can maintain balance while walking. We did this test at different speeds (20%, 50%, 80% and 100% of Nao's full speed) and we saw that the faster the NAO walks the more stable its movement is. Obviously a heavy weight has an impact on NAOs stability. On the other the hand the extra battery attached to the robot lasts for two or three hours when both boards are running at full capacity, while NAOs battery lasts only half an hour when NAO is walking, or one hour and a half if the robot is idle.

To test the quality of the selected RGB-D cameras (fig. 4), we compared their point clouds. With this aim, we first implemented a "plane detection" with PCL and C++, as a ROS node. This node extracts from the original point cloud, the majority plane's point cloud (typically the ground plane) and the rest of the point cloud and publish them as two ROS topics. They can be visualized on real time with the Rviz tool. Nevertheless, the data that the ROS wrapper of Creative Senz3D send does not work with PCL. Hence in order to compare the point clouds we pointed the cameras to the floor and, using Rviz, we took two captures from the data that send the ROS wrappers of both cameras. In figure 5 we can see that the point cloud of Creative Senz3D is noisier than the point cloud of Asus Xtion Pro Live.

We developed a little program called "Mimo Nao" [24]. This program detects and tracks the movement of the arms of a person, so that this movement can be replicated by Nao. Figure 6 shows a capture from our software. The way our software works can be described in the next stages: first the NITE library is used to filter all users but the

Fig. 6 Capture from the Nao's human imitation software.

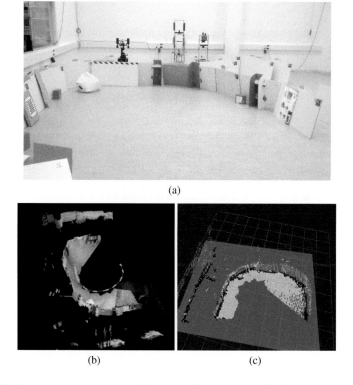

Fig. 7 RTABmap test: a) Environment, b) Point cloud in the middle of the Nao's trajectory, c) Final occupancy grid.

one in front of the camera. Second, the coordinates of the user limbs are detected and projected into the camera space and, finally, the process described in [3] is followed to compute the value of the angles made by the user joints, translate them into NAOs space, and send a ROS message with this information to the robot NAO.

Finally, we have tried to create a map with the Asus Xtion and the Odroid U3 mounted on Nao while it is walking on the environment. To make the map we used the Asus Xtion like a fake laser and we also used visual odometry from RTABmap algorithm. The Odroid XU3 is able to process 11 FPS and the Odroid U3 is able to process 4 FPS. Our test was very simple, it consisted on describing an arch with Nao around our robotic laboratory (fig. 7(a)). The captures obtained in one of these tests are shown in the figures 7(b) and 7(c). We recorded some videos which can be seen in [24].

6   Conclusions and Future Work

In this work we have integrated both a RGB-D camera and a mini-PC into a humanoid Nao robot with two objectives in mind: First we endowed Nao with an RGB-D camera (Asus Xtion Pro Live), so that it can perceive the environment. We also tested the Creative Senz3D camera for short distances, but it was discarded due to a lack of support for ARM architectures and because the data retrieved by this camera is very noisy.

Our second objective was an attempt to improve the computational power of our robot. In this case, we attached an ARM embedded board on its back. Both Odroid U3 and XU3 were tested with very good results, but U3 is preferred due to its low power requirements. For simplicity, we have also included a dedicated external battery to power the embedded board.

To test the Nao's performance regarding computational autonomy and 3D perception, we have implemented two basic tasks: The first one is a simple application that detects and tracks in real time the body and arms of a person standing in front of the robot. Our application is able to make the robot Nao mimic the human movements. The second task we have carried out is a SLAM using the RTABmap algorithm while Nao walks in the environment. An additional purpose of this second task was to prove the stability of Nao's walking with the additional accessories (embedded board, camera, battery and assemblies). As it has been discussed, there are several limitations: on one hand, some ROS wrappers are not fully functional on ARM boards (RTABmap, Creative Senz3D, NaoQi) To solve this, an easy solution might be using an x86 embedded board, for example the MeegoPad T02 (a clone of the new Intel Computer Stick). Another option could be fixing each wrapper. Finally, a better Nao's balancing control will be required when Nao carries the Odroid XU3 and its battery.

The mimic application we have presented and tested in this paper, will be used to get our robot being able to learn by demonstration, that is, any person can teach a simple behaviour to Nao (grasp, walk). The new Nao's 3D perception capabilities will allow it to detect relevant objects for a task: for example a pointer and a board to play games [8]. At this point, a better support of Creative Senz3D or a new RGB-D camera for short distances will be welcome.

We will expand the basic SLAM application in several ways: active exploration, autonomous map creation, indoor localization and scene recognition. We will study how to integrate human salience models in the SLAM data.

Acknowledgements This work was supported by the research grant TIN2012-32262 (FEDER cofunded) and by the Galician Government (Xunta de Galicia) under the Consolidation Program of Competitive Reference Groups: GRC2014/030 (FEDER cofunded) and GRC2013/055 (FEDER cofunded).

References

1. Tomic, T., Schmid, K., Lutz, P., Domel, A., Kassecker, M., Mair, E., Lynne Grixa, I., Ruess, F., Suppa, M., Burschka, D.: Toward a fully autonomous uav. research platform for indoor and outdoor urban search and rescue. IEEE Robotics & Automation Magazine, pp. 46–56 (2012)
2. Chacón, J.D.: Visual-Topological Mapping: An approach for indoor robot navigation. Master Thesis. Univ. of Groningen (2011)
3. Rodriguez, I., Astigarraga, A., Jauregi, E., Ruiz, T., Lazkano, E.: Humanizing NAO robot teleoperation using ROS. IEEE-RAS Int. Conf. on Humanoid Robots, pp. 179–186 (2014)
4. Wen, S., Othman, K.M., Rad, A.B., Zhang, Y., Zhao, Y.: Indoor SLAM Using Laser and Camera with Closed-Loop Controller for NAO Humanoid Robot. Abstract and Applied Analysis, vol. 2014 (2014)
5. Tjernberg, I.: Indoor Visual Localization of the NAO Platform. Master Thesis, KTH Computer Science and Comunication (2013)
6. Labb, M., Michaud, F.: Online Global Loop Closure Detection for Large-Scale Multi-Session Graph-Based SLAM. IEEE/RSJ Int. Conf. on Intelligent Robots and Systems (IROS 2014), pp. 2661–2666 (2014)
7. Chatzilygeroudis, K., Sako, D., Synodinos, A., Aspragathos, N.: Navigation of humanoid robot NAO in unknown space. Diploma Thesis. Univ. Patras (2014)
8. Calvo-Varela, L., Regueiro, C.V., Canzobre, D.S., Iglesias, R.: Development of a Nao humanoid robot able to play Tic-Tac-Toe game on a tactile tablet. Second Iberian Robotics Conference ROBOT 2015 (2015)
9. Hardkernel's official website. http://www.hardkernel.com/main/main.php. Last visit: June 30, 2015
10. Anker's official website. http://www.ianker.com/. Last visit: June 30, 2015
11. ROS install in Nao's embedded system tutorial. http://wiki.ros.org/nao/Tutorials/Installation/local. Last visit: June 30, 2015
12. ROS official website. http://ros.org/. Last visit: July 07 2015
13. ROS Nao wrapper website. http://wiki.ros.org/nao. Last visit: July 07, 2015
14. NaoQi official documentation. http://doc.aldebaran.com/1-14/dev/naoqi/index.html. Last visit: July 07, 2015
15. Microsoft Kinect official website. https://msdn.microsoft.com/en-us/library/jj131033.aspx. Last visit: July 07, 2015
16. Asus Xtion Pro Live official website. http://www.asus.com/US/Multimedia/Xtion_PRO_LIVE/. Last visit: July 07, 2015
17. SoftKinetic DS325 official website. http://www.softkinetic.com/Products/DepthSense Cameras. Last visit: July 07, 2015
18. Openni2 official website. http://structure.io/openni. Last visit: July 07, 2015

19. Openni2 ROS wrapper website. http://wiki.ros.org/openni2_launch. Last visit: july 07, 2015
20. SoftKinetic official website. http://www.softkinetic.com/. Last visit: July 07, 2015
21. SoftKinetic SDK ROS wrapper website. https://github.com/Myzhar/ros_depthsense_camera. Last visit: July 07, 2015
22. NITE official website. http://openni.ru/files/nite/. Last visit: July 07, 2015
23. PCL official website. http://pointclouds.org/ Last visit: July 07, 2015
24. Wiki CITIUS. http://wiki.citius.usc.es/. Last visit: July 07, 2015

Interpreting Manipulation Actions: From Language to Execution

Bao-Anh Dang-Vu, Oliver Porges and Máximo A. Roa

Abstract Processing natural language instructions for execution of robotic tasks has been regarded as a means to make more intuitive the interaction with robots. This paper is focused on the applications of natural language processing in manipulation, specifically on the problem of recovering from the instruction the information missing for the manipulation planning, which has been traditionally assumed to be available for instance via pre-computed grasps or pre-labeled objects. The proposed approach includes a clustering process that discriminates areas on the object that can be used for different types of tasks (therefore providing valuable information for the grasp planning process), the extraction and consideration of task information and grasp constraints for solving the manipulation problem, and the use of an integrated grasp and motion planning that avoids relying on a predefined grasp database.

Keywords Manipulation · Grasp planning · Natural language processing

1 Introduction

Teaching and interacting with robots is shifting from being a highly specialized task requiring specific knowledge, to a more natural interaction that allows agnostic users to instruct the robot in the same way as they would instruct a human apprentice. Such capability requires the ability to process natural language (NL) instructions coming from a human. For instance, a simple instruction like "Bring me a glass of water" requires a decomposition into a set of simpler subtasks (look for the glass, pick it up, verify if it already contains water, go to the faucet, open the faucet, pour water onto the glass, close the faucet, move the glass to an inferred goal position), and in turn each

B.-A. Dang-Vu · O. Porges · M.A. Roa(✉)
Institute of Robotics and Mechatronics,
German Aerospace Center (DLR), 82234 Wessling, Germany
e-mail: {Bao-Anh.Dang-Vu,Oliver.Porges,Maximo.roa}@dlr.de
http://www.robotic.dlr.de

© Springer International Publishing Switzerland 2016
L.P. Reis et al. (eds.), *Robot 2015: Second Iberian Robotics Conference*,
Advances in Intelligent Systems and Computing 417,
DOI: 10.1007/978-3-319-27146-0_14

one of the subtasks requires inferring the objects involved in the action (glass, faucet), and the required action (pick up, hold, open, move). Environmental constraints are tacitly embedded into the instructions; for instance, if the glass already contains water, then the instruction is simply translated into a pick and place command.

To successfully execute the tasks, the robot must be able to recognize the objects in the environment, and infer and understand the spatial relations between them. For instance, if the glass must be filled with water, the robot should recognize that it needs to go to the kitchen, a physical location in a home environment [10]. Physical spaces define then a subproblem of navigation, which can also be modified via spatial constraints as in "do not go to the toilet", or "stay away from the stairs". Prepositions embedded into the NL are often used to express these spatial relations, and have been considered either through predefined, hard-coded models of relational primitives, or through probabilistic inference of the spatial meaning that they enclose [14]. Typical spatial prepositions include, among others: to, from, along, across, through, toward, past, into, onto, out of, and via. Learning and probabilistic reasoning have been also used, for instance, for the grounding and semantic interpretation of phrases that lead to the definition of navigational paths for service robots [4], or for the creation of an extended vocabulary for tabletop manipulation scenarios [6].

Natural instruction of robotic manipulation problems has become an active research topic. Typical everyday activities for humans involve a large amount of inferred and possibly hidden knowledge about the way the activity should be performed, what objects are required to solve a specific task, the relation between them, and the affordances that each one of those objects allows [10]. Several works have tackled the grounding of NL commands to robot instructions for these cases. The concept of Probabilistic Robot Action Cores (PRAC) was introduced to encode action-specific knowledge in a probabilistic knowledge database, which helps to solve ambiguity and underspecification of the NL commands [10]. Inclusion of environment and task context and variation of language has also been considered by learning from example sequences performed in an online simulated game [9]. There, a learning model based on an energy function that encodes the validity of pre- and post-conditions, length of the instructions and consistency, was used to translate sequences of NL commands into sequences of instructions appropriate for the environment and task at hand. Learning the spatial semantics of manipulation actions can also be obtained through the analysis of the evolution of spatial relations between objects [15].

This paper is also focused on the problem of performing manipulation tasks with instructions given in natural language (Fig. 1). Different to previous works, our aim is to exploit the information contained in the NL expression to deduce the appropriate constraints for simple skills, in this case grasping and manipulating an object. For instance, if the object is an electric driller, a user could specify a pick and place action on the object, which only requires a stable grasp on it, or could suggest drilling a hole on some other material, which necessarily translates into constraints for grasping and activating the object to perform the intended task. Therefore, given a NL expression we parse the sentence and exploit its syntactical structure to determine actions and semantic relations between the objects. Action-specific knowledge is extracted from the instruction, not only to deduce the current

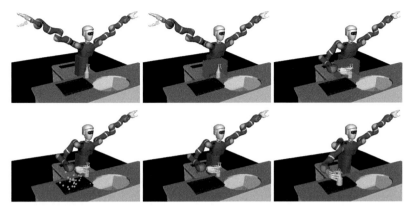

Fig. 1 Sequence of planned actions from the instruction "put the spray on top of the tray," using the DLR robot SpaceJustin: a) initial configuration; b) potential approach directions, obtained with a capability map; c) initial grasp: d) sampling the goal location; e) object being transported; f) object in final location.

location and goal region for the manipulated object, but also to understand how the object should be grasped, or if there are special restrictions on the way it should be manipulated. Naturally, this knowledge must be complemented with a suitable understanding of the object-specific affordances. A semantic clustering method is used to obtain regions that could be employed with different purposes on the object. The desired goal, the task constraints and the grasp restrictions are considered in an integrated grasp and motion planner that derives feasible trajectories for performing the manipulation action.

After this introduction, the paper is organized as follows. Section 2 describes the overall architecture of the planning system. Section 3 presents the framework used to analyze the NL instruction and extract constraints for the instantiation of the manipulation task. Section 4 explains the inclusion of these constraints on the motion planner. Finally, Section 5 concludes the paper.

2 Overall Concept

A simple human-to-human instruction, like "cut the bread", involves hidden knowledge that a service robot should infer: *cut* is an action that requires an agent object, a knife, which is a tool and has a proper way to be used, and *bread* is a passive object, the object that is acted upon to perform the action. The goal of our system is to interpret the meaning coming from an input sentence in order to send appropriate commands to the robot to perform the grasping or manipulation skill. First, the natural language text is processed using a statistical parser. The syntactic structure is exploited in order to determine action verbs and semantic relations between the objects. Possibly hidden relations should be inferred also at this level. Then, the action-specific knowledge is deduced, i.e. what to do to which objects in a particular

situation. The passive object that is acted upon is deduced by analyzing the preposi-
tional relations (with, from, to, into) that modify the corresponding action verb. The
action verb describes a range of tasks that we simplify into three categories: pick and
place tasks (direct manipulation), tool usage, or device usage. We make here a subtle
distinction between tool and device by considering that a device contains a trigger
that should be pressed/moved to generate the intended action (e.g. an electric driller),
while the tool is just an object with some affordance that allows its usage to perform
some particular task (e.g. a knife). This distinction is critical for the grasp planning
process, as a device requires an active finger capable of triggering the desired action
while the other fingers grab the device. The objects, nouns and prepositions in the
NL instruction help also to decide the spatial relations, i.e., to identify an initial and
final pose for an object. For instance, the sentence "Take the mug that is behind the
plate and put it on the tray" provides some hints on where the mug can be found,
indicates the goal position, and defines a pick and place action. A vision module (not
explained in this paper) identifies the required objects and estimates its location, and
provides information to infer possible goal positions for the grasping or manipulation
action.

To complement the knowledge coming from the NL instruction, the system is
supported by an object database that includes the semantic knowledge of the objects.
The functional parts of the objects could be manually labeled, or can also be obtained
in a more systematic way through an active clustering process, as proposed here. In
any case, one object in the database, for instance a spoon, has a pre-stored geometrical
interpretation of its parts: a handle, where the object should be grasped, and a bowl,
the part required to perform some action. Also, the database includes predefined
tasks or action verbs associated to the object: pick and place, serve, mix.

With all the previous elements in consideration, a planner is executed to create a
grasp and a path –or a sequence of paths– for the arm and end effector to successfully
accomplish the task. The task specification defines grasp constraints on the object,
and the initial and final location defines the starting and goal configuration for the
path planner. To simultaneously consider these restrictions, an integrated grasp and
motion planner decides the best grasp on the object and the collision-free path that
allows the successful execution of the task [5].

The architecture of the overall system is presented in Fig. 2. This paper is focused on the NL processing system, and is presented in two parts: 1) Extraction of
constraints: spatial relations, current and desired object location, role of the object,
desired task (Section 3), and 2) use of the action-specific knowledge to plan the
motion: constrained areas and specifications of required contact points for grasping,
and integrated grasp and motion planning (Section 4). Note that the generation of
long sequences of actions, for instance for making a pancake [1], considerations on
the performance of the NL processing like generality, templating and instantiation of
actions or parsing success, and an extensive evaluation of the system performance are
out of the scope of the paper; the paper is devoted to the description of the proposed
architecture for extracting the task-related knowledge from the NL instruction, in
order to provide parameters for the generation of the plan required for executing the
manipulation task.

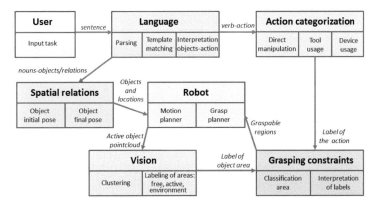

Fig. 2 Architecture of the system.

3 Natural Language Processing: From Task Specification to Description of Constraints

Processing the natural language instruction not only provides the definition of the task, but also includes important clues on the intended task, mainly:

– the role of the objects: an object can be active, like a tool/device, or passive, i.e., modified by the action of some other object. Note that a given object can be passive or active, depending on the provided NL instruction.
– information on the spatial relations: defines for instance possible reference locations for the objects, or an intended goal area for a pick and place task.

To perform the NL processing we use the Natural Language Toolkit, NLTK [3], a Python-based suite of libraries and programs for processing free text. The general approach for the language processing is summarized in Fig. 3. The syntactic parser

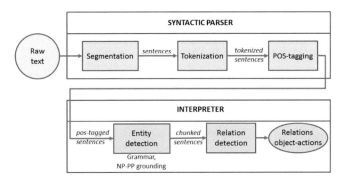

Fig. 3 Language processing approach.

is used directly from NLTK. However, since NLTK does not include a semantic analyzer (interpreter), we developed our own semantic module, specifically suited for manipulation instructions.

3.1 Language Processing: Syntactic Parsing

To correctly interpret a command such as "Put the bottle on the tray" or "Use the screwdriver", the system must ground and interpret the sentence and figure out the relations and states of the objects. The planning system is managed by an interface that allows the user to type free sentences (Fig. 2). The input is then a sentence in a free-form natural language, which can contain information about the objects through their names or properties (e.g., "the bottle", or "the red bottle behind the plate"), and description of their spatial relations with respect to other objects (e.g., "to the left of"). The syntactic parser module outputs the interpretation of the command as the action type (verb), the identity of the objects (nouns), and the spatial location (prepositions). The prepositions considered in this work for the grasping and manipulation skills are: above, behind, below, close to, far from, in front of, inside of, on, to the left of, to the right of, and under. The sentence is segmented and tokenized, i.e. divided in words, punctuation signs, etc., and POS (Part-Of-Speech)-tagged, i.e. tokens are marked with their corresponding word type based on the token itself and the context of the token.

3.2 Grammar: Language Grounding

The arbitrary structure of the natural language is reduced to a structure based on clausal decomposition. The clause C is defined by a tuple $C = (a, [role], r)$ containing the action a described by the verb, the set [role] of *object-role* that defines the object to move as active and the referenced object (if present) as passive, and a relationship r between objects, deduced from the spatial relations (e.g., "left", "from").

In order to parse and describe the structure of the sentences, we use a grammar G that identifies the verb (V), noun (N), preposition (PP), noun-phrase (NP $\rightarrow$ N PP), and spatial relation (SP $\rightarrow$ NP PP). For example, for the sentence:

$$\underbrace{\text{"Put the bottle on the left of tray"}}$$

$$C = (a=\text{put}, [\text{role}]=\{\text{bottle:active, tray:passive}\}, r=\text{"on the left"})$$

the use of the grammar generates the tree shown in Fig. 4.

The parse-tree is analyzed to find nodes of action-verb type that form a sentence, and each sentence is used to find a clause. The action-verb type retrieved by the clause can have multiple meanings in different contexts. For example, the

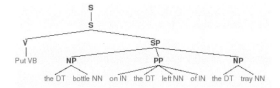

Fig. 4 Parsing: "Put the bottle on the left of tray".

action-verb "pick" has the same meaning as "take", and both belong to the same group of actions that require direct manipulation of the object (pick and place). Therefore, we perform a disambiguation step in order to classify each action-verb into specific categories, similar to the procedure of [16]. To understand the different meanings, we reduce the dimensionality of the set of action verbs by semantic clustering. We divide the actions into two categories: "direct manipulation" such as pick, hold, place, where the goal is to pick and move objects directly, and "tool/device use" such as screw or cut (e.g. non pick and place), where the object must enter in contact with the environment to perform some action.

The meaning of each action-verb is obtained using the corpus reader WordNet [8] that provides cognitive synonyms, or synsets, of a word. The synsets of "pick" are generated in this way, and the synsets for tool/device use are manually entered. During execution, the retrieved action-verb is compared to the list of synsets in order to classify the action into one of the two categories already defined.

3.3 Spatial Relations

The deduction of spatial relations depends on the vision module, to consider the location of objects from the point of view of the robot, and on the prepositions present in the sentence, to interpret the spatial relations between the objects.

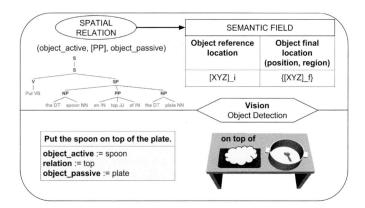

Fig. 5 Interpretation of spatial relations.

The spatial relations r are described by prepositions such as "to the left of" or "near to". To interpret those relations in the environment, we use a semantic field model, sketched in Fig. 5.

The semantic field of a spatial relation is represented as a multivariate Gaussian distribution over 3D points that fall into the described area of the preposition, relative to the passive object, and assigns weight values to the points depending on the meaning of the preposition. For example, to reason about some active object being "to the left" of a passive object (a tray in the example of Fig. 6), the spatial region to the left of the passive object is considered (which depends on the current location of the robot), and then we compute the multivariate Gaussian where its mean is defined by the extents of the passive object.

Fig. 6 Spatial language interpretation with semantic field depending on the point of view, for the instruction "Put the bottle to the left of the tray". The robot is assumed to be located perpendicular to the page.

4 Planning: From Constraints to Semantic Grasps

The actions and spatial relations found from the NL define the constraints for the grasp and motion planning problem. The approach is supported by an object database that contains information on the functionality. Previous works dealing with the addition of such functional information to object databases rely on manual input from a human [7]. Here, this process is replaced by a clustering and annotation process that simplifies the generation of differentiated areas on the objects. This functional information and the task information coming from the NL processing are used to generate desirable areas or constraints for grasping the object. Finally, these constraints, along with the initial location of the object and the goal region defined by the sentence, are used as input to an integrated grasp and motion planner that generates a feasible trajectory for the robot.

4.1 Semantic Clustering and Annotations of Objects

For utilizing an object, a robot must be aware of the affordances of the object, and the implications that they have in terms of restrictions on the object. For instance, a glass can be grasped in different ways, but if the glass is needed to pour a liquid into it, then the top part of the glass must be free to allow the pouring action. More complex objects are classified into the categories of tools/devices; these objects are meant to actively perform some action on another object or the environment. The semantic difference considered here to define a device is that it requires the actuation of a finger on some mobile part (button, lever, trigger) to initiate its expected action. Examples of devices are a driller or an electric screwdriver, while knives or pencils would be tools.

Natural language instructions, originally meant for humans, often do not explicitly specify the part of the object that should be considered for using the tool. For instance, the instruction "Use the spray bottle to clean the table" does not indicate the location of the spray trigger. Thus, we identify the functions of the object through symbolic attributes defined with three labels:

- Action area $\mathcal{A}$: area that the hand should interact with in order to perform an action.
- Environment area $\mathcal{E}$: area meant for the interaction with the environment, which cannot be touched or blocked (forbidden area).
- Free area $\mathcal{F}$: area where no constraints are applied, and therefore is free to be used by the gripper/multifingered hand to grasp the object.

For example, a driller is defined with three different parts: a trigger that must be touched in order to activate it, a main body that can be grasped, and a forbidden region that interacts with a screw (Fig. 7). For an object described with a pointcloud, each point belongs to one of the described categories, and the whole object $\mathcal{O}$ is the union of the three sets: $\mathcal{O} = \mathcal{A} \bigcup \mathcal{E} \bigcup \mathcal{F}$.

To achieve a clustering of the object, we use the Locally Convex Connected Patches (LCCP) method [13], which gives a rough clustering that approximately conforms to object boundaries. The clustering can be further refined in order to have the three distinct parts. The annotation of the object is then validated by a user; this validation is required only once for each object in the database. Methods to make this process in a completely autonomous way are currently under investigation.

Fig. 7 Semantic clustering of a spray bottle and a driller: the action area $\mathcal{A}$ is displayed as blue dots, the environment area $\mathcal{E}$ as purple dots, and the remaining gray zone corresponds to the free area $\mathcal{F}$.

4.2 Semantic Grasps

The generation of the sets defined in the previous section helps to identify the graspable areas on the object, required for the grasp planner. The grasp planner used in this work is based on the concept of reachable Independent Contact Regions (rICR), i.e. regions (patches) that are reachable for the current hand pose and that guarantee a force closure grasp when each finger is (independently) located inside its corresponding contact region on the object surface [12].

The kinematics of the hand is considered by (offline) computing the reachable workspace for each one of the fingers. First, reachable points on the object surface are obtained by computing the collision points between the pointcloud of the object and the workspace for each finger. This leads to the reachable regions $\mathcal{R}$:

$$\mathcal{R} = \{\{xyz\}_{i=1,\cdots,\text{nFingers}}\}$$

Then, the labeled areas on the object (action, free and environment area) are used to obtain a subset of the reachable points that satisfies the constraints of the task. If a is a pick and place task, the reachable points in $\mathcal{F}$ can be used for grasping

$$\mathcal{G}_1 = \left\{\{\mathcal{R}\bigcap\mathcal{F}\} - \{\mathcal{A}\bigcup\mathcal{E}\}\right\}$$

If a is a task using a tool, the points used for grasping must include also points from $\mathcal{A}$, to be able to effectively use the object

$$\mathcal{G}_2 = \left\{\{\mathcal{R}\bigcap\{\mathcal{F}\bigcup\mathcal{A}\}\} - \mathcal{E}\right\}$$

These graspable areas $\mathcal{G}$ are the ones considered in the grasp planning process, as illustrated in Fig. 8 for a spray bottle considering a device usage or a simple pick and place task.

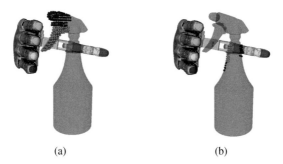

(a)   (b)

Fig. 8 Graspable regions for different tasks: a) The use of the spray bottle requires at least one finger to be located on the action area $\mathcal{A}$ (in blue), while the environment area $\mathcal{E}$ (in purple) should be free; b) the action area $\mathcal{A}$ is removed from the reachable regions, thus creating a graspable area $\mathcal{G}_2$ for a pick and place task, that avoids possible activation of the device while it is being transported.

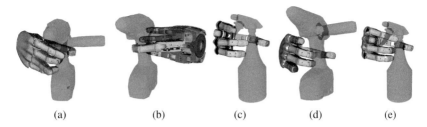

| (a) | (b) | (c) | (d) | (e) |

Fig. 9 Constrained grasps for action (a, b and c), and for pick and place (d and e).

Fig. 9 illustrates different grasps for a driller and a spray bottle, for use of the device and for pick and place tasks. Note that qualitatively different grasps can be obtained for using the driller, and still they meet the constraint of having one finger on the action button. In the examples of grasps for a pick and place task, the action area is avoided even though the hand is in its vicinity.

4.3 Integrated Grasp and Motion Planner

Traditionally, grasp and arm motion planning are considered as separate tasks. This approach generally relies on a set of precomputed grasps (grasp database) for finding a grasp suitable for the current situation, which greatly limits the number of grasp possibilities that can be tried, i.e. there is a low adaptability to the environment as no new grasps can be explored, even if they mean only a slight change of pose with respect to one of the predefined grasps. The feasibility of the grasps is evaluated for the given scenario, and only grasps that have a corresponding inverse kinematics (IK) solution for the arm are considered in later stages. One feasible grasp defines one goal configuration for the robot. Then, given the initial and final arm configuration, a collision-free path for the arm is searched using some path planning method. If no path is found a new grasp is chosen, until a path is obtained or until the complete database has been explored and no solution is found [2].

A better adaptability is obtained when using an integrated grasp and motion planner that only requires the initial configuration of the arm and the pose of the target object to simultaneously plan a good hand pose and arm trajectory to grasp the object [5]. The planner relies on the computation of independent contact regions to look for the best possible grasp; the planner is here adapted so that the obtained region for each finger is a subset of the actual graspable area $\mathcal{G}$. The potential grasp poses are obtained by using a capability map, i.e. an offline computed representation of the reachable and dexterous workspace for the robotic manipulator [11]. Once a goal grasp has been identified, the corresponding collision-free motion for the arm and hand is obtained by using bidirectional RRTs (Rapidly-exploring Random Trees). The integrated planner can be sequentially called as required; for instance, for pick and place tasks an initial plan is required to approach and grasp the object, and a second plan is used to move the object from the initial to the final configuration.

5 Final Comments

This paper presented a system to process a natural language instruction, with the main focus being how to retrieve the information required for instantiating and executing a grasp or manipulation skill. Given a task description in natural language, the system grounds the sentences in order to find spatial relations between objects, and to discover the meaning of the task itself. The system relies on an associated object database that contains information on the possible usage (action verbs) and the geometrical regions of the object, which are obtained through an automatic clustering process whose annotations are later validated by a user. This allows the generation of grasp constraints that are inherent to the object to be used, and therefore independent of the end effector of the particular application. The semantic knowledge of the object and the task allows the reduction of the search space explored during the grasp planning process. A vision module, not considered in this paper, additionally provides information on the location of objects and goal locations, as specified in the NL instruction. Once the manipulation skill is completely specified, including identification and role of the involved objects, desired usage of the object and current and goal locations, an integrated grasp and motion planner generates the collision-free trajectory required for the skill execution.

The described pipeline was implemented on top of OpenRAVE , and preliminary tests of the approach in simulation were successful for different instructions for table-top scenarios. Some examples of the tested instructions are: "take the spray", "clean with the spray", "put the spray on top of the tray", "use the drill", "grab the drill", "put the drill to the right of the spray". Fig. 1 shows a sequence of snapshots identifying different steps in the generation of the plan for the instruction "put the spray on top of the tray", executed for the DLR robot SpaceJustin, using two five-fingered DLR/HIT hand II as end effectors. The planning time varies from a few seconds to about a minute for more complex problems. The main factor that influences the planning time is not the NL processing module, but the relative location and reachability of the required objects. Next steps for the development of this system include the expansion of the vocabulary that the NL processor considers, the development of an automatic clustering algorithm that avoids the need for user intervention to verify the annotations generated by the system, an extensive and quantitative study of the performance of the system, and the integration of the system and evaluation of the success rate on a real robotic system.

Acknowledgment The research leading to these results has received partial funding through the project DPI2013-40882-P.

References

1. Beetz, M., Kresse, U.K.I., Maldonado, A., Mosenlechner, L., Pangercic, D., Ruhr, T., Tenorth, M.: Robotic roommates making pancakes. In: Proc. IEEE-RAS Int. Conf. on Humanoid Robots, pp. 529–536 (2011)
2. Berenson, D., Diankov, R., Nishikawi, K., Kagami, S., Kuffner, J.: Grasp planning in complex scenes. In: Proc. IEEE-RAS Int. Conf. on Humanoid Robots, pp. 42–48 (2007)
3. Bird, S., Loper, E.: NLTK: the natural language toolkit. In: Proc. ACL (Association for Computational Linguistics) on Interactive presentation sessions (2004)
4. Fasola, J., Mataric, M.J.: Using semantic fields to model dynamic spatial relations in a robot architecture for natural language instruction of service robots. In: Proc. IEEE/RSJ Int. Conf. on Intelligent Robots and Systems, pp. 143–150 (2013)
5. Fontanals, J., Dang-Vu, B., Porges, O., Rosell, J., Roa, M.A.: Integrated grasp and motion planning using independent contact regions. In: Proc. IEEE-RAS Int. Conf. on Humanoid Robots, pp. 887–893 (2014)
6. Guadarrama, S., Riano, L., Golland, D., Gouhring, D., Jia, Y., Klein, D., Abbeel, P., Darrell, T.: Grounding spatial relations for human-robot interaction. In: Proc. IEEE/RSJ Int. Conf. on Intelligent Robots and Systems, pp. 1640–1647 (2013)
7. Leidner, D., Borst, C., Hirzinger, G.: Things are made for what they are: solving manipulation tasks by using functional object classes. In: Proc. IEEE-RAS Int. Conf. on Humanoid Robots, pp. 429–435 (2012)
8. Miller, G.A.: Wordnet: a lexical database for English. Communications of the ACM **38**(11), 39–41 (1995)
9. Misra, D., Sung, J., Lee, K., Saxena, A.: Tell me Dave: Context-sensitive grounding of natural language to mobile manipulation instructions. In: Proc. Robotics: Science and Systems, RSS (2014)
10. Nyga, D., Beetz, M.: Everything robots always wanted to know about housework (but were afraid to ask). In: Proc. IEEE/RSJ Int. Conf. on Intelligent Robots and Systems, pp. 243–250 (2012)
11. Porges, O., Stouraitis, T., Borst, C., Roa, M.A.: Reachability and capability analysis for manipulation tasks. In: Armada, M., Sanfeliu, A., Ferre, M. (eds.) ROBOT2013: First Iberian Robotics Conference, vol. 253, pp. 703–718. Advances in Intelligent Systems and Computing. Springer (2014)
12. Roa, M.A., Hertkorn, K., Borst, C., Hirzinger, G.: Reachable independent contact regions for precision grasps. In: Proc. IEEE Int. Conf. on Robotics and Automation, pp. 5337–5343 (2011)
13. Stein, S.C., Schoeler, M., Papon, J., Worgotter, F.: Object partitioning using local convexity. In: Proc. IEEE Conf. on Computer Vision and Pattern Recognition (CVPR), pp. 304–311 (2014)
14. Tellex, S., Kollar, T., Dickerson, S., Walter, M.R., Banerjee, A., Teller, S., Roy, N.: Understanding natural language commands for robotic navigation and mobile manipulation. In: Proc. Nat. Conf. on Artificial Intelligence - AAAI (2011)
15. Zampogiannis, K., Yang, Y., Fermueller, C., Aloimonos, Y.: Learning the spatial semantics of manipulation actions through preposition grounding. In: Proc. IEEE Int. Conf. on Robotics and Automation, pp. 1389–1396 (2015)
16. Zoliner, R., Pardowitz, M., Knoop, S., Dillmann, R.: Towards cognitive robots: Building hierarchical task representations of manipulations from human demonstration. In: Proc. IEEE Int. Conf. on Robotics and Automation, pp. 1535–1540 (2005)

Multi-robot Planning Using Robot-Dependent Reachability Maps

Tiago Pereira, Manuela Veloso and António Moreira

Abstract In this paper we present a new concept of robot-dependent reachability map (RDReachMap) for mobile platforms. In heterogeneous multi-robot systems, the reachability limit of robots motion and actuation must be considered when assigning tasks to them. We created an algorithm that generates those reachability maps, separating regions that can be covered by a robot from the unreachable ones, using morphological operations. Our method is dependent on the robot position, and is parameterized with the robot's size and actuation radius. For this purpose we introduce a new technique, the partial morphological closing operation. The algorithm was tested both in simulated and real environment maps. We also present a common problem of multi robot routing, which we solve with a planner that uses our reachability maps in order to generate valid plans. We contribute a heuristic that generates paths for two robots using the reachability concept.

Keywords Multi-robot systems · Map representation · Reachability · Planning and scheduling · Morphological operations · Coordination

T. Pereira(✉) · M. Veloso
Carnegie Mellon University, Pittsburgh, PA 15213, USA
e-mail: tpereira@cmu.edu, mmv@cs.cmu.edu

T. Pereira · A. Moreira
FEUP - Faculty of Engineering, University of Porto, Porto, Portugal
e-mail: {tiago.raul,amoreira}@fe.up.pt

T. Pereira · A. Moreira
INESC TEC - INESC Technology and Science, Porto, Portugal

This work is financed by the ERDF European Regional Development Fund through the COMPETE Programme (operational programme for competitiveness), by National Funds through the FCT Fundao para a Cincia e a Tecnologia (Portuguese Foundation for Science and Technology) within project FCOMP-01-0124-FEDER-037281, and under Carnegie Mellon-Portugal Program grant SFRH/BD/52158/2013.

© Springer International Publishing Switzerland 2016   189
L.P. Reis et al. (eds.), *Robot 2015: Second Iberian Robotics Conference*,
Advances in Intelligent Systems and Computing 417,
DOI: 10.1007/978-3-319-27146-0_15

1 Introduction

Multi-robot systems have been the focus of research for some time, because they can often accomplish tasks more efficiently than a single robot and are more resilient to failures [12]. Common applications of multi-robot systems are exploration, search and rescue, and scheduling of service robots.

The planning problems are usually solved using optimization techniques, such as the ones used in the multiple vehicle routing and the multiple traveling salesman problems [1]. Generally, one wants to distribute tasks among robots.

When planning, it is important to consider heterogeneity and the reachability limit of each robot. We have nowadays a high diversity of robot configurations, and should be able to use that in our favor [7]. Furthermore, in multi-robot systems, robots must not ignore their own heterogeneity, and should be able to plan together considering their different capabilities.

Although multi-robot problems have already been studied regarding the distribution of tasks, there is less work on how to determine if those tasks are feasible for a robot. Therefore, we investigated the geometric and spacial properties of robots in the world, their impact on which positions are achievable, and how it affects task assignment in multi-robot scenarios.

One simple example is when we want to coordinate two robots to go to a goal position to execute a task. Assuming we have circular robots, a bigger robot has more limited motion capabilities, and in principle cannot reach as many places as a smaller robot. This simple example shows it is important to consider physical characteristics, i.e. heterogeneity, when planning. Moreover, coordination of heterogeneous robots should not be hard-coded. Robots should be able to objectively represent their heterogeneity, having their own representation of the environment that is dependent on their physical characteristics. This representation would be useful not only for the robots themselves, but also to share information with teammates, and allow more optimal coordination.

One common way of representing the accessibility of a map considering the robot shape is the configuration space [13]. However, robots need also take into account their other capabilities, like actuation. And for that purpose, the configuration space does not give enough information.

As an example, we can consider a vacuum cleaning robot. We show in Figure 1 a simulated environment designed to test the many specifications of our problem, such as corridors, and rooms connected through openings of different shapes and sizes. In Figure 1(a) we have a map with black representing walls and obstacles. And in Figure1(b), we have the configuration space which inflates the walls and obstacles for some robot size. White represents where the robot center is allowed to be, and not what parts of the environment it can clean.

In Figure 1(c), assuming the robot is in the center of the image, we show what we introduce as *reachability map*, with white representing regions in the environment the robot can clean. Therefore, it adds to the configuration space (regions where the

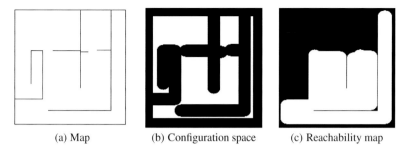

(a) Map   (b) Configuration space   (c) Reachability map

Fig. 1 Difference between the configuration space and the actuation space for a vacuum cleaning robot.

robot can be) the information on the actuation capabilities, in this case, its cleaning reachability. As expected, the robot cannot reach corners.

Our problem is then to determine 2D reachability maps, representing the points in the environment that can be actuated by a robot from some position that is reachable from the initial position.

To solve the problem of constructing a 2D reachability map, we propose a methodology that uses algorithms borrowed from image processing. Image processing is a common framework in robotics, but used preferably for perception tasks. Here, in a new intersection of image processing and map transformation, we use it to achieve extraction of higher level information from SLAM based grid maps. These occupancy grid maps can be obtained with, for example, rao-blackwellized particle filters [10].

In the next section we discuss related work, and then we present in depth our algorithm to obtain the Robot-Dependent Reachability Map (RDReachMap). We show results in a real map obtained from SLAM. We also contribute an heuristic for multi-robot planning for multiple goals, using the reachability maps. Finally, we present our conclusions and the directions for future work.

2   Related Work

We focus our work on heterogeneous multi-robot scenarios, creating the concept of reachability maps to extract information that is useful for coordination.

Multi-robot systems have been a research subject for some time, because robots can merge their measurements to produce more precise maps, also allowing for faster and more robust task execution. [3]

Various map representations can be used, such as occupancy grids, geometric features, or topology. We combine in our approach both lower level (occupancy grid maps) and higher level (reachability maps) data. Comparable approaches have been used before, using manifolds for the map representation [11].

Regarding cooperation, it is accomplished through exchange of information. Robots can communicate their reachability maps so that they choose where each should go in an efficient way. Related works on the literature use information sharing among teammates to improve the robot beliefs, complementing the different perceptions of each one [4]. They also show how important it can be to coordinate robots using transfers in order to achieve an efficient solution of the pickup and delivery problems [5].

It was also shown by [2] how to do multi robot path planning, with highly dynamic domains. However, they do not consider feasibility of goals before planning. Another alternative approach to coordination has been suggested by [6], using market-based approaches, and auctions, in domains ranging from mapping and exploration to robot soccer. However, none of these approaches determine feasibility of goals for each robot, as we do in this work.

Regarding heterogeneous multi-robot systems, not only terrestrial mobile robots of different sizes and characteristic, but also aerial robots, have been used for the purpose of rescuing missions [15]. In this example, they use the different sensing capabilities of each one to coordinate the robots.

Regarding the use of techniques borrowed from the image processing field, there are more examples, such as automatically extracting topology from an occupancy grid [8]. Finally, [16] uses reachability to coordinate robots with humans in shared workspaces.

3 RDReachMap

Considering maps are discrete representations of the environment, we found a duality between images and maps, because both of them are a discrete sampling of the world. Therefore, our main idea is to perform purely geometric operations on the map in order to obtain coverage, using image oriented algorithms.

The first step is to transform an occupancy grid map M (with probabilities of occupation in each cell) into a binary map of free and obstacle cells, using a given threshold. Then we have a image-like map, I, and we can extract the reachability coverage considering a given robot radius.

We assume robots have circular shape. While there are many robots with a circular footprint, there are others with different shapes. However, as we want to find the reachable regions, which can be seen from different positions and orientations that can be reached from the initial position, we need a rotation-invariant model of the robot. Moreover, many robots have a close to circular shape. Assuming circular robots with omni-directional motion, the actuator model is irrelevant, and only the maximum range of actuation is needed.

3.1 Morphological Operations

We create a simulation test map, which we show in Figure 2(a). Given a grid of positions G (discretized environment), a black and white binary image I (one pixel per grid point) as a binary map representation, and a structuring element R (robot), the morphological operation dilation is defined as

$$I \oplus R = \{z \in G \mid R_z \cap I \neq \emptyset\} = \bigcup_{z \in R} I_z \qquad (1)$$

where $R_z = \{p \in G \mid p = r + z, r \in R\}, \forall z \in G$.

Applying the dilation operation to black points in the image, the algorithm inflates the obstacles of the map by the robot size, achieving the configuration space. In Figure 2(b), we show in white the free parts of the configuration space, C_{free}. The same kind of approach is used in most motion planning algorithms, inflating objects in order to find a path that minimizes some cost. Indeed, inflation is the solution used, for example, in the ROS navigation package[14].

However, the configuration space shows where the robot center can be, not giving any information on its covering capabilities.

Considering A to be a structuring element representing the actuation capabilities (also circular), the morphological operation erosion is defined as

$$I \ominus A = \{z \in G \mid A_z \subseteq I\} = \bigcap_{z \in A} I_{-z} \qquad (2)$$

We can apply the erosion morphological operation to the configuration space, with a given actuation radius (size of A). The combination of erosion after dilation $(I \oplus R \ominus A)$ is called in computer vision morphological closing operation [18].

As we show in Figure 2, if using the same radius for the erosion operation (actuation size) as the size of the structuring element for the dilation, we obtain the original map with rounded corners. Going back to the vacuum cleaner example, the "closed" map implies the robot can reach all positions, except the square corners, because of its circular shape.

However, Figure 2(c) implies all regions can be vacuum cleaned, which is not true. The configuration space has disconnected regions, meaning the robot cannot transverse the entire environment. The disconnected parts do not show up after the morphological closing, meaning the topology is changed. In Figure 2(b), a robot in the bottom cross cannot reach the upper cross. However, in Figure 2(c), there is a feasible path between the two crosses, even tough the robot cannot start from one cross and clean in the other cross position.

The lack of disconnected components after the closing operation happens because it returns the reachability for any initial position of the robot. Therefore, the map after the morphological closing does not serve our purpose, because it represents reachability from any initial position of the robot in the configuration space. And for the planning purposes, we need to obtain the reachability for the current initial position of each robot.

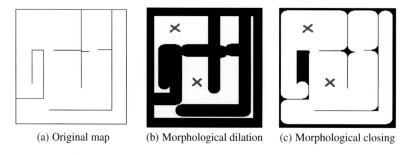

(a) Original map   (b) Morphological dilation   (c) Morphological closing

Fig. 2 Example of morphological operations in a map to obtain the (b) configuration and (c) actuation spaces. Configuration space shows feasible points for the robot center, and the actuation space shows in white what the robot can actuate.

3.2 Partial Morphological Closing

Although the morphological closing does not yield the desired reachability map for a given initial position, it is still interesting, because it represents a very similar concept, which is overall reachability for any initial position. Therefore, in order to use this technique to achieve our goal, we introduce the concept of partial morphological closing.

After applying dilation to the original map in order to get the configuration space, we obtain a set of disconnected regions. Going over the image, we cluster those regions using the "Flood Fill" technique, labeling C_{free}. There are more complex methods for labeling, with better performances [9]. Flood Fill proved fast enough for our requirements. This operation is equivalent to the morphological operation propagation (conditional successive dilations).

The algorithm finds the label of the robot's initial position x_0, and uses that region as the only reachable part of the configuration space, $Reach(x_0, C_{free})$.

For each image, with $Reach(x_0, C_{free})$ as white pixels and the rest as black, we can now apply the erosion operation $Reach(x_0, C_{free}) \ominus A$, and finish the partial morphological closing. The result obtained, in Figure 3, shows the reachability maps for three regions corresponding to three different initial positions.

In the end of the procedure we get the robot-dependent reachability map (RDReachMap) that closes small corridors, transforming them into obstacles. It also separates regions, and is able to detect all points of the world that can be actuated from any position in a feasible path from the initial location. Thus we call the result a reachability map, because it shows all the points that are reachable to the robot's actuators, given an initial position.

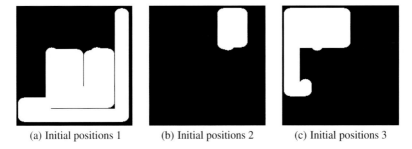

| (a) Initial positions 1 | (b) Initial positions 2 | (c) Initial positions 3 |

Fig. 3 Reachability map as a result of partial morphological closing operation, shown for the three sets of possible initial positions.

3.3   Overall Algorithm

In this section we present the algorithm that we use to get the reachability map. The main idea of RDReachMap is to use the closing morphological operation to eliminate from the map regions that are not reachable. This technique is enough to erase from the final map corridors whose width is smaller than the circumscribing radius of the robot. However, in order to get the reachability map, our algorithm then uses the concept of partial morphological closing. The full algorithm is shown in Algorithm 1.

Algorithm 1. RDReachMap: Creating reachability maps from grid maps

---
Require: Grid Maps G

1: **for** each G **do**
2:    Initialization: B&W Image I from M
3:    Dilation: inflated image C^{free} from I
4:    Labeling: Reachable set $Reach(x_0, C_{free})$ from C^{free} and initial position x_0
5:    Partial Closing: Reachability map $RM = Reach(x_0, C_{free}) \ominus A$
6: **end for**
7: **return** RM

3.4   Topology Analysis for Different Robot and Actuation Radii

We now show the result of our algorithm in both simulated and real environments. The real world map was obtain by a robot doing SLAM inside a building.

Although we used static maps, this work can be used on online SLAM and exploration, because it is an algorithm that only depends on the current map, and allows for semi-explored regions. RDReachMap always considers unexplored regions as free cells when extracting the reachability map. The optimistic assumption makes the planner always send robots to explore unkown regions.

In Figure 4 we show reachability maps for different robot sizes. For test purposes, we defined the robot size as a number of cells in the grid map.

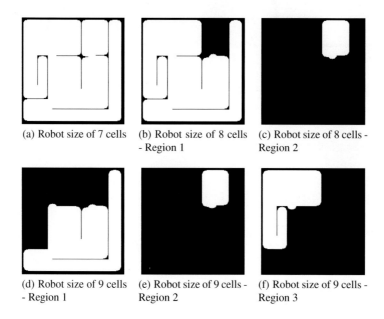

(a) Robot size of 7 cells   (b) Robot size of 8 cells   (c) Robot size of 8 cells -
                             - Region 1   Region 2

(d) Robot size of 9 cells   (e) Robot size of 9 cells - (f) Robot size of 9 cells -
- Region 1   Region 2   Region 3

Fig. 4 Reachability maps for different sizes of robots. Size of map is 200×200 cells.

Depending on the size of the regions, corridors start being blocked when the robot size increases. Figures 4(b) and (c) show a scenario where only a room was separated from the others compared to the smaller robot case in Figure 4(a), where all regions were accessible, except for corners of the environment. Therefore, either the robot is in one region and the rest of the environment is unreachable, or the robot is in the other region and then the first one is unreachable. In figures 4(d) to 4(f) more rooms are separated from each other, due to the increase in robot size.

Figure 5 shows the reachability map of a real world map. Again, as the robot size increases, more and more regions will be added to the reachability map, reducing the number of reachable places at each point of the map.

If the actuation size is the same as the robot, then it is a special case because the total morphological closing yields the correct reachability.

It is important to notice that RDReachMap can only be used for actuation radii smaller then the robot size, otherwise the algorithm does not work properly. However, it is reasonable to assume equal size because most times the actuation range is given by the robot size, as with the vacuum cleaning robot.

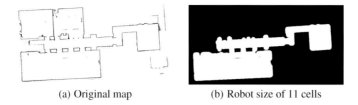

(a) Original map   (b) Robot size of 11 cells

Fig. 5 Reachability in a real world map

4   Coordination with Reachability Maps

The purpose of RDReachMap is to extract information that can be used to improve the efficiency of coordination. Knowing a priori the reachability, the combinatorial cost of finding solutions in multi-robot scenarios is reduced.

As an example, when doing search and rescue, or when doing SLAM, robots need to explore the environment. One possibility is to use the concept of exploration frontiers, and robots need to coordinate in order to efficiently distribute tasks [17]. When using RDReachMap, frontiers in unreachable places can be automatically given to smaller robots. But in this case, more than helping efficiency, RDReachMap is an essential feature, because it is needed to determine which goals each robot can accomplish.

4.1   Multiple Vehicle Routing

As an example, we use a scenario with two robots in the environment (with different reachability for each one), and a set of goals randomly distributed over the environment. The objective is to have a centralized planner, with access to reachability maps, distributing tasks to each robot.

We want to minimize the total time the robots need to get to all goals. Being p a possible solution combining the paths of both robot, $t_r(p)$ is the time it takes for robot r to complete its part of the path in solution p. Therefore, the idea is to find the route that minimizes the maximum time of execution of all robots. The optimal exploration route is given by

$$p^* = \arg\min_{p}\{\max_{r}\left(t_r(p)\right)\} \tag{3}$$

We assume that robots travel at the same speed. So, we transform the time minimization problem into a distance minimization. This is a multi traveling salesman problem, and with the combinatorial explosion, it is impossible to calculate the optimal solution efficiently. One possible solution to the problem is to use a greedy heuristic, where the nearest goal is added to the correspondent robot path. That solution is often very poor, getting stuck in local minima. Therefore, we propose a different heuristic.

Heuristic. We know the route for a robot includes with certainty all the goals only reachable to it. So, our planner can have a broader view of where the robots are moving in the future. With our cost function, we have at the same time to minimize the cost (total distance) of each path, and minimize the difference between the cost of paths of all robots. An intuitive heuristic is to add goals to one robot path if its distance to that path is much less than the distance to the other robot's path, thus not only minimizing distance to one path, but also maximizing the distance to the other robot current path.

In our problem we have a robot that is bigger (R_1) and can only reach some small set of regions (S_1), and another (R_2) that can reach a bigger set, including the regions only accessible to smaller robots ($S_2 \cup S_1$). We start to allocate the certain goals, which are the ones only accessible to R_2 (goals in S_2). Then in order to capture the non-local perspective of the problem, we search for the one with the maximum distance to the initial position. We are using this first goal as an initial estimate of the path it will have to do.

Furthermore, we add more goals to the paths. In order to choose where to add goal position in the current path of robots, we analyze the increase in cost (total distance) given by the additional size of the path when adding the goal. So, if we add goal g between points i and j and being $d(.,.)$ the distance function calculated with A* between two points, the additional cost of adding goal position g between i and j is

$$c(g, i, j) = d(i, g) + d(g, j) - d(i.j) \tag{4}$$

We can also add it to the end of the path, where the cost is only going to be $d(j, g)$. Goals are added at points where cost is minimized, and we choose a goal that minimizes the adding cost. Goals in S_2 are added until all of them are in the path of R_2 (the smaller robot that is the only to reach goals in S_2).

Then, we need to choose a goal for R_1 from S_1, also giving a good initial estimate of the global path. We want robots to move apart from each other for efficient exploration. Thus our metric now will return the goal which has the bigger distance to both robots, and we use the sum for this purpose.

In the final step, goals in S_1 are added to both robots' paths, until there are no more goals left. We want to start adding goals to some path if we know it is clear it should go for that robot and not the other one (minimizing uncertainty of attribution). Therefore, we compute for every goal left, the cost of adding a goal to both robot paths. Being d_1 the distance to path of R_1, and d_2 for R_2, the cost of adding the goal to R_1 path is $c_1 = d_1 - d_2$, and vice versa for R_2. This means we are trying to minimize the distance of the added goal to the current path, and at the same time, maximize the distance of that goal to the path of the other robot. We choose two goals, one for each robot, that minimize the cost for each one. However, we always add goals incrementally.

Algorithm 2. Heuristic for creating paths

Require: Grid Reachability Maps from 2 Robots
1: Separate Goals in categories, S_1 and S_2, using reachability maps
2: Choose goal from S_2 with maximum distance to R_2 (smaller robot) initial position
3: Allocate goals S_2 greedily to R_2
4: Choose first goal from S_1 to R_1 maximizing distance to both robots.
5: **while** Goals in S_1 **do**
6:     Sort goals, minimizing uncertainty of attribution
7:     Choose goal minimizing length difference of the two robots paths.
8: **end while**
9: **return** Path p

As we want to make the total distance for both robots more similar, we choose from the previous two goals the one that makes the total path distance for both robots more similar. In Algorithm 2 we show the overall heuristic. The same algorithm can be used for more robots, extrapolating the strategy of separating the regions that are reachable only to certain robots.

4.2   *Results*

In Figure 6, we use the simulated map presented in Figure 1(a). It shows in black the regions only reachable to one of the robots, R_2. We show the result of our route planning for two robots using reachability maps.

In this example scenario, when using 10 goals, the heuristic took around 3.5 seconds, and the brute force 27.5 seconds, both yielding the same paths for each

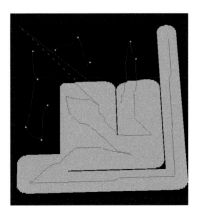

Fig. 6 Paths in multi-robot planning using reachability maps. Blue path for robot with smaller reachability, red for one with more reachability. Crosses are robot initial positions, with circles representing the robot sizes. Green balls are goal positions.

robot. The path costs for each robot were approximately 432 and 480 (units are grid cells). For 28 goals, the total distance for each robot was approximately 622 and 624 using the heuristic (no groundtruth, too many goals for brute-force), taking around 23.5 seconds to compute.

5 Conclusion

In this paper we presented RDReachMap, an algorithm that creates robot-dependent reachability maps. RDReachMap is able to create maps separating reachable regions from the unreachable ones, using only image processing operations. We introduced the concept of partial morphological closing in order to solve the reachability problem. We tested our algorithm both in a simulated environment and a real world map, showing the influence of the size of the robot on the resulting reachability map.

We showed the importance of using the concept of reachability for planning. When robots work in cooperation, it is important that they can execute tasks according with their own characteristics. Robot size and actuation radius are some of these characteristics, and in our work we built a framework so that robots can cooperate taking that into account. We presented a common multi vehicle routing problem, and we solved it contributing a heuristic that uses reachability to try to minimize the maximum time of operation.

As future work, we want to extend our framework to 3D maps. Moreover, we want to consider the heterogeneity of robots not only in their size, but also in their shape. The goal is to drop in the future the circular robot restriction. We want to use these concepts of reachability to be able to coordinate a fully heterogeneous multi-robot team.

References

1. Bektas, T.: The multiple traveling salesman problem: an overview of formulations and solution procedures. Omega **34**(3), 209–219 (2006)
2. Bruce, J., Veloso, M.: Real-time randomized path planning for robot navigation. In: IEEE/RSJ International Conference on Intelligent Robots and Systems, 2002, vol. 3, pp. 2383–2388. IEEE (2002)
3. Burgard, W., Moors, M., Stachniss, C., Schneider, F.: Coordinated multi-robot exploration. In: IEEE Transactions on Robotics (2005)
4. Coltin, B., Liemhetcharat, S., Meriçli, C., Tay, J., Veloso, M.: Multi-humanoid world modeling in standard platform robot soccer. In: 2010 10th IEEE-RAS International Conference on Humanoid Robots (Humanoids), pp. 424–429. IEEE (2010)
5. Coltin, B., Veloso, M.: Scheduling for transfers in pickup and delivery problems with very large neighborhood search. In: Twenty-Eighth AAAI Conference on Artificial Intelligence (2014)
6. Dias, M.B., Zlot, R., Kalra, N., Stentz, A.: Market-based multirobot coordination: A survey and analysis. Proceedings of the IEEE **94**(7), 1257–1270 (2006)
7. Dorigo, M., et al.: Swarmanoid: A novel concept for the study of heterogeneous robotic swarms. IEEE Robotics and Automation Magazine **20**(4), 60–71 (2013)

8. Fabrizi, E., Saffiotti, A.: Extracting topology-based maps from gridmaps. In: Proceedings of the IEEE International Conference on Robotics and Automation, ICRA2000, vol. 3, pp. 2972–2978. IEEE (2000)
9. Grana, C., Borghesani, D., Cucchiara, R.: Fast block based connected components labeling. In: 2009 16th IEEE International Conference on Image Processing (ICIP), pp. 4061–4064. IEEE (2009)
10. Grisetti, G., Stachniss, C., Burgard, W.: Improved techniques for grid mapping with rao-blackwellized particle filters. IEEE Transactions on Robotics $23(1)$, 34–46 (2007)
11. Howard, A.: Multi-robot mapping using manifold representations. In: Proceedings of the 2004 IEEE International Conference on Robotics and Automation, ICRA 2004, vol. 4, pp. 4198–4203. IEEE (2004)
12. Howard, A., Parker, L., Sukhatme, G.: Experiments with a large heterogeneous mobile robot team: Exploration, mapping, deployment and detection. The International Journal of Robotics Research $25(5–6)$, 431–447 (2006)
13. LaValle, S.M.: Planning algorithms. Cambridge University Press (2006)
14. Lu, D.V., Smart, W.D.: Towards more efficient navigation for robots and humans. In: 2013 IEEE/RSJ International Conference on Intelligent Robots and Systems (IROS), pp. 1707–1713. IEEE (2013)
15. Luo, C., Espinosa, A., Pranantha, D., De Gloria, A.: Multi-robot search and rescue team. In: 2011 IEEE International Symposium on Safety, Security, and Rescue Robotics (SSRR) (2011)
16. Pandey, A.K., Alami, R.: Mightability maps: a perceptual level decisional framework for co-operative and competitive human-robot interaction. In: 2010 IEEE/RSJ International Conference on Intelligent Robots and Systems (IROS), pp. 5842–5848. IEEE (2010)
17. Pereira, T., Moreira, A.P., Veloso, M.: Coordination for multi-robot exploration using topological maps. In: CONTROLO2014–Proceedings of the 11th Portuguese Conference on Automatic Control, pp. 515–524. Springer (2015)
18. Soille, P.: Morphological image analysis: principles and applications. Springer-Verlag, New York (2003)

Development of a Nao Humanoid Robot Able to Play Tic-Tac-Toe Game on a Tactile Tablet

Luis Calvo-Varela, Carlos V. Regueiro,
David S. Canzobre and Roberto Iglesias

Abstract This paper describes the challenges that involve playing with the Nao humanoid robot on a tablet. For that purpose, an inverse kinematic solver that allows the robot to move it's limbs, and a computer vision algorithm that allows the robot to understand the items displayed on the tablet, are needed. The presented solution uses NAOqi's Cartesian Control and OpenCV's Hough Transform respectively. To overcome the lack of force and tactile sensors on Nao's hand, we propose a touch movement based on visual feedback. As an initial approach, we chose the Tic-Tac-Toe game and we introduced interaction mechanisms to make it more pleasant and enjoyable, with the objective of creating a template for HRI and machine learning integration. The experimental results show the robustness of the proposed architecture.

Keywords Nao humanoid robot · Computer vision · Inverse Kinematic Solver · Tic-Tac-Toe game · Human Robot Interaction

1 Introduction

The domain of robotics has evolved greatly thanks to the progress of technology and computing in recent years, achieving automatons capable of solving arduous tasks, which have allowed them to become more present in our society.

One of the most promising variants are the social or personal robots, i.e., those who interact with human beings. Among them we can find the humanoid robots,

L. Calvo-Varela · D.S. Canzobre · R. Iglesias
CITIUS (Centro Singular de Investigación En Tecnoloxías da Información),
15782 Santiago de Compostela, Spain
e-mail: {luis.calvo,david.sanchez.canzobre,roberto.iglesias.rodriguez}@usc.es

C.V. Regueiro(✉)
Department of Electronics and Systems, Universidade da Coruña, 15071 Coruña, Spain
e-mail: cvazquez@udc.es

© Springer International Publishing Switzerland 2016
L.P. Reis et al. (eds.), *Robot 2015: Second Iberian Robotics Conference*,
Advances in Intelligent Systems and Computing 417,
DOI: 10.1007/978-3-319-27146-0_16

i.e., robots with human physiognomy able to mimic our behaviour to a certain degree, which have great acceptance.

The aim of this paper is to implement a system capable of allowing one of these humanoid robots, Nao [5], to interact dynamically and autonomously with human beings through the act of playing. To be more specific, we want our Nao robot being able to play a board game, displayed on a tablet, using a digital pointer. The results described in this paper will provide a basics for a wider topic: human-robot interaction and robot learning from these games. We are interested on applications able to learn and adapt to the user, i.e., games that offer a progressive level of difficulty and evolve at the same speed as the human user (which can be helpful for children, or a a tool in therapy, etc.).

We chose the Tic-tac-toe for its simplicity. It allowed us to create a basic template for robotic board gaming, transferable to other activities like drawing, with or without a tablet.

Playing a game with a Nao humanoid robot involves three main challenges:

- First, the robot has to recognize the board, that is, understand all the items displayed on the tablet application, even under partial occlusion. This can be solved using Computer Vision techniques.
- Second, the robot has to move one of it's arms (or both) dynamically to interact with the tablet and play the game. One intelligible way to solve this is through Inverse Kinematics.
- Third, include visual feedback to overcome the limitation of identifying the end of the arm movement.

This paper is structured as follows: Section 2 presents the related work. In Section 3, the architecture and implementation of the system that allows Nao to play a board game is described. Section 4 shows and discuss the experimental results. Finally, in Section 5, conclusions and future work are presented.

2   Related Work

There are few works that cover the problem of robot playing, specially with Nao humanoid robots using a tablet. Poddigue and Roos [1] compare several Inverse Kinematics methods, more precisely, "traditional" methods based on the Jacobian inverse technique, FABRIK (an heuristic algorithm) and some neural networks. The test domain is the movement of the arm of a Nao humanoid robot playing a tic-tac-toe game. The robot has to draw a cross or circle on a white board displayed in front of him. The game state is analysed using the robot camera and OpenCV algorithms.

In another work, Poddigue [2] focus on the challenges that needed to be solved to play the game itself: drawing coordination and figures recognition. The methods used to detect the board and map it to a game state belong to the OpenCV image processing library: probabilistic Hough lines and Hough circles transform. The IK algorithm used to move the Nao's arm needs a mapping from pixel coordinates to cartesian coordinates. This is done by triangulation between known dimensions, with the help of the camera's focal length. The chosen tic-tac-toe logic algorithm was Minimax.

Mingueza [3] presents the design and implementation of a gaming module that also allows the Nao robot to play tic-tac-toe, and which includes a basic level of interaction: voice synthesis and bumper confirmation. The techniques selected include learning by demonstration through Nao's Choregraphe application [10] (Nao movements and behaviours), computer vision algorithms through OpenCV image processing (detection of the gaming board elements by histogram comparison) and heuristic searching using the AIMA library (game algorithm and logic by a combination of Minimax and Alpha-Beta pruning). Our architecture not only expands the interaction capabilities of Nao, but also provides a movement based on visual feedback, allowing Nao to interact with a digital tablet.

3   System Architecture

The system architecture is implemented on the ROS robotics framework [6], as a "de facto" standard. Following ROS principles, the system is distributed among several independent nodes (fig. 1). This modularity increases the robustness of the final application and code reusability.

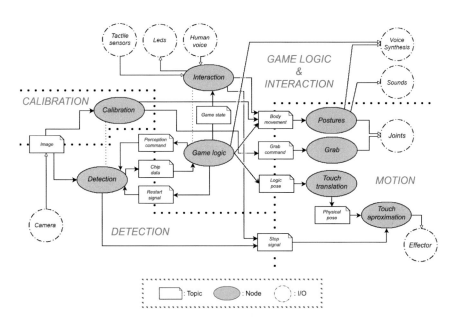

Fig. 1   Global system architecture to play tic-tac-toe game with Nao humanoid robot.

As we can see in figure 1, the system is composed by four modules: detection, calibration, motion and logic. We will explain each one of them in detail.

3.1 Detection

Nao has to be able to understand all the items displayed on the tablet that are related to the game itself. We will use the images captured by Nao's lowest camera to obtain all the relevant information, through several image transformations. The chosen computer vision library to do so is OpenCV [7].

The Detection module has three stages: game board, lines and chips. The result of them all is the current game logic state.

Board Detection. The first stage involves segmenting the game board on the image (fig. 2(a)) from all the other objects, so that the robot can analyse the items displayed on it. It is also important to carry out a perspective transformation, so that the items are not distorted due to the way the tablet is presented to the robot. The technique used to solve this problem is called Homography [11], and the process consists in three steps:

1. Obtain the four corners coordinates from the white background of the game app and measure it's width and height in pixels.
2. Use this data to calculate the homography matrix, which give us the perspective transformation between two planes.
3. Use the homography matrix to warp the perspective of the original image, obtaining only the undistorted board white background (fig. 2(b)).

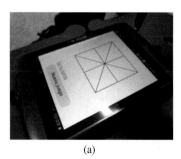 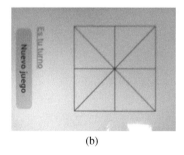

(a)   (b)

Fig. 2 Game board homography transformation: a) Original image, b) Undistorted and segmented image of the game board.

Line Detection. On the second stage, the game board is analysed to find all the playing positions (see fig. 3). Given the already undistorted image (fig. 2(b)) as the source of information, the techniques used to solve this problem are the Canny[12] Edge detector and the Hough[13] Line Transform, following the next process:

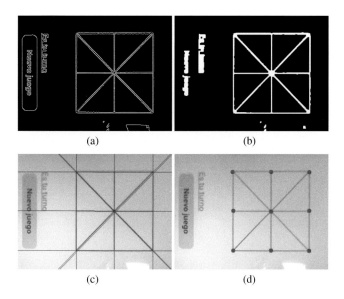

(a)   (b)

(c)   (d)

Fig. 3 Game positions detection through lines estimation: a) Canny edge detector, b) Morphological transformations, c) Hough lines estimation, d) Valid points.

1. Apply the Canny Edge detector to the undistorted image (fig. 3(a)).
2. Delete undesired edges with morphological transformations (fig. 3(b)).
3. Apply the Hough Line Transform function to estimate the board lines (fig. 3(c)).
4. Calculate the intersection points of all the estimated lines and classify them as valid, provided that some conditions (such as proximity to white colour) are fulfilled (fig. 3(d)).

Chip Detection. The last stage is responsible for detecting where and when a game chip appears on the tablet. This is essential to project the physical game into a logical game. This detection is relatively simple, given the undistorted image (fig. 2(b)) to which we apply both colour segmentation techniques and Hough Circles Transform. In this case the process involves four steps:

1. Change the colour space from BGR to HSV (fig. 4(a)), where the colour tone is represented independently to the brightness.
2. Threshold the HSV image (fig. 4(b)) with the colour range of each chip (see section 3.3), obtaining one mask for each colour.

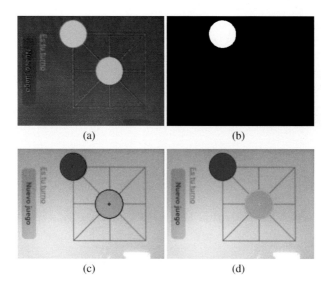

(a)   (b)

(c)   (d)

Fig. 4 Game chip detection: a) Undistorted image in HSV, b) Colour mask (blue), c) Hough Circles Transform, d) Valid chips.

3. Apply the Hough Circle Transform to estimate the board circles (fig. 4(c)).
4. Validate each detected circle (counting the number of coloured pixels inside it) and assign it to a logic position on the board (fig. 4(d)).

3.2 Motion

At this point, Nao is capable of translating the tablet content to a game state. After deciding it's next logical movement, it has to move it's arm to interact with the tablet. For that task we need an Inverse Kinematic Solver (IKS), which allow us to move the robot arm in terms of cartesian coordinates, by providing the objective coordinates that the end effector of the arm (hand) has to reach.

As IKS we have chosen the Cartesian Control tool [8] integrated in Nao's own sdk, NAOqi [9]. We have tried another alternatives, like MoveIt, but nowadays it is not completely operational on the Nao robot.

We have classified the Nao movements in three categories, as they involve different approximations and difficulties: grab, posture and touch.

Grab. Before starting the game, Nao needs to grab the tactile pointer (fig. 5(a)). This is done by controlling the hand joint directly instead of using an IKS. It is also not autonomous, i.e., the robot needs help from a human to grab it.

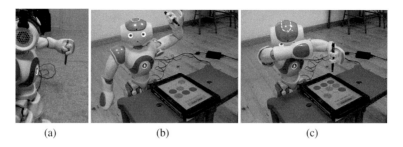

(a)   (b)   (c)

Fig. 5 Simple movements: a) Grab of tactile pointer, b) Victory posture, c) Defeat posture.

We have also programmed an autonomous grab, but it is limited by the pointer detection and it's relative position to the robot frame. A monocular vision algorithm (as is the one from the Perception module) does not provide us enough precision and robustness. We have also tried some RGB-D cameras [4].

Postures. Nao adopts a game posture each time before it moves it's arm (we call it the "Stand" posture), where it holds the pointer perpendicular to the tablet (and the whole arm is parallel to the tablet). This posture is achieved by moving the arm's joints directly, as in the grab case, and was designed by moving the arm through Nao's Choregraphe application.

This "Stand" posture is needed to restore the initial arm configuration, since the IKS distorts the angles of the arm on each movement. Not restoring the posture would result in different and unexpected behaviours each time Nao wants to touch the tablet.

On the other hand, we have also implemented some "outcome" postures, as a visual and intuitive feedback that depends on the outcome of the game. The victory and defeat postures can be seen on figures 5(b)) and 5(c), respectively.

Touch. The process of touching the tablet is critical for playing games, and Nao's limitations make this kind of movement one of the most difficult. Nao does not include force sensors in it's arms or tactile sensors in it's hands palms, therefore it cannot know when the tactile pointer is touching a surface. The precision of the arm movements and the visual perception algorithms are not enough to carry out an "open-loop" technique.

To solve this dilemma we decided to use as the stop signal for the touch movement the appearance of Nao's turn game chip. Hence, we implemented a touch movement based on visual feedback.

The touch movement is divided in two stages: first, Nao moves it's hand to a intermediate position perpendicular to the final position and adjusts it's wrist (that is, the hand's orientation). Then, the hand descends slowly in the Z plane with the same orientation until Nao detects the touch "visually" (a new chip appears on the board). The detailed description of this process is:

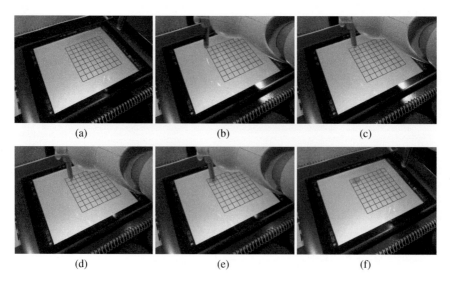

Fig. 6 Touch movement with visual feedback: a) Initial state on the "Stand" posture, b) arm before wrist adjustment, c) arm after wrist adjustment, d) arm before touch, e) arm after touch (see chip), f) arm returns to initial state.

1. Nao compares it's hand actual position with the destination position, so it can calculate the distance it has to move the hand in terms of each space axis (fig. 6(a)).
2. Movement of the hand in the XY plane, parallel to the tablet plane, followed by the movement of the hand in the Z axis (75%) (fig. 6(b)). We call this the "basic approach".
3. Adjustment of the WristYaw joint angle, to make the hand's orientation perpendicular to the board (fig. 6(c)).
4. Movement of the hand in the Z axis (>25%) in small intervals. We call this the "precise approach" (figs. 6(d) and 6(e)).
5. When a stop signal is received (fig. 6(e)), the hand raises, adopting the "Stand" posture afterwards (fig. 6(f)).

3.3   Calibration

The presented system needs to be calibrated before Nao can play with the human player. The first stage consists in placing all the items in a convenient way, as Nao's arm has a very limited reach.

The second stage implies the calibration of all the visual algorithms. Usually, it only has to be done once on each environment:

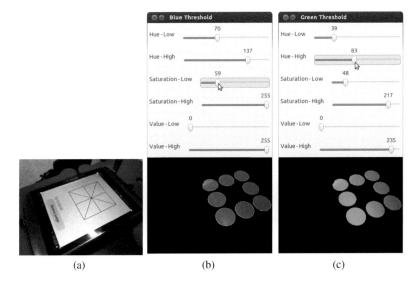

Fig. 7 Detection calibration: a) Board identification (green dots), b) Blue colour range, c) Green colour range.

- It begins with the Homography calibration (fig. 7(a)), where the user has to double click on a window the four corners of the white background from the game application.
- Then, the game logic positions are registered through the line detection process. This is only done once because we assume that the tablet will remain in the same position during the whole game.
- And finally, all the game chips colours are calibrated with a window that shows, in real time, the effects of changing the values of each variable that defines a colour in the HSV space (figs. 7(b) and 7(c)). This step is really important, given the nature of the methods used and the colour itself, extremely dependent of the lighting conditions.

3.4 Interaction and Game Logic

A very simple and heuristic Tic-Tac-Toe algorithm was designed. It follows the next hierarchy: first it attempts to win, then it attempts to prevent the rival from winning or, finally, it follows a predefined order of movements. It also has memory, allowing the occlusion of the tablet. The logic is kept simple because our next objective is to add an algorithm capable of allowing Nao to learn on it's own how to play a board game, and to model human behaviour so it can adapt the level of difficulty according to the performance of the opponent.

Finally, the system includes several interaction mechanisms between the robot and the human player in order to make the game more intuitive and attractive. The objective is for the human being to be able to play with Nao without the need of any external resources (PC, monitor, keyboard or mouse) and to include HRI learning mechanisms to help engaging with humans.

We have integrated four types of interaction based on Nao's capabilities:

1. **Voice synthesis and recognition:** Nao is capable of speaking and recognising words in both English and Spanish. These capabilities are used to indicate the game state (example: "It's your turn") and to start a new game cycle if desired (example: "I want to start a new game"), respectively.
2. **Tactile:** We use two tactile sensors from Nao's head for security reasons, as a counter measure to stop the touch movement and to indicate the start of a new game when each implemented feature (item detection and voice recognition) do not work as intended.
3. **Leds:** They adopt the chips colour from the current player to indicate their turn.
4. **Outcome behaviours:** Nao reacts to each possible game outcome with a predefined behaviour, like is crying when losing.

4 Experimental Tests

As we have mentioned before, the game is presented to the robot on a digital tablet and the robot uses a digital pointer to play, both regular devices that were not modified for this specific application at all. However, the tablet Tic-Tac-Toe application was created for this test, so we could change the items parameters easily. The tablet position is shown to us by the robot by pointing where the board center position should be, given the hand reachable area.

The Tic-Tac-Toe game with Nao was tested under different circumstances, i.e., different human players and lighting conditions, and has been extensively used on monthly robotics demonstrations in our center [14], usually with young people and with very positive reactions. In particular, we carried out 15 demos, with and average of 3 active participants per demo and 2 games per participant.

Overall, the system works properly, but it presents some limitations:

- The vision module detection is really dependent of the lighting conditions, specially for the lines, since the chip detection can be calibrated before each game (although it shouldn't be too precise or it could harm the game performance as well). It also requires for the tablet to be in the same position through the whole game.
- The touch module allows Nao to play in every board position, but given the degrees of freedom (DOF) of Nao's wrist (only one), the nature of the selected IKS and the variability of the pointer position inside the hand due to previous touches, the touch can present some inclination.

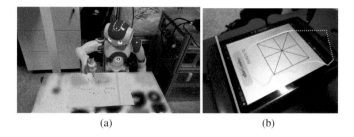

(a)   (b)

Fig. 8 Reachable area of Nao's right hand (Nao's center represented with a green line): a) Reachable points on the table plane, b) Efective reachable area supperposed to the board game.

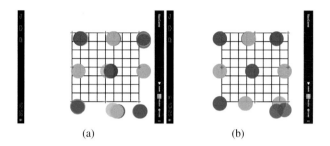

(a)   (b)

Fig. 9 Repeteability of Nao's arm movement (5 touches on each target position, marked by a red cross): a) right arm, b) Left arm.

– The game logic and interaction module works as intended as long as the human player follows the stated rules, making the game more pleasant and enjoyable to the player and removing the need to interact with the robot through other devices.

4.1 *Reach, Precision and Repeatability of the Arm's Movement*

Another test domain for the system was the measure of the arm's movement reach, precision and repeatability. For that purpose, an eight by eight cell board was added to the tablet application, which showed a translucent circle each time the screen was touched.

The designed touch movement has a limited reach (see fig. 8), given the arm size, number of joints and gaming position. Nonetheless, it is possible to reach the four corners of a square board with a side of 7.2 cm at a height of 31 cm, if the aforementioned board is centered on the arm's reachable area.

Apart from the pointer's position variability inside the hand, which can be solved by making it thicker, the touch precision is highly conditioned by the selected IKS and the lack of multiple DOF on Nao's wrists, which induce an offset to the touch point in the direction of the forearm, towards Nao (see fig. 9). This can be solved by applying an opposite offset to the desired touch destination.

Now, the repeatability of the touch movement is greatly consistent (fig. 9), only disturbed by anomalies in the joint motors due to quick movements of the arm, or just as a consequence of heavy use. This translates to almost identical touches among iterations. Nevertheless, we can note that several times, the pointer can move inside Nao's hand, which produces sudden error increments (example: bottom right target position on figure 9(b)).

5 Conclusions

In this paper we have presented a system architecture that allows Nao to play a board game on a tablet, specifically the well known Tic-Tac-Toe game. As long as the system is correctly calibrated, the Nao robot can interact autonomously with people through board games: not only it is able to understand all the game items through OpenCV's computer vision algorithms, but it also can decide movements and reach the tablet through the Cartesian Control's IKS. To overcome Nao's arms and hands limitations (there are not force nor tactile sensors in them), we propose a touch movement with visual feedback (the detection of a new chip on the board). To make the game more pleasant and attractive to the human being, we have implemented several interaction mechanisms based on Nao's capabilities. Experimental results show the robustness of the proposed architecture.

Regarding future work, our next aim would be to improve the design to overcome some of the current limitations, and thus make the system more robust and efficient. Firstly, we need to optimize the arm movements using an IKS that includes planning. MoveIt's algorithm shows promising results, but it is still not completely functional on a Nao robot. Secondly, the use of a stereo camera or a RGB-D camera [4] would allow us to achieve a better estimate of the position of the items, thus enhancing the system autonomy. Thirdly, we are finalising a new behaviour that detects polylines on a sheet of paper and allows Nao to replicate them. Lastly, we are involved on a project to add machine learning to the game logic behaviour, so the robot can learn how to play a specific board game on it's own by simply interacting with an user. Furthermore, this way of interacting with humans could also be learnt, so that the robot adapts to the specific characteristics of the user. We have achieved encouraging results so far with reinforcement learning.

Acknowledgment This work was supported by the research grant TIN2012-32262 (FEDER co-funded) and by the Galician Government (Xunta de Galicia) under the Consolidation Program of Competitive Reference Groups: GRC2014/030 (FEDER cofunded) and GRC2013/055 (FEDER cofunded).

References

1. Poddigue, R., Roos, N.: A NAO robot playing tic-tac-toe, comparing alternative methods for Inverse Kinematics. In: Proceedings of the 25th Benelux Conference on Artificial Intelligence, BNAIC 2013, November 7–8, 2013, Delft, The Netherlands (2013)
2. Poddigue, R.: Playing Tic-tac-toe with the NAO humanoid robot. Maastricht University, Department of Knowledge Engineering (2013)
3. L. Mingueza, N.: Desarrollo de un sistema de juego al Tres en Raya para el robot NAO H25. Master thesis, Universidad Carlos III de Madrid. Departamento de Informática (2013)
4. Canzobre, D.S., Regueiro, C.V., Calvo-Varela, L., Iglesias, R.: Integration of 3-D perception and autonomous computation on a Nao humanoid robot. ROBOT 2015 (accepted)
5. Who is Nao?. https://www.aldebaran.com/en/humanoid-robot/nao-robot Last Visit: July 07, 2015
6. ROS. http://www.ros.org/ Last Visit: July 07, 2015
7. OpenCV. http://opencv.org/ Last Visit: July 07, 2015
8. Cartesian control. http://doc.aldebaran.com/2-1/naoqi/motion/control-cartesian.html. Last Visit: July 07, 2015
9. NAOqi SDK. https://www.aldebaran.com/en/robotics-solutions/robot-software/development.html. Last Visit: July 07, 2015
10. Choregraphe. http://doc.aldebaran.com/2-1/software/choregraphe/choregraphe_overview.html. Last Visit: July 07, 2015
11. Smith, J., Belongie, S.: CSE 252: Computer Vision II. Lecture 4, Planar Scenes and Homography (2004)
12. Kalra, P.K.: Canny Edge Detection. CSL783, Digital Image Processing (2009)
13. De Rezende, A.: Hough Transform. MC851, Implementation Projects in Computing - Computer Vision (2013)
14. Wiki CiTIUS. https://wiki.citius.usc.es/demostradores-nao. Last visit: july 07, 2015

Cooperative Adaptive Cruise Control for a Convoy of Three Pioneer Robots

F.M. Navas Matos, E.J. Molinos Vicente, A. Llamazares Llamazares,
Manuel Ocaña Miguel and V. Milanés Montero

Abstract This paper deals with the stability analysis of a Cooperative Adaptive Cruise Control system (CACC) in a convoy of vehicles, in order to solve traffic congestion. The differences between the Adaptive Cruise Control (ACC) and its cooperative version will be studied from the string stability point of view, highlighting the advantages of adding Vehicle-to-Vehicle (V2V) communications. The system will be tested in simulation using MATLAB/Simulink for a convoy of three vehicles and, after that, the control system will be translated to real experimentation for a convoy of three Pioneer robots.

Keywords ACC · CACC · V2V · String stability

1 Introduction

Road transport has significantly absorbed people and good transportation in recent years. According to the European Commission, road transport was 44.9% of the total goods transport and 82.4% of the passenger transport by powered two-wheelers and buses in the European Union (EU) [1]. These figures point out to an increment in the number of road transport vehicles, leading to more congested roads.

The Intelligent Transportation Systems (ITS) domain tries to find more efficient and greener solution to this kind of problems by developing advanced aiding systems that can help drivers in their daily life. Related to traffic congestion, intelligent longitudinal speed control is a suitable aiding system to improve traffic flow in highways by avoiding consecutive speed changes on the part of the driver.

First commercially available system for improving traffic congestion was the Cruise Control (CC) system [2]. It acts over the throttle pedal, allowing the ve-

F.M.N. Matos(✉) · E.J.M. Vicente ·
A.L. Llamazares · M.O. Miguel
RobeSafe, University of Alcalá, Madrid, Spain
e-mail: franiti2107@gmail.com,
    {eduardo.molinos,angel.llamazares,mocana}@depeca.uah.es

V.M. Montero
INRIA, Paris, France
e-mail: vicente.milanes@inria.fr

© Springer International Publishing Switzerland 2016
L.P. Reis et al. (eds.), *Robot 2015: Second Iberian Robotics Conference*,
Advances in Intelligent Systems and Computing 417,
DOI: 10.1007/978-3-319-27146-0_17

hicle to follow a set speed. Next already commercially available system was the Adaptive CC (ACC) where brake control feature was added, allowing the vehicle to keep a safety distance with a preceding vehicle by using a radar/lidar front sensor [3]. Current research work is focused on adding vehicle-to-vehicle (V2V) communications to ACC system, leading to the development of Cooperative ACC (CACC) systems. CACC systems dramatically reduce vehicle response time to speed changes on the part of preceding vehicles in the lane, allowing to significantly reduce inter-vehicle safety distance [4]. Simulation results indicate that highway capacity can be doubled by adding communications to the already available ACC systems [5].

This paper presents a comparative study between ACC and CACC systems from the string stability point of view. To this end, a longitudinal dynamic vehicle model is developed for analyzing stability responses for both control algorithms. Simulation results show up the benefit of adding communication to car-following control systems. Finally, the CACC control algorithm is tested on three real Pioneer robots equipped with frontal Hokuyo laser. Real experiments validate theoretical results by demonstrating string stability when adding V2V communications to the car-following control algorithm. The robotic platform uses a low-speed profile. Low-speed tests for ACC and CACC systems have not been carried out yet because of the complex dynamic in such conditions. This work presents the behavior of a robotic CACC system for low-speeds.

2 Problem Formulation

2.1 Vehicle Following Objective

There are two different spacing strategies: constant intervehicle distance or constant head time [6]. The first one is only justified for consumption reduction in heavy vehicles, in other cases is preferable a velocity-dependent strategy, because provides more security. So, constant head time has been chosen.

The vehicles inside the platoon are interconnected according to the vehicle following objective. Each vehicle has to follow its predecessor while maintaining a distance dependent on the velocity of the vehicle and a constant head time. The head time represents the time that it will take the vehicle to reach the position of its predecessor. Thus, the desired spacing ($d_{r,i}$) between two consecutive vehicles is:

$$d_{r,i} = r_i + h_{d,i} v_i \qquad (1)$$

where i is the vehicle index; r_i is a constant term that forms the gap between consecutive vehicles at standstill; $h_{d,i}$ is the constant head time; and v_i is the vehicle velocity. This desired distance will be compared with the real distance between vehicles (d_i) obtained by laser measurement. Thus, the local control objective can be defined as the error regulation to zero, as follows:

$$e_i = d_i - d_{r,i} \tag{2}$$

2.2 String Stability

String stability can be defined as the attenuation of disturbances along the vehicle platoon, and it is evaluated through different signals as amplification of distance error [7] [8] [9] or velocity, acceleration and absolute position [10] [11] [12] as the vehicle index i increases within the string.

Besides, it is necessary to study the string stability in frequency domain. This property can be defined according to [6] by the following stability transfer function:

$$SS_{\Delta_i}(s) = \frac{\Delta_i(s)}{\Delta_{i-1}(s)}, i \geq 1 \tag{3}$$

where $\Delta_i(s)$ is the signal of interest to study the string stability. The string stability requirement can be interpreted through H-infinity norm of the string stability transfer function (4), representing the maximal amplification of perturbations along the string.

$$\left\| SS_{\Delta_i}(jw) \right\|_\infty = \sup_{w \in \mathbb{R}} \left\| SS_{\Delta_i}(jw) \right\|_\infty \tag{4}$$

In accordance with (3) and (4), the condition for string stability will be:

$$\left\| SS_{\Delta_i}(jw) \right\|_\infty \leq 1, \forall w, i \geq 1 \tag{5}$$

2.3 Longitudinal Dynamics Model

This section details the longitudinal vehicle and robot dynamics models. They will be used in MATLAB/Simulink to obtain the corresponding simulation results. Firstly, the longitudinal vehicle model will be used to obtain a CACC control algorithm. It will allow establishing the necessary knowledgebase to its adaptation to the experimental set-up composed by a string of three Pioneer robots. Before the experimental set-up, the correct adaptation of the control system will be tested using the longitudinal robot dynamics model. Both models will be used to obtain the feedforward filter proposed to the cooperative part of the adaptive cruise controller.

The chosen longitudinal vehicle dynamics model is widely used in the literature [13] [14] [15]. This vehicle model can be represented by the transfer function:

$$G_i(s) = \frac{Q_i(s)}{U_i(s)} = \frac{1}{s^2(\eta_i s + 1)} e^{-\tau_{a,i} s} \tag{6}$$

where $Q_i(s)$ is the absolute position of the vehicle i; $U_i(s)$ is the commanded acceleration; η_i represents the internal actuator dynamics; and $\tau_{a,i}$ is a constant actuation delay between the desired acceleration and the commanded one.

On the other hand, the longitudinal Pioneer robot dynamics model has been obtained by identification techniques [16]. The model in (7) corresponds to a P3-DX Pioneer robot. It also relates absolute position and commanded acceleration.

$$G_i(s) = \frac{s+5}{s^2(s+5)(s+4.1)} \tag{7}$$

3   Adaptive Cruise Control

Once the basic concepts and the dynamics model have been defined, we proceed to the development of the ACC system. In first place, the control structure will be shown. After that, a string stability analysis to the developed control system will be carried out; control gains will be chosen in order to ensure the stability at any speed.

The adopted values will be verified through the results corresponding to a platoon composed by three vehicles, and for the string composed by three Pioneer P3-DX robots. It will be done using Simulink and the models described in (6) and (7).

3.1   *Control Structure*

There is a string of three vehicles where each one has to follow its predecessor at a distance imposed by the spacing strategy. This distance is compared with the laser measurement, obtaining the error shown in (2). To regulate the error to zero, we use a Proportional-Derivative (PD) controller, obtaining the acceleration signal to apply to the longitudinal vehicle dynamics model $G_i(s)$. The control scheme is shown in Fig. [1]. PD transfer function is given in the Laplace domain as follows:

$$U_i(s) = \omega_{c,i}(\omega_{c,i} + s)E_i(s) \tag{8}$$

where $\omega_{c,i}$ is the bandwidth of the controller and is chosen under comfort and not-saturation condition, that is, $\omega_{c,i} < \omega_{g,i} = 1/\eta_i$ [14], where $\omega_{g,i}$ is the vehicle's bandwidth.

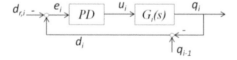

Fig. 1 ACC Control Scheme. Where q_i is the absolute position of vehicle i, q_{i-1} is the absolute position of vehicle $i-1$, d_i is the real relative distance, $d_{r,i}$ is the desired relative distance and u_i is the commanded acceleration.

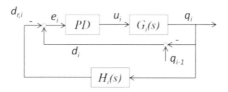

Fig. 2 ACC Control Scheme with Spacing Policy.

Now, if the spacing strategy is included in Fig. [1] to obtain the desired distance, the scheme changes as seen in Fig. [2]. $H_i(s)$ is the spacing policy transfer function corresponding to (1) $H_i(s) = 1 + h_{d,i}s$. The closed loop transfer function of this scheme, which relates absolute position of the vehicle i with its predecessor, results:

$$T_i = \frac{H_i G_i PD}{1 + H_i G_i PD} \tag{9}$$

3.2 String Stability Analysis

String stability will be studied through absolute position signals. According to stability transfer function in (3), we need a transfer function which relates q_i and q_{i-1}. This is the closed loop transfer function obtained in (9).

Assuming $\tau_i = 0$ and a full bandwidth situation, ideal vehicle dynamics results, yielding $G_i(s) = s^{-2}$. This assumption makes easier the string stability analysis. The consequences of this assumption with respect to practice will be evaluated later on. Taking into account PD and $H_i(s)$ in (9) with the string stability condition (5) shows that string stability can be guaranteed if (extended justification in [6]):

$$h_{d,i} \leq h_{d,i,\min} = \sqrt{2}\omega_{c,i}^{-1}, i > 1 \tag{10}$$

To enhance the congestion problem, we will be interested in the minimum value of head time, to pack the driving vehicles together as tightly as possible.

3.3 MATLAB/Simulink Validation. Vehicles

In this subsection, the explained control structure is validated. With this aim, a heterogeneous traffic string of three vehicles is modeled in MATLAB/Simulink, obtaining the signal shown in Fig. [3], corresponding to the parameters of Table [1]. The bandwidth of the underlying ACC controller is taken based on speed response and passenger comfort. The head time is calculated according to (10), so the minimum head time that makes the system stable is 2.82s, which is approximated to 3s.

Table 1 Parameters ACC system. MATLAB/Simulink.

Vehicle i	$\eta_i(s/rad)$	$\tau_i(s)$	$\omega_{c,i}(rad/s)$	$h_{d,i}(s)$
1	0.1	0	0.5	3
2	0.5	0.1	0.5	3
3	0.4	0.3	0.5	3

The velocity signal shown in Fig. [3] validates the calculated head time to ensure string stability, because there is not amplification of the signal. Notice how there is a reference vehicle 0, which gives its absolute position as reference to the vehicle 1; thus all vehicles are equipped with ACC system.

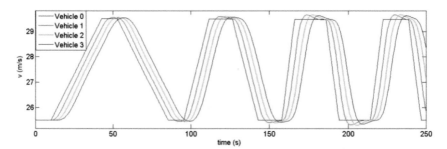

Fig. 3 Velocity Simulink results for a convoy of three heterogeneous vehicles equipped with ACC system. Head time 3 s. MATLAB/Simulink.

3.4 MATLAB/Simulink Validation. Robots

Once the ACC system has been validated for a platooning of three vehicles, we use the same control structure with the P3DX model shown in (7). Notice how the reference velocity is adapted. The obtained velocity signal (see Fig. [4]) through MATLAB/Simulink is also satisfactory.

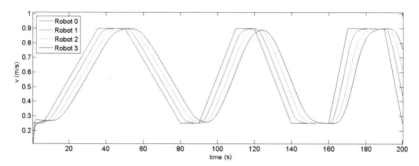

Fig. 4 Velocity Simulink results for a convoy of three P3DX robots equipped with ACC system. Head time 3 s. MATLAB/Simulink.

4   Cooperative Adaptive Cruise Control

The cooperative version of ACC system includes wireless communication with its predecessor. Specifically, the acceleration $\ddot{q}_{i-1}$ will be sent through wireless communication. This data will be used in the feedforward control for obtaining a better stability result.

4.1   Control Structure

The ACC system is extended to the cooperative version by adding the wireless information from the preceding vehicle to the control structure showed in Fig. [2], leading to a new control scheme according to   Fig. [5]. Notice how a constant delay is taken into account to send the data; its transfer function is $D_i(s)$ (11), with a constant delay θ_i.

$$D_i(s) = e^{-\theta_i s}, i > 1 \tag{11}$$

The predecessor's acceleration will be the input of a feedforward filter $F_i(s)$, which design is adopted to improve the string stability of the system [17]. The feedforward filter has the following transfer function:

$$F_i(s) = \frac{1}{H_i(s)G_i(s)s^2}, i > 1 \tag{12}$$

Thus, to obtain the feedforward filter is necessary to know the head time and the longitudinal dynamics model.

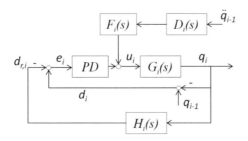

Fig. 5 Control Structure CACC system with Spacing Policy.

Now, the closed loop transfer function which relates the absolute position of vehicle i with the vehicle $i-1$ changes:

$$T_i = \frac{G_i F_i D_i s^2 + G_i PD}{1 + H_i G_i PD}, i > 1 \tag{13}$$

4.2 String Stability Analysis

For the cooperative case, the string stability analysis is also studied with absolute position. The transfer function which relates q_i and q_{i-1} is the closed loop function in (13). Considering substitution of $F_i(s)$ in (13), the string stability transfer function is:

$$SS_i = \frac{D_i + H_i G_i PD}{H_i(1 + H_i G_i PD)}, i > 1 \tag{14}$$

The benefit to the congestion problem with the inclusion of wireless information is clearly seen in the not-delay ideal situation. In this case, $D_i = 1$, so (14) results:

$$SS_i = \frac{1}{H_i}, i > 1 \tag{15}$$

In this case the string stability is guaranteed for any positive value of head time, improving the congestion problem.

The real case presents variable delays imposed by the wireless communication. If a communication delay $\theta_i > 0s$ is taken into account, the analytical derivation of the minimal required headway time $h_{d,i,\min}$ becomes rather complex and does not provide additional insight.

The worst case would occur where delays tend to infinite (i.e. wireless link is lost), downgrading the system to the regular ACC system.

4.3 MATLAB/Simulink Validation. Vehicles

The process to validate the designed control structure is analog to the ACC case. Table [2] shows the simulation parameters where a constant delay has been considered. Additionally, the head time has been halved to show the benefits with respect to ACC systems from the traffic congestion problem point of view.

Table 2 Parameters CACC system. MATLAB/Simulink.

Vehicle i	$\eta_i\,(s\,/\,rad)$	$\tau_i(s)$	$\omega_{c,i}\,(rad\,/\,s)$	$h_{d,i}(s)$	$\theta_i(s)$
1	0.1	0	0.5	1.5	0
2	0.5	0.1	0.5	1.5	0.2
3	0.4	0.3	0.5	1.5	0.2

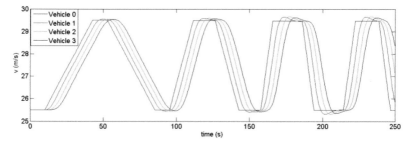

Fig. 6 Velocity Simulink results for a convoy of three heterogeneous vehicles equipped with CACC system. Head time 1.5 s. MATLAB/Simulink.

The velocity result corresponding to these parameters is shown in Fig. [6]. The system keeps stability with the head time halved compared with the ACC experiment.

4.4 MATLAB/Simulink Validation. Robots

Once the CACC system is validated for a convoy of three vehicles, we proceed to validate the system with a convoy of three P3DX robots, taking into account the necessary change in the feedforward filter according to equation (7). The velocity signal shown in Fig. [7] indicates that string stability is ensured, because there is not amplification of perturbations. In fact, the overshoot that appears in v_1 is not amplified along the string of robots in longer simulations. This overshoot doesn't affect to the system, because is too small to be able to cause a collision.

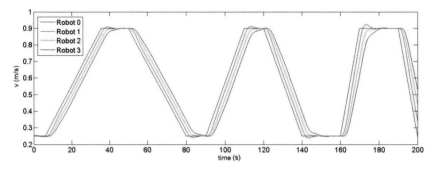

Fig. 7 Velocity Simulink results for a convoy of three P3DX robots equipped with CACC system. Head time 1.5 s. MATLAB/Simulink.

5 Experimental Results

With the experimental results corresponding to a string composed by two P3-AT and one P3-DX robots, a real experiment was carried out in order to validate the simulation results. It will also allow to carry out the fine tuning of the controller gains when moving the system to the robotic development platform.

5.1 Experimental Set-up

The string is composed by a total of three robots. The leader is a P3-AT robot, which receives relative position to a reference robot 0 (i.e. virtual robot leader). The second one, is another P3-AT, equipped with a Hokuyo URG-04LX laser to measure d_i, and which receives the acceleration signal from the first one using a Wi-Fi link. The last robot of the convoy is a P3-DX, with laser and which receives information from the second one.

Notice how each robot needs a computer to execute the client application developed in C language to obtain the control system in Player. The scheme of the experimental set-up is shown in Fig. [8].

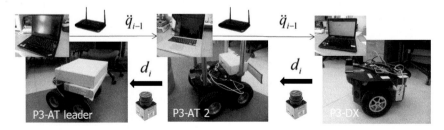

Fig. 8 Experimental Set-up.

5.2   Results and Discussion

The experimental results for both systems, ACC and CACC, correspond to $h_{d,i} = 3s$ and $h_{d,i} = 1.5s$ respectively. Besides, the value of the delay is unknown, because is directly imposed by the Wi-Fi link. The velocity signals corresponding to each system are shown in Fig. [9] and [10].

In both figures, a large overshoot can be observed, this is due to the start of each robot, and its spacing policy strategy, that has to be fulfilled as soon as possible, so maximum acceleration is adopted to reach the predecessor robot.

One can appreciate how velocity signals are not able to keep the constant head time during the deceleration phase. This failure is caused by a random error in the driver responsible for obtaining velocity measurement.

In spite of this, these signals secure the string stability of the system. The velocity signal is not amplified along the robots' string. Comparing the results of the ACC system in Fig. [9] and the CACC system in Fig. [10] clearly illustrates the potential of the proposed design.

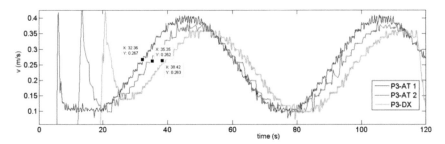

Fig. 9 Velocity experimental result for a convoy of two P3-AT and one P3-DX robots with ACC system. Head time 3 s. Player.

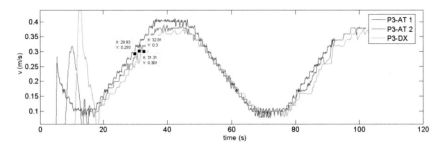

Fig. 10 Velocity experimental result for a convoy of two P3-AT and one P3-DX robots with CACC system. Head time 1.5 s. Player.

6   Conclusions and Future Work

This paper carries out a comprehensive study of the cooperative adaptive cruise control system in comparison with the commercially available ACC system. First, longitudinal vehicle dynamic model was identified and CACC algorithm was developed from a ACC system. The control algorithm was validated by Simulink's results. A study for a string with three vehicles was carried out to establish the minimum levels of head time to guarantee the string stability.

Then, the same minimum values were used to validate the control system for a string of three Pioneer Robots in MATLAB/Simulink. The control system's adaptation was proved.

The experimental results corresponding to a P3AT-P3AT-P3DX string validates the assumptions made in simulation and its correct adaptation to the robotic platform Player.

Thus, the V2V communication's addition to share acceleration with the predecessor reduces traffic congestion problem by reducing head time while ensuring string stability.

Despite the random error found in Player's driver, the string stability is ensured. In the future, this error will be mitigated by either velocity laser's correction or changing to a robotic platform without this error, like ROS.

Another interesting work will be the string stability analysis taking into account other wireless imperfections, as packet loss, ZOH and multihop between vehicles. Experimental results depending on what kind of wireless communication (Zigbee, WiFI and Wimax) will be also explored.

Acknowledgments This work has been partially found by ABS4SOW a Ministerio de Economía y Competitividad (TIN2014-56633-C3-3-R) project and RoboCity2030-III-CM a CAM (P2013/MIT2748) project.

References

1. Comission, E: EU transport in figures. Statistical pocketbook 2014 (2014)
2. Ioannou, P., Chien, C.: Autonomous intelligent cruise control. IEEE Trans. on Vehicular Technology **42(4)** November 1993
3. Adaptive Cruise Control published by SAE International, ISBN 978-0-7680-1792-2, p. 488 (2006)
4. Milanes, V., Shladover, S., Spring, J., Nowakowski, C., Kawazoe, H., Nakamura, M.: Cooperative adaptive cruise control in real traffic situations. IEEE Transactions on Intelligent Transportation Systems **15**(1), 296–305 (2014)
5. Shladover, S., Su, D., Lu, X.: Impacts of cooperative adaptive cruise control on freeway traffic flow. 91st Transportation Research Board Annual Meeting (2012)
6. Naus, G., Vugts, R., Ploeg, J., van de Molegraft, R., Steinbuch, M.: Cooperative adaptive cruise control and design and experiments. In: American Control Conference (2010)

7. Yanakiev, D., Kanellakopoulos, I.: Nonlinear spacing policies for automated heavy-duty vehicles. IEEE Trans. Veh. Technol. **47**(4), 1365–1377 (1998)
8. Sheikholeslam, S., Desoer, C.: Longitudinal control of a platoon of vehicles with no communication of lead vehicle information: A system-level study. IEEE Trans. Veh. Technol **42**(4), 546–554 (1993)
9. Swaroop, D., Hedrick, J., Chien, C., Ioannou, P.A.: A comparison of spacing and headway control strategy for automatically controlled vehicles. Veh. Syst. Dyn. **23**(8), 597–625 (1994)
10. Khatir, M. Davison, E.: Decentralized control of a large platoon of vehicles using non-identical controllers. In: *Proc. Amer. Control Conf.*, pp. 2769–2776 (2004)
11. Huppe, X., Lafontaine, J., Beauregard, M., Michaud, F.: Guidance and control of a platoon of vehicles adapted to changing environment conditions. Proc. IEEE Int. Conf. Syst. **4**, 3091–3096 (2003)
12. Shaw, E., Hedrick, J.: String stability analysis for heterogeneous vehicle strings. In: *Proc. Amer. Control Conf.*, pp. 3118–3125 (2007)
13. Liu, X., Goldsmith, A., Mahal, S., Hedrick, J.: Effects of communication delay on string stability in vehicle platoons. Proc. IEEE Intl. Conf. Intell. Trans. Systems **15**(1), 625–630 (2001)
14. Ploeg, J., Scheepers, B., Nunen, E., van deWouw, N., Nijmeijer, H.: Design and experimental evaluation of cooperative adaptice cruise control. In: *Proc. IEEE Intell. Transp. Syst.Conf.*, pp. 260–265 (2011)
15. Naus, G., Ploeg, J., de Molengraft, M.V., Heemels, W., Steinbuch, M.: Design and implementation of parameterized adaptive cruise control: An explicit model predictive control approach. Control Engineering Practice **18**, 882–892 (2010)
16. Santos, H., Jr., M.M., Espinosa, F.: Adaptive self-triggered control of a remotely operated robot. Technical report, Electronics Department, University of Alcalá (Spain) and INCAS3, Assen (The Netherlands) and University of Groningen (The Netherlands) (2014)
17. Naus, J., Vugts, P., Ploeg, J., van de Molengraft, J., Steinbuch, M.: String-stable cacc design and experimental validation: A frequency-domain approach. IEEE Transactions on vehicular technology **59**(9), 4268–4279 (2010)

Design and Development of a Biological Inspired Flying Robot

Micael T.L. Vieira, Manuel F. Silva and Fernando J. Ferreira

Abstract This paper describes the design and development of a biologically inspired flying robot prototype (a machine able to fly by beating its wings, as birds do). For its implementation, the flight of biological beings was analysed, as well as the techniques involved in ornithopter's construction. Some parameters adopted by biological beings to maintain a stable flight were studied, and the prototype was designed based on these values. To conclude the project, an ornithopter was built, aiming to perform a stabilized flight and some preliminary experiments were performed to check if its behaviour meets the design expectations.

Keywords Robotics · Ornithopter · Flight · Biological inspiration

1 Introduction

One active line of research and development in robotics is the development of biologically inspired robots. Whether robots that use legs, fins or wings as a means to implement locomotion, the idea is to acquire knowledge of biological beings, whose evolution took place over millions of years, and utilize the knowledge thus acquired to implement the same methods of locomotion (or at least use the biologically inspiration) on the machines being developed. It is believed that in this way we are able to develop machines with capabilities similar to those of biological beings in terms of locomotion capacity and energy efficiency.

The pioneer in the study of flying machines was Leonardo Da Vinci, around 1490, but it took a long way until these machines could be successfully implemented. The first projects in the late nineteenth century achieved successfully stabilized flight

M.T.L. Vieira · M.F. Silva(✉) · F.J. Ferreira
ISEP-IPP - School of Engineering of the Polytechnic of Porto and INESC-TEC,
Rua Dr. António Bernardino de Almeida, Porto, Portugal
e-mail: {1081510,mss,fjf}@isep.ipp.pt
http://ave.dee.isep.ipp.pt/mss

© Springer International Publishing Switzerland 2016   231
L.P. Reis et al. (eds.), *Robot 2015: Second Iberian Robotics Conference*,
Advances in Intelligent Systems and Computing 417,
DOI: 10.1007/978-3-319-27146-0_18

by using the elastic energy exerted by a rubber band to promote the oscillatory motion of the wings. After this, projects started being developed with the use of steam and compressed air, human motors, internal combustion engines and, more recently, electric motors [18].

Currently there are several ornithopters, such as SmartBird [4], Sean Kinkade Park Hawk [6], Cybird P1 [5] and robot Raven [2]. These ornithopters are designed with a lightweight structure and, at the same time, with high resistance to the efforts exerted during flight and possible impacts due to crashes. The preferred material for the structure of this type of machines is carbon fiber, ensuring the stated requirements. To achieve a stabilized flight and ensure the necessary lifting force, these machines rely on an interrelation between the wings and the tail. The wings, usually built in ripstop nylon and Dacron tapes, are responsible for obtaining the thrust forces while the tail is responsible for the orientation of the machine (interfering with the direction and inclination). For ensuring the different inclinations of the wing along the movement cycle, these designs divide the wing into two zones: the first is fixed to the transmission rods and is constantly taut, while the second is composed by small carbon rods that provide linearity to the first zone, but makes the bending element of the wing and gets the desired angles of the wing in a passive way. Concerning the electronics, most of these robots use a brushless motor and an Electronic Speed Controller (ESC), a lithium polymer (LiPo) battery for powering the system, two servomotors to actuate the tail and a receiver that sends the signals received from a radio controller (RC) into the machine motors controller.

After this brief introduction, in the sequel, section two introduces the flight principles of biological beings. Next, sections three and four are devoted to describing the ornithopter design and its implementation, respectively. Section five presents several tests that were performed in order to validate the prototype, and the results obtained. Finally, section six summarizes the main conclusions of the work and states some ideas for future improvements of the ornithopter.

2 The Flight of Biological Beings

There are four forces involved in the act of flight: lift (F_L), thrust (F_T), drag and gravity. The vertical component of lift force is the opposite force from the own mass in the way that, depending on the desired movement of the bird, one should be higher or lower than another. The thrust force is responsible for promoting speed to the biological being and counteracts the drag force exerted on it.

Linton [10] argues that F_L is dependent on the dimensions of the wings, flight speed and its angle of attack (α). An increased α promotes a greater F_L (increases the lift coefficient (C_L)) but only until a certain value, called critical angle (around $15°$ in the case of most birds). Above this value is originated turbulence, increasing the drag force which forces the loss of speed necessary to the animal flight. The maximum C_L usually is between 1,5 and 1,7 for birds and corresponds to the value of critical angle.

The higher the aspect ratio of the wing (AR), the lower the critical angle but, on the contrary, there is provided a higher C_L. The wing's AR varies between 4,8 for small birds and 15 for albatrosses [7], and depends on the total length between the tips of each wing (b) and the wing area (S_W), according to:

$$AR = \frac{b^2}{S_W}. \tag{1}$$

Relating the Newtonian formulas developed by Linton [9], it's possible to describe that C_L is related with AR and α, by the following expression:

$$C_L = \frac{2\pi \times \alpha}{(1 + \frac{2}{AR})}. \tag{2}$$

For an animal to move in a fluid at a constant speed, it was defined by Tayler *et al*. [16] that it is necessary to achieve a Strouhal number (σ) between 0,2 and 0,4. This parameter relates the amplitude A_0 (relative to the length between the tip of a single wing from its highest point to the lowest point, measured in meters), with the wing beat frequency (f), for a given velocity v:

$$\sigma = \frac{A_0 \times f}{v}. \tag{3}$$

F_T is caused by the undulations made by the wing membrane. When the wing moves upwardly it obtains a positive α and when moving downwardly a negative α. This angle ranges from $0°$ at the wing root up to about $22°$ or $-22°$ at the wing tip. In short, F_T at the wing root is almost zero and at the tip is generated its maximum value (negative or positive). Based on this, Linton [10] does an integration along the length of the wing to obtain F_T. Then, using another integration that can consolidate the angle of incidence of the bird (β), with the value of α acquired along the wing, Linton explains how F_L is calculated. Since both forces are made when the wings are in the down-stroke, and at the up-stroke aren't produced any forces, the average F_T and F_L are given by:

$$F_{T\ MEAN} = \frac{1}{6} \times \frac{2\pi}{(1 + 2 \times AR)} \times S_w \times \rho \times v^2 \times \sigma^2, \tag{4}$$

$$F_{L\ MEAN} = \frac{1}{4} \times \frac{2\pi}{(1 + 2 \times AR)} \times S_w \times \rho \times v^2 \times \beta. \tag{5}$$

2.1 Biomechanics of Animal Flight

The wings of biological beings are usually led by a membrane called humerus. This membrane is in contact with the pectoralis and supracoracoideus muscles which emit the necessary contractions for moving it. The pectoralis is the main responsible for the down-stroke. The supracoracoideus, lying on the inner side of the pectoralis, is linked to a tendon that is linked into the top of humerus. The contraction of this muscle allows the tendon to bring the member up [17]. Figure 1 demonstrates how these muscles promote certain wing motions. Most of the projects studied use this structure, replacing the humerus for carbon rods.

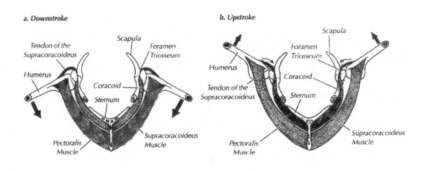

Fig. 1 Muscles involved in the mechanics of the wing [11].

To promote a certain initial speed, essential for the flight, biological beings have legs that give a boost at the time of takeoff. Thus, the energy efforts produced in the wing muscles are reduced. Furthermore, to reduce the efforts of the wings, two types of flight are used when the bird is not promoting wingbeat: the glide and the limited flight [17]. To make curves, birds use asymmetric contractions in the pectoralis, being the side which makes the greater contraction the one to which the bird rolls. This movement creates a momentum about the longitudinal axis that goes from the head to the tail of the bird. This maneuver, known as roll, suggests the rotation of the body. The tail of the bird provides stability in flight and influences the lift and drag forces. With a determined inclination, it can promote better values of α to obtain the desired lift motion, and a big drag caused by a long tail can promote better horizontal stability.

2.2 The Wings

According to The Cornell Lab of Ornithology, there are four types of birds wings [12]: passive soaring, active soaring, elliptical and high-speed. Each one presents some specific flight characteristics [3].

An important parameter in the design of the wing is the wing loading (W_L), reflecting the mass supported per area unit [15]. Usually W_L varies between 1 and 20 kg/m^2, being estimated that the maximum permissible value for birds is 25 kg/m^2 [8]. The fact that the W_L value of a bird is greater than another indicates that this will have higher top speed but needs a larger distance for landing and taking-off and performs maneouvers with greater difficulty.

3   Prototype Design

3.1   Animal's Wings Study and Ornithopter Reference Parameters

The first point to be established for the implementation of an ornithopter is to decide on what type of wings to use, and its parameters. For this project were chosen elliptical wings (similar to the ones of bats and sparrows). This feature reduces the torque required to handle the wing, uses a small AR and provides greater control of the flight. To check this information, the data of 7 birds [13][14] and of 23 bats [1], all having elliptical wings, was collected to determine if AR lies close to the minimum value of 4,8. In fact, the bats provide a mean value of 6,3 while the seven birds promote an average value of 8,6. Calculations were also conducted to verify if W_L was in the range of [1; 20] kg/m^2. Bats obtained an average of 0,69 kg/m^2 and the seven birds an average of 6,02 kg/m^2. It can be concluded that this range does not fit to the bats but it is suitable for all birds involved in Pennycuick's studies [13][14]. The sample of 7 birds and 23 bats have an average W_L of 0,97 kg/m^2. This value is similar to the values used in other projects, such as Sean Kinkade Park Hawk (1,65 kg/m^2) [6] and Cybird P1 (0,82 kg/m^2) [5]. For this reason, it appears that bats are the source of inspiration for the wings of these projects.

Considering these factors, it is possible to find the existence of some interdependencies between some variables involved in flight, using the bats as a sample. Figure 2(a) shows a close correlation of 0,92 between bats W_L and their mass. Bullen and Mckenzie [1] found correlations that vary between 0,748 and 0,905 correlating f with the mass of bats, and these three expressions are different for different flight speeds. In conjunction with the other studies mentioned, it is possible to verify that f is related to the mass of the bats as it can be seen in Figure 2(b). To obtain a stabilized flight, σ must be within a certain range of values. This leads to an understanding that the amplitude is interconnected to f, as reviewed in Figure 2(c) where a correlation of 0,903 is obtained. It is also possible to draw a correlation between the wingspan and S_W. Although the area being dependent on the wingspan value, Figure 2(d) shows that the wings always follow the same elliptical shape.

With the study conducted on the flight of bats, it is possible to define the parameters that the ornithopter must present depending on its mass. Using SolidWorks 2013 CAD software, a machine was designed with 0,593 kg of mass. This value reflects the reference parameters shown in Table 1.

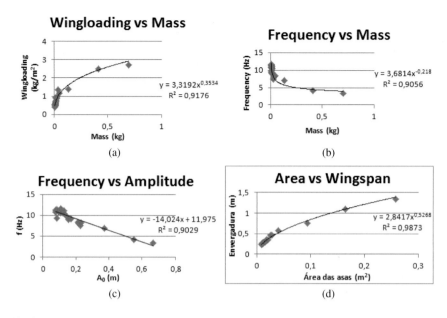

Fig. 2 Correlation between: (a) W_L and the mass of bats, (b) f and the mass of bats, (c) f and wing amplitude on bats, and (d) S_W and wingspan of bat wings.

Table 1 Parameters for flight reference

Parameter	Value	Unit	Expression
Mass	0,593	kg	-
Wing Loading (W_L)	2,760	kg/m^2	$3,3192 \cdot m^{0,3534}$
Wing Area (S_W)	0,215	m^2	$\dfrac{m}{W_L}$
Wingspan (b)	1,264	m	$2,8417 \cdot S_W^{0,5268}$
Aspect Ratio (AR)	7,436	-	$\dfrac{b^2}{S_W}$
Wingbeat Frequency (f)	4,126	Hz	$3,6814 \cdot m^{-0,218}$
Amplitude (A_0)	0,560	m	$\dfrac{f - 11,975}{-14,024}$
Amplitude Angle	52,56	degree	$Arcsen(2 \cdot (\frac{A0}{2})/(\frac{b}{2}))$

3.2 Ornithopter Electronic Architecture

The choice of the electronic components was exercised in parallel with the design of the machine, since they influence its weight. The power is supplied by a two cells

LiPo battery, adequate when using brushless motors, and offering a low weight, with 1300 mAh capacity and a voltage of 7.4 V. The motors are two servomotors HD-1160A (for the tail orientation), and a brushless motor Sunnysky Angel A2212, with 800 KV (in charge of the wings cyclic motion). The brushless motor provides the required torque, but to get voltage regulation from 7.4 V to 5 V, conversion from direct to alternate current, and transformation from the PWM signal into the rotational speed of the motor, the Turnigy Plush 30 A ESC is used. For communication is used a RC controller pre-programmed with a 6 channel receiver. This receiver operates at 5 V, powered by the ESC.

The ornithopter must glide when no torque is provided to the wings. For this reason, a magnet is implemented in the last gear wheel and is used a Hall effect sensor to ensure that the wings stop in the desired position. An Arduino Nano is used for this function, in order to process the impulses provided by the sensor, acting as an intermediary between the receiver and the ESC, to stop the brushless motor in the desired position.

3.3   Ornithopter Mechanical Design

The wings are constructed of ripstop nylon, together with adhesive and non-adhesive Dacron tapes. As the wings perform cyclic movements which are not linear with the machine, ball joins are used to transmit power through a perpendicular axis connection. Thus, both the rear and front carbon rods can play the desired motion for the wings. The wings are designed to promote an amplitude of 0,56 m with an angle of 55,2° (28,4° upward and 23,8° downward). The carbon rods used for the transmission of power have 3 mm diameter and offer a much higher yield stress than the stress theoretically imposed to them.

The torque reduction system uses Delrin gear wheels with a module of 0,8 and a reduction of 13,3:1, from the motor to the last wheel. This system is based on the main frame and two pieces that are always at the same distance from it, with the help of threaded tabs. The motor rotational direction is important since the wings should provide faster speed during the descent movement than on the ascent. Here are made 179° to make the down-stroke and 181° for the up-stroke. Figure 3(a) shows how this influences the rotation direction when using the perpendicular axis system. The brushless motor's imposed torque is 0,046 Nm. To overcome this torque is expected a current of 4,13 A. To obtain the desired $f = 4, 1$ Hz is calculated a current of 4,4 A, being the motor maximum allowed current 10 A.

To complement the locking system, performed by the Hall effect sensor, this component is incorporated into a mechanical locking arm with a spring in the last geared wheel to prevent the gliding flight position if the motor gives up to the imposed torque. Figure 3(b) depicts how this system is implemented. Once the sensor indicates the stop position of the wings, the wheel built-in screw aligns with the locking arm due to the tension spring action. At this moment, as the machine will make a downward

M.T.L. Vieira et al.

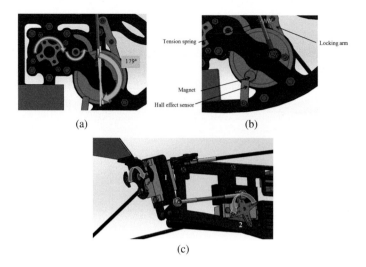

(a)   (b)

(c)

Fig. 3 Transmission orientation (a), braking system (b) and servomotors' motions (c).

movement imposed by gravity, the wings will move up by wind action and at this time the bolt is fixed in the arm.

Figure 3(c) shows how the servomotors operate to promote the tail movements. Servomotor 1 is responsible for the direction made by the tail. It's function is to move the vertical component of the force exerted by the tail to one side, creating a moment along the longitudinal axis. This torque forms a rolling movement required to make curves during the flight. The intermediate part between this servomotor and the tail should ensure a certain inclination in order to promote a career in bowl and keep the angle of incidence promoted by another servomotor. Servomotor 2 is responsible for setting the inclination of the first servomotor and obtain the desired angles of attack of the machine.

The tail surface area is $1,31$ dm^2 and the servomotors provide a maximum torque of $2,0$ kgcm. These calculations are difficult to perform accurately since they depend on the drag coefficient exerted by the air resistance, a parameter normally obtained by laboratory experiments. However, calculations were made for an extreme situation where the drag coefficient takes the value of $0,67$ (value obtained by iterating) and the air velocity is $11,5$ m/s. In this particular case, it is determined that the maximum area for the tail is $3,8$ dm^2, for the maximum torque exerted by the servo-elevator.

The main frame is responsible for supporting all components involved in the ornithopter and is designed to allow their easy assembly. In its front is placed the torque reduction box, in the middle the electronic components, at the rear the servomotors for actuating the tail, and above the wings. The center of mass (CoM) of the machine should be between 1/3 and 1/4 of the length of the wing, along the longitudinal axis. The main challenge in the design of this part is reflected by the placing of the components in a forward position and the comparison of this point with the wing position

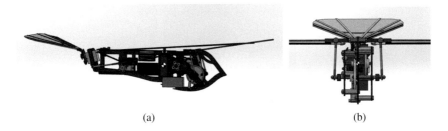

(a)   (b)

Fig. 4 Lateral (a) and frontal view of the ornithopter (b).

on the machine. This CoM is theoretically at 0,319 of the length of the wing. Within this 2 mm thick frame, the CoM lies at 1,2 mm from one of the surfaces. Figures 4(a) and 4(b) show the position of this CoM.

4   Implementation of the Prototype

Ripstop nylon was used for the wings and tail construction, being fixed to the mechanical structure through Dacron tapes and 3M VHB adhesive double-sided tape. Using the VHB double sided tape, initially the non-adhesive Dacron tapes (blue tapes in Figure 5(a)) are incorporated into the wing to create "bags" where the carbon rods are inserted. After are incorporated the small carbon rods, with resort to the double sided VHB adhesive tape and covered with small strips of ripstop nylon to create the waving zone of the wing. In a final phase is used the Dacron adhesive tape (black tape shown in Figure 5(b)) to reinforce the fixing of the non-adhesive tapes. The wings are made separately and coupled to one another, at the end. On this connection, non-adhesive Dacron tape is used with the double sided VHB tape to create a "bag" in the center of the wing which functions as the backbone of a human being (see in Figure 5(c)). Figure 6(a) identifies the part that is inserted on this central pocket. This area also uses the adhesive Dacron tape to reinforce the connection between the wings. The same construction method is used on the tail. Four 1 mm diameter rods are used to ensure that the tail is always stretched in the vertical direction (from Figure 6(b) perspective) and another bent rod from one side to another, which is fixed in the adhesive double sided tape layers of main rods (located on the sides).

The use of adhesives also brings an advantage to this project. In the case of the wings and tail, a strong connection with some ductility between the rods and respective coupling pieces is required. For these connections is used SikaForce 7752 FRW L60. For the fixation of the bearings used at the transmission box its used Loctite 603, which ensures radial fixation between the contact parts.

The material used for the mechanical structural parts (in Figure 6(c)) is aluminium 5083, being processed though laser and water jet cutting and bending.

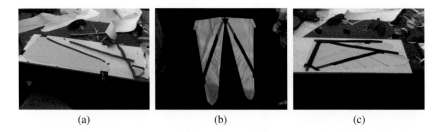

Fig. 5 Using the non-adhesive (a) and adhesive Dacron tape (b), and interconnected wings (c).

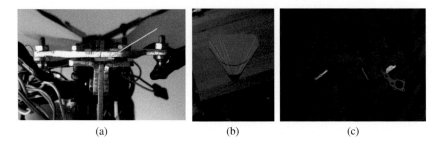

Fig. 6 Aluminium part which is inserted in the central bag of the wing (a), tail construction (b) and aluminium parts obtained by cutting and bending (c).

The mechanical locking arm needs to be in a fixed position, but also to rotate free. Using a flapped bushing, this part is placed in the desired position by use of washers, being obtained a permanent fixation by the screw which passes through the center of the bushing. The sensor involved in the locking system takes a few microseconds to send the signal. Therefore, its position should be adjustable to ensure that the wing stops immediately after the passage of the retaining bolt through the locking arm.

To control the ornithopter, a program was developed for the Arduino Nano to function as an intermediary between the receiver, the Hall effect sensor and the ESC (connected to the motor). The values coming from the RC controller are read by the Arduino controller, every 20 ms, to compute the action to perform. The PWM signal transmitted from the RC controller takes values between 1,1 and 1,7 ms (it was checked that the machine will display a maximum f of 5 Hz at 1,7 ms) and, in this case, when the fader is below 1,35 ms is indicated to the ESC to obtain a signal of 1,40 ms until the Hall effect sensor reaches a value other than 0, point in which the motor is commanded to stop. After obtaining the desired operation for the machine, it was fully assembled (Figure 7).

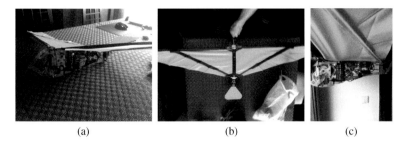

(a)   (b)   (c)

Fig. 7 Different views of the prototype's final version.

5 Tests Performed and Results Achieved

Initially was determined the total mass of the machine. Since the ripstop nylon density is 56 grams/m^2 and the wing area is 0,215 m^2, a mass of 12 grams was expected for this component. However, the wings and tail expected masses differ from the real mass. After being built, it was verified that it weighs about 45 grams. This is due to the incorporation of the 1 mm diameter rods at the waving zone, the Dacron tapes and the VHB double sided adhesive tape. This factor explains the discrepancy between the expected mass of 0,593 kg and the real value of 0,647 kg. However, the values determined for this weight, such as f and S_W, are not subject to large differences from the measurements in the implemented prototype.

For determining the generated F_T, the ornithopter was suspended from a rope, aligned with its CoM, with two ropes tied laterally to ensure that F_T is given only in the forward direction. Another rope was tied from the back of the machine to a dynamometer. This one uses a wooden rod to help the alignment between the dynamometer and the ornithopter. The values measured by the dynamometer fluctuate, since the ornithopter only develops F_T when the wings are moving downwards, but the recorded peak values are between 1 and 2 N, for a $f = 5$ Hz and an amplitude of 0,55 m. These values are close to the theoretical ones ($F_T = 1, 31 N$) computed through the expression given by Linton [10].

To understand whether the machine can exert the F_L necessary for sustained flight, was used a rope aligned with its CoM. This test consisted on throwing the ornithopter in a circular path (the center is the rope top attachment point), and check if it can practice a stabilized flight. Due to the stiff behaviour of the rope, tensions exerted by it eventually influence the stability of the ornithopter. Therefore, the experiment was repeated using an elastic rope, which does not guarantee a circular path and eventually also influences the flight forces. Measuring instruments were not used for F_L quantification but the experiment concluded that the machine could not exert the force necessary for its elevation.

6 Conclusions and Future Developments

This paper described the design and development of an ornithopter. Its construction demonstrated the existence of details not noticeable in the design phase. The mechanical behaviour of the structure behaves as expected. The servomotors transmit the forces necessary to stabilize the tail in a fixed position and the brushless motor ensures a $f = 5$ Hz. For the in-suspension test, it is possible to understand that the CoM lies on the expected point. The 3 mm diameter rods provide good flexural strength and ensure the torque transmission. The electronics works as expected and is well dimensioned. The limitation of this system is the transmission of the PWM signal between 1,1 and 1,7 ms. With a larger investment, it is possible to get RC controllers complying with an expected range of 1 to 2 ms. Thus it can be obtained a larger F_T and secure higher speed, propitious to increasing F_L.

Based on the tests performed, it is concluded that the expected value of F_T is within the experimental results, and that it depends on f. According to Linton, this information is confirmed, since he states that this force depends on f, amplitude and dimensions of the wing, while F_L depends on the flight speed, the wing dimensions and α. For this reason, an issue that can help in the design of an ornithopter is to construct wings with good surface area, always taking into account the values for σ and A_R. Thus, in those conditions, the necessary forces for the ornithopter to fly can be generated with lower rates of speed or f.

Acknowledgments This work is financed by the ERDF-European Regional Development Fund through the COMPETE Programme (operational programme for competitiveness) and by National Funds through the FCT-Fundação para a Ciência e a Tecnologia (Portuguese Foundation for Science and Technology) within project "FCOMP-01-0124-FEDER-037281".

References

1. Bullen, R.D., McKenzie, N.L.: Scaling bat wingbeat frequency and amplitude **205**(17), 2615–2626 (2002)
2. Center, M.R.: Pioneering flight of 'robo raven' is major breakthrough for micro air vehicles. http://www.robotics.umd.edu/news/news_story.php?id=7337 (last acessed on July 17, 2015)
3. Education, A.D.: Bird wings. http://www.acsedu.co.uk/Info/Environment/Environmental-Science/Bird-Wings.aspx (last acessed on July 17, 2015)
4. Fisher, M.: Bird flight deciphered. http://www.festo.com/cms/en_corp/11369.htm (last acessed on July 17, 2015)
5. Groups, R.: The ornithopter zone cybird p1 review. http://www.rcgroups.com/forums/showthread.php?t=822285 (last acessed on July 17, 2015)
6. Groups, R.: Park hawk. http://www.rcgroups.com/forums/showthread.php?t=189849 (last acessed on July 17, 2015)
7. Henderson, C.L.: Birds in Flight: The Art and Science of How Birds Fly. 1st edn. Voyageur Press, October 2008
8. Hub, S.L.: Wing loading. http://sciencelearn.org.nz/Contexts/Flight/Science-Ideas-and-Concepts/Wing-loading (last acessed on July 17, 2015)
9. Linton, J.O.: The physics of flight: I Fixed and rotating wings. Physics Education **42**, 351–357 (2007)

10. Linton, J.O.: The physics of flight: II Flapping wings. Physics Education **42**, 358–364 (2007)
11. North, J.: Hummingbird - muscling. In: Discussion of challenge question 4. http://www.learner. org/jnorth/spring2002/species/humm/Update032802.html (last acessed on July 17, 2015)
12. of Ornithology, T.C.L.: Four common wing shapes in birds. http://www.birds.cornell.edu/ education/kids/books/wingshapes (last acessed on July 17, 2015)
13. Pennycuick, C.: Wingbeat frequency of birds in steady cruising flight: new data and improved predictions. Journal of Experimental Biology **199**, 1613–1618 (1996)
14. Pennycuick, C.: Speeds and wingbeat frequencies of migrating birds compared with calculated benchmarks. Journal of Experimental Biology **204**, 3283–3294 (2001)
15. Stern, A.A., Kunz, T.H., Bhatt, S.S.: Seasonal wing loading and the ontogeny of flight in phyllostomus hastatus (chiroptera : phyllostomidae). Journal of Mammalogy **78**(4), 1199–1209 (1997)
16. Taylor, G.K., Nudds, R.L., Thomas, A.L.R.: Flying and swimming animals cruise at a strouhal number tuned for high power efficiency. Nature **425**, 707–711 (2003)
17. Tobalske, B.W.: Biomechanics of bird flight **210**(18), 3135–3146 (2007)
18. Zone, O.: Full history of ornithopters. http://www.ornithopter.org/history.full.shtml (last acessed on July 17, 2015)

UBRISTES: UAV-Based Building Rehabilitation with Visible and Thermal Infrared Remote Sensing

Adrian Carrio, Jesús Pestana, Jose-Luis Sanchez-Lopez,
Ramon Suarez-Fernandez, Pascual Campoy, Ricardo Tendero,
María García-De-Viedma, Beatriz González-Rodrigo,
Javier Bonatti, Juan Gregorio Rejas-Ayuga,
Rubén Martínez-Marín and Miguel Marchamalo-Sacristán

Abstract Building inspection is a critical issue for designing rehabilitation projects, which are recently gaining importance for environmental and energy efficiency reasons. Image sensors on-board unmanned aerial vehicles are a powerful tool for building inspection, given the diversity and complexity of façades and materials, and mainly, their vertical disposition. The UBRISTES (UAV-based Building Rehabilitation with vISible and ThErmal infrared remote Sensing) system is proposed as an effective solution for façade inspection in urban areas, validating a method for the simultaneous acquisition of visible and thermal aerial imaging applied to the detection of the main types of façade anomalies/pathologies, and showcasing its possibilities using a first principles analysis. Two public buildings have been considered for evaluating the proposed system. UBRISTES is ready to use in building

A. Carrio(✉) · J. Pestana · J.-L. Sanchez-Lopez · R. Suarez-Fernandez · P. Campoy
Computer Vision Group, Centre for Automation and Robotics, CSIC-UPM, Madrid, Spain
e-mail: {adrian.carrio,jesus.pestana,jl.sanchez}@upm.es
http://www.vision4uav.eu

J. Pestana
Aerial Vision Group, University of Technology (TU Graz), Graz, Austria

P. Campoy
Robotics Institute, University of Technology (TU Delft), Delft, The Netherlands

R. Tendero · M. García-De-Viedma
ETS de Edificación, Technical University of Madrid (UPM), Madrid, Spain

B. González-Rodrigo
ETS de Ingeniería Civil, Technical University of Madrid (UPM), Madrid, Spain

J. Bonatti
Centro de Investigación En Ciencias Atómicas, Nucleares Y Moleculares (CICANUM),
University of Costa Rica (UCR), San José, Costa Rica

J.G. Rejas-Ayuga · R. Martínez-Marín · M. Marchamalo-Sacristán
Laboratorio de Topografía Y Geomática, ETSI de Caminos, Canales Y Puertos,
Technical University of Madrid (UPM), Madrid, Spain

© Springer International Publishing Switzerland 2016   245
L.P. Reis et al. (eds.), *Robot 2015: Second Iberian Robotics Conference*,
Advances in Intelligent Systems and Computing 417,
DOI: 10.1007/978-3-319-27146-0_19

inspection and has been proved as a useful tool in the design of rehabilitation projects for inaccessible, complex building structures in the context of energy efficiency.

Keywords Building inspection · Computer vision · Façade pathologies · Thermal loss · UAV

1 Introduction

Building rehabilitation is a promising discipline in Architecture and Civil Engineering, as many buildings are approaching the end of their life cycles while the improvement of energy efficiency in urban areas is triggering building renovation and rehabilitation for economic and environmental reasons, and also as required by European Directives, such as EPBD (Energy Performance of Buildings Directive) 2002/91/EC [5] and EPBD 2010/31/UE [7].

Rehabilitation is a complex field of work, as there is no general solution for all buildings, but rather technical solutions that should be chosen for each building according to its specificities [4]. Energy audits in urban areas get more difficult as the architecture gets denser and more complex. Even in the same building, different energy performances may be observed in different parts of it, depending mainly on its materials and façade typologies [9].

Building inspection is a key process in the design of rehabilitation projects. In particular, building thermography is a method of indicating and representing the temperature distribution over a part of the surface of a building envelope [3]. Infrared thermography can be used in various fields of building rehabilitation to assess construction conditions and detect anomalies (i.e. humidity, infiltrations, thermal bridges, etc.). The sensed thermal infrared (TIR) radiation is a function of the surface temperature, the characteristics of the surface, the ambient conditions, and the sensor itself. Therefore, obtaining temperature measurements involves using the information contained in the thermal images, corresponding to a digital signal proportional to the intensity of the sensed radiation, together with a number of parameters including surface emissivity, distance between sensor and surface, air temperature, air humidity, and others. Furthermore, different effects should be considered when analysing thermal images for building inspection. For example, heat flow can lead to either warming up or cooling the building surfaces due to different mechanisms such as conductive differences, thermal bridges and air filtrations. Also surface humidity often reduces the surface temperatures due to evaporative cooling [23].

The main disadvantage when capturing thermal images of building façades, comes from the lack of accessibility and shooting angle necessary to limit the effect of reflected temperature on the captured images (recommended angle of incidence $\leq$ 30°). In urban environments it is possible to collect images correctly from the first floors at street level, but in tall buildings, the effect of reflected temperature (TRFL) increases upwards as a function of the shooting angle, and it becomes difficult to account only for the thermal energy radiated by the surfaces.

Nowadays the opportunity arises to exploit the complementarity of the information provided by TIR images captured at street level and those captured from Unmanned Aerial Vehicles (UAVs). TIR cameras have been developed and have expanded rapidly in the last years due to the development of digital imaging technology and bolometers, bringing low cost, lightweight, high-quality thermal cameras into the market. Microbolometers are the dominant uncooled IR detector technology with more than 95% of the market in 2010 [19]. These recent developments in thermal image sensors widen the possibilities for UAV applications in remote sensing. Also the development of UAVs in the last decade for many different applications has powered the use of onboard TIR image sensors for building inspection purposes, proposed as a mid-term challenge by González-Aguilera et al. [10]. Eschmann et al. [6] also included this kind of design in the framework of promising developments in the non-destructive testing (NDT) domain. The integration of these sensors allows for the inspection of inaccessible places, ensuring a close enough and orthogonal camera position in an inexpensive way. Similarly, it is possible to take pictures of structures or façades from above, by changing the orientation of the on-board camera(s). For these reasons, remote sensing analysis in urban areas for monitoring and management purposes presents challenging problems, mainly due to the diversity and complexity of materials and façade typologies [18].

The present paper presents the UBRISTES system, a UAV-based solution with an integrated TIR sensor for pre-rehabilitation building inspection, validating the usage of visible (VIS) and TIR UAV monitoring to detect construction pathologies in façades and demonstrating its potential.

The remainder of this paper is organized as follows. Firstly, a review of the related works is presented. Secondly, a description of the system including the selected UAV platform, the sensors and how they are integrated is provided. Thirdly, a methodology for evaluating the system is presented. Fourthly, the evaluation results are discussed. Finally, conclusions are presented.

2 Related Works

Research based on performing thermographies from UAVs has been sparse in the past, but is recently gaining impetus. Martinez-de-Dios et al. [14] proposed the use of UAVs for passive building thermography and to detect heat losses originating from windows. Later, the work by Iwaszczuk et al. [11] focused on matching VIS and infrared airborne imagery to an existing building 3D model. Eschmann et al [6] were able to obtain highly detailed mosaics of building façades using an octocopter UAV. Although they mentioned the possibility of using their algorithms to acquire thermal images, they only showed results using visual imagery. In this direction, but using ground-based imagery, González-Aguilera et al. [10] applied a modern 3D reconstruction image processing pipeline to thermographies to obtain a thermographic model of a façade. Recent work by Yahyanejad et al. [24] proposed novel image feature descriptors to match images in different spectra, and showed their performance for image mosaicing using visual and thermal imagery taken from multiple quadrotors.

A description of the main methodologies to obtain thermographies from an area of interest including the different types of surveys, characteristics and related bibliography has been published by Fox et al. [8].

Another field that is getting increased attention is the georeferenciation of imagery taken from UAVs. In this direction, Lagüela et al. [12] developed a methodology for the automatic extraction of building geometry directly from aerial oblique thermographic imagery. The obtained geometric and thermal 3D models are usually inserted into a complete Geographic Information System (GIS). The geo-referenciation of imagery acquired from UAVs is an active field of research but it is often focused on visual imagery. For instance, Rumpler et al. [21] and Maurer et al. [15] have focused on matching point clouds with publicly available data sources. In this manner, the acquired images are geo-referenced. In more recent research, Rumpler et al. [22] have also explored the use fiducial markers to increase the automation and the quality of the obtained 3D reconstructions.

3   UBRISTES System Description

3.1   UAV Platform

The proposed UAV platform is an AscTec Pelican quadrotor [1], equipped with an AscTec Atomboard (Figure 2) as on-board computer. This quadrotor is commonly used in research laboratories because of its high payload capacity (up to 650 g), flight performance and adequate structure design that allows to integrate a number of additional sensors. Furthermore, its small size (651 x 651 x 188 mm) makes it suitable and less dangerous for urban operations, as it can fly closer to façades and in narrow spaces.

3.2   Sensors

Thermoteknix MicroCAM 384M (Fig. 1a) is a passive long-wave infrared (LWIR) sensor with a spectral range of 8 to 12 μm. 7.5 mm optical lens were used, providing a field of view (FOV) of 65° x 51°. The thermal sensitivity of this camera is 80mK, and it can operate at a maximum frame rate of 60 frames per second. Thermal images with a resolution of 384 x 288 pixels can be acquired, with the TIR radiation being codified into an analog video output. The MicroCAM 384M TIR image sensor has been integrated on-board the AscTec Pelican, as shown in Fig. 2.

Street level images were taken with a calibrated Flir B335 LWIR camera (Fig. 1b), in order to obtain accurate temperature measurements. This hand-held camera has a spectral range of 7.5 to 13 μm. With a field of view (FOV) of 25° x 19°, it features a 320 x 240 pixel resolution and a thermal sensitivity of 50mK. The scene temperature range supported is $-20°C$ to $120°C$ while temperature measurements have an accuracy of $\pm 2°C$.

(a) Thermoteknix MicroCAM 384M.   (b) Flir B335 LWIR camera.

Fig. 1 Thermal sensors integrated

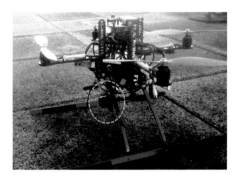

Fig. 2 AscTec Pelican quadrotor with the onboard MicroCAM 384M TIR image sensor highlighted in green.

Street level images in the VIS spectrum were captured as well using an IDS UI-3250CP USB 3.0 camera, with a sensor size of 1/1.8" and a maximum image resolution of 1600x1200 pixels. A 2.9 mm focal length camera lens was used.

3.3   UAV Integration Details

The data acquisition software was implemented using the Robot Operating System middleware [16]. The software set-up allows to configure the camera on runtime. TIR images from the MicroCAM 384M can be acquired and saved into the on-board computer at 30 fps and the battery on-board allows a flight time of approximately 10 minutes, enough for the inspection of most building façades.

4 Evaluation Methodology

Two buildings, both academic centres in the Technical University of Madrid, were selected for evaluating the UBRISTES system: ETSI Caminos (Civil Engineering) main building and ETSI Industriales Automation and Robotics building. ETSI Caminos main building was built in 1963 mainly with concrete (structure and façade) and carries linear narrow lines of windows with poor insulation. ETSI Industriales Automation and Robotics building was a pre-existent building from the 1930's with thick brick façade. In the beginning of the XXI Century, a new insulated roof was built and new insulated windows were installed. The façades selected for monitoring were north facing (ETSI Caminos) and east facing (ETSI Industriales) to avoid noise from solar radiation.

During our evaluation, the quadrotor was remotely piloted by a human operator. Three types of potential façade pathologies were monitored: dampness, degradation and thermal bridges. These pathologies could not be automatically detected in the collected imagery due to the enormous complexity of the problem. However, the mentioned pathologies were previously estimated via visual assessment and evaluated through image analysis by experts. Indoor and outdoor temperatures were continuously monitored during image capture.

A room was warmed up during the night in order to enhance the temperature changes related to thermal bridges, since only drastic changes in the surface temperature may allow the detection of pathologies in these areas.

4.1 Image Analysis

Outgoing surface radiation is formed by the portion emitted by an object (which depends on its temperature), the portion reflected from ambient sources and the emission from the atmosphere. Only the radiation emitted by an object provides information about its temperature, and the rest of the radiation must be filtered to obtain correct temperature measurements.

Data processing techniques have been used to retrieve information from thermal data. They usually include geometric correction, radiometric correction, and a number of analysis (image enhancement, image arithmetics and statistical analysis). The choice of techniques depends on the image quality and the required output. The main function of the different processes applied to the acquired data is to produce a single perfectly co-recorded, multi-sourced file. This is aimed at undertaking a posterior analysis for the purpose of spatially correlating the elements of monitoring interest (walls, windows, structures, materials, etc.) with the superficial temperature at which they were recorded.

Flir B335 images were processed with Flir's software in order to obtain the radiometric (temperature) information, including the ambient temperature and humidity during the tests. The emissivity and reflected temperature were also inputs in the image software.

MicroCAM 384M images were analyzed using the image processing package ENVI from ITT [2] (also available from other packages as ERDAS-Imagine or PCI Geomatics). No radiometric correction was applied to the MicroCAM images. For geometric correction we selected ground control points from the images captured with the Flir B335 camera. The thin plate spline method (a method for geometric correction and geo-reference of images based on a set of interpolating bi-dimensional polynomials) was applied to a number of ground control points in the original image.

5 Results and Discussion

In this section the results of the proposed methodology are presented. Firstly, a description of two radiometric effects observed in the MicroCAM 384M TIR camera is presented, together with the proposed solutions. Then an example of image analysis for active humidity detection is discussed. Finally, examples of the assessment of thermal losses in different types of façades and slabs are presented.

5.1 Radiometric Effects

In the MicroCAM 384M TIR camera images two radiometric effects were observed. One consisted of an unusually brighter region in the image edges. The second one was a local defocussing effect in some image regions. Both effects were caused by problems in the instrument during the experiments, and require specific solutions. For the brighter band, a normalization function has been computed as the mean of the ND (radiometric or digital value) for each image column in the area of study and the effects have been removed by applying a Minimum Noise Fraction (MNF) transform to the output multi-source file. The defocussing has been minimized using a standard high pass filter. The same strategy has been followed to build a single image per band, mosaicing the individual frames. No atmospheric correction was applied. As a final step, a textural analysis was applied using state-of-art filters designed for enhancing image structures (i.e. Sobel filter).

5.2 Image Analysis for Humidity Detection

We carried out a pattern recognition analysis. With the Flir and the MicroCAM thermal channels we calculated a thermal index [17], profiting from the separability between the spectral sensitivities of both sensors. We generated image convolutions using a median filter, which were used afterwards to make a ratio between the 8 μm

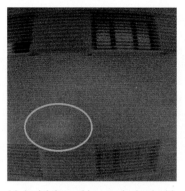 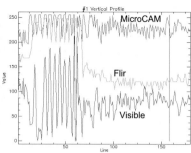

(a) Aerial thermal image obtained with the UBRISTES system and enhanced with morphological operations for visualization purposes. The green ellipse indicates a humidity area in the façade.

(b) Multisource vertical profile of the humidity area: red (MicroCAM), Green (Flir) and Blue (Visible).

Fig. 3 Analysis of a humidity area in a façade.

and 12 μm wavelengths (spectral range), weight corrected by the ratio between each channel's gain. We established thresholds on the resultant variable for highlighting detected pixels as possible anomalies, as shown in Figure 3.

5.3 Active Humidity Detection Using VIS-TIR Imagery

The UBRISTES system allowed for acquiring both buildings' façades in the VIS and TIR spectra. Image filtering, analysis and visual assessment of the images was carried out, identifying potential pathologies in the VIS and TIR channels. These potential pathologies were validated with a second fieldwork visit to both buildings.

Figure 4 presents an example of combined humidity detection, using VIS and TIR images. The combined VIS and TIR monitoring allows for the identification of humidity-induced pathologies. Two zones were detected in visual inspection (Fig. 4a), whereas thermal images showed that one presented a colder area (Fig. 4b), probably caused by the presence of humidities, while the other did not (Fig. 4c). Profiles show the normalized pixel intensity in the TIR channel across both humidities (Figs. 4d and 4e).

5.4 Thermal Loss Assessment Using VIS-TIR Imagery

Figure 5 shows the evaluation of thermal losses in two cases, one related to a poorly-insulated window and the second related to an air-conditioned room with low external insulation. Air conditioning conductions show important thermal losses. Similar thermal losses were identified with hand-held thermography in the Instituto Eduardo

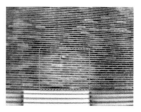

(a) Visible image indicating the presence of two possible humidity areas in the façade, highlighted in green.

(b) Aerial thermal image captured with the MicroCAM 384M showing an active humidity, highlighted in green.

(c) Aerial thermal image captured with the Micro-CAM 384M showing a non-active humidity, highlighted in green.

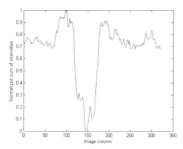
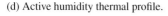
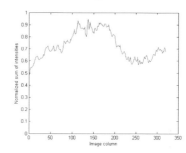

(d) Active humidity thermal profile.

(e) Non-active humidity thermal profile.

Fig. 4 Humidity analysis in the ETSI Industriales building (UPM).

Torroja (CSIC) building, built in 1953 with a brick façade and poorly-insulated windows [13].

Figure 6 shows the detection of thermal losses in vertical front walls in façades. The aerial image captured by the UBRISTES system (Fig. 6c) shows a vertical loss, highlighted in red, in the junction between the front wall and the concrete façade.

Figures 7b and 7c show the presence of thermal anomalies related to slabs in a façade, highlighted in green. Junctions of slabs and façades are usually vulnerable to thermal losses, as façade insulation may be lost if the construction process is not accurate.

ETSI Caminos building images (concrete façade) show a constant temperature pattern over the whole surface, not related with thermal bridges or other pathologies, but with façade texture, conditioned by each of the concrete finishing surfaces depending on the form-work tables with which it was made and the lack of flatness. As a result, a distortion is appreciated in a continuous surface due to small changes in the orientation of each form-work table and hence small temperature changes in the reflected IR are registered by the sensor.

Analyzed buildings presented typical pathologies common to other 20th century buildings in Spain, as the IETCC building [13]. Furthermore, as whole districts in our cities built in the postwar era (1945 - 1965) are approaching the end-of-life of

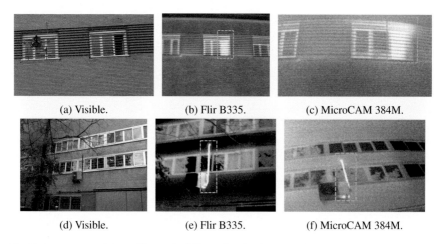

(a) Visible.   (b) Flir B335.   (c) MicroCAM 384M.

(d) Visible.   (e) Flir B335.   (f) MicroCAM 384M.

Fig. 5 Assessment of thermal losses in different façade elements. A poorly insulated window and an air-conditioning conduction, highlighted in green.

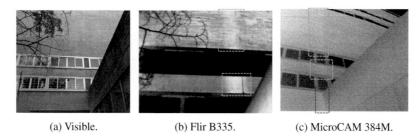

(a) Visible.   (b) Flir B335.   (c) MicroCAM 384M.

Fig. 6 Assessment of thermal losses, highlighted in green, in a vertical front wall. A thermal loss in a vertical junction is also shown in the MicroCAM 384M image, highlighted in red.

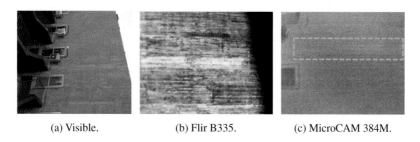

(a) Visible.   (b) Flir B335.   (c) MicroCAM 384M.

Fig. 7 Assessment of a thermal loss, highlighted in green, in a façade slab.

their buildings and infrastructures, inspection for rehabilitation should be carried out at neighbourhood level [20]. This will require the combination of monitoring techniques, including satellite, airborne and UAV remote sensing.

6   Conclusions

The UBRISTES system is a powerful tool for obtaining close aerial imagery buildings under inspection. The acquired imagery effectively fills the gap between high altitude imagery and ground-based imagery.

The evaluation experiments, in which ground-level VIS images, and aerial and ground-level TIR images have been compared, confirm that combined VIS-TIR UAV monitoring can enable logging and tracing façades spectral data with high spatial resolution. The exploratory data analysis, which has focused on the study of space-thermal correlation between the surfaces and materials, can help in identifying anomalies and pathologies on façades.

Multispectral UAV monitoring, with both VIS and TIR image sensors, allows to reach inaccessible parts of the buildings, obtaining global homogeneous models that can be analysed off-line with enough time and detail. These global models ensure objective diagnoses even in most complex building and city environments. Measurement of areas and perimeters of the affected areas is also possible, with an accuracy that depends on the image resolution and geometric correction.

The effects of reflected temperature observed during the experiments also demonstrate that these studies are more reliable in untextured, homogeneous surfaces.

The UBRISTES system is ready to use in building inspection and has been proved as a useful tool in the design of rehabilitation projects for inaccesible, complex building structures in the context of energy efficiency.

Acknowledgments The authors would like to thank the Consejo Superior de Investigaciones Científicas (CSIC) of Spain for the JAE-Predoctoral scholarships of two of the authors and their research stays, and the Spanish Ministry of Science MICYT DPI2010-20751-C02-01 for project funding, as well as the Spanish Ministry for Education, Culture and Sports for funding the international research stay of one of the authors.

References

1. Ascending Techonologies. http://www.asctec.de/en/, accessed: Feb 16, 2015
2. Exelis Visual Information Solutions. http://www.exelisvis.com/, accessed: February 16, 2015
3. AENOR 2000. EN-13187- Thermal performance of buildings. Qualitative detection of thermal irregularities in building envelopes. Infrared method (ISO 6781:1983 modified). AENOR - Spanish Association for Standardisation and Certification, 2000
4. Charlot-Valdieu, C., Outrequin, P.: An approach and a tool for setting sustainable energy retrofitting strategies referring to the 2010 EP. Informes de la Construcción **63**(Extra) 2011
5. EC European Commission et al: Council Directive 2002/91/EC of 16 December 2002 on the energy performance of buildings. Official Journal of the European Communities **1**, 65–71 (2003)

6. Eschmann, C., Kuo, C.M., Kuo, C.H., Boller, C.: High-resolution multisensor infrastructure inspection with unmanned aircraft systems. ISPRS-International Archives of the Photogrammetry, Remote Sensing and Spatial Information Sciences **1**(2), 125–129 (2013)
7. Parliament European Union and of the Council. Directive 2010/31/EU on the energy performance of buildings (2010)
8. Fox, M., Coley, D., Goodhew, S., de Wilde, P.: Thermography methodologies for detecting energy related building defects. Renewable and Sustainable Energy Reviews **40**, 296–310 (2014)
9. García-Navarro, J., González-Díaz, M.J., Valdivieso, M.: Estudio PRECOST&E: evaluación de los costes constructivos y consumos energéticos derivados de la calificación energética en un edificio de viviendas situado en Madrid. Informes de la Construcción **66**(535), e026 (2014)
10. González-Aguilera, D., Lagüela, S., Rodríguez-Gonzálvez, P., Hernández-López, D.: Image-based thermographic modeling for assessing energy efficiency of buildings façades. Energy and Buildings **65**, 29–36 (2013)
11. Iwaszczuk, D., Hoegner, L., Stilla, U.: Matching of 3D building models with IR images for texture extraction. In Urban Remote Sensing Event (JURSE), 2011 Joint, pp. 25–28. IEEE (2011)
12. Lagüela, S., Díaz-Vilariño, L., Roca, D., Armesto, J.: Aerial oblique thermographic imagery for the generation of building 3D models to complement Geographic Information Systems. The 12th International Conference on Quantitative Infrared Thermography **25** (2014)
13. Martín-Consuegra, F., Oteiza, I., Alonso, C., Cuerdo-Vilches, T., Frutos, B.: Análisis y propuesta de mejoras para la eficiencia energética del edificio principal del Instituto cc Eduardo Torroja-CSIC. Informes de la Construcción **66**(536), e043 (2014)
14. Martinez-De-Dios, J.R., Ollero, A.: Automatic detection of windows thermal heat losses in buildings using UAVs. In: World Automation Congress, WAC 2006, pp. 1–6. IEEE (2006)
15. Maurer, M., Rumpler, M., Wendel, A., Hoppe, C., Irschara, A., Bischof, H.: Geo-referenced 3d reconstruction: fusing public geographic data and aerial imagery. In: 2012 IEEE International Conference on Robotics and Automation (ICRA), pp 3557–3558. IEEE (2012)
16. Quigley, M., Conley, K., Gerkey, B., Faust, J., Foote, T., Leibs, J., Wheeler, R., Ng, A.Y.: ROS: an open-source robot operating system. In: ICRA workshop on open source software, vol. 3, p. 5 (2009)
17. Rejas, J.G., Burillo, F., López, R., Cano, M.A., Sáiz, M.E., Farjas, M., Mostaza, T., Zancajo, J.J.: Teledetección pasiva y activa en arqueología. Caso de estudio de la ciudad celtíbera de Segeda. In: XIII Congreso de la Asociación Española de Teledetección. Calatayud, Spain. September, pp 23–26 (2009)
18. Rejas, J.G., Martínez-Marín, R., Malpica, J.A.: Hyperspectral remote sensing application for semi-urban areas monitoring. In: Urban Remote Sensing Joint Event 2007, pp. 1–5, April 2007
19. Rogalski, A.: History of infrared detectors. Opto-Electronics Review **20**(3), 279–308 (2012)
20. Rubio del Val, J.: Rehabilitación Urbana en España (1989–2010). Barreras actuales y sugerencias para su eliminación. Informes de la Construcción **63**, 5–20 (2011)
21. Rumpler, M., Irschara, A., Wendel, A., Bischof, H.: Rapid 3d city model approximation from publicly available geographic data sources and georeferenced aerial images. In: Computer vision winter workshop (CVWW) (2012)
22. Rumplera, M., Daftryab, S., Tscharfc, A., Prettenthalera, R., Hoppea, C., Mayerc, G., Bischofa, H.: Automated End-to-End Workflow for Precise and Geo-accurate Reconstructions using Fiducial Markers. ISPRS Annals of Photogrammetry, Remote Sensing and Spatial Information Sciences **1**, 135–142 (2014)
23. Vollmer, M., Möllmann, K.-P.: Infrared thermal imaging: fundamentals, research and applications. John Wiley & Sons (2010)
24. Yahyanejad, S., Rinner, B.: A fast and mobile system for registration of low-altitude visual and thermal aerial images using multiple small-scale UAVs. ISPRS Journal of Photogrammetry and Remote Sensing (2014)

An Adaptive Multi-resolution State Lattice Approach for Motion Planning with Uncertainty

A. González-Sieira, Manuel Mucientes and Alberto Bugarín

Abstract In this paper we present a reliable motion planner that takes into account the kinematic restrictions, the shape of the robot and the motion uncertainty along the path. Our approach is based on a state lattice that predicts the uncertainty along the paths and obtains the one which minimizes both the probability of collision and the cost. The uncertainty model takes into account the stochasticity in motion and observations and the corrective effect of using a Linear Quadratic Gaussian controller. Moreover, we introduce an adaptive multi-resolution lattice that selects the most adequate resolution for each area of the map based on its complexity. Experimental results, for several environments and robot shapes, show the reliability of the planner and the effectiveness of the multi-resolution approach for decreasing the complexity of the search.

Keywords Motion planning · State lattice · Multi-resolution

1 Introduction

Motion plans must consider the uncertainty due to the inaccuracy of the motion model and the measurements. As the uncertainty depends on the current state and control commands, different paths will have different uncertainties. Safety and accuracy are critical issues, specially to guarantee the integrity of the robot and to ensure

A. González-Sieira(✉) · M. Mucientes · A. Bugarín
Centro de Investigación En Tecnoloxías da Información (CiTIUS),
Universidade de Santiago de Compostela, Santiago de Compostela, Spain
e-mail: {adrian.gonzalez,manuel.mucientes,alberto.bugarin.diz}@usc.es

A. Bugarín—This work was supported by the Spanish Ministry of Economy and Competitiveness under projects TIN2011-22935, TIN2011-29827-C02-02 and TIN2014-56633-C3-1-R, and the Galician Ministry of Education under the projects EM2014/012 and CN2012/151. A. González-Sieira is supported by a FPU grant (ref. AP2012-5712) from the Spanish Ministry of Education, Culture and Sports.

© Springer International Publishing Switzerland 2016
L.P. Reis et al. (eds.), *Robot 2015: Second Iberian Robotics Conference*,
Advances in Intelligent Systems and Computing 417,
DOI: 10.1007/978-3-319-27146-0_20

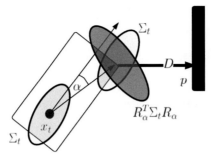

Fig. 1 To estimate accurately the probability of collision of the robot with an obstacle it is required to know the PDF at the point of the robot shape closest to the obstacle (p). The simple translation of the covariance (Σ_t) from the center of rotation (x_t) to p generates a PDF completely different from the real one —$R_\alpha^T \Sigma_t R_\alpha$, in blue. More details are given in Sec. 3.3.

the successful completion of the task. Stochastic sampling methods —like probabilistic roadmaps (PRM) [6] and rapidly-exploring randomized trees (RRT) [7]—, and deterministic sampling techniques —like state lattices [13]—, have proven to be successful approaches in the field of motion planning. With the aim to reduce the search space and obtain solutions faster, several state lattice approaches rely on multi-resolution planning techniques [12]. These proposals adjust the resolution of the state lattice, using for example high resolutions near the start and goal states, and lower resolutions in other areas of the environment.

Most planners assume, that a full knowledge about the states and possible actions is available, leaving the uncertainty management to be solved by feedback controllers that execute the paths. However, in this way planners do not take into account hazardous states, e.g. when a small deviation could cause a collision. Also, the robustness of a solution lies exclusively in the ability of the controller to follow the path with precision. To solve these drawbacks, several authors have proposed different solutions for stochastic sampling motion planners [2][15]. These approaches model the uncertainty at planning time with *a-priori* probability density functions (PDFs) of the states along the paths generated by the planner. These distributions allow to evaluate the candidate paths and select the one that minimizes the probability of collision.

In this paper we present a motion planner based on a state lattice that models the uncertainty with realistic PDFs that take into account the shape of the robot, and the controller that executes the paths. Also, the proposal uses a novel multi-resolution technique to reduce the complexity of the search. The planner is able to obtain safe and optimal paths, i.e., it maximizes the probability of success while minimizes the length of the path —its traversal time. The probability of collision is evaluated in a realistic way, taking into account the uncertainty in the state and the real shape of the robot, instead of making an approximation that underestimates the probability of collision (Fig. 1). Our approach handles stochasticity in both controls and observations and takes into account the corrective effect of using a Linear Quadratic Gaussian (LQG) controller. The prediction of the PDFs along the paths is done in a realistic way without making assumptions about the measurements —i.e. we do not use maximum likelihood observations. We introduce an adaptive approach for multi-resolution planning that groups the actions used to connect the lattice according to their length and maneuvering complexity. As each group of actions allows different kind of motions, each one defines a resolution for the state lattice —where the connectivity and the distances between neighbors change. For each region of the map, our proposal estimates the maneuvering complexity needed to avoid the obstacles and drive the robot to the goal. Depending on this estimation, it automatically selects the minimum resolution that can be used in each region and, therefore, the planner can obtain solutions faster and without a significant loss in the optimality of the path.

2   Related Work

Successful approaches in the field of motion planning are based in the combination of a search algorithm and random sampling —probabilistic roadmaps (PRM) [6] and rapidly-exploring randomized trees (RRT) [7]— or deterministic sampling techniques —state lattices [13]. Although random sampling methods provide a very efficient way for the exploration of the state space, the advantage of state lattices is the possibility to generate offline the set of control actions, instead of using the motion model to connect the random samples in planning time.

Motion planning with uncertainty has raised an increased attention in the last years. Stochasticity in motions arises from different sources: noise in the controls and measurements, partial information about the state and uncertainty in the map. Some approaches only take into account uncertainty in controls as [10], which presents a planner that avoids rough terrain to reduce the differences between the predictions and the result of the controls. In [1] the probability of collision avoidance is maximized using a Markov Decision Process (MDP) approach. Sensor uncertainty can be managed using Partially Observable MPDs (POMPD), as in [4], but this has scalability issues [11].

Van den Berg et. al. [15] propose an algorithm (LQG-MP) based on RRT which takes into account both the uncertainty associated to the control and the measurements and minimizes the probability of collision, but due to the stochasticity of RRT the generated paths may be non smooth. This is addressed via Kalman smoothing but it requires additional processing after planning. It also presents an approach to use LQG-MP in a roadmap with a search algorithm, but with the limitation of splitting the LQG control in the individual trajectories instead of using the same for the whole path. Bry et. al. [2] present a similar approach that uses a user-defined threshold of tolerance to the risk to define a chance-constrained search. Both methods use a bounding disk and a point to represent the robot, respectively. The safety and feasibility of the solutions obtained by the planners may be affected using these approximations, specially in cases where the shape of the robot is significantly different.

The application of multi-resolution planning techniques over state lattices was introduced by Pivtoraiko et. al. [12]. The approach uses a low resolution to represent the state space, except in three defined regions of high resolution: the initial and goal regions, and the area centered on the robot. This significantly increases the search performance, but as the robot moves along the path the resolution of the lattice changes, requiring to replan to keep the solution updated.

3 Planning with Uncertainty

Our motion planner obtains the path that minimizes the probability of collision and the time needed to reach the goal. The state space $\mathcal{X}$ is sampled using a state lattice, a deterministic and regular sampling that obtains a set of lattice states $x_t \in \mathcal{X}^{lat} \subset \mathcal{X}$. These states are connected by a finite set of actions $\mathcal{U}$, also called motion primitives, which are extracted from the vehicle dynamics. $\mathcal{U}$ is generated offline, and in this paper we have used an iterative optimization technique based on Newton-Raphson to obtain that set [3].

The motion (f) and measurement (h) models are assumed to be linear or locally linearizable:

$$\begin{aligned} x_t &= f(x_{t-1}, u_t, m_t), \quad m_t \sim (0, M_t) \\ z_t &= h(x_t, n_t), \quad n_t \sim (0, N_t) \end{aligned} \tag{1}$$

where $x_t \in \mathcal{X} = \mathcal{X}^{free} \cup \mathcal{X}^{obs}$ is a state of the robot, $u_t \in \mathcal{U}$ is a control, z_t is the measurement, m_t and n_t are the random motion and observation noises, and $\mathcal{X}^{obs}$ is the set of states in which there are obstacles.

Each action belonging to $\mathcal{U}$ is composed by a set of control commands ($u^{a:b}$) that drive the vehicle from $x^a \in \mathcal{X}^{lat}$ to $x^b \in \mathcal{X}^{lat}$ in a time $t^{a:b}$:

$$\begin{aligned} u^{a:b} &= \left(u_1^{a:b}, u_2^{a:b}, ..., u_{t^{a:b}}^{a:b}\right) \\ x^{a:b} &= \left(x_1^{a:b}, x_2^{a:b}, ..., x_{t^{a:b}}^{a:b}\right) \\ x_1^{a:b} &= x^a, x_{t^{a:b}}^{a:b} = x^b \\ x_t^{a:b} &= f(x_{t-1}^{a:b}, u_t^{a:b}, 0), \forall t \in [1, t^{a:b}] \end{aligned} \tag{2}$$

where the intermediate states ($x^{a:b}$) are obtained from the motion model (f) in absence of process noise.

Because of the regularity of the lattice, the primitives are position-independent, and the same commands can be used to connect every pair of states in $\mathcal{X}^{lat}$ equally arranged. Also, as these actions are extracted from the vehicle dynamics, it is clear that the lattice observes the kinematic restrictions. It is straightforward to translate this representation to a directed weighted graph, where the nodes are the states in $\mathcal{X}^{lat}$ and the arcs are the motion primitives $\mathcal{U}$. The solution to a planning problem is obtained executing a discrete search algorithm over the graph structure.

Algorithm 1 Path planning algorithm main loop

1: **while** solution not found **do**
2:    select arg $\min_{x^a \in \mathcal{X}^{lat}} (cost(x^{0:a}) + \epsilon \cdot e(x^a))$
3:    **for all** $x^b \in successors(x^a)$ **do**
4:       $x^{a:b} = uncertaintyPrediction(x^a, x^b, u^{a:b}, t^{a:b})$
5:       $cost(x^{0:b}) = cost(x^{0:a}) + cost(x^{a:b})$
6:    **end for**
7: **end while**

3.1 Search Algorithm

Our proposal relies on the search algorithm Anytime Dynamic A* (AD*)[9]*. The algorithm is able to obtain sub-optimal bounded solutions varying an heuristic inflation parameter (ϵ). Although AD* typically performs a backwards search, our proposal uses a forward variant to allow the estimation of the PDFs along the generated paths.

An overall approach of the operations done by the motion planning algorithm is summarized in Alg. 1. Iteratively, the most promising state (x^a) is selected. This is the state that minimizes the aggregated cost from the initial state —$cost(x^{0:a})$—, and the estimated cost to goal —$\epsilon \cdot e(x^a)$— scaled by ϵ. The heuristic function —$e(x^a)$— is calculated using the mean of the PDFs and combines two values: the cost of the path with kinematic constraints under free space —stored in a Heuristic Look-Up Table [5]— and the cost of the path regardless the motion model but with information about the environment —calculated in a 8-connected 2D grid [8].

Then, the motion planner (Alg. 1) propagates the uncertainty from x^a to its successors according to Alg. 2 (Sec. 3.2). Finally, the outgoing actions of x^a are evaluated to calculate the cost of the path to each successor (Sec. 3.3).

The planner estimates the PDFs for each state along the generated paths. As both the noises and the prior probability follow a Gaussian distribution, and the motion and measurement models can be locally approximated with linear functions, the PDFs are efficiently estimated using EKF-based methods.

The prediction of the PDFs at planning time using an EKF assumes that the observations are those with the maximum likelihood. This underestimates the uncertainty and also ignores the corrective effect of using a feedback controller. To solve these drawbacks, our planner uses the EKF-based method presented in [2]. This method poses that the system will execute the paths using a LQR controller, so the true state and the estimated one are dependent reciprocally. Combining the LQR control policy and the EKF allows to analyze their joint evolution as functions of each other, so the prediction of the PDFs can be done without making assumptions about the observations in planning time.

---

* We used the implementation included in Hipster4j [14].

Algorithm 2 *uncertaintyPrediction*$(x^a, x^b, u^{a:b}, t^{a:b})$

1: $\bar{x}^0 = \bar{x}^a$; $\Sigma^0 = \Sigma^a$; $\Lambda^0 = \Lambda^a$; $x^{a:b} = \emptyset$
2: **for all** $t \in [1, t^{a:b}]$ **do**
3: $\bar{\Sigma}_t = A_t \Sigma_{t-1} A_t^T + M_t$
4: $K_t = \bar{\Sigma}_t H_t^T (H_t \bar{\Sigma}_t H_t^T + N_t)^{-1}$
5: $\Lambda_t = (A_t + B_t L_t) \Lambda_{t-1} (A_t + B_t L_t)^T + K_t H_t \bar{\Sigma}_t$
6: $\bar{x}_t = \tilde{x}_t = f(\bar{x}_{t-1}^{a:b}, u_t^{a:b}, 0)$
7: $\tilde{\Sigma}_t = (I - K_t H_t) \bar{\Sigma}_t$
8: $\Sigma_t = \tilde{\Sigma}_t + \Lambda_t$
9: $x^{a:b} = \{x^{a:b} \cup x_t \sim \mathcal{N}(\bar{x}_t, \Sigma_t)\}$
10: **end for**
11: **return** $x^{a:b}$

3.2 Uncertainty Prediction

Alg. 2 details the estimation of the PDFs along a trajectory $x^{a:b}$. The calculation starts with the PDF at the initial state of the trajectory ($\mathcal{N}(\bar{x}^a, \Sigma^a)$), and iteratively calculates the PDFs of the intermediate states along the motion primitive that connects them. L_t, A_t, B_t and H_t represent respectively the gain of the LQG controller, the Jacobians of the motion model and the Jacobian of the measurement model. The PDFs are calculated based on the state estimations given by an EKF —lines 3, 4, 6 and 7— which return the distribution $P(x_t|\tilde{x}_t) = \mathcal{N}(\tilde{x}_t, \tilde{\Sigma}_t)$, where $\bar{x}_t$ and $\bar{\Sigma}_t$ are the mean and covariance of the predicted state of the EKF, and $\tilde{x}_t$ and $\tilde{\Sigma}_t$ are the mean and covariance of the true state of the EKF. The algorithm also calculates $P(\tilde{x}_t) = \mathcal{N}(\bar{x}_t, \Lambda_t)$, which models the uncertainty due to the estimation of the state without the real observations —lines 4 and 5. These two distributions are used to obtain the joint distribution of the real state of the robot and the true state of the EKF, $P(x_t, \tilde{x}_t) = P(x_t|\tilde{x}_t)P(\tilde{x}_t)$. This distribution allows to obtain the PDF of the real state, $P(x_t) = \mathcal{N}(\bar{x}_t, \tilde{\Sigma}_t + \Lambda_t)$, which can be used by the motion planner as:

$$x_t \sim \mathcal{N}(\bar{x}_t, \Sigma_t) \tag{3}$$

3.3 Path Evaluation

The planner minimizes the probability of collision and the minimum length (represented by the traversal time). The cost function returns a vector with three elements: a safety cost (c_s) related to the probability of traversing the path without collisions, the traversal time of the path ($t^{a:b}$), and the uncertainty at the goal (Σ^b). Each element of the cost defines an objective to minimize in the search. To prioritize the different objectives, the algorithm compares the cost of two paths in the following way:

$$\begin{aligned}
cost(x^{0:a}) < cost(x^{0:b}) \Leftrightarrow (c_s^{0:a} &< c_s^{0:b}) \vee \\
(c_s^{0:a} = c_s^{0:b} \wedge t^{0:a} &< t^{0:b}) \vee \\
(c_s^{0:a} = c_s^{0:b} \wedge t^{0:a} = t^{0:b} \wedge \Sigma^a &< \Sigma^b)
\end{aligned} \tag{4}$$

Algorithm 3 $cost(x^{a:b})$

1: $c_s = 0$
2: **for all** $t = 1 : t^{a:b}$ **do**
3:   $d_m = \infty$
4:   **for all** obstacles in confidence region **do**
5:     $D =$ distance between vehicle at $x_t^{a:b}$ and obstacle
6:     $d_m = \min(d_m, D^T (R_\alpha^T \Sigma_t R_\alpha)^{-1} D)$
7:   **end for**
8:   $c_s = c_s - \log(\Gamma(1, d_m/2))$
9: **end for**
10: **return** $\left[c_s, t^{a:b}, \Sigma^b\right]$

We developed a new method to obtain the probability of successfully traversing a path without collisions. The method avoids the sampling of the PDFs —which allows to our planner to have a deterministic behavior— and takes into account the shape of the vehicle. Alg. 3 describes the procedure to evaluate the cost of the path between the states x^a and x^b, $x^{a:b}$. Given a PDF estimated by our planner, we use an efficient approximation of the probability of collision based on the number of the standard deviations that the vehicle can afford before colliding at the stage t of the path [15]. The algorithm obtains from the PDF $x_t \sim N(\bar{x}_t, \Sigma_t)$ the eigenvalues and eigenvectors of Σ_t, and calculates the 2D region of the map for which the PDF has an accumulated probability close to the 100%. Then, it obtains the minimum Mahalanobis distance (d_m) between the vehicle and the obstacles. d_m is calculated taking into account the real robot shape —as D is the vector and the border of the robot with minimum d_m (Fig. 1)— and the relative angle α between the point in the border and the heading of the vehicle —which requires the transformation of the covariance with matrix R_α, the Jacobian of the transformation function of the point in the border respect to the state. For a multivariate Gaussian distribution, a lower bound for the probability of a sample being within m standard deviations is given by $\Gamma(n/2, m^2/2)$, being Γ the regularized Gamma function, n the dimensions of the distribution and m the number of standard deviations. The probability of successfully traversing the path is the accumulation of all the probabilities along the intermediate states (line 8).

4 Adaptive Multi-resolution Planning

The computational efficiency of the motion planner is very dependent on the resolution of the state lattice. Lower resolutions result in a faster search, but decreasing the diversity of maneuvers at higher resolutions. However, in some cases low resolutions could generate paths with higher costs. Multi-resolution planning techniques can manage this trade-off between computational efficiency and performance. Our proposal uses an adaptive selection of the resolution depending on the complexity of each area of the environment. Those regions that require complex maneuvers should

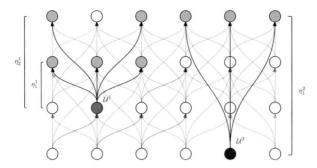

Fig. 2 A multi-resolution state lattice. $\mathcal{U}$ is divided in $\mathcal{U}^1$ and $\mathcal{U}^2$ to define two resolutions. Applying $\mathcal{U}^1$ on the gray state generates the successors in blue, and applying $\mathcal{U}^2$ on the black one generates the successors in orange.

be explored with a higher resolution, whereas if simple motions can drive the vehicle to the goal, the resolution could be reduced without a significant loss of performance.

The aim of the proposed adaptive multi-resolution technique is to represent the state lattice with low resolution in those regions where the absence of obstacles allows to drive the vehicle to the goal using long and simple motion primitives. When the search algorithm —Alg. 1, line 3— selects the next state to be expanded ($x^a \in \mathcal{X}^{lat}$), it must obtain its neighborhood. The neighborhood of x^a depends on its resolution. Alg. 4 describes the generation of the successors of a state x^a through the selection its resolution. $\mathcal{U}$ contains the motion primitives sets for the γ different resolutions of the lattice. Each set —$\mathcal{U}^k$, the lower the k the higher the corresponding resolution— has several motion primitives subsets for different neighborhoods ($\mathcal{U}(\eta_j^k)$), where η_j^k is the j-th neighborhood for the k-th resolution (see Fig. 2 for an example).

The resolution of a state x^a is obtained by searching in a 2D grid —where the positions match the states belonging to the highest resolution of the lattice— the path in the direction given by the heading. This allows to check the presence of obstacles in the moving direction —Alg. 4, lines 1-3. Given the path for the 2D grid, the method counts the number of complex maneuvers for each of the resolutions using the closer

---

Algorithm 4 *successors*(x^a)

Require: $\mathcal{U} = \{\mathcal{U}^1, ..., \mathcal{U}^\gamma\}$
 where $\mathcal{U}^k = \{\mathcal{U}(\eta_1^k), ..., \mathcal{U}(\eta_\delta^k)\}$
1: $x^{a:g} = path2D(x^a, x^g)$
2: **for all** $i = 2 : k$ **do**
3:   $\beta = \sharp complexMove(x^{a:g}, \eta_1^i)$
4:   **if** $\beta > 0$ **then**
5:     **return** $\{x^{b_1}, x^{b_2}, ..., x^{b_n}\}$ given by $\mathcal{U}^{i-1}$
6:   **end if**
7:   **return** $\{x^{b_1}, x^{b_2}, ..., x^{b_n}\}$ given by $\mathcal{U}^k$
8: **end for**

neighborhood for that resolution —it checks if there is a complex maneuver (based on the turning angle) in the next η_1^i steps of the 2D grid path. The algorithm ends when there is a complex maneuver at a given resolution. Finally, it selects the motion primitives of the previous resolution, and generates the neighbor states of x^a using the corresponding primitives ($\mathcal{U}^{i-1}$).

This approach allows to work with several resolutions which are automatically selected depending on the estimated motions. The 2D search is an optimistic approximation to the real path, as it does not take into account the motion model or the real shape. This may lead to pick a resolution higher than necessary. Nevertheless, the opposite is not possible; this ensures that the adaptive selection of the resolution does not underestimate the complexity of the maneuvers in that area of the path. Our proposal reuses the 2D heuristic described in Sec. 3.1. This search is executed once before the planning process, and whenever a change in the environment occurs. Therefore, the adaptive multi-resolution technique does not require any new operations at planning time.

5   Experimental Results

This proposal has been tested with a differential drive robot, which state is a 5-dimensional variable described by the pose at its center, the linear and the angular speeds: $x_t = (x_x, x_y, x_\theta, x_v, x_\omega)^T$. Commands are described by the linear and the angular velocity controls, $u_t = (u_v, u_\omega)^T$. The trajectories of the motion primitives ($\mathcal{U}$) were generated for neighborhood distances: 1, 2, 4 and 8 —each unit is 0.5 m— resulting in trajectories between 0.5 m and 4 m long. The primitives have been divided in two resolutions ($\gamma = 2$) as seen in Fig. 3: high ($\mathcal{U}_H$) and low ($\mathcal{U}_L$) containing trajectories up to 2 m and longer, respectively.

We tested different configurations for the planner: with and without multi-resolution —MR and SR—, with and without estimating the covariance at the border of the robot —TC and NC—, two different shapes —S1 and S2— and two uncertainty

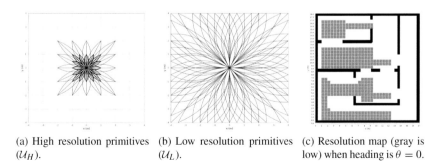

(a) High resolution primitives ($\mathcal{U}_H$).   (b) Low resolution primitives ($\mathcal{U}_L$).   (c) Resolution map (gray is low) when heading is $\theta = 0$.

Fig. 3 Different sets of actions ($\mathcal{U}$) are used for each resolution.

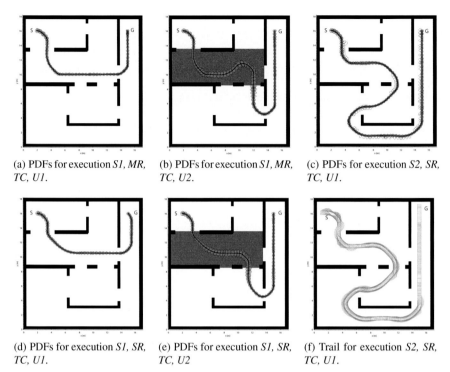

(a) PDFs for execution *S1, MR, TC, U1*.

(b) PDFs for execution *S1, MR, TC, U2*.

(c) PDFs for execution *S2, SR, TC, U1*.

(d) PDFs for execution *S1, SR, TC, U1*.

(e) PDFs for execution *S1, SR, TC, U2*

(f) Trail for execution *S2, SR, TC, U1*.

Fig. 4 Planning results for different shapes and uncertainty conditions. Executions of the planned paths are in blue, the PDFs of the rotation center in black, and their transformation to the border of the robot in red. Regions with higher uncertainty in observations are in gray—lower localization precision. Environment is $17\,m \times 18\,m$.

conditions —*U1* and *U2*. Detailed results are given in Table 1: number of expansions of the search algorithm, number of nodes evaluated (insertions), planning time, cost of the solution path (traversal time), probability of collision estimated by the planner and the real one. This one was obtained simulating the execution of the planned path 100 times using a LQG controller and introducing stochastic noise in controls and observations with covariances:

$$M = diag\left(0.1^2, \frac{\pi^2}{60^2}\right), N = diag\left(0.1^2, 0.1^2, \frac{\pi^2}{60^2}\right), N' = diag\left(0.3^2, 0.3^2, \frac{\pi^2}{9^2}\right)$$

A comparison of the planning results for different uncertainty conditions and robot shapes is shown in Fig. 4. Performing several simulations for the planned path also allows to validate the good performance of the a-priori PDFs (in black) to the paths obtained in execution time (in blue). Translating the PDFs from the rotation center to the border of the shape does not reflect the real distribution in that point. Figure 4(c) stands out those situations in which the PDF estimated at the border provides a more realistic estimation of probability of collision, while the last two rows in

Table 1 Summary of the tests. *MR* and *SR* mean multi and single-resolution, while *TC* indicate the estimation of the PDF at the border of the robot. *U1* and *U2* are different uncertainty conditions. *S1* dimensions are 0.5x0.5 *m* and *S2* 2.5x0.5 *m*.

Execution	Expansions	Insertions	Time (s)	p_{col} (est)	Cost (s)	p_{col} (real)
S1, MR, TC, U1	422	2326	13	0.0	50.82	0.0
S1, SR, TC, U1	1566	8356	40	0.0	49.50	0.0
S1, MR, TC, U2	1164	5376	32	0.0	77.88	0.0
S1, SR, TC, U2	3591	19489	39	0.0	71.28	0.0
S2, MR, TC, U1	1990	9627	97	8.099E-9	117.81	0.0
S2, SR, TC, U1	4285	23068	180	1.404E-12	111.87	0.0
S2, SR, NC, U1	4423	23433	519	1.738E-11	113.19	0.0

Table 1 compare the results of the planner with and without using the correct PDFs at the border. Our method to predict the real PDFs at the border achieves a better performance when turning near obstacles at the sides of the robot. This results in a lower cost of the solution and a better estimation of the probability of collision along the planned path.

The experiments also show the reliability of our planner to work with several robot shapes. While *S1* can safely cross the first door (Fig. 4(a)), for *S2* this becomes unfeasible and the planner selects the longest —but safest path in the environment (Fig. 4(c)). The coverage of shape *S2* when executing the maximum likelihood path in this experiment is shown in Fig. 4(f). Moreover, Fig. 4(a) and Fig. 4(b) also show how the planner selects a different path for shape *S1* to avoid crossing the first door because of the high probability of collision when adding uncertainty to the environment.

Fig. 3(c) shows the resolution assigned to each region of the environment when the robot heading is $\theta = 0$. The complexity in each state x^a is estimated counting the number of turnings of the 8 first steps —maximum neighborhood distance in the set of primitives $\mathcal{U}$— of the 2D path in the direction given by the heading. Figs. 4(a), 4(b), 4(d) and 4(e) compare the paths obtained with and without multi-resolution. The adaptive selection of the resolution decreases the average neighborhood of the lattice and reduces the branching factor of the search —related to the number of insertions. It also decreases the operations required to propagate the uncertainty along the candidate paths. As table 1 shows, our multi-resolution approach obtains a significant reduction in the number of nodes managed by the search algorithm —62% in expansions and 66% in insertions, on average—, while the cost of the planned paths increases only in a 5.6%. Also, it does not affect to the estimation of the probability of collision, which is similar to the one obtained from the executions of the paths.

6   Conclusions

We introduced an adaptive multi-resolution state lattice approach for motion planning. The proposal selects automatically the resolution of each state according to the estimated complexity of the maneuvers in the next steps. The planner also predicts at planning time the motion uncertainty due to the imprecision in controls and observations and takes into account the corrective effect of executing the paths with a LQG controller. Moreover, the algorithm evaluates the probability of successfully traversing the path by taking into account the real shape of the robot, i.e., the PDF is calculated for the point of the robot closer to the obstacles. The experiments show a reduction in the number of explored states due to the adaptive multi-resolution technique. Also, the prediction of the PDFs along the paths has proved to be a very good approximation to the real PDFs. As future work, we expect to use the information of a multi-resolution map in the selection of the lattice resolution in order to achieve a better reduction of the search space. This will significantly reduce the planning time.

References

1. Alterovitz, R., Siméon, T., Goldberg, K.: The stochastic motion roadmap: a sampling framework for planning with Markov motion uncertainty. In: Robotics: Science and Systems, pp. 246–253 (2007)
2. Bry, A., Roy, N.: Rapidly-exploring random belief trees for motion planning under uncertainty. In: IEEE International Conference on Robotics and Automation (ICRA), pp. 723–730 (2011)
3. González-Sieira, A., Mucientes, M., Bugarín, A.: Anytime motion replanning in state lattices for wheeled robots. In: Workshop on Physical Agents (WAF), pp. 217–224 (2012)
4. Kaelbling, L.P., Littman, M.L., Cassandra, A.R.: Planning and acting in partially observable stochastic domains. Artificial intelligence 101(1), 99–134 (1998)
5. Knepper, R., Kelly, A.: High performance state lattice planning using heuristic look-up tables. In: IEEE/RSJ International Conference on Intelligent Robots and Systems (IROS), pp. 3375–3380 (2006)
6. LaValle, S.M.: Planning Algorithms. Cambridge University Press, Cambridge (2006)
7. LaValle, S.M., Kuffner, J.J.: Randomized kinodynamic planning. The International Journal of Robotics Research 20(5), 378–400 (2001)
8. Likhachev, M., Ferguson, D.: Planning Long Dynamically Feasible Maneuvers for Autonomous Vehicles. The International Journal of Robotics Research 28(8), 933–945 (2009)
9. Likhachev, M., Ferguson, D., Gordon, G., Stentz, A., Thrun, S.: Anytime dynamic A*: an anytime, replanning algorithm. In: Proceedings of the International Conference on Automated Planning and Scheduling (ICAPS), pp. 262–271 (2005)
10. Melchior, N.A., Simmons, R.: Particle RRT for path planning with uncertainty. In: IEEE International Conference on Robotics and Automation (ICRA), pp. 1617–1624 (2007)
11. Papadimitriou, C.H., Tsitsiklis, J.N.: The complexity of Markov decision processes. Mathematics of operations research 12(3), 441–450 (1987)
12. Pivtoraiko, M., Kelly, A.: Differentially constrained motion replanning using state lattices with graduated fidelity. In: IEEE/RSJ International Conference on Intelligent Robots and Systems (IROS), pp. 2611–2616 (2008)
13. Pivtoraiko, M., Knepper, R.A., Kelly, A.: Differentially constrained mobile robot motion planning in state lattices. Journal of Field Robotics 26(3), 308–333 (2009)
14. Rodriguez-Mier, P., Gonzalez-Sieira, A., Mucientes, M., Lama, M., Bugarin, A.: Hipster: An open source java library for heuristic search. In: 2014 9th Iberian Conference on Information Systems and Technologies (CISTI), pp. 1–6. IEEE (2014)
15. Van Den Berg, J., Abbeel, P., Goldberg, K.: LQG-MP: Optimized path planning for robots with motion uncertainty and imperfect state information. The International Journal of Robotics Research 30(7), 895–913 (2011)

Improving Teleoperation with Vibration Force Feedback and Anti-collision Methods

André Casqueiro, Diogo Ruivo, Alexandra Moutinho and Jorge Martins

Abstract This paper presents a two folded solution to facilitate and improve the tele-operation of unmanned vehicles in unknown scenarios. The first part of the solution regards increasing the operators perception of the vehicle surroundings by means of a new vibration feedback transmitted by a haptic controller. The second part concerns the implementation of new anti-collision methods that take into account both vehicle and environment constraints through a spatial representation of the allowed vehicle velocities. The solution was tested and validated by 28 subjects tele-operating an omnidirectional ground vehicle through an unseen maze. The experiment results show a reduction of the human operator workload and of the time taken to complete the task. The vibration feedback was compared by the subjects with other haptic feedbacks in an experiment to identify the direction of a single obstacle, outperforming these in the effective indication of the presence and direction of the existing obstacle.

Keywords Haptic feedback · Obstacle avoidance · Deconfliction · Teleoperation · Holonomic vehicle · Planar 2D LIDAR

1 Introduction

Nowadays, mobile robots have an increasingly importance in high risk tasks where the presence of humans can be hazardous or impractical due to distance, scale or other environmental barriers. Such applications include land demining [1], search and rescue [2], maintenance of nuclear plants [3] and space exploration [4].

A. Casqueiro · D. Ruivo
Instituto Superior Técnico, Universidade de Lisboa, Lisbon, Portugal
e-mail: {andre.casqueiro,diogo.ruivo}@tecnico.ulisboa.pt

A. Moutinho(✉) · J. Martins
IDMEC/LAETA, Instituto Superior Técnico, Universidade de Lisboa,
Av. Rovisco Pais, 1049-001 Lisbon, Portugal
e-mail: {alexandra.moutinho,jorgemartins}@tecnico.ulisboa.pt

© Springer International Publishing Switzerland 2016
L.P. Reis et al. (eds.), *Robot 2015: Second Iberian Robotics Conference*,
Advances in Intelligent Systems and Computing 417,
DOI: 10.1007/978-3-319-27146-0_21

In the last years, a significant progress has been made in the control systems of mobile robots. However, the complexity and uncertainty of many tasks require the sophisticated cognitive capabilities of a human operator. Ideally, in teleoperation, the operator safely controls the vehicle at a distance while perceiving the task environment and making real time decisions. Therefore, having a complete awareness of the vehicle surroundings is vital to guarantee the success of the task.

Several studies have shown that the use of haptic feedback increases the transparency in teleoperation [5–8]. During the process of teleoperation, the user main focus should be on the mission. By using an anti-collision system it is possible to reduce the attention needed on the working environment, hence reducing the workload and increasing performance [7, 8].

The key contribution of this paper is the implementation of a collision-free haptic teleoperation scheme for a holonomic vehicle. This scheme includes a novel haptic feedback based on vibration and a collision avoidance system that adapts the desired velocity in accordance to the surroundings of the vehicle. The referred methods were implemented and then evaluated by 28 subjects in two different experiments.

The rest of the paper is organized as follows. In Section 2, the teleoperation architecture is introduced. The obstacle identification, the anti-collision systems and the haptic interfaces are explained. The implementation and evaluation of the referred methods and respective experimental results are detailed in Section 3. Finally, concluding remarks are provided in Section 4.

2 Bilateral Teleoperation Scheme

This section introduces the structure of the proposed teleoperation system. Posteriorly, the main methods used for the anti-collision systems and haptic feedback are briefly explained.

2.1 Architecture of the Teleoperation Loop

The scheme of the bilateral haptic teleoperation considered is represented in Fig. 1. The operator interacts with the system through the master device, a haptic device capable of transmitting to the user a force related to the slave device's state and perception of the environment. The master controller is in charge of mapping the user command to a reference signal for the slave device, an unmanned vehicle being teleoperated. The master controller is also responsible for converting the slave output into an appropriate force feedback for the user.

The slave controller, composed of the onboard microcontroller and sensors, adapts the master reference signal received according to the state and surroundings of the slave device. By controlling the vehicle locally, the slave controller guarantees the

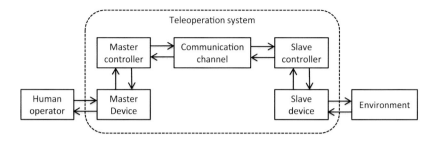

Fig. 1 Generic architecture of bilateral teleoperation

stability of the vehicle and ensures a collision-free navigation, even in the absence of commands from the master.

2.2 Workspace Mapping

Contrary to classical bilateral teleoperation systems, in haptic teleoperation of mobile robots the master and the slave do not share a similar workspace. The master device workspace is bounded by its physical limitations. On the other hand, the slave moves in an almost unbounded environment. Consequently, direct mapping of the master position to a slave position is not proper for most scenarios.

In order to overcome the difference of workspaces, a rate control is used, i.e., the position of the master is mapped to a reference velocity of the slave. This mapping is described by

$$v_{s_r}(t) = \lambda q_m(t) \tag{1}$$

with v_{s_r} the reference velocity for the slave, q_m the position of the master device end effector and λ a diagonal scaling matrix.

2.3 Obstacle Detection

An anti-collision method is dependent on some perception of the environment and the capacity to detect existing obstacles. In this work the vehicle under operation is assumed to have a horizontal scanner installed. Within its field of view and for a defined sampling angle, the scanner returns a set of distances to all obstacles detected. This information then needs to be transformed into features to be used by the anti-collision system.

The features correspond to the most influential edges and vertexes of the obstacles and their respective direction relative to the vehicle's body frame of reference.

Although the Obstacle Detection problem is not new, the approach taken to select the features is not conventional. This is due to: i) no distinction is being made between static and moving obstacles, ii) one object could return multiple features, and iii) the simplicity of the features needed in the anti-collision methods.

The process of computing these features starts by transferring the scanned points from the sensor's frame of reference to the vehicle's frame of reference. The points are then grouped to form obstacles, by selecting the consecutive distances below a certain threshold, the Observation Radius R_o (see Fig. 2). The identified obstacles undergo a line simplification algorithm (Douglas-Peucker Algorithm [9]), in order to obtain a simplified version of the obstacle's contour, corresponding to a set of line segments per obstacle. The final step is to calculate the closest point of each line segment to the origin (location of vehicle). This point is only relevant if it is between the segment edges or if it is an edge point selected from two segments. In this case, the feature vector (see Fig. 2a), defined by a distance and direction, is stored.

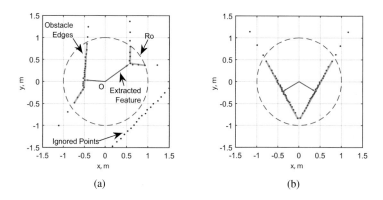

(a)   (b)

Fig. 2 Two obstacle edge detection examples. O is the origin (location of vehicle). R_o is the Observation Radius.

2.4   Anti-collision Methods

Two Anti-Collision Methods were devised to make the slave deal with the obstacles. The first is an Avoidance System that ignores the user's command if it means getting closer to an obstacle than allowed. The second is a Deconfliction System which, like the Avoidance System, does not allow the vehicle to get too close to an obstacle, but still takes into account the user's commands. It redirects the vehicle in a direction that is both safe and most similar to the desired one.

The presented anti-collision methods take a new approach to the collision avoidance problem, taking into account both vehicle and environment constraints through a spatial representation of the allowed vehicle velocities.

In both anti-collision methods the vehicle velocity reference must be inside an allowed velocity space. This space is built taking into account the vehicle's physical constraints and is adjusted (clipped) during the operation depending on the obstacles encountered. As an example, consider that the initial velocity space for a holonomic ground vehicle consists of all points inside a circle of radius V_{Max}, the maximum speed allowed for the vehicle (see Fig. 3).

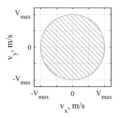

Fig. 3 Initial velocity space

The velocity space clipping takes into account the information collected during the Obstacle Detection phase. The reference speed of the vehicle in the direction of the obstacle feature is set to have its value depending on the distance to the obstacle, as seen in Fig. 4. If the obstacle is in the safe zone then there are no restrictions. If the obstacle is in the warning zone its speed is reduced to zero when at a distance R_d (Danger Radius). If the obstacle passes this point into the danger zone, its velocity is set in the opposite direction of the obstacle, achieving the maximum value when it reaches the critical zone, defined by the Critical Radius, R_c.

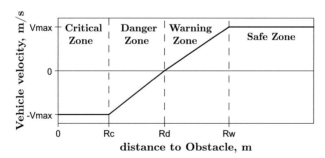

Fig. 4 Relation between Velocity and Distance

The clipping induced on the velocity space by an obstacle approaching the vehicle from north-east is seen in Fig. 5.

During the clipping operation the velocity space may be cropped multiple times, which could lead to an empty velocity space. The Anti-Collision Methods proposed and discussed in the following should guarantee that the vehicle does not reach such a

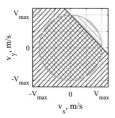 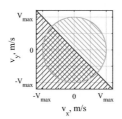 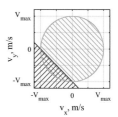

(a) Obstacle inside the Warn- (b) Obstacle in the Danger Ra- (c) Obstacle inside the Danger
ing Zone   dius   Zone

Fig. 5 Three imposed velocities from same direction. Resulting allowed velocity space is the intersection between the two areas.

solution for static environments, or environments where moving obstacles are slower than the considered vehicle.

Avoidance System. The Avoidance System (AS) goal is to limit the operator input velocity if it means getting closer than allowed to an obstacle. This is done by analyzing the desired velocity regarding the velocity space.

Consider Fig. 6. If the desired velocity (blue arrow) is inside the allowed velocity space (Fig. 6a), then it is accepted and passed to the vehicle. If it is outside the allowed velocity space, there are two cases to consider. If the desired velocity crosses the allowed velocity space, its magnitude is decreased until it is in the allowed space (red arrow in Fig. 6b). If it does not cross the allowed velocity space, the desired velocity is ignored and substituted by the closest velocity to the origin of the velocity space (no velocity) that belongs to the allowed velocity space (red arrow in Fig. 6c).

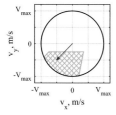 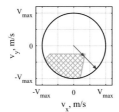 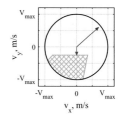

(a) Desired velocity inside al- (b) Desired velocity crossing (c) Desired velocity outside
lowed velocity space   allowed velocity space   allowed velocity space

Fig. 6 Avoidance Scenarios. Circle defines velocity space. Shaded area: allowed velocity space. Blue arrow: desired velocity. Red arrow: corrected velocity.

Deconfliction System. Unlike the Avoidance System, the Deconfliction System (DS) does not deny the user's desired velocity, rather, it transforms it if needed. When the desired velocity is inside the allowed velocity space no transformation occurs (Fig. 7a). If the desired velocity is outside the allowed velocity space it is substituted by the closest velocity in the allowed space (Fig. 7b and Fig. 7c). This allows the vehicle to keep moving in a direction that corresponds to a compromise between the operator's desired velocity and what is imposed by the environment where the vehicle is working.

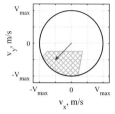 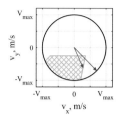 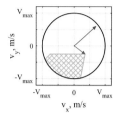

(a) Desired velocity inside allowed velocity space

(b) Desired velocity crossing allowed velocity space

(c) Desired velocity outside allowed velocity space

Fig. 7 Deconfliction Scenarios. Circle defines velocity space. Shaded area: allowed velocity space. Blue arrow: desired velocity. Red arrow: corrected velocity.

2.5 Haptic Feedback

After the description of how the teleoperation is organized and how the obstacle detection and avoidance is done in the slaves controller, this section presents some solutions for the bilateral interaction between the user and the master system, i.e., how to haptically transmit information referent to the state of the slave and its surroundings to the user.

In haptic teleoperation, the master system includes a haptic device capable of transmitting to the user a force feedback. This force is composed of two components: a local force and an obstacle driven force. The local force acts as a spring-damper system that pushes the end effector to its center of operation. For the obstacle driven force, three different force feedback methods were studied and compared: a standard force feedback, a stiffness feedback and a novel vibration feedback.

In the force feedback case [10], a force is applied to the end effector pointing away from the obstacle. The intensity of this force is proportional to the distance to the obstacle. This force may be interpreted as a force offset (Fig. 8a), shifting the neutral position of the master device. When moving in the direction of an obstacle, the force exerted by operator in the end effector has to increase to maintain it in the same position. If the operator releases the end effector, the slave vehicle will move away from the obstacle until it reaches a safety distance, since the end effector is pushed to a non-neutral position. In the presence of an obstacle, this method requires the operator to counter the force feedback to maintain a neutral position, increasing the workload.

The stiffness feedback [10] is interpreted as an extra spring to the spring-damper system which stiffness increases with the proximity to an obstacle. As in the preceding case, the operator needs to increase the applied force to the end effector in order to maintain the reference velocity when heading to an obstacle. In this case, no force feedback is given to the operator when the end effector is in a neutral position (Fig. 8a) or when the slave vehicle is moving away from an obstacle. If released, the end effector moves to the neutral position.

Finally the vibration feedback is designed as a pulsated stiffness feedback which frequency increases with the proximity to obstacles. Both methods mentioned before, require the user to counteract the feedback forces, possibly increasing the workload of the task. By taking advantage of the vibrotactile sensing, it is possible to intuitively transmit information of the slave surrounding without demanding high physical efforts from the operator. Moreover, small vibrations seem easier to perceive than small increases in force, therefore improving the human perception of obstacles through the master device.

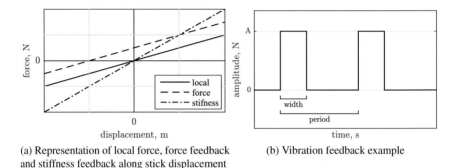

(a) Representation of local force, force feedback and stiffness feedback along stick displacement

(b) Vibration feedback example

Fig. 8 Haptic Feedback Representation

Figure 8b provides an example of vibration feedback and respective pulse parameters to be adjusted: width, period, amplitude and direction. For the duration of the pulse (width), a value around 0.02s is recommended. In case a smaller value is used, the user might be unable to perceive the vibration feedback. On the other side, higher width values will make the user feel small kicks instead of a simple vibration, causing discomfort. The amplitude should be represented as a decreasing monotonic function of the distance to the obstacle. The amplitude is maximum when the obstacle is inside the danger zone and null when the obstacle is in the safe zone. The direction should point away from the obstacle.

3 Experiments

Two experiments were conducted to evaluate the different types of haptic feedback and the effectiveness of both anti-collision systems presented. The main goal of the first experiment was to identify which haptic feedback provides the operators/subjects with a better perception of a single obstacle. In the second experiment, subjects had to teleoperate an omnidirectional vehicle through an unseen obstacle-laden course, using different haptic feedback and anti-collision systems. Workload and task completion time were the performance parameters evaluated.

3.1 Haptic Experiment

The first experiment consists of a series of rounds where the subject has to guess the direction of an obstacle by interacting with the master device. Figure 9a presents the visual interface of this experiment. In each round, the direction of the obstacle is randomly selected between eight possible directions. A round ends when the user makes a guess, clicking on the respective button. Holding the master device, the subject feels the haptic feedback and tries to identify where the obstacle is located. The three haptic feedbacks previously described were used: force feedback, stiffness feedback and vibration feedback. As time passes during a round, the distance to the obstacle reduces at a steady rate, increasing the force feedback, and making the identification of the obstacle location easier.

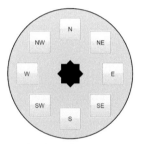

(a) Visual interface, one button   (b) Haptic Device, Novint Falcon
per direction

Fig. 9 Haptic experiment

Twenty eight subjects with little to no teleoperation experience participated in the experiment. The subjects were informed of the objective of the task and about the presence of different haptic feedbacks. However, the subjects were not aware of how many different feedbacks were used, or how they worked. The master device used was the 3 DoF fully-actuated haptic joystick Novint Falcon, Fig. 9b. Each subject executed eighteen rounds corresponding to six rounds for each kind of haptic feedback. The haptic feedback used as well as the obstacle direction were randomly selected in each round. The performance evaluation of each haptic feedback was based on the number of correct guesses and on the time taken to make a decision.

The main results of the haptic experiment are shown in Fig. 10. A full-factorial ANOVA test was applied. In average, subjects correctly answered 67.2% of the times using vibration feedback, against 37.5% and 45.8% for force and stiffness feedbacks respectively (Fig. 10a). The respective average response times for vibration, force and stiffness feedbacks are 13.48s, 16.98s and 18.33s (Fig. 10b).

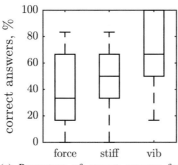
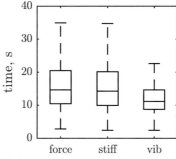

(a) Percentage of correct answers for each haptic feedback. ANOVA test ($p < 0.001$).

(b) Average response time for each haptic feedback. ANOVA test ($p < 0.01$).

Fig. 10 Main results of haptic experiment conducted on 28 subjects for each type of feedback. Boxes = 25th and 75th percentiles; bars = min and max values.

3.2 Assisted Teleoperation Experiment

Setup. In the assisted teleoperation experiment, the subject has to teleoperate the vehicle inside a small unknown maze and find the exit (Fig. 11a). During this experience the operator is unable to see the vehicle or the maze. The only information available is the haptic feedback from the master device and a top view of the obstacles detected by the onboard sensor (Fig. 11b).

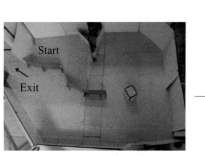
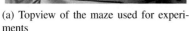
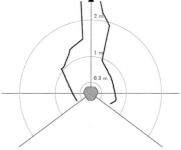

(a) Topview of the maze used for experiments

(b) Visual interface

Fig. 11 Maze Task

In this setup (Fig. 9b), the master device is composed of a desktop computer running Matlab. The computer is connected to the user's input device with feedback capabilities, Novint Falcon, and connects with the slave wirelessly through a xBee.

The slave is a 3 wheeled holonomic vehicle (OMNI-ANT, Fig. 3.2). Its central processing unit is an Arduino Mega ADK that connects to the master with a xBee, perceives the environment with a planar LiDar scanner (Hokuyo URG-04LX-UG01), and controls the motors via a MD25 motor controller board.

Fig. 12 Slave vehicle, OMNI-ANT

Each subject executes two sets of trials, each with four rounds, one for each feedback (no feedback, basic force feedback, stiffness feedback and vibration feedback). Regarding the collision avoidance methods, the first trial uses the avoidance system, while the second trial uses the deconfiction system. The different feedbacks appear in a randomly generated order for each subject. After each round, the subject was asked to rate his/her workload using the NASA TLX rating scale [11]. In the end, each subject completed a small questionnaire. The performance of the teleoperation system was based on the task duration, task workload and questionnaire results.

Results. A first analysis compares the overall results using the avoidance system and the deconfliction system. A significant decrease in the duration of the task when the deconfliction system is used is confirmed in the results of Fig. 13a, with a reduction in the average completion time from 46.94 s to 34.30 s. The NASA TLX scores (Fig. 13b) indicate that the deconfliction system resulted in less workload, as expected.

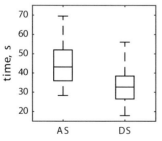

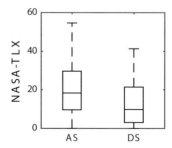

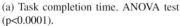

(a) Task completion time. ANOVA test (p<0.0001).

(b) NASA-TLX scores. ANOVA test (p<0.001).

Fig. 13 Comparison results between avoidance system and deconfliction system. Box = 25th and 75th percentiles; bars = min and max values.

The comparison of the results between each type of haptic feedback shows a slight increase in workload for the force feedback. However, the difference in workload and completion time between vibration feedback, stiffness feedback and no feedback was statistically insignificant.

In the end of the experiment, subjects were asked to try to identify the different kinds of haptic feedback and differentiate their favorite (Fig. 14). Although having a lower score, the stiffness feedback was placed in second most of the times, whereas the vibration and force feedback switched between first and third favorite choices. Some subjects either liked all solutions with feedback or were unable to identify the different feedback solutions. It is important to notice that there were no collisions during these experiments.

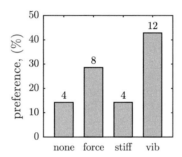

Fig. 14 Feedback type preference

4 Conclusions

This paper presents a generic haptic teleoperation scheme for unmanned water, ground or air vehicles. A novel vibration haptic feedback enables the user to intuitively perceive the vehicle surroundings, decreasing the task workload. The proposed deconfliction system for holonomic vehicles autonomously adapts the reference velocity permitting a smooth collision-free navigation.

Experiments showed the successful teleoperation of a holonomic vehicle using a haptic joystick. The experimental results reveal a reduction of both human operator workload and respective time taken to complete the task when using the deconfliction system. Regarding the haptic vibration feedback, results support that the developed vibration interface is capable of more effectively transmitting information regarding the presence and direction of an obstacle than existing haptic feedbacks. Still, the influence of the haptic vibration feedback in assisting the teleoperation is reduced when anti-collision methods are implemented as well as when visual feedback is available. Future work includes the implementation and evaluation of the described teleoperation system in multirotor platforms.

Acknowledgments This work was supported by Fundação para a Ciência e a Tecnologia (FCT), through IDMEC, under LAETA UID/EMS/50022/2013.

References

1. Havlik, S.: Land Robotic Vehicles for Demining. In: Maki K. Habib (Ed.) Humanitarian Demining. InTech (2008)
2. Nourbakhsh, I.R., et al.: Human-robot teaming for search and rescue. IEEE Pervasive Computing **4**(1), 72–79 (2005)
3. Kim, K., Lee, H., Park, J., Yang, M.: Robotic contamination cleaning system. In: IEEE/RSJ International Conference on Intelligent Robots and Systems, vol. 2, pp. 1874–1879 (2002)
4. Schenker, P.S., et al.: Planetary rover developments supporting Mars exploration, sample return and future human-robotic colonization. Autonomous Robots **14**(2-3), 103–126 (2003)
5. Hokayem, P.F., Spong, M.W.: Bilateral teleoperation: An historical survey. Automatica **42**(12), 2035–2057 (2006)
6. Diolaiti, N., Melchiorri, C.: Teleoperation of a mobile robot through haptic feedback. In: IEEE International Workshop Haptic Virtual Environments and Their Applications, HAVE 2002, pp. 67–72. IEEE (2002)
7. Lee, S., et al.: Haptic teleoperation of a mobile robot: A user study. Presence: Teleoperators and Virtual Environments **14**(3), 345–365 (2005)
8. Omari, S., Hua, M.D., Ducard, G., Hamel, T.: Bilateral haptic teleoperation of an industrial multirotor UAV. In: Gearing Up and Accelerating Cross-fertilization between Academic and Industrial Robotics Research in Europe, pp. 301–320. Springer International Publishing (2014)
9. Douglas, D.H., Peucker, T.K.: Algorithms for the reduction of the number of points required to represent a digitized line or its caricature. Cartographica: The International Journal for Geographic Information and Geovisualization **10**(2), 112–122 (1973)
10. Lam, T.M., Mulder, M., van Paassen, M.M.: Haptic Feedback for UAV Tele-operation - force offset and spring load modification. In: IEEE International Conference on Systems, Man and Cybernetics 2006, vol. 2, pp. 1618–1623, October 8–11, 2006
11. Hart, S.G., Staveland, L.E.: Development of NASA-TLX (Task Load Index): Results of empirical and theoretical research. In: Hancock, P.A., Meshkati, N. (eds.) Human Mental Workload, pp. 139–183. Elsevier Science Publishers, North-Holland (1998)

Human-Aware Navigation Using External Omnidirectional Cameras

André Mateus, Pedro Miraldo, Pedro U. Lima and João Sequeira

Abstract If robots are to invade our homes and offices, they will have to interact more naturally with humans. Natural interaction will certainly include the ability of robots to plan their motion, accounting for the social norms enforced. In this paper we propose a novel solution for Human-Aware Navigation resorting to external omnidirectional static cameras, used to implement a vision-based person tracking system. The proposed solution was tested in a typical domestic indoor scenario in four different experiments. The results show that the robot is able to cope with human-aware constraints, defined after common proxemics rules.

1 Introduction

In the last few years, robotics is becoming focused on Human-Robot Interaction and on its role in social environments. When people think of a robot interacting with a person, what comes to mind is a robot that can speak with her or hand over some object. However, the motion itself is of great importance in a social context (e.g. when a robot is requested to fetch an item), or simply when a normal navigation behavior needs to be adjusted according to proxemics rules, so it does not disturb people. The study of robot navigation in the presence of people is called Human-Aware Navigation.

Most approaches in the literature used only sensors on-board the robot. Even though those sensors bring the advantage of context-independence, most robust people tracking methods are based on computer vision and computationally expensive. Hence not suited for on-board computational devices [2]. In this paper, a Networked

A. Mateus(✉) · P. Miraldo · P.U. Lima · J. Sequeira
Institute for Systems and Robotics (LARSyS), Instituto Superior Técnico,
Universidade de Lisboa, Torre Norte - 7 Piso Av.Rovisco Pais, 1, 1049-001 Lisbon, Portugal
e-mail: andre.mateus@tecnico.ulisboa.pt

This work was supported by EU-funded projects [FP7-ICT-2011-9-601033] MOnarCH, [FP7-ICT-601012] RoCKIn, and FCT project [UID/EEA/50009/2013].

© Springer International Publishing Switzerland 2016   283
L.P. Reis et al. (eds.), *Robot 2015: Second Iberian Robotics Conference*,
Advances in Intelligent Systems and Computing 417,
DOI: 10.1007/978-3-319-27146-0_22

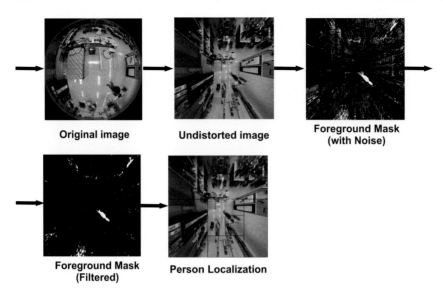

Fig. 1 Representation of the steps used for detecting people. The first picture corresponds to the original image (distorted image). The second is the undistorted image. Third and fourth correspond to background subtraction and foreground filtering steps, respectively. The last image corresponds to the final detection (the blue ellipse identifies the person on the image).

Robot Systems have been used to overcome the limitations of on-board sensing. Particularly, external omnidirectional fisheye cameras were mounted on the ceiling. This setup ensures a broader perception of the environment, capable of seeing both humans and robots at the same time. In addition, an increase in processing power is achieved (computational effort can be distributed through external devices), which gives the robot extra computation time that can be used in other tasks (e.g finding objects).

In 2007, Sisbot et al. [19] proposed the Human-Aware Navigation Planner (HANP). Their work is focused on human comfort, which is addressed by three criteria: preventing personal space invasions; navigating in the humans' field of view (FOV); and preventing sudden appearances in the FOV of humans. Those criteria are modeled as cost functions in a 2D cost-map and path planning is performed with A^*. Even though HANP accounts for replanning if people move, it does not adapt their personal space during the motion. With that in mind, two extensions to HANP were proposed:

- a prediction cost function which, by increasing the cost in front of a moving human, decreases the probability of the robot entering that area [11];
- the concept of compatible paths, which says that two paths are compatible if both agents can follow their paths (reaching the goal position), without any deadlocks [10].

Kirby at [8] proposed a new solution, which differ from HANP on the considered constraints and their formulation. Instead of focusing simply on human comfort, constraints concerning social rules (e.g. overtake people from the left) and low-level human navigation behavior (e.g. face direction of movement) are also taken into account.

Another important issue related with human comfort, in a social context, is the interference with humans interacting with other humans and/or objects. This issue is tackled in [16] where, besides considering proxemics and the back space of a person, a constraint is included to model the space between interacting entities.

Other important work was presented in [15], a framework for planning a smooth path through a set of milestones. Those are added, deleted and/or modified, based on the static and dynamic components of the environment.

In 2013, Kruse et al. [12] defined the three goals for Human-Aware Navigation as: human comfort (e.g. space that people keep from each other in different contexts, known as the theory of proxemics [6], and velocity that robots navigate close to humans [18]); respect social rules; and mimic low-level human behavior.

In this paper, we propose a solution for Human-Aware Navigation resorting external cameras, for person state estimation. A cost-map is computed by combining several constraints associated with Human-Aware Navigation. Each time the robot receives a new goal, it computes a path on that cost-map. When compared with state-of-the-art approaches, the main contributions are:

– the use of static external camera sensors, increasing the field of view (allowing to see the robots and people at the same time and not requiring the re-estimation of the camera pose, thus eliminating one cause of error), Sec. 2; and
– standardization of human-aware constraints (collection of a set of constraints from different state-of-the-art methods and reformulation of some of them), Sec. 3.

The solution was tested in simulated and realist scenarios in four different experiments. The results (Sec. 4) show that the proposed solution fulfills the respective goals.

2 Vision-Based Person Tracking

One of the most important steps of Human-Aware Navigation is person tracking. That problem is solved by a vision-based tracking system, consisting of two steps: people detection, including posture (identify if people are seated, standing or if it is just noise, e.g. a robot); and person tracking with a Kalman Filter. The goal is to get a setup that is capable of seeing both the robot and people, at same time. For that purpose, omnidirectional fisheye cameras mounted on the ceiling are used.

2.1 People Detection Including Posture

The goal of the detector is to be as fast as possible, for a good real-time tracker. With this is mind, our solution first finds blobs of interests on the image; associates these blobs to posture states (also filters some errors due to noise); and finally tracks those blobs in the world frame.

Each time a new image is received, the respective undistorted image is computed (the method/model used is proposed at [17]). Moving objects in the scene are detected using background subtraction. In addition, since resulting foreground is very noisy, morphological operations are employed (opening and closing). Blobs are then extracted and, from the resulting contours, those with an area bigger than a threshold are fitted onto an ellipse (in our experiments we used 3000 pixels as threshold). In addition, one needs to ensure that the localization only takes into account fully detected people, i.e., the fitted ellipses cover people shapes from head to toe. As a result, ellipses that have none of the limits of its major axis inside a predefined interest region are discarded. An example of the application of each step is shown in Fig. 1.

Regarding person posture classification, let us assume that a person's silhouette looks similar when she appears at the same location multiples times (this is verified for different people as long as their heights do not differ much, e.g. children and adults). Using this assumption, a Support Vector Machine (SVM), particularly a C-SVM, [3], was used to classify the fitted ellipses in three possible classes. Those classes are associated with person postures: standing, seated, or noise. Videos of people walking and sitting in the scene were recorded and used for training. The robot was also taken into account in the training, and classified as noise.

For each frame, three features of the ellipses were retrieved:

– (u, v) position of their center on the image;
– size of major axis; and
– size of minor axis.

The features of the ellipses are collected into vectors and classified by the SVM. If ellipses are classified as noise (e.g. a robot), they are discarded. Otherwise, each ellipse is considered to be a person and is going to be localized in the world. Examples of classification using C-SVM are shown in Fig. 2.

The location of a standing person is considered to be her feet. These correspond to the point in the contours closest to the image center. This is true for every point on the image, except when the person is exactly below the camera.

If an ellipse is classified as a sitting person, its center is checked to assess on which of the seating areas the person is (e.g. couch and chairs). If the ellipse is not on any of them, it will be discarded. Otherwise, its position is set to a default for that sitting region. We chose to associate this default position, because feet are often occluded when a person is seated, which will significantly affect the estimation of localization.

To conclude this section, one needs to associate the coordinates of the persons to the world frame (same as the robot). For that, we used the following assumptions:

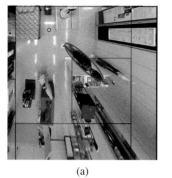 

(a)   (b)

Fig. 2 Results of people detection (including posture) on zoomed images. Blue ellipse on Fig. (a) means that the classification is standing person (the center of the yellow circle is the point that is assumed to be the feet. Green ellipse on Fig. (b) means that the person is seated. A video of this results is sent as supplementary material.

– cameras are on the ceiling and vertically align with the ground plane (z-axis are aligned);
– only people and the robot are moving on the scene;
– they are all moving on the same plane (floor);

For each detection, we compute the inverse projection of the respective pixel and its 3D coordinates, in the camera reference frame (which can be easily obtained by intersecting the 3D projection ray with the 3D floor plane). Then, a homography is applied to transform coordinates of this point from camera to world coordinate systems, [7], concluding the localization step.

2.2 Tracking

Let the person state (position and velocity on the ground plane), at time k, be denoted by $\mathbf{x}_k = (x_k, y_k, v_k^x, v_k^y)$. Taking the position measurements from people detection scheme (described in the previous subsection), one can track the person using a Kalman Filter. The observation equation is

$$\mathbf{z}_k = \begin{bmatrix} 1 & 0 & 0 & 0 \\ 0 & 1 & 0 & 0 \end{bmatrix} \mathbf{x}_k + \mathbf{v}, \tag{1}$$

where $\mathbf{z}_k$ is the measurement vector, $\mathbf{v} \sim \mathcal{N}(\mathbf{0}, \mathbf{R})$ (normal distribution at $\mathbf{0}$ with covariance matrix equal to $\mathbf{R}$). Humans tend to walk at a constant velocity, [1], hence a constant velocity model was assumed. Then, the state equation is

$$\mathbf{x}_k = \begin{bmatrix} 1 & 0 & \Delta t & 0 \\ 0 & 1 & 0 & \Delta t \\ 0 & 0 & 1 & 0 \\ 0 & 0 & 0 & 1 \end{bmatrix} \mathbf{x}_{k-1} + \mathbf{w}, \tag{2}$$

where $\mathbf{w} \sim \mathcal{N}(\mathbf{0}, \mathbf{Q})$ and Δt is the time difference between messages received from the people detector. $\mathbf{R}$ and $\mathbf{Q}$ are the measurement and process noise covariance matrices.

For each iteration, there are four possibilities (note that we are assuming that only one person is being tracked):

1. a person is not being tracked;
2. a person is being tracked and a new detection was received;
3. a person is being tracked and more than one detection was received;
4. a person is being tracked and no detection was received.

In the first case, if there is only one detection, the tracker is initialized. Otherwise (none or more than one detection was sent) that message is discarded. For the second case, the filter predicts a new state and then corrects it based on the measurement. If there is more than one detection, the measurement closest to the last estimated state is used to correct the filter's prediction. Finally, if there are no detections the estimated state is the prediction from the filter, without correction.

3 Human-Aware Path Planning

To be as Human-Aware as possible, the three goals of the field (namely human comfort; social rules respect; and naturalness) were considered. As a result, following constraints were taken into account:

1. Take least effort path (naturalness);
2. Keep a distance from static obstacles (naturalness);
3. Respect personal spaces (human comfort);
4. Avoid navigating behind sitting humans
   (human comfort);
5. Do not interfere with human-object interactions
   (human comfort);
6. Overtake people by the left (social rule);

These constraints are based on previous approaches. However, some of them are reformulated (constraints 4 and 5) in order to be better integrated in our approach and to standardize their formulation. Next, we fully described the formulations used for each constraint.

The first is addressed by the path planner. In this work A^* was used, ensuring a minimum cost path as long as the heuristic is admissible. The total cost of a node is given by the sum of the cost of reaching that node, with the heuristic cost. The latter was considered to be the euclidean distance from the node to the goal position. Since the environment is dynamic, planner computes a path periodically.

The second constraint prevents the robot from passing too close to obstacles. This is solved by attributing a high cost to areas surrounding the obstacles, [4].

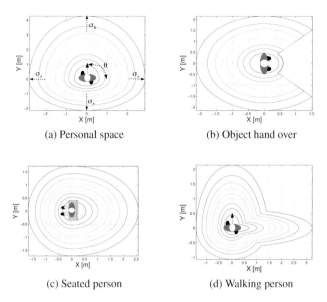

(a) Personal space   (b) Object hand over

(c) Seated person   (d) Walking person

Fig. 3 Representation of cost functions associated with different people posture: Fig. (a) represents the cost function for the personal space of a person walking in the y direction at 1 [m/s]; Fig. (b) shows the cost function of a person standing, oriented in the x direction, during an object hand over; Fig. (c) represents the cost function for the case were a person is seated; and Fig. (d) shows the total cost function of a walking person, including the social rule of overtaking her by the left and her personal space.

The third constraint accounts for personal space. We consider three different situations: when a person is standing; walking; or seated. For the case of a person walking, we used the formulation proposed in [8], which the authors call asymmetric Gaussian

$$f = asymGauss(x, y, \theta, \sigma_h, \sigma_s, \sigma_r),\tag{3}$$

where

- θ - orientation of the function;
- σ_h - variance in the θ direction;
- σ_r - variance in the $\theta - \pi$ direction;
- σ_s - variance in the $\theta \pm \frac{\pi}{2}$ direction.

A graphical representation of these parameters is shown in Fig. 3(a). Then personal space of a walking person was modulated as

$$f_1 = asymGauss(x, y, \theta_p, \beta, \frac{2}{3}\beta, \frac{1}{2}\beta), \text{ where } \beta = \max(v, 0.8),\tag{4}$$

where θ_p is the person's orientation and v her speed. A graphical representation of the person walking along y–axis direction with a velocity of 1 [m/s] is presented in Fig. 3(a).

Regarding a walking person, it makes sense for personal space in front to be larger than in the back (to ensure the robot does not pass in front of the person, decreasing risk of collision). On the other hand, if a person is standing and we consider personal space defined using previous formulation, the robot may pass behind too close the person, causing discomfort. Thus, for this case we suggest that personal space be modulated as a circular Gaussian,

$$f_2 = e^{-\left(\frac{(x-x_p)^2}{2\sigma_x^2} + \frac{(y-y_p)^2}{2\sigma_y^2}\right)}, \tag{5}$$

where (x_p, y_p) is the person position, σ_x and σ_x are the standard deviation in the x and y direction respectively. This formulation was also considered for a seated person.

If an object hand over is required, the robot should be able to enter the personal space, to be at "arm's length". However, the robot cannot be allowed to enter from a random direction, instead it should only be allowed to approach a person from the front, [9]. Thus, our solution is to open the region in front of the person 45 degrees, to a distance of 0.6[m]. Personal space, in a hand over scenario, is depicted in Figure 3(b). Constraint 4 concerns preventing discomfort from passing behind a seated person. We reformulated this problem with an asymmetric Gaussian

$$f_3 = asymGauss(x, y, \theta_p - pi, 1.2, 0.8, 0.006). \tag{6}$$

The fifth constraint prevents the robot from interfering with a person interacting with an object. It is represented by an interaction set modelled as a circle

$$f_4 = \begin{cases} \alpha & \text{if } (x - x_c)^2 + (y - y_c)^2 \leq r \\ 0 & \text{otherwise} \end{cases}. \tag{7}$$

Its center is the middle position of the interacting entities, (x_c, y_c), and the radius r is half the distance between the entities (only one-on-one interactions are considered). α is an importance factor, which varies from 0 to 1. Constraint 6 represents the social rule of overtaking people by the left (considered only for walking persons). This constraint is also represented using an asymmetric Gaussian

$$f_5 = asymGauss(x, y, \theta_p - \frac{\pi}{2}, 1.5, 0.3, 0.0075), \tag{8}$$

For the three possible postures of a person (standing, seated and walking), there are two where more than one cost function is applied and they must be combined. Since the main goal of the framework is to maximize the comfort of the humans, the cost functions are combined by taking the maximum cost value attributed to each point in

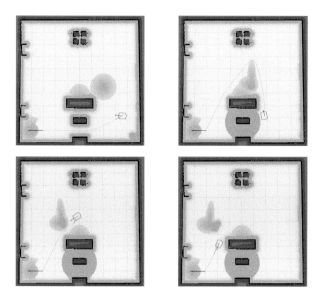

Fig. 4 Evaluation of the proposed navigation system using simulated environments. The environment was created using Gazebo and results are shown in rviz (ROS package). The robot is requested to hand over an object to a person who across the room. However it needs to pass between a seated person and a TV. As it starts moving the TV is turned on. To prevent interference, the robot replans around that area. However, there is yet another person, who is walking behind the couch, who must be taken into account.

space. The first case of multiple cost functions affecting the same space, is a seated person, which personal space is given by

$$f_6 = \max\left(f_2, f_3\right),\tag{9}$$

this cost function is depicted in Fig. 3(c). The second case concerns a walking person, where the personal space must be combined with the respective social rule (a person should be overtaken by the left),

$$f_7 = \max\left(f_1, f_5\right).\tag{10}$$

A graphical representation of this cost function is shown in Fig. 3(d).

4 Experiments

To evaluate the proposed framework, five experiments were defined. They were conducted both on simulation, Sec. 4.1, and using a mobile platform [14], in a typical domestic indoor scenario, with fisheye omnidirectional cameras on the ceiling, Sec. 4.2.

The proposed system was implemented as an extension of ROS navigation stack, [4]. The cost functions described in the previous section were implemented as plug-ins to the cost-map layered structure [13]. The trajectory controller used was the Trajectory Rollout algorithm, [5]. The implementation of computer vision algorithm was based on OpenCV. Based on the frame rate of the vision-based person tracker, a frequency of 1 [Hz] was used for replanning the paths.

The simulation environment was Gazebo. For the vision-based person tracking, a computer with an Intel Core i5-2430M, with 6GB of RAM (external CPU) was used. For navigation components (reconfiguring cost-maps, path path planning, and trajectory execution), we used an Intel Core i7-3770T with 8GB of RAM (on-board CPU).

The experiments performed were:

Experiment 1: The robot is navigating in free space and a standing person appears on the robot's path. The goal is to see if the robot is able to replan its path, without violating personal space.

Experiment 2: The robot is navigating when encounters a slow walking person, which it must overtake. The goal is to verify if it respects constraints 3 and 6, Sec. 3.

Experiment 3: A person is seated on a couch, watching TV, and the robot wants to go across the room. The goal is to test if the robot respects constraints 4 and 5, Sec. 3.

Experiment 4: The robot navigates towards a person, to hand over some object. The goal is to verify the modification of the personal space, constraint 3, Sec. 3.

Experiment 5: Combines the setup of Experiments 2, 3, and 4. The robot is requested to hand over some object to a person across the room. As it starts moving, the TV is turned on, and the robot replans its path. However, another person is going across the room behind the seated person, who must be overtaken for it to reach its goal.

Next, we present experiments results using both simulated and realistic environments. Videos of the experiments can be seen at `http://tinyurl.com/HANEOCExp` (youtube playlist).

4.1 Results on a Simulated Environment

The goal of these experiments (1 to 4) is to evaluate the proposed navigation system (note that some of the proposed constraints were reformulated and thus needed to be evaluated). The experiments were conducted in a simulated room with some furniture. Three people were placed in the environment, one sitting on a couch and the other two standing. For the first experiment, there were no people in the environment at first, only after the robot started moving, a person was added in its path. Several runs of each experiment were conducted to ensure the robot's behaviour was the desired.

During those runs the trajectory controller parameters were fine tuned. Finally, an experiment integrating all the cost functions was conceived (Experiment 5).

When the robot receives the action, the personal space of the target opens in front, and the robot starts moving. However, the TV is turned on, as the robot is reaching the area between a seated person and the TV. Since the importance given to the interaction area was 1 the area was marked as forbidden and the robot needs to replan around. Meanwhile another person is walking slowly behind the seated person, thus the robot plans a path to overtake her. A dark grey area continues on the map after the person started moving. This is were the person was before and it was marked as an obstacle, which is cleared as soon as the robot understands from the laser scans that the obstacle is no longer there. Since the robot could only pass on the person's right, the path is longer (to satisfy Constraint 6). Given the longer path the robot is not able to overtake the person, so it passes behind her when there is enough space behind the person. Finally passing the moving person the robot is free to reach its goal. Pictures of the run in question are presented in Fig. 4.

4.2   Real Experiments in an Indoor Domestic Scenario

In this subsection, we validate the complete framework using real experiments, in a typical indoor domestic scenarios. Experiments 1 and 4 were performed 10 times, with fixed start and goal positions. The computed paths and the position of the person in each run are presented in Fig. 5(a) and 5(d), respectively. Experiment 3 was performed 10 times with the TV off and 10 times with the TV on. The paths were recorded and plotted in Fig. 5(c). Finally, Experiment 2 was performed 3 times. The results are shown in Fig. 5(b).

Throughout the experiments, the robot displayed a similar behaviour to the simulation in terms of trajectory execution in most cases. However, the parameters of the cost functions (4), (5), (6), and (8) needed to be adjusted. The values presented previously were derived empirically, taking into account: the values in the literature; the space restrictions of the real scenario, and our intuition of comfort distances. Even though the final distances are similar to previous works, an effort should be made to further validate those parameters, by conducting user studies.

Computational performance of the overall system is mainly constrained by the performance of the vision-based person tracker. When a person is moving slowly, standing and/or seated, the system is able to keep track of the person, as can be seen by the high repeatability of the execution of Experiments 1, 3 and 4. However, for a person walking at faster pace, the system may not show the same performance, Fig. 5(b). As the person moves faster and faster, the vision system may take longer to estimate the person state (position and velocity) than she takes to take another step, thus the estimate may be delayed when compared with the true position of the person. The limiting factor in the vision-based tracker is the slow background subtraction step (high resolution images were used).

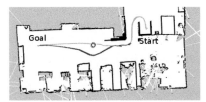 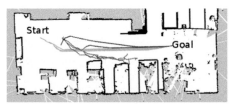

(a) Results for Experiment 1. Green lines represent paths computed before the person entered the scene. Blue lines represent the final path (which was replaned when person interfere with robots path). Person location is represented by red circles.

(b) Results for Experiment 2. Paths computed before the person started walking are in green. Blue lines show the paths computed after the person started walking. The estimated person's path is shown as red.

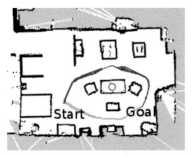 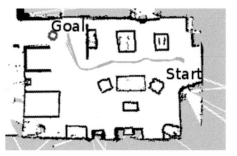

(c) Results for Experiment 3. Green lines show robot's final paths computed with TV off. Blue lines show the final robot's paths with TV on. Red circles identify person location.

(d) Results for Experiment 4. Green lines show robot's paths to reach the person for object hand over. Person location in each run are shown in red circles.

Fig. 5 Results of the real experiments in a realistic domestic indoor scenario, using a mobile robot, [14].

5 Conclusions

This work addresses the problem of Human-Aware Navigation for a robot in a social context. The proposed framework combines a networked robotics environment, with specific motion constraints representing the human-aware concerns. A new framework was presented, which increases the field of view by using external omnidirectional cameras and attempts to standardize human-aware constraints. Five experiments were conducted in simulation and four in a real scenario to assess the system performance. On simulation, the navigation was shown to satisfy all human-aware constraints. Regarding the real experiments, when adding the people detector and tracking, successful experiments were also achieved, even though a limitation was identified.

As future work, we are planning on fusing data from external and on-board cameras, and extend the tracking system to account for multiple people. Finally, conduct user studies to evaluate how people feel about the robot and validate cost functions parameters.

References

1. Bitgood, S., Dukes, S.: Not Another Step! Economy of Movement and Pedestrian Choice Point Behavior in Shopping Malls. Environment and Behavior **38**(3), 394–405 (2006)
2. Brooks, A., Williams, S.: Tracking people with networks of heterogeneous sensors. In: Proceedings of the Australasian Conference on Robotics and Automation (ACRA), pp. 1–7 (2003)
3. Chang, C.C., Lin, C.J.: LIBSVM: A Library for Support Vector Machines. ACM Trans. Intelligent Systems and Technology (TIST) **2**(3), 27 (2011)
4. Marder Eppstein, E., Berger, E., Foote, T., Gerkey, B., Konolige, K.: The office marathon: robust navigation in an indoor office environment. In: IEEE Proc. Int'l Conf. Robotics and Automation (ICRA), pp. 300–307 (2010)
5. Gerkey, B., Konolige, K.: Planning and control in unstructured terrain. In: IEEE Proc. Int'l Conf, Robotics and Automation (ICRA) Workshop on Path Planning on Costmaps (2008)
6. Hall, E.: The Hidden Dimension: Man's Use of Space in Public and Private. Bodley Head (1969)
7. Hartley, R.I., Zisserman, A.: Multiple View Geometry in Computer Vision, 2nd edn. Cambridge University Press (2004)
8. Kirby, R.: Social Robot Navigation. Ph.D. thesis, Robotics Institute, Carnegie Mellon University (2010)
9. Koay, K.L., Sisbot, E.A., Syrdal, D.S., Walters, M.L., Dautenhahn, K., Alami, R.: Exploratory study of a robot approaching a person in the context of handing over an object. In: AAAI Spring Symposium: Multidisciplinary Collaboration for Socially Assistive Robotics, pp. 18–24 (2007)
10. Kruse, T., Basili, P., Glasauer, S., Kirsch, A.: Legible robot navigation in the proximity of moving humans. In: IEEE Workshop on Advanced Robotics and its Social Impacts (ARSO), pp. 83–88 (2012)
11. Kruse, T., Kirsch, A., Sisbot, E.A., Alami, R.: Exploiting human cooperation in human-centered robot navigation. In: IEEE Int'l Symposium on Robot and Human Interactive Communication (RO-MAN), pp. 192–197 (2010)
12. Kruse, T., Pandey, A.K., Alami, R., Kirsch, A.: Human-aware robot navigation: A survey. Robotics and Autonomous Systems **61**(12), 1726–1743 (2013)
13. Lu, D.V., Hershberger, D., Smart, W.D.: Layered costmaps for context-sensitive navigation. In: IEEE/RSJ Proc. Int'l Conf. Intelligent Robots and Systems (IROS), pp. 709–715 (2014)
14. Messias, J., Ventura, R., Lima, P., Sequeira, J., Alvito, P., Marques, C., Carriço, P.: A robotic platform for edutainment activities in a pediatric hospital. In: IEEE Int'l Conf. Autonomous Robot Systems and Competitions (ICARSC), pp. 193–198 (2014)
15. Pandey, A., Alami, R.: A framework towards a socially aware mobile robot motion in human-centered dynamic environment. In: IEEE/RSJ Proc. Int'l Conf. Intelligent Robots and Systems (IROS), pp. 5855–5860 (2010)
16. Scandolo, L., Fraichard, T.: An anthropomorphic navigation scheme for dynamic scenarios. In: IEEE Proc. Int'l Conf. on Robotics and Automation (ICRA), pp. 809–814 (2011)
17. Scaramuzza, D., Martinelli, A., Siegwart, R.: A toolbox for easily calibrating omnidirectional cameras. In: IEEE/RSJ Int'l Conf. Intelligent Robots and Systems (IROS), pp. 5695–5701 (2006)
18. Shi, D., Collins Jr., E.G., Goldiez, B., Donate, A., Liu, X., Dunlap, D.: Human-aware robot motion planning with velocity constraints. In: Int'l Symposium Collaborative Technologies and Systems (CTS), pp. 490–497 (2008)
19. Sisbot, E.A., Luis, F.M.U., Alami, R., Simeon, T.: A Human Aware Mobile Robot Motion Planner. IEEE Trans. Robotics **23**(5), 874–883 (2007)

Motion Descriptor for Human Gesture Recognition in Low Resolution Images

António Ferreira, Guilherme Silva, André Dias,
Alfredo Martins and Aurélio Campilho

Abstract A great variety of human gesture recognition methods exist in the literature, yet there is still a lack of solutions to encompass some of the challenges imposed by real life scenarios. In this document, a gesture recognition for robotic search and rescue missions in the high seas is presented. The method aims to identify shipwrecked people by recognizing the hand waving gesture sign.

We introduce a novel motion descriptor, through which high recognition accuracy can be achieved even for low resolution images. The method can be simultaneously applied to rigid object characterization, hence object and gesture recognition can be performed simultaneously.

The descriptor has a simple implementation and is invariant to scale and gesture speed. Tests, preformed on a maritime dataset of thermal images, proved the descriptor ability to reach a meaningful representation for very low resolution objects. Recognition rates with 96.3% of accuracy were achieved.

Keywords Motion descriptor · Gesture descriptor · Search and rescue · Robotics · Gesture recognition

1 Introduction

Human-computer interaction is a multidisciplinary field that aims to develop communication channels between humans and machines. In this context, human gesture

A. Ferreira(✉) · G. Silva · A. Dias · A. Martins
INESC TEC - INESC Technology and Science and ISEP/IPP - School of Engineering,
Polytechnic Institute of Porto, Porto, Portugal
e-mail: ajbf@inescporto.pt

A. Campilho
INESC TEC - INESC Technology and Science and FEUP - Faculty of Engineering,
University of Porto, Porto, Portugal

© Springer International Publishing Switzerland 2016
L.P. Reis et al. (eds.), *Robot 2015: Second Iberian Robotics Conference*,
Advances in Intelligent Systems and Computing 417,
DOI: 10.1007/978-3-319-27146-0_23

recognition is a key research topic with various branches. Applications range from real-time interaction with entertainment devices to robotic perception.

In robotic applications, gesture recognition is mostly associated with video analysis. The large variety and variability of gestures, the large number of degrees of freedom involved, among variations in view point and scale, constitute the main challenges for this type of problems.

Methods fall in two main categories [1]: some make use of traditional computer vision techniques and decouple the spatial from the temporal aspects, while others build upon the spatio-temporal relationships to achieve an integrated gesture description.

Generally, the first class of methods, that often involve feature tracking or optical flow calculation [2], preform weakly [3] [4] in case of low aperture, occlusions, sudden motion changes, low resolution video, changes in appearance, between other perturbations. The second class of methods blend spatial and temporal characteristics, to achieve a description for the entire motion, and hence retain a continuous representation of each behavior sequence.

In [3], a spatial-time shape, composed of a 2D object silhouette sequence, is organized chronologically to encode each behavior as a 3D space-time shape. A target behavior can be easily classified, by comparing its spatial-time shape with the ones from previously learned behaviors, using a simple nearest-neighbor search. Robustness to partial occlusions, non-rigid deformations, silhouette imperfections, and motion variability is reported.

A similar notion, called motion history gradients, is explored in [5]. This time, silhouette sequences are encoded in a single 2D image (motion history image), by a successive layering process. Every time a new silhouette arrives, the brightness level of existing silhouettes decrease, and the new silhouette is blended at full brightness. Gestures are identified through a gradient analysis on the motion history image. Fast computation and high recognition rates are highlighted features by the authors. Nevertheless, low performance is expected for low resolution images, small amplitude movements and varying motion speeds.

Both low resolution and noise aspects are considered in [2], where direction histograms of optical flow are computed in 8 regions around the object silhouette center. By extracting the optical flow direction in several image regions, the motion of each human body part is approximately estimated. A feature vector is achieved through concatenation of all normalized region histograms. Each element, form the feature vector, corresponds to a bin of a particular histogram. Feature vector noise, introduced by the optical flow method over time, is smoothed using a moving average filter. Although, the efficient noise reduction stage, motion recognition results show low discriminant capability by the motion descriptor. Gestures with similar body posture and motion are easily misclassified.

Our main contribution consists on a simple gesture descriptor, devoted to human gesture recognition in low resolution images. A polar histogram is introduced to build and store a motion signature for each gesture. The gesture descriptor was also applied to rigid object characterization, enabling simultaneous object and gesture recognition using the same methodology. Perturbation aspects like gesture speed variations, scale and low resolution are efficiently tackled.

Gesture classification can be achieved in several different ways. A nearest neighbor (NN) search, using a similarity measure like the Sum of Square Differences (SSD), as well as Support Vector Machines (SVM), proved to give acceptable results. The method was applied to a proprietary dataset of thermal images, collected by a water surface robot, for detection of shipwrecked victims asking for help in the water. In the presence of a large number of false positives, gesture analysis is used to identify victims in a disaster scenario.

Auxiliary image processing routines were needed to achieve a fully functional system. A significant image section depicts the sky and not the water surface. To focus the processing resources in the most important area – the water surface – the horizon line was detected, and only the image portion below the horizon is taken into account for target detection and gesture recognition. Additionally, gesture descriptors are built along several video frames, so target identity must be preserved. For that purpose, a Kalman Filter based multi-target tracking scheme was implemented.

This document is organized as follows: Section 2 details the steps involved in the construction of the gesture descriptor, as well as, the technique for horizon line detection in maritime scenarios. Results are presented in Section 3, followed by results discussion on Section 4. Section 5 provides some final conclusions.

2   Methods

In search and rescue operations of shipwrecked victims, the area of interest is the water surface, as the image portion associated with the sky does not bring any valuable information. Therefore, a horizon line detection operation is carried out at first, allowing the gesture recognition method to focus just on the interest area. The gesture descriptor is then built by combining a set of polar histograms, constructed along a sequence of frames for each detected object. Gesture recognition is finally performed by comparing the gesture descriptors with a previously constructed dataset of gestures.

2.1   Horizon Line Detection

During daylight and with clear sky, the horizon line in maritime scenarios can be clearly identified. Figure 1 shows the process developed for that purpose. First, a Gaussian filter is applied to smooth the image and reduce the edge detector response. Edges are readily detected through the Canny edge detector, followed by a line detection stage using the Hough transform. Outliers are rejected in a first stage by establishing a minimal length. In the thermal images available, the sky is represented by a high intensity value, while lower intensities are associated with the sea water.

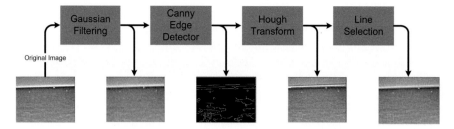

Fig. 1 Representation of the processing pipeline for horizon line detection.

For that reason, the horizon line is finally selected by analyzing the mean intensity difference between pixels above and below the line. This analysis is centered on the pixels from both ends of each line. Considering (x_r, y_r) the coordinates of the right end point of a line, the intensity analysis is performed through the inequation below.

$$\frac{\sum_{i=1}^{N} I(x_r, y_r - i) - \sum_{i=1}^{N} I(x_r, y_r + i)}{N} > th \qquad (1)$$

Where N defines the number of pixels to consider and th is the intensity threshold. The line is selected when expression (1) is satisfied and the same happens for the left end point of the line.

2.2 Motion Descriptor

The new motion descriptor detailed here is directed for body human gesture encoding. It builds upon the idea that gestures occur in the vicinity of a person's center of mass. It also assumes that each gesture can be properly described by measuring the silhouette occupancy rate around the center of mass. This method was designed to be immune to gesture speed and frame rate invariant, so each body gesture type should have a distinct motion pattern, as the descriptor does not differentiate between faster and slower movements. It does not differentiate also movements in the reverse direction.

Similarly to [5], the motion descriptor is built in a layering manner. Nevertheless, instead of overlaying the silhouettes directly, polar histograms are generated for each frame and combined at the end, to obtain a histogram descriptor for the complete gesture. In this way, the method does not rely on gradient analysis, being a remarkable advantage in case of low resolution and noisy images, in which gradients become inconsistent.

Single Frame Polar Histograms. The single frame polar histograms are built using the structure depicted in figure 2. Each video frame should be preprocessed to retrieve the silhouette of the object of interest. In the developed maritime experiment, silhouettes are extracted from thermal images using a simple threshold technique.

A set of circles is then arranged around the silhouettes center of mass. In the next step, each circle is divided into equal size angular regions to define the various histogram bins. Each bin counts the number of silhouette intersections with the circles, inside the respective bin region. For the example illustrated in figure 2, the intersections counting for bin 1 are highlighted by the red circles. In this case the first bin will have a count of 2.

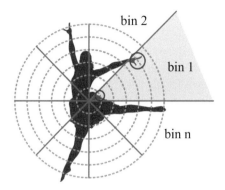

Fig. 2 Method for building the polar descriptor for a single frame. Different ring sizes, arranged around the center of mass, are divided into angular sections to form the polar histogram bins. Each bin counts the number of rings intersected by the silhouette.

Parameters like the amount of histogram bins, number of circles and respective sizes, are crucial to achieve an acceptable object characterization performance. A large number of parameter combinations can be used, however, simpler implementation and adequate performance can be attained by following some guidelines established here.

It was defined that the perimeter of the smallest circle should be equal to the number of histogram bins. This means that there is a one-to-one relationship between the number of bins and the points from the smallest circle. Furthermore, from one circle to the next, the perimeter should double. All these rules are very convenient, ensuring that each bin is obtained through an integer partition of the circle points, hence sub-pixel operations are avoided.

To achieve scale invariance, the total number of circles should be dynamically adjusted, to cover the object area almost completely. The largest circle must have enough size to enclose almost the entire silhouette, although it should also maintain contact with the object, as shown in figure 2. From the descriptor point of view, there are no disadvantages if the outer circle is too big and cannot intersect the silhouette. The only downside is related with the computational effort, spent in the computation of a circle that does not contribute to the histogram building process.

After building the polar histogram for a given frame, it should be normalized so that all bins become in the range [0, 1]. This will provide immunity with respect to the amount of circles used for each frame. Nonetheless, when little object size variations are expected, this step can be neglected, causing meaningless impact on the final gesture descriptor.

The individual polar histogram encodes the silhouette shape for only one frame. For static object classes with low similarity, this type of representation can also be used for object recognition. Moreover, since individual polar histograms are fully compatible with the gesture descriptor detained next, both encoding methods can be applied simultaneous to gesture and object recognition.

Gesture Descriptor. The methodology used to generate the gesture descriptor is presented in figure 3. Individual polar histograms are collected from as many frames as necessary to represent, at least, an entire gesture cycle. Individual polar histograms are normalized and summed together. The result is normalized in order to make the descriptor invariant to the number of frames, frame-rate and variations in gesture speed.

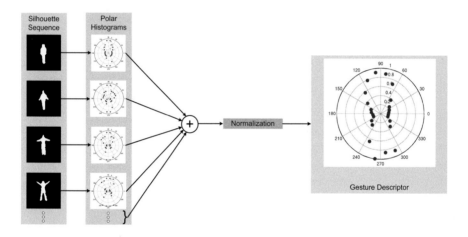

Fig. 3 Illustration of the steps used to build the gesture descriptor from the individual polar histograms.

Online Gesture Descriptor. The motivation for this work came from the need for a gesture recognition technique oriented to very low resolution images. Our target application consists of shipwrecked victim detection, based on hand waving recognition, in the context of robotic search and rescue missions. Figure 4 illustrates the integration of the gesture descriptor in a multi-target tracking framework.

In the first stage, targets of interest are tracked over the video stream. Targets are detected, below the horizon line, by a simple image thresholding operation. As expected, this process is characterized by a high false positive rate.

All targets are tracked within a single Kalman Filter. A constant velocity model is used for prediction and data association is performed using a nearest neighbor

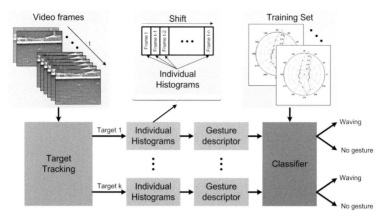

Fig. 4 Overview of the framework for gesture recognition in a multi-target configuration

criterion. New observations, that do not belong to current targets, generate new targets. So the Kalman Filter vectors and matrices expand in order to accommodate new targets. Whenever a target reaches a high uncertainty level in position, due to the lack of observations, it is removed from the Kalman Filter.

A set of polar histograms is then associated with each target. The histogram set, stores the polar histograms built over the last set of frames. As a new frames arrives, new polar histograms are built and stored, while the oldest ones are deleted from the set. For each new frame, an updated gesture descriptor is obtained by combining all stored histograms as detailed previously. After that, the gesture descriptor is used for classification, by comparison with gesture examples in the training set.

Gesture Classification. After collecting some examples, gesture classification can be attempted in several ways. In this work, multiclass classification is accomplished using Support Vector Machines. The k-Nearest Neighbor (k-NN) classifier is also tested, following a Sum of Squared Differences (SSD) metric:

$$SSD = \sum_{i \in bins} \left(H_{ref}(i) - H_{target}(i) \right)^2 \qquad (2)$$

Where H_{ref} is an histogram from the training set, that describes a given gesture, and H_{target} is the histogram for the gesture we want to classify.

In the multi-target problem version, the training set can be composed by only one gesture class. In this case, one-class classification can be accomplished through the k-NN classifier, by measuring the similarity to the training set according to the SSD metric. A threshold applied to the SSD measure could be used to validate the test example within the class.

3   Results

A maritime dataset of thermal images was used to evaluate the different method components. The dataset was collected during the Portuguese Navy Robotics Exercises (REX) at Alfeite Naval Base in 2014. The images, with a resolution of 320x240 pixels, display a maritime disaster scene, with various objects floating in the water, containing shipwrecked people and robotic rescue boats. The low image resolution originates target representations with little detail.

3.1   Descriptor Evaluation

As stated earlier, the gesture descriptor is fully compatible with the individual polar histograms. In this experiment both representation were combined in order to

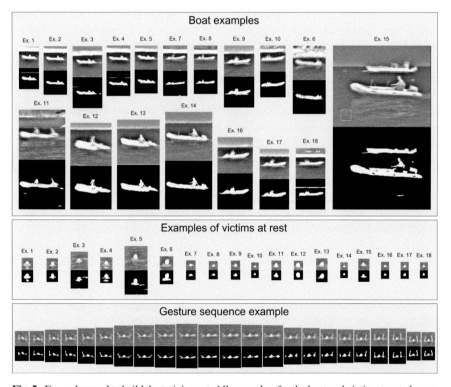

Fig. 5 Examples used to build the training set. All examples, for the boat and victim at rest classes, are shown. The original image patches and the segmented silhouettes are shown for each example. In the gesture sequence box, the data used to build one gesture example is depicted. Similar sequences were used to obtain the remaining 17 gesture examples.

assess their discriminant capability, attempting gesture and rigid object recognition simultaneously. Through the individual polar histograms, two classes – boat and shipwrecked victims at rest – were encoded. On the other hand, the gesture descriptor was applied to sequences of shipwrecked victims performing the waving gesture. An example of a gesture sequence is depicted in figure 5. Figure 5 also illustrates all examples used to characterize the boat and resting victim classes. The dataset contains 54 examples in total, 18 examples per class.

All individual polar histograms and gesture descriptors were generated with a fixed number of 12 bins. Scale invariant representations were obtained by adjusting automatically the number of rings, for individual polar histogram building, according to the object silhouette size, as illustrated in figure 6. Immunity, with respect to the amount of circles, is achieved through normalization of each individual polar histogram.

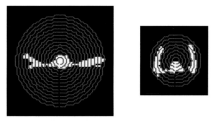

Fig. 6 Dynamic number of rings depending on silhouette size.

Based on the training set, classification was performed following a leave-one-out cross-validation strategy, using SVMs (with radial basis kernel) and k-NN classifiers. In a leave-one-out cross-validation scheme, only one example is used for validation at a time, while the remaining examples are used for training.

The classification accuracy was evaluated for both classifiers. For the k-NN case, different numbers of neighbors were tested. Table 1 summarizes the results. By analyzing table 1, it becomes clear that the best performance is achieved for the SVM and the k-NN (k=1) classifiers. Confusion matrices, depicted in figure 7, were generated for those two cases.

Table 1 Classification accuracy achieved by each classifier following the leave-one-out procedure.

Classifier	SVM	k-NN (k=1)	k-NN (k=3)	k-NN (k=5)	k-NN (k=7)
Accuracy	96.3%	96.3%	92.6%	92.6%	90.7%

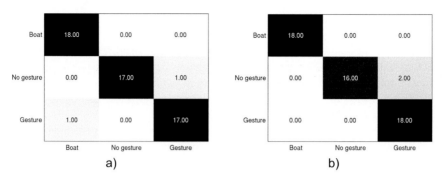

Fig. 7 Confusion matrices for the a) SVM classifier and b) k-NN classifier with k=1. Columns represent the predicted classes and rows represent the actual class

3.2 Video Sequence Experiment

In this experiment, one-class gesture recognition was performed during a video sequence of 571 frames. A training set composed by the same 18 hand waving descriptors from the first experiment was used. The target video shows two shipwrecked victims standing still, in some moments, or performing the hand waving gesture in the

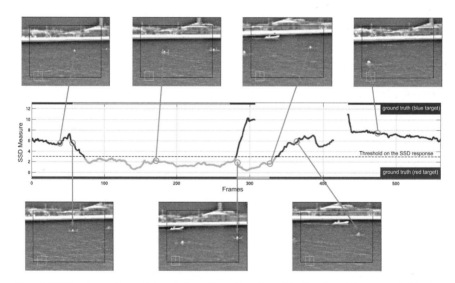

Fig. 8 SSD response for each target along the entire video. The first target is represented by the blue line and the second one by the red line. The line colors changes to green whenever the waving gesture is identified. The true action for the blue target is represented by the color of the top line marked as "ground truth (blue target)". The ground truth for the second target is given by the line in the bottom. Once again, blue and red represent no gesture and green denotes waving. Some video frames were added to better illustrate the real gesture being preformed at different instants.

other instants. The similarity between the descriptor, computed for each victim, and the training set descriptors is used to identify the instants in which victims execute the gesture. The SSD metric is used as similarity measure.

Figure 8 characterizes the gesture recognition results obtained for the entire target video length. The gesture classification was attained by the k-NN method with k=1 using the SSD metric. In this case, the gesture descriptor was built using the last 30 frame polar histograms. Waving gesture was considered whenever the mean SSD value, computed by averaging SSD responses relative to all training set examples, falls below a threshold of value 3. This threshold value was experimentally determined, by comparing the similarity response with the ground truth, using another video sequence.

4 Discussion

Despite the poor quality of the dataset silhouettes, illustrated in figure 5, accuracy levels of 96.3% were achieved for the multiclass classification experiment. This performance level corresponds to only 2 wrong classifications out of 54 attempts. The high accuracy level was accomplished for a dataset with wide scale variations, proving the methods ability to deal with this perturbation.

Both SVM and k-NN with k=1 achieve the same accuracy. Increasing the number of neighbors considered in the k-NN classifier leads to a worst performance. This indicates some difficulty to properly describe each class, which may be related to the low quality of the silhouettes used. In figure 5, some inconsistencies can be readily identified. Boat example 15 contains not only the object of interest, but also another boat in the background. In boat examples 1, 3, 4 and 18 the horizon line was not properly removed. Bad segmented silhouettes can also be identified in victim and waving examples.

Confusion matrices in figure 7 were generated for the SVM and k-NN (k=1) cases. It is possible to verify that all boat examples are correctly classified. The k-NN classifier also identifies correctly all waving gestures. In the SVM case, one waving gesture is misclassified as belonging to the boat class, and one victim at rest is confused with the waving gesture. For the k-NN case, two victims at rest are assigned to the waving gesture class. Again, the high variability that characterizes the three classes, along with the relatively small number of training examples, justify this results.

Results from the application in the multi-target framework give also valuable information about the descriptor performance. To build the descriptor, a set of past polar histograms is used. This generates a memory effect causing some delay in gesture recognition. This effect can be seen for the blue target. The waving gesture is detected later than the gesture starting instant marked by the ground truth. Due to the same cause, gestures continue to be identified after they finish. This phenom occurs for both targets, as they keep being identified after the finishing instant defined by the ground truth.

Apart from this, the immediate recognition of the waving gesture as soon as the red target enters the scene is surprising. This happened, since the set of polar histograms was empty and the person entered the scene already performing the gesture. Such behavior, along with the high noise level from our silhouette segmentation, led to the gesture being recognized by considering only one frame.

5 Conclusion

The gesture descriptor presented here is directed to applications that deal with very low resolution images. The descriptor is invariant to gesture speed and scale variations, has a simple implementation and can be applied to rigid object characterization, enabling object and gesture recognition simultaneous within the same framework.

The descriptor is built by accumulating information along the video sequence. Despite considering only the recent past, this causes the method to present a delayed recognition reaction. Nevertheless, an high recognition accuracy can be achieved even for targets with very low resolution.

Acknowledgment This work is financed by the ERDF – European Regional Development Fund through the COMPETE Programme (operational programme for competitiveness) and by National Funds through the FCT – Fundação para a Ciência e a Tecnologia (Portuguese Foundation for Science and Technology) within project ≪FCOMP-01-0124-FEDER-037281≫.

References

1. Kaâniche, M., Bremond, F.: Recognizing Gestures by Learning Local Motion Signatures of HOG Descriptors. IEEE Transactions on Pattern Analysis and Machine Intelligence **34**, 2247–2258 (2012)
2. Lertniphonphan, K., Aramvith, S., Chalidabhongse, T.H.: Human action recognition using direction histograms of optical flow. In: 2011 11th International Symposium on Communications and Information Technologies (ISCIT), pp. 574–579, October 2011
3. Gorelick, L., Blank, M., Shechtman, E., Irani, M., Basri, R.: Actions as space-time shapes. IEEE Trans. Pattern Anal. Mach. Intell. **29**(12), 2247–2253 (2007)
4. Laptev, I.: On space-time interest points. Int. J. Comput. Vision **64**(2–3), 107–123 (2005)
5. Bradski, G., Davis, J.: Motion segmentation and pose recognition with motion history gradients. In: Fifth IEEE Workshop on Applications of Computer Vision, pp. 238–244 (2000)

Genome Variations

Effects on the Robustness of Neuroevolved Control for Swarm Robotics Systems

Pedro Romano, Luís Nunes, Anders Lyhne Christensen,
Miguel Duarte and Sancho Moura Oliveira

Abstract Manual design of self-organized behavioral control for swarms of robots is a complex task. Neuroevolution has proved a viable alternative given its capacity to automatically synthesize controllers. In this paper, we introduce the concept of Genome Variations (GV) in the neuroevolution of behavioral control for robotic swarms. In an evolutionary setup with GV, a slight mutation is applied to the evolving neural network parameters before they are copied to the robots in a swarm. The genome variation is individual to each robot, thereby generating a slightly heterogeneous swarm. GV represents a novel approach to the evolution of robust behaviors, expected to generate more stable and robust individual controllers, and benefit swarm behaviors that can deal with small heterogeneities in the behavior of other members in the swarm. We conduct experiments using an aggregation task, and compare the evolved solutions to solutions evolved under ideal, noise-free conditions, and to solutions evolved with traditional sensor noise.

Keywords Neuroevolution · Robot controllers · Genome Variations · Swarm robotics · Robustness · Heterogeneity

1 Introduction

Synthesizing control for swarms of robots is a challenging process since the individual rules that govern the behavior of each robot are non-trivial to discover and implement [10]. In the research field of evolutionary robotics (ER), neuroevolution (NE) has been successfully applied to the synthesis of self-organized behavioral control for swarms of robots [4]. Given the set of sensor inputs, the controller, an artificial neural network (ANN), determines the value of the outputs that will be sent to the

P. Romano · L. Nunes(✉) · A.L. Christensen · M. Duarte · S.M. Oliveira
Instituto de Telecomunicações, Instituto Universitário de Lisboa (ISCTE-IUL), Lisboa, Portugal
e-mail: luis.nunes@iscte.pt

© Springer International Publishing Switzerland 2016   309
L.P. Reis et al. (eds.), *Robot 2015: Second Iberian Robotics Conference,*
Advances in Intelligent Systems and Computing 417,
DOI: 10.1007/978-3-319-27146-0_24

actuators. The weights, and sometimes the topology, of the ANN are evolved by an evolutionary algorithm (EA) to optimize a given fitness function. The evolution of controllers is usually conducted in simulation due to its time-intensive nature, and, after evolution ends, the highest-performing controllers are transferred to real robotic hardware. Transferring controllers from simulation to real robots often leads to a decrease in performance due to differences between the simulated and real world, an issue known as "the reality gap" [17].

In the evolution of homogeneous swarms, the same controller is cloned for every robot [27], which means that each genome must encode the behavior for all situations and all team members will react similarly in similar conditions. This is different from biological swarms, where each individual has a different genome and slightly different characteristics (physical, behavioral, etc.). The question raised is "do these slight differences in swarm members play a part in the swarms' capabilities to solve problems?". Potter et al. [22] show that behavior homogeneity is a disadvantage when more complex swarm behaviors are needed and specialization becomes a necessity. Even though certain types of ANN-based controllers, such as continuous-time recurrent neural networks (CTRNN) [3], have been shown to enable the emergence of dynamic task-allocation under specific conditions [11], and even in some cases the appearance of different roles [2, 23], such controllers tend to display homogeneous behaviors.

This paper introduces the concept of *Genome Variations* (GV), a novel technique that is based on mutating the individual controllers of each robot prior to evaluation. Variations are obtained by adding noise to the weights of the artificial neural networks controlling the robots. The use of GV during training may provide more stable and robust solutions for swarm controllers in coordinated tasks and help bridge the reality gap. There are several potential benefits of varying swarm controllers' genomes during evolution, namely: (i) increased tolerance to minor, but unavoidable, hardware differences between different robots, (ii) increased robustness to environmental noise and to (iii) slight heterogeneities in swarm behavior. Increased tolerance to hardware differences [24], noisy sensors [19] and varying environmental conditions, is expected since a small bias in the input has very similar effects in terms of the internal computations from a change in an input weight or a slight bias in the readings from a sensor. If the GV solution is impervious to small variations in the weights, then it should also be able to tolerate small changes in the input, whether these may be caused by the environment or by the robot's hardware.

GV should also increase robustness to small differences in behaviors of other swarm members, since candidate solutions are exposed to slightly heterogeneous swarms during training. It may be advantageous to have slight behavioral differences in different individuals to break ties that can block the solution to a problem (e.g. two robots both insisting on manipulating the same object). Another way in which GV may be advantageous is by forcing the robots to adapt to different behaviors from teammates, whether these are simply small performance differences or robots experiencing faults.

In this paper, we evaluate the effects of GV in swarms of robots in an aggregation task [25, 26]. We chose an aggregation task, since it is a fundamental behavior in many collective systems found in nature [6]. The ability for a group of individuals to aggregate is a precursor to many collective behaviors since being within sensory range of each other is a paramount condition to coordinate collective swarm behaviors. Also, successful aggregation requires the combined use of different skills that are common with other problems (e.g. distributed search, coordinated movement or cooperation).

We assess the performance of controllers evolved in three different scenarios, an ideal evolutionary scenario with no noise (NF), which is used as a baseline, one with sensor noise (SN), and a Genome Variations scenario. The highest performing solution evolved in each scenario is tested in all three setups (NF, SN and GV).

2   Related Work

Variation is a fundamental part of the evolutionary process, since new solutions are based on previously successful solutions by applying evolutionary operators such as mutation and recombination [1]. However, the use of variation for other purposes in the evolution of controllers for swarms of robots remains unexplored. In this section, we present a discussion of work related to variation, robustness and generation of heterogeneity in swarm robotics.

The need to create robust solutions in simulation was identified early on in the field of evolutionary robotics (ER). Cliff et al. [8] use the lowest fitness score obtained, as opposed to the mean, obtained in multiple evaluation samples of the evolved individuals in a simulated visually-guided robot navigation problem. The authors adopt the selection method "to encourage robustness, remembering that there is noise at all levels in the system". Miglino et al. [19] and later Jakobi [17] build on the ideas in [8] and apply noise to simulated sensors, actuators, and environmental elements to create controllers that are more capable of crossing the reality gap.

As other authors attempted alternatives to improve solution robustness that did not involve noise, to compensate for the overhead that noisy conditions impose on training in simulation [21], Gomez and Miikkulainen [16] report that "the appropriate use of noise during evolution can improve transfer [of behaviors to real world scenarios] significantly by compensating for inaccuracy in the simulator". The "appropriate use" refers to adding sensor noise and trajectory noise. Authors report significant improvements in transferred behavior performance.

Lehman et al. [18] introduced the concept of *reactivity* and showed that it is as effective in addressing the robustness problem as training with noise, while being computationally more efficient. Reactivity promotes the learning agent to use behaviors where there are dependencies in the magnitude of changes in a robot's sensors and actuators, i.e. if there is a large change in the sensors, there should be some large change in the actuators. Reactivity should not be confused with the concept of a *reactive agent* usually associated with the subsumption architecture [5].

In the evolution of heterogeneous multirobot teams and swarms, either co-evolution [14, 22], or approaches in which whole teams are encoded in a single genome [22] are typically used. Potter et al. [22] show that as tasks become more difficult, heterogeneity and specialization become more important, although there is a trade-off in learning speed. Nitschke et al. [20] study the evolution of collective behaviors, with a focus on facilitating specialization and team heterogeneity. The authors use genotypic and behavioral difference metrics to group genomes in different sub-populations in order to regulate inter-population recombination. Gomes et al. [14] propose Hyb-CCEA to facilitate the evolution of heterogeneous, cooperative multi-agent systems using co-evolution and a method for merging and splitting sub-teams.

Silva et al. [25] apply online evolution [13] to swarms of robots by continuously generating controllers that can adapt to periodic changes during task execution. Shang et al. [24] report a variant of EA for creating heterogeneity by adapting to each robot's specific hardware: robots compensate for hardware differences by individually re-training to adapt to the specificities of their hardware before re-insertion in the team. D'Ambrosio et al. [9] see each element of the team as a combination of policies, creating a continuum between homogeneity and heterogeneity. As in our approach, all agents have the same base-genome, but these agents also possess a "policy geometry" that makes the behavior of the agent depend on its position and provides for the necessary diversity, as well as keeping the homogeneity of the basic genome.

In the work discussed above, no attempt was made to generate heterogeneity in team or swarm behavior simply by introducing genome variations before evaluation. The most common approach to achieve robustness is by introducing noise during training at different levels of the simulation (sensors, actuators, etc.). Also, the focus of heterogeneity is on task specialization, which generally requires separate evolution of individuals or sub-teams, not on the effects of small genome heterogeneities, as seen in biological swarms and in our approach.

3   Methodology

Artificial evolution is typically an iterative process composed of different stages. The stages can be divided into: population generation, genome evaluation and genome selection. Iterations are usually referred to as *generations*. The first generation is commonly composed of a number of randomly-generated genomes. During evaluation, the fitness of each genome is typically sampled several times under different initial conditions to get a proper estimate of the genome's quality. In each sample, the genome is decoded into a controller that is copied to each robot, and the performance of the swarm with respect to the target task is assessed. In the final stage of the iterative process, a set of genomes is selected for reproduction based on the fitness scores obtained. The selected genomes are used as the seed for the next generation. The process typically continues for a fixed number of generations, until stagnation is detected, or until a solution of sufficient quality has been evolved.

In this study, the robotic controllers that are decoded from genomes are CTRNNs [3]. Genomes are composed of the values of the weights, bias and recurrent connections of a fixed-size CTRNN (in this case: 8 x 5 x 2, with delayed self-connections at each node). In the evaluation phase, the fitness of each genome is sampled 30 times in order to assess its fitness. The final fitness score corresponds to the mean fitness score obtained in the 30 samples. After all genomes of a generation have been evaluated, we use a simple elitist selection strategy that retains the five highest-performing controllers from a generation and also uses these elite genomes to create the new genomes for the next generation (a total of 100 genomes). A new genome is created by mutating one of the elite genomes. The evolutionary process continues until the desired number of generations is reached (400).

In the case of conventional evolution with ANN-based controllers for swarms of robots, an individual's genome is decoded into an ANN, which is copied to each robot in the swarm. With the GV approach studied in this paper, the genome assigned to each robot in the swarm prior to evaluation is not an exact clone of the genome under evolution. Instead, a slight variation of the base-genome is copied to each robot. GV is applied to the base-genome in each evaluation sample by adding a uniformly distributed random value (noise) in a defined range, to a certain portion of the ANN weights, creating an heterogeneous swarm. Different ranges for both the variation magnitude and portion of weights varied are tested on the experiments conducted in this study (see Section 4). GV is only used when assessing the performance of a genome. A different random variation is applied to the ANN copied to each robot, thereby generating a different heterogeneous swarm in every sample from the same base genome.

For our experiments, we use JBotEvolver [12], an open-source simulation platform and neuroevolution framework.

4 Experiments

In this section, we compare our GV approach, neuroevolution with (SN) and without sensor noise (NF). Adding noise to simulated sensors during evolution has been shown to promote robust behaviors that are more capable of crossing "the reality gap" [19], so we will replicate evolution with sensor noise to compare with the robustness of the GV evolution in different scenarios. In the following experiments, we study GV with different settings for both the magnitude of variations and the proportion of ANN parameters varied.

The task used in these experiments is standard swarm robotics aggregation [26]. Our aggregation task is performed in a bounded arena, where 15 robots should find each other and aggregate, i.e. form a connected cluster with the greatest number of robots possible. A robot is considered in a cluster if it is within 0.25 m of at least one other robot from that cluster.

Robots can sense each other with four sensors dispersed evenly on their body, and can also sense walls, using four sensors that are also dispersed evenly on their body. All sensors have an opening angle of 90°, providing each robot with omnidirectional sensing, and they have a range of 1 m. All sensory readings are linearized to the interval [0, 1] before being fed to the controller. The value of the actual reading of the sensor is linearly proportional to the distance at which the object is sensed. In this task, the size of the arena is 4 m in odd samples and 8 m in even samples. The size of the arena is varied to prevent overfitting of solutions to a particular size. Controllers are scored based on the largest cluster of robots in the environment, according to the following equation inspired by [15]:

$$Fitness = \frac{\sum_{t=0}^{T} \left(max \left(\sum_{r=0}^{R} (C_{rt}) \right)^2 \right)}{T} \tag{1}$$

where T is the number of control cycles in the sample (3000), R the number of robots (15), and C_{rt} is the number of robots clustered with robot r at control cycle t.

Throughout the following experiments we have three scenarios: Noise-Free (NF), Sensor Noise (SN), and Genome Variations (GV), that will be detailed in Sections 4.1, 4.2 and 4.3, respectively. We assess the performance of genomes evolved in each scenario in all test setups (NF, SN and GV).

The noise-free NF scenario will be the baseline. In the SN scenario (Section 4.2), each sensor in each robot is assigned with an offset uniformly distributed within the range $[-0.1, 0.1]$, and a random noise on each of its sensor readings, also in the same range. Sensor offset noise is kept constant through the sample's lifetime while sensor reading noise is random for each reading. Total noise in the SN environment has an amplitude lower than the 0.4 suggested by [17]. In the SN scenario, all genomes are evolved in a noisy environment. In the GV scenario (Section 4.3), GV is used during evolution. We study three different variants of the GV scenario, each with different parameters: (i) $[-0.125, 0.125]$ variation on 12.5% of the ANN weights, (ii) $[-0.25, 0.25]$ variation on 25% of the ANN weights, iii) $[-0.5, 0.5]$ variation on 50% of the ANN weights. Post-evolution we assess the performance of the highest scoring controllers from each of the scenarios in all three test setups: NF, SN, and GV.

We are interested in assessing how the different solutions perform, and in comparing the robustness and self-organization capabilities of controllers evolved in the different scenarios. Five evolutionary runs are conducted in each scenario (NF, SN, and three GV variants) with a total of 400 generations per run. Each generation is composed of 100 genomes, which are evaluated in 30 samples each. After the evolutionary process ends, the highest-performing genome of the last generation is post-evaluated in 100 samples in the different test scenarios considered for each experiment.

4.1 Noise-Free Neuroevolution

The first set of experiments serves as a baseline: we conducted a total of five evolutionary runs in the noise-free (NF) scenario. We then assessed the performance of the

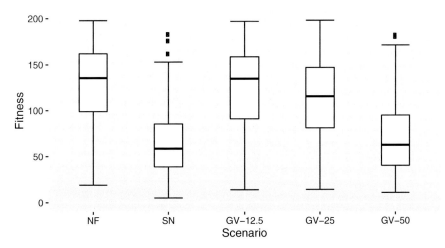

Fig. 1 Results for NF evolved controller in the three different scenarios: NF, SN and GV with [−0.125, 0.125] variation on 12.5% of the ANN weights (GV-12.5), GV with [−0.25, 0.25] variation on 25% of the ANN weights (GV-25) and GV with [−0.5, 0.5] variation on 50% of the ANN weights (GV-50). Mean fitness obtained in 100 post-evaluation trials with the highest-scoring controller for each run (five runs), resulting in a total of 500 observations per box.

highest-scoring genome evolved in each run in all test setups, see results in Figure 1. Notice that swarms are not being evolved with SN or GV, only in the post-evolution assessment is SN or GV introduced.

Controllers evolved in the NF scenario suffer from a significant fitness drop when exposed to SN. Recall that in the GV test setups, we use a different variation for each robot in the swarm, creating an heterogeneous swarm. Still, since controllers were not evolved with GV, performance degrades as the GV interval increases (GV-12.5, GV-25 and GV-50). This provides an indication of the amount of variation the solutions can cope with, before performance degrades, namely between 0.25 and 0.5, in GV-25 and GV-50.

4.2 Sensor Noise Environment

In the second set of experiments, we conducted five evolutionary runs in the SN scenario, and assessed the performance of the highest-scoring genome evolved in each run in all test setups, see results in Figure 2. Details on how SN is applied can be found in the beginning Section 4.

Controllers evolved in a SN environment have a good behavior in both NF and SN test setup. The GV used in this test (heterogeneous swarms) maintains a performance on par with NF and SN trained up to GV-25. Performance in tests with high GV intervals (GV-50) degrades but performance is still higher than the noise-free-evolved

316 P. Romano et al.

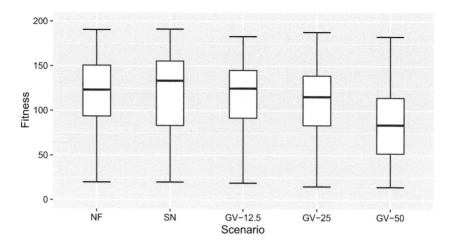

Fig. 2 Results for SN evolved controller in the three different scenarios: NF, SN and GV. In this setup noise was added to random sensors in each robot. The GV scenarios have $[-0.125, 0.125]$ variation on 12.5% of the ANN weights (GV-12.5), GV with $[-0.25, 0.25]$ variation on 25% of the ANN weights (GV-25) and GV with $[-0.5, 0.5]$ variation on 50% of the ANN weights (GV-50). Mean fitness obtained in 100 post-evaluation trials with the highest-scoring controller for each run (five runs), resulting in a total of 500 observations per box.

solutions in the SN test (SN in Figure 1). It can be expected since SN evolution should enable the genome to cope with noise in the input up to $[-0.1, 0.1] \pm [-0.1, 0.1]$.

In an additional set of experiments (not shown), GV evolved controllers were tested in a setup that had both SN and GV simultaneously. These experiments tested homogeneous versus heterogeneous swarms in SN scenarios, but we chose not to include them since they were not significantly different from the results presented in Figure 2.

4.3 Genome Variations

In the third set of experiments, we conducted five evolutionary runs in each of the three GV scenarios, and assessed the performance of the highest-scoring genome evolved in each run in all test setups, see results in Figure 3.

When using a small variation interval (GV-12.5), GV evolved solutions are similar to the solutions evolved in the NF scenario. In this set of experiments, both NF and GV-12.5, evolved behaviors in which the robots aggregate with search-like patterns moving freely inside the environment. GV assessments with larger variation intervals, GV-25 and GV-50, sometimes evolved a behavior in which the robots group

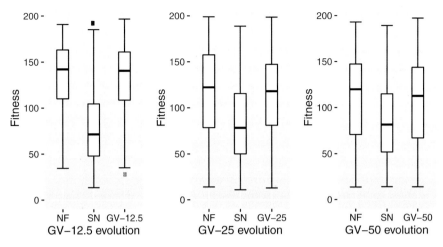

Fig. 3 Results for GV-12.5, GV-25 and GV-50 evolved controller in the three different scenarios: NF, SN and GV (heterogeneous swarms). In GV tests, the variation interval is the same as the one used on the evolution of that specific controller. Mean fitness obtained in 100 post-evaluation trials with the highest-scoring controller for each run (five runs), resulting in a total of 500 observations per box.

by following the walls, turning corners, and eventually finding each other. All the behaviors evolved in the GV scenario solved the aggregation task in the NF test setup.

While a $[-0.125, 0.125]$ variation on 12.5% of the weights has almost no impact in the fitness, larger variations in more weights lead to solutions with an increasingly worse fitness. GV-evolved solutions generally required more time to form a group (around 1700 control cycles on NF evolved controllers and 2040 on GV-25 — tests done in the NF test scenario, small arena). Apart from the quantitative evaluation shown above, a qualitative evaluation of the solutions shows that the solutions evolved with GV with a low degree of variation, appear, to the naked eye, to be equal to the ones evolved in the noise-free NF setup. Also, robots in GV-evolved solutions with larger variation intervals lack the smoothness of movements that NF and narrow-variation GV's have.

As seen in Section 4.1, when tested in SN setup, controllers evolved in the NF scenario have a significant drop in fitness when exposed to SN. Surprisingly, the fitness drop of GV-evolved controllers in presence of SN is comparable to that of noise-free-evolved controllers in this case. As mentioned above, larger GV intervals tend to evolve different behaviors. The behavior where the robots aggregate by following the walls and finding each other near them is probably a more robust strategy in a SN environment.

5   Conclusions and Future Work

In this paper, we introduced Genome Variations, a technique in which small mutations are applied to the evolving neural networks before they are copied to each robot in a swarm. The variation is introduced only in the evaluation of the swarm, and it is individual to each robot, leading to slight behavioral heterogeneity in the swarm. GV was expected to provide a certain degree of robustness whether by resembling the effects of sensor noise, and/or by promoting behaviors robust enough to cope with slightly different peer behaviors during evolution.

Experiments in the aggregation task only partially confirm the hypothesis. GV does indeed appear to produce more robust behaviors than those evolved in noise-free scenarios, but robustness is not comparable to solutions evolved with sensor noise, nor does the resulting mild degree of heterogeneity seem to have any significant impact on the robots' behavior in this task.

Future work will focus on increasing the number of evolutionary runs and include additional tasks to provide a sufficient basis to determine the significance of any performance differences. We intend to determine whether GV can provide some form of generic robustness that is less expensive to train than tolerance to each specific type of noise applied to sensors, actuators, or to the environment. We will also try to use NEAT (more reactive and flexible than CTRNNs). NEAT will also allow an interesting analysis of the different types of networks evolved with each setup (number of nodes, connections, recurrent connections). Also, we plan to assess the impact of GV on swarm fault-tolerance and its potential advantages on behavior transfer from simulation to real-robot scenarios.

Acknowledgements This work was supported by Fundação para a Ciência e a Tecnologia (FCT) under the grants, SFRH/BD/76438/2011, UID/EEA/50008/2013, and EXPL/EEI-AUT/0329/2013, integrated in the CORATAM and HANCAD projects [7].

References

1. Bäck, T., Schwefel, H.P.: An overview of evolutionary algorithms for parameter optimization. Evolutionary Computation **1**(1), 1–23 (1993)
2. Baldassarre, G., Nolfi, S., Parisi, D.: Evolving mobile robots able to display collective behaviors. Artificial Life **9**(3), 255–267 (2003)
3. Beer, R.D., Gallagher, J.C.: Evolving Dynamical Neural Networks for Adaptive Behavior. Adaptive Behavior **1**(1), 91–122 (1992)
4. Bongard, J.C.: Evolutionary robotics. Communications of the ACM **56**(8), 74–83 (2013)
5. Brooks, R.A.: Elephants don't play chess. Robotics and Autonomous Systems **6**(1), 3–15 (1990)
6. Camazine, S., Deneubourg, J.L., Franks, N.R., Sneyd, J., Theraulaz, G., Bonabeau, E.: Self-organisation in biological systems. Princeton University Press (2001)
7. Christensen, A.L., Oliveira, S., Postolache, O., de Oliveira, M.J., Sargento, S., Santana, P., Nunes, L., Velez, F., Sebastião, P., Costa, V., Duarte, M., Gomes, J., Rodrigues, T., Silva, F.: Design of communication and control for swarms of aquatic surface drones. In: Proceedings of the International Conference on Agents and Artificial Intelligence (ICAART), pp. 548–555. SCITEPRESS, Lisbon (2015)

8. Cliff, D., Husbands, P., Harvey, I.: Evolving visually guided robots. In: Proceedings of the Second International Conference on Simulation of Adaptive Behavior (SAB), pp. 374–383. MIT Press, Cambridge (1993)

9. D'Ambrosio, D.B., Stanley, K.O.: Scalable multiagent learning through indirect encoding of policy geometry. Evolutionary Intelligence **6**(1), 1–26 (2013)

10. Dorigo, M., Trianni, V., Şahin, E., Groß, R., Labella, T., Baldassarre, G., Nolfi, S., Deneubourg, J., Mondada, F., Floreano, D., Gambardella, L.M.: Evolving self-organizing behaviors for a swarm-bot. Autonomous Robots **17**(2), 223–245 (2004)

11. Duarte, M., Oliveira, S., Christensen, A.L.: Towards artificial evolution of complex behavior observed in insect colonies. In: Proceedings of the Portuguese Conference on Artificial Intelligence (EPIA), pp. 153–167. Springer, Berlin (2011)

12. Duarte, M., Silva, F., Rodrigues, T., Oliveira, S.M., Christensen, A.L.: JBotEvolver: a versatile simulation platform for evolutionary robotics. In: Proceedings of the International Conference on the Synthesis & Simulation of Living Systems (ALIFE), pp. 210–211. MIT Press, Cambridge (2014)

13. Ficici, S.G., Watson, R.A., Pollack, J.B.: Embodied evolution: a response to challenges in evolutionary robotics. In: Proceedings of the European Workshop on Learning Robots, pp. 14–22. Citeseer (1999)

14. Gomes, J., Mariano, P., Christensen, A.L.: Cooperative coevolution of partially heterogeneous multiagent systems. In: Proceedings of the International Conference on Autonomous Agents and Multiagent Systems, pp. 297–305. International Foundation for Autonomous Agents and Multiagent Systems (2015)

15. Gomes, J., Urbano, P., Christensen, A.L.: Evolution of swarm robotics systems with novelty search. Swarm Intelligence **7**(2–3), 115–144 (2013)

16. Gomez, F.J., Miikkulainen, R.: Transfer of neuroevolved controllers in unstable domains. In: Proceedings of Genetic and Evolutionary Computation Conference (GECCO), pp. 957–968. Springer, Berlin (2004)

17. Jakobi, N.: Evolutionary robotics and the radical envelope-of-noise hypothesis. Adaptive Behavior **6**(2), 325–368 (1997)

18. Lehman, J., Risi, S., D'Ambrosio, D.B., Stanley, K.O.: Rewarding reactivity to evolve robust controllers without multiple trials or noise. In: Proceedings of the Thirteenth International Conference on Artificial Life (ALIFE), pp. 379–386. MIT Press, Cambridge (2012)

19. Miglino, O., Lund, H.H., Nolfi, S.: Evolving mobile robots in simulated and real environments. Artificial Life **2**(4), 417–434 (1995)

20. Nitschke, G.S., Eiben, A.E., Schut, M.C.: Evolving team behaviors with specialization. Genetic Programming and Evolvable Machines **13**(4), 493–536 (2012)

21. Paenke, I., Branke, J., Jin, Y.: Efficient search for robust solutions by means of evolutionary algorithms and fitness approximation. IEEE Transactions on Evolutionary Computation **10**(4), 405–420 (2006)

22. Potter, M.A., Meeden, L.A., Schultz, A.C.: Heterogeneity in the coevolved behaviors of mobile robots: the emergence of specialists. In: Proceedings of the International Joint Conference on Artificial Intelligence (IJCAI), pp. 1337–1343. Citeseer (2001)

23. Quinn, M., Smith, L., Mayley, G., Husbands, P.: Evolving controllers for a homogeneous system of physical robots: structured cooperation with minimal sensors. Philosophical Transactions. Series A, Mathematical, Physical, and Engineering Sciences **361**(1811), 2321–2343 (2003)

24. Shang, B., Crowder, R., Zauner, K.P.: Simulation of hardware variations in swarm robots. In: Proceedings of the IEEE International Conference on Systems, Man, and Cybernetics, pp. 4066–4071. IEEE Press, Piscataway (2013)

25. Silva, F., Urbano, P., Correia, L., Christensen, A.L.: odNEAT: An Algorithm for Decentralised Online Evolution of Robotic Controllers. Evolutionary Computation **23**(3), 421–449 (2015)

26. Trianni, V., Nolfi, S., Dorigo, M.: Cooperative hole avoidance in a swarm-bot. Robotics and Autonomous Systems **54**(2), 97–103 (2006)

27. Waibel, M., Keller, L., Floreano, D.: Genetic team composition and level of selection in the evolution of cooperation. IEEE Transactions on Evolutionary Computation **13**(3), 648–660 (2009)

Adaptive Sampling Using an Unsupervised Learning of GMMs Applied to a Fleet of AUVs with CTD Measurements

Abdolrahman Khoshrou, A. Pedro Aguiar and Fernando Lobo Pereira

Abstract This paper addresses the problem of real-time adaptive sampling using a coordinated fleet of Autonomous Underwater Vehicles (AUVs). The system setup consists of one leader AUV and one or more follower AUVs, all equipped with conductivity, temperature and depth (CTD) sensor devices and capable of running in real-time an on-line unsupervised learning computer algorithm that uses and updates Gaussian Mixture Models (GMMs) to model the CTD data that is being acquired in real-time. The path to be traced by the leader is predefined. The followers path will depend on the CTD data. More precisely, during each resurfacing of the AUVs (and this has to be done in a coordinated fashion), every follower AUV receives the GMM hypothesis of the leader and computes the *variational distance* error between its own GMM and the received one. This error, that provides a notion of how different is the CTD data of each follower from the leader, is used to reconfigure the formation by scaling the distance between the AUVs in the formation (making a zoom-in and zoom-out), in order to improve the efficiency of the CTD data acquisition in a given region. The simulation results show the feasibility of the proposed strategy in uniform and more complex environments.

Keywords Adaptive sampling · AUV · Gaussian mixture models · Unsupervised learning

1 Introduction

Over the last decade, there has been a flurry of activity in the development of autonomous marine robotic vehicles to improve the means available for ocean exploration and exploitation. An Autonomous Underwater Vehicle (AUV) is a robot

A. Khoshrou · A.P. Aguiar(✉) · F.L. Pereira
Research Center for Systems and Technologies, Faculty of Engineering,
University of Porto (FEUP), Rua Dr. Roberto Frias, 4200-465 Porto, Portugal
e-mail: {a.khoshrou,pedro.aguiar,flp}@fe.up.pt

© Springer International Publishing Switzerland 2016   321
L.P. Reis et al. (eds.), *Robot 2015: Second Iberian Robotics Conference*,
Advances in Intelligent Systems and Computing 417,
DOI: 10.1007/978-3-319-27146-0_25

designed to operate underwater, and it is typically a free swimming body without an "umbilical cord" attached to the support vessel, to which is launched. The vehicle is programmed to complete a particular mission and uses its on-board sensors to estimate its states in order to complete the mission objectives. A particular scenario where AUVs can play an important role is in the automatic acquisition of marine environmental data. In this case, one or more AUVs acting in cooperation are programmed to survey a given region. To this end, an important problem that has to be addressed is the sampling motion control strategy, that is, the high-level software system that is running on a computer system at the AUV that decides based on on-board sensors where (and in some cases when) to acquire environmental data.

In the literature, it is possible to find several interesting works that have proposed adaptive sampling strategies for marine systems. In [2], following the approach proposed first in [10], a robotic ocean sampling network is developed where AUVs move on their own, collecting data on the ocean physics and biology to gain a better understanding of ecosystem and ocean climate change. The algorithm used is based on artificial potentials and *virtual leaders*. Artificial potentials define the interaction control forces between neighbouring vehicles and are designed in this context to manipulate the formation and geometry of the fleet. A virtual leader is a moving reference point, at the center of the formation, that attracts or repels the neighbouring vehicles by means of additional artificial potentials. Ogren et al. in [4], describe a stable control strategy for a group of vehicles to direct and reconfigure cooperatively based on changes in acquired measurements. In [5], the authors propose a cooperation algorithm for adaptive oceanographic sampling taking into account range communication constraints. For a fleet of AUVs with static topology, a distributed dynamic programming algorithm is applied to solve the global optimization problem of maximizing the oceanographic sampling area coverage. In [13], a map is used to decrease the uncertainty of the sampling by considering the correlations among the ocean values. Smith et al. in [25], present a near-real time path planning solution for tracking and sampling an evolving ocean feature using a Regional Ocean Model System (ROMS). In [26] a model-predictive, local trajectory planning algorithm for a fleet of Unmanned Surface Vehicles (USVs) is proposed. The reported work aims at increasing resolution and focused sampling by considering the International Regulations for Prevention of Collisions at Sea (COLREGs).

In this paper, we use a slightly modified version of the proposed unsupervised algorithm in [15] to develop an adaptive sampling strategy to obtain relevant conductivity, temperature and depth (CTD) information of a given area using a coordinated fleet of AUVs. In the proposed setup, a leader AUV is tasked to acquire CTD data by running a set of user-defined mission instructions like for example following a desired path profile, and the rest of the fleet (the follower AUVs) will follow the leader closely with a desired formation that will adaptively change according to the CTD data are being acquired. More precisely, each AUV is in charge of running in real-time an unsupervised learning of GMMs algorithm that is fed by the CTD data. Each time the vehicles resurface (and this is done in a coordinated fashion), the leader AUV broadcast its currently estimated parameters of the GMM, and the followers based on this and their own estimated GMM compute the variational distance error between them. This error, which provides a notion of how different is the CTD measurements of each follower with respect to the leader, is then used to update the next formation configuration, which typically scales the distance between the AUVs in the formation (making a zoom-in and zoom-out), in order to improve the efficiency of data acquisition in a given region. It is important to stress that the proposed approach does not need to have prior knowledge of the dynamics of the environment due to the fact that the AUVs are running an unsupervised motion learning algorithm. Further, the communication data transmitted between the AUVs is very small (only the parameters of the GMM are transmitted) compared with the CTD data. This enables in the future to consider approaches where the communications can be done underwater (avoiding the resurface phase of the AUVs). The rest of this paper is organized as follows. Section 2 provides some background knowledge in GMMs and an overview of the on-line unsupervised learning algorithm. Section 3 proposes the motion control design architecture. The applicability of the proposed adaptive sampling strategy to obtain CTD information of a given area using a coordinated fleet of AUVs is investigated in Section 4. Conclusions are provided in Section 5.

2   Preliminaries and Background

2.1   Variational Distance between GMMs

The Gaussian mixture model defined in this work reflects the CTD feature distributions of an environment by a linear combination of Gaussian densities. Given two probability density functions $f(x)$ and $g(x)$, one can apply the Kullback-Leibler (KL) divergence, which is a widely used tool in statistics and pattern recognition, to measure the similarity between them as follows:

$$D(f\|g) = \int f(x)\log\frac{f(x)}{g(x)}\,dx \tag{1}$$

Note that unlike two single Gaussian distributions, there is no closed form expression to measure the similarity between two GMMs. Thus, in this work we use the *variational* approximation [20] to measure the dissimilarity between GMM hypotheses, that is,

$$\zeta = D_{variational}(f\|g) = \sum_a w_a \log\frac{\sum_{a'} w_{a'}e^{-D(f_a\|f_{a'})}}{\sum_b w_b e^{-D(f_a\|g_b)}} \tag{2}$$

where w_a and $w_{a'}$ are the mixing weights of the components in $f(x)$ and w_b is the mixing weights of the components in $g(x)$.

2.2 On-line Unsupervised Learning of Gaussian Mixture Models

The present paper borrows from the unsupervised learning of Gaussian mixture models in the presence of dynamic environments algorithm proposed in [15]. It is worth mentioning that the probability density function for a K-component mixture of d-dimensional Gaussian distributions is a linear combination of the individual probability density functions given by

$$p(\mathbf{y}|\theta) = \sum_{k=1}^{K} w^{[k]} p^{[k]}(\mathbf{y}|\theta^{[k]}), \qquad (3)$$

where $\mathbf{y}$ represents one particular outcome of the random variable distributed according to the mixture, $\theta = \{\theta^{[1]}, \ldots, \theta^{[K]}\}$ is composed by the parameters of each Gaussian components and $w = \{w^{[1]}, \ldots, w^{[K]}\}$ is the mixing weight set that satisfies

$$\sum_{k=1}^{K} w^{[k]} = 1, \quad w^{[k]} > 0, \qquad (4)$$

Knowing the parameters of the kth component $\theta^{[k]} = \{\mu^{[k]}, \Sigma^{[k]}\}$ and its mixing weight $w^{[k]}$, the probability density of that component can be obtained as follow

$$p^{[k]}(\mathbf{y}|\theta^{[k]}) = \frac{1}{(2\pi)^{d/2}|\Sigma^{[k]}|^{1/2}} \exp\left(-\frac{1}{2}(\mathbf{y} - \mu^{[k]})^T (\Sigma^{[k]})^{-1}(\mathbf{y} - \mu^{[k]})\right). \qquad (5)$$

The original algorithm in [15], that addresses the problem of on-line unsupervised learning of Gaussian mixture models in the presence of uncertain dynamic environments, consists of four major stages that are performed each time a new sample becomes available: 1) Updating all the current hypotheses; 2) Postulating a new hypothesis; 3) Checking if Gaussian mixture reduction is possible; and 4) Keeping a limited number of hypotheses and discarding the rest. Each hypothesis consists of particular values for the parameters of the GMM, the number of points of each component, the log-likelihood and the description length of the GMM. As in [15], the minimum description length principle is used to determine the complexity of the model in an on-line fashion.

There is, however, some important modifications to the algorithm presented in [15] that were done in this work, which is basically related to the method of calculating the log-likelihood of each GMM. The difficulty in calculating the likelihood of a GMM hypothesis $p(Y|\theta)$, is that Y, the set of all observations, is not available in real-time scenario and is being discarded after updating the GMM. Therefore, instead, we compute the expected likelihood of the same number of data points and, hence, the expected description length of the GMM hypotheses that are being calculated.

More precisely, consider two Gaussian components $g_1(Y) \sim \mathcal{N}(Y; \theta^{[g_1]})$, $g_2(Y) \sim \mathcal{N}(Y; \theta^{[g_2]})$ with the parameters $\theta^{[i]} \equiv \{\mu^{[i]}, \Sigma^{[i]}\}, i \in \{g_1, g_2\}$ and $w^{[g_1]} + w^{[g_2]} = 1$.

Let g_{12} be the moment preserving merge of these two components, see [19]. The expected likelihood of N_1 points drawn from the former and N_2 points from the latter given model $w^{[g_1]}g_1(Y) + w^{[g_2]}g_2(Y)$ is

$$\mathbb{E}[p(Y|\theta^{[g_{12}]})] = \left(\int g_1(Y)(w^{[g_1]}g_1(Y) + w^{[g_2]}g_2(Y))dY \right)^{N_1} \left(\int g_2(Y)(w^{[g_1]}g_1(Y) + w^{[g_2]}g_2(Y))dY \right)^{N_2} \tag{6}$$

where the integrals of type $\int p^i(y)p^j(y)dy$ are recognized as the Bhattacharyya distance, which for Gaussian distributions can be computed as

$$d_B(p^i, p^j) = \int p^i(y)p^j(y)dy = \frac{\exp(-C/2)}{(2\pi)^{(d/2)}|\Sigma^i \Sigma^j \Sigma^{ij}|^{1/2}} \tag{7}$$

where

$$\Sigma^{ij} = (\Sigma^{i-1} + \Sigma^{j-1})^{-1} \quad , \quad \mu^{ij} = \Sigma^{ij}(\Sigma^{i-1}\mu^i + \Sigma^{j-1}\mu^j)$$
$$C = \mu^i \Sigma^{i-1}\mu^{iT} + \mu^j \Sigma^{j-1}\mu^{jT} - \mu^{ij}\Sigma^{ij-1}\mu^{ijT}$$

After calculating the expected log-likelihood, the expected description length can be obtained similar to what is done in [15].

3   The Adaptive Sampling Strategy

We address the problem of real-time adaptive sampling using a coordinated fleet of AUVs. To this end, we provide a motion control algorithm that includes a modified version of the unsupervised learning of Gaussian Mixture Models (GMMs) in [15]. Fig. 1 illustrates the concept. The system setup consist of one leader AUV and two follower AUVs. All vehicles are equipped with CTD sensor devices. The leader moves according to a predefined path, while is acquiring CTD data. Simultaneously, it constructs a GMM profile by running the unsupervised learning of GMM algorithm.

The objective is to execute a coordinated formation maneuver, by driving and maintaining the other vehicles, the followers, at a desired distance with respect to the leader that is a function of the dissimilarity of the CTD profiles of the leader and the followers. This dissimilarity measure is the *variational distance* between two GMM profiles. To make the scheme robust to fault of underwater communications, which is very prone to happen, in this approach we assume that the followers only on the surface can receive the GMM hypothesis of the leader to calculate the error (dissimilarity measure). Note that on the surface, the vehicles can also use GPS to correct their navigation errors. The adopted control strategy consist in generating adequate speed and heading commands so as to drive the follower to the desired

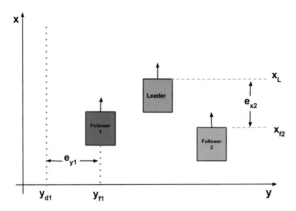

Fig. 1 Schematic representation of the setup

position along time. These commands are used as input to the local inner loops for yaw and speed control.

3.1 Path Following

As it can be seen in Fig. 1, the vehicles are required to follow segments of line, which for simplicity and without loss of generality are perpendicular to the y axis. During each resurfacing, every follower separately regulates the desired linear velocity u_d and heading ψ_d. These are then fed to inner loop controllers specific to the vehicle. A simplified version of the kinematic model of an underwater vehicle can be described as follows

$$\begin{bmatrix} \dot{x} \\ \dot{y} \\ \dot{\psi} \end{bmatrix} = \begin{bmatrix} u\cos\psi - \upsilon\sin\psi \\ u\sin\psi - \upsilon\cos\psi \\ r \end{bmatrix} \tag{8}$$

where u is the surge velocity, υ is the sway velocity, r is the angular velocity and ψ is the heading. Using the path-following algorithm in [27], we have

$$e_y = y - y_d \Rightarrow \dot{e}_y = \dot{y} - \dot{y}_d = u\sin\psi - 0 = uU \tag{9}$$

where $U = \sin\psi$, and y_d define the desired position of the line to be followed by the AUV, which will correspond to the desired distance from the leader AUV (see Fig. 1). Therefore, if $U = \sin\psi$ we obtain $\dot{e}_y = -k_1 e_y$, which implies that the error e_y will converge exponentially to zero. The desired heading angle can be obtained as

$$\psi_d = \sin^{-1}(sat(U)) \tag{10}$$

where

$$sat(U) = \begin{cases} U & \text{if } |U| < \epsilon \\ \epsilon & \text{if } U > \epsilon \\ -\epsilon & \text{if } U < -\epsilon \end{cases} \qquad (11)$$

and $\epsilon \in (0, 1)$. A possibility for the inner-loop in heading can be derived by noting that

$$\tilde{\psi} = \psi - \psi_d \Rightarrow \dot{\tilde{\psi}} = r - \dot{\psi}_d \qquad (12)$$

Thus, if $r = -k_\psi \tilde{\psi} = -k_\psi (\psi - \psi_d)$, $k_\psi > 0$ it can be concluded that ψ will converge to ψ_d for ψ_d constant. Note that it is possible to show, using Lyapunov-based analysis tools, that the above nonlinear control law yields convergence of the cross track error to zero if the actual vehicle heading equals the desired heading reference ψ_d. The work in [27] also shows that "similar behaviour" is obtained when the dynamics of the heading autopilot (inner loop) and the sideslip of the vehicle are taken into account.

3.2 Coordinated Formation

For simplicity we consider a formation in line parallel with the y axis as illustrated in Fig. 1. Every follower has to keep the coordinated formation with respect to the leader. To do so, during each resurfacing the vehicle must correct the navigation errors and accelerate (or decelerate) to position itself along the leader. To achieve this goal, a PI controller is used

$$\dot{\xi} = e_x, \, u_f = -k_u \, e_x - k_2 \xi \qquad (13)$$

where e_x is the difference between the position of the leader and each follower along the x axis (see Fig. 1).

Fig. 2 shows a schematic representation of the subsystems running on the leader AUV. As mentioned above, its task is to follow some desired predefined path while acquiring CTD data. By running the proposed unsupervised learning of GMMs, a real-time CTD profile for the leader is being obtained. The AUV real-time mission planner is a high-level decision system that monitors all system states and issues the speed and heading commands to the low-level thruster controllers (AUV guidance and navigation block).

A scheme of a follower AUV is represented in Fig. 3. To implement this strategy, we propose the following steps to compute the error or dissimilarity measure:

– The CTD data is modelled as a Gaussian Mixture Model (GMM) and each vehicle runs an on-line unsupervised learning algorithm to estimate the GMM parameters. In the proposed approach, the number of components of the GMM is not fixed

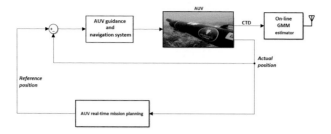

Fig. 2 Schematic representation of the leader

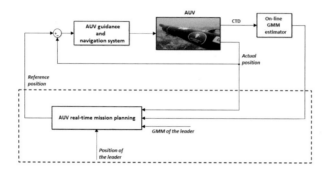

Fig. 3 Schematic representation of the follower

and can change over time, which provides more flexibility to deal with uncertain dynamic environments.

- During each resurfacing, the fleet use the available GPS signals to correct the navigation errors as suggested in [17]. Suitable velocity command can be issued to have a coordinated formation with respect to the last seen position of the leader.
- Also, on the surface, every follower learn the position of the leader and its GMM hypothesis. Then, together with its own estimated GMM parameters, compute an error that measures the dissimilarity of the two GMM probability density functions, based on Equation 2.

The desired position y_d for each follower corresponds to the case when variational distance ζ becomes equal to some reference value ζ_r

$$e_\zeta = (\zeta - \zeta_r) \; / \; \zeta_r$$
$$y_d = y_f + \lambda \, e_\zeta \tag{14}$$

where λ can be positive or negative, depending on which direction we want to guide the follower. y_d is the desired position in $x - y$ plane and y_f is the current position of the follower. The intuition behind the normalization of e_ζ between $[0, 1]$ is to moderate the changes in the formation with respect to the value of ζ specially at the beginning of the experiment (expansion and contraction of the formation to be more dependent on user-defined value of λ than unknown value of ζ).

4 Simulation Results

This section illustrates the real-time adaptive sampling strategy. The task of the leader vehicle is to follow some predefined path profile while is acquiring CTD data. The aim of the each follower is to keep a coordinated formation with respect to the leader by keeping a desired distance from it. The performance of the algorithm that we propose is demonstrated in uniform and changeable environments, which were reconstructed based on real CTD data provided by [24]. The simulation parameters are shown in Table 1. Extensive simulations, with $\zeta_r = 4$, were performed to evaluate the effectiveness of the presented approach. In this section we present two different scenarios. Worth noting that in the simulations some limits were put for the acceleration of the vehicles and expansion of the formation, to address the various operational constraints associated with real world applications, and to maintain the group structure: 1) attraction to the leader up to a maximum distance, 2) repulsion from the leader when it is too close, and 3) alignment or velocity matching with the leader. Every follower makes its own decision based on the difference or distance between its GMM and the leader. Fig. 4(a) shows the temperature dynamics of a reconstructed uniform environment, along the x axis. The top layer represents the near surface and the lowest one is $12m$ beneath the surface. A fleet of three vehicles, the leader is in the middle, is tasked to do as follows: The leader follows a straight line along the x axis. The followers have to expand the formation to find the regions where the CTD data is desirably different from the leader. Fig. 4(c) represents the overall performance of the fleet, from the top view, and the position of the vehicles during resurfacing intervals in $x - y$ plane. Fig. 4(e), compatible with Fig. 4(c) and based on Equation 14, confirms that the followers reached to the desirable environment, by expanding the formation at the beginning of the experiment. For the second scenario, we assume that the environment on the left side of the leader is not uniform and changes at some point in $x - y$ frame, Fig. 4(b). Like the previous case, the followers learn their path toward desired position according to Equation 14. Fig. 4(f) shows the variations of the dissimilarity measures of the left vehicle, followed by entering into a new environment as can be seen in Fig. 4(d).

Table 1 Simulation parameters

ϵ	k_1	k_ψ	k_u	k_2
0.8	1	1.5	1.2	1

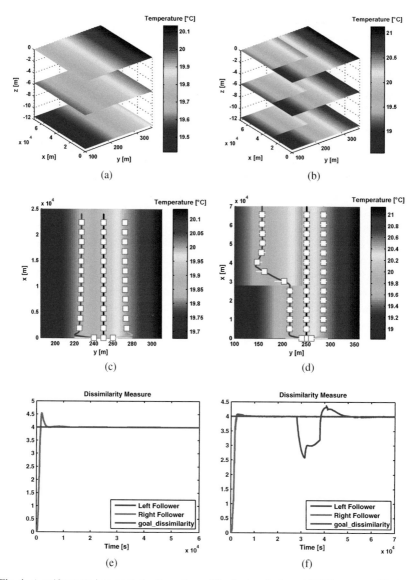

Fig. 4 A uniform environment, the formation of the fleet and relevant GMM error profile (a,c,e). A more complex scenario, the formation of the fleet and relevant GMM error profile (b,d,f).

5 Conclusion

This paper demonstrated the applicability of a modified version of the proposed unsupervised learning algorithm in [15] to solve a CTD adaptive sampling problem for a fleet of Autonomous Underwater Vehicles (AUVs). The proposed setup consists of one leader AUV and one or more follower AUV(s), all equipped with CTD sensor devices, and capable of running in real-time the proposed on-line unsupervised learning computer algorithm that uses and updates Gaussian Mixture Models to model the acquired CTD data in real-time. For the sake of simplicity, we assumed that all vehicles followed the same saw-tooth path in z axis. The path to be traced by the leader is predefined. Every follower, which necessarily does not need to know the path profile of the leader AUV, updates its mission plan during each resurfacing. The practical implementation requires a careful tuning of the gain parameters. Factors such as, how fast the vehicles can place themselves in the desired position or the greatness of the displacement based on the dissimilarity measure (λ), can be decisive in the fleet manoeuvrability, since unreasonable values can deteriorate the overall performance of the model. Future work will address the implementation and application of the proposed algorithms in real scenarios.

References

1. Runnalls, A.R.: Kullback-Leibler approach to Gaussian mixture reduction. IEEE Transactions on Aerospace and Electronic Systems **43**(3), 989–999 (2007)
2. Fiorelli, E., Leonard, N.E., Bhatta, P., Paley, D.A., Bachmayer, R.: IEEE Journal of DM Fratantoni Oceanic Engineering **31**(4), 935–948 (2006)
3. Grnwald, P.D.: The minimum description length principle. MIT press (2007)
4. Ogren, P., Fiorelli, E., Leonard, N.E.: Cooperative control of mobile sensor networks: Adaptive gradient climbing in a distributed environment. IEEE Transactions on Automatic Control **49**(8), 1292–1302 (2004)
5. Alvarez, A., et al.: Folaga: a low-cost autonomous underwater vehicle combining glider and AUV capabilities. Ocean Engineering **36**(1), 24–38 (2009)
6. Figueiredo, M.A.T., Jain, A.K.: Unsupervised learning of finite mixture models. IEEE Transactions on Pattern Analysis and Machine Intelligence **24**(3), 381–396 (2002)
7. Bishop, C.M.: Pattern recognition and machine learning, vol. 4(4). Springer, New York (2006)
8. Li, D., Xu, L., Goodman, E.: On-line EM variants for multivariate normal mixture model in background learning and moving foreground detection. Journal of Mathematical Imaging and Vision **48**(1), 114–133 (2014)
9. Titterington, D.M.: Recursive parameter estimation using incomplete data. Journal of the Royal Statistical Society. Series B (Methodological), 257–267 (1984)
10. Leonard, N.E., Fiorelli, E.: Virtual leaders, artificial potentials and coordinated control of groups. In: Proceedings of the 40th IEEE Conference on Decision and Control, 2001, vol. 3, pp. 2968–2973. IEEE (2001)
11. Carroll, K.P., McClaran, S.R., Nelson, E.L., Barnett, D.M., Friesen, D.K., William, G.N.: AUV path planning: an A* approach to path planning with consideration of variable vehicle speeds and multiple, overlapping, time-dependent exclusion zones. In: Proceedings of the 1992 Symposium on Autonomous Underwater Vehicle Technology, AUV 1992, pp. 79–84, June 2–3, 1992
12. Cover, T.M., Thomas, J.A.: Elements of Information Theory. John Wiley & Sons, November 28, 2012
13. Yilmaz, N.K., et al.: Path planning of autonomous underwater vehicles for adaptive sampling using mixed integer linear programming. IEEE Journal of Oceanic Engineering **33**(4), 522–537 (2008)
14. Eickstedt, D.P., Benjamin, M.R., Curcio, J.: Behavior based adaptive control for autonomous oceanographic sampling. In: IEEE International Conference on Robotics and Automation (2007)

15. Khoshrou, A., Aguiar, A.P.: Unsupervised learning of gaussian mixture models in the presence of dynamic environments. In: CONTROLO 2014–Proceedings of the 11th Portuguese Conference on Automatic Control. Springer International Publishing (2015)

16. Dempster, A.P., Laird, N.M., Rubin, D.B.: Maximum likelihood from incomplete data via the EM algorithm. Journal of the royal statistical society. Series B (methodological), 1–38 (1977)

17. Yun, X., et al.: Testing and evaluation of an integrated GPS/INS system for small AUV navigation. IEEE Journal of Oceanic Engineering **24**(3), 396–404 (1999)

18. Zivkovic, Z., van der Heijden, F.: Recursive unsupervised learning of finite mixture models. IEEE Transactions on Pattern Analysis and Machine Intelligence **26**(5), 651–656 (2004)

19. Arandjelovic, O., Cipolla, R.: Incremental learning of temporally-coherent gaussian mixture models. Society of Manufacturing Engineers (SME) Technical Papers, 1–1 (2006)

20. Hershey, J.R., Olsen, P.A.: Approximating the Kullback Leibler divergence between Gaussian mixture models. In: IEEE International Conference on Acoustics, Speech and Signal Processing, ICASSP 2007, vol. 4. IEEE (2007)

21. Aguiar, A.P., Hespanha, J.P.: Trajectory-tracking and path-following of underactuated autonomous vehicles with parametric modeling uncertainty. IEEE Transactions on Automatic Control **52**(8), 1362–1379 (2007)

22. Bahr, A., Leonard, J.J., Fallon, M.F.: Cooperative localization for autonomous underwater vehicles. The International Journal of Robotics Research **28**(6), 714–728 (2009)

23. Woolsey, C.A.: Review of marine control systems: Guidance, navigation, and control of ships, rigs and underwater vehicles. Journal of Guidance, Control, and Dynamics **28**(3), 574–575 (2005)

24. Laboratrio de Sistemas e Tecnologia Subaqutica. http://lsts.fe.up.pt/

25. Smith, R.N., et al.: Autonomous underwater vehicle trajectory design coupled with predictive ocean models: a case study. In: 2010 IEEE International Conference on Robotics and Automation (ICRA). IEEE (2010)

26. Svec, P., et al.: Dynamics-aware target following for an autonomous surface vehicle operating under COLREGs in civilian traffic. In: 2013 IEEE/RSJ International Conference on Intelligent Robots and Systems (IROS). IEEE (2013)

27. Maurya, P, Aguiar, A.P., Pascoal, A.M.: Marine vehicle path following using inner-outer loop control. In: 8th IFAC International Conference on Manoeuvring and Control of Marine Craft 2009, September 16–18, 2009, Guaruja-Brazil, p. 6. IFAC International Conference (2009)

Part II
Agricultural Robotics and Field Automation

Stability Analysis of an Articulated Agri-Robot Under Different Central Joint Conditions

R. Vidoni, G. Carabin, A. Gasparetto and F. Mazzetto

Abstract In hilly terrains, the exploitation of (semi-)autonomous systems able to travel nimbly and safely on different terrains and perform agricultural operations is still far from reality.

In this perspective, the articulated 4-wheeled system, that shows an optimal steering capacity and the possibility to adapt to uneven terrains thanks to a passive degree of freedom on the central joint, is one of the most promising mobile wheeled-robot architectures. In this work, the instability of this robotic platform is evaluated in the two different conditions, i.e. phase I and phase II [1], and the effect of blocking the passive DoF of the central joint investigated in order to highlight possible stabilizing conditions and best manoeuvring practices for overturning avoidance. In order to do so, a quasi-static model of the robotic platform has been developed and implemented in a Matlab™ simulator thanks to which the different conditions have been studied.

Keywords Articulated-robot · Stability · Agricultural robotics

1 Introduction

Agricultural robotics and autonomous systems for planting, weeding, fruit picking and monitoring have been studied since many years ([2]) and now, smart, cheap and miniaturized sensors and controllers could allow the development of new efficient mechatronic applications that can speed up the race towards the agricultural automation both in the management of field processes ([3]) and in the safety of machines

R. Vidoni(✉) · G. Carabin · F. Mazzetto
Faculty of Science and Technology, Free University of Bozen-Bolzano,
Piazza Universita, 39100 Bolzano, Italy
e-mail: {renato.vidoni,giovanni.carabin,fabrizio.mazzetto}@unibz.it

A. Gasparetto
DIEGM - University of Udine, Via delle Scienze, 33100 Udine, Italy
e-mail: gasparetto@uniud.it

© Springer International Publishing Switzerland 2016
L.P. Reis et al. (eds.), *Robot 2015: Second Iberian Robotics Conference*,
Advances in Intelligent Systems and Computing 417,
DOI: 10.1007/978-3-319-27146-0_26

335

operating on slopes ([4]; [5]) areas. The latter is strongly related to the configuration of the mobile robot platform, the choice of which is directly dependent on the working environment ([2, 6, 7]).

The vehicle stability is highly affected by the terrain condition and slope ([8]; [9]; [10]) and effective, safe and self-stabilizing systems are still not on-board. Then, mobile terrain platforms for agri-, hilly-, mountain- applications, either human-driven or (semi-)autonomous, are rare.

In [11], it has been highlighted how a versatile robotic platform that could be the solution for easily moving and turning on different slopes and between rows, e.g. vineyards, is the articulated-frame one. Its central joint is made of two (yaw and roll) degrees of freedom (DoFs), one actuated (yaw) to steer and the other passive (roll), to allow the system to adapt to the terrain. It has a smaller external turning radii with respect to vehicles with a conventional configuration [11]. This platform shows two different possible overturning manners ([12]; [1]): the classical stability condition (type II instability) related to the quadrilateral polygon made of the four wheel contacts, and a second critical stability condition created by the passive roll DoF (type I instability). In literature, the Thype I instability has been firstly defined by [12]. In order to study an anti-overturning mechatronic system able to both forecast and prevent critical configurations in a mobile robot, it is very important to model and simulate the system instabilities together with to inverstigate the effect of the passive DoF and its blocking by means of, for example, a mechanical brake.

In this work, the Guzzomi's kinematic and (quasi-)static model has been firstly revised (Sections 2). Then, in Section 3 the instability phases are discussed and, in Sections 4, a Matlab™ emulator and the stability maps for the two different phases computed. Then, in Section 5, the stabilizing effect of the passive DoF blocking has been investigated and evaluated.

2 Model of the Articulated Robot

2.1 Model Assumptions

Under the following basic hypothesis ([1]):

- the roll DoF of the articulated joint is considered frictionless;
- since the robot speed is going to be slow in practical activities, the dynamic effects can be ignored;
- the robot does not slide down the slope, due to a non-limiting coefficient of friction between surface and tyres;
- tyres are considered stiff, so the contact surfaces result in discrete points (not areas);
- the joint mass is negligible, so it does not affect the dynamic behaviour.

the kinematic and quasi-static model can be described.

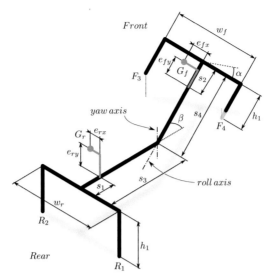

Fig. 1 Kinematic model.

Table 1 Main parameters of the kinematic model.

G_r	CoG of the rear part	e_{rx}	Rear CoG x distance from rear midplane
G_f	CoG of the front part	e_{ry}	Rear CoG y height above roll axis
R_1	Contact point between rear wheel 1 and surface	s_2	Front CoG distance from front axle
R_2	Contact point between rear wheel 2 and surface	e_{fx}	Front CoG x distance from front midplane
F_3	Contact point between front wheel 3 and surface	e_{fy}	Front CoG y height above roll axis
F_4	Contact point between front wheel 4 and surface	w_r	Rear track width
α	Roll angle between rear and front part	w_f	Front track width
β	Yaw angle between rear and front part	h_1	Roll axis height from ground
s_3	Distance from rear axle to central joint	m_r	Rear mass
s_4	Distance from front axle to central joint	m_f	Front mass
s_1	Rear CoG distance from rear axle		

Given the model in Fig. 1, the articulated robot can be explained: a front "f" and a rear "r" parts are connected by a 2 DoF joint which is made of a first revolute DoF, i.e. the yaw β angle, and of a second passive revolute DoF, i.e. the roll α angle. In such a manner the articulated chassis can maintain the four wheels in contact with the substrate even in case of uneven terrains.

In table 1 the geometric parameters of the model shown in Fig. 1 are explained. In order to study the system configurations, the robot is supposed to travel a circle on a sloped surface, which slope ϑ. The robot position related to the maximum slope direction is named φ, β sets the trajectory followed by the robot and α describes the surface conformation ($\alpha = 0$ implies a plane surface), see Fig. 2.

A global coordinate system $(x_0 \ y_0 \ z_0)$ and two local ones $(x_1 \ y_1 \ z_1)$ and $(x_2 \ y_2 \ z_2)$, rigidly attached on the rear and front robot parts respectively, are defined. Then, the matrix $\boldsymbol{R}_1^0$ that describes the rotation from the global system to the rear local one is $\boldsymbol{R}_1^0 = \boldsymbol{R}(\vartheta)\boldsymbol{R}(\varphi)$, i.e. the product of the elementary rotations around the current z and y axis. The rotation matrix $\boldsymbol{R}_2^0$ becomes $\boldsymbol{R}_2^0 = \boldsymbol{R}_1^0 \boldsymbol{R}(\beta)\boldsymbol{R}(\alpha)$ with $\boldsymbol{R}(\beta)$ and $\boldsymbol{R}(\alpha)$ elementary rotation matrices around the current y and z axis, respectively.

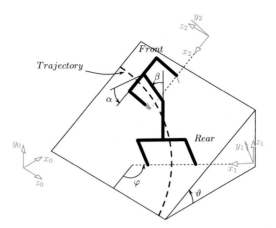

Fig. 2 Robot orientation angles and reference systems.

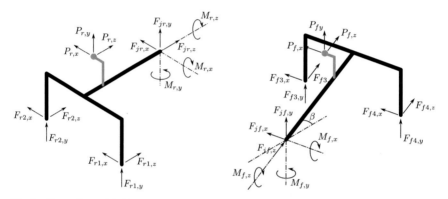

Fig. 3 Dynamic model: (a) rear part, (b) front part.

In order to define and develop a dynamic (quasi-static) model, forces and moments that act on the model have to be evaluated. By referring to Fig. 3, two weight forces P_r and P_f, respectively on the rear and front CoG are present. These are counteracted by the four reaction forces F_{r1}, F_{r2}, F_{f3} and F_{f4}, sum of the force normal to the plane and the friction force parallel to the plane. Through the central joint, the forces (F_{jr} and F_{jf}) and moments (M_r e M_f) are exchanged. The two weight forces P_r and P_f, and, due to the absence of friction, the moment $M_{f,z}$ are known.

Since the four normal forces $F_{r1,y}$, $F_{r2,y}$, $F_{f3,y}$ and $F_{f4,y}$ acting on the wheels are needed to study the stability, it is desirable to reduce the system dimension to improve the computational speed with respect to solve the whole system made of 23 equations. By considering the relations of forces and torques in the joint and the fact that all the forces F_{r1}, F_{r2}, F_{f3} and F_{f4} have a vertical direction with respect to the global reference system, with some reformulations the system can be simplified and, from the initial 23 unknowns, reduced to only 6: F_{r1}, F_{r2}, F_{f3}, F_{f4}, $M_{f,x}$ and $M_{f,y}$.

In such a manner, six equilibrium equations can be written:

$$F_{r1} + F_{r2} + F_{f3} + F_{f4} = P_r + P_f \tag{1}$$

$$M_{f,x} = F_{f3}\left(k_{fy}s_4 + k_{fz}h_1\right) + F_{f4}\left(k_{fy}s_4 + k_{fz}h_1\right) + P_f\left[k_{fy}(s_4 - s_2) - k_{fz}e_{fy}\right] \tag{2}$$

$$M_{f,y} = F_{f3}\left(-k_{fx}s_4 + k_{fz}\frac{w_f}{2}\right) + F_{f4}\left(-k_{fx}s_4 - k_{fz}\frac{w_f}{2}\right) + P_f\left[-k_{fx}(s_4 - s_2) + k_{fz}e_{fy}\right] \tag{3}$$

$$M_{f,z} = F_{f3}\left(-k_{fx}h_1 - k_{fy}\frac{w_f}{2}\right) + F_{f4}\left(-k_{fx}h_1 + k_{fy}\frac{w_f}{2}\right) + P_f\left(k_{fx}e_{fy} - k_{fy}e_{fx}\right) = 0 \tag{4}$$

$$M_{r,x} = -\cos\alpha\cos\beta\, M_{f,x} + \sin\alpha\cos\beta\, M_{f,y}$$
$$= F_{r1}\left(-k_{ry}s_3 + k_{rz}h_1\right) + F_{r2}\left(-k_{ry}s_3 + k_{rz}h_1\right) + P_r\left[-k_{ry}(s_3 - s_1) - k_{rz}e_{ry}\right] \tag{5}$$

$$M_{r,z} = \cos\alpha\sin\beta\, M_{f,x} - \sin\alpha\sin\beta\, M_{f,y}$$
$$= F_{r1}\left(-k_{rx}h_1 + k_{ry}\frac{w_r}{2}\right) + F_{r2}\left(-k_{rx}h_1 - k_{ry}\frac{w_r}{2}\right) + P_r\left(k_{rx}e_{ry} - k_{ry}e_{rx}\right) \tag{6}$$

in the 6 unknowns F_{r1}, F_{r2}, F_{f3}, F_{f4}, $M_{f,x}$ and $M_{f,y}$.

3 Instability Phases

The instability of an articulated robot can be subdivided in phase I and II ([12] and [1]). By increasing the slope (in a quasi-static condition), the force distribution on the four wheels changes according to the configuration and system properties. The articulated system is stable until one of the four reaction forces falls to zero. After that, the roll moment equilibrium is not satisfied, one wheel loses the contact and one part of the robot starts to roll, i.e. the phase I instability occurs. To detect the phase I instability limit condition, the system of six equations has to be solved for every configuration in terms of slope (ϑ angle), robot placement (φ angle), robot trajectory (β angle) and terrain conformation (α angle).

The phase I instability creates a roll motion in a part of the robot. This motion stops when the joint reaches its mechanical limit (also a brake can stop the motion in an intermediate position) and the robot chassis becomes a unique rigid body. In this condition the instability occurs only when the CoG projection point falls out from the equilibrium polygon made of the wheel contact points, i.e. the phase II instability.

4 Numerical Implementation

The overall stability model has been implemented in a Matlab™ environment. Its outputs are a matrix with the stability limits and an instability map where, for a given robot configuration, i.e. α and β angles, the following information are shown:

– plane slope limits for phase I instability ($\vartheta_{lim,I}$);
– tyre that looses the contact with the terrain in phase I instability;

Table 2 Robot emulator geometric and physical parameters.

s_1 [mm]	s_2 [mm]	s_3 [mm]	s_4 [mm]	w_r [mm]	w_f [mm]	h_1 [mm]	e_{rx} [mm]	e_{fx} [mm]	e_{ry} [mm]	e_{fy} [mm]	m_r [kg]	m_f [kg]
26	55	200	200	240	180	94	0	0	20	14	1,34	1,84

– plane slope limits for phase II instability ($\vartheta_{lim,II}$);
– angular margin $\Delta\vartheta = \vartheta_{lim,II} - \vartheta_{lim,I}$, that will be gained if the joint is blocked.

The simulator finds the solution by an iterative algorithm based on the bisection method, one for the phase I and one for the phase II instability. In that manner, by the superimposition of the two instabilities limits, it is possible to evaluate the possible stability enhancement of the joint's passive DoF blocking. Thus, the idea is to evaluate the stabilizing effect of a blocking action when approaching the phase I instability and find out some directives and "best driving practices" for the articulated robotic platform. In table 2, the simulated robot parameters referred to a real emulator, see figure 4, are listed.

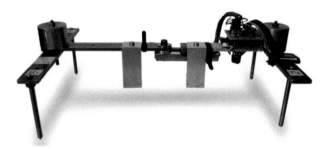

Fig. 4 Emulator of the articulated robotic platform.

5   Simulations and Results

The stability map is the main output of the Matlab simulator and the first considered case is the one of a robot travelling along a straight line ($\beta = 0°$) and a terrain with a regular surface ($\alpha = 0°$), figure 5.

The phase I instability curve is always below the phase II curve, except for some φ values in which they are overlapped. So, unless of these points, the strategy of blocking the passive roll DoF would guarantee an extra margin to the robot instability. More in details, this $\Delta\vartheta$ safety angular margin changes with φ and presents local maxima when φ is about 60, 120, 244 and 295° (i.e. when the robot goes uphill or descends with an oblique trajectory). It has to be underlined how these maxima correspond to $\vartheta_{lim,I}$ values that are high; this means that, in this case, the configurations where the blocking action could give the greatest stabilizing effect are when the slope condition is very critical. However, if the phase I minima are evaluated, i.e. φ equal to 0 and 180°, is it possible to achieve an extra slope margin of about 4°. In particular, practical conditions where φ goes from 0° to 75°, the extra slope margin is always over 4°. There is no extra slope margin for φ equal to 95°, 228°, 270°, and 310°, due to overlapping of phase I and phase II instability curves.

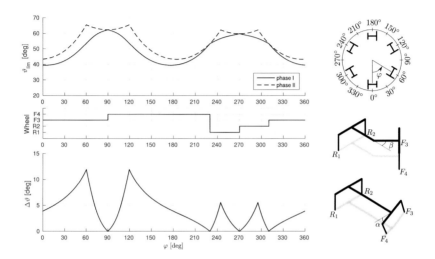

Fig. 5   Stability map related to $\beta = 0°$ and $\alpha = 0°$.

The considered case, in which the robot travels sideways the slope ($\varphi = 0°$ and $\varphi = 180°$), is one of the most critical and common practical cases. However, it is extremely important to consider the possible stabilizing effect if also the β and α angles change, e.g. when the robotic system goes out to a row of wines in a hill and starts turning up in order to go in to the next one or with an uneven sloped terrain.

In figures 6 and 7, the stability map for a regular surface terrain ($\alpha = 0°$) and two robot steering cases, i.e. upstream turning with $\beta = 20°$ and $\beta = 45°$ are presented while in figure 8 the phase I limits for the three different cases are shown.

It can be seen how, the variation of the turning angle β influences both the minimum value of the instability slope angle and its map position along the φ axis: if β increases (upstream turning), the minimum moves from $\varphi = 0°$ to a lower value, and decreases its magnitude; in the downstream turning condition, i.e. $\varphi > 180°$, the contrary occurs.

By considering uneven terrains, i.e. $\alpha \neq 0°$, other important considerations can be made. Figure 9 shows the phase I stability angle for $\beta = 0°$ and considering three different α values. In this case, α influences only the magnitude of the local minima, and not its position along φ axis.

Now, looking at the case in which $\varphi = 0°$ (similar to $\varphi = 180°$ due to the fact that the two COGs locations are almost on the midplane), the correlation between the

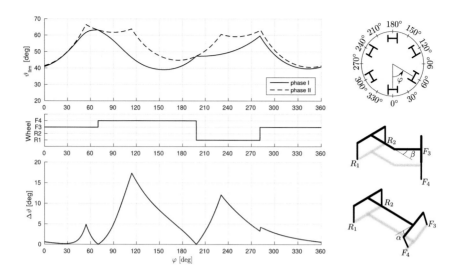

Fig. 6 Stability map related to $\beta = 20°$ and $\alpha = 0°$.

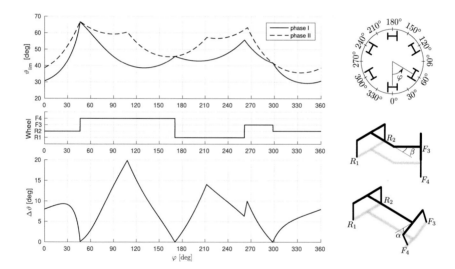

Fig. 7 Stability map related to $\beta = 45°$ and $\alpha = 0°$.

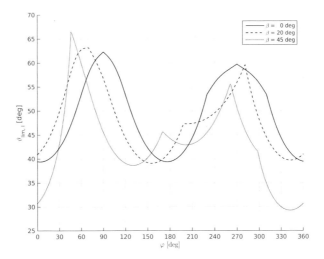

Fig. 8 Stability map for different robot turn levels (β angle), and $\alpha = 0°$.

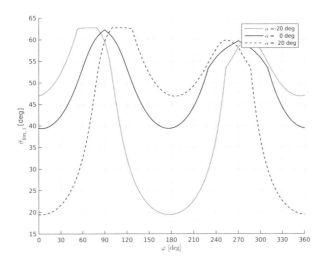

Fig. 9 Stability map for different surface conformation (α angle), and $\beta = 0°$.

angle β and the stability is shown in figure 10. If the robot is in a stable condition with $\beta = 0°$ and starts turning downstream (negative values of β), the phase I stability angle increases; so it should not be necessary to activate the joint brake, since the robot would already be in a stable condition. On the contrary, if the robot starts turning upstream, the phase I stability value increases until a maximum point and, after that,

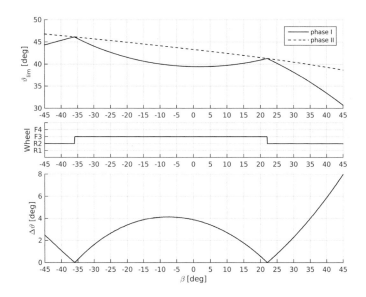

Fig. 10 Stability vs β angle, at $\varphi = 0°$ and $\alpha = 0°$.

it fast decreases. In the first turning phase, blocking the joint would not be necessary due to the positive curve slope; indeed, by increasing the turning angle, increases also the stability. After the point of maximum, it would be necessary (negative curve slope) to block the passive DoF in order to increase the stability.

There is another motivation to block the joint between the interval $\beta = [0, 24]°$, i.e. upstream motion up to the stability curve slope changing: if the robot is near to an unstable condition and β is inner this interval and the operator wants to move in a safer condition, he instinctively reduces the steering angle by a counter-steering manoeuvre. In this way, the robot tends to roll-over, since in this range the phase I instability angle limit decreases with β. Otherwise, with the passive DoF blocked, it does not occur, i.e. the curve slope is negative.

In figure 11 the case of $\varphi = 0°$ and $\beta = 0°$ is considered. When α has positive values, i.e. the robot front part is in a more sloped condition than the rear one, both the phase I and II stability limit angles decrease. However, the phase II angle shows a lower slope, thus blocking the passive DoF would give the possibility to gain an extra angular margin. On the contrary, when α is negative both the phase I and II stability limit angles increase up to the value of $-8°$. After that, the phase I stability limit angle remains constant due to the fact that the phase I instability critical condition goes to the robot rear part, while the phase II instability limit angle still increases thus allowing a more stable condition if the passive DoF is blocked.

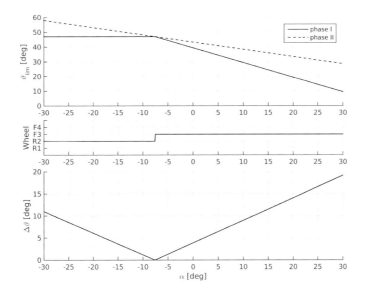

Fig. 11 Stability vs α angle, at $\varphi = 0°$ and $\beta = 0°$.

6 Conclusions

In this work, an articulated 4-wheeled robotic platform suitable for side-slope agricultural activities has been evaluated in its stability conditions. First of all the kinematic and (quasi-)static model has been revised and the two different instability conditions, i.e. phase I and phase II, evaluated. These results have been implemented in a Matlab simulator which gives as output the stability maps and roll-over limits. Then, by considering the fact that this platform shows an optimal steering capacity and the possibility to adapt to uneven terrains thanks to a passive degree of freedom on its central joint, the most critical conditions have been investigated. By focusing on the possibility to block the passive DoF of the central joint, its possible stabilizing effect and best manoeuvring practices for overturning avoidance have been studied and highlighted. Future work will cover experimental tests on a real robotic prototype and/or articulated tractor.

References

1. Baker, V., Guzzomi, A.L.: A model and comparison of 4-wheel-drive fixed-chassis tractor rollover during phase I. Biosystems Engineering **116**(2), 179–189 (2013)
2. Billingsley, J., Visala, A., Dunn, M.: Robotics in agriculture and forestry. In: Siciliano, B., Khatib, O. (eds.) Springer Handbook of Robotics, pp. 1064–1077 (2008)

3. Fumagalli, M., Acutis, M., Mazzetto, F., Vidotto, F., Sali, G., Bechini, L.: A methodology for designing and evaluating alternative cropping systems: application on dairy and arable farms. Ecological Indicators **23**, 189–201 (2012)
4. Khot, L.R., Tang, L., Steward, B.L., Han, S.: Sensor fusion for improving the estimation of roll and pitch for an agricultural sprayer. Biosystems Engineering **101**(1), 13–20 (2008)
5. Kise, M., Zhang, Q.: Sensor-in-the-loop tractor stability control: look-ahead attitude prediction and field tests. Computers and Electronics in Agricultur **52**(1–2), 107–118 (2006)
6. Lee, J.-H., Park, J.B., Lee, B.H.: Turnover prevention of a mobile robot on uneven terrain using the concept of stability space. Robotica **27**(5), 641–652 (2009)
7. Agheli, M., Nestinger, S.: Study of the foot force stability margin for multilegged/wheeled robots under dynamic situations. In: Proceedings of the 8th IEEE/ASME International Conference on Mechatronics and Embedded Systems and Applications, pp. 99–104 (2012)
8. Gravalos, I., Gialamas, T., Loutridis, S., Moshou, D., Kateris, D., Xyradakis, P., Tsiropoulos, Z.: An experimental study on the impact of the rear track width on the stability of agricultural tractors using a test bench. Journal of Terramechanics **48**(4), 319–323 (2011)
9. Mazzetto, F., Bietresato, M., Vidoni, R.: Development of a dynamic stability simulator for articulated and conventional tractors useful for real-time safety devices. Applied Mechanics and Materials **394**, 546–553 (2013)
10. Previati, G., Gobbi, M., Mastinu, G.: Mathematical models for farm tractor rollover prediction. International Journal of Vehicle Design **64**(2–3–4), 280–303 (2014)
11. Vidoni, R., Bietresato, M., Gasparetto, A., Mazzetto, F.: Evaluation and stability comparison of different vehicle configurations for robotic agricultural operations on side-slopes. Biosystems Engineering **129**, 197–211 (2015)
12. Guzzomi, A.L.: A revised kineto-static model for phase I tractor rollover. Biosystems Engineering **113**(1), 65–75 (2012)

A Path Planning Application for a Mountain Vineyard Autonomous Robot

Olga Contente, Nuno Lau, Francisco Morgado and Raul Morais

Abstract Coverage path planning (CPP) is a fundamental agricultural field task required for autonomous navigation systems. It is also important for resource management, increasingly demanding in terms of reducing costs and environmental polluting agents as well as increasing productivity. Additional problems arise when this task involves irregular agricultural terrains where the crop follows non-uniform configurations and extends over steep rocky slopes. For mountain vineyards, finding the optimal path to cover a restricted set of terraces, some of them with dead ends and with other constraints due to terrain morphology, is a great challenge. The problem involves other variables to be taken into account such as speed, direction and orientation of the vehicle, fuel consumption and tank capacities for chemical products. This article presents a decision graph-based approach, to solve a Rural Postman Coverage like problem using A* and Dijkstra algorithms simultaneously to find the optimal sequence of terraces that defines a selected partial coverage area of the vineyard. The decision structure is supported by a graph that contains all the information of the Digital Terrain Model (DTM) of the vineyard. In this first approach, optimality considers distance, cost and time requirements. The optimal solution was represented in a graphical user OpenGL application developed to support the path planning

O. Contente(✉)
ESTGV, DEMGI, Politechnic Institute of Viseu, Viseu, Portugal
e-mail: ocont@estv.ipv.pt

N. Lau
DETI, IEETA, Aveiro University, Aveiro, Portugal
e-mail: nunolau@ua.pt

F. Morgado
ESTGV, DI, Politechnic Institute of Viseu, Viseu, Portugal
e-mail: fmorgado@estv.pt

R. Morais
Polo INESC TEC, UTAD University, Vila Real, Portugal
e-mail: rmorais@utad.pt

© Springer International Publishing Switzerland 2016
L.P. Reis et al. (eds.), *Robot 2015: Second Iberian Robotics Conference*,
Advances in Intelligent Systems and Computing 417,
DOI: 10.1007/978-3-319-27146-0_27

347

process. Based on the results, it was possible to prove the applicability of this approach for any vineyards which extend like routes. Near optimal solutions based on other specific criteria could also be considered for future work.

Keywords Coverage path planning · Path planning · Mountain vineyards

1 Introduction

With the widespread use of automated machines in agricultural tasks to replace humans hard work (such as preparing the crop, seeding, fertilizing, plant care and harvesting), intelligent path planning techniques, developed along with other study areas, have been adopted to address agriculture issues and to increase productivity. For autonomous agricultural machines, or tractors with auto-guidance systems, the field plots are planned with the help of human know-how, normally keeping the vehicle on-lane based on accurate satellite positioning, with or without curved lines or headland driving support. For those cases, path planning for coverage of the entire crop field refers to copy a realized path, throughout all the uniform field, and paste it apart from the width of the implement. Recent developments in path planning methods for large scale agricultural productions, for full coverage using different optimization techniques and for single or multiple synchronized vehicles which move through the entire field, are reviewed in the literature [1]. For agricultural field machines, the greedy algorithms can be applied when the goal is to find a route as efficiently as possible, while executing the required condition of covering the whole field [2].

The characteristics of some vineyards' terrain can be one of the major limitations for automated agriculture and consequently the need to develop specific path planning systems arises. Due to its unique characteristics, the Douro Demarcated Region (DDR) (a UNESCO World Heritage Site and the oldest Wine Demarcated Region in the World) poses very specific challenges, mainly due to its unique topographic profile, pronounced climatic variations and the complex characteristics of its soil [3]. The DDR is located in Northeast Portugal and consists mostly of steep hills and narrow valleys that flatten out in plateaus. The Douro River digs deep into the mountains to form its bed and the vineyards are planted on terraces fashioned from the steep rocky slopes and supported by hundreds of kilometers of dry stone wall. Within the application of precision viticulture (PV) in the DDR region, addressing the scarcity problem of human resource and pursuing the production of excellent quality wine with minimal costs, developing autonomous vineyard navigation systems requires the use of a very specific path planning strategy. The vineyard under study is part of the Bateiras property, Ervedosa do Douro, São João da Pesqueira, Viseu, Portugal, Fig.1. The prior knowledge of the vineyard 3D terrain map [4] as well as the navigation tracks (obtained, for instance, during spraying [5]) and the traffic lines (obtained from the skeletonization of the vineyard routes [6]) are undoubtedly a huge aid to define the autonomous navigation path planning strategy for this kind of vineyard. This article presents a decision graph-based approach, to solve a Rural

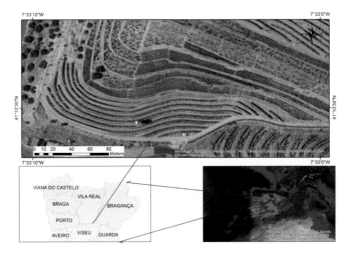

Fig. 1 Localization of the vineyard under study

Postman Coverage like problem using A* and Dijkstra algorithms simultaneously to find the optimal sequence of terraces that defines a selected partial coverage area of the vineyard. The decision structure is supported by a graph that contains all the information of the Digital Terrain Model (DTM) of the vineyard.

The paper begins by introduction our motivation and goals. Section 2 presents a brief overview on coverage path planning methods for agriculture robotics. The problem definition and considerations for this particular study are pointed out in Section 3. Section 4 explains the objective behind the development of the Vineyard Viewer and Path Planing application and its potential. Section 5 describes the methods used in our experimental setup. The results are presented and discussed in Section 6. Conclusions are drawn and suggestions for future work are presented in section 7.

2 Related Work

Generally speaking, the aim of Coverage Path Planning (CPP) is to determine a path that passes over all of the points of interest, at least once, optimizing some criteria, while avoiding obstacles. A recent survey on coverage path planning robotic methods, reflecting advances in the last ten years, can be seen in [1]. The survey presents a new qualitative classification of methods, different from the earlier one presented by Choset [7], based on the different approaches, however, applied, on on-line or off-line classes of algorithms and showing how environments are handled. In addition, it points out different works framed in such method categories.

For agricultural mobile robots, solving collision-free CPP for known environments usually involves the subdivision of the area to cover in small polygonal regions, to find the best coverage path for each cell and then aggregate the results.

In fields where row crops remain unchanged over a decade, for instance, off-line methods can be considered (based on field map information). Environments with configurations similar to street or road networks, or even mountain vineyard crop configuration, that can be represented as a graph, optimality is addressed by assigning a cost to each edge of the graph and either looking for the optimal solution (when no time constraints exists) or for an approximate one (otherwise). Various optimization techniques have been adopted to extract an exact or approximate solution that solves the CPP problem, such as neural networks [8], genetic [9, 10], fuzzy, ant colony [11] and artificial potential field algorithms or a combination of them [12, 13]. The selection of an appropriate algorithm is very important, for both path planning itself and mapping or modeling the environment.

Some CPP approaches for agricultural applications, such as the traveling salesman like (TSP) problem, can be indicated here. Sorensen et al. [14] presented a method for optimizing vehicle routes by defining relevant field information as a graph and formulating the routing problem as the Chinese Postman Problem or as the Rural Postman Problem, when a headland graph is considered. To find the solution, a heuristic approach was proposed based on NP-Hard combinatorial optimization problem consideration. Using a spraying operation the method was tested, assigning distance costs to all edges and generating a route that minimizes the overall cost.

Ryerson and Zhang [15], proposed a genetic algorithm based approach to solve the CPP. Considering a grid representation of the field, the solution comprises determining the optimal sequence of cells that forms the route taking into account the minimal distance and maximum coverage criteria. Using B-patterns [16] as fieldwork patterns representing the results of a combinatorial optimization process, different methods were tested. Bochtis el al [17] proposed a multi traveling salesman problem for planning a fleet to combine harvesters operating in a field. B-patterns are used for mission planning of an autonomous tractor for area coverage operations such as grass mowing, seeding and spraying. Bakhtiari et al. [18] used the B-patterns as the result of an ant colony optimization (which minimizes operational criteria such as time, non-working distance traveled, fuel consumption among others) to generate optimal routes for field area coverage.

A planning method for simulating field operations in agricultural fields with multiple obstacle areas that uses an ant colony algorithmic optimization, as an approximate approach was also presented by Zhou et al, [11]. The planning method involves a first stage where the field area and the in-field objects are represented as a geometrical graph (including the obstacle and the field work-track representation), a second stage where a decomposition of the field into block areas (with no obstacles) takes place (implying the division of the previous work-tracks) and a final stage where the optimal traversal sequence of blocks is obtained, considering field parameters as well as machine ones.

Oksanen and Visala [2], developed a split and merge path planning algorithm to evaluate the optimal routes and their efficiency in terms of energy consumption needed to operate in Finnish complex shape crop fields. This study, considering fields with obstacles and the need of refuel the machine, includes the combination of two greedy algorithms (an off-line and an on-line algorithm). These are used to divide the coverage region into sub-regions (with driving lines side-by-side and parallel to each other), to select the sequence of those sub-regions that has the minimum headland turns and to generate a path that covers each sub-regions taking into account the desired working direction. It assumes the vehicle's pose is precisely known and the localization problem does not exist. The field boundaries are known and considered invariable. As the authors said, although neither of the two algorithms alone or combined can solve the CPP problem optimally, they constitute a further contribution to solve the routing problem for agricultural operations. A practical approach to multi-path planning for a fleet of heterogeneous and autonomous vehicles, during a treatment operation is proposed in [19], which considers the combinational optimization of criteria such as distance, input cost and time. Total coverage is considered for a parallel sown field like rows involving the determination of the bests track permutation and transitions for refueling the vehicles tank's as the solution. The simulated annealing method is used to solve the problem as rapidly as possible. The author concludes the result depends on the features of the fleet vehicles and on the optimization criteria used.

3 Problem Definition

3.1 Practical Considerations

As previously mentioned, the mountain vineyard rows, follow the curl of the hill forming 3D configuration parcels with nonuniform contour. The rows remain unchanged over the years, unless, for instance, the land subsides due to heavy rainfall. As result an off-line path planning approach is used. For simplification, because less steep terraces often have a narrow width, U-turn maneuvers are not allowed. Especially in dead-end terraces the solution is to return in the reverse gear. Some maneuvers are allowed on paths (that usually surrounds the vineyard and have a steeper slope). However paths cannot be considered as headlands because some of them have vineyards planted in one of its sides. The transitions, as crossover vineyard areas between terraces and between terraces and paths, were defined to behold a maneuver place, like the ones shown in Fig. 2. In these places changes in direction are allowed as well as the forward and backward maneuvers. To simplify U-turn are not allowed anywhere on the vineyard.

A simplified geometric representation of the terraces, paths and transitions was also used. The navigation routes are identified as the skeletons of both path and terraces. Both are limited by their ends (three sub-figures above in Fig. 3). When a path allows access to a terrace, a transition is defined (three sub-figures below in Fig. 3). As the skeleton does not overlap the path, this is divided in three parts: one before (Skl_1, in pink), one between (Skl_2, in white) and one after the transition (Skl_3, in green). Consequently resulting in two more pair of ends. For transition definition, it was used the representation of three 1D curves: the inner part of the skeleton

Fig. 2 Changing direction through different transitions: (a) and (b) terraces to terrace transitions, (c) terrace to path transition, (d) in a headland path transition

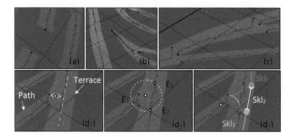

Fig. 3 Vineyard top-down view including: (a),(b) and (c) Skeletons ends for different configurations terraces; (d1), (d2) and (d3) Steps involving the transition definition, its control points and skeleton partition.

subdivision and two Bézier curves which links each end of the subdivided skeleton to the end of the terrace skeleton that borders the transition. So, with the new Digital Terrain Model (DTM) of the vineyard (where the ends of the skeleton, sub-skeleton and Béziers curves where added to the original geographic information) a graph of nodes (end points) and edges (hereinafter tracks) can be built. Optimal path planning based on the map of the vineyard is proposed.

Our problem can be defined as similar to the Rural Postman Problem (RPP) which, given a graph $G = (N, E_S \subset E)$ where N represents the set of node's, E the set of undirected edges and E_S is a subset of E, finds the minimum cost tour on G that visits each edge $e \in E_S$. The RPP routing problem seeks a tour that traverses a required subset of the graph edges using the remaining edges as travel links. However it fits further the Windy Rural Postman Problem (WRPP) which seeks a route for an directed graph where the cost of each edge changes depending on the direction of traversal. Classified as a NP-Hard problem, it might be suitable to be approached by Dijkstra or A* methods, that uses a best-first search and finds a least-cost path from a given initial node to one target node.

3.2   The Problem Formulation

Considering the directed graph $G = (N, E)$ where G is named the map graph of the vineyard, $N = \{n_1, ..., n_N\}$ the set of nodes of the graph which represents the track-ends and $E = \{p_1, ..., p_P\}$ the set of the directed edges which represents the totality of the map tracks. Let S be the set of the n_S tracks selected for coverage ($S \subset E$) and U the set of all different permutations, u_i, obtained using all elements of S, where $i \in \{1, ..., n_U\}$ and n_U is the number of permutations. Considering also two special track-ends, e_{Start} and e_{Goal} which belong to N (e_{Start} defined as the start and e_{Target} as the target points respectively).

The problem consists of finding the optimal permutation of tracks $u_i \in U$, to ensure the coverage of S at least once. Because tracks may be covered, in one or two directions, at least once, the edges are considered directed. The cost criteria definition may involves parameters like distance, task time, type of task, type of maneuvers or variation in transition slope. The estimation of the time spent in maneuvers do not consider turning angles and dynamic vehicle constrains in this phase of the work. The objective function assigns the optimization criteria such as the distance traveled (Eq. 1), the input cost (Eq. 2) and time (Eq. 3) required to complete the task. $d(p_j)$, $c(p_j)$ and $t(p_j)$ are the distance, cost and time spent traveling the selected tracks p_j, where $j \in \{1, ..., n_S\}$. The input costs refers to funds spent, for instance, on chemical product and fuel. The slope of each track greatly influence the fuel consumption and indirectly the input costs. A linear combination of the individual objective function was adopted for finding the optimal solution, $u^* : \Phi(u^*) = Min(\Phi(u_i)), \forall u_i \in U$. Each individual function, settled for each edge by a searching in the map graph G was defined by the following expressions:

$$\Phi_d(u_i) = \sum_{j=1}^{n_S} d(p_j) \tag{1}$$

$$\Phi_c(u_i) = \sum_{j=1}^{n_S} c(p_j) \tag{2}$$

$$\Phi_t(u_i) = \sum_{j=1}^{n_S} t(p_j) \tag{3}$$

Mathematically, the problem can be expressed using the following expression:

$$\Phi(u^*) = Min\{\Phi(u_i)\} = Min(\alpha_d \times \Phi_d(u_i) + \alpha_c \times \Phi_c(u_i) + \alpha_t \times \Phi_t(u_i)), \forall u_i \in U \tag{4}$$

Where α_d, α_c, α_t are the relative weights assigned for a vineyard operation task, for distance, input cost and time, respectively and u^* the optimal permutation.

An Heap's algorithm implementation was used to find the set all different permutations of the selected tracks, U. To find, on the map graph G the best path between the start point e_{Start} and the closer track-end of the first track p_j of each permutation element u_i it was used the A* algorithm. This was also used to compute the best path between the closer track-end of the last track of each permutation element and the target point e_{Target}. The set of permutations, U, is used as a guide to the Dijkstra algorithm to find, step by step from the start point, the possible expansion of tracks to find the best navigation path for the task do be done which ends in the target point e_{Target}.

4 Vineyard Viewer and Path Planning Application

The Vineyard Viewer and Path Planning application (VVPP) will be used to assist an autonomous vineyard navigation system. It was developed in C#, with OpenGL facilities and enables 3D vineyard details to be viewed. The DTM of the vineyard (comprising terraces and paths position and vineyard characteristics), the boundary points of each terrace and each path, the tractor route taken previously (during a spraying operation), and the skeletons of each terrace and path can be loaded to the application. The selected elements drawn in different colors can be seen in the viewer. Vineyard information was identified by a letter followed by a number. The letters P and C were used for the terraces and paths respectively. Both were followed by a number which represents their relative position from the southeast corner of the vineyard. Information about previous developed functionalities and how to access them can be consulted in previous authors publications [5, 6]. Recent developments includes the options associated to the path planning tool described below. In the vertical menu it is possible to select the path or terraces to be covered and the start and the goal points. The loading process of the skeletons is automatically started by selecting the suitable algorithm (through a Thinning or Geometric skeletonization approach). By selecting the Limit Points or Transition check-boxes it is possible to hide (Fig. 4 (a)) or show (Fig. 4(b)) the limit points of the skeletons or the transitions' geometric configuration. All the skeleton and transition end points are considered nodes of the vineyard map graph that supports the decision-making system.

In the GraphViewer environment the node and associated edge can be found, Fig. 5 (a). Going back to the VVPP, the terraces and paths to be covered are selected using the mouse key. As shown in Fig. 5 (b) the P7, P8 and P9 terraces were selected

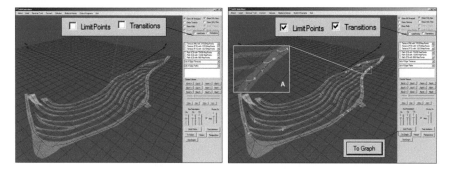

Fig. 4 Vineyard skeletons: (a) without ends (b) with ends and transitions

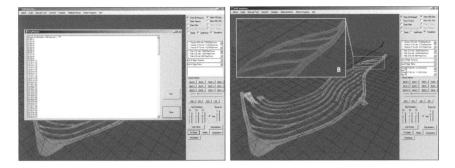

Fig. 5 (a) Graph-viewer (b) Set of selected terraces to be covered

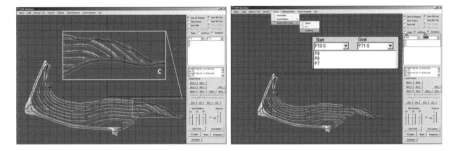

Fig. 6 (a) Selection for combined algorithm application (b) Set of selected terraces to be covered, in perspective

to be covered. Through a vineyard top view (using the Perspective option) a 2D spot of the covered area can be seen. In the vertical menu, the start and the goal points can be selected from the range of possible end points. The Heap's bottom start the process of finding all possible permutation of the selected path ends [20]. After generating the decision graph structure, when the Calculus/Decision Point Graph/Combined horizontal menu option is selected, Fig. 6 (b), the path planning system automatically opens the Graph Viewer. If the Run button which has become available is selected, the path planning system starts the combined Dijkstra and A* new algorithm, optimizing the previously set forth objective function. The resulting optimal sequence is both displayed there and in the VVPP interface.

5 Description and Planning of the Experience

Two procedures were done, one following a specific branch crushing task performed by an expert driver and other developing the new algorithm and applying it for the mentioned task.

5.1 Branch Crushing After Pruning Grapevines

During a manual pruning the branches are usually left on the floor. They are later decomposed and absorbed but when mowed and crushed they form a mat that facilitates the passage of both human and vehicle. In fact this task was done using a New Holland TCE 50 tractor equipped by a Becchio-Mandrille (Series TSE) flail mower implement and covered all the vineyard. However, the present study only considered the crushing of three terraces. As shown in Fig. 7 (a) the set implement must be lifted when maneuvering in transition sites or when backwards to position itself for repeating the task. During branches crushing the implement must be in down position, Fig. 7 (b).

Fig. 7 Tractor with: (a) Mower up (b) Mower down

5.2 Path Planning Implementation

As initial information for this work, the skeletons of all the terraces and path of the vineyard where clearly defined and their ends were computed and considered as decision points. However, the definition of the transition led to the emergence of more points to be added as decision making. So tracks and track-end were considered in the graph map updating step. The VVPP application start and goal points were selected as well as the set of tracks to be covered. Then the permutation step returns the information needed to establish the edges of the decision graph. In this first test of the path planning method only distance is considered. So, the objective function was then adapted. The results, in terms of the best chosen permutation appear in a graphical color representation as explained in results section.

6 Results

The tour done by the expert driver, Fig. 8 (a), calculated by summing the length of the following tracks, can be expressed as:

$$(P_{10}, P_{91}, P_9', P_9, P_{81}, P_8', P_8, P_{71}, P_7', P_7, P_7') \qquad (5)$$

The total route resulting from the planner, Fig. 8 (b), calculated by summing the length of the following tracks, can be expressed as:

$$(C_{2-1}, P_9, P_9', C_{2-2}, P_8, P_8', C_{2-3}, P_7,' P_7) \tag{6}$$

Where the $'$ refers the backward travel. As results the cost of route obtained using the path planning system is much smaller.

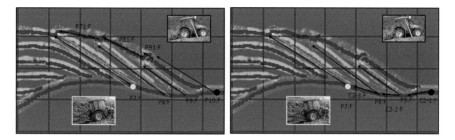

Fig. 8 Tractor route: (a) Experimental (b) Path Planning

7 Conclusion

This article presents a decision graph-based approach, to solve a Rural Postman Coverage like problem using A* and Dijkstra algorithms simultaneously to find the optimal sequence of terraces that defines a selected partial coverage area of the vineyard. The decision structure is supported by a graph that contains all the information of the vineyard DTM. Despite having a more general approach considering distance, cost and time parameters, the simulation was only carried out using distance optimization. The results obtained from a comparative analyses between the path planning and a human expert tractor driver during a cane crushing operation reveals the great importance of path planning. However, to validate the system, the inclusion of the remaining tuning parameters is critical, especially the ground slope parameter. To validate this approach, it will also need to be extended including a wider coverage and including tests for other operations in mountain vineyards. Several of the VVPP's features can also be improved. Another aspect, initially not included, is the implementation of dynamic path planning to change the number of times a track could be covered, for instance. Although the study only considers the coverage of three terraces, the driver found that the existing amount of wicker in one of the terraces requires a second pass. Issues like these will require the reformulation of the optimization problem. This approach can be applied to other row like vineyards. Near optimal solutions based on other specific criteria could also be considered for future work.

References

1. Galaceran, E., Carreras, M.: A Survey on coverage path planning for robotics. Robotics and Autonomous Systems **61**(12), 1258–1276 (2013)
2. Oksanen, T., Visala, A.: Coverage path planning algorithm for agricultural field machines. Journal of Field Robotics **26**(8), 651–668 (2009)
3. Andresen, T., Bianchi, F., Curado, M.J.: The Alto Douro Wine Region Geeenway. Landscape and Urban Planning **68**(2–3), 289–303 (2004)
4. Contente, O., Aranha, J., Martinho, J., Ferreira, P., Lau, N., Morais, R.: 3D digital maps for vineyard autonomous robot navigation. In: Advances in Artificial Intelligence - Proceedings EPIA- XVI Portuguese Conference on Artificial Intelligence (2009)
5. Contente, O., Aranha, J., Martinho, J., Morgado, J., Reis, M., Ferreira, P., Morais, R., Lau, N.: 3D map and DGPS validation for a vineyard autonomous navigation system. In: CONTROLO 2014 - Proc. of the 11th Port. Conf. on Autom. Control. LNEE, vol. 321, pp. 617–625. Springer, Heidelberg (2015)
6. Contente, O., Lau, N., Morgado J., Morais, R.: Vineyard skeletonization for autonomous robot navigation. In: 2015 IEEE International Conference on Autonomous Robot Systems and Competition (ICARSC), pp. 50–55 (2015)
7. Choset, H.: Coverage for Robotics a Survey of Recent Results. Annals of Mathematics and Artificial Intelligence **31**(1–4), 113–126 (2001)
8. Yang, S.X., Luo, C.: A Neural Network Approach to Complete Coverage Path Planning. IEEE Transactions on System, Man and Cybernetics, Part B: Cybernetics **34**(1), 718–724 (2004)
9. Hameed, I.A., Bochtis, D., Sorensen, C.A.: An Optimized Field Coverage Planning Approach for Navigation of Agricultural Robots in Fields Involving Obstacle Areas. International Journal of Advanced Robotic Systems **10**(231), 1–9 (2013)
10. Hameed, I.A.: Intelligent Coverage Path Planning for Agricultural Robots and Autonomous Machines on Three-Dimensional Terrain. Journal of Intelligent and Robotic Systems **74**(3–4), 965–983 (2014)
11. Zhou, K., Jensena, A.L., Srensena, C.G., Busatob, P., Bothtisa, D.D.: Agricultural operations planning in fileds with multiple obstacle areas. Computers and Electronics in Agriculture **109**, 12–22 (2014)
12. Noguchi, N., Terao, H.: Path planning of an agricultural mobile robot by neural network and genetic algorithm. Computers and Electronics in Agriculture **18**(2–3), 187–204 (1997)
13. Garcia, M.A., Montiela, O., Castillo, O., Seplveda, R., Melinb, P.: Path planning for autonomous mobile robot navigation with ant colony optimization and fuzzy cost function evaluation. Applied Soft Computing **9**(3), 1102–1110 (2009)
14. Bochtis, D.D., Vougioukas, S.G., Griepentrog, H.W.: A Mission Planner for an Autonomous Tractor. Transactions of the ASABE **52**(5), 1429–1440 (2009)
15. Ryerson, A.E.F., Zhang, Q.: Vehicle Path Planning for Complete Field Coverage Using Genetic Algorithms. Agricultural Engineering International: the CIGR E-Journal, 9 (2007)
16. Bochtis, D.D.: Planning and control of a fleet of agricultural machines for optimal management of field operations. Greece: Aristotle University. PH. D. Thesis (2008)
17. Bochtis, D.D., Vougioukas, S.G.: Minimising the non-working distance travelled by machines operating in a headland field pattern. Biosystems Engineering **101**(1), 1–12 (2008)
18. Bakhtiari, A.A., Navid, H., Mehri, J., Bochtis, D.D: Optimal route planning of agricultural field operations using ant colony optimization. Agricultural Engineering International: CIGR Journal 13(4), 1–16 (2011)
19. Conesa-Muoz, J., Bengochea-Guevara, J.M., Andujar, D., Ribeiro, A.: Efficient distribution of a fleet of heterogeneous vehicles in agriculture: a practical approach to multi-path planning. In: 2015 IEEE International Conference on Autonomous Robot Systems and Competition (ICARS), pp. 56–61 (2015)
20. Sedgewick, R.: Permutation Generation Methods. ACM Computing Surveys **9**(2), 137–164 (1997)

Agricultural Wireless Sensor Mapping for Robot Localization

Marcos Duarte, Filipe Neves dos Santos, Armando Sousa and Raul Morais

Abstract Crop monitoring and harvesting by ground robots in steep slope vineyards is an intrinsically complex challenge, due to two main reasons: harsh conditions of the terrain and reduced time availability and unstable localization accuracy of the Global Positioning System (GPS). In this paper the use of agricultural wireless sensors as artificial landmarks for robot localization is explored. The Received Signal Strength Indication (RSSI), of Bluetooth (BT) based sensors/technology, has been characterized for distance estimation. Based on this characterization, a mapping procedure based on Histogram Mapping concept was evaluated. The results allow us to conclude that agricultural wireless sensors can be used to support the robot localization procedures in critical moments (GPS blockage) and to create redundant localization information.

1 Introduction

Crop monitoring and harvesting by robots remains a complex challenge, particularly due its low efficiency, accuracy, and robustness on sensing, perception and interpretation of the agricultural environment [1]. The strategic European research agenda for robotics [2] states that robots can improve agriculture efficiency and competitiveness. But there are still very few available commercial robots for agricultural applications [1].

M. Duarte · F.N. dos Santos(✉) · A. Sousa
INESC TEC - INESC Technology and Science (formerly INESC Porto) and
Faculty of Engineering, University of Porto, Porto, Portugal
e-mail: {ei10007,asousa}@fe.up.pt, fbsantos@inesctec.pt

R. Morais
INESC TEC - INESC Technology and Science (formerly INESC Porto),
Universidade de Trás-os-Montes e Alto Douro, UTAD, Vila Real, Portugal
e-mail: rmorais@utad.pt

© Springer International Publishing Switzerland 2016
L.P. Reis et al. (eds.), *Robot 2015: Second Iberian Robotics Conference*,
Advances in Intelligent Systems and Computing 417,
DOI: 10.1007/978-3-319-27146-0_28

Fig. 1 A typical steep slope Terraced Vineyard in the Douro region of Portugal.

In Europe space we can identify two ongoing research and development projects to deploy monitoring robots on flat vineyards: the VineRobot [3], and Vinbot [4]. However, there are other kinds of vineyards that are not built on flat terrains but on steep slope hills which is a complex environment for the machinery and for the robotic algorithms (such as localization, mapping and path planning). These called steep slope vineyards exist in Portugal in the Douro region - an UNESCO heritage place - Fig. 1, and in other regions of five European countries. As these crops are not built on flat terrain but in steep hills, the development of robotic localization and mapping module that is accurate, reliable and all-time-available is a challenge because [5]:

— the GPS availability and accuracy are largely reduced due to the signal blockage or multi-reflection;
— the dead-reckoning systems (for example odometry and inertial measurement systems) accuracy is drastically reduced due to the harsh conditions of the terrain.

In [5], a localization system based on natural vineyard features, such as the trunks and poles of the vine, is proposed. This system can be seen as an alternative/complement to GPS-based localization systems. However, localization systems based on natural feature detectors are not all the time reliable, because the detector accuracy depends largely on the environment conditions (light, fog and dust) and for that reason the use of artificial beacons is advisable to increase the accuracy, reliability and availability of these localization systems.

The latest techniques for precision agriculture are applying static sensors to the fields for continuously monitoring important variables, such as temperature, humidity and conductivity. The use of wires to connect these sensors to the base station is not always easy or an option and for that reason wireless sensors are a common approach. These sensors inserted in Wireless Sensor Network (WSN) can be used as artificial beacons and they have an enormous potential to increase the accuracy of the robot location estimation.

Various international standards have been established for WSN applications in the past decades. Among them, the standards for wirelless LAN IEEE802.11b ("WiFi"),

IEEE 802.15.1 (BT) and IEEE 802.15.4 (ZigBee) are the most commonly used [6]. In this work, BT version 4.0 also known as Bluetooth Smart or Bluetooth Low Energy (BLE) based technologies were selected as it provides some features like [7]: "Ultra-low peak, average and idle mode power consumption; ability to run for years on standard coin-cell batteries; lower implementation costs; multi-vendor interoperability; enhanced range". Particularly, the beacons manufactured by Kontakt[1] were used to explore the wireless sensor signals in order to improve and increase redundancy of robot localization estimation. In the tests, the robot used was AGROB V14 system defined in [5].

In this paper, in section 2 a global overview about localizing and mapping considering RSSI-based distance estimations is given. In section 3 the RSSI-based distance estimation using *iBeacon* artifacts is evaluated. In section 4 our beacons mapping procedure (BMP) is described. In section 5 the realized tests are detailed and the obtained results are presented. Paper conclusions are presented in section 6.

2   Localization and Mapping Based on RSSI

Localization can be realized considering several techniques, among them lateration and angulation are the most common techniques. However, a distance and/or an angle measurement between the robot and beacons is required for the robot location to be observable. In Radio Frequency based beacons it is possible to estimate the distance and angle between a beacon and a receiver, by considering three solutions [8]:

− Received Signal Strength (RSS) to distance (path loss models)
− RSS to Angle of Arrival (AoA) (directional antenna models)
− Time-of-Flight to distance(ToF) (Speed of light)

RSS is defined as the voltage measured by a receivers RSSI circuit. Most of the time, RSS is equivalently reported as measured power (the squared magnitude of the signal strength). RSS can be used for acoustic, Radio Frequency (RF), or other signals. Agricultural wireless sensors naturally communicate with neighboring sensors/base-stations, so the RSS of RF signals can be measured by each receiver during normal data communication without presenting additional bandwidth or energy requirements. RSSI is available in a large percentage of commercial receivers, these measurements are relatively inexpensive and simple to implement in hardware.

As the GPS signals are not available in indoor scenarios, the RSS based localization has become an important and popular topic of localization research [9]. However, RSS measurements are notoriously unpredictable. When we consider this RSS measurements for a robust localization system, their sources of error must be precisely modeled. In optimal conditions (open-spaces and non-noisy) the signal power decays proportional to d^{-2}, where d is the distance between the transmitter and receiver.

---

[1] Available in http://www.kontakt.io

In real conditions, the signal reflection is the most common source of errors. Reflection generates multiple signals with different amplitudes and phases at the receiver antenna, and these signals add constructively or destructively as a function of the frequency, causing frequency-selective fading, in [8] several methods are presented to minimize these errors and increase the distance estimation accuracy.

After the best estimation of the distance between the receiver and beacon, the robot/receiver location can be estimated or corrected. However, for 2D localization purposes, the number of measurements must be enough (at least 3) and the beacons position well known. Nevertheless most of times these agricultural wireless sensors are installed without being georeferenciated, which is a requirement for this distance to be useful for robot location estimation. Here emerges a question: *how can a robotic mapping procedure for artificial beacons (agricultural wireless sensors) be developed?*

The process of mapping beacons can be solved considering the robot location known or unknown. When it is unknown we must formalize a mapping procedure considering Simultaneous Localization and Mapping (SLAM) based techniques. In [10], a FastSLAM based algorithm is used for mapping seventy Radio-Frequency IDentification (RFID) tags, which are spread through an indoor environment. After the RFID tags mapping stage they have evaluated the robot location considering the RFID distance measurements and ignoring the odometry information, with this work the authors have concluded that the robot was able to localize itself with the same magnitude of accuracy and reduce the computational demands for the global localization of a moving mobile robot.

Despite the fact of SLAM based techniques being the most appropriated for situations when the environment and robot location are unknown, in outdoor robots and particularly in agricultural robots, we can consider that the robot location is always know. In this particular work, of applying robots to steep slope vineyard, we will consider that the robot position is always known during the beacons mapping procedure. This can be considered true, because the mapping procedure will occur only when GPS signals are available or the robot location is observable by other means with an acceptable accuracy.

Knowing the robot position, to be possible to estimate the beacon position in a 2D space, at least 3 distance measurements from different observations points are required. This beacon location estimation is obtained from the circle interception points/spaces. However:

- **To the distance measurements is always associated an uncertainty** which makes the circle boundaries less defined and instead of a single interception point one or several likelihood interception regions are obtained; and,
- **Only one receiver will be available**, which means that at each instant only one distance measurement will be obtained per beacon.

The number of beacon distance measurements can be solved by considering distinct moments and robot locations. With minimum distance measurements satisfied, the beacon location can be estimated. However as several uncertainties are associated to the distance measurements and robot position, the used circle/ellipsoid equation interception solvers are intractable. Instead of finding a single interception point our problem now is to find the most likely interception region. In this way three filters used in robot localization problem can be considered for the beacons mapping procedure [11]:

− Kalman filter - continuous tracking representation;
− Histogram filters - Grid based approaches; and,
− Particle filters;

The Histogram based filters makes a discretization of the 2/3D space into a set of cells and assign to each cell a value with the probability of the robot/object/beacon being localized in that cell. In contrast particle filters spread randomly a large number of particles through the space and to each particle is associated a probability value of that particle being closer to the robot/object/beacon, in each iteration these probability values are updated and those particle with the lowest probability are deleted and new ones are generated in the neighborhood of the most likelihood particles. Due to the non-linearity of the beacons mapping problem, Histogram filters and Particle filters are the most adequate for this problem.

3   iBeacons RSSI Based Distance Evaluation

The BT based sensors called *iBeacons*[2] are simple emitters that advertises a packet with a certain frequency. They are developed using BT version 4.0 or "Bluetooth Low Energy" (BLE) that specifies the use of Generic Attribute Profile (GATT) built on top of the Attribute Protocol (ATT)[12]. This defines common operations and a framework for use of stored and transported data. GATT has two types of roles:

− server - stores the data transported over ATT and accepts requests, commands and GATT client confirmations;
− client - sends requests, commands and confirmations trough ATT to GATT servers.

Thus, GATT is divided into Attributes, which in turn contain Services with certain Characteristics. These have a Value and a Description. Among others [13], some key features provided by these *iBeacons* are:

− "TX Power Level" - sets the transmission signal strength based on which GATT clients will get RSSI;
− "Advertising Interval" - sets the period advertising packets are emitted.

To characterize the RSSI's and distances of the *iBeacons* some outdoor tests have been made based on the description of Android Beacon Library[14]. On an empty football field, one put two wooden benches with approximately 0.7 m high and separate from each other for needed and known distances of 0.25, 0.5, 1, 2, 3, 4, 5, 6, 7, 8, 9, 10, 12, 14, 16, 18, 20 and 25 m. Then put the *iBeacon* (GATT servers)

---

[2] Devices that complies to certain adopted specifications created by Apple Inc.. See https://developer.apple.com/ibeacon/.

in a bench and in the other an Android mobile device with an application (GATT clients) that measures the RSSI from the *iBeacon*.

In this particular case, the beacons used were the ones from Kontakt and the mobile device a Samsung GT-I9505. The beacons were set to transmit the advertising packet[3] each 350 ms and its TxPower to a moderate value[4] of −12 dBm. On the mobile side, the application developed records the RSSI readings also in intervals of 350 ms during a period of 2 minutes. That takes to 342 samples of RSSI that are saved to a text file to future analysis. In Fig. 2 are examples of histograms based on data recorded by the application[5].

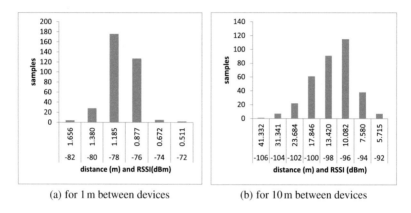

(a) for 1 m between devices   (b) for 10 m between devices

Fig. 2 Example histograms of recorded RSSI samples.

Accordingly to the model described in the Android Beacon Library [14]

$$d = A * \left(\frac{RSSI}{Ref\,Power} \right)^{B} + C \,. \tag{1}$$

In 1, d is the distance in meters that we want to obtain, $RSSI$ is the one received from the *iBeacon* by the mobile application, $Ref\,Power$ is the reference received RSSI for the configured TxPower at the distance of 1 m between devices[6], A, B and C are constants.

With that data, each distance has a corresponding RSSI obtained with a mean value of the 342 samples measured in our field tests. Then, a fitting curve is constructed

---

[3] Advertising interval can be set to a value between 20 ms and 10,240 ms.[13].

[4] Transmission power can be set to −30, −20, −16, −12, −8, −4, 0 or 4 dBm.[13].

[5] To improve readability the values of measured RSSI were merged using a interval of 2 dBm instead of 1 dBm.

[6] RefPower can be extracted from the advertising packet or can be measured for calibration of an *iBeacon* with the specific mobile device. The calibration can give more accuracy to the function of distance estimation.

and a function for converting RSSI to distance is derived using 1 and the nonlinear least squares method to the pairs of distance and mean RSSI.

The function derived:

$$d = 0.3534536 * \left(\frac{RSSI}{Ref\,Power} \right)^{14.8393466} + 0.7566785 \;. \tag{2}$$

Using 2, a graph of the distances corresponding to the mean RSSI for each distance can be constructed (Fig. 3). Reading the generated graph, one can see that the error on converting from RSSI to distance increases with the distance.

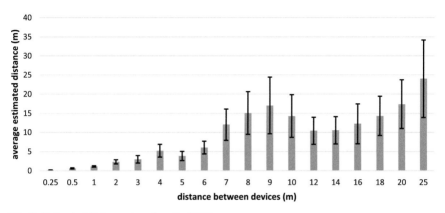

Fig. 3 Estimated distances based on the RSSI.

4 Beacons Mapping Procedure

Due to their simplicity and accuracy, a grid based filter was selected for our BMP. An occupancy grid map is a classical metric map representation [11] and is based on the idea that occupancy grids represent the map as a field of random variables, arranged in an evenly spaced grid. They assign to each (x, y) (or even in 3D (x, y, z)) coordinate a occupancy value which specifies whether or not a location is occupied with an object.

Our BMP algorithm at the first beacon distance observation, generates a likelihood grid map (Fig. 4) with its center located at the robot position with the next constrain:

$$Scale * NumberCells = max_{BeaconDistance} \;. \tag{3}$$

where $max_{BeaconDistance}$ is maximum observable beacon distance, $Scale$ the scale value for longitude and latitude axis, and $NumberCells$ the horizontal and vertical resolution.

M. Duarte et al.

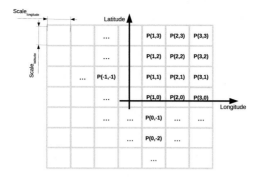

Fig. 4 Likelihood grid map for beacon existing probability region description.

The gold standard of any occupancy grid mapping algorithm is to calculate the posterior over maps given the data:

$$p(m|z_{1:t}x_{1:t}) \,. \tag{4}$$

where m is the map, $z_{1:t}$ the set of all measurements up to time t, and $x_{1:t}$ is the set of robot poses from time 1 to t.

Based on these concepts and considering that the errors associated to beacons measurements and robot localization follows a normal distribution it is time to formalize a cell probability value as a function of distance observations and robot position estimation. So from a beacon distance observation z_t^{beacon} at time t - in the BMP algorithm - the probability of existing a beacon at the cell $c(x, y)$ is given by:

$$P_t(x, y) = \frac{1}{\sigma_{RB}\sqrt{2\pi}} e^{-\left(d_{RB}-z_t^{beacon}\right)^2/2\sigma_{RB}^2} \,. \tag{5}$$

where, d_{RB} is the distance between the robot location and the cell gridmap center, and σ_{RB} is the sum of uncertainties associated to the robot position estimation and beacon distance estimation.

For the fusion of several distance measurements at different time instants, the probability value at each grid map cell $P(x, y)$ is given by:

$$P(x, y) = \frac{P_t(x, y) * P_{t-1}(x, y) * P_{t-2}(x, y) * \dots}{\sum_{t=0}^{T} \sum_{i=0}^{X} \sum_{j=0}^{Y} P_t(i, j)} \,. \tag{6}$$

Based on these probabilities values, the beacon position estimation is simply given by:

$$B_{x,y} = \sum_{i=0}^{X} \sum_{j=0}^{Y} C_{i,j} * P(i, j) \,. \tag{7}$$

where $B_{x,y}$ is bi-dimensional vector that represents the beacon position and $C_{i,j}$ is bi-dimensional vector that represents the cell position center.

5 Tests and Results

For BMP algorithm test we have selected the AGROB V14 platform. AGROB V14 is small and cost effective outdoor robot for application on steep slope vineyard monitoring tasks. The robot is built on top of a commercial radio-controlled model (RC-model) Traxxas E-Maxx, a 1/10 scale 4WD Electric Monster Truck, Fig. 5. To this RC-mode are attached two tiny computers (one UDOO quadcore version, and one RaspberryPi version B), with Advanced RISC Machine (ARM) processors. This robot has two processing units running Linux (Ubuntu 12.04) with the robotic operating system (ROS-groovy) on top. More details about AGROB V14 can be found in [5].

(a) At steep slope vineyard   (b) Beacon mapping test site

Fig. 5 AGROB V14 robot at the vineyard and test site. Red dots are the places where the beacons were deployed.

To make the robot able to acquire distance measurements to these beacons, an Android application was developed and deployed into a smartphone (Samsung GT-I9505), which was attached to the robot. The Android application running in the mobile device starts reading the RSSI of every visible beacon and sends the real-time information every 1 s via User Datagram Protocol (UDP) to the AGROB V14 robot (via the wireless interface).

In Fig. 5 is shown in the left the AGROB V14 operating in a steep slope vineyard, and in the right the robot at the beacon mapping test site. As shown in Fig. 5 and in 6, ten *iBeacons* were deployed over ten park benches, which are spaced by 12 meters longitudinally and 6 meters transversely. To illustrate their positions, the minor ID of each *iBeacon* has been used. Robot estimated trajectories and beacons position estimation were drawn using Google Maps API.

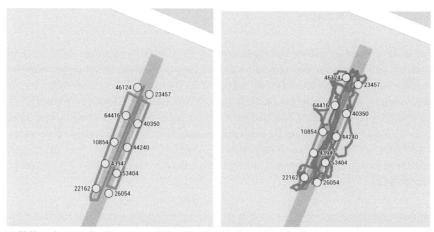

(a) Yellow dots are the *iBeacons* and blue line the (b) Green line is the robot trajectory extracted planned trajectory to be realized by the robot.   from the GPS receiver estimations.

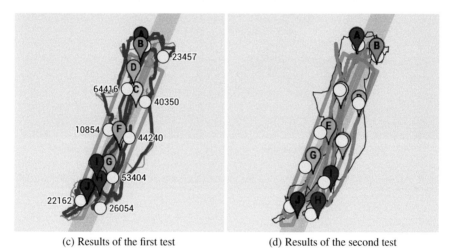

(c) Results of the first test   (d) Results of the second test

Fig. 6 The obtained test results.

The data acquired by the robot at the test site has been recorded and stored in a ROSbag file[7]. Using this file two tests were realized. On the first one, the robot location estimation was obtained from the fusion of GPS and odometry velocity, by means of a simple particle filter. In Fig. 6.(a) is shown the resulted estimated robot trajectory (Red bold line) and beacons position (colorful pinpoints). From this figure is possible to verify a better estimation of the robot trajectory (in contrast to GPS estimation (Green thin line)).

---

[7] The file can be accessed at http://hyselam.com/agrob.html.

On the second, the robot location estimation was obtained from the fusion of GPS, odometry velocity and considering the mapped beacons and beacons distance observations, by means of a simple particle filter. In Fig. 6.(b) is shown the estimated robot trajectory (Green bold line) and beacons position (colorful pinpoints). In contrast to the first test, from this figure is possible to verify a better estimation of the robot trajectory and better beacons position estimation. When the robot position estimation was realized considering the beacons distance observations (and their beacons position estimation) has improved significantly the robot trajectory estimation and consequently the beacons position estimation, as can be seen in 6.

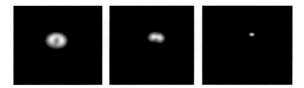

Fig. 7 Grid map with resolution of 0.25 meters/pixel for *iBeacon* with minor ID 64416 at 10, 230, 460 seconds instant.

Fig. 7 shows the obtained likelihood grid map for the *iBeacon* with minor ID 64416 at three different instants. These likelihood grid maps were obtained during the second test and it is possible to verify the evolution of the beacon position estimation uncertainty. The possible beacon position has evolved from a circular region, to two possible regions and ends in likelihood single spot.

On the second test, the beacon position estimations has in average an error of 1.5meters. The main origin of this error is associated to the low accuracy of robot position estimation, which is drifted from the original trajectory as can be seen in Fig. 6.(b). However, if we analyze the relative error positioning of the beacons (distance between beacons) the average error is 0.43meters, near to the likelihood grid-map resolution, which is 0.25meters/pixel. This means that beacon mapping accuracy is highly dependent on the robot position estimation accuracy, as expected.

6 Conclusion

In this work, was shown that the use of agricultural wireless sensor signals can help to improve and increase redundancy of robot location estimation, even when beacons mapping procedure is occurring, as show in section 5. Of course there is the need of doing more and repeated tests to have enough data to help the support of these results.

Besides that the BMP algorithm is easily implementable and integrable to other localization and mapping procedures. As future work we will integrate agricultural wireless sensors and BMP into the hybrid-map based localization system proposed in [5], in order to evaluate the system localization robustness in situations of GPS absence.

Acknowledgment This work is financed by the ERDF European Regional Development Fund through the COMPETE Programme (operational programme for competitiveness) and by National Funds through the FCT - Fundacão para a Ciência e a Tecnologia (Portuguese Foundation for Science and Technology) within project FCOMP-01-0124-FEDER-037281.

References

1. Bac, C.W., van Henten, E.J., Hemming, J., Edan, Y.: Harvesting Robots for High-value Crops: State-of-the-art Review and Challenges Ahead. Journal of Field Robotics **31**(6), 888–911 (2014)
2. euRobotics. Strategic research agenda for robotics in Europe. Draft 0v42 (2013). http://ec.europa.eu/research/industrial_technologies/pdf/robotics-ppp-roadmap_en.pdf
3. VineRobot - FP7 project. http://www.vinerobot.eu/
4. Vinbot - FP7 project. http://vinbot.eu/?lang=pt
5. Dos Santos, F.N., Sobreira, H., Campos, D., Morais, R., Moreira, A.P., Contente, O.: Towards a reliable monitoring robot for mountain vineyards. In: 2015 IEEE International Conference on Autonomous Robot Systems and Competitions (ICARSC), pp. 37–43, April 2015
6. Ning, W.: 13 Worksite Management for Precision Agricultural. Agricultural Automation: Fundamentals and Practices, 343 (2013)
7. Bluetooth: The Low Energy Technology Behind Bluetooth Smart. http://www.bluetooth.com/Pages/low-energy-tech-info.aspx
8. Patwari, N., Ash, J.N., Kyperountas, S., Hero, A.O., Moses, R.L., Correal, N.S.: Locating the nodes: cooperative localization in wireless sensor networks. IEEE Signal Processing Magazine **22**(4), 54–69 (2005)
9. Honkavirta, V., Perala, T., Ali-Loytty, S., Piche, R.: A comparative survey of WLAN location fingerprinting methods. In: 6th Workshop on Positioning, Navigation and Communication, WPNC 2009. IEEE (2009)
10. Hahnel, D., Burgard, W., Fox, D., Fishkin, K., Philipose, M.: Mapping and localization with RFID technology. In: Proceedings of the 2004 IEEE International Conference on Robotics and Automation, ICRA 2004, vol. 1, pp. 1015–1020. IEEE (2004)
11. Sebastian, T., Burgard, W., Fox, D.: Probabilistic robotics. MIT press (2005)
12. Bluetooth S.I.G.: Specification of the Bluetooth System, pp. 201–203 (2010). https://www.bluetooth.org/docman/handlers/downloaddoc.ashx?doc_id=229737
13. Kontakt: Kontakt.io Beacon Datasheet v2.0 (2014). http://docs.kontakt.io/beacon/kontakt-beacon-v2.pdf
14. Android Beacon Library - Distance Estimates. http://altbeacon.github.io/android-beacon-library/distance-calculations.html

Crop Row Detection in Maize for Developing Navigation Algorithms Under Changing Plant Growth Stages

David Reiser, Garrido Miguel, Manuel Vázquez Arellano, Hans W. Griepentrog and Dimitris S. Paraforos

Abstract To develop robust algorithms for agricultural navigation, different growth stages of the plants have to be considered. For fast validation and repeatable testing of algorithms, a dataset was recorded by a 4 wheeled robot, equipped with a frame of different sensors and was guided through maize rows. The robot position was simultaneously tracked by a total station, to get precise reference of the sensor data. The plant position and parameters were measured for comparing the sensor values. A horizontal laser scanner and corresponding total station data was recorded for 7 times over a period of 6 weeks. It was used to check the performance of a common RANSAC row algorithm. Results showed the best heading detection at a mean growth height of 0.268 m.

Keywords Ground-truth · Reference · Algorithms · RANSAC · Total station · LIDAR · Plant position · Growth status · Row navigation

1 Introduction

Autonomous robots can have a key role in increasing sustainability and resource efficiency in food production for future world population [1]. Therefore the

D. Reiser(✉) · M.V. Arellano · H.W. Griepentrog · D.S. Paraforos
Institute of Agricultural Engineering, University of Hohenheim,
Garbenstr. 9, 70599 Stuttgart, Germany
e-mail: {dreiser,manuel_vazquez,hw.griepentrog,d.paraforos}@uni-hohenheim.de

G. Miguel
Laboratorio de Propiedades Físicas (LPF)-TAGRALIA,
Technical University of Madrid, 28040, Madrid, Spain
e-mail: miguel.garrido.izard@gmail.com

© Springer International Publishing Switzerland 2016
L. Paulo Reis et al. (eds.), *Robot 2015: Second Iberian Robotics Conference*,
Advances in Intelligent Systems and Computing 417,
DOI: 10.1007/978-3-319-27146-0_29

navigation must be planed precisely and be robust enough to deal with the changing conditions on a field. But this requires, that the machines know where the crop plants are and that they don´t get destroyed by the vehicle. As most of the current crops are planted in row structures, detecting these rows is one of the basic needs for the autonomous navigation of robots in semi-structured agricultural environments. Many researches had been conducted on detecting this line structures by camera images ([2],[3]), light detection and ranging (LIDAR) laser scanner data ([4],[5],[6]), or other types of sensors. Nevertheless, precise line detection, relying on noisy sensor data, is still a challenging task for a computer algorithm due to the inherent uncertainty in the environment [6]. Humans can detect objects and shapes because of experience rather than a formal mathematical definition, like a computer algorithm does [7]. The environment has a countless number of variables influencing the sensors, making it hard to get the right information out of the values [8]. Aside from that, plants on the field are changing their shape rapidly, making object recognition even more challenging. First the plants are growing and, second, the conditions are changing. Therefore, there is a necessity for calibration of the algorithms before the robot is able to perform the task autonomously [6].

Also, weather and lighting conditions can already produce big changes in the results. This is especially problematic for image analysis, where alternate and discontinuous luminance usual affects the outcome [7].

To deal with these uncertainties, researchers have used simulated datasets [9], artificial plants ([9], [10]) or recorded datasets ([1],[3],[6]) to evaluate their algorithms. Since a simulation is always an approximate model of the environment, it will never cover all possibilities [8]. When recording data, the question is of how to refer to the algorithm performance. One option is to set the crop row manually [6]. The precise sensor value recording of the same plants over different growth stages, can be a good way for the later evaluation of navigation algorithms. In order to understand how algorithms behave under changing conditions in a field, it is necessary to know the pictured objects and how the sensors react on them. Therefore, it is important to know the correct plant position and parameters. In order to achieve that, the plant parameters must be mapped and referenced in every new test.

The aim of this paper is to show how algorithm analyzing could be improved using precise referenced sensor data, especially when the same plants can be investigated with the same sensors over different growth stages. For that purpose, the data set of a horizontal LIDAR is used. With the help of a highly accurate total station, all sensor data sets can be converted into the same reference frame, in order to obtain comparable results. This approach is tested by the performance evaluation of a common random sample consensus (RANSAC) line fitting algorithm [11]. The RANSAC algorithm has the advantage of being fast and robust against outliers, resulting in advanced performance when dealing with noisy sensor data. Therefore the performance at different growth stages can be precisely evaluated, by using the same reference.

2 Materials and Methods

2.1 Hardware and Sensors

A small 4-wheel autonomous robot with differential steering was used to move the sensors with manual control through the crop rows (see Fig. 1). The size of the robot platform was 500 x 600 x 1100 mm. The weight of the robot is 50 kg and it is equipped with four motors with a total power of 200 W. Maximum driving speed is 0.8 m/s and the maximum static motor torque is 4 x 2,9 Nm. The robot system is equipped with wheel encoders, a VN-100 Inertial Measurement Unit (IMU) (VectorNav, Dallas, USA) and a LMS111 2D-LIDAR laser scanner (SICK, Waldkirch, Germany). The laser scanner was mounted horizontally at the front of the robot at a height of 0.2 m above the ground level. All other mounted sensors had not been used in this paper.

For evaluating the precise position of the robot, the SPS930 Universal Total Station (Trimble, Sunnyvale, USA) was utilized. The total station was tracking a Trimble MT900 Machine Target Prism, which was mounted on top of the robot at a height of 1.07 m in order to guarantee always line of sight to the total station (see Fig. 1).

The robot is controlled by an embedded computer, equipped with i3-Quadcore processor with 3.3 GHz, 4 GB RAM and SSD Hard drive. For energy supply, two 12V/48Ah batteries are providing an operating time of around 4-5 h, depending on the load torque, task and additional weight of equipment placed on the robot platform. The total station data was sent to a Yuma 2 Rugged Tablet Computer (Trimble, Sunnyvale, USA); this tablet is equipped with an Intel Atom CPU N2600 dual-core processor with 1.6 GHz, 4 GB RAM, SSD Hard drive, and a self-sufficient battery. Connectivity to the SPS930 total station is provided by an internal 100mW radio antenna at the 2.4 GHz (IEEE 802.11) range. The Yuma 2 Rugged Tablet Computer was connected to the robot computer via serial RS232 interface for continuous data exchange.

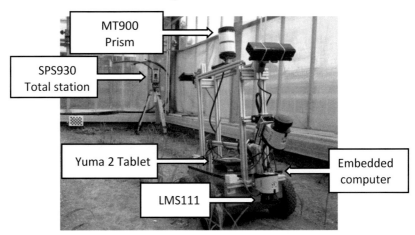

Fig. 1 Robot platform, equipped with the sensors and the reference prism

2.2 Software

The robot computer runs by Ubuntu 14.04 and use the Robot Operating System (ROS-Indigo) middleware for the data recording. The software components had been programmed in a combination of C++ and Python programming languages.

The Trimble Yuma 2 Tablet was running under Windows 7 Professional and executed the Trimble SCS900 Site Controller Software Version 3.4.0. The Trimble software includes an easy-to-use graphical interface for fast calibration and point measurement. It also has the option to export the actual prism position via serial RS232 interface. The tablet was placed on the robot and the serial output was directly used by the ROS system to refer the robot position to the total station coordinate frame. The prism position data was time stamped, according to the computer system time, together with the sensor data. The data flow diagram can be seen in Fig. 2.

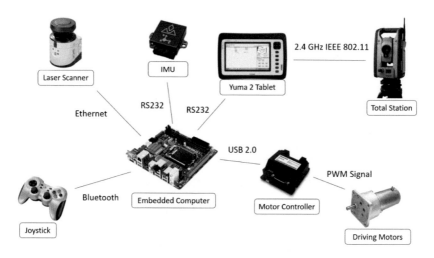

Fig. 2 Data flow diagram of the robot sensor setup

2.3 Referencing & Data Acquisition

To create a relative coordinate frame in the greenhouse, five fixed attachments for the MT1000 Prism had been placed on predefined positions. The accurate position of these five points could be located by just screwing the prism on the attachments for every new field test. The absolute point distances were stored during the first setup. Before every subsequent data acquisition, these points were measured once again by the total station in order to recalibrate the system to the first total station position with the original Cartesian coordinate frame. After every test, the inaccuracy in the static measurement was evaluated by reassessing each of these fixed points. The shift between the first reference points and the actual measurement was evaluated by the Trimble SPS software and was in all tests below 4 mm for all three dimensions. The total station was always placed almost at the same position inside the greenhouse, which lies around the zero point of the coordinate frame.

After plant emergence, the stem position was measured with the aid of pendulum hanging from a tripod; the MT1000 Prism was attached at the center of the tripod. It was assumed that the center of each plant stem's position will not change during the period of growth. Consequently, each of these points was used as the reference position of each individual plant.

Due to the robot rigid body frame that was carrying the sensors, a static transformation between the prism and the sensor positions was performed. The three-dimensional orientation of the robot in space was evaluated by the IMU, which was placed at the center of the robot and on the same axis as the prism. As the orientation of the prism could not be evaluated by the total station, the position of the prism was fused with the IMU data to transform the laser scans to the greenhouse frame.

This procedure allowed to directly transform the recorded sensor data into the same coordinate frame, and even to assign them to single plant positions. In Fig. 3a the test environment with the moving robot is presented and in Fig. 3b the corresponding sensor value visualization of all attached sensors in the ROS environment is illustrated. The blue points correspond to the horizontal laser scanner data.

(a)   (b)

Fig. 3 (a) the greenhouse environment and (b) a visualisation of the transformed sensor data in the ROS visualisation environment

The sensor files were separated at a size of 4 GB. The timestamp was according to the robot embedded computer system time, with a resolution of one millisecond. The LIDAR data was collected with an average of 25 Hz and a resolution of 0.5 degree. The total station updated the data with 15 Hz. The IMU data was transmitted with 40 Hz. Linear interpolation was used to fuse robot position and the sensor data before transforming it to the global coordinates of the greenhouse frame.

2.4   RANSAC Algorithm

The RANSAC is a commonly used algorithm for evaluating plane parameters in noisy point cloud data [12], [13]. But also for row or line estimation in image analysis [1], and 2D-LIDAR data [14], [15]. To evaluate the general algorithm behavior, the RANSAC was chosen due to its previously mentioned robustness against outliers. The implemented RANSAC algorithm is part of the Point Cloud Library [16] and was integrated in the ROS environment, for direct analysis of the published scans. To get always precise reference of the extracted lines, the LIDAR data was first transformed to the robot body frame and then to the overall greenhouse frame.

As the distance between maize crop rows is 0.75 m, this parameter was used to filter roughly the row area with the known robot position with a rectangle. The resulting point cloud was separated to have for each crop row an individual point cloud. This was done by using the known robot position and robot direction. The RANSAC was then applied to each distinct point cloud. The maximum iteration limit was set to the input point number and the maximal distance range for the line to 0.5 m. These parameters were fixed for all performed line fittings.

3   Experiments

Five rows of maize were planted with a length of 5.2 m each. The row spacing was defined according to common agricultural practice to 0.75 m, with 41 plants per row. The maize was planted in a greenhouse to be independent of external weather conditions. The measured positions of the plants, total station and reference points can be found in Fig. 4. After every data acquisition, the height, stem width and leaf numbers of each single plant had been measured. This was done manually with a measuring tape and a sliding caliper.

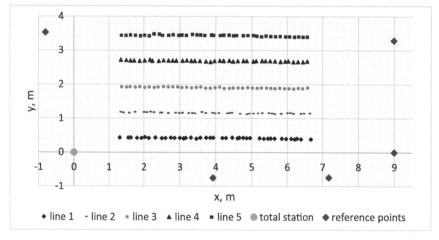

Fig. 4 Manually measured plant positions, reference points and total station position

The ideal line parameters were evaluated by the plant germination positions, measured with the total station. Because this line has no outliers, the least square algorithm can result in the most accurate line fitting. As reference, a 2D line in the XY-plane was estimated using the equation $f(x) = ax + b$. The residual r of every data point $P_i(x_i, y_i)$ can be described with:

$$r_i = f(x_i) - a * x_i - b \tag{1}$$

Using the least square estimation the best line fit can be estimated as:

$$\min_{a\,b} \sum_{i=1}^{n} r_i^2 \tag{2}$$

This was adapted to the emerging points of the plants, resulting in the line parameters presented in Table 1.

Table 1 Listing of the line parameters

line number	line equation in [m]	intersection point with last row in [m]
1	$f(x) = -0.0062x + 0.4441$	-
2	$f(x) = -0.0079x + 1.1913$	$P\ (439.53, -2.28)$
3	$f(x) = -0.0085x + 1.9414$	$P\ (1250.17, 8.69)$
4	$f(x) = -0.0068x + 2.7186$	$P\ (-457.18, 5.83)$
5	$f(x) = -0.0101x + 3.4811$	$P\ (231.06, 1.15)$

As it can be seen in the intersection points, shown in table 1, the row with the smallest angle difference between the lines is the path between crop row 2 and 3; here, the intersection point had the longest distance to the row center. The best performance was expected from the most parallel lines for evaluating the row detection algorithm. So crop row line 2 and 3 were selected. The sensor data recording took place from 23.04.15 until 1.06.15 in Stuttgart Hohenheim. In total 7 tests were performed. In every test, the robot was driven by a remote joystick with a constant speed through the crop rows. The average speed was around 0.05 m/s in order to acquire a high data density. Both rows were recorded twice, once for each travel direction. For all 7 test days, each travel direction was evaluated, resulting in a total of 14 recorded and analyzed datasets. The laser scanner data was used, when the robot reference prism was in an area between 2 to 5 m in the x direction. For filtering purposes, the scans were first transformed to Cartesian coordinates and then to the greenhouse coordinate frame. The reflected points of the robot vehicle were removed and the point cloud was separated as described above. By doing this, the RANSAC could be performed for each crop line separately. Each of the line fittings was addressed directly to one single point cloud set without using any prior knowledge about the last dataset or the robot position. Scans with less than three points in the line area were ignored. In total 10277 different laser scans were evaluated. In Table 2 the number of analyzed scans per line are presented.

Table 2 Numbers of analyzed scans per line

Test Number:	Date	Days after seeding	Analyzed scans line 3	Analyzed scan line 2
1	23.04.2015	28	67	688
2	27.04.2015	32	601	990
3	30.04.2015	35	910	1041
4	05.05.2015	40	897	941
5	13.05.2015	48	754	763
6	18.05.2015	53	653	652
7	01.06.2015	67	660	660

The difference between ideal line and the algorithm solved line, was evaluated with the help of the Root Mean Square Error (RMSE), defined by the following equation:

$$RMSE = \sqrt{\frac{\sum_{i=1}^{n}(\delta - \beta_i)^2}{n}} \tag{3}$$

With δ as ideal line parameters and β_i as the resolved algorithm parameters at the scan i.

4   Results and Discussion

As the greenhouse soil was not homogeneous, there was a huge diversity in growth status. For tracking the crop development, the highest point of each plant was measured and the mean value was evaluated for each crop row. The variability is expressed by the standard deviation of all 41 plants heights per crop row. The results are shown in Fig. 5. The average height of the plants at line 3 had been lower than at line 2. 48 days after seeding, most of the plants reached the level of 0.2 m height. At all tests afterwards the number of analyzed scans had been almost the same for both sides (see Table 2). As in the first two tests the average plant height of line 3 was below the height of the laser scanner, the RANSAC algorithm for line 3 detected points, just when the vehicle was turned downwards, because of uneven ground. The absolute mean value for the height of the line 3 was 0.47 m while for line 2 the mean value was 0.65 m. The tallest plant reached 0.82 m at line 3 and 0.86 m at line 2. For later growth stages the standard deviation was increased. Along with the height, the numbers of leaves, covering the row, were also increased. This caused limited sight of view for the LIDAR.

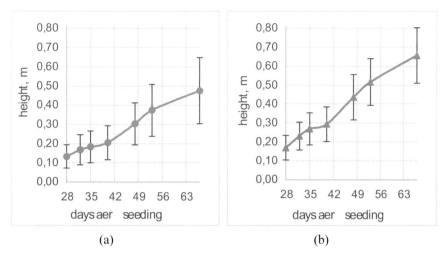

Fig. 5 Mean growth status of the plants for (a) line 3 and (b) line 2

For evaluating the change between crop row and RANSAC output, the RMSE for every data set was estimated. The results for the row position (Fig. 6a) showed a higher error in the first tests for all mean values below 0.2 m. For all other heights, the RMSE value fluctuated at a value around 0.1 m. The best matches were after 35 and 67 days at line 2 with an RMSE value of 0.07 m and 0.06 m, respectively.

Also the heading error (Fig 6b) had a higher value at the first tests with low mean plant height. For both lines a local minima could be detected after a 35 days (see Fig. 6b). As it could be seen in Table 2, this was the first test with almost equal number of detected lines out of the scans. With the growth of the plants, the precision decreased back again. Only the last measurement of line 3 did performed better than the first minima of the same line. A reason for this could be the inhomogeneous growth of the crop plants. In total the RANSAC performed better at line 2 than in line 3. Reasons for that could be the more homogenous growth of the plants, which is expressed by the standard deviation of the two lines (see Fig. 5). Especially line 2 had almost constant RMSE values between 35 and 53 days after seeding.

To better understand the evaluated error of the real line parameters, the direct RANSAC output is shown in Fig. 7. For evaluating the values, two tests after 35 days and two tests after 67 days are visualized. For both test days, the robot moved through the row in each direction once. 200 measurements were evaluated and compared in the following graph (see Fig. 7). The RANSAC heading output after 35 days is shown in Fig 7a and the output after 67 days can be seen in Fig 7b. The inclination parameter of line 2 is -0.0079, which under ideal conditions should be the same like the computed RANSAC parameter. The nearest heading to this theoretical value can be seen at 35 days after seeding (see Fig 6b). After 67 days, the computed values increased and produced out of these a bigger shift to the reference value. At the end of the row a higher shift can be noticed. A reason for this could be, the lower number of points, detectable at the end of a row, to balance the outliers. After 67 days this performance was worse compared to the values after 35 days as illustrated in Fig. 7b. First this could be reasoned, by the high amount of leaves hanging into the laser scan area. These leaves blocked the detection of the stem positions that were necessary to evaluate the line orientation.

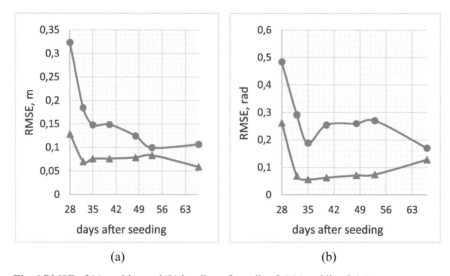

(a)   (b)

Fig. 6 RMSE of (a) position and (b) heading of row line 2 (▲) and line 3 (●).

For every direction the robot moved, a static shift of the heading was observed (see Fig. 7). This can be explained by some reflections of leaves, which caused a shift of the detected line to the middle of the row. This effect was stronger after 67 days and caused a narrow detection of the plant stems. The minimal reached RMSE was 0.05 rad for the line detection with the RANSAC after 37 days. After 67 days this value increased.

In worst cases the noise could be much higher than 5 degrees (0.1 rad) compared to the real value. This can cause problems on line following, especially when there is not enough space between the vehicle and the rows. The failure rate could be seen in many cases of the evaluated data. A part of the analyzed error could also be resulted by the inaccuracy of the LIDAR measurements.

For getting a RANSAC algorithm robust for navigation, this heading uncertainty must be compensated. Higher algorithm robustness could be accomplished using a Kalman filter. When the growth status is known, the heading error could also be decreased by a static offset, which must be evaluated before starting the line following. Filtering for outliers or mean filter methods could also bring better results.

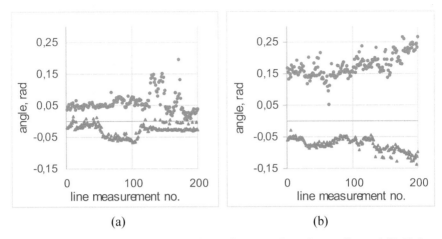

(a)   (b)

Fig. 7 Heading values of the RANSAC for (a) line 2, 35 days after seeding and (b) 67 days after seeding. Robot movement in positive x direction (orange triangles) and in negative x direction (blue dots).

The results of the RANSAC showed a high variability in the dataset, with different outcomes of the algorithm. So it could be assumed, that the variability in the dataset brings additional options for testing the robustness of line following algorithms for different growth states. The experimental setup allowed to detect a heading offset, which is dependent on the growth status of the crop plants. It could be shown that this offset is dependent on the sensor position and the movement of the robot. Also the best height for the line detection with the given laser position could be evaluated. This was at a mean height of 0.265 m of the plants.

5   Conclusions

The results of the collected data set showed high precision and good referenced sensor data for all measured growth stages. The application of a RANSAC algorithm for line detection to the horizontal laser data showed high diversity in heading and positioning. The smallest heading error was detected 35 days after seeding and at an average plant height of 0.268 m. After that, the error increased and brought a higher RMSE value to the detection. Also a drift dependent on the travel direction of the robot was observed, which was caused by leaves inside the row. This effect increased with the growth of the plants. In many cases of the given data set, the deviation of the line heading was higher than 5 degrees. This would cause problems for precise row navigation. The position error was for most cases acceptable. For line following applications in maize with a RANSAC algorithm, robust filtering of the laser data and algorithm results should be considered. In total the approach was helpful in order to evaluate some basic problems of outdoor line detection with LIDARs and a RANSAC algorithm. Aside of that, the accurate reference of the heading difference could be evaluated.

Acknowledgments The project is conducted at the Max-Eyth Endowed Chair (Instrumentation & Test Engineering) at Hohenheim University (Stuttgart, Germany), which is partly grant funded by the DLG e.V.

References

1. English, A., Ross, P., Ball, D., Corke, P.: Vision based guidance for robot navigation in agriculture. In: IEEE International Conference on Robotics and Automation (ICRA). pp. 1693–1698 (2014)
2. Marchant, J., Brivot, R.: Real-Time Tracking of Plant Rows Using a Hough Transform. Real-Time Imaging **1**, 363–371 (1995)
3. Jiang, G., Zhao, C., Si, Y.: A machine vision based crop rows detection for agricultural robots. In: Proceedings of the 2010 International Conference on Wavelet Analysis and Pattern Recognition, pp. 11–14 (2010)
4. Hansen, S., Bayramoglu, E., Andersen, J.C., Ravn, O., Andersen, N.A., Poulsen, N.K.: Derivative free Kalman filtering used for orchard navigation. In: 13th international Conference on Information Fusion (2010)
5. Barawid, O.C., Mizushima, A., Ishii, K., Noguchi, N.: Development of an Autonomous Navigation System using a Two-dimensional Laser Scanner in an Orchard Application. Biosyst. Eng. **96**, 139–149 (2007)
6. Hiremath, S.A., van der Heijden, G.W.A.M., van Evert, F.K., Stein, A., ter Braak, C.J.F.: Laser range finder model for autonomous navigation of a robot in a maize field using a particle filter. Comput. Electron. Agric. **100**, 41–50 (2014)
7. Papari, G., Petkov, N.: Edge and line oriented contour detection: State of the art. Image Vis. Comput. **29**, 79–103 (2011)
8. Russell, S.J., Norvig, P.: Artificial Intelligence: A modern approach. Ptrentice-Hall, Englewood Cliffs (1995)
9. Weiss, U., Biber, P.: Plant detection and mapping for agricultural robots using a 3D LIDAR sensor. Rob. Auton. Syst. **59**, 265–273 (2011)
10. Bochtis, D., Griepentrog, H.W., Vougioukas, S., Busato, P., Berruto, R., Zhou, K.: Route planning for orchard operations. Comput. Electron. Agric. **113**, 51–60 (2015)
11. Fischler, M.A., Bolles, R.C.: Random Sample Consensus: A Paradigm for Model Fitting with Applications to Image Analysis and Automated Cartography. Commun. ACM **24**, 381–395 (1981)
12. Choi, S., Park, J., Byun, J., Yu, W.: Robust ground plane detection from 3D point clouds. In: 14th International Conference on Control, Automation and Systems (ICCAS 2014), pp. 1076–1081 (2014)
13. Weiss, U., Biber, P., Laible, S., Bohlmann, K., Zell, A.: Plant species classification using a 3D LIDAR sensor and machine learning. In: Proc. - 9th Int. Conf. Mach. Learn. Appl. ICMLA 2010, pp. 339–345 (2010)
14. Zhang, J., Maeta, S., Bergerman, M., Singh, S.: Mapping orchards for autonomous navigation. In: ASABE Annual International Meeting, pp. 1–9 (2014)
15. Marden, S., Whitty, M.: GPS-free localisation and navigation of an unmanned ground vehicle for yield forecasting in a vineyard. In: Proceedings of the 13th International Conference IAS-13 (2014)
16. Rusu, R.B., Cousins, S.: 3D is here: point cloud library. In: IEEE Int. Conf. Robot. Autom., pp. 1–4 (2011)

Part III
Autonomous Driving and Driver Assistance Systems

Stereo Visual Odometry for Urban Vehicles Using Ground Features

Arturo de la Escalera, Ebroul Izquierdo, David Martín,
Fernando García and José María Armingol

Abstract Autonomous vehicles rely on the accurate estimation of their pose, speed and direction of travel to perform basic navigation tasks. Although GPSs are very useful, they have some drawbacks in urban applications. Visual odometry is an alternative or complementary method, because it uses a sensor already available in many vehicles for other tasks and provides the ego motion of the vehicle with enough accuracy. In this paper, a new method is proposed that detects and tracks features available on the surface of the ground, due to the texture of the road or street and road markings. This way it is assured only static points are taking into account in order to obtain the relative movement between images. A Kalman filter is used taking into account the Ackermann steering restrictions. Some results in real urban environments are shown in order to demonstrate the good performance of the algorithm.

Keywords Autonomous vehicles · Visual odometry

1 Introduction

Vehicle localization is a fundamental task in autonomous vehicle navigation. It relies on accurate estimation of pose, speed and direction of travel to achieve basic tasks including mapping, obstacles avoidance and path following. Nowadays, many vehicles rely on GPS-based systems for estimating their ego motion. Although GPSs are very useful, they have some drawbacks: the price of the equipment is still high for centimeter accuracy. Moreover, above all for urban applications, the shortcomings of the system are clearer, since there is no direct

A. de la Escalera(✉) · D. Martín · F. García · J.M. Armingol
Universidad Carlos III de Madrid, Leganés, Spain
e-mail: {escalera,dmgomez,fegarcia,armingol}@ing.uc3m.es

E. Izquierdo
Queen Mary University London, London, UK
e-mail: ebroul.izquierdo@qmul.ac.uk

© Springer International Publishing Switzerland 2016
L. Paulo Reis et al. (eds.), *Robot 2015: Second Iberian Robotics Conference*,
Advances in Intelligent Systems and Computing 417,
DOI: 10.1007/978-3-319-27146-0_30

line of sight to one or numerous satellites because of the presence of a building or a tree canopy, the urban canyon effect or the absence of the signal for an important task as driving along tunnels. Other sensors available are low-cost IMUs; however, though they are fast, they have a measurement bias and therefore needs frequent corrections. Several solutions can be proposed to solve this problem, such as the use of maps or odometry provided by the vehicle wheels. The first one needs a continuous updating of the maps to be useful and the second lacks enough precision for several applications. That is why another sensor is needed and here is where digital cameras can play an important role. On one hand because, as it will be shown, they are useful for obtaining the vehicle's ego motion and, on the other hand, because nowadays they are already used for other tasks such as pedestrian, traffic sign or road lane detection [1], accordingly it is a sensor that can be applied for multiple assignments. Visual Odometry (VO) estimates the ego motion of a camera or a set of cameras mounted on a vehicle using only the visual information provided by them. The term is related to the wheel odometry used in robotics and was formulated in 2004 by Nister [2]. Usually, VO algorithms have three steps:

1. Detect points of interest (POI) in every image and find a matching between two consecutives ones.
2. Find and remove the wrong matches.
3. Estimate the relative movement of the cameras.

This can be done using monocular or stereo cameras and assuming planar or non-planar motion model. A tutorial on VO can be found in [3-4]. In [5] a stereo system is presented, where it estimates the rigid body motion that best describes the transformation among the sets of 3D points acquired in consecutive frames. Optical flow and stereo disparity are computed to minimize the re-projection error of tracked feature points. Instead of performing this task using only consecutive frames, they use the whole history of the tracked features to compute the motion of the camera. The camera motion is estimated in [6] using a quaternion and RANSAC [7] for outlier removal and a Two-stage Local Binocular Bundle Adjustment for optimizing the results. In [8] the rotation and translation between consecutive poses are obtained, minimizing the distance of the correspondent point projections. They take into account that farther 3D points have higher uncertainty and RANSAC for outliers and the pose estimation is constrained by temporal flow. A persistent map containing 3D landmarks localized in a global frame is presented in [9]. They distinguish automatically some frames, used to update the landmark map, which serves for ego-localization. The other frames are used to track the landmarks and to localize the camera with respect to the map. Other sensors, like lasers, has been used in [10-11].

Urban environments are highly dynamic, so the case of a static scene cannot be assumed. Moreover, these surroundings are highly cluttered with frequent occlusions. Consequently, there are some specific difficulties any method has to face:

• Detected POI can belong to the moving objects and, as a consequence, the camera motion estimation would be erroneous.
• Due to the ego motion and occlusions, some detected POI in an image are not detected in the next one, and vice versa, but this can lead to an erroneous matching.

The novelty of the proposed algorithm is related to the previous difficulties. In order to detect only points belonging to the static part of the scene, the stereo system detects and tracks features available on the surface of the ground due to the texture of the road or street and road markings. Unlike other approaches that assume that only the yaw angle changes, estimations of the camera roll, pitch and height are obtained for every image.

The rest of the paper describes the algorithm. The features are going to be detected in a virtual bird-view image. In order to do this, section 2 explains how the extrinsic parameters of the stereo camera are obtained for every image. Section 3 explains how the features are matched and the relative movement of consecutive images is found. The Kalman filter is explained in section 4 and the results in real driving situations are shown in section 5. Finally, the conclusions are presented.

2 Continuous Extrinsic Parameters Estimation

The first block of the algorithm is to find the extrinsic parameters of the stereo camera. Other approaches find the initial position of the cameras and assume that only the yaw angle changes. Although this is valid for several domains, it is not practical in urban applications due to the change in the extrinsic parameters because of the vehicle movements, the effect of the shock absorbers and driving on uneven road surfaces. The road is assumed to be flat up to a near distance and the plane of the road is found. Besides the application for visual odometry, finding the plane is also useful for other tasks of the vehicle as obstacle detection.

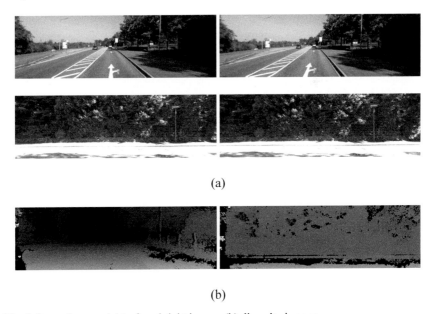

(a)

(b)

Fig. 1 Stereo images. (a) Left and right images (b) disparity images

2.1 Obtaining the 3D Point Cloud

As shown in Fig. 1, the changes in illumination inside the images, the lack of texture in many objects and the presence of repetitive patterns in others are the three main problems in order to obtain 3D points from stereo images taken in urban environments. Accordingly, stereo local methods, although being fast, are not reliable and at least a semi-global method has to be used. A popular one in vehicle applications, which is used here, is [12]. Not all the provided 3D points are needed, as the information is going to be used to detect the plane of the road in front of the vehicle. So, from all the points in the 3D cloud, only those points between a minimum and a maximum distance and within a certain width are taking into account. Moreover, they are normalized within a grid. This way, although there are much more points per square meter in the nearest distances, there is not a bias towards them when the road plane is obtained.

2.2 Stereo Extrinsic Parameters.

As shown in Fig 2, the relationship between road, P_r, and image, P_c, coordinates is defined by a rotation matrix, R_{cr}, and translation vector, T_{cr}:

$$P_r = R_{cr}P_c + T_{cr} \tag{1}$$

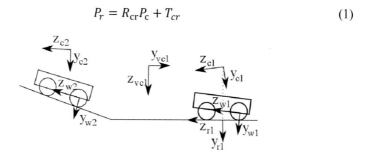

Fig. 2 Different reference axes used in the algorithm

If θ is the yaw, ψ the roll and ϕ the pitch of the camera, one of the Tait-Bryan matrixes is:

$$R_{cr} = \begin{pmatrix} C\theta C\psi & S\theta S\phi - C\theta S\psi C\phi & S\psi C\phi + C\theta S\psi S\phi \\ S\psi & C\psi C\phi & -C\psi S\phi \\ -S\theta C\psi & C\theta S\phi + S\theta S\psi C\phi & C\theta C\phi - S\theta S\psi S\phi \end{pmatrix} \tag{2}$$

As the pixels belonging to the road have nil height and any yaw angle, equation (2) can be simplified to:

$$\begin{bmatrix} x_r \\ 0 \\ z_r \end{bmatrix} = \begin{pmatrix} C\psi & -S\psi C\phi & S\psi S\phi \\ S\psi & C\psi C\phi & -C\psi S\phi \\ 0 & S\phi & C\phi \end{pmatrix} \begin{bmatrix} x_c \\ y_c \\ z_c \end{bmatrix} + \begin{bmatrix} 0 \\ h \\ 0 \end{bmatrix} \tag{3}$$

So, the plane equation is:

$$0 = S\psi\, x_c + C\psi C\phi\, y_c - C\psi S\phi\, z_c + h \tag{4}$$

The road in front of the vehicle is assumed to be flat and is defined by the plane:

$$a\, x_c + b\, y_c + c\, z_c + d = 0 \tag{5}$$

From the point cloud, the plane of the road is obtained with the Sample Consensus Model Perpendicular Plane method [13]. Thus, the roll, pitch and height of the camera are calculated from the obtained plane:

$$\psi = \mathrm{asin}(a)$$

$$\phi = atan\frac{-c}{b} \tag{6}$$

$$h = d$$

The results form a sequence of 1.200 images and are shown in Fig. 3. The nominal height of the camera is 1.65m while the median and mean is the same, -1.649m. The pitch angle has a median value of -1.049° and a mean of -1.050° while for the roll angle are 0.132° and 0.130° respectively.

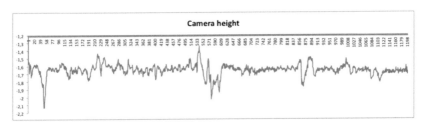

(a)

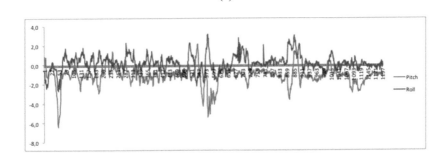

(b)

Fig. 3 Extrinsic parameters (a) camera height (b) pitch and roll angles

2.3 Virtual Camera

A virtual camera with the same intrinsic parameters, K, as the stereo system is used in order to "capture" the image where the features on the road are going to be looked for (Fig 2). This camera is looking perpendicular to the road and captures a defined area of it: the same area that it has been used for the road plane detection. In order to obtain the homography, which relates both images, the relationship between virtual camera coordinates, p_{vc}, and camera coordinates, p_c, is needed. The pin-hole model is:

$$p_c = KP_c \tag{7}$$

and from equation (1)

$$P_r = R_{cr}K^{-1}p_c + T_{cr} \tag{8}$$

$$p_c = KR_{cr}^{-1}(P_r - T_{cr}) \tag{9}$$

Similarly, the virtual camera:

$$p_{vc} = KR_{vcr}^{-1}(P_r - T_{vcr}) \tag{10}$$

Therefore, the homography, H, is obtained using the projection on both images, obtained from equations (9) and (10), of four points on the road.

$$p_{vc} = Hp_c \tag{11}$$

Some examples can be seen in Fig. 4.

(a)

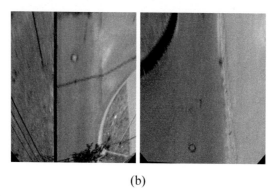

(b)

Fig. 4 Birdview images. (a) Original images captured by the stereo cameras (b) Birdview images

3   Feature Detection and Matching

3.1   Road Features Detection

6000 SIFT features [14] are detected on the bird-view images. Some features correspond to textured areas of the road, but others are related to the projection of objects like cars, pedestrians, buildings, etc. Hence, whether the points belong to the ground, or not, has to be checked. From equation (11) the corresponding pixel of the feature in the bird-view image can be obtained:

$$p_c = (u \quad v)^t = H^{-1}p_{vc} \tag{12}$$

Knowing the image coordinates, the disparity image, *ImaDisp*, and the intrinsic parameters, it is possible to obtain the real 3D coordinates of that point:

$$disp = ImaDisp(u, v) \tag{13}$$

$$z_c = fD/disp \tag{14}$$

$$x_c = z_c(u - u_0)/f \tag{15}$$

$$y_c = z_c(v - v_0)/f \tag{16}$$

where f is the focal length, D the stereo baseline, and (u_0, v_0) the image centre. Finally, the distance to the plane is obtained:

$$dist = |ax_c + by_c + cz_c + d| \tag{17}$$

If it is less than a threshold the feature belongs to a point on the ground and is kept, otherwise it is discharged. In Fig. 5, some examples can be seen. Green features are on the road plane, the red ones do not belong to the plane and the white ones represent features where there is not stereo information available.

Fig. 5 Road features detection

3.2   First Speed Filter

The previous process is done for every image and a matching between features of two consecutive images is done. Many errors can be expected from the matching as many features are being taken into account and the road surface has a minimum texture to detect them and not much information for their description. Although the algorithm for detecting the vehicle's displacement can deal with outliers, it will work better and faster if there are as minimum errors as possible. Therefore a previous filter is done where:

- Those matched features whose module is higher than a threshold, 130km/h, are discharged since the vehicle is not allowed to drive over that speed.
- From the rest, the median of their module is obtained and all the matched features with a speed below 75% and higher than 125% of the median are also discharged

3.3   Camera Displacement

The camera displacement is obtained using a Perspective-n-Point (PnP) algorithm [15][16]. It estimates the camera displacement given a set of 3D points, P_{c2} and their corresponding image projections, p_{c1}. It minimizes the re-projection error, that is, the sum of squared distances between the observed projections p_{c1} and the projected P_{c2}:

$$p_{c1} = K(R|T)P_{c2} \qquad (18)$$

RANSAC is used, in the implementation of the algorithm, in order to be robust against outliers. In Fig. 6, all the matched features between consecutive images can be seen. In blue are the ones discharged by the first filter and in green the ones whose displacement is given by the algorithm and in red the ones considered as outliers.

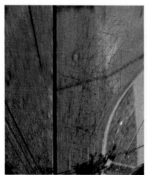

Fig. 6 Matched features. In blue are the ones discharged by the first filter and in green the ones whose displacement is given by the algorithm and in red the ones considered as outliers

4   Kalman Filter

4.1   *Ackerman Constraint*

From equation (18) the rotation and translation between the two camera poses are obtained. But if some dynamic restriction is going to be applied, the rotation and translation of the rear wheel axis is needed. The relationship between wheel, P_w, and image, P_c, coordinates is defined by a rotation matrix, R_{cw}, and a translation vector, T_{cw}:

$$P_w = R_{cw}P_c + T_{cw} \qquad (19)$$

At time 2, the 3D values of the detected SIFT features on the bird-view image are obtained. So, from the previous equation and equation (7) the rotation and translation between the wheels are obtained:

$$P_{w1} = R_{w1}P_{w2} + T_{w1} \qquad (20)$$

$$T_w = (I - R_{cw}RR_{cw}^{-1})T_{cw} + R_{cw}T \qquad (21)$$

$$R_{w1} = R_{cw}RR_{cw}^{-1} \qquad (22)$$

From equation (2) the three angles are:

$$\psi = \operatorname{asin}(R_w(1,0))$$

$$\phi = \operatorname{atan}\left(\frac{-R_w(1,2)}{R_w(1,1)}\right) \qquad (23)$$

$$\theta = \operatorname{atan2}\left(\frac{-R_w(2,0)}{\cos(\psi)}, \frac{-R_w(0,0)}{\cos(\psi)}\right)$$

Due to the Ackerman steering, the trajectory is a circumference (Fig. 7), so some cinematic restrictions can be applied. What is the relationship between the rotation and translation obtained from the camera and the vehicle movement? Applying trigonometric rules, it can be deduced that is:

$$T = \begin{bmatrix} \rho \sin(\theta/2) \\ 0 \\ \rho \cos(\theta/2) \end{bmatrix} \qquad (24)$$

when ρ is the displacement.

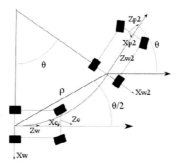

Fig. 7 Ackerman constraints

4.2 Kalman

The state vector is $\left[v_k\ \omega_k\ \dot{z}_k \dot{\psi}_k \dot{\phi}_k\right]$, where v_k and ω_k are the linear and angular speed, $\dot{z}_k$ is the vertical speed and $\dot{\psi}_k$ and $\dot{\phi}_k$ the roll and yaw speed at instant k. The linear and angular speeds are supposed to be constant and the pitch, roll and height speeds are nil.

$$\begin{aligned} v_k &= v_{k-1} + r_1 \\ \omega_k &= \omega_{k-1} + r_2 \\ \dot{z}_k &= 0 + r_3 \\ \dot{\psi}_k &= 0 + r_4 \\ \dot{\phi}_k &= 0 + r_5 \end{aligned} \tag{25}$$

The measurements are obtained from the values from (21) and (23):

$$\hat{v}_k = \frac{\sqrt{T_{x,w}*T_{x,w}+T_{z,w}*T_{z,w}}}{\Delta t} \tag{26}$$

$$\hat{\omega}_k = \frac{\theta}{\Delta t} \tag{27}$$

$$\hat{\dot{z}}_k = \frac{T_{z,w}}{\Delta t} \tag{28}$$

$$\hat{\dot{\psi}}_k = \frac{\psi}{\Delta t} \tag{29}$$

$$\hat{\dot{\phi}}_k = \frac{\phi}{\Delta t} \tag{30}$$

The prediction is compared with the measurements and if the Normalized Innovation Squared is greater than a threshold, the prediction is used as the new state, if not, the filter is updated.

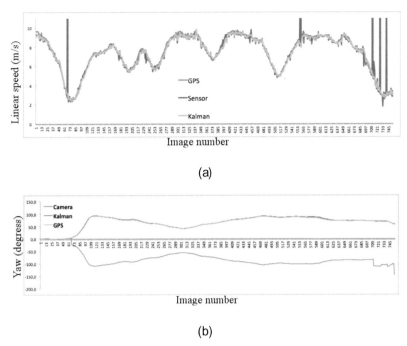

(a)

(b)

Fig. 8 Linear speed and yaw angle profiles

5   Results

The results are based on sequences of the KITTI Vision Benchmark Suite [17-18]. The stereo cameras for this benchmark were place on the middle of the vehicle's roof. Because the presented algorithm looks for features on the road, the cameras are not placed on the best place, on the vehicle's windshield. Because of the camera placement, the minimum distance that the camera captures is a bit far, 6m but as the sequences have a ground truth using a centimeter GPS, they are useful to show if the method is valid or not. In Fig. 8 the linear speed and the yaw angle along a sequence of 558m are shown. It shows how the Kalman filter follows the real speed and yaw of the vehicle despites the occasional errors of the measurements.

In Fig. 9 two sequences are shown. The first one has a final error of only 0.5m after 558 m (0.1%), while the second one has an error of 17.3 m after 917.78m (1.9 %).

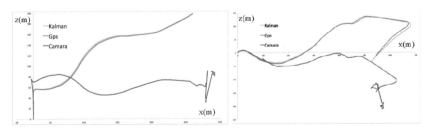

Fig. 9 Trajectory followed by the vehicle along two trajectories

6   Conclusions

A new method for VO using stereo vision is proposed, which detects and tracks features available on the surface of the ground. This way, it is assured only static points are taking into account in order to obtain the relative movement between images. A Kalman filter is used taking into account the Ackermann steering restrictions. The results in real urban environments show the algorithm is able to estimate the linear and angular speeds of the vehicle with high accuracy. Although VO is different than SLAM (Simultaneous Localization And Mapping), the results show its ability to follow the real trajectory drove by the vehicle along long paths with a minimum linear and angular error.

Acknowledgements This work was supported by the Spanish Government through the CICYT project TRA2013-48314-C3-1-R.

References

1. Martín, D., García, F., Musleh, B., Olmeda, D., Peláez, G., Marín, P., Ponz, A., Rodríguez, C., Al-Kaff, A., de la Escalera, A., Armingol, J.M.: IVVI 2.0: An Intelligent Vehicle based on Computational Perception. Expert Systems with Applications **41**, 7927–7944 (2014)
2. Nister, D., Naroditsky, O., Bergen, J.: Visual odometry. In: IEEE Computer Society Conference on Computer Vision and Pattern Recognition, pp. 652–659. IEEE Press, New York (2004)
3. Scaramuzza, D., Fraundorfer, F.: Visual Odometry Part I: The First 30 Years and Fundamentals. IEEE Robotics & Automation Magazine **18**, 80–92 (2011)
4. Fraundorfer, F., Scaramuzza, D.: Visual Odometry Part II: Matching, Robustness, Optimization, and Applications. IEEE Robotics & Automation Magazine **19**, 78–90 (2012)
5. Badino, H., Yamamoto, A., Kanade, T.: Visual odometry by multi-frame feature integration. In: IEEE International Conference on Computer Vision Workshops, pp. 222–229. IEEE Press, New York (2013)
6. Lu, W., Xiang, Z., Liu, J.: High-performance visual odometry with two-stage local binocular high-performance visual odometry with two-stage local binocular BA and GPU. In: IEEE Intelligent Vehicles Symposium, pp. 1107–1112. IEEE Press, New York (2013)
7. Fischler, M.A., Bolles, R.C.: Random Sample Consensus: A Paradigm for Model Fitting with Applications to Image Analysis and Automated Cartography. Comm. of the ACM **24**, 381–395 (1981)
8. Bellavia, F., Fanfani, M., Pazzaglia, F., Colombo, C.: Robust selective stereo SLAM without loop closure and bundle adjustment. In: Image Analysis and Processing – ICIAP 2013. LNCS, vol. 8156, pp. 462–471. Springer, Heidelberg (2013)
9. Sanfourche, M., Vittori, V., Le Besnerais, G.: eVO: a realtime embedded stereo odometry for MAV applications. In: IEEE/RSJ International Conference on Intelligent Robots and Systems, pp. 2107–2114. IEEE Press, New York (2013)

10. Bosse, M., Zlot, R.: Map matching and data association for large-scale two-dimensional laser scan-based SLAM. The International Journal of Robotics Research **27**, 667–691 (2008)
11. Almeida, J., Santos, V.M.: Real time egomotion of a nonholonomic vehicle using LIDAR measurements. Journal of Field Robotics **30**, 129–141 (2013)
12. Hirschmuller, H.: Stereo Processing by Semiglobal Matching and Mutual Information. IEEE T on PAMI **30**, 328–341 (2008)
13. Point Cloud Library (PCL). http://pointclouds.org
14. Lowe, D.G.: Distinctive image features from scale-invariant keypoints. International Journal of Computer Vision **60**, 91–110 (2004)
15. Hesch, J.A., Roumeliotis, S.I.: A direct least-squares (DLS) method for PnP. In: IEEE International Conference on Computer Vision, pp. 383–390. IEEE Press, New York (2011)
16. Open source Computer Vision. http://Opencv.org
17. Geiger, A., Lenz, P., Urtasun, R.: Are we ready for autonomous driving? the KITTI vision benchmark suite. In: IEEE Conference on Computer Vision and Pattern Recognition, pp. 3354–3361. IEEE Press, New York (2012)
18. Geiger, A., Lenz, P., Stiller, C., Urtasun, R.: Vision meets Robotics: The KITTI Dataset. International Journal of Robotics Research **32**, 1231–1237 (2013)

Recognizing Traffic Signs Using a Practical Deep Neural Network

Hamed H. Aghdam, Elnaz J. Heravi and Domenec Puig

Abstract Convolutional Neural Networks (CNNs) surpassed the human performance on the German Traffic Sign Benchmark competition. Both the winner and the runner-up teams trained CNNs to recognize 43 traffic signs. However, both networks are not computationally efficient since they have many free parameters and they use highly computational activation functions. In this paper, we propose a new architecture that reduces the number of the parameters 27% and 22% compared with the two networks. Furthermore, our network uses Leaky Rectified Linear Units (Leaky ReLU) activation function. Compared with 10 multiplications in the hyperbolic tangent and rectified sigmoid activation functions utilized in the two networks, Leaky ReLU needs only one multiplication which makes it computationally much more efficient than the two other functions. Our experiments on the German Traffic Sign Benchmark dataset shows 0.6% improvement on the best reported classification accuracy while it reduces the overall number of parameters and the number of multiplications 85% and 88%, respectively, compared with the winner network in the competition. Finally, we inspect the behaviour of the network by visualizing the classification score as a function of partial occlusion. The visualization shows that our CNN learns the pictograph of the signs and it ignores the shape and color information.

Keywords Convolutional Neural Networks · Traffic sign recognition · Visualization

1 Introduction

Traffic sign recognition is one of the major tasks in Advanced Driver Assistant Systems (ADAS) and it is an indispensable part of the autonomous cars. One of the

H.H. Aghdam(✉) · E.J. Heravi · D. Puig
Intelligent Robotic and Computer Vision Group, Department of Computer
Engineering and Mathematics, University Rovira i Virgili, Tarragona, Spain
e-mail: {hamed.habibi,elnaz.jahani,domenec.puig}@urv.cat

© Springer International Publishing Switzerland 2016
L.P. Reis et al. (eds.), *Robot 2015: Second Iberian Robotics Conference,*
Advances in Intelligent Systems and Computing 417,
DOI: 10.1007/978-3-319-27146-0_31

important characteristics of traffic signs is their design simplicity which facilitates their recognition for a human driver. First, they have a simple shape such as circle, triangle, polygon or rectangle. Second, they are usually composed of basic colors such as red, green, blue, black, white and yellow. Finally, the meaning of traffic sign is acquired using the pictograph in the center. Even though the design is clear and discriminative for a human driver, there are challenging problems in real world applications such as shadow, distance, weather condition and age of the sign that need to be addressed in the traffic sign recognition systems.

Convolutional Neural Networks (CNNs) have shown a great success on challenging datasets such as CIFAR [7] and ImageNet [13, 16] and, recently, they suppressed the human performance on ImageNet dataset [5]. Moreover, CNNs beat the human performance on a practical dataset called German Traffic Sign Benchmark [14]. To be more specific, the winner algorithm proposed by Cireşan *et.al.*[3] computes the average score of 25 CNNs with the same architecture trained on 5 variations of the original dataset. In addition, the second place was also awarded to another CNN proposed by Sermanet and Lecun[12]. In contrast to the winner, the second algorithm uses only one CNN to recognize the traffic signs. In fact, the data-augmentation procedure utilized by both teams is almost identical and the only difference is the architecture of the networks. A single CNN of the winner team requires optimizing 1, 543, 443 parameters. However, the runner-up team trains a network with 1, 437, 791 parameters. Last not the least, the first network uses the hyperbolic tangent activation function and the second network utilizes the rectified sigmoid activation function. Despite their high accuracies, both networks are not computationally efficient and they require a large amount of multiplications on the hardware. As we discuss in Section 3, their utilized activation functions are not efficient since they need more than ten multiplications to calculate the output of a single neuron. While the both networks have mainly concentrated on reducing the classification error, yet, the practical applications require that the network consumes as few CPU cycles as possible in order to reduce the run-time of the network.

Contribution: In this paper, we design a new CNN which is *more accurate* than the previously proposed networks in recognizing traffic signs and it is *computationally much more efficient*. To achieve this, we use Leaky Rectified Linear Units (ReLU) [9] activation function. In addition to the favourable mathematical properties of Leaky ReLUs, they are also computationally very efficient since they calculate the output using only one comparison operator and one multiplication. Furthermore, beside the common convolutional, pooling and fully-connected layers, we add a trainable linear transformation layer to the network which is directly applied on the input image to enhance the illumination of the network. Finally, we visualize our CNN and show that it learns to classify traffic signs using their pictographs and it ignores the shape and the color information.

The rest of this paper is organized as follows: Section 2 reviews the state-of-art and Section 3 explains the proposed network architecture. Then, Section 4 shows the experimental results on German Traffic Sign Benchmark and it mentions the computational differences between the networks. Finally, Section 5 concludes the paper.

2   Related Work

Traffic sign recognition has been extensively studied and some impressive results on uncontrolled environments have been reported. In early works, a traffic sign was considered as a rigid and well-defined object and their image were stored in the database. Then, the new input image was compared with the all templates in the database to find the best matching. The methods based on template matching usually differ in terms of similarity measure or template selection. In general, these methods are not stable and accurate in uncontrolled environments. For more detail the reader can refer to [11] and [10].

Moreover, traditional classification approaches have achieved accurate results on practical databases. Stallkamp *et.al.* [15] achieved 95% classification accuracy on their database, called German Traffic Sign Benchmark database, by extracting the HOG features and classifying the images into 43 classes using the linear discriminant analysis. Zaklouta and Stanciulescu [18, 19] extracted the same HOG features on the same database and classified them using the random forest model. They could increase the performance up to 97.2%.

Liu *et.al.* [8] extracted the SIFT features after transforming the image to the log-polar coordinate system and found the visual words using k-means clustering. Then, the feature vectors were obtained using a sparse coding method. Finally, the traffic signs were recognized using a support vector machine.

Different from the previous approaches, Wang *et.al.* [17] employed a two step classification. In the first step, the input image is classified into 5 super-classes using HOG features and a support vector machine. In the second stage, the final classification is done using HOG and support vector machine after doing perspective adjustment on the image taking into account the information from the super class. Recently, we also recognized traffic signs using visual attributes and Bayesian network [1].

Deep networks outperformed the human performance by classifying more than 99% of the images, correctly. Cireşan *et.al.* [2, 3] developed a bank of 7-layer deep networks whose inputs are transformed versions of the input image. The input image is classified by computing the average score of 25 CNNs. In addition, Sermanet and LeCun [12] proposed a 7-layer deep network for recognizing the traffic signs and obtained 99% accuracy in their experiments.

3   Proposed Network

Fig. 1 illustrates the proposed network architecture. Our network consists of a transformation layer, 3 convolution-pooling layers and two fully connected layers with a dropout layer [6] in between. The network accepts a 48×48 RGB image and classify it into one of the 43 traffic sign classes in the dataset. It should be noted that we have reduced the number of the parameters by dividing the two middle convolution-pooling layers into two groups.

More specifically, the transformation layer applies an element-wise linear trans-formation $f_c(x) = a_c x + b_c$ on c^{th} channel of the input image where a and b are trainable parameters and x is the input value. Consequently, the transformation layer does not change the number of the channels. Next, the image is processed using 100 filters of size 7×7. The notation $C(w, y, z)$ indicates a convolution layer containing y filters of size $z \times z$ that are applied on the input with w channels. Then, the output of the convolution is passed through a Leaky ReLU[9] layer and fetched into the pooling layer where a MAX-pooling is applied on the 3×3 window with stride 2. Indeed, there is a Leaky ReLU layer after each convolution layer and after the first fully-connected layer.

In general, a $C(w, y, z)$ layer contains $w \times y \times z \times z$ parameters. In fact, the second convolution layer accepts a 100-channel input and applies 150 filters of size 4×4. Using this configuration, the number of the parameters in the second convolution layer would be 240, 000. We halve the number of the parameters in the second layer by dividing the input channels into two equal parts and fetch each part into a layer including two separate $C(50, 75, 4)$ convolution-pooling units. Similarly, the third convolution-pooling layer halves the number of the parameters using two $C(75, 125, 4)$ units instead of one $C(150, 250, 4)$ unit. Our proposed architecture is collectively parametrized by 1, 123, 449 weights and biases which is 27% and 22% reduction in the number of the parameters compared with the networks proposed in [3] and [12], respectively. Table 1 compares the computational complexity of the three networks.

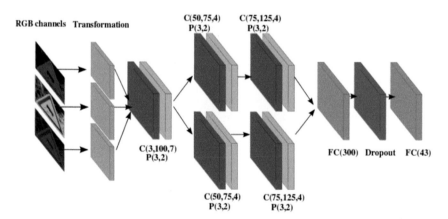

Fig. 1 The network architecture. Orange, red, green, blue and purple indicate the transformation, convolution, pooling, fully connected and dropout layers, respectively.

Complexity: Complexity of CNNs can be computed from two perspectives including *number of the parameters* and *number of the operations*. We previously compared the number of the parameters in our CNN and the two other CNNs. Note that activation functions utilized in the three networks do not have trainable parameters. As the

Table 1 Comparing the computational complexity of the proposed network with [3] and [12]

	Cireşan et.al. [3]	Sermanet et.al. [12]	Proposed architecture
Number of parameters	1,543,443	1,437,791	1,123,449
Activation function	10 multiplications	10 multiplications	1 multiplication

results, they do not change the overall number of the parameters. However, number of operations depends also on the activation function. Cireşan et.al. [3] have used the hyperbolic activation function which is defined as $tanh(x) = \frac{e^x - e^{-x}}{e^x - e^{-x}} = \frac{e^{2x} - 1}{e^{2x} + 1}$. Even with an efficient implementation of exponentiation e^x, it still requires many multiplications. Note that x is a floating point number since it is the weighted sum of the input to the neuron.

An efficient way to calculate e^x is as follows: First, write $x = x_{int} + r$, where x_{int} is the nearest integer to x and r is a real number in range $[-0.5 \ldots 0.5]$ which gives us $e^x = e^{x_{int}} \times e^r$. Second, multiply the number e by itself x_{int} times. The multiplication can be done quite efficiently. To further increase efficiency, various integer powers of e can be calculated once and stored in a lookup table. Finally, e^r can be estimated using polynomial $e^r = 1 + x + \frac{x^2}{2} + \frac{x^3}{6} + \frac{x^4}{24} + \frac{x^5}{120}$ with estimation error $+0.00003$. Consequently, calculating $tanh(x)$ needs $[x] + 5$ multiplications and 5 divisions. Although division needs more computations, yet, we assuming that division and multiplication needs the same amount of CPU cycles. Therefore, $tanh(x)$ can be computed using $[x] + 10$ multiplications. In contrary, a Leaky ReLU function needs only one multiplication in the worst case. Assuming $x \in [-0.5 \ldots 0.5]$ then $tanh(x)$ can be calculated using 10 multiplications. Based on this, the total number of multiplications in the network proposed in [3] and our network are $128, 308, 800$ and $76, 527, 012$, respectively, which is 40% reduction in the number of multiplications when we consider a single network. As we describe in the next section, we outperform the results of the winner CNN by computing the average score of 5 CNNs. Compared with the winner network that uses ensemble of 25 networks, our ensemble of 5 CNNs reduces the number of multiplications 0.88%.

4 Experiments

We applied our architecture on the German Traffic Sign Recognition Benchmark [14]. The original color images contain one traffic sign each and they vary in size from 15×15 to 250×250 pixels. The training set consists of $39, 209$ images and the test set contains $12, 630$ images. We crop all the images and process only the image within the bounding box (The bounding box is part of the annotation). Then, we resize (downsample or upsample) the all images to 48×48 pixels. Finally, the mean image is obtained over the whole dataset and it is subtracted from each image in the dataset since previous studies[4] show that subtracting the mean image increases the performance of the network.

Data Augmentation: To reduce the effect of over-fitting, we augment the original dataset by applying 12 transformations on the images as follows: 1) *Translation*: We perturb the images in positions [2, 2], [2, −2], [−2, 2] and [−2, −2]. 2) *Smoothing*: The images are smoothed using a 5 × 5 Gaussian filter and a 5 × 5 median filter. 3) *HSV manipulation*: The saturation of images are modified by transforming the images into the HSV color space and scaling the saturation component by factors 0.9 and 1.1. In another transformation, the value component is scaled by factor 0.7. 4) *PCA*: First, the matrix of principal components $P \in \mathbb{R}^{3\times3}$ of the pixels of the training dataset is computed. Then, pixels of the input image are projected onto the principal components to obtain the coefficient vector b. Next, a 3 × 1 vector v is randomly generated in range [0, 0.5] and the image is reconstructed by back-projecting the pixels onto RGB color space using the new coefficient vector $b' = b - b \odot v$ where $\odot$ is an element-wise multiplication operator. 5) *Other*: The dataset is further augmented by applying histogram equalization transformation and sharpening the images.

Results: Each function in the transformation layer is implemented using a convolution layer with 1 × 1 kernel and a bias parameter. The network is trained using mini-batch stochastic gradient descent (batch size=50) with exponential weight annealing. We fixed the learning rate to 0.02, momentum to 0.9, L2 regularization to $1e - 5$, annealing parameter to 0.9998, dropout ratio to 0.5 and the negative slope of the Leaky ReLU to 0.01. The network is trained 5 times and their classification accuracies are calculated. It is worth mentioning that [14] has only reported the classification accuracy and it is the only way to compare our results with [3] and [12]. In this paper, we report two different results based the accuracy of a single network and the accuracy obtained by calculating the average score of 5 networks.

Table 2 Classification accuracy of the single network.

Trial	Top 1 acc. (%)	Top 2 acc. (%)
1	99.05	99.69
2	99.01	99.65
3	98.79	99.52
4	98.67	99.53
5	98.86	99.59
average	98.87 ± 0.14	99.59 ± 0.04
Human	98.84	NA

   Table 2 shows the result of training 5 networks with the same architecture and different initializations. The average classification accuracy of the 5 trials is 98.87% which is just above the average human performance reported in [14]. In addition, the standard deviation of the classification accuracy is small which shows that the proposed architecture trains the networks with very close accuracies. We argue that this stability is the result of reduction in the number of the parameters and regularizing the network using a dropout layer. Moreover, we observe that the top-2[1] accuracy is very close in all trials and their standard deviation is 0.04. In other words, although the difference in top-1 accuracy of the Trial 1 and Trial 4 is 0.38%, notwithstanding, the same difference for top-2 accuracy is 0.16%. This implies that some cases are always within the top-2 results. In other words, there are images that have been classified correctly in Trial 1 but they have been misclassified in Trial 4 (or vice versa). As a consequence, if we fuse the scores of two networks the classification accuracy might increase.

Table 3 Comparing the results with CNNs in [3] [14] and [12]

CNN	accuracy(%)
Single network (best) [3]	98.80
Single network (avg.) [3]	98.52
Multi-scale CNN (official)[14]	98.31
Multi-scale CNN (best) [12]	98.97
Proposed CNN (best)	99.05
Proposed CNN (avg.)	98.87
Committee of 25 CNNs [3][14]	99.46
Ensemble of 5 proposed CNNs	**99.51**

Based on this observation, we calculated the average score of the 5 networks to classify the input image. As it is shown in Table 3, the overall accuracy of the network increases 0.6% by this way. Furthermore, the proposed method has established a new record compared with the winner network reported in the competition [14]. Besides, we observe that the results of the single network has outperformed the two other CNNs. It is worth mentioning that the winner architecture calculates the output by computing the average score of 25 CNNs which collectively indicate a model with $25 \times 1,543,443 = 38,586,075$ parameters. Taking into account the fact that we only compute the average score of 5 networks with $1,123,449$ parameters each, our model requires $5 \times 1,123,449 = 5,617,245$ parameters which is 85% reduction in the number of the parameters.

Table 4 Effect of the linear transformation layer in our proposed CNN.

Trial	Top 1 acc(%)	Top 2 acc(%)
1	98.46	99.22
2	98.10	99.35
3	98.74	99.50
4	98.66	99.46
5	98.80	99.50
average	98.55 ± 0.25	99.40 ± 0.11

---

[1] The percent of the samples which are always within the top 2 classification scores.

Effect of Linear Transformation: We investigated the effect of the linear transformation layer by removing the layer from the network and repeating the training 5 times. Table 4 shows the results. We observe that by ignoring the transformation layer the average accuracy of the network drops 0.3%. Moreover, the stability of the training procedure drops by 44%. We inspected the database, manually, and realized that there are images with poor illumination. In fact, the transformation layer enhances the illumination of the input image by multiplying the whole pixels with a constant factor and adding an intercept to the result. With this goal, the transformation layer learns the parameters of the linear transformation such that it increases the classification accuracy. Fig. 2 illustrates the output of the network up to the first pooling layer. We observe that the input image suffers from a poor illumination. However, applying the linear transformation on the image enhances the illumination and, consequently, the subsequent layers represent the image properly so it is classified correctly.

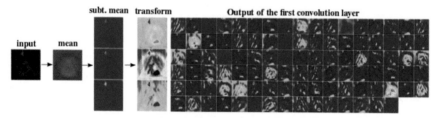

Fig. 2 Visualization of the transformation and the first convolution layers.

Misclassified Images: We computed the class-specific *precision* and *recall* (Table 5). Besides, Fig. 3 illustrates the incorrectly classified traffic signs. The number below each image shows the predicted class label. For presentation purposes, all images were scaled to a fixed size. First, we observe that there are 6 cases where the images are incorrectly classified as class 42 while the true label is 6. We note that all these cases are low-resolution images with a poor illumination. Moreover, class 42 is distinguishable from class 6 using the fine differences in the middle. However, rescaling a poorly illuminated low resolution image to 48×48 pixels causes some artifacts on the image. As the result, the network is not able to discriminate these two classes. For this reason, class 42 has the lowest precision due to high false-positive rate. Similarly, class 6 has the lowest recall because of high false-negative rate. In addition, by inspecting the rest of the misclassified images, we realize that the wrong classification is mainly due to occlusion or low-quality of the images.

Visualizing: It is important to understand the behaviour of the CNN. In particular, we are interested in finding those parts of the images where they have the greatest impact on the classification. To achieve this goal, we synthetically occlude the image using a mask and compute the classification score of the image. We slide the mask all over the image and record the classification score of the CNNs as a function of the position of the mask. Then, the resulting matrix is illustrated using a heatmap plot. Our goal is to assess the behaviour of the network when the area under mask

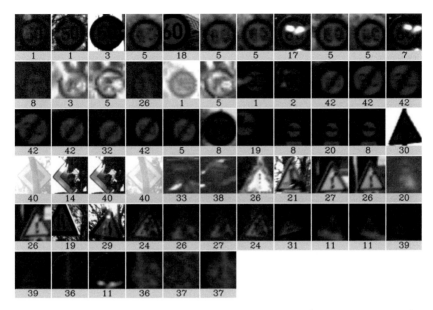

Fig. 3 Incorrectly classified images. The number below each image shows the predicted class label. Actual class labels are illustrated in Table 5

Table 5 Class specific precision and recall obtained by our network. Bottom images show corresponding class number of each traffic sign.

Class	precision	recall	Class	precision	recall	Class	precision	recall
0	1.00	1.00	15	1.00	1.00	30	0.99	0.99
1	0.99	1.00	16	1.00	1.00	31	1.00	0.99
2	1.00	1.00	17	1.00	1.00	32	0.98	1.00
3	1.00	0.98	18	1.00	0.98	33	1.00	1.00
4	1.00	0.99	19	0.97	1.00	34	1.00	1.00
5	0.99	1.00	20	0.98	1.00	35	1.00	1.00
6	1.00	0.95	21	0.99	1.00	36	0.98	1.00
7	1.00	1.00	22	1.00	1.00	37	0.97	1.00
8	0.99	1.00	23	1.00	1.00	38	1.00	1.00
9	1.00	1.00	24	0.98	0.98	39	0.98	0.98
10	1.00	0.99	25	1.00	0.99	40	0.97	0.97
11	0.99	1.00	26	0.97	1.00	41	1.00	1.00
12	1.00	0.99	27	0.97	1.00	42	0.94	1.00
13	1.00	1.00	28	1.00	1.00			
14	1.00	1.00	29	0.99	1.00			

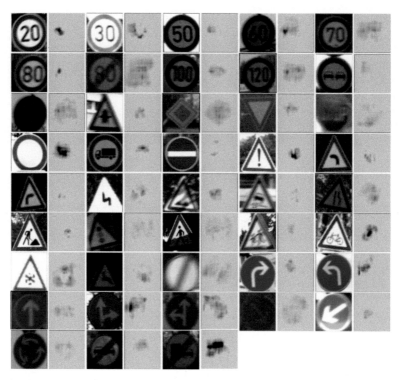

Fig. 4 Visualization of the CNN. The darker color, the smaller class score.

maximally changes. For this purpose, we complement the pixels under the mask and calculate the network score. Fig. 4 shows the heatmap of each traffic sign.

We observe that the score of the CNN drops on pictographs. In contrast, dramatic changes on the margin of the traffic signs do not affect the classification score. These findings suggest that our CNN learns the pictographs and it ignores the shape and color of the traffic signs. One of the interesting findings is related to "speed limit 20"(class 0) and " speed limit 120"(class 8) signs. In the one hand, the CNN recognizes the class 0 by learning the appearance of digit 2 in the image. This is implied by the fact the the score of this class drops when the digit 2 is obscured by a mask. On the other hand, CNNs are translation invariant. Hence, our CNNs could also recognize the digit 2 in class 8. However, to discriminate class 8, the CNNs learns the digits 1, 2 and 0 jointly. For this reason, any small change on these digits causes the dramatic reduction in the classification score. The same intuition is applied on sign "speed limit 100". In general, this visualization technique helps us to find which parts of the image are important for classification of each traffic sign.

As it is shown in Fig. 3, there are four signs from class 12 ("priority road" sign) which have been incorrectly classified. The common point in these four signs is that the middle part of the sign has been occluded. Moreover, we observe in Fig. 4 that any occlusion in the middle part of the "priority road"sign reduces the classification score dramatically. Consequently, the CNN classifies the four images, incorrectly.

5 Conclusion

In this paper, we proposed a new Convolutional Neural Network (CNN) for the traffic sign recognition problem with fewer number of parameters and better classification accuracy. Although two different networks has been previously proposed for this problem, notwithstanding, they are not computationally efficient for two reasons. First, the networks have been designed with many parameters. More specifically, these networks are defined by $1,543,443$ and $1,437,791$ parameters. Second, the activation functions of these networks need many multiplication operations in order to compute the activation of a single input. Taking into account the overall number of times that activation functions are called during one forward pass, they consume a considerable amount of CPU/GPU cycles. To alleviate these two problem, we utilized Leaky Rectified Linear Units (ReLU) as the activation function that only needs one comparison and one multiplication to compute the output. Then, we divided the two middle convolution/pooling layers into two separate parts so we could halve the number of the parameters in the middle layers. Using this architecture, our network reduces the number of the parameters to $1,123,449$ which is 27% and 22% reduction compared with the two previously proposed networks. Most importantly, the proposed architecture improved the classification accuracy and established a new record compared with the winner network reported in the German Traffic Sign Benchmark competition. Specifically, we could increase the classification accuracy up to 99.51%. It should be noted that our model computes the average score of 5 networks which collectively shows a model with $5,617,245$ parameters. Comparing with the winner model which computes the average score of 25 networks with $1,543,443$ parameters each ($25 \times 1,543,443 = 38,586,075$), our method reduces the overall number of the parameters and the number of multiplications 85% and 88%, respectively. Finally, we visualized our network and showed that the CNN learns the pictographs of the traffic signs and it ignores the shape and the color information.

Acknowledgements The authors are grateful for the support granted by Generalitat de Catalunya's Agència de Gestió d'Ajuts Universitaris i de Recerca (AGAUR) through FI-DGR 2015 fellowship.

References

1. Aghdam, H.H., Heravi, E.J., Puig, D.: A unified framework for coarse-to-fine recognition of traffic signs using bayesian network and visual attributes. In: Proceedings of the 10th International Conference on Computer Vision Theory and Applications, pp. 87–96 (2015)
2. Ciresan, D., Meier, U., Masci, J., Schmidhuber, J.: A committee of neural networks for traffic sign classification. In: The 2011 International Joint Conference on Neural Networks (IJCNN), pp. 1918–1921, July 2011
3. Cirean, D., Meier, U., Masci, J., Schmidhuber, J.: Multi-column deep neural network for traffic sign classification. Neural Networks **32**, 333–338 (2012). http://dx.doi.org/10.1016/j.neunet.2012.02.023
4. Coates, A., Ng, A.Y.: Learning feature representations with K-means. Lecture Notes in Computer Science (including subseries Lecture Notes in Artificial Intelligence and Lecture Notes in Bioinformatics) LECTU, vol. 7700, pp. 561–580 (2012)

5. He, K., Zhang, X., Ren, S., Sun, J.: Delving Deep into Rectifiers: Surpassing Human-Level Performance on ImageNet Classification
6. Hinton, G.E., Srivastava, N., Krizhevsky, A., Sutskever, I., Salakhutdinov, R.R.: Improving neural networks by preventing co-adaptation of feature detectors, pp. 1–18 (2012). arXiv: 1207.0580, http://arxiv.org/abs/1207.0580
7. Lin, M., Chen, Q., Yan, S.: Network In Network, p. 10 (2013). arXiv preprint http://arxiv.org/abs/1312.4400
8. Liu, H., Liu, Y., Sun, F.: Traffic sign recognition using group sparse coding. Information Sciences **266**, 75–89 (2014)
9. Maas, A., Hannun, A., Ng, A.: Rectifier nonlinearities improve neural network acoustic models. ICML Workshop on Deep Learning **28** (2013). http://www.stanford.edu/~awni/papers/relu_hybrid_icml2013_final.pdf
10. Paclik, P., Novovicova, J., Duin, R.P.W.: Building road sign classifiers using trainable similarity measure. IEEE Transactions on Intelligent Transportation Systems **7**(3), 309–321 (2006)
11. Piccioli, G., Micheli, E.D., Parodi, P., Campani, M.: A robust method for road sign detection and recognition (1996)
12. Sermanet, P., Lecun, Y.: Traffic sign recognition with multi-scale convolutional networks. In: Proceedings of the International Joint Conference on Neural Networks, pp. 2809–2813 (2011)
13. Simonyan, K., Zisserman, A.: Very Deep Convolutional Networks for Large-Scale Image Recognition, pp. 1–13 (2015)
14. Stallkamp, J., Schlipsing, M., Salmen, J., Igel, C.: Man vs. computer: Benchmarking machine learning algorithms for traffic sign recognition. Neural Networks **32**, 323–332 (2012). http://dx.doi.org/10.1016/j.neunet.2012.02.016
15. Stallkamp, J., Schlipsing, M., Salmen, J., Igel, C.: Man vs. computer: Benchmarking machine learning algorithms for traffic sign recognition. Neural Networks **32**, 323–332 (2012). selected Papers from IJCNN 2011
16. Szegedy, C., Reed, S., Sermanet, P., Vanhoucke, V., Rabinovich, A.: Going deeper with convolutions, pp. 1–12 (2014)
17. Wang, G., Ren, G., Wu, Z., Zhao, Y., Jiang, L.: A hierarchical method for traffic sign classification with support vector machines. In: The 2013 International Joint Conference on Neural Networks (IJCNN), pp. 1–6, August 2013
18. Zaklouta, F., Stanciulescu, B.: Warning traffic sign recognition using a hog-based k-d tree. In: 2011 IEEE Intelligent Vehicles Symposium (IV), pp. 1019–1024, June 2011
19. Zaklouta, F., Stanciulescu, B.: Real-time traffic sign recognition in three stages. Robotics and Autonomous Systems **62**(1), 16–24 (2014). new Boundaries of Robotics

Particle Filter SLAM on FPGA: A Case Study on Robot@Factory Competition

Biruk G. Sileshi, Juan Oliver, R. Toledo, Jose Gonçalves and Pedro Costa

Abstract Particle filters are sequential Monte Carlo estimation methods with applications in the field of mobile robotics for performing tasks such as tracking, simultaneous localization and mapping (SLAM) and navigation, by dealing with the uncertainties and/or noise generated by the sensors as well as with the intrinsic uncertainties of the environment. This work presents a field programmable gate arrays (FPGA) implementation of a particle filter applied to SLAM problem based on a low cost Neato XV-11 laser scanner sensor. Post processing is performed on data provided by a realistic simulation of a differential robot, equipped with a hacked Neato XV-11 laser scanner, that navigates in the Robot@Factory competition maze. The robot was simulated using SimTwo, which is a realistic simulation software that can support several types of robots. The simulator provides the robot ground truth, odometry and the laser scanner data. The results achieved from this study confirmed the possible use such low cost laser scanner for different robotics applications.

Keywords Particle filter · SLAM · FPGA · Laser scanner

B.G. Sileshi(✉) · J. Oliver
MISE Department, Univ. Autònoma de Barcelona, Bellaterra, Barcelona, Spain
e-mail: {birukgetachew.sileshi,joan.oliver}@uab.cat

R. Toledo
CC Department, Computer Vision Center, Univ. Autònoma de Barcelona,
Bellaterra, Barcelona, Spain
e-mail: ricardo.toledo@uab.cat

J. Gonçalves · P. Costa
INESC-TEC, Porto, Portugal
e-mail: goncalves@ipb.pt, paco@fe.up.pt

J. Gonçalves
ESTiG IPB, Bragança, Portugal

P. Costa
FEUP UP, Porto, Portugal

© Springer International Publishing Switzerland 2016   411
L.P. Reis et al. (eds.), *Robot 2015: Second Iberian Robotics Conference*,
Advances in Intelligent Systems and Computing 417,
DOI: 10.1007/978-3-319-27146-0_32

1 Introduction

In mobile robotic applications the most common tasks comprise localization, mapping, Simultaneous Localization and Mapping (SLAM), navigation and obstacles avoidance. In order to perform them efficiently, the robot needs to sense, calculate the distances to the obstacles and build the map for robot navigation. To achieve that, laser scanners are widely used in mobile robotics localization systems [1]. However, the high price tag in the most commonly used laser scanners, SICK LMS 200 and Hokuyo URG-04LX [2], has been a major drawback for many hobbyist and educational robotics practitioners which results in the need for alternative low cost laser scanners.

A domestic vacuum cleaner robot, the Neato XV-11 [3] shown in Fig. 1 (left), has included an alternative low cost 360 degree laser scanner [4]. The laser scanner can be removed from the robot Fig. 1 (right) in order to allow robotics practitioners to use it in their projects. However, as a primary step towards facilitating the development of robotic solutions, mimicking the behavior of the laser scanner based on its model in a virtual simulation environment is required. Therefore, in this study using a SimTwo [6] realistic simulations software that supports several types of robots, a simulation is performed with a robot. The robot is equipped with a Hacked Neato XV-11 laser scanner and navigates in a Robot@Factory competition maze [5]. Based on the actuator [7] and laser scanner models [8] used in the simulator, the simulation finally provides the robot odometry and laser scanner data information. The odometry and laser data are used for post processing in order to verify the performance and possible use of such alternative low cost laser scanner in robotic applications.

SLAM is one of the applications where laser scanner data is primarily used by mobile robots to make a map and navigate in an environment. The use of robust and accurate estimation methods in SLAM is crucial for optimizing the high uncertainties and/or noise generated by such low cost laser scanner. Particle filter, which is based on sequential Monte Carlo and recursive Bayesian estimation, is one of the state estimation methods that provide accurate estimations and addresses systems that are non-linear and non-Gaussian sensor models [10, 11]. But, the disadvantages

Fig. 1 Neato XV-11.

of the complex algorithm architecture, enormous computations and the low speed solutions have restricted the particle filter use in most real-time systems. However, an efficient implementation of the particle filter algorithm can be achieved using flexible hardware/software platforms, such as field programmable gate arrays (FPGA). In this paper an approach for the implementation of the particle filter to the SLAM application based on low cost laser scanner is performed on an FPGA platform.

The main contributions outlined in this work are:

1) Verifying the possible use of low cost laser scanner for robotic application based on an FPGA implementation.

2) Proposing simple data structure for map representation in SLAM to avoid the expensive map copying process during the resampling step of the particle filter.

3) Proposing an observation model which takes into account the error in the measurement and matching between laser scans and map data.

The rest of the paper is organized as follows. In Section 2 a general background information on particle filtering algorithm is presented, followed by Section 3, where a description to the application of particle filter to the SLAM problem is presented. Section 4 is a discussion about the probabilistic sensor measurement model and the occupancy grid map representation of the robots environment. A discussion on the FPGA implementation and the achieved results are given in Section 5 and 6 respectively. Finally Section 7 provides the conclusions.

2 Particle Filtering Algorithm

Particle filters are Bayesian non-linear filtering methods for recursive estimation of the state x_t from noisy measurement z_t, where the initial target state is assumed to have a known probability density function $p(x_0)$. The estimate of the unknown state x_t is performed based on the sequence of all available measurements $Z_{1:t} = z_i, i = 1, ..., t$ up to time t via the posterior distribution $p(x_t|Z_t)$.

The posterior distribution is obtained with the two recursive prediction and update steps. The prediction step is based on the probabilistic model of the transitional density, $p(x_t|x_{t-1})$, and the pdf, $p(x_{t-1}|Z_{t-1})$, at time $t-1$ to obtain the prediction (prior) density of the state at time t by:

$$p(x_t|z_{t-1}) = \int p(x_t|x_{t-1})p(x_{t-1}|z_{t-1})dx_{t-1} \qquad (1)$$

In the update step, the prior pdf is updated while the measurement z_t becomes available at time step t by:

$$p(x_t|z_t) = \frac{p(z_t|x_t)p(x_t|z_{t-1})}{p(z_t|z_{t-1})} \qquad (2)$$

where the normalizing constant $p(z_t|Z_{t-1})$ is :

$$p(z_t|Z_{t-1}) = \int p(z_t|x_t)p(x_t|Z_{t-1})dx_t \qquad (3)$$

Particle filters apply a Monte Carlo integral approximation method for the representation of the posterior pdf with a random measure composed of discrete set of samples (particles) $\{x_t^i\}_{i=1}^N$ drawn from a proposal distribution $q(x_t|x_{t-1}, z_t)$ with their corresponding weights $\{w_t^i\}_{i=1}^N$.

The approximation to the posterior density is given by [12].

$$p(x_t|Z_t) \approx \sum_{i=1}^N w_t^i \delta(x_t - x_t^i) \qquad (4)$$

Where $\delta(.)$ is a Dirac Delta function.

In the most widely adopted variant of the particle filter, Sampling Importance Resampling particle filter (SIRF), three steps are used for propagating the particle set $S_t = \{x_t^i, w_t^i\}_{i=1}^N$ at time t, from the set S_{t-1} at the previous time $t - 1$. The three steps are sampling, importance weight and resampling steps. In the sampling step new particles are drawn from a prior density $p(x_t^i|x_{t-1}^i)$ which is defined by a system equation $x_t = f_t(x_{t-1}, v_{t-1})$. Where f_t is possibly non-linear function of the state x_{t-1} and process noise v_{t-1} at time $t - 1$. In the importance weight step, weight is assigned to individual particles according to the measurement z_t by:

$$w_t^i \propto w_{t-1}^i p(z_t|x_t^i) \qquad (5)$$

where, $p(z_t|x_t^i)$ is the measurement likelihood.

In the resampling step, new set of particles are sampled from the set $\{x_t^i, w_t^i\}_{i=1}^N$ according to their weights, and the resulting set of particles are used in the next time step to predict the posterior probability. Resampling helps to avoid the particle degeneracy problem and improves the estimation of states by concentrating particles into domains of higher posterior probability [10].

3   The SLAM Application

In this work it is considered a SLAM problem for the application of the particle filter. SLAM is the problem of localization and building a map of a given environment simultaneously [13]. It is given by a joint posterior probability density distribution $p(x_t, m|z_{0:t}, u_{0:t}, x_0)$ of the map (m) and the robot states (x_t) at time t given the observations $(z_{0:t})$ and control inputs $(u_{0:t})$ upto and including time t together with the initial state of the robot (x_0).

A recursive Bayes solution for the computation of the joint probability density function is used by starting with an estimate for the distribution at time $t - 1$, $p(x_{t-1}, m|z_{0:t-1}, u_{0:t-1})$, and following a control u_t and observation z_t. Therefore, the SLAM algorithm is implemented with the two standard recursive state update and measurement update steps. Where the recursion is a function of the robot motion model $p(x_t|x_{t-1}, u_t)$ and an observation model $p(z_t|x_t, m)$. The two recursive state and measurement updates are given by [14]:

State Update:

$$p(x_t, m|z_{0:t-1}, u_{0:t-1}, x_0) =$$

$$\int p(x_t|x_{t-1}, u_t)p(x_{t-1}, m|z_{0:t-1}, u_{0:t-1}, x_0)dx_{t-1} \qquad (6)$$

Measurement Update

$$p(x_t, m|z_{0:t}, u_{0:t}, x_0) = \frac{p(z_t|x_t, m)p(x_t, m|z_{0:t-1}, u_{0:t-1}, x_0)}{p(z_t|z_{t-1}, U_{0:t})} \qquad (7)$$

For the analytical solution of the state update and measurement update equations 6 and 7, a particle filter based approach based on Monte Carlo Localization (MCL) and Occupancy grid mapping (OGM) is used [15, 16]. This approach to SLAM, shown in Fig. 7, is known as the grid based Fast SLAM. In the MCL step, a given number of particles are used for the prediction of the trajectory (pose) of a robot, and the occupancy grid map represents the map of the environment constructed from a measurement data in respect to the individual particles generated in the MCL step. The occupancy grid map is a representation of the environment by a fine grained metric grids of cells and estimate the probability of any cell being occupied depending on the sensor readings. It provides information about the presence and absence (free-space) of objects in the environment.

In the basic grid based Fast SLAM algorithm every particle is required to maintain its individual map [15]. This requires a huge amount of memory and it also needs

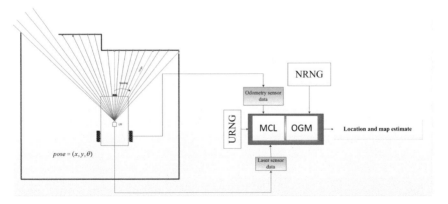

Fig. 2 A 2D SLAM for a robot equipped with odometry and laser range finder(LRF) sensors based on MCL and OGM.

the copying of the individual particle maps during the resampling step of the particle filter [17]. As opposed to the basic grid based Fast SLAM algorithm, this work uses a different data structure to maintain the particle maps. The proposed approach maintains a single global occupancy grid map for all the particles, where each grid cell of the global map maintains the occupancy likelihood information of all the particles. In general, if there are N particles and the map is $M \times M$, the proposed approach stores $N \times M \times M$ scalars, which is identical to storing N different $M \times M$ maps as in the basic grid based Fast SLAM algorithm. However, the advantage of the proposed approach is in avoiding the requirement of copying the particle maps during the resampling step by maintaining the index of the particle in each grid cell.

4   Observation Model and Map Update

An observation model is used to compute the probability of the laser measurement at time t and assign weight for the individual particles based on the current observation of the robot to environment. The observation model considered in this work takes into account the error model of the sensor and a scan matching to determine how well the current scan matches the occupied points in the map.

The laser range finder measurement error is modeled by a Gaussian distribution (equation 8) with mean z_t^n of the actual laser measurement at time t, where n is the index of the laser beams, and standard deviation given by the function $\delta(z_t^n)$. Therefore, the probability that the range measurement is returned by a grid cell m_k along the path of the laser beam is given by:

$$p_o(m_k^n|z_t^n) = \frac{1}{\delta(z_t^n)\sqrt{2\pi}}e^{\frac{-(m_k^n - z_t^n)^2}{2\delta^2(z_t^n)}} \tag{8}$$

The standard deviation function $\delta(z_t^n)$ is specific to a given laser scanner and can be obtained from experiments. For the specific laser adopted in this work, the function is given by equation 9, as presented in [8].

$$\delta(z_t^n) = 0.0001524e^{(1.3103z_t^n)} \tag{9}$$

As occupancy grid map provide information about occupied and unoccupied grid cells, the probability of any path that a laser takes through a series of grid cells has to be calculated. The probability of the complete laser path is computed by taking the product of the probability that the laser is passing through unoccupied grid cells, and the probability that it will be stopped by an obstacle in the grid cell that terminates the path. For grid cells m_k along the path of the laser, the probability that the laser measurement will have traveled through unoccupied grid cells and stopped by an obstacle at grid cell m_k is given by [17]:

$$p(m_k = occupied) = p_o(m^k) \prod_{j=1}^{k-1} \left(1 - p_o(m^j)\right) \qquad (10)$$

The scan matching between the current laser scan points and the occupied points in the map is obtained by the function $fsm(Z_t^n)$ given by equation 11. The current laser scan points which fall in unoccupied or unexplored grid cells contribute to small values to the final value of the scan matching function $fsm(Z_t)$, when compared to the occupied points in the evaluation of scan matching. Therefore, this function is maximized while the current laser scan points match well with the occupied points in the map.

$$fsm(Z_t^n) = \sum_{n=1}^{L} p(m_{t-1}|z_t^n) \qquad (11)$$

The total probability of the laser scan is obtained by taking the product of the $fsm(Z_t^n)$ and sum of the $p(m_k = occupied)$ for all laser beams at time t, as shown in equation 12.

$$fsm(Z_t) \sum_{n=1}^{L} \left(p_o(m^k) \prod_{j=1}^{k-1} \left(1 - p_o(m^j)\right) \right) \qquad (12)$$

The map update is performed by tracing all the grid cells which lies along the path of the laser beams and updating their corresponding occupancy likelihood. However, the occupancy likelihood of those grid cells that are outside the laser's beam range of view remain unchanged. The trace of the grid cells along the path of the laser beam is performed based on the a Bresenham's line tracing algorithm [18] and the occupancy likelihood of each grid cell is read and updated accordingly. Those grid cells that lies between the starting point and end point of the laser beam corresponds to unoccupied points and their occupancy likelihoods are updated by a certain factor f_{free} and for the grid cell that corresponds to the end point of the laser beam, its occupancy likelihood is updated by incrementing its previous value by a factor of $f_{occupied}$. The value $f_{free} = 0.1$ and $f_{occupied} = 0.1$ are heuristic parameter determined from extensive tests.

5   FPGA Implementation and Performance Evaluation

The global architecture for the implementation of SLAM on the FPGA is shown in Fig. 3. The architecture shows the major computational steps of the particle filter for the grid based Fast SLAM application. The FPGA implementation is based on a hardware/software (HW/SW) co-design approach, where the software parts of the algorithm are executed on the embedded Microblaze soft core processor and communicate with a peripheral PF hardware acceleration module.

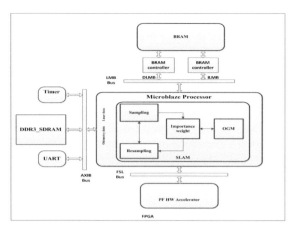

Fig. 3 FPGA architecture for particle filter SLAM implementation

In the architecture, the sampling module is responsible for generating particles by applying a probabilistic motion model of the robot and odometry sensor data. The importance weight module is responsible for the evaluation of the weights of the particles using the laser scanner measurement data and the occupancy grid map data. Finally, the resampling module conducts the evaluation of the replication factors of the particles based on their weight obtained from the importance weight module. For its operation this module requires a uniform random number generation.

The PF hardware acceleration module shown in Fig. 4 accelerates the computational demanding steps based on a custom COordinate Rotation DIgital Computer (CORDIC) and random number generator cores, which are connected to the Microblaze soft-core processor through a dedicated one-to-one communication bus (Fast simplex Link) for fast streaming of data. Individual CORDIC hardware modules are assigned for the evaluation of different functions in the sampling, $f_S()$, and importance weight, $f_I()$, steps and are configured through their $config$ port. Uniform (URN) and Gaussian random (GRN) numbers are provided to the resampling and sampling steps of the particle filter respectively by the random number generator (RNG) module. The design of the RNG module is based on Tausworthe [20] and Ziggurat [21] algorithms. The details on the hardware design of these modules is presented in the previous work [19]. For verification, a realistic simulation generated log data from odometry and laser scanner is stored in the DDR3_SDRAM memory and used in the computation of new particles and the evaluation of their weights in the sampling and importance weight steps respectively.

The Xilinx version 14.6 Integrated Software Environment (ISE), Embedded Development Kit (EDK) and Software Development Kit (SDK) are used for the design and implementation of the system architecture on Xilinx Kintex-7 KC705 FPGA device running at 100 MHz. The output was demonstrated via UART on hyperterminal of a computer. The design of the hardware modules is written in VHDL language, where the different variables are represented as fixed point numbers with

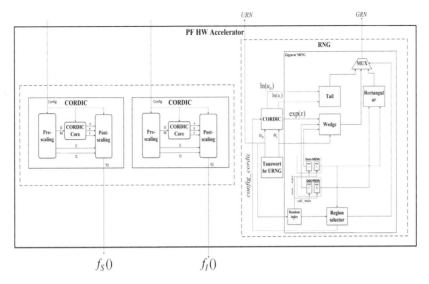

Fig. 4 Internal structure of the PF HW Acceleration module[19]

$Q_{(}8 : 24)$ format (i.e. 8 bits for the integer part and 24 bits for the fractional part). For the different variables in the software part of the algorithm a 32 bit floating point representation is used by enabling the floating point unit (FPU) of the Microblaze processor.

6   Results

In Fig. 5 (left) is shown a simulated Robot Factory competition maze environment, where a robot is guided to navigate. While the robot is navigating in the maze, odometry and laser scanner data are collected for post processing in the SLAM algorithm on the FPGA. The raw collected data is first processed to calibrate the odometry, and synchronization between the calibrated odometry data and laser data is performed. The final data is used for performing the SLAM algorithm on the FPGA, with the objective of constructing a probabilistic occupancy grid map of the robot's maze environment while estimating the trajectory of the robot simultaneously. Fig. 5 (right) shows the FPGA implementation result for the occupancy grid map generated based on the odometry and laser log data. The size of the occupancy grid map is $4meter \times 4meter$ with a grid resolution of 0.01 meter. The gray and white colors in the map correspond to unexplored and free space of the robot environment respectively. The boundary between the free and unexplored space defines the existence of obstacle in the robot environment.

For qualitative analysis of the performance of the implementation, the plot of the trajectory of the robot obtained from odometry, ground truth and particle filter

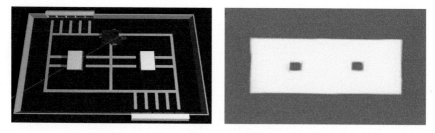

Fig. 5 A simulation maze environment (left) and the occupancy grid map (right) constructed based on the laser model on the FPGA.

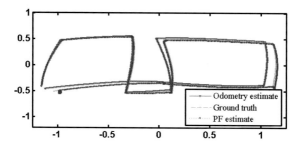

Fig. 6 Robot trajectories for the odometry estimate, ground truth and the particle filter estimated path

estimate are provided in Fig. 6. The results in Fig. 6 shows that the particle filter estimated path of the robot is close to the actual ground truth path of the robot, which confirms the good performance of the implementation. Besides the qualitative analysis, a quantitative evaluation of the performance of the particle filter to the pose estimation of the robot is performed by evaluating the error between the ground truth and the particle filter estimated pose (Fig. 7), and the root mean squared error (RMSE) metrics (Fig. 8). The pose error shown in Fig. 7 is obtained by taking the difference of the sample mean of the particle filter with 100 particles and the ground truth data. The RMSE for the pose is calculated by averaging 50 independent runs of the algorithm at each time step. The results in Fig. 7 and 8 show that initially there is a relatively higher rate in the increase of the pose error with time, which can be attributed to the fact that the robot is trying to localize it self with in its map. However, after certain point in time the robot is able to acquire some knowledge to localize itself based on the sensor measurement data and the map.

The hardware resource utilization of the whole HW/SW system architecture (Fig. 3) and the PF acceleration module (Fig. 4) is given in Table 1, where very few of the available FPGA resources is used for the implementation. From the synthesis results a maximum clock frequency of 1426.228 MHz is achieved for the PF HW acceleration module. Which implies with a 100 MHz clock frequency of the Xilinx Kintex-7 development board used in this work, the PF acceleration module is able to provides the computational results every 142.62 nano seconds, which is much higher

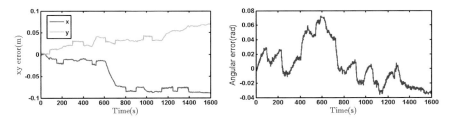

Fig. 7 Translational (left) and angular (right) estimation errors

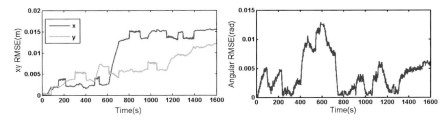

Fig. 8 RMSE errors for pose estimation, translational (left) and angular (right)

Table 1 Resource utilization for PF hardware acceleration module and the HW/SW co-design system on Xilinx Kintex-7 KC705 FPGA

Resources	PF HW Accelerator	HW/SW System	Available
Slice registers	101 (0%)	9406 (2%)	407600
Slice LUTs	4216 (2%)	12246 (6%)	203800
DSP48E1s	12 (1%)	23 (2%)	840
Block RAMB36E1	2 (0%)	42 (9%)	445

than update rate of the laser scanner which is 10 Hz. The computational speedup of the hardware/software implementation over the embedded software implementation for each step of the particle filter computations is provided in our previous work [19], where 140×, 14.87× and 19.36× speedup is achieved in the sampling, importance weight and resampling steps respectively.

7 Conclusion and Future Work

The Neato XV-11 laser scanner is a very low cost alternative, when compared to the current available laser scanners. Using data based on the model of such low cost laser scanner, this work presented an FPGA implementation of particle filter for a SLAM application. The achieved results demonstrate a good performance in the estimation and the potential of such very low cost laser range finder in many robotic application areas. As future work, the parallelization of the particle filter computations and its application to real time robotic system with the use of such low cost laser scanner is expected.

Acknowledgements The authors would like to thank the Universitat Autònoma de Barcelona for the financial support to conduct this research work in collaboration with the Polytechnic Institute of Bragança (IPB). The authors also express their gratitude to the IPB for inviting Mr. Biruk Getachew Sileshi to perform this work as as an invited researcher for the duration of the research work. The authors would also acknowledge the project *TIN2014-56919-C3-2-R DETECCION Y MONITOREO DE INCENDIOS USANDO IMAGENES MULTIESPECTRALES* for the support provided to this work.

References

1. Arras, K.O., Tomatis, N.: Improving robustness and precision in mobile robot localization by using laser range finding and monocular vision. In: 1999 Third European Workshop on Advanced Mobile Robots, (Eurobot 1999), pp. 177–185 (1999)
2. Alwan, M., Wagner, M.B., Wasson, G., Sheth, P.: Characterization of infrared range-finder PBS-03JN for 2-D mapping. In: Proceedings of the 2005 IEEE International Conference on Robotics and Automation, ICRA 2005, pp. 3936–3941, April 18–22, 2005
3. Hacking the Neato XV-11 (2015). https://xv11hacking.wikispaces.com/
4. Neato Robotics XV-11 Tear-down (2015). https://www.sparkfun.com/news/490
5. Gonçalves, J., Lima, J., Costa, P., Moreira, A.: Manufacturing education and training resorting to a new mobile robot competition. In: Ferry Cruise Conference on Flexible Automation Intelligent Manufacturing (Faim 2012), Helsinki-Stockholm-Helsinki, June 10–13, 2012
6. Costa, P., Gonçalves, J., Lima, J., Malheiros, P.: SimTwo realistic simulator: A tool for the development and validation of robot software. International Journal of Theory and Applications of Mathematics Computer Science (2011)
7. Gonçalves, J., Lima, J., Costa, P., Moreira, A.: Modeling and simulation of the EMG30 geared motor with encoder resorting to simtwo: the official robot@factory simulator. In: Flexible Automation Intelligent Manufacturing (Faim 2013), Porto, July 28, 2013
8. Campos, D., Santos, J., Gonçalves, J., Costa, P.: Modeling and simulation of a hacked neato XV-11 laser scanner. In: Second Iberian Robotics Conference ROBOT 2015 (2015)
9. Kalman, R.E.: A new approach to linear filtering and prediction problems. Trans. ASME Journal of Basic Engineering **82**, 35–45 (1960)
10. Arulampalam, M.S., Maskell, S., Gordon, N., Clapp, T.: A tutorial on particle filters for online nonlinear/non-Gaussian Bayesian tracking. IEEE Transactions on Signal Processing **50**(2), 174–188 (2002)
11. Gustafsson, F.: Particle filter theory and practice with positioning applications. IEEE Aerospace and Electronic Systems Magazine **25**(7), 53–82 (2010)
12. Ristic, B., Arulampalam, S., Gordon, N.J.: Beyond the Kalman Filter: Particle Filters for Tracking Applications. Artech House Publishers, Norwood (2004)
13. Durrant-Whyte, H., Bailey, T.: Simultaneous localization and mapping: part I. IEEE Robotics and Automation Magazine **13**(2), 99–110 (2006)
14. Doucet, A., de Freitas, N., Murphy, K., Russell, S.: Rao-Blackwellized particle filtering for dynamic Bayesian networks. In: Proceedings of the 2000 Conference on Uncertainty in Artificial Intelligence (2000)
15. Thrun, S., Burgard, W., Fox, D.: Probabilistic Robotics (Intelligent Robotics and Autonomous Agents series). Intelligent robotics and autonomous agents. The MIT Press, August 2005
16. Dellaert, F., Fox, D., Burgard, W., Thrun, S.: Monte carlo localization for mobile robots. In: Proceedings of the 1999 IEEE International Conference on Robotics and Automation, vol. 2, pp. 1322–1328 (1999)

17. Eliazar, A.I., Parr, R.: DP-SLAM 2.0. In: Proceedings of the 2004 IEEE International Conference on Robotics and Automation ICRA 2004, vol. 2, pp. 1314–1320, April 26–May 1, 2004
18. Bresenham, J.E.: Algorithm for computer control of a digital plotter. IBM Systems Journal **4**(1), 25–30 (1965)
19. Sileshi, B.G., Ferrer, C., Oliver, J.: Hardware/software co-design of particle filter in grid based fast-SLAM algorithm. In: 2014 International Conference on Embedded Systems and Applications (ESA), pp. 24–27, July 2014
20. L'Ecuyer, P.: Maximally equidistributed combined Tausworthe generators. Mathematics of Computation **65**(213), 203–213 (1996)
21. Marsaglia, G., Tsang, W.W.: The ziggurat method for generating random variables. Journal of Statistical Software **5**, 1–7 (2000)

Modeling and Simulation of a Hacked Neato XV-11 Laser Scanner

Daniel Campos, Joana Santos, José Gonçalves and Paulo Costa

Abstract Laser scanners are widely used in mobile robotics localization systems but, despite the enormous potential of its use, their high price tag is a major drawback, mainly for hobbyist and educational robotics practitioners that usually have a reduced budget. This paper presentes the modeling and simulation of a hacked Neato XV-11 Laser Scanner, having as motivation the fact that it is a very low cost alternative, when compared with the current available laser scanners. The modeling of a hacked Neato XV-11 Laser Scanner allows its realistic simulation and provides valuable information that can promote the development of better designs of robot localization systems based on this sensor. The sensor simulation was developed using SimTwo, which is a realistic simulation software that can support several types of robots.

Keywords Laser Scanner · Modeling · Simulation · Neato XV-11

1 Introduction

In mobile robotics applications the most common tasks comprise mapping, localization, navigation and obstacle avoidance. In order to perform them efficiently, the robot needs to sense, calculate the distances to the obstacles and to build the map for robot navigation.

D. Campos · J. Santos · J. Gonçalves(✉) · P. Costa
INESC-TEC, Porto, Portugal
e-mail: {danielfbcampos,joana.r.santos}@inesctec.pt

J. Gonçalves
Department of Electrical Engineering, Polytechnic Institute of Bragança, Bragança, Portugal
e-mail: goncalves@ipb.pt

P. Costa
Department of Electrical Engineering and Computers,
Faculty of Engineering of the University of Porto, Porto, Portugal
e-mail: paco@fe.up.pt

© Springer International Publishing Switzerland 2016
L.P. Reis et al. (eds.), *Robot 2015: Second Iberian Robotics Conference*,
Advances in Intelligent Systems and Computing 417,
DOI: 10.1007/978-3-319-27146-0_33

425

To achieve that, laser scanners are widely used in mobile robotics localization systems [1][2] but, despite the enormous potential of its use, their high price tag is a major drawback, mainly for hobbyist and educational robotics practitioners that usually have a reduced budget. The Neato XV-11, shown in Figure 1, is a robot sold to vacuum domestic rooms[3], that includes a low cost 360 degree laser distance scanner. The laser scanner can be removed from the XV-11, allowing robotics practitioners to use it in their projects, being a very low cost alternative [10].

A comparison between three laser rangefinders (URG-04LX, XV-11 laser scanner, Kinect derived) was developed [8], the XV-11 laser demonstrated to be reasonable accurate and precise with the more competitive cost. In [11] Neato XV-11 was used for Simultaneous Localization and Mapping, being modeled and simulated using V-Rep software with satisfactory results.

Fig. 1 Neato XV-11 [10]

In this paper, the laser scanner was modeled and simulated concerning the parameters: distance and color. The laser scanner hardware approach differs from the previous work [13], because the motor laser scanner is now controlled in closed loop. The presented approach is a much more reliable hardware implementation, when compared with the typical open loop approaches, being the presented model and its simulation specific for the presented hack. Simulation has established itself as an important tool in the mobile robotics field. The ability to test and develop robotic solutions in a virtual environment is widely used in research as well as in education, allowing the rapid development of robot software. The laser scanner sensor was

simulated using SimTwo, shown in Figure 2, which is a realistic simulation software that can support several types of robots. Its original purpose was the simulation
of mobile robots that can have wheels or legs, although industrial robots, conveyor
belts and lighter-than-air vehicles can also be defined. Basically any type of terrestrial robot definable with rotational joints and/or wheels can be simulated in this
software [4] [7] [6]. This paper is structured as follows. Section 2 describes the main
features of the Hacked Neato XV-11 Laser Scanner. In Section 3 and 4 the laser
scanner modeling and its simulation are described, respectively. Section 5 presents a
possible calibration of the laser scanner. Finally, Section 6 presents the conclusions
and future work.

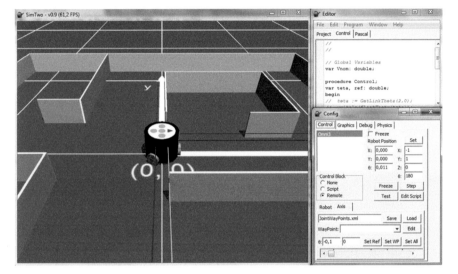

Fig. 2 SimTwo Realistic Simulator.

2   Hacked Neato XV-11 Laser Scanner Description

As described by *Konolige et al.* in [8], the Neato XV-11 laser scanner, shown in Figure
3a, is a low-cost laser scanner equipped with features like eye-safe, fully functional
in standard indoor lighting conditions and some outdoor conditions, it is small sized
and has a low power consumption. Instead of using time of flight measurement, like
the more expensive laser scanners, it uses triangulation to determine the distance
to the target, using a fixed-angle laser, a CMOS imager and a DSP for subpixel
interpolation [9].

The sensor establishes a serial communication with a 115200 bps baudrate, sending data up to about 5 Hz. Its power consumption without motor is relatively low:
~145 mA @ 3.3 V, which is a very important factor in order to increase the autonomy
of a mobile robot with its power based only on the on-board batteries.

It provides a 360° range of measurements, with an angular resolution of 1°, with its range from 0.2 m up to 6 m with an error inferior to 0.03 m. When the laser scanner is removed from the Neato XV-11 robot, its motor has to be controlled by the user, being necessary to be powered with 3.0 V continuous voltage (~60 mA), in order to produce a turn rate of 240 rpm. Typically it is used a voltage regulator to obtain the 3.0 V. Although this approach is the most popular, it is not the most efficient, because it is an open loop control, being observed oscillations in the motor velocity. An alternative to the referred approach is the use of the turn rate information contained in the data to close the loop [3] [8].

In this project the motor was controlled in closed loop. To control and to obtain measurements of the hacked Neato laser scanner, it was used an Arduino Due, which provides the 3.3 V requested by the laser scanner and can establish the needed serial communication. The data packet sent by the sensor is composed by a start header, an index byte, the motor speed (V_n), the laser measured data and a checksum.

Using the received motor speed data the control loop is closed by calculating the error relative to the speed (V), needed to maintain the laser frequency up to 5 Hz (5 Hz @ 240 rpm). Posteriorly the error is passed by an integrative like filter, resulting in a PWM control signal, which actuates on a N-Channel Mosfet powering the motor. In Figure 3b it is shown the control loop diagram.

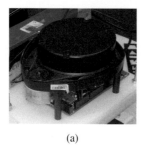
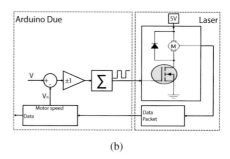

(a)   (b)

Fig. 3 (a) Hacked Neato XV-11 laser scanner. (b) Laser scanner motor closed loop control.

3 Laser Scanner Modeling

3.1 Experimental Setup

To model the hacked Neato XV-11 Laser Scanner an experimental setup was developed in order to obtain several measurement datasets. The data was obtained with the goal of extracting information about the sensor minimum and maximum ranges and its measurement error.

Initially it was printed an A4 sheet with a target with three areas, the center area white and the two sides areas in black, shown in Figure 4a, as a way to also assert the color influence in the measure chain. It also can be seen in Figure 4a the robot prototype. This has a square shape with the laser(hacked from Neato XV-11) centered in the front side.

To ensure that the laser angle stayed in the same position, the prototype was placed parallel and against a wall, distanced from the corner 6.2 m. This way the laser remained static during the experiments, while the target moved perpendicular to the wall and parallel to the prototype front. As for the target positioning, it was centered with the laser scanner.

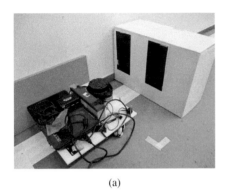 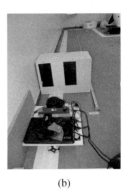

(a)   (b)

Fig. 4 (a) Alignment between robot and target. (b) Experimental setup layout example.

During the experiment it were obtained 44 datasets with different distances from the target to the laser scanner. The measurements were taken from 0.15 to 6 m, with the step sizes listed in Table 1.

Table 1 Measure datasets distance and respective step size

Distances (m)	Step size (m)
0.15 - 0.6	0.05
0.6 - 2	0.1
2 - 6	0.2

In each dataset it were obtained 128 samples of the 360° scans. An example of a dataset extraction experimental layout is shown in Figure 4b.

3.2 Laser Scanner Model

The datasets obtained via the experimental setup previously described, show the minimum measure range is lower than the described in [8], being of 0.15 m instead of the 0.2 m. Nevertheless, despite the use of close loop control of the motor to ensure the rotation speed to obtain the frequency required by the laser scanner, measurements above 5 m suffer frequent data loss, which results on inconclusive data to support the model, so it could not be considered in the model development. The maximum range described in [9] is 6 m ,despite sometimes data being retrieved suffer from frequent missed measures, reducing the usable range to 5 m.

For each dataset, 128 samples were taken and the mean value and standard deviation of each dataset were calculated in order to obtain the laser scanner model and the noise estimation to apply in the simulator.

It was observed that obstacle color did not influenced the measured distances, this becomes one advantage when compared with laser scanners based on different technologies [12].

In Figure 5a are shown the mean values of the samples of each dataset relative to the real distance. As it can be seen in Figure 5b the laser scanner measurements tend to increase the error with the distance, reaching values up to 0.54 m at 5 m measurements to the object. This brings the need for sensor calibration, which will be presented at Section 5, in order to retrieve a more accurate measure of the real distance to the obstacle.

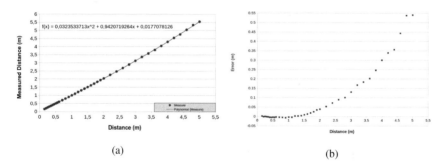

(a)   (b)

Fig. 5 (a) Distance measured by the laser scanner. (b) Experimental setup layout example.

Using a second order polynomial regression, as shown in Figure 5a represented by a red line, a viable function is obtained to model the sensor measurements (D_l) relative to the real distance (D). This way the model can be defined by equation 1, to apply posteriorly onto a simulator, in order to retrieve the Neato XV-11 laser scanner distances.

$$D_l(D) = a_1 * D^2 + a_2 * D + a_3 \tag{1}$$

with
$a_1 = 0.0323533713$
$a_2 = 0.9420719264$
$a_3 = 0.0177078126$

The standard deviation of each dataset, shown in Figure 6a, has with an exponential behavior, increasing with the distance, getting more noticeable after 2.4 m, and reaching values up to 0.072 m at an obstacle distance of 5 m. The standard deviation obtained for each distance is applied to model the sensor noise.

In Figure 6b is shown the noise using a logarithmic scale, being applied an exponential regression to the standard deviation values (σ), obtaining an approach of the measures noise, represented by the Equation 2.

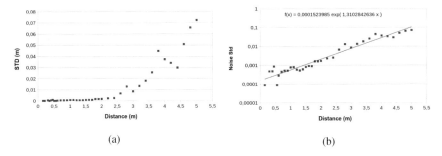

(a)   (b)

Fig. 6 (a) Standard deviation from each dataset. (b) Measurement noise standard deviation.

$$\sigma(D) = b_1 * e^{b_2 * D} \tag{2}$$

with
$b_1 = 0.0001523985$
$b_2 = 1.3102842636$

To display a more accurate model of the laser scanner, the measurements obtained by the equation 1 should be disturbed by Gaussian white noise, with a standard deviation given by the equation 2.

An example of the laser scanner measure histogram is shown in Figure 7a. In order to demonstrate that sensor provides data with a gaussian probability distribution it was used the normal probability plot, which is a graphical technique for assessing whether or not a data set is approximately normally distributed. The data is plotted against a theoretical normal distribution in such a way that the points should form an approximate straight line. Departures from this straight line indicate departures from normality [5]. The points in Figure 7b form a nearly linear pattern, which indicates that the normal distribution is a good model for this data set. The effect of the discretization on the measurement is also observable.

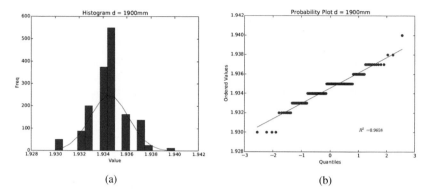

(a)   (b)

Fig. 7 (a) Laser scanner measure probability distribution. (b) Normal probability plot.

4   Laser Scanner Simulation

4.1   SimTwo Simulator

SimTwo is a simulator that allows the modulation and test in three-dimensional environments and using different scenarios. It can simulate several types of robots, like industrial manipulators, conveyors and mobile robots. Also makes a realistic approach allowing to model surface friction, mass and shapes as well as models which approximates and captures non-linear elements that can be used to represent sensors and actuators.

Due to the possibility of developing simulations of different robotic applications that use a laser scanner, SimTwo presents itself as a potential simulator candidate to implement the hacked Neato XV-11 laser scanner model. This way the XV-11 can be modeled and tested on a simulation environment to demonstrate the achieved results.

This simulator allows the creation of graphical scenarios using *XML - eXtensible Markup Language*, existing several, already developed, description tags. One of them is a two-dimensional laser scanner, which can retrieve ideal measures, by using the tag presented in Algorithm 1.

In Figure 8 it is shown an example of a Neato laser scanner applied on an omni-directional robot, applied to determine the robot localization.

4.2   Laser Scanner Simulated in SimTwo

In Section 3, the hacked Neato XV-11 laser scanner model and its standard deviation equations are described, becoming possible the simulation in SimTwo.

Algorithm 1. XML Ideal laser scanner tag

```
<sensors>
  <ranger2d>
    <ID value='ranger2d'/>
    <period value='0.1'/>
    <beam length='6' initial_width='0.01'
          final_width='0.015'/>
    <pos x='0.1' y='0' z='0.09'/>
    <rot_deg x='0' y='0' z='0'/>
    <tag value='0'/>
    <beam angle='360' rays='360'/>
    <noise stdev='0.0' stdev_p='0'
           offset='0' gain='1'/>
    <color_rgb r='255' g='0' b='0'/>
  </ranger2d>
</sensors>
```

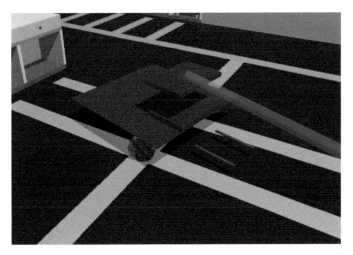

Fig. 8 Neato XV-11 laser scanner model in SimTwo.

As shown in the block diagram presented in Figure 9, the simulator, SimTwo, provides the ideal sensor measurements (D) needed to obtain the measurements model (D_l) of the laser scanner and the standard deviation (σ).

The standard deviation is used in the definition of the Gaussian white noise (N) which will be added to the measurements D_l, obtaining a simulated laser scanner data (D_s) similar to the real sensor.

The definition of the laser scanner model in SimTwo is made by accessing the ideal measurements obtained by the laser scanner tag described in the Algorithm 1, *ranger2d*, and to each beam apply Equations 1 and 2, in order to obtain the distance value and standard deviation, respectively. To achieve that, the following Pascal function can be applied onto the SimTwo control loop script, as shown in Algorithm 2.

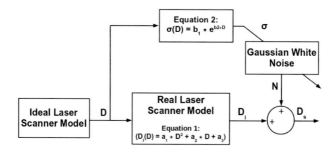

Fig. 9 Block diagram of the laser scanner integration on SimTwo.

Algorithm 2. Pascal function implementing the model

```
function LaserModel(Robot, Laser:Integer;
               a1, a2, a3, b1, b2 : double): Matrix;
var
  D, N, sigma, Dl : double;
  Laser_scan, Ds : Matrix;
  i : integer;
begin
  Laser_scan := GetSensorValues(Robot, Laser);

  Ds := Mzeros(MNumRows(Laser_scan),
             MNumCols(Laser_scan));

  for i:= 0 to MNumRows(Laser_scan)-1 do begin
    D := Mgetv(Laser_scan,i,0);
    Dl := a1*D*D + a2*D + a3;
    sigma := b1*exp(b2*D);
    N := RandG(0,sigma);
    Msetv(Ds, i, 0, Dl + N);
  end;

  Result := Ds;
end;
```

---

5 Laser Scanner Calibration

As seen in Section 3, the error for the hacked Neato XV-11 laser scanner increases when the distance to the object increases. So in order to increase the laser measurement accuracy a non linear function was used as a calibration step, using the data obtained through the experimental setup.

This way it was developed a function, which is the inverse of Equation 1, to convert the sensor measurements to more accurate distances (D_{cal}). The laser calibration is shown in equation 3.

$$D_{cal}(D_l) = c_1 * D_l^2 + c_2 * D_l + c_3 \tag{3}$$

with
$c_1 = -0.0242594$
$c_2 = 1.03703951$
$c_3 = -0.0086895$

After applying the calibration equation to the laser data the measurement error
was reduced, From the one that was shown in Figure 5b, the error was reduced to a
maximum of 0.035 m, as shown in Figure 10.

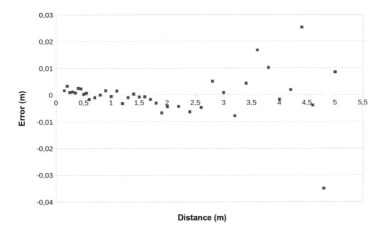

Fig. 10 Distance error after applying the calibration.

6 Conclusions and Future Work

Neato XV-11 is a robot that includes a low cost 360° laser scanner, this sensor can
be extracted from the robot, allowing robotics practitioners to use it in their projects.
The Neato XV-11 laser scanner is a very low cost alternative, when compared to the
current available laser scanners.

   In this paper was presented a study concerning with the accuracy the hacking, modeling and simulation of the Neato XV-11 laser scanner. The modeling of a hacked
Neato XV-11 Laser Scanner allows realistic simulations providing valuable information that can promote the development of better robot localization systems based
on this sensor. The laser was simulated in the SimTwo realistic simulator.

As future work the authors intend to evaluate and characterize localization systems based on the Neato XV-11 Laser Scanner, using simulation and real robots.

Acknowledgements This work is financed by the ERDF – European Regional Development Fund through the COMPETE Programme (operational programme for competitiveness) and by National Funds through the FCT – Fundaçãão para a Ciência e a Tecnologia (Portuguese Foundation for Science and Technology) within project «*FCOMP-01-0124-FEDER-037281*».

References

1. Arras, K., Tomatis, N.: Improving robustness and precision in mobile robot localization by using laser range finding and monocular vision. In: 1999 Third European Workshop on Advanced Mobile Robots, Eurobot 1999, pp. 177–185. IEEE (1999)
2. Surmann, H., Nuchter, A., Joachim, H.: An autonomous mobile robot with a 3D laser range finder for 3D exploration and digitalization of indoor environments. Robotics and Autonomous Systems **45**(3), 181–198 (2003)
3. Hacking the Neato XV-11 (2015). https://xv11hacking.wikispaces.com/
4. Costa, P., Gonçalves, J., Lima, J., Malheiros, P.: SimTwo realistic simulator: A tool for the development and validation of robot software. International Journal of Theory and Applications of Mathematics Computer Science (2011)
5. Chambers, J., Cleveland, W., Kleiner, B., Tukey, P.: Graphical Methods for Data Analysis. Wadsworth (1983)
6. Browning, B., Tryzelaar, E.: UberSim: a realistic simulation engine for robotsoccer. In: Proceedings of Autonomous Agents and Multi-Agent Systems (2003)
7. Michel, O.: $Webots^{TM}$: Professional Mobile Robot Simulation. International Journal of Advanced Robotic Systems **1**(1) (2004). ISSN 1729–8806
8. Konolige, K., Augenbraun, J., Donaldson, N., Fiebig, C., Shah, P.: A low-cost laser distance sensor. In: IEEE International Conference on Robotics and Automation, ICRA 2008, pp. 3002–3008, May 19–23, 2008
9. Shah, P., Konolige, K., Augenbraun, J., Donaldson, N., Fiebig, C., Liu, Y., Khan, H., Pinzarrone, J., Salinas, L., Tang, H., Taylor, R.: Distance sensor system and method, US 2010/0030380 A1 (2010)
10. Neato Robotics XV-11 Tear-down (2015). https://www.sparkfun.com/news/490
11. Bajracharya, S.: BreezySLAM: A Simple, efficient, cross-platform Python package for Simultaneous Localization and Mapping. Washington Lee university (2014)
12. Lima, J., Goncalves, J., Costa, Paulo., Moreira, A.: Modeling and simulation of a Laser Scanner: An industrial application case study. FAIM (2013)
13. Lima, J., Goncalves, J., Costa, P.: Modeling of a low cost laser scanner sensor. In: Proceedings of the 11th Portuguese Conference on Automatic Control, CONTROLO 2014 (2014)

Appearance Based Vehicle Detection by Radar-Stereo Vision Integration

Marko Obrvan, Josip Ćesić and Ivan Petrović

Abstract This paper proposes a novel method for appearance based vehicle detection by employing stereo vision system and radar units. In the vein of utilizing advanced driver assistance systems, detection and tracking of moving objects or particularly vehicles, represents an essential task. For the merits of such application, it has often been suggested to combine multiple sensors with complementary modalities. In accordance, in this work we utilize a stereo vision and two radar units, and fuse the corresponding modalities at the level of detection. Firstly, the algorithm executes the detection procedure based on stereo image solely, generating the information about vehicles' position. Secondly, the final unique list of vehicles is obtained by overlapping the radar readings with the preliminary list obtained by stereo system. The stereo vision–based detection procedure consists of (i) edge processing plugging in also the information about disparity map, (ii) shape based vehicles' contour extraction and (iii) preliminary vehicles' positions generation. Since the radar readings are examined by overlapping them with the list obtained by stereo vision, the proposed algorithm can be considered as high level fusion approach. We analyze the performance of the proposed algorithm by performing the real-world experiment in highly dynamic urban environment, under significant illumination influences caused by sunny weather.

Keywords Vehicle detection · Stereo vision · Radar · Fusion

M. Obrvan · J. Ćesić(✉) · I. Petrović
Faculty of Electrical Engineering and Computing, Departement of Control and Computer
Engineering, University of Zagreb, Unska 3, Zagreb, Croatia
e-mail: josip.cesic@fer.hr

I. Petrović—This work has been supported by the European Regional Development Fund under the project Advanced technologies in power plants and rail vehicles.

L.P. Reis et al. (eds.), *Robot 2015: Second Iberian Robotics Conference*,
Advances in Intelligent Systems and Computing 417,
DOI: 10.1007/978-3-319-27146-0_34

437

1 Introduction

Accurately comprehending the environment under various conditions is an essential task in nearly any computer vision application. Among the applications of interest, the research community has engaged significant resources in developing advanced driver assistance systems (ADAS). These applications involve adaptive cruise control, collision avoidance, lane change assistance, traffic sign recognition, parking assistance, and many others foremostly aiming towards fully autonomous driving. ADAS have been in the focus of the research for a few decades, intending to enhance the safety and reduce the possibility of a human error as a cause of road accidents [1]. Considering the employment of ADAS applications, the detection and tracking of moving objects (DATMO) in the surrounding of a vehicle represents an essential task. DATMO allows platform to be aware of dynamic objects around it and predicting future behaviour of tracked entities. Since the robustness of such application under various environmental conditions represents a pertinent feature, it has become clear there exists no such sensing system that could solely deliver full information required for adequate quality of ADAS applications [2]. Hence, the combination of multi-modality sensors is further discussed.

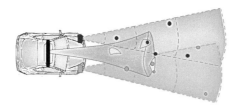

Fig. 1 Experimental platform equipped with a stereo system (green and blue) and two radar units (orange).

ADAS usually rely on utilization of different sensing systems, and accordingly great deal of works rely on employment of either (i) vision-based systems, (ii) millimeter-wave radars or (iii) lidar units. Seeking for the sensing system being able to produce accurate measurements of relative speed or deliver the information regarding the distance to an object, one comes to the possibilities of utilizing either lidar or radar sensor unit. Regarding the robustness to a weather conditions, radar units are much more robust to rain, fog, snow, and other appearances that may cause significant troubles when employing a laser range sensor units, but produce significant size of a clutter as a drawback. On the other hand, lidar sensors can detect the occupancy area of an object due to higher lateral resolution much better than radars, while radars have much lower average price with respect to lidars. As opposed to radars, vision based sensing systems have much wider field of view and more accurate lateral information, and therefore provide an effective supplement to radar-based road scene analysis. As a result, a stereo vision sensor can provide target detection with high azimuth resolution, while usually bringing enough information for identification and classification of objects in the surrounding, whereas radar provides a measurement of an accurate range and relative speed information. Therefore, in this work we utilize the integration of radar units and stereo vision for carrying out the moving objects detection task.

Employing multiple sensors and consequently exploiting their different modalities thus extracting comparative advantages, inherently requires to perform the fusion between sensing systems at appropriate levels. To endow an optimal gain, the fusion can take place at three levels; (i) before objects detection (low level) [3, 4], (ii) at objects detection level (fused list of objects) [5, 6] or (iii) at track level (updating the states of objects in the list for each sensor system) [7, 8].

In this work, we propose a novel algorithm which relies on fusion of complementary modalities of radar units and stereo vision. The fusion is performed at the objects detection level, providing the unique list of moving objects in the surrounding of a vehicle at every discrete time step. The algorithm firstly processes the stereo image based on appearances, and afterwards incorporates the radar readings by overlapping the sets of detections of both sensors. The stereo vision–based detection algorithm consists of

(i) *Edge processing step*. It strongly relies on careful edge filtering approach, as well as the information obtained by utilizing the disparity map computation.
(ii) *Vehicles' contour extraction*. Relying on the assumption the edges belonging to a vehicle are either situated vertically or horizontally within the image, the contours are carefully segmented.
(iii) *Dimensionality check*. If some inappropriate dimensions show up, this step is to ensure the stereo vision–generated clutter is significantly smaller than radar–based one, thus further exploiting the complementary multi-modalities.

In our application, we have utilized *Bumblebee XB3* stereo vision system and two *Continental SRR 208* short range radar units, as is illustrated in Fig. 1.

2   Related Work

Hereafter, we will focus our overview of recent results in the field of detection of moving objects by considering the results relevant for the nature of our particular application. It involves the detection methods focused on integration of radars and stereo vision systems, but alongside we also provide the brief description of methods that rely solely on vision based sensing. The reasoning for reviewing it stems from the fact that our approach relies on fusion that takes place at the object detection level. Apparently, since the primal hypotheses of detections are based on mere stereo vision, it is relevant to consider such single sensor algorithms to some extent as well.

Vision-Based Detection. Respecting the nature of the taken approach for utilization of detection task, an algorithm either relies on appearances at a single time step, or on motion over more frames. These two are often referred to as (i) appearance and (ii) motion based detection approaches [2]. The moving objects detection approaches may also be divided by the environment for operating in, hence they are optimized for either indoor [9] or outdoor [10–13] operation.

The work in [11] relies on perceiving the environment by tracking a temporal motion of detected corners based on pyramidal Lucas-Kanade algorithm. The scene flow computation based detection is also an often employed motion based approach [12, 13]. In [12], for every interest point, disparity and optical flow values are known and consequently, scene flow is calculated. Adjacent interest points describing a similar scene flow are considered to belong to one rigid object. In [13] authors partially decouple the depth estimation from the motion estimation, thus endowing many practical advantages. The dense scene flow estimation is a common approach for extraction of moving objects in indoor environments [9].

As opposed to mono vision systems that rely on HOG SVM [14], Random forest of local experts [15] and others, the appearance based stereo vision approaches are not as widespread. Still, the work in [16] employs the detection procedure based on appearances in disparity space, where both clustering and extraction of moving objects are performed. In [17], the objects are extracted based on computation of stereo and optical flow, thus employing the scene flow information. Upon this, the reliable depth information and the motion of investigated scene are exploited. Another similar approach, given in [10], combines depth-based and optical-flow-based clustering with active learning-based method, which is usually employed in mono vision systems.

Radar-Vision Integration Based Detection. An instance of low level integration of radar and vision was utilized in [4]. Therein, a method is based on fitting the model of a contour of a vehicle to both stereo depth image and radar readings, and has revealed to result in satisfactory quality of detection. Another low level integration algorithm was presented in [18]. In particular, that work provides a strategy that actively introduces the radar based depth information into the image segmentation step. The idea is to split an edge map of a binocular image into edge layers corresponding to different target depth information so that different layers contain the edge pixels of targets at different depth ranges.

Many authors have emphasized the complementarity of stereo vision and radar [4, 6, 19], but have also polemicized the cost of between affordability, complexity and usefulness of mono vision with respect to binocular vision. Therefore, in [19] and [6] the authors propose a potential solution to this problem by using of motion stereo which recovers the three dimensional structure information from motion itself, by employing monocular vision. In [19] the objects are firstly detected with the radar, while vision based feature points are carefully selected and tracked through frames. Alongside, the distance is estimated from motion in the image, employing motion stereo technique. A similar approach is used in [6], where the camera is mainly used to refine the vehicles' boundaries detected by radar, for discarding those detections that might correspond to false positives. In that work, instead of tracking features through frames, an inertial sensor is used in order to improve the ego-motion estimation.

In [20], the authors have proposed an approach to a vehicle detection by using radar observations for localizing areas of interest in images. Afterwards, the objects

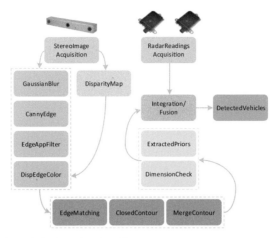

Fig. 2 The flow chart of the detection algorithm.

are extracted mainly based on vertical objects' symmetry. Nevertheless, significant differences wrt [20] are met in a fusion strategy utilized through three level fusion approach consisted of: a radar-vision point alignment, region searching for potential target detection, and real obstacle determination with adaptive thresholding algorithm.A completely independent detection procedures for two sensor types were used in [5]. Herein, a complete software architecture for a frontal object perception task is presented, where both camera and radar sensors were processed separately delivering two lists of objects. These are then merged employing a Dempster-Shafer theory.

Since the detection procedures often serve as input for tracking applications, we direct the interested reader to refer the works presented in [8] and [7]. Both these are oriented towards efficient state estimation and tracking of multiple moving objects in the surrounding of a vehicle, while not discussing the detection procedures into details.

3   Appearance Based Detection Procedure

As previously introduced, the entire detection process relies on detecting the preliminary assumptions about vehicles' location in the surrounding of a platform by employing the stereo vision system. Afterwards, such generated data is overlapped with the radar readings, and the unique list is hence obtained. The flow chart of the proposed algorithm is given in Fig. 2. The entire algorithm can be partitioned by purpose into a few fragments. In particular, this involves (i) edge processing, (ii) vehicles' contour extraction, (iii) shape related dimensionality check and (iv) radar readings integration. Obviously, the first three steps correspond to stereo image processing, while the radar readings endow the algorithm at the very last step. We now proceed with detailed descriptions of each segment of this algorithm, starting with edge processing.

3.1 Edge Processing

Gaussian Blur. Before any further processing, Gaussian blurring is applied over the stereo image acquired with *Bumblebee XB3*. Blurring itself is resource demanding, since relying on usage of $N \times N$ kernel, $N = 2n + 1$, $n \in \mathbb{N}_0$, for each pixel. However, it is an indispensable step. By employing this step, some insignificant or noise–generated edges are removed. Since the detector utilized further in this work, i.e. Canny edge, the size of the kernel is set to 3×3. By this, most of edges remain uninfluenced by the filter, while large monotonic are smoothed. After kernel is applied over the image, segments of road with peaks in color will no longer be detected by edge detector. Still, true objects appearing in traffic related environments such as lanes, sings or shadows, could not be filtered by only blurring the image.

Canny Edge Detection. For detecting edges in blurred image, we have employed the popular Canny Edge Detector [21]. Its main features can be expressed via following:

– *Low error rate*. Only existing edges can be detected–no false positives.
– *Good localization*. The distance between real edge and detected edge pixel tends to be minimized.
– *Minimal response*. Maximally one detector response per edge.

Good localization of Canny detector is backbone of algorithm, since there is need for overlapping disparity image with edges during different steps of detection for finding matching disparity of edges. Bad localization could lead to obtaining wrong disparity, which may later lead to wrong mapping in 3D. Since most of filters will use 3D information, it is of our huge interest to obtain good corresponding disparity of each edge. From now on, when referring to edge image processing, we consider the left image only, rather than repeating the same procedure for both images.

Edge Filtering. After applying Canny edge detector for edge extraction, significant noise is still present in the edge image, mostly in form of small and sharp edges. Hence, here we provide an algorithm for edge image filtering based on shape of edges. The procedure starts with scanning through image, detecting and removing small edges, while keeping larger ones and smoothing them. This can be summarized in the following:

– Remove small/insignificant edges from the edge image,
– Remove all insignificant intersections and
– Smooth significant edges.

Let us define P and P^e, as the sets of all pixels and edge pixels in edge image, respectively. The pseudo code of the filtering approach is presented in Algorithm 1, which involves the recursion *findLongestPath()* presented in Algorithm 2. The illustrating effect of applying edge filter is presented in Fig. 3. The most left image represent the original image, while all succeeding images show the results after detecting and deleting edges with low pixel length. The most right image shows the output of filter after small intersecting edge is removed from largest edge, whose length is greater than threshold.

Algorithm 1. Edge Filtering

0: **Require:** P, P^e, l_{thrsh}
0: **Initialization:** $P^{e'} = \emptyset$
1: **Iterate** through each P
2: $l = findLongestPath(P_i, l)$
3: **if** $l < l_{thrsh}$ **then**
4:    delete edge and restore deleted intersections in P^e
5: **else**
6:    mark edge as visited and add edge pixels to P^e
7:    mark intersections as unvisited so other edges can walk over them
8:    save intersection so they can be restored later
9: **end if**
10: **return** $P^{e'}$

Algorithm 2. Find Longest Path

0: **Require:** P_i, P^e
1: **if** P_i belongs to P^e and P^e not visited **then**
2:    **find** all neighbours $P^{e'}$
3:    **for** for each $P^{e'}$ **do**
4:       **while** exists unvisited $P^{e''}$ neighbour of $P^{e'}$ **do**
5:          increase edge length for corresponding direction $P^{e'}$
6:          move to $P^{e''}$ and mark $P^{e'}$ as visited
7:          **if** $P^{e'}$ is intersection **then**
8:             $l+ = findLongestPath(P_i^{e'}, l)$
9:             save intersection pixel $P_i^{e'}$
10:          **end if**
11:       **end while**
12:    **end for**
13:    discard all but longest edge and its directions from current iteration
14:    **return** l
15: **end if**

Fig. 3 The edge filtering result. From the left, first image is initial scene. Firstly, top left closed edge is detected and removed due to length. Secondly, right most edge is removed. Thirdly, small edge intersecting significant edge is removed. Result is noise free scene with smooth edge.

Edge Colouring. Disparity map plays vital role in determining position of detected edges in 3D scene. However, we utilize the sparse disparity map computation approach, and hence there are some points in disparity map that do not contain disparity information. Furthermore, one edge can match different disparity values in disparity map. Inconsistency in disparity map is especially strong at borders of two areas, usually background and foreground, where features for image comparisons are strong, but due to inaccuracy of disparity computation, both background and foreground

M. Obrvan et al.

could be consolidated under same disparity value. In most cases background object are labelled with foreground object disparity, since closer object naturally generate more robust features during stereo matching. In order to correctly join the disparity information with each edge, we have employed the following algorithm. Let's define following:

- Let E be a union of all valid edges $e_0, ..., e_n$,
- Let p_{ik} be pixel of edge e_k in range $p_0, ..., p_z$,
- Let d_{ik} be value in disparity map overlapping with p_{ik} coordinate,
- Let D_{ek} be histogram of all disparities within edge e_k with values in range $1, ..., 255$.

$$\forall p_{ik} \in e_k \wedge \exists(d_{ik}! = 0) \rightarrow D_{ek}(d_{ik}) + 1. \tag{1}$$

Algorithm walks over all pixels p_{ik} of edge e_k and checks for disparity existence at coordinates d_{ik}. If the one exists, i.e. not equal to zero, increase the number of its appearances in histogram D_{ek}. Once the histogram is created, disparity with maximum number of occurrences is considered as disparity for edge e_k. After disparity is determined, corresponding edge e_k is coloured with the same color as disparity value shifted in grayscale space. Since the disparity can range in $-1, ..., 127$ where -1 stands for no disparity detection, it is converted to grayscale range $0, ..., 128$ unsigned character, where 0 stands for no disparity detected. Now we can proceed with the contour extraction.

3.2 Contour Extraction

Edge Matching. After colouring edges upon the corresponding disparity, there exists just enough information for matching edges. There are three parameters used for extracting contours:

- Length of edge being grouped,
- Disparity match with neighbouring edge,
- Distance between edges in pixels.

The algorithm itself can be divided into four steps:

1. **Find** leftmost and rightmost point of edge,
2. **Find** length of current edge,
3. **Check if** there are any neighbouring edges within the range of leftmost and rightmost pixel,
4. **If** a neighbour is detected
   then compare disparities via following:

   (a) **If** disparities are within bounds
       then connect edges,
   (b) **If** disparities are over higher bound
       then stop searching,

Else continue searching and decide if eventually connecting edges based on succeeding detections. In case there is no other detections, connect edges.

Since determined length of edges greatly affects ability of forming closed contours during edge matching, the edges with larger length are supposed to search for their pairs further away than the short edges are. In our case, we have set the searching range as one third of length of an edge in pixels. Search path is simply computed as verticals in leftmost and rightmost pixel. Once they are obtained, all neighbours could be found as intersections between vertical and any other edge in computed range. Since every edge possesses corresponding disparity information, its 3D position can be determined. Once it is obtained, 3D position of an edge while its neighbours can be used for deciding whether connecting edges or not.

Closing Contours and Merging. Once edge matching is performed, we have used an openCV implementation for finding all contours in the image (OpenCV function called *findContours()*). Function parameters are set to retrieve all the contours and reconstruct a full hierarchy of nested contours. Using obtained information about contours, the closed contours as candidates for objects in the scene are extracted, based on their hierarchy level. During this step, most of the contours are discarded based either on their too small or too large area. Furthermore, most of nested and open contours are discarded, while clusters of overlapping contours are merged, providing us with small set of candidates that need to be further considered.

3.3 Dimensionality Check

At this point, each significant contour in the scene possesses the bounding box. It is also possible to join the disparity information to each bounding box. There are two possible ways to obtain this disparity–either from any sample point belonging to particular contour, or by looking for disparity values within bounding box and finding best matches. In this case we have utilized the latter.

Choosing a sample point on a contour might yield better results for an area in its surrounding, but since edges with different disparities are eventually merged together, this could cause inaccuracy in the approach. It is considered being more relevant to assume the rectangles are bounding boxes for objects, hence the area around its center has the average disparity value of entire object. Using detected disparity, all the means for calculating 3D position of bounding rectangles are determined.

We have chosen four key points for calculating depth, lateral translation and height of an object with respect to the center of stereo system, i.e. top left corner, bottom left corner, top right corner and central point. Three corners are used to calculate rectangle's width and height in 3D world. Based on the assumed dimensions of vehicles, it is possible to discard all bounding boxes with unrealistic heights and widths. Since camera is well calibrated, it is satisfactory to assume 3D coordinates of ground level and ignore all detections that deviate much from it. However, very few contours should be discarded this way, since the algorithm needs to be as robust to platform movements as possible, hence the threshold parameters need to be chosen carefully.

Once all false detections are discarded, remaining rectangles are considered as being primers for radar detections. Their 3D positions are calculated as the middle point of a rectangle. Obtained coordinates represent input points for radar-stereo vision integration step.

3.4 Radar Integration

The integration is performed such that the candidates from mere stereo vision–based detection are matched with radar readings. If match is established, we can assume with great certainty that object is correctly detected (blue rectangle box in 4). If match is not found, data is not discarded (green rectangle in 4), but rather considered as weak detection. Due to the complementarity of the used sensor systems, it is even more appropriate to keep both detections and involve such information into a multi-target tracking estimation framework. However, this topic is out of scope of the paper and as such represents a challenging topic for further research.

Considering the synchronisation aspect, we have employed a synchronization tool from the Robot Operating System [22], where all the algorithms were implemented and tested in. The calibration of the system involving several additional sensor systems is the topic of the future work.

3.5 Experimental Results

The experiments are performed in a challenging highly dynamic urban environment under very challenging illumination conditions (strong sun). The results for several consecutive frames are given in Fig. 4. Therein, the results of the detection procedure are given for four consecutive frames. The first row represents the original images, the second row provides outputs from Canny Edge Detector, the third row represents the disparity images, the fourth row provides edge maps after utilizing previously presented edge filtering algorithm based on disparity map, the fifth row are edge images connected based on disparity of neighbouring edges, while the last row gives the final detections of vehicles. Considering the results, some mistakes can be noticed. Firstly, it can be seen that at some points the radar readings did not overlap with the stereo generated primers (time frame 1 and 2) due to radar misdetections. Secondly, at the fourth frame it can also be seen that a false negative has appeared since a car was situated very close to a passing tram. Thirdly, a few false alarms were caused by the tram itself. However, the results of the detection procedure seem to be very promising for incorporating this algorithm as part of detection and tracking procedure, which is also in the focus of our future work.

Experiments were running at 2.6 Ghz $i7$ processor with turbo 3.5 Ghz, while only one core was used during computation. The execution was lasting for between 120 and 170 ms, mostly depending on the complexity of a scene. Furthermore, the algorithm is yet not fully optimized and some needless floating point operations could be filtered out from the code. It is also possible to achieve even greater speed-ups by enabling hyper-threading and simultaneously detecting and removing edges, which turned out to be a time critical part of the algorithm, taking up from 30 and 70 ms.

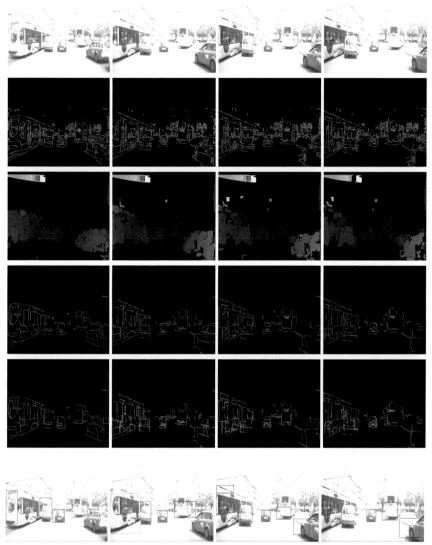

Fig. 4 Vehicle detection procedure over four consecutive frames. Row 1 represents the original images. Row 2 provides outputs from Canny Edge Detector created over original images. Row 3 are disparity images. Row 4 represents edge maps after edge filtering. Row 5 are edge images, where edges are connected based on disparity of neighbouring edges, which serves as template for finding contours. Hence the candidates for bounding rectangles are obtained, and afterwards checked if fitting the expected vehicle dimensions. Finally, the rectangles are drawn over remaining closed contours and images in row 6 are obtained.

4 Conclusion

In this paper we have presented an algorithm for appearance based vehicle detection by employing stereo vision system and two short range radar units mounted on a platform vehicle. The detection procedure was independent for two types of sensors, whence the fusion can be considered as performed at the level of detection. The stereo vision–based detection procedure consisted of (i) Canny edge detection procedure and processing based on disparity map information, (ii) shape based vehicles' contour handling and extraction and (iii) primer vehicles' positions generation. The radar integration was performed by overlapping the sets of stereo vision–based primers and radar readings, thus providing weak and strong assumptions about the positions of vehicles. The results showed that the proposed algorithm is capable of reliably performing the detection procedure based on appearances, without tracking them over time, hence representing a useful segment for endowing its future incorporation within the entire detection and tracking procedure.

References

1. Dickmanns, E.: The development of machine vision for road vehicles in the last decade. In: Intelligent Vehicle Symposium (IV), pp. 268–281. IEEE (2002)
2. Sivaraman, S., Trivedi, M.M.: Looking at vehicles on the road: A survey of vision-based vehicle detection, tracking, and behavior analysis. IEEE Transactions on Intelligent Transportation Systems 14(4), 1773–1795 (2013)
3. Wang, T., Zheng, N., Xin, J., Ma, Z.: Integrating millimeter wave radar with a monocular vision sensor for on-road obstacle detection applications. Sensors 11(9), 8992–9008 (2011)
4. Wu, S., Decker, S., Chang, P., Camus, T., Eledath, J.: Collision sensing by stereo vision and radar sensor fusion. IEEE Transactions on Intelligent Transportation Systems 10(4), 606–614 (2009)
5. Chavez-Garcia, R.O., Vu, T.-D., Burlet, J., Aycard, O.: Frontal object perception using radar and mono-vision. In: Intelligent Vehicles Symposium (IV), pp. 159–164. IEEE (2012)
6. Bertozzi, M., Bombini, L., Cerri, P., Medici, P., Antonello, P.C., Miglietta, M.: Obstacle detection and classification fusing radar and vision. In: Intelligent Vehicles Symposium (IV), pp. 608–613. IEEE (2008)
7. Liu, F., Sparbert, J., Stiller, C.: IMMPDA vehicle tracking system using asynchronous sensor fusion of radar and vision. In: Intelligent Vehicles Symposium (IV), pp. 168–173. IEEE (2008)
8. Richter, E., Schubert, R., Wanielik, G.: Radar and vision-based data fusion - advanced filtering techniques for a multi-object vehicle tracking system. In: Intelligent Vehicles Symposium (IV), pp. 120–125. IEEE (2008)
9. Herbst, E., Ren, X., Fox, D.: RGB-D flow: dense 3-D motion estimation using color and depth. In: Int. Conf. on Rob. and Autom. (ICRA), pp. 2276–2282 (2013)
10. Ohn-bar, E., Sivaraman, S., Trivedi, M.: Partially occluded vehicle recognition and tracking in 3D. In: Intell. Veh. Symposium (IV), pp. 1350–1355 (2013)
11. Lefaudeux, B., Nashashibi, F.: Real-time visual perception: detection and localisation of static and moving objects from a moving stereo rig. In: Intelligent Transportation Systems Conference (ITSC), pp. 522–527 (2012)
12. Lenz, P., Ziegler, J., Geiger, A., Roser, M.: Sparse scene flow segmentation for moving object detection in urban environments. In: Intelligent Vehicles Symposium (IV), pp. 926–932 (2011)

13. Wedel, A., Brox, T., Vaudrey, T., Rabe, C., Franke, U., Cremers, D.: Stereoscopic scene flow computation for 3D motion understanding. International Journal of Computer Vision **95**(1), 29–51 (2011)
14. Cao, X., Wu, C., Yan, P., Li, X.: Linear svm classification using boosting hog features for vehicle detection in low-altitude airborne videos. In: IEEE International Conference on Image Processing (ICIP), pp. 2421–2424 (2011)
15. Marin, J., Vazquez, D., Lopez, A., Amores, J., Leibe, B.: Random forests of local experts for pedestrian detection. In: 2013 IEEE International Conference on Computer Vision (ICCV), pp. 2592–2599 (2013)
16. Broggi, A., Cappalunga, A., Caraffi, C., Cattani, S., Ghidoni, S., Grisleri, P., Porta, P.P., Posterli, M., Zani, P.: TerraMax vision at the Urban challenge 2007. IEEE Trans. on Intell. Transp. Systems **11**(1), 194–205 (2010)
17. Barrois, B., Hristova, S., Wohler, C., Kummert, F., Hermes, C.: 3D pose estimation of vehicles using a stereo camera. In: Intelligent Vehicles Symposium (IV), pp. 267–272 (2009)
18. Fang, Y., Masaki, I., Horn, B.: Depth-Based Target Segmentation for Intelligent Vehicles: Fusion of Radar and Binocular Stereo. IEEE Transactions on Intelligent Transportation Systems **3**(3), 196–202 (2002)
19. Kato, T., Ninomiya, Y., Masaki, I.: An Obstacle Detection Method by Fusion of Radar and Motion Stereo. IEEE Transactions on Intelligent Transportation Systems **3**(3), 182–187 (2002)
20. Alessandretti, G., Broggi, A., Cerri, P.: Vehicle and guard rail detection using radar and vision data fusion. IEEE Trans. on Intelligent Transp. Systems **8**(1), 95–105 (2007)
21. Canny, J.: A Computational Approach to Edge Detection. IEEE Transactions on Pattern Analysis and Machine Intelligence **PAMI-8**(6), 679–698 (1986)
22. Quigley, M., Conley, K., Gerkey, B.P., Faust, J., Foote, T., Leibs, J., Wheeler, R., Ng, A.Y.: Ros: an open-source robot operating system (2009)

Vision-Based Pose Recognition, Application for Monocular Robot Navigation

Martin Dörfler, Libor Přeučil and Miroslav Kulich

Abstract This paper presents improvements made to previous method for monocular teach-and-repeat navigation of mobile robots. The method is based on recording the position of image features in camera image, and moving the robot so their position matches during the recall. The method has shown good reliability, though requires odometry to perform well. This paper targets improvements of the method by replacement of a simple odometry by visual pose recognition approach. Thus, localization becomes independent of preceding pose computation. This prevents accumulation of error during the run of the algorithm.

A pose recognition method based on angle differences is presented herein. The substitution of odometry implies necessary adjustments to the aforementioned method to be used. Suitability of the method for pose recognition is evaluated experimentally. The method has shown to be feasible for the nav task, although the achieved accuracy is lower than the original method.

Keywords Pose recognition · Vision-based · Robust image features · Monocular Localization and Navigation

M. Dörfler(✉)
Department of Cybernetics, Faculty of Electrical Engineering,
Czech Technical University in Prague, Technicka 2, 166 27 Prague 6, Czech Republic
e-mail: dorflmar@fel.cvut.cz

L. Přeučil · M. Kulich
Czech Institute of Informatics, Robotics, and Cybernetics,
Czech Technical University in Prague, Zikova 1903/4, 166 36 Prague 6, Czech Republic
e-mail: {preucil,kulich}@ciirc.cvut.cz

L. Přeučil and M. Kulich—This research was supported by the Grant Agency of the Czech Republic (GACR) with the grant no. 15-22731S entitled "Symbolic Regression for Reinforcement Learning in Continuous Spaces" and Technology Agency of the Czech Republic under the project no. TE01020197 "Centre for Applied Cybernetics".

L.P. Reis et al. (eds.), *Robot 2015: Second Iberian Robotics Conference*,
Advances in Intelligent Systems and Computing 417,
DOI: 10.1007/978-3-319-27146-0_35

451

1   Introduction

With increasing computational power making real-time image processing possible, multiple navigation methods based on vision have been investigated in the last two decades.

One of the most popular approaches is monocular SLAM (monoSLAM). A feature-based variant of monoSLAM extracts discrete feature observations from consecutive images and matches them in order to determine full pose of the camera and all the features themselves [5, 9, 17]. Dense monocular SLAM approaches work with raw images, model the environment as a dense surface and align these complete images to determine the camera pose [8].

Another stream takes advantage of a pre-learnt map built during a human-guided teleoperated drive. Afterwards, the robot is controlled by a kind of image-based visual servoing in the autonomous navigation mode, i.e. difference between the expected and current image is to be minimized [3].

For example, Bekris et al. [2] present a local control law for a homing strategy exploiting bearings of at least three salient features in a panoramic image of the scene. Determination of the actual control then relies on measuring the angle between pairs of features. Long-range homing is managed by extracting milestone positions on the trajectory with partially overlapping sets of observable features. The milestones are then traversed sequentially.

Another approach was used by Diosi et al. [6]. Path is stored as a sequence of reference images. Localization is performed by computing local geometry between current image and nearest reference frames. Navigation is simpler, relying on comparing expected and actual positions of detected landmarks in the current view and next reference image in sequence.

Similar approach is used SURFnav proposed by Krajník et al. [11]. In contrast to previous method, no attempt to reconstruct local geometry is performed. Learned trajectory is split into a sequence of linear segments. A lateral deviation during forward motion along a particular segment is reduced via comparison of 2D positions of currently observed features with positions of features expected to be visible in the pre-learnt map. A longitudinal position within the segment is determined from odometry.

Using these more relaxed constraints sacrifices localization of the robot, without impacting the ability to navigate the path. In return, it is possible to perform faster computation and use less reference frames, leading to sparser map and thus smaller representation. This makes the method suitable for use on resource-constrained on-board computers of smaller robots.

It has been proven, that the position error for closed polygonal trajectories does not diverge [10] with the following weaknesses:

1. The salient feature existence, their stability and recognition/matching may not be sufficient if displacement of the robot is not known with sufficient precision and confidence.
2. The limited relative dead-reckoning precision, which accumulates robot positioning errors along the trajectory may raise above the limits if the robot trajectory

does not exhibit sufficient curvature to keep the odometry error within certain bounds in all degrees of freedom.

Consequently, the improper robot positioning may lead to weaker determination of correspondences between subsequent observations (frames) in feature matching. This brings the method [10] to less precise performance as the trajectory is followed with lower precision. This may even lead to complete failure if the robot looses ability to observe features in the previous frame completely. In this aspect, the robustness of the robot dead-reckoning system stands crucial for stable performance of feature-based visual servoing approaches (*FB-nav*).

Nitsche et al. [15] overcome dead-reckoning precision by employing computationally efficient Monte Carlo localization to estimate the robot position within a segment relative to the segment start. Nevertheless, their approach still depends on odometry measurements.

The herein introduced approach addresses adjustment of positioning errors of the robot on the fly, using the same salient features only as applied in the *FB-nav* method itself (or a similar one, to limit the computational intensity of the process) without a need for odometry. Primarily, as the relative dead-reckoning error typically grows proportionally over time, or the path driven, a periodic (or even a steady) corrective process shall be elaborated. This keeps the positioning deviation within specific bounds and thus avoids failure of the method.

The paper is structured as follows: The next section explains the principles of the presented method and motivation for our approach. Section 3 deals with the algorithm itself. Necessary changes to *FB-nav* are presented, followed by explanation of the method we replace the dead-reckoning with, and technical details. The last part details the performed experiment and the results.

2   Motivation

The aim of the work is to elaborate a method for robust visual odometer, that may serve as a localization of a mobile robot along its trajectory. The method relies on visual features that are used for robot navigation along the given trajectory through application of [11]. The features used are natural salient properties of the observed scene described by suitable robust image descriptors [14].

As a regular odometry suffers from accumulative errors with non-zero mean, the overall error displacement of the robot grows without limits with the path driven. This featuring may cause limitation on performance of the used navigation method, that determines the robot heading control and requires guidance of the robot based on odometry for a certain portion of time, or path. Erroneous odometry may cause a complete failure of the method. Therefore, a procedure that refines the robot position steadily along the trajectory it drives is foreseen.

Normally, the SURFnav method allows the robot to drive along a linear path segment under assumption that the positioning error on the segment is constrained since position re-calibration can be only accomplished at the next turn — a rapid change of the robot heading along a piecewise linear trajectory.

In the cases, where the path segment is oversized, the dead-reckoning error may rise high and the robot may loose its capability to observe the desired scene features at all, causing the method to fail. Therefore, limiting of the robot displacement error on the fly is desired. Our approach is an extension of the original SURFnav approach and relies on permanent observation of salient features by omni-directional camera along the robot path, providing their observation angles only. Our contribution shows that the obtained directional information can then be processed to compute robot displacement errors and to suggest the necessary correction of its position without a need for any further calibration.

Moreover, we show prospective properties of the method, that appears incrementally stable i.e. it is capable to bring the robot into the correct position (= the previous position, where the robot is expected to be with respect to its preceding occurrence in this location) if the correction step is primarily finite and under specific limits, imposed by the constrains of the camera, manoeuvering capabilities, etc.

2.1 Problem Specification

Generally the *FB-nav* methods [10, 11], irrespective of the type of image feature being used, are based on extraction of those stable image regions from either omni- or forward looking camera. Knowing the assignment of particular features to a specific observation position (comprising robot heading and coordinates), correspondences of these within subsequent frames can be established and used for robot heading adjustment. In other words, the method, dead-reckoning serves two necessary roles:

1. To determine expected features to be observed from each position on the path.
2. To decide about a path segment end point and which is to be continued by the next path segment. In a typical case, this situation is accompanied by a change in direction.

For the first point, dead-reckoning is not strictly necessary. Expected features can be determined based on segment rather than precise position. Resulting larger feature set increases computational costs of feature matching, but should not hamper the function of the method. Shorter path segments can be used to keep the costs from becoming unbearable.

The key issue is the detection of the segment endings. Relying on the dead-reckoning is very unsafe in real cases. Therefore, replacing the end-of-segment detection by visual means allows to omit the dead-reckoning process completely and decouples the results on a segment from previous path steps and accumulative errors along the direction of the movement.

3 Method Description

3.1 Approach

To determine whether the end of a segment has been reached or passed, the method based on angle differences is used. In a way similar to the original method, the first step is to employ robust features in the camera image to detect landmarks in the scene. As the feature detection has already been performed by the *FB-nav* method in each step, this data can be reused. For this step, angles between the landmark directions are exploited instead of the directions themselves.

The core idea of the method is based on a simple geometric intuition: when observing a pair of landmarks, the difference of observation angles between them appears larger from closer distance and smaller, if they are more far away. Therefore by comparing views to the same landmarks from two diverse positions, it can be determined which one is located closer to the pair of landmarks. Considering a convex region from which the observations are done, while preserving their ordering in the view, the afore rule can be applied: if starting at a distant position and moving towards a landmark pair, the distance will decrease and the observations will turn more similar to a closer view case.

Expanding this idea for multiple landmarks, it becomes possible to state whether the motion direction is forwards or backwards with respect to each landmark pair. Depending on the use case, this piecewise information can be fused to estimate direction to the other position, or to perform voting to ascertain whether the motion appears forwards or backwards in a single specified direction.

In this work, the latter approach is addressed. For each segment of the *FB-nav* recorded path, the visual information of the segment end is described in terms of features observed. This information enables to later determinate whether the robot is located near the recorded position and subsequently navigate precisely to the same location.

The method used is based on comparison of angular measurements from the two positions. In both the recorded and current position, the image of surrounding environment is acquired. The matching landmarks are detected in these images, and identified using either SURF [13], SIFT [1], ORB [16] or BRISK [12] descriptors. The angle sizes between the landmarks are compared. Difference in each single angle yields only limited information regarding the relative position, but with sufficient number and good distribution of detected landmarks, an approximate vector towards the next position can be gained.

Since the robot used can only move on ground, we can simplify the computation and project all landmarks onto a plane, performing the computation completely in 2D.

3.2   Geometry of Single Angle

According to the generalized Thales Theorem, all the points observing two landmarks under a given angle are located on a circumference segment. Figure 1 gives an overview of the situation.

Distance d of observation point from the center and distance v of the center of the circumference from the line connecting the landmarks are given by following equations. s denotes scalar distance of the landmarks, α is the observation angle.

$$v = \frac{s}{2 \tan \alpha} \tag{1}$$

$$d = \frac{s}{2 \sin \alpha}, \tag{2}$$

If the landmark positions are known and fixed, the parameters of the circumference containing all possible observation points are dependent only on the observation angle. A smaller angle corresponds to a larger circumference and vice versa. Thus, we can state that moving towards the center of the circumference means a shift towards positions with a greater observation angle, while moving away means the opposite.

When the position of the landmarks is unknown, the information about landmark positions is limited to the corresponding direction vectors and angle α between them. In such case, the direction vector towards the center cannot be computed. However, it can be approximated by the axis of the angle α. As shown in [7], the error of this estimation is bounded by Eq. 3. The tightness of the bound is dependent on the angle size only. Angles with sufficiently small error can thus be selected.

$$\beta \leq \frac{\pi}{2} - \frac{\alpha}{2} \tag{3}$$

3.3   Estimation

From each angle, relative position of the two viewpoints can be estimated in the direction of the angle axis. Without knowing the respective landmark distance distance, only the direction and orientation can be computed, not absolute scale. In the next step, these partial estimates need to be aggregated into a single correction vector. As the goal is limited to distinguishing whether the target position at the end of a segment is located still ahead or behind the robot, a simplified aggregation by voting can be used.

For each pair of observed landmarks, consider the angle between them. Let a be the direction of the angle axis. Compare the angle with the corresponding one recorded at the target location in terms of size. Result determines the direction on the axis vector. The landmark pair contributes to voting according to Eq. 4 with the

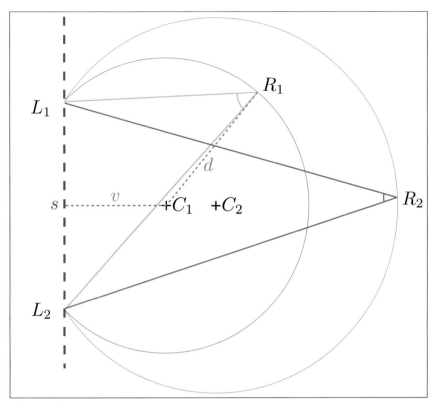

Fig. 1 Two landmarks observed under different angles. The points L_1, L_2 are landmarks, R_1, R_2 are two observing positions, while the points C_1, C_2 are the respective circumference centers. The corresponding d and v are also shown for the point R_1.

magnitude equal to the angle between the vectors representing angle axis a and the direction of robot movement r. The contribution is weighted by maximum error β_a, calculated by Eq. 3.

$$C = \sum_a \frac{\frac{\pi}{2} - |a - r|}{\max(\beta_a, 1)} \tag{4}$$

If the sum of contributions C is positive, the target is considered to be still ahead of the robot. The negative C signifies the opposite, and is interpreted as passing the end of the current path segment.

3.4 Order-Based Filtration

Nevertheless, the method exhibits a considerable weakness. While the influence of noised landmark positions has little effect on the final method performance, the

Fig. 2 Unwrapped image from the robot camera.

method remains highly sensitive to mismatches of landmark pairs. As the landmarks are handled pairwise in the computation, even small number of mismatches can introduce a large systemic error. To counteract this fact, a method to detect and discard outliers is necessary.

Assuming that viewpoints of two subsequent images are nearby, the geometry of the scene will be similar in the both. Therefore, horizontal order of majority of landmarks will be the preserved. Any deviations from the previous rule can be caused either by mismatching of features between the two frames, or by landmarks whose distance from the observer is significantly dissimilar to others. Neither of those is favorable to the computation, and all such landmarks can be disregarded.

The filtration process is iterative. Landmark ordering is considered pairwise. Landmark pairs that have different ordering between the two images are considered in conflict. In each cycle, landmark contained in greatest number of conflicting pairs is removed. Process is repeated until conflicts are eliminated.

Although only a small number of detected points is necessary to perform direction estimation, effect of any erroneous detection is significant in such small sets. To increase robustness, a lower bound on a number of detected matches can be set and any result calculated from lesser amount of matches is not considered valid.

4 Experimental Results

4.1 Experiment Setup

The proposed method has been validated in a real environment. The mobile robot used is Evolutionary Robotics, model ER1 fitted with a laptop PCB camera with catadioptric lens. Video stream with resolution 2048x1536 was recorded, and the omnidirectional image was unwrapped to 3111x241 pixels before processing (Fig. 2).

The camera placement in a sufficient height above the top of the robot prevents observing of the robot body. The robot takes the advantage of a reliable dead-reckoning system used in the experiment as the ground truth. Nevertheless, the robot design is tailored for indoor use only.

The experiment was performed in the indoor office-type of environment, see Fig. 3. The experimental scene has primarily been set as static with minor variation in shape and structure. The variations due to presence of the operator and other pedestrians have not been observed as imposing any substantial influence on the experimental results.

Fig. 3 The area of the experiment. The red line shows the trajectory of the robot.

The robot was manually driven multiple times along a trajectory consisting of several straight-line segments (Fig. 3). To test the error of the segment-end detection, each segment was required to be recorded past the end (therefore, past the point when the dead-reckoning would terminate it). For this reason, the original method was not used for the data gathering itself.

The first run was considered for a training, during which the endings of segments used by *FB-nav* in its learning phase are saved as a collection of observation angles. During the following runs, segment ends are detected by the presented method. A success level has been calculated as a distance between the closest position to a previously saved target point and the point indicated by the method as the end of a segment (as explained in Fig. 4).

4.2 Results

Localization was performed on the recorded experimental data. Various image feature detectors (SIFT, SURF, ORB, BRISK) were used in the first step of the computation. The OpenCV implementation [4] of all these detectors was used. The images underwent compensation of the camera embedded distortion from use of the catadioptric lens and thus potentially contain deformations. Consequently, the suitability

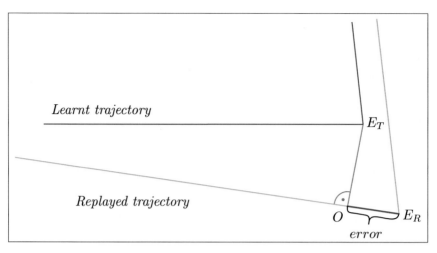

Fig. 4 The error is measured as a distance of the proposed segment end E_R from closest point O on the replayed trajectory.

of feature detectors was in question. Results obtained of diverse features are reported side-by-side for comparison.

Principal results of the experiment are in the Table 1, which shows errors in calculation of the segment end. For each feature type, the five point statistics is presented. As observed herein, the error in finding the segment ending is mostly in the order of centimeters. Even this exhibits less precision than afforded by a dead-reckoning approach, this error is not accumulative thus does not increase beyond any limits over the run time of the system.

Considering the size of robot and scale of the environment, the precision in average case is sufficient for the task of segment ending detection in the *FB-nav* method. The results for ORB and BRISK detectors are less reliable.

The efficiency of any feature-based method is conditioned by existence of a sufficient amount of detected features along the recorded path. To investigate limits of the presented method in this parameter, comparison based on the number of detected features was also performed. Fig. 5 details relation between the number of successfully matched image features in the vicinity of the target point, and the error in the segment end detection.

Table 1 Error in segment termination (five point statistic)

	SIFT	SURF	ORB	BRISK
Average difference (cm)	9.5	8.9	13	13
Median difference (cm)	7.2	6	7.5	10
Standard deviation	6.4	6.6	13	9.3
Maximum difference (cm)	23	26	50	38
Minimum difference (cm)	1.8	1.8	0	1.4

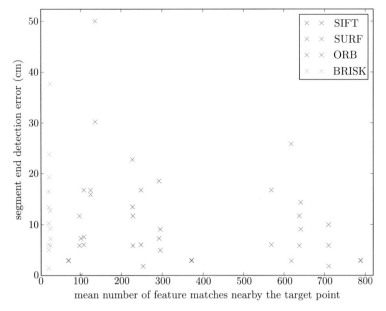

Fig. 5 Segment end detection error, plotted against number of detected features. Higher number of features is a slight advantage.

We can see that detectors yielding higher amount of features have lower median error as shown in Table 1. However, the advantage that higher amount of features convey is not significant. We conclude that the method is not dependent on a high number of features with corresponding high computational costs. Past certain necessary minimum, additional features are of a limited benefit.

Also important is the fact how large vicinity of the target contains sufficient feature matches for navigation. Figure 6 shows the relation of distance from the target to the number of matched features. The number of detected features is highest at the target location and declines further afield. Note, that these values are dependent on the experiment location, visibility and a distance to landmarks.

Contrary to expectations, ORB and BRISK features were considerably less successful in matching than the other ones, with greater fallout caused by distance from target. This might be caused by deformation of the image imposed by unwrapping the image from omni-directional camera. Broad invariance of SURF feature detector brings minor advantages in this case.

These results also explain greater variance of error of endpoint detection based on ORB and BRISK-based features. If the target point neighborhood containing sufficient amount of features is smaller, even small deviations in the trajectory can have substantial impact.

By selecting a desirable amount of features for localization, the afore data can also be used to estimate the maximum admissible length of a path segment. Longer segments run risk of the robot losing localization in the run in a given environment.

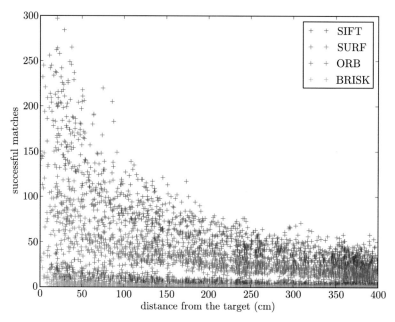

Fig. 6 The number of matches in relation to a distance from the target location

In this case, the SURF detector enables segments of 2 meters, SIFT half that value, and other detectors require even shorter segments. These estimates necessarily vary between scenes.

5 Conclusion

The presented approach builds a novel *FB-nav* method as a generalization of the original *SURFnav* approach, deriving corrections of the robot bearing from forward-looking camera via usage of robust image features. This work aims to improve its performance by removal of the dependency on a reliable dead-reckoning system. Classical dead-reckoning approaches suffer accumulative errors and may fail completely under certain constraints (skid-control or aerial robots, wheel slippage, etc.) loosing their precision substantially. Hereby we suggest a novel method reusing the same modality – the same image features as used for the navigation in its previous outfit.

Angles between observed locations of such image feature-based landmarks are partial information that is readily available through most scenes. Although insufficient for absolute triangulation, this information is satisfactory for direction estimation, under certain constraints. In a limited form, it is able to support decisions, whether a pre-learned location has been achieved or allow judgements on a relative distance of these features for each particular observation point.

In this paper, a method using this information is presented. We show, that such method is able to replace dead-reckoning in the task of segment end detection, thus upgrading *FB-nav* to fully visual navigation. Experiments on real data verify feasibility of the presented approach.

References

1. Bay, H., Ess, A., Tuytelaars, T., Van Gool, L.: Speeded-up robust features (SURF). Comput. Vis. Image Underst. **110**(3), 346–359 (2008)
2. Bekris, K.E., Argyros, A.A., Kavraki, L.E.: Exploiting Panoramic Vision for Angle-Based Robot Navigation, pp. 229–251. Springer (2006)
3. Blanc, G., Mezouar, Y., Martinet, P.: Indoor navigation of a wheeled mobile robot along visual routes. In: Proceedings of the 2005 IEEE International Conference on Robotics and Automation, ICRA 2005, pp. 3354–3359, April 2005
4. Bradski, G.: The OpenCV Library. Dr. Dobb's Journal of Software Tools (2000)
5. Davison, A.J., Reid, I.D., Molton, N.D., Stasse, O.: Monoslam: Real-time single camera slam. IEEE Trans. Pattern Anal. Mach. Intell. **29**(6), 1052–1067 (2007)
6. Diosi, A., Remazeilles, A., Segvic, S., Chaumette, F.: Outdoor visual path following experiments. In: IEEE/RSJ International Conference on Intelligent Robots and Systems, IROS 2007, pp. 4265–4270, October 2007
7. Dörfler, M., Přeučil, L.: Position correction using angular differences. In: 18th International Student Conference on Electrical Engineering, POSTER 2014. Czech Technical University, Prague, October 2014
8. Engel, J., Schöps, T., Cremers, D.: LSD-SLAM: large-scale direct monocular SLAM. In: Fleet, D., Pajdla, T., Schiele, B., Tuytelaars, T. (eds.) ECCV 2014, Part II. LNCS, vol. 8690, pp. 834–849. Springer, Heidelberg (2014)
9. Klein, G., Murray, D.: Parallel tracking and mapping for small AR workspaces. In: Proc. Sixth IEEE and ACM International Symposium on Mixed and Augmented Reality (ISMAR 2007), Nara, Japan, November 2007
10. Krajník, T., Faigl, J., Vonásek, V., Košnar, K., Kulich, M., Přeučil, L.: Simple yet stable bearing-only navigation. Journal of Field Robotics **27**(5), 511–533 (2010)
11. Krajník, T., Přeučil, L.: A simple visual navigation system with convergence property. In: Bruyninckx, H., Přeučil, L., Kulich, M. (eds.) European Robotics Symposium 2008, Springer Tracts in Advanced Robotics, vol. 44, pp. 283–292. Springer, Berlin Heidelberg (2008)
12. Leutenegger, S., Chli, M., Siegwart, R.: BRISK: binary robust invariant scalable keypoints. In: 2011 IEEE International Conference on Computer Vision (ICCV), pp. 2548–2555, November 2011
13. Lowe, D.G.: Distinctive image features from scale-invariant keypoints. Int. J. Comput. Vision **60**(2), 91–110 (2004)
14. Mikolajczyk, K., Schmid, C.: A performance evaluation of local descriptors. IEEE Transactions on Pattern Analysis and Machine Intelligence **27**(10), 1615–1630 (2005)
15. Nitsche, M., Pire, T., Krajník, T., Kulich, M., Mejail, M.: Monte carlo localization for teach-and-repeat feature-based navigation. In: Mistry, M., Leonardis, A., Witkowski, M., Melhuish, C. (eds.) TAROS 2014. LNCS, vol. 8717, pp. 13–24. Springer, Heidelberg (2014)
16. Rublee, E., Rabaud, V., Konolige, K., Bradski, G.: ORB: an efficient alternative to sift or surf. In: 2011 IEEE International Conference on Computer Vision (ICCV), pp. 2564–2571, November 2011
17. Strasdat, H., Montiel, J.M.M., Davison, A.: Scale drift-aware large scale monocular slam. In: Proceedings of Robotics: Science and Systems, Zaragoza, Spain, June 2010

Two-Stage Static/Dynamic Environment Modeling Using Voxel Representation

Alireza Asvadi, Paulo Peixoto and Urbano Nunes

Abstract Perception is the process by which an intelligent system translates sensory data into an understanding of the world around it. Perception of dynamic environments is one of the key components for intelligent vehicles to operate in real-world environments. This paper proposes a method for static/dynamic modeling of the environment surrounding a vehicle. The proposed system comprises two main modules: (i) a module which estimates the ground surface using a piecewise surface fitting algorithm, and (ii) a voxel-based static/dynamic model of the vehicle's surrounding environment using discriminative analysis. The proposed method is evaluated using KITTI dataset. Experimental results demonstrate the applicability of the proposed method.

Keywords Velodyne perception · Motion detection · Dynamic environment · Piecewise surface · Voxel representation

1 Introduction

Intelligent vehicles have seen a lot of progress recently. An intelligent vehicle is generally composed by three main modules: perception, planning and control. The perception module builds an internal model of the environment using sensor data, while the planning module performs reasoning and makes decisions for future actions based on the current environment's model. Finally the control module is responsible for translating actions into commands to the vehicle's actuators.

Today, the perception system of an intelligent vehicle is able to sense and interpret surrounding environment in 3D using sensors such as stereo vision [1], [2] and 3D

A. Asvadi(✉) · P. Peixoto · U. Nunes
Department of Electrical and Computer Engineering, Institute of Systems and Robotics,
University of Coimbra, Coimbra, Portugal
e-mail: {asvadi,peixoto,urbano}@isr.uc.pt

© Springer International Publishing Switzerland 2016
L.P. Reis et al. (eds.), *Robot 2015: Second Iberian Robotics Conference*,
Advances in Intelligent Systems and Computing 417,
DOI: 10.1007/978-3-319-27146-0_36

laser scanners [3], [4]. The data acquired by 3D sensors needs to be processed to build a 3D internal representation of the environment surrounding the vehicle. Awareness of moving objects is one of the key components for the perception of the environment. By detecting, tracking and analyzing the moving objects present in the scene, an intelligent vehicle can make a prediction about objects' locations and behaviors and plan for next actions.

In this paper, we propose a voxel-based representation of the dynamic/static three-dimensional environment surrounding a moving vehicle equipped with a Velodyne Lidar and an Inertial Navigation System (GPS/IMU). The main contributions of the present work are: 1)- a piecewise surface fitting method for ground estimation and object/ground separation; 2)- a voxel-based discriminative static/dynamic environment modeling. The output of the proposed method is a voxel representation of the environment surrounding a moving vehicle where voxels are classified as being static or dynamic (see Fig. 1).

Fig. 1 Top picture shows an image of a given frame from KITTI dataset and the result of the proposed method projected onto it. Static/dynamic voxels are shown with green and red colors respectively. The bottom picture shows the corresponding dynamic and static cells in surrounding environment of the intelligent vehicle in 3D.

The remaining part of this paper is organized as follows. Section 2 describes the related state of the art. Section 3 describes the proposed voxel-based static/dynamic environment modeling by introducing a piecewise surface fitting algorithm. Experimental results are presented in Section 4 and Section 5 brings some concluding remarks.

2   Related Work

The perception of a 3D dynamic environment captured by a moving vehicle requires a 3D sensor and an ego-motion estimation mechanism. The representation of the environment is another important issue for 3D perception of dynamic environment. Pfeiffer and Franke [5] used stereo vision system for acquiring 3D data and visual odometry for ego-motion estimation. They proposed the Stixel representation, consisting on sets of thin, vertically oriented rectangles used for the representation of the environment. Stixels are segmented based on the motion, spatial and shape constraints and tracked using a 6D-vision Kalman filter framework that is a framework for the simultaneous estimation of 3D-position and 3D-motion. Asvadi et al. [6] used 2.5D elevation grids to build the environment representation using as input data from a Velodyne laser scanner and GPS/IMU localization system. They combined 2.5D grids with localization data to build a local 2.5D map. In every frame, using a robust spatial reasoning, the last generated 2.5D grid was compared with the local 2.5D map to detect the 2.5D motion grid. Motion grids are grouped to provide an object-level representation of the scene. Next, they applied data association and Kalman filtering for tracking grouped motion grids. Broggi et al. [7] used setero vision as a 3D sensor. They estimated ego-motion using visual odometry and used it to distinguish between stationary and moving objects. A color-space segmentation of voxels that are above the ground plane is also performed. Voxels with similar features are grouped together. Next, the center of mass of each cluster is computed and Kalman filtering is applied to estimate their velocity and position. Azim and Aycard [8] proposed a method based on the inconsistencies between observation and local grid maps represented by an Octomap [9]. An Octomap is a 3D occupancy grid with an octree structure. Next, they segmented objects using density based spatial clustering. Finally, Global Nearest Neighborhood (GNN) data association, and Kalman filter for tracking, and Adaboost classifier for object classification are used. They used data from a Velodyne laser scanner, and they estimated ego motion using odometry and scan matching. The summary of the aforementioned methods is shown in Table 1. In comparison with these methods, the work presented in this paper contributes with a piecewise surface

Table 1 Some recent related work on 2.5D/3D perception of dynamic environments surrounding a vehicle.

Reference	Representation	Approach for perception of dynamics
Pfeiffer and Franke, 2010 [5]	Stixel (2.5D vertical bars in depth image)	Segmentation of stixels based on motion, spatial and shape constraints using graph cut algorithm
Asvadi et al., 2015 [6]	2.5D elevation grid	Build a local 2.5D map and compare the last generated 2.5D grid with the local map
Broggi et al., 2013 [7]	3D voxel grid	Distinguish stationary/moving objects using ego-motion estimation and color-space segmentation of voxels
Azim and Aycard, 2014 [8]	Octomap	Inconsistencies on map and density based spatial clustering

fitting and a discriminative voxel-based static/ dynamic environment modeling, with voxels around the vehicle being classified as static or dynamic.

3   Proposed Method

In this section, we present an algorithm for voxel-based representation of the static/dynamic environment surrounding a vehicle equipped with a Velodyne Lidar and an Inertial Navigation System (GPS/IMU). First, at every time step, point clouds from the m last sensor measurements are integrated. Next, a piecewise surface fitting algorithm is applied on the integrated points to estimate the ground parameters (see Fig. 4-ii). Piecewise surface parameters are used for removing ground points from a point cloud. Voxelization process is performed by quantizing and counting the total number of points that falls into each cell. A voxel-based representation of the integrated point clouds and the last frame are used to build the static/dynamic model of the surrounding vehicle environment.

3.1   Point Clouds Integration and Ground Estimation

In this section, we present an algorithm for fitting a piecewise surface model on a set of registered and integrated point clouds to estimate the ground geometry. Fig. 2 shows the architecture of the algorithm. In every frame, by going over m previous scans (m is the number of scans to integrate and n is the number of current scan), point clouds Pts are loaded and transformed into the current coordinates of the vehicle. The integrated point clouds $IPts$ are cropped to a region inside the local grid, which is an area covering 5 to 15 meters ahead of the vehicle, 10 meters on the left and right sides of the vehicle, with 2 meters in height. This procedure is summarized in Algorithm 1. Point clouds are integrated for two purposes: 1)- for robust estimation of ground parameters; 2)- to extract the static model of the environment from it, as explained in the next section.

Those points that belong to the ground can cause false detections in undulated roads and slow the process of building voxel grid. To address this problem, the ground is cut into stripes according to the car orientation and a piecewise surface fitting

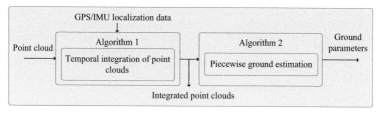

Fig. 2 Point clouds integration and ground estimation.

Algorithm 1. Temporal integration of point clouds.

Inputs: Point cloud of a single scan $Pts(i)$; localization and pose data of the vehicle in Euclidean space given by GPS/IMU measurements $LaP(i)$; number of integrated point clouds m.

Output: Integrated point clouds $IPts$.

  start

    for $i : n - m$ to n

      $IPts$ ← Transform $Pts(i)$ using $LoP(i)$ and $LoP(n)$

    end

    $IPts$ ← Remove outliers from $IPts$

  end

---

method is proposed to estimate the ground stripes' parameters. Ground estimation process starts with the stripe closest to the host vehicle. Because in closer regions, point clouds are denser and measured with less localization errors, therefore stripes are estimated with more confidence. As a pre-process, points with a height greater than 1 meter in the first stripe (closest stripe to the host vehicle) are considered as outliers and rejected. The result is a filtered point cloud in the stripe region $IPts_f$. Next, piecewise surface fitting is performed on the inlier points of the stripe's region by fitting a plane using a least square method. Estimation of the next stripe is started from endmost edge of the first stripe in the vehicle movement direction. A similar approach is used for computing the next stripes (See Fig. 3). Every stripe parameter is checked for acceptance. The validation process starts from the closest stripe to the farthest stripe. If the slope difference between previous plane's stripe and current stripe is less than 15 degrees, it is considered as a valid value and a new plane is initialized, or else it is considered the two stripes are the same ground plane and the parameters from the previous stripe are used instead. Every stripe's length is selected as 5 meters in the vehicle movement direction. The process is shown in Algorithm 2. These parameters are used in the next step, to exclude ground's points while building the voxel grid.

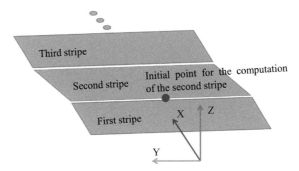

Fig. 3 Piecewise ground estimation using stripes. Vectors show the pose of the vehicle. Initial point for estimation of the second stripe is shown in red.

Algorithm 2. Piecewise ground estimation.

Inputs: Integrated point clouds $IPts$; number of stripes k.
Output: Surface stripes parameter matrix Prm.
   start
      for surface stripe's number i: 1 to k
         $IPts_f(i)$ ← Reject outliers from $IPts$
         $Prm(i)$ ← Least square fit on $IPts_f(i)$.
         if $\Delta(Prm(i), Prm(i-1)) > threshold$
            $Prm(i)$ ← $Prm(i-1)$
         end if
      end for
   end

3.2 Voxel-based Static/Dynamic Modeling of the Environment

Surface parameters from the previous step are used for removing the ground points from two groups of point clouds: 1)- integrated point clouds; 2)- point clouds from single scans. For removing the ground points, the distance between a point cloud and piecewise surfaces is computed, and all points under the surface are removed. To make the approach more robust against undulated roads, points with heights lower than 20 cm from the surface are also rejected. Next, the point cloud is voxelized and sent to the next module for building the static/dynamic model of the environment.

Voxelization. Voxel grids dicretize the space into equally-sized cubic volumes or voxels, where each voxel contains information about the space it's representing. Grids are a memory-efficient dense representation with no dependency to predefined features which allow them to provide detailed representation of complex environments. These attributes made the voxel grid an efficient tool for integrating temporal data and representing the surrounding environments of intelligent vehicles in 3D. Here, the voxelization process is performed by quantizing end-points of the beams, counting the total number of points that falls into each grid cell, and storing a list of occupied voxels. This simplified model drastically speeds up the process at the cost of discarding information about free and unknown spaces. The size of a voxel in each dimension was chosen to be equal to 10 cm. The selected size provides enough number of voxels to represent objects and keeps the discretization errors low. Fig. 4 shows a voxel grid with estimated piecewise surfaces. Two previous groups of point clouds (e.i. integrated point clouds with removed ground points and point clouds from single scans with removed ground points) are voxelized and used for static/dynamic modeling of the environment. The voxlization results are the integrated grid voxels $IGrd$ which is computed from integrated point clouds with rejected ground points and voxel grids from single scans with removed ground points $Grd[i]$.

Fig. 4 From top to bottom: i) RGB image of the scene, ii) piecewise surface fitting and voxel representation of objects. Notice the curvature of the ground that makes it impossible to model using only one surface. Red lines show corresponding locations on images.

Static/Dynamic Modeling. The main idea behind this section is that since, a moving object occupies different voxels along time, a Velodyne Lidar captures and saves more data in the static voxels in comparison with voxels belong to moving objects. Therefore, the voxel values for the static parts of the environment are greater. To realize this concept, a two-stage process is proposed (Fig. 5). First stage provides a rough estimation of static/dynamic voxels by using a simple subtraction mechanism. Second stage further refines the results using a discriminative analysis on the 2D histograms computed from the output of the first stage.

In the first stage, integrated voxel grid $IGrd$ is compared with each of the last m grids $Grd[i], \{i : n - m, .., n\}$ to remove the dynamic voxels of the integrated voxel grid $IGrd$. By comparing each pairs of $IGrd$ and $Grd[i], \{i : n - m, .., n\}$, those voxels belonging to the integrated grid that have the same value with each of corresponding voxels in $Grd[i], \{i : n - m, .., n\}$ are removed. It means those voxels of $IGrd$ have been seen only for one time and therefore, more likely belong to a moving object (dynamic cells). The process is shown in Algorithm 3. Using this approach we filter out dynamic voxels and build a rough stationary model of

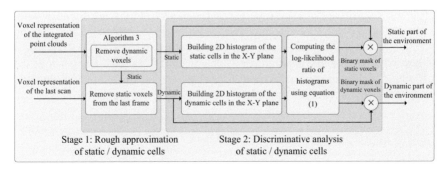

Fig. 5 Two-stage discriminative static/dynamic environment modeling.

Algorithm 3. Procedure of removing the dynamic voxels.

Inputs: Integrated voxel grid $IGrd$; Point cloud of a single scan $Pts(i)$; localization and pose data given by GPS/IMU $LaP(i)$.

Output: Integrated voxel grid with removed dynamic cells.

   start

      for $i : n - m$ to n

         $Pts(i) \leftarrow$ Transform $Pts(i)$ using $LoP(i)$ and $LoP(n)$

         $Pts(i) \leftarrow$ Remove outliers from $Pts(i)$

         $Grd[i] \leftarrow$ Voxelization of $Pts(i)$.

         Compare $IGrd$ with $Grd[i]$; remove dynamic cells of $IGrd$.

      end

   end

---

the environment. Next, the stationary model is used to remove static voxels from the voxel representation of the last scan using a simple subtraction operation. The outputs of the first stage are roughly approximated static and dynamic grids that are inputed to the second stage. The output of stage 1 is inaccurate because some parts of very slowly moving objects may have seen more than once in the same voxels and therefore may wrongly be inserted into the static model of the environment (See Fig. 6-i). To remove false detections in the second stage, we assume that all voxels in every column of the x-y plane have the same state (static or dynamic). Based on this assumption, we proposed to build the 2D histogram of the cells in the X-Y plane. Histograms are built for both approximated static/dynamic cells from stage 1. The log-likelihood ratio of 2D histograms of the approximated dynamic and static cells are employed to determine the binary mask for the dynamic voxels. It is computed by:

$$L_i = log \frac{max\{h_d(i), \delta\}}{max\{h_s(i), \delta\}} \tag{1}$$

where δ is a small value (we set it to 1) that prevents dividing by zero or taking the log of zero. h_d is the 2D histogram computed from the approximated dynamic grid of the stage 1, and h_s is the 2D histogram computed from the approximated static grid of stage 1. Cells belonging to the dynamic part have higher values in the computed log-likelihood ratio, static cells have negative values and cells that are shared by both, the dynamic and static, tend towards zero (See Fig. 6-ii). By applying a thresholding operation to L_i, a 2D binary mask of dynamic cells is obtained. The 2D mask is applied to all levels of the approximated dynamic gird from stage 1 to generate the final output. A similar approach is used for computing the binary mask of the static voxels. The outputs of stage 2 are voxels labeled as static or dynamic (See Fig. 6-iii).

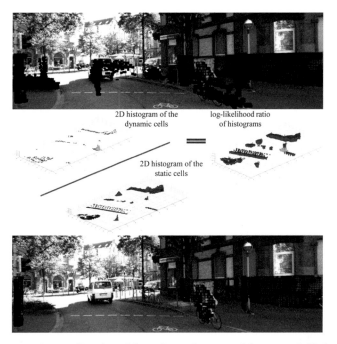

Fig. 6 From top to bottom: i) static and dynamic voxels outputted from stage 1, ii) the process of computing the binary mask of the dynamic voxels, iii) static and dynamic voxels outputted from stage 2 after discriminative analysis. For a better visualization, the voxel grids outputted from stages 1 and 2 were projected and displayed onto the RGB image. Static/dynamic voxels are shown with green and red colors respectively.

4   Experimental Results

The presented algorithm was tested on the KITTI dataset [10]. The proposed method is currently implemented in MATLAB. The ground-truth for the task of static/dynamic environment modeling or motion detection is not available yet, therefore we have performed a qualitative evaluation. In the following sections, we describe the dataset and present experimental results.

4.1   Dataset

The KITTI dataset was captured using a Velodyne 3D laser scanner and a high-precision GPS/IMU inertial navigation system. The Velodyne HDL-64E spins at 10 frames per second with vertical resolution of 64 layers and angular resolution of 0.09 degree. The maximum recording range is 120 m. The inertial navigation system is a OXTS RT3003 inertial and GPS system with 100 Hz speed of recording data and a resolution of 0.02m / 0.1 degree.

474   A. Asvadi et al.

4.2 Evaluation

In order to evaluate the performance of the proposed algorithm, a variety of challenging sequences were used. The eight most representative sequences are summarized
in Fig. 7, in which each row corresponds to one sequence. The proposed method
detects and classifies dynamic/static voxels around moving vehicles when they get
into the local perception field.

Fig. 7 Sample screenshots of the results obtained in the considered sequences. Static/dynamic
voxels are shown in green and red colors respectively. Each row represents one sequence. Left to
right we see the results obtained in different time instants.

In the first sequence, our method detects a cyclist and a car (while they are in
the perception field) as dynamic objects within the scene and models the walls and
stopped cars as part of the static model of the environment. The second sequence
shows a downtown area, where the proposed method successfully models moving
pedestrians and cyclists as dynamic part of the environment. The third sequence shows
a cross section scenario, our method models almost all passing pedestrians as dynamic voxels.Sequence number 4 shows another challenging downtown scenario with
moving objects range from pedestrian, groups of pedestrian to cyclist. The proposed
method successfully models moving objects as a dynamic part of the environment.

4.3 Computational Analysis

There is a compromise between the computational cost and performance of the
static/dynamic modeling of the proposed method. Increasing the number of the

integrated scans, will lead to a better removal of the dynamic cells and generates a stronger static/dynamic model. However, it adds an additional computational cost and makes the method slower. On the other hands, less integrated scans make the environment model weaker. The proposed method has a stable results when the number of integrated frames are more than 10. The number of integrated scans of the proposed algorithm is set as 20. The computational cost of the proposed method depends on the size of the local grid, the size of a voxel, the number of integrating frames, and the number of non-empty voxels, since only non-empty voxels are indexed and processed. We performed the experiment on the first scenario of the previous section with a fixed sized local grid. The scenario has in average nearly 1% non-empty voxels. The size of a voxel and the number of integrating point clouds are two key parameters that correspond with spatial and temporal properties of the proposed algorithm respectively, and directly impact on the computational cost of the method. The experiment was carried out with a quad core 3.4 GHz processor with 8 GB RAM under MATLAB R2013a. The average speed of the proposed algorithm (frames per second) together with the value of each parameter (voxel size and number of integrating frames) are reported in Fig. 8. As it can be seen the number of integrated frames has the greatest impact on the computational cost of the proposed method. The proposed method configured with the defaults parameters works at $1.05\,fps$.

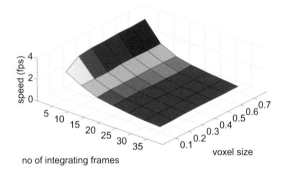

Fig. 8 Computational analysis of the proposed method.

5   Concluding Remarks and Future Work

The 3D perception of dynamic environment is one of the key components for intelligent vehicles to operate in real-world environments. In this paper, we propose a voxel-based representation of the dynamic/static three-dimensional environment surrounding a moving vehicle equipped with a Velodyne Lidar and an Inertial Navigation System (GPS/IMU). A piecewise surface fitting algorithm is proposed to estimates the ground surface and remove the ground points from the point clouds. A discriminative voxel-based static/dynamic environment modeling is proposed with voxels classified into the static and dynamic classes. The proposed method was evaluated using KITTI dataset and experimental results demonstrate the applicability of the proposed method.

We propose two new directions for the future work. First, the color information from the image can be incorporated to provide a more robust static/dynamic environment modeling. Second, the dynamic model of the environment can be investigated for object detection and tracking purposes.

Acknowledgments This work has been supported by the FCT project "AMS-HMI2012 - RECI/ EEIAUT/0181/2012" and project "ProjB-Diagnosis and Assisted Mobility - Centro-07-ST24- FEDER- 002028" with FEDER funding, programs QREN and COMPETE.

References

1. Laugier, C., Paromtchik, I.E., Perrollaz, M., Yong, M.Y., Yoder, J.-D., Tay, C., Mekhnacha, K., Negre, A.: Probabilistic analysis of dynamic scenes and collision risks assessment to improve driving safety. IEEE Intelligent Transportation Systems Magazine 3(4), 4–19 (2011)
2. Ziegler, J., Bender, P., Schreiber, M., Lategahn, H., et al.: Making bertha drive? - an autonomous journey on a historic route. IEEE Intelligent Transportation Systems Magazine 6(2), 8–20 (2014)
3. Urmson, C., Anhalt, J., Bagnell, D., Baker, C., Bittner, R., Clark, M.N., et al.: Autonomous driving in urban environments: Boss and the urban challenge. Journal of Field Robotics 25(8), 425–466 (2008)
4. Montemerlo, M., Becker, J., Bhat, S., Dahlkamp, H., Dolgov, D., Ettinger, S., Haehnel, D., et al.: Junior: The stanford entry in the urban challenge. Journal of Field Robotics 25(9), 569–597 (2008)
5. Pfeiffer, D., Franke, U.: Efficient representation of traffic scenes by means of dynamic stixels. In: IEEE Intelligent Vehicles Symposium (IV), pp. 217–224 (2010)
6. Asvadi, A., Peixoto, P., Nunes, U.: Detection and tracking of moving objects using 2.5D motion grids. In: 18th International IEEE Conference on Intelligent Transportation Systems (ITSC) (2015)
7. Broggi, A., Cattani, S., Patander, M., Sabbatelli, M., Zani, P.: A full-3D voxel-based dynamic obstacle detection for urban scenario using stereo vision. In: 16th International IEEE Conference on Intelligent Transportation Systems (ITSC) (2013)
8. Azim, A., Aycard, O.: Layer-based supervised classification of moving objects in outdoor dynamic environment using 3D laser scanner. In: IEEE Intelligent Vehicles Symposium (IV), pp. 1408–1414 (2014)
9. Hornung, A., Wurm, K.M., Bennewitz, M., Stachniss, C., Burgard, W.: OctoMap: An efficient probabilistic 3D mapping framework based on octrees. Autonomous Robots 34(3), 189–206 (2013)
10. Geiger, A., Lenz, P., Stiller, C., Urtasun, R.: Vision meets robotics: The kitti dataset. The International Journal of Robotics Research 32(11), 1231–1237 (2013)

Automatic Calibration of Multiple LIDAR Sensors Using a Moving Sphere as Target

Marcelo Pereira, Vitor Santos and Paulo Dias

Abstract The number of LIDAR sensors installed in robotic vehicles has been increasing, which is a situation that reinforces the concern of sensor calibration. Most calibration systems rely on manual or semi-automatic interactive procedures, but fully automatic methods are still missing due to the variability of the nearby objects with the point of view. However, if some simple objects could be detected and identified automatically by all the sensors from several points of view, then automatic calibration would be possible on the fly. This is indeed feasible if a ball is placed in motion in front of the set of uncalibrated sensors allowing them to detect its center along the successive positions. This set of centers generates a point cloud per sensor, which, by using segmentation and fitting techniques, allows the calculation of the rigid body transformation between all pairs of sensors. This paper proposes and describes such a method with encouraging preliminary results.

Keywords Point cloud · 3D data fitting · Rigid body transformation

1 Introduction

Many vehicles with autonomous navigation capabilities, and also many advanced drivers assistance systems, rely on LIDAR based devices. Moreover, most of the

M. Pereira(✉) · V. Santos
Department of Mechanical Engineering, DEM, University of Aveiro, Aveiro, Portugal
e-mail: {marcelopereira,vitor}@ua.pt

P. Dias
Department of Electronics, Telecommunications and Informatics,
DETI, University of Aveiro, Aveiro, Portugal
e-mail: paulo.dias@ua.pt

V. Santos · P. Dias
Institute of Electronics and Informatics Engineering of Aveiro,
IEETA, University of Aveiro, Aveiro, Portugal

© Springer International Publishing Switzerland 2016
L.P. Reis et al. (eds.), *Robot 2015: Second Iberian Robotics Conference*,
Advances in Intelligent Systems and Computing 417,
DOI: 10.1007/978-3-319-27146-0_37

developed systems use multiple sensors simultaneously, sometimes even combining different types (1D, 2D, n×2D, 3D). LIDAR sensors provide reasonably reliable and robust data in many circumstances, but when there is more than one sensor in the same setup, a calibration procedure must take place so the data from the different sensors can be combined or fused in a common reference frame. In order to solve that necessity, this paper presents a new extrinsic calibration method. Differently from the majority of existing methods, which are manual or semi-automatic, the proposed method is fully automatic, having no requirement of manual measurements or manual correspondences in the data from the devices. Instead of using those approaches, a ball is used as a calibration target allowing the detection of its center by the different sensors. To estimate the transformation between the sensors, at least three 3D point correspondences between the sensors are needed. These points can be estimated during ball motion, which furthermore increases the accuracy of the method as a significant number of points can be estimated during that movement.

1.1 Context of the Work

This work is part of the ATLASCAR project [1], carried out at the University of Aveiro, whose main purpose is the research and development of solutions on autonomous driving and Advanced Driver Assistance System (ADAS). For that purpose, and besides an extensive intervention on many fronts, a common Ford Escort car was equipped with a rich set of sensors dedicated mainly to the perception of the surrounding environment (Fig. 1).

The car is equipped with several exteroceptive sensors, namely a stereo camera, a 3D LIDAR, a foveated vision unit and additional planar laser range finders. The planar lasers already installed on the car are two Sick LMS151, and the 3D laser is actually a Sick LMS200 in a rotating configuration. There is also a Hokuyo laser

Fig. 1 AtlasCar, a Ford Escort adapted for autonomous driving capabilities. On the right, the three LIDAR sensors used in this study.

range finder to extract road profiles ahead of the car. Additionally, a new multi-layer laser (sick LD-MRS 400001) is to be installed in the car and, finally, a SwissRanger 3D TOF camera is also occasionally used in some experiments and contexts. The sensors that are part of the study in this paper are illustrated on the right side of Figure 1, and Table 1 presents some of their main characteristics.

1.2 Related Work

Over the past several years, a number of proposed solutions for the calibration between a camera and a laser were introduced [2–5], including some automatic on-line calibration solutions as presented in [6]. However, the studies for the specific problem of calibration between LIDAR sensors haven't yet reached the ideal solution. Previous works on the ATLASCAR project used a technique that also uses a calibration target, but it requires manual input from the user[7]. Other authors developed algorithms to calibrate one [8] or two [9] single-beam LIDARs within the body frame; both methods use approaches that rely on establishing feature correspondences between the individual observations by preparing the environment with laser-reflective tape, which additionally requires an intensity threshold for correspondences and some initial parameters. And a more recent method [10] is based on the observation of perpendicular planes; this calibration process is constrained by imposing co-planarity and perpendicularity constrains on the line segments extracted by the different laser scanners; despite being an automatic calibration method, it also requires some initial conditions, and does not provide the versatility of the method presented in this paper.

Table 1 Properties of the main LIDAR based sensors onboard the ATLASCAR

	Sick LD-MRS400001	Sick LMS151	SwissRanger sr4000
Scan planes	4, with full vertical aperture of 3.2°	1	
Field of view	2 scan planes: 85° 2 scan planes: 110°	270°	43.6°(h) × 34.6°(v)
Scanning frequency	12.5 Hz/25 Hz/50 Hz	25 Hz/50 Hz	
Angular resolution	0.125°/0.25°/0.5°	0.25°/0.5°	
Operating range	0.5 m - 250 m	0.5 m - 50 m	0.1 m - 5.0 m
Statistical error	± 100 mm	± 12 mm	±10 mm

2 Proposed Solution

The problem consists of estimating the rigid transformation between three different sensors with respect to a reference one; the proposed solution is to use a ball as the calibration target; the only restriction of the calibration target is about its size (diameter). The size of the calibration target is related with the angular resolution of

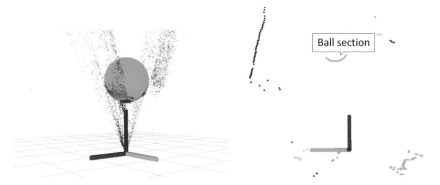

Fig. 2 Detection in a Swissranger cloud of points.   **Fig. 3** Result of the segmentation on sensor
Sick LMS151.

the sensors used; after some empirical experiments, it was observed that, in order to obtain feasible results, the ball must have a diameter large enough for the sensors to have at least 8 measurements on the target at 5 m.

The approach used to obtain the calibration between all the devices is achieved in three stages. First, each sensor must detect the center of the ball; then, the ball is placed in motion in front of the sensors allowing them to detect its center along successive positions, creating at the same time three point clouds (since there are three sensors); the condition to consider a new point for each point cloud is to require that each new point is separated from the previous point some minimum distance. Finally, one sensor is chosen to be the reference frame and the remainder are calibrated relatively to it one at a time by using an algorithm of the Point Cloud Library (PCL) [11].

2.1 Sphere Center Detection

The method to find the center of the ball depends on the type of the range data. The method for 2D data is based in finding circular arcs, taking advantage on the particularity that any planar section of the ball is a circle. Thus, the process to detect the center of the ball is divided in the following sequence: segmentation of the laser scan, detection of circle and calculation of its properties (center coordinates and radius), and calculation of the center of the ball given the properties of the target (diameter). In case of 3D data, an algorithm from PCL is used, as will be further explained with more detail.

It is important to mention that in 2D scans, due to the symmetry of the ball, there is the ambiguity relatively to which of the hemispheres belongs the detected circle, since every section of the ball has a symmetric section relatively to the hemisphere, which gives two solutions for the center of the ball (one above and another below the detected section). Consequently, some a priori information about the position of the sensor relatively to the ball is required.

2.1.1   Sphere Center Detection in 3D Data

The ball recognition in the SwissRanger is achieved using the PCL segmentation capabilities, namely the sample consensus module. The Random Sample Consensus (RANSAC) [12] is the algorithm applied in the point cloud from the SwissRanger, which is an iterative method used to estimate parameters of a mathematical model from a set of data containing outliers. In this case, the model is a sphere, so the resulting parameters are the coordinates of the center of the ball and its radius. Figure 2 shows the application of the RANSAC algorithm on the ball detection.

2.1.2   Segmentation

Segmentation is a very important part before the calibration process. The main goal is to cluster the point cloud in sub sets of points which have high probability to belong to the same object through detected discontinuities in the laser data sequence, which are called break-points.

Several method are available to perform 2D point cloud segmentation. Based on the work from Coimbra [13], the Spatial Nearest Neighbour (SNN) was used since it appears to be the most consistent over different tested scenes.

The SNN is a recursive algorithm where the distance between a scanned point and all the other points that are not yet assigned to a cluster is computed. If that distance is smaller than a certain threshold the points are assigned to that cluster. Figure 3 shows the result of the segmentation in a scan and the cluster related to the ball. The only variable in this algorithm is the threshold value, so it's expected that for a higher D_{th} the result will be bigger clusters, and for a smaller D_{th} the result will be smaller clusters.

2.1.3   Circle Detection

The method used for circle detection is inspired on a work developed for line, arc/circle and leg detection from laser scan data [14]. The circle detection makes use of a technique named Internal Angle Variance (IAV). This technique uses the trigonometric properties of the arcs: every point in an arc has congruent angles (angles that have the same amplitude) in respect to the extremes. This property can be verified in Figure 4. Let P_1 and P_4 be the extremes of the arc, and P_2 and P_3 random points belonging to the same arc. Then $\angle P_1 P_2 P_4 = \angle P_1 P_3 P_4$ because both angles measure one-half of $\angle P_1 O P_4$.

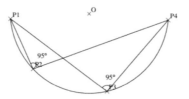

Fig. 4 Congruent angles of points on an arc in respect to the extremes.

The detection of circles involves calculating the mean of the aperture angle ($\bar{m}$) between the extreme points and the remaining points of a cluster, as well as the standard deviation (σ). To verify a positive detection at the beginning, values of standard deviation smaller than 8.6° and values of mean aperture between 90° and 135° were used, considering that the scan covers approximately half a circle. However, those values are dependent of two factors: the error associated to the sensors, and how much of the circle the scan covers. Thus, after analyzing the results empirically, the values were adjusted to standard deviations values under 5° and values of mean aperture between 105° and 138° for the sick LMS151, and 10° and between 110° and 135°, respectively for the sick LD-MRS400001. These adjustments allowed to obtain the best results avoiding the detection of false circles. Considering that a segment S has n points (P), say $S = \{P_1, P_2, ..., P_n\}$, the mean and standard deviation of the angle are calculated as follows: $\bar{m} = \frac{1}{n-2} \sum_{i=2}^{n-1} \angle P_1 P_i P_n$ and $\sigma = \sqrt{\frac{1}{n-2} \sum_{i=2}^{n-1} (\angle P_1 P_i P_n - \bar{m})^2}$.

2.1.4   Calculation of the Circle Properties

The calculation of the center and radius of the circle uses the method of least squares to find the circle that best fits the points. Given a finite set of points in $\mathbb{R}^2$, say $\{(x_i, y_i) | 0 \leq i < N\}$, first calculate their mean values by $\bar{x} = \frac{1}{N} \sum_i x_i$ and $\bar{y} = \frac{1}{N} \sum_i y_i$. Let $u_i = x_i - \bar{x}$, $v_i = y_i - \bar{y}$ for $0 \leq i < N$, and defining $S_u = \sum_i u_i$, $S_{uu} = \sum_i u_i^2$, etc. The problem is to solve first in (u, v) coordinates, and then transform back to (x, y).

Considering that the circle has center (u_c, v_c) and radius R, the main goal is to minimize $S = \sum_i g(u_i, v_i)^2$, where $g(u, v) = (u - u_c)^2 + (v - v_c)^2 - \alpha$, and $\alpha = R^2$. To do that, it is necessary to differentiate $S(\alpha, u_c, v_c)$, giving the following solutions:

$$u_c S_{uu} + v_c S_{uv} = \frac{1}{2}(S_{uuu} + S_{uvv}) \tag{1}$$

$$u_c S_{uv} + v_c S_{vv} = \frac{1}{2}(S_{vvv} + S_{vuu}) \tag{2}$$

Solving equations (1) and (2) simultaneously gives (u_c, v_c). Then the center (x_c, y_c) of the circle in the original coordinate system allows to write $(x_c, y_c) = (u_c, v_c) + (\bar{x}, \bar{y})$, being $\alpha = u_c^2 + v_c^2 + \frac{S_{uu} + S_{vv}}{N}$. The result of the combination of circle detection with the properties of the circle is illustrated on Figure 5.

Fig. 5 Result of the circle detection in real data from the sensor Sick LMS151.

2.1.5 Calculation of the Center of the Ball

After knowing all the properties of the circle, it is possible to calculate the center of the ball through trigonometric relations as shown in Figure 6, where R is the radius of the ball, R' the radius of the circle and d the distance between the center of the circle and the center of the ball, with $d = \sqrt{R^2 - R'^2}$. Taking into account the ambiguity for the 2D lasers mentioned in subsection 2.1, and considering that the center of the circle has (x_c, y_c) coordinates, the coordinates of the center of the ball for this laser are defined as follows: $(X_c, Y_c, Z_c) = (x_c, y_c, \pm d)$. Figure 7 illustrates an example of the ball detection in a 2D scan.

In order to simplify the calculations in the 3D multi-layer laser, the coordinates of the circle center are transformed into the XY plane for each layer; then for each layer, the center of the ball is calculated in the same way as in the 2D laser. After this, the centers of the ball are transformed back into the respective original plane.

At this point, for each layer, the center of the ball was calculated, which means that there are as many centers as layers. Thus, statistically, the center of the ball is obtained by calculating the mean of all the centers. Considering that there are n layers and the different circles have $(x_{c_1}, y_{c_1}, z_{c_1}; \ldots; x_{c_n}, y_{c_n}, z_{c_n})$

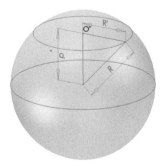

Fig. 6 Example of the cross-section of the sphere at a distance d from the sphere's center.

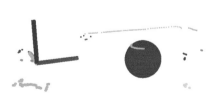

Fig. 7 Detection of the ball in a 2D scan from the Sick LMS151.

coordinates, the center mean coordinates are defined as follows: $(X_c, Y_c, Z_c) = (\frac{1}{n} \sum_{i=1}^{n} x_{c_i}, \frac{1}{n} \sum_{i=1}^{n} y_{c_i}, \frac{1}{n} \sum_{i=1}^{n} z_{c_i})$

2.2 Calibration

The 3D transformation between the cloud of points generated from the ball centers in each sensor are estimated using an Absolute Orientation algorithm. The implementation used is available in PCL, which is based on the Singular Value Decomposition (SVD) method proposed in [15, 16] to estimate the 3D translation and rotation transformation between two clouds of points.

3 Software Architecture

Evaluation tests have been realized in order to estimate the behavior and robustness of the presented method in real conditions. The method has been implemented in C++ on the ROS platform.

ROS [17] is a development environment specifically for applications in robotics, usually used in large projects that, due to its modular architecture, allows reduce the projects complexity by splitting a large project into smaller modules, each with a specific application. Also, those modules can be used for other applications and not only in a single project. Figure 8 shows all the processes that run during the

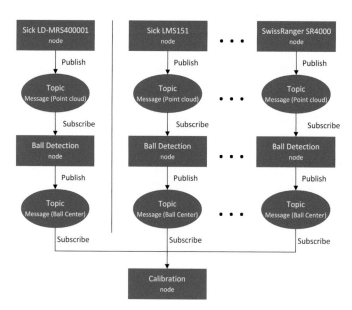

Fig. 8 Calibration process scheme.

calibration and what each of them is publishing and subscribing. Three processes are shown (one for each sensor), but this is scalable for more sensors.

4   Results

Preliminary experiments were carried out. In this stage, the interest was especially in: i) evaluating the consistency in the ball detection in each sensor; and ii) the consistency in the 3D transformation estimation depending on the number of points used. Even though these experiments do not provide absolute validation of the results are, nevertheless, a good preliminary evaluation of the general quality of the method.

The errors related to those experiments weren't evaluated due the difficulty to obtain a good ground-truth that would allow the comparison of results. However, future tests are planned to evaluate and prove the efficiency of the process. These tests will place the ball in a static known distance from the chosen sensor defined as the reference, and apply the resulting transformations of the calibration on the remainder sensors. Hence, if the transformation is correct, the detected center of the ball by each sensor must have the same coordinates.

4.1   Ball Detection

In the ball detection test, and in similar conditions, the ball was placed statically at different distances from the sensors and, for each position, 500 samples of coordinates of the ball center were acquired. From these samples, a mean and a standard deviation to this mean were calculated. Figure 9 shows the standard deviations of the measurements of the three sensors used on this test.

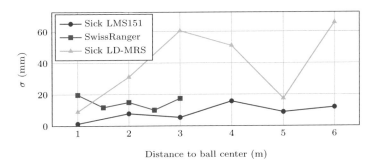

Fig. 9 Standard deviation of ball center detection for several distances.

The results show that the variation verified is consistent with the error associated to each sensor as shown in Table 1 (below 10 cm for the Sick LD-MRS, and about 1 cm for the remaining sensors). In the case of the SwissRanger a greater variation is verified for the closest and furthest distances; it may be due to the proximity of the ball and a lower density of points for further distances, but this is only a suspicion that needs a more detailed study. Nonetheless, the standard deviation in the detection of the ball is not larger than the error associated to the standalone sensors, indicating that the method does not introduce new measurement errors. For the Sick LD-MRS, even with results in accordance with its associated error, it is difficult to evaluate the interference of the distance of the ball on its detection, which can be possibly explained by the fluctuations in the data provided by the sensor; for example, it is possible to have a set of 500 measures of a static object at 10 m, where 80% of the measures have a deviation of ±0.01 m and the remainder 20% have a deviation of ±0.08 m. On the other hand, for the exact same conditions, if a new set of 500 measures is obtained, a ratio of 50%/50% may be found instead.

4.2 Transformation Estimation

In the second test, a calibration between all the sensors was performed using always the same setup; this was done by using different sizes of the point clouds (number of points), where each point of the cloud corresponds to a different position of the ball along its motion in front of the sensors. Thus, for each size of the point clouds, a set of 20 calibrations was performed with the respective matrix of the estimated rigid transformation. The analysis of results compares the translation and rotation; the translation is obtained directly from the matrix of rigid associated geometric transformation; considering the rotation matrix (R) from calibration and R_x, R_y and R_z the generic rotations around each axis, R is defined as $R = R_x(Roll)R_y(Pitch)R_z(Yaw)$, which allows to be solved and obtain the Euler angles ($Roll$, $Pitch$, Yaw). Then, as in the first test, a mean and a standard deviation of the translation and Euler angles are calculated, with the difference that the Euler angles are analyzed individually. Figure 10 shows the results for the translation, and Figure 11 presents the results for the Euler angles.

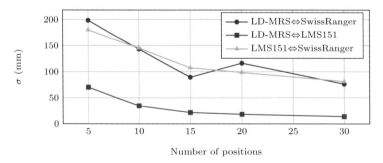

Fig. 10 Standard deviation of the translation between the three sensor pairs.

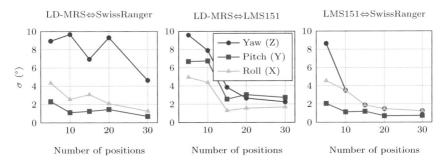

Fig. 11 Standard deviation of the Euler angles between the three sensor pairs.

Figure 12 shows the setup used for the calibration and the estimated positions of the sensors after being calibrated. As expected, the standard deviation decreases with the number of points, and stabilizes at around 20 points; for this test the minimum distance between consecutive points was 10 cm. The mean variation from point clouds of 20 points is lower than 10 cm and about 4°or 5°, which, once again, is in the range of the standalone sensors.

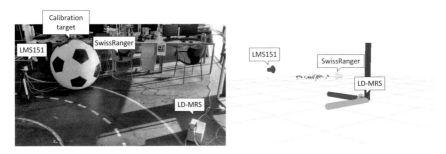

Fig. 12 Layout of the sensors on then scene and positions detected by the calibration process.

Since the tests are limited by the SwissRanger limited maximum range, new calibration tests for the Sick sensors are planned; these new tests are intended to evaluate the influence of the distance of the ball to the sensors.

5 Conclusions and Future Work

This paper proposes a new automatic calibration process capable of calibrating different 2D/3D sensors by using a ball as calibration target. The only initial parameter required concerns the position of the 2D sensor relatively to the ball (whether it's above or below the equator of the ball).

The proposed method presents good initial results for real tests, which is proved also by the diversity of sensors that it supports. However, for the calibration between the two sensors with the lowest measurement uncertainties (Sick LMS151 and Swiss Ranger) better results were expected, which is an issue for future work and study.

The future work involves studies in outdoors, and also the ball in motion in more complete 3D paths, possibly hopping, to solve all ambiguities for any positions of the sensors requiring no a priori information on hemisphere relative position. Additionally, the future work will introduce cameras, making possible a full automatic calibration of all sensors present on the ATLASCAR. The calibration of the cameras will follow the same methodology; through the intrinsic parameters of the cameras and an algorithm able to identify the ball, calculate its center in the frame of each camera. After that, the process is the same used for the sensors.

Aknowledgments This work was partially funded by National Funds through FCT - Foundation for Science and Technology, in the context of the projects UID/CEC/00127/2013 and Incentivo/EEI/UI0127/2014.

References

1. Santos, V., Almeida, J., Avila, E., Gameiro, D., Oliveira, M., Pascoal, R., Sabino, R., Stein, P.: ATLASCAR - technologies for a computer assisted driving system on board a common automobile. In: 13th International IEEE Conference on Intelligent Transportation Systems (ITSC2010), Madeira Island, Portugal (2010)
2. Bok, Y., Choi, D., Kweon, I.: Generalized laser three-point algorithm for motion estimation of camera-laser fusion system. In: IEEE International Conference on Robotics and Automation (ICRA), pp. 2880–2887 (2013)
3. Rodriguez F., S.A., Fremont, V., Bonnifait, P.: Extrinsic calibration between a multi-layer lidar and a camera. In: IEEE International Conference on Multisensor Fusion and Integration for Intelligent Systems, pp. 214–219 (2008)
4. Scaramuzza, D., Harit, A., Siegwart, R.: Extrinsic self calibration of a camera and a 3d laser range finder from natural scenes. In: IEEE/RSJ International Conference on Intelligent Robots and Systems (ICRA), pp. 4164–4169 (2007)
5. Zhang, Q., Pless, R.: Extrinsic calibration of a camera and laser range finder (improves camera calibration). In: Proceedings of the IEEE/RSJ International Conference on Intelligent Robots and Systems, vol. 3, pp. 2301–2306 (2004)
6. Levinson, J., Thrun, S.: Automatic online calibration of cameras and lasers. In: RoboticsScience and Systems (RSS) (2013)
7. Almeida, M., Dias, P., Oliveira, M., Santos, V.: 3D-2D laser range finder calibration using a conic based geometry shape. In: Campilho, A., Kamel, M. (eds.) Image Analysis and Recognition, pp. 312–319. Springer Verlag (2012)
8. Underwood, J., Hill, A., Scheding, S.: Calibration of range sensor pose on mobile platforms. In: IEEE/RSJ International Conference on Intelligent Robots and Systems, pp. 3866–3871 (2007)
9. Chao, G., Spletzer, J.: On-line calibration of multiple lidars on a mobile vehicle platform. In: IEEE International Conference on Robotics and Automation (ICRA), pp. 279–284 (2010)
10. Fernandez-Moral, E., Gonzalez-Jiménez, J., Arévalo, V.: Extrinsic calibration of 2D laser rangefinders from perpendicular plane observations. In: The International Journal of Robotics Research (2015)
11. Rusu, R.B., Cousins, S.: 3D is here: point cloud library (PCL). In: IEEE International Conference on Robotics and Automation (ICRA), pp. 1–4 (2011)

12. Fischler, M.A., Bolles, R.C.: Random sample consensus: a paradigm for model fitting with applications to image analysis and automated cartography. Commun. ACM **24**(6), 381–395 (1981)
13. Coimbra, D.: LIDAR Target Detection and Segmentation in Road Environment. Masters' thesis, Universidade de Aveiro (2013)
14. Xavier, J., Pacheco, M., Castro, D., Ruano, A.: Fast line, arc/circle and leg detection from laser scan data in a player driver. In: Proceedings of the 2005 IEEE International Conference on Robotics and Automation (ICRA), pp. 3930–3935 (2005)
15. Arun, K.S., Huang, T.S., Blostein, S.D.: Least-squares fitting of two 3-D point sets. IEEE Transactions on Pattern Analysis and Machine Intelligence (PAMI) **9**(5), 698–700 (1987)
16. Horn, B.K.P.: Closed-form solution of absolute orientation using unit quaternions. Journal of the Optical Society of America A **4**(4), 629–642 (1987)
17. Quigley, M., Berger, E., Ng, A.Y.: STAIR: hardware and software architecture. In: AAAI 2007 Robotics Workshop (2007)

Pedestrian Pose Estimation Using Stereo Perception

Jorge Almeida and Vitor Santos

Abstract This paper presents an algorithm to perform pedestrian pose estimation using a stereo vision system in the Advanced Driver Assistance Systems (ADAS) context. The proposed approach isolates the pedestrian point cloud and extracts the pedestrian pose using a visibility based pedestrian 3D model. The model accurately predicts possible self occlusions and uses them as an integrated part of the detection. The algorithm creates multiple pose hypotheses that are scored and sorted using a scheme reminiscent of the Monte Carlo techniques. The technique performs a hierarchical search of the body pose from the head position to the lower limbs. In the context of road safety, it is important that the algorithm is able to perceive the pedestrian pose as quickly as possible to potentially avoid dangerous situations, the pedestrian pose will allow to better predict the pedestrian intentions. To this end, a single pedestrian model is used to detect all pertinent poses and the algorithm is able to extract the pedestrian pose based on a single stereo depth point cloud and minimal orientation information. The algorithm was tested against data captured with an industry standard motion capture system. Accurate results were obtained, the algorithm is able to correctly estimate the pedestrian pose with acceptable accuracy. The use of stereo setup allows the algorithm to be used in many varied contexts ranging from the proposed ADAS context to surveillance or even human-computer interaction.

1 Introduction

Pedestrians are one of the most vulnerable and unpredictable road users. The pedestrians ability to suddenly start motion or change direction can create a dangerous

J. Almeida(✉) · V. Santos
Department of Mechanical Engineering, University of Aveiro, Aveiro, Portugal
e-mail: {almeida.j,vitor}@ua.pt

V. Santos
IEETA, University of Aveiro, Aveiro, Portugal

© Springer International Publishing Switzerland 2016   491
L.P. Reis et al. (eds.), *Robot 2015: Second Iberian Robotics Conference*,
Advances in Intelligent Systems and Computing 417,
DOI: 10.1007/978-3-319-27146-0_38

situation in hundreds of milliseconds. In the Advanced Driver Assistance Systems (ADAS) context, the prediction of the pedestrians' intentions could potentially prevent accidents and possible injuries. For instance, the detection of the pedestrian intent to either cross a road at a crosswalk or to stop. Systems that are able to perceive pedestrian motion as soon as possible will improve safety for road users. In [1], the authors studied how humans detect the intentions of pedestrians to cross the road. The authors presented the participants videos of pedestrians crossing in natural traffic situations. The authors conclude that parameters of body language, such as legs or head movements, are indispensable for a consistent behavior prediction. Pedestrian trajectories alone are not sufficient to a correct and robust prediction. In this context, estimation of the pedestrian pose is of crucial importance to achieve a fast response system.

In this work, a technique to estimate the body pose of pedestrians is presented; with this estimation a subsequent system could potentially interpreter the poses to perform motion recognition. To achieve the proposed goal, the system must not depend on any previous manual initialization step or on a multi-frame tracking system. As such, the proposed system is able to estimate the pose from a single frame and minimal prior orientation information. The system performs a hierarchical top down geometrical search on a segmented pedestrian point cloud using an anthropomorphic constrained sampling scheme to detect body parts and limbs.

The human body pose estimation is a complex task with a large number of possible applications; robust interactive human body tracking has applications including gaming, humancomputer interaction, security, telepresence, and health care [2]. The problem is made complex due to the high dimensional search-space, frequent ambiguities between poses and high number of local minima. A great deal of research has been dedicated to detect human poses based only on monocular vision systems (survey in [3]); this is an especially ill-posed problem due to fact that many different poses present the same image projection. In this work, the use of a stereo system is proposed. This system provides dense 3D point clouds by using a state of the art stereo matching algorithm. The extra information provided by the point cloud, depth information, relieves many of the ambiguities in pose compared to the monocular system. Existing high performance depth-based systems are mostly dependent on structured light sensors [2]; theses sensors provide high precision, frame rate and definition, but are not suited to work on outdoors environments due to saturation of the sensor and range limits, therefore are not applicable in the ADAS context. The stereo setup is still the most attractive approach in the ADAS context given its low cost and low complexity, especially compared to active laser systems [4].

In section 2 the related work in markerless pose detection is presented. The proposed system is described in 3 with body parts detections in 3.4. The experimental results are presented in section 4 and final conclusions are presented in section 5.

2 Related Work

Previous work on markerless detection and tracking of a human body pose has been primarily focused in the use of intensity images, as stated above. In [3] the authors provide a survey of the different techniques used. The authors mark the distinction between model-based (generative) and model-free (discriminative) approaches, with the model-based methods using *a priori* information of the human body.

In [5], the authors propose a generic model for human detection and articulated pose estimation. The authors train detectors for anatomically defined body parts, which are then used as the likelihood in a generative model. The authors employ a flexible kinematic tree prior using pictorial structures on the configuration of body parts. In [6], the authors expand the previous work to include evidences from multiple frames. They model the temporal prior as a hierarchical Gaussian Process Latent Variable Model (hGPLVM) combined with Hidden Markov Model (HMM) to extend pedestrian tracklets. Their approach generates bottom-up evidence from 2D body models and so it constitutes a hybrid generative/discriminative approach.

The work proposed in [7] treats pose estimation as a nonlinear regression problem and proposes to estimate body poses directly from silhouette images. They employ a discriminative learning approach of body parts and embedded the algorithm in a tracking framework to facilitate disambiguation between poses. The absence of a previous model makes their technique easily adapted to different people, appearances or representations of 3D body poses.

Current monocular systems suffer from pose ambiguity problems due to the limitations of data used. These systems employ tracking architectures to solve pose ambiguity but the tracking implies the need to use multiple frames increasing the response time of these systems.

Work has also been performed using multiple monocular cameras to help with pose ambiguity. In [8], the authors propose to perform 3D human upper body pose estimation using multiple camera views. Their system creates multiple 3D pose hypotheses on a single view using a probabilistic hierarchical shape matching algorithm. These hypotheses are re-projected into other camera views and are then ranked according to their likelihood. Their system also applies a tracking mechanism integrating a motion model and observations in a maximum-likelihood approach. The need of multiple points of view severely limits the applicability of these systems.

Recently, the introduction of real-time depth cameras simplified greatly the pose estimation problem, when compared to monocular systems. The work presented in [9] makes use of a time-of-flight camera to estimate human body pose at video frame rates. The authors take a bottom-up approach to detect the body pose, starting with an interest point detector with a subsequent classification system.

Stereo has been previously applied to estimate human body pose, [10, 11]. In [12] the authors treat the pose tracking problem as a registration of two 3D point sets. The authors integrate Iterative Closest Point (ICP) with an unscented Kalman filter to yield a registration algorithm capable of tracking articulated bodies. In [13], the authors propose a system that uses stereo vision and a skin color filter. The skin color filter is used as a segmentation method to extract the point cloud belonging to the human body. The approach uses multiple models in different poses and computes an error metric to identify the correct pose. The work was performed in indoor environments and focused on upper body poses. The algorithm proposed in [14] also makes use of a variant of the ICP algorithm to match a simplified human model. The authors apply a Kalman filter based tracking architecture with a subsequent pose classification based on HMMs. All the proposed systems are either based on tracking algorithms or are not applicable in the ADAS context.

In the topic of predicting pedestrians' intentions in the ADAS context, the work by [15] presents a system that is able to predict if a pedestrian, walking towards the road curbside, will cross the road or stop. Asides from classification, the system uses dense optical flow from a stereo camera, with egomotion compensation, to obtain motion clues for the pedestrian upper torso and legs. A dimensional reduction using Principal Component Analysis (PCA) is applied to create Histogram of Orientation Motion (HOM) features. The current motion is matched to the database using Quaternion-based Rotationally Invariant Longest Common Subsequence (QRLCS) similarity metric.

On the same topic, the work by [16] presents a system that allows to detect early the intention of a pedestrian to cross a road lane. This system uses the body language as an early indicator of a crossing intent. Their system uses an infrastructure monocular vision system to extract Motion Contour Histogram of Oriented Gradients (MCHOG) feature descriptor. They apply a linear Support Vector Machine (SVM) system to identify the point when the pedestrian starts to enter the lane.

Both of these works would benefit from a more accurate and complete perception of the pedestrian motion. With additional detail the pedestrians' intentions could be inferred more accurately and also sooner. The use of stereo vision makes possible pose estimation in outdoors environments. The system is less susceptible to pose ambiguity, a serious problem in monocular systems, and performs well in outdoors environments with the desirable range. The proposed systems focus attention in the pertinent poses in ADAS context, especial attention is given to the legs pose. Previous works do not focus on this problem neither present a solution with the required characteristics; a solution that works in outdoors environments capable of, quickly and without initialization, estimate the pose of the human lower limbs during a normal walking cycle.

3 Stereo Pose Estimation

Human body poses are obtained using 3D point clouds from a stereo camera, as shown in 1. The pose estimation is performed using a method that compares the visibility of the point cloud from the stereo camera with the expected visibility from a pose hypothesis.

The visibility at each point is defined as one of three possible values: free space, occupied or occluded. A free space classification indicates that a point is visible from the camera point of view but is not occupied. A occupied point is visible from the camera and occupied by a 3D point. Finally an occluded point is a point that is not visible by the camera because there is an occupied point in front.

A dense voxel cloud is created overlapping the extracted pedestrian point cloud. A set of 3D rays interests this dense cloud, the intercepted voxels for each ray are classified according to their visibility using the pedestrian point cloud as the blocking element. After classification, this dense voxel cloud will be the base element for calculating the score of different hypotheses.

For each body part hypothesis, a set of 3D rays is used to calculate the visibility. The hypothesis score is calculated by comparing the classification of the points intercepted by the rays and the corresponding classification of the original dense voxel cloud.

When calculating the visibility of body parts hypotheses, previous detected body parts are used as blocking elements, for instance: the first detected leg will occlude the hypotheses for the second leg. This method allows to estimate the position of the occluded leg.

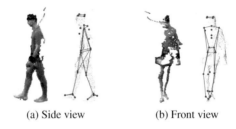

(a) Side view   (b) Front view

Fig. 1 Example of an estimated pose. On the left the segmented pedestrian point cloud, on the right the estimated pose. The arms are not detected.

This work uses data from an industry standard motion capture system as ground truth. The motion capture system provides millimeter accurate position of a set of infrared reflective markers, visible on figure 1. To establish a direct comparison, a set of virtual markers, matching the motion capture markers, is used by the pose estimation algorithm.

3.1 Preprocessing

To extract the pedestrian point cloud three steps are applied: ground plane estimation, background subtraction and Euclidean clustering. The ground place estimation uses the RANSAC algorithm and helps to remove points near the feet. The background subtraction algorithm removes most of the points not belonging to the pedestrian. Finally, the resulting points from the two previous steps are clustered according to Euclidean distance between them and a specified threshold, the largest cluster is assumed to be the pedestrian.

This pedestrian extraction scheme works well in the dataset used, but in a more complex scenario some other state-of-the-art pedestrian detection algorithm could be used to segment the pedestrian point cloud. The developed algorithm does not require a perfect segmentation of the pedestrian from the background.

3.2 Visibility Calculation

The pose estimation algorithm here proposed assumes that a point cloud, comprised mostly of points belonging to a single pedestrian, was previously obtained. It is also assumed that the pedestrian is in an upright pose, a common assumption in the pedestrian detection context.

As stated before, ray tracing is used to calculate which voxels are either free, occupied or occluded, figure 2. The algorithm defines a set of rays using the original pedestrian cloud and the sensor position. For each ray, the intercepted voxels are classified. The end result is a dense voxel cloud in which each voxel contains the above classification, $V_{pedestrian}$. This process is repeated for the pose hypotheses. Each body part pose hypothesis consists of a 3D model of the part, section 3.3, in a hypothesis pose. For each hypothesis the visibility is calculated. A score is obtained comparing the visibility of the hypothesis with the visibility of the original cloud.

In figure 2 two torso samples are presented. Each sample represents the same 3D model but in a different pose. The left hypothesis has a much larger area visible to the sensor and, as such, a much larger occluded volume. The left sample is aligned with the pedestrian, therefore the visibility will be very similar. The right sample will score a much higher value that the left sample.

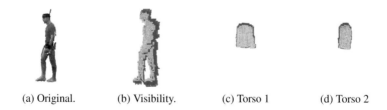

(a) Original.   (b) Visibility.   (c) Torso 1   (d) Torso 2

Fig. 2 Visibility dense voxel cloud representation. On the left, the original point cloud and the visibility calculated with the cloud. On the right, two different samples used to detect the torso orientation. Occupied voxels are represented as yellow squares, occluded voxels are colored blue. Empty voxels are not represented but used to score the sample.

Let $V = \{v_1, \dots, v_N\}$ represent all the voxels in the hypothesis, the score of each hypothesis Ψ_i is calculated as the sum, equation (2), of the score of every voxel, equation (1).

$$\forall v \in V, s(v) = \begin{cases} P1 & \Leftarrow (v = v_{\text{pedestrian}}) \wedge (v = \text{free}) \\ P2 & \Leftarrow (v = v_{\text{pedestrian}}) \wedge (v = \text{occluded}) \\ P3 & \Leftarrow (v = v_{\text{pedestrian}}) \wedge (v = \text{occupied}) \\ P4 & \Leftarrow (v = \text{occluded}) \wedge (v_{\text{pedestrian}} = \text{occupied}) \\ P4 & \Leftarrow (v = \text{occupied}) \wedge (v_{\text{pedestrian}} = \text{occluded}) \\ 0 & \Leftarrow \text{otherwise} \end{cases} \tag{1}$$

$$\Psi_i = \frac{\sum\limits_{n=1}^{N} s(v_n)}{N} \tag{2}$$

The different weights ($P1$, ..., $P4$) in equation (1) allow the algorithm to compensate for the different percentage of voxels with each classification.

Several performance optimizations were applied. The ray tracing can be very computationally expensive, as such, it is only performed once, for the $V_{\text{pedestrian}}$ cloud. The rays and the intercepted voxels positions are reused for each pose hypotheses. The samples, after transformation, are geometrically aligned to the $V_{\text{pedestrian}}$ cloud to allow the reuse of the rays. The geometric alignment of the samples also allows for a very fast indexing of the two clouds, avoiding the need for expensive nearest neighbor searches.

Ray tracing is not performed for each point in the pedestrian cloud. The rays are created starting in the sensor position and defining a square angular grid with a specific vertical and horizontal resolution, R_V and R_H respectively. The grid limits are defined from the point cloud, as to avoid unnecessary rays. The vertical and horizontal resolutions are key parameters of the algorithm. A more refined grid will account for greater detail, with the limit of the sensor own angular resolution, while a more coarse grid will correspond to lower number of rays improving computational performance.

3.3 3D Model

The proposed algorithm compares the visibility of a pose hypothesis with the visibility of the current pedestrian point cloud. To this end, a realistic geometric 3D model of a pedestrian is used. The 3D model defines the shape that will be used to calculate the visibility of each different pose hypothesis. The method is hierarchical and sequential, the first body part to be detected is the torso, followed by the head and upper legs, and finally the lower legs. As such, the 3D model was segmented into different body parts for individual use.

Let $\mathcal{P} = \{p_1, \dots, p_N\}$ represent the pedestrian point cloud with N points. The overall bounding box of $\mathcal{P}$ provides a rough approximation to the pedestrian height. The height approximation allows to estimate the size of the different body parts. The original 3D model is scaled to fit this measurement.

3.4 Detecting Body Parts

The first body part to be detected is the torso. The torso pose is extracted in three steps.

The pivot position is directly defined from the centroid position and a penetration factor. The penetration factor is used to correct the centroid in the sensor direction, placing the torso pivot inside the body and not at the surface.

The second step estimates the torso orientation θ_{torso} in the vertical direction $\hat{z}$. To this end, a set of samples is created with different orientation angles. Each sample is scored and a graphic, such as figure 3, is obtained. From this graphic, it is clearly visible that, there are two main peaks with 180° offset. The two peaks appear due to the fact that the torso shape is similar on the front and back, leading to pose ambiguity. To solve this ambiguity more information is required. In the proposed method, the θ_{torso} maximum closest to the previous estimated orientation is used.

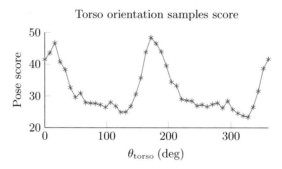

Fig. 3 Torso orientation samples score. The two peaks are created by the ambiguity between the front and back of the torso. The algorithm is able to correctly estimate the correct orientation using the peak closest to the previous orientation.

The third step estimates the torso forward inclination ϕ_{torso}, the rotation on the axis perpendicular to the vertical direction and the direction derived from the torso orientation $\hat{\phi} = \hat{z} \times \hat{\theta}$. This rotation is especially important when the pedestrian is moving quickly or running.

The head pose is estimated after the torso pose. The head pivot is directly derived from the torso pose and a set of samples is created to detect the head rotation θ_{head} in the vertical axis $\hat{z}$.

After estimation of the head pose, the legs positions are estimated. The algorithm starts by identifying which leg is more exposed to the sensor as a function of its predicted distance. This distance is based on the hip distance using the torso pose. The pose of the leg more exposed is the first to be estimated. Each leg is segmented in two parts, the upper leg and the lower leg. The upper leg comprises the distance from the hip to the knee, and the lower leg the distance from the knee to the foot.

The upper leg samples are created using two degrees of freedom, rotation on the $\hat{\phi}$ axis and rotation on the $\hat{\theta}$ axis. The upper leg pivots on the hip joint, defined by the torso pose. A set of samples is created by composed rotation of the two degrees of freedom. The samples are scored using the method described above. The lower leg samples pivots on the knee joint and rotates on the two same axes. All rotations are limited by anthropomorphic constrains.

The second leg pose is only estimated after the first. The first leg pose will influence the visibility of the second leg. The first leg will be used as an obstacle when calculating the visibility for the second leg. This method allows to estimate the position of the leg even when it is occluded. The created samples will reflect the fact that there is an obstacle in front and samples that are occluded will be correctly classified.

4 Results

The proposed algorithm was compared to a high precision industry standard motion capture system. The test trial consisted of a simulated pedestrian road crossing, figure 4. In the trial, several pedestrian trajectories were obtained. The test was composed of pedestrian trajectories parallel to the sensor, perpendicular and at an angle. The test contained trajectories where the pedestrian stopped at the simulated road entrance, and also trajectories where the pedestrian runs. The trial consisted of a total of 1588 frames, of witch 1053 were used. Frames where the pedestrian was not fully visible in the stereo camera were discarded. Also, the motion capture system was not always able to acquire all markers, in a frame, if a specific maker was not found the pose estimation marker was discarded.

(a) Walking perpendicular. (b) Walking parallel. (c) Running. (d) Walk and stop.

Fig. 4 Sample images from the trial. The images present some of the several different trajectories used. The running trajectories were affected by the weak lighting conditions of the laboratory that led to some blurry images.

Quantitative results were obtained. A direct comparison was made possible by defining virtual markers analogous to the motion capture markers, on the 3D body parts. Figure 5 presents the histogram of the Euclidean distance from the motion capture markers to the pose estimation markers for the whole trial. The markers placement on the 3D body parts affects the results. Incorrect placement will appear as error on figure 5, an attempt to minimize this error was made. Table 1 presents the parameters values used in the trial.

Table 1 Parameters used in the test trial.

Parameter	Value
$P1$	10
$P2$	50
$P3$	100
$P4$	1
R_V	1.5°
R_H	0.5°

As can be observed, a large percentage, 72%, of the results are under 0.1m, and 94% of results are under 0.2m. The person's self occlusion presents some serious challenges, typically only one shoulder is visible and legs frequently occlude each other. The proposed method allows to estimate the person's orientation even with high occlusions. Given the hierarchical nature of the method, lower body parts suffer from errors in the upper parts. To account for this fact, lower body parts' samples are created with broader limits that would otherwise be necessary. Figure 6 presents the results for pose orientation. This orientation is calculated using the shoulders markers projected on the $X - Y$ plane. The figure presents a histogram of the body orientation error of the algorithm.

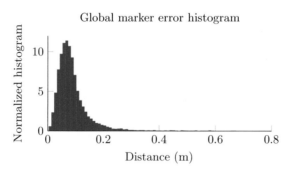

Fig. 5 Histogram of the euclidean distance between each marker of the pose estimation and the motion capture system.

The pose orientation is estimated with good accuracy. The largest errors occur when the pedestrian runs. The stereo setup used, performed poorly on low light conditions, such as the motion capture laboratory. Fast movements cause the image to become blurred due to the large exposure time. This in turn, decreased the quality of the stereo algorithm.

The stereo data used is of good quality but, nevertheless, presents some pronounced noise; the stereo noise presents the main limitation to the accuracy of the proposed approach.

The lack of a strong prior in our algorithm presents some advantages, but also disadvantages. With a good prior, the search space for each body part could be dramatically reduced, thus improving estimation accuracy. The current proposal could

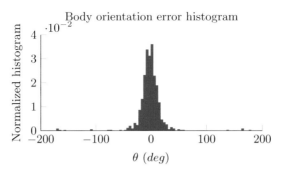

Fig. 6 Histogram of the body orientation error for the trial.

be expanded to use such a tracker. The presented algorithm, as is, could be used to initialize the tracker and also to recover from failure.

5 Conclusions

An algorithm capable of detecting human poses using stereo point clouds was presented. The algorithm is able to estimate poses using single point clouds and minimal motion orientation, used to relieve ambiguity between left and right poses. The proposed approach uses a hierarchical visibility based pose estimation algorithm. The algorithm focuses attention on the legs position, the legs motion will provide cues on the early intention of pedestrians trying to enter or cross a road.

The algorithm was tested with millimeter accurate industry motion capture data of a pedestrian simulating a possible pedestrian road crossing. Results presented show the potential of the algorithm to correctly recover poses even with noisy stereo data. The stereo setup presents some serious advantages over traditional monocular systems or even structured light systems. The point cloud data presents much less pose ambiguity than a monocular system and has the advantage of working in outdoors environments at long ranges.

Our proposed algorithm does not require any pose initialization or an elaborate pose tracking algorithm. This presents an obvious advantage by allowing the estimation of the pose of a pedestrian entering the scene without the need of a long multi-frame tracking system that would delay any conclusion. Nevertheless, the posterior application of a tracking algorithm would improve computational performance as well as performance under occlusion. The proposed algorithm could be used in the initialization step of the tracker or to recover from failure.

Future work will be focused on the implementation of a probabilistic pose tracker and finally on a system integrating the pose detection with the estimation of the pedestrians intentions in an advanced pedestrian safety system.

Acknowledgments This work was supported by the Portuguese foundation for Science and Technology (FCT) under grant SFRH/BD/73181/2010 and partially funded in the context of the projects UID/CEC/00127/2013 and Incentivo/EEI/UI0127/2014.

References

1. Schmidt, S., Färber, B.: Pedestrians at the kerb recognising the action intentions of humans. Transportation Research Part F: Traffic Psychology and Behaviour **12**(4), 300–310 (2009)
2. Shotton, J., Sharp, T., Kipman, A., Fitzgibbon, A., Finocchio, M., Blake, A., Cook, M., Moore, R.: Real-time human pose recognition in parts from single depth images. Commun. ACM **56**(1), 116–124 (2013)
3. Poppe, R.: Vision-based human motion analysis: An overview. Computer Vision and Image Understanding **108**(1–2), 4–18 (2007)
4. Geronimo, D., Lopez, A., Sappa, A., Graf, T.: Survey of pedestrian detection for advanced driver assistance systems. IEEE Trans. Pattern Anal. Machine Intell. **32**(7), 1239–1258 (2010)
5. Andriluka, M., Roth, S., Schiele, B.: Pictorial structures revisited: people detection and articulated pose estimation. In: IEEE Conference on Computer Vision and Pattern Recognition, (CVPR), pp. 1014–1021 (2009)
6. Andriluka, M., Roth, S., Schiele, B.: Monocular 3D pose estimation and tracking by detection. In: IEEE Conference on Computer Vision and Pattern Recognition, (CVPR), pp. 623–630 (2010)
7. Agarwal, A., Triggs, B.: Recovering 3D human pose from monocular images. IEEE Trans. Pattern Anal. Machine Intell. **28**(1), 44–58 (2006)
8. Hofmann, M., Gavrila, D.M.: Multi-view 3D human pose estimation in complex environment. International Journal of Computer Vision **96**(1), 103–124 (2012)
9. Plagemann, C., Ganapathi, V., Koller, D., Thrun, S.: Real-time identification and localization of body parts from depth images. In: IEEE International Conference on Robotics and Automation, (ICRA), pp. 3108–3113 (2010)
10. Urtasun, R., Fua, P.: 3D human body tracking using deterministic temporal motion models. In: Pajdla, T., Matas, J.G. (eds.) Computer Vision, (ECCV). LNCS, vol. 3023, pp. 92–106. Springer, Heidelberg (2004)
11. Yang, H.D., Lee, S.W.: Reconstruction of 3D human body pose from stereo image sequences based on top-down learning. Pattern Recognition **40**(11), 3120–3131 (2007)
12. Ziegler, J., Nickel, K., Stiefelhagen, R.: Tracking of the articulated upper body on multi-view stereo image sequences. In: IEEE Computer Society Conference on Computer Vision and Pattern Recognition, (CVPR), vol. 1, pp. 774–781 (2006)
13. Muhlbauer, Q., Kuhnlenz, K., Buss, M.: A model-based algorithm to estimate body poses using stereo vision. In: IEEE International Symposium on Robot and Human Interactive Communication, (RO-MAN), pp. 285–290 (2008)
14. Pellegrini, S., Iocchi, L.: Human posture tracking and classification through stereo vision and 3D model matching. J. Image Video Process. 2008, 7:1–7:12, January 2008
15. Keller, C.G., Hermes, C., Gavrila, D.M.: Will the pedestrian cross? probabilistic path prediction based on learned motion features. In: Mester, R., Felsberg, M. (eds.) Pattern Recognition. LNCS, vol. 6835, pp. 386–395. Springer, Heidelberg (2011)
16. Kohler, S., Goldhammer, M., Bauer, S., Doll, K., Brunsmann, U., Dietmayer, K.: Early detection of the pedestrian's intention to cross the street. In: IEEE Conference on Intelligent Transportation Systems, ITSC, pp. 1759–1764 (2012)

Scene Representations for Autonomous Driving: An Approach Based on Polygonal Primitives

Miguel Oliveira, Vítor Santos, Angel D. Sappa and Paulo Dias

Abstract In this paper, we present a novel methodology to compute a 3D scene representation. The algorithm uses macro scale polygonal primitives to model the scene. This means that the representation of the scene is given as a list of large scale polygons that describe the geometric structure of the environment. Results show that the approach is capable of producing accurate descriptions of the scene. In addition, the algorithm is very efficient when compared to other techniques.

Keywords Scene reconstruction · Point cloud · Autonomous vehicles

1 Introduction

Recent research in the fields of pattern recognition suggest that the usage of 3D sensors improves the effectiveness of perception, "since it supports good situation aware-

M. Oliveira(✉)
Instituto de Engenharia de Sistemas e Computadores,
Tecnologia e Ciência, R. Dr. Roberto Frias, 465, 4200 Porto, Portugal
e-mail: miguel.r.oliveira@inesctec.pt

V. Santos · P. Dias
Institute of Electronics and Telematics Engineering of Aveiro,
Campus Universitario de Santiago, 3800 Aveiro, Portugal

V. Santos
Department of Mechanical Engineering, University of Aveiro,
Campus Universitario de Santiago, 3800 Aveiro, Portugal

A.D. Sappa
Computer Vision Center, Edificio O, Campus UAB, Bellaterra, 08193 Barcelona, Spain

A.D. Sappa
Facultad de Ingeniería Eléctrica y Computación (FIEC), Escuela Superior Politécnica del Litoral
(ESPOL), Campus Gustavo Galindo, Km 30.5 Vía Perimetral, 09-01-5863, Guayaquil, Ecuador

© Springer International Publishing Switzerland 2016
L.P. Reis et al. (eds.), *Robot 2015: Second Iberian Robotics Conference*,
Advances in Intelligent Systems and Computing 417,
DOI: 10.1007/978-3-319-27146-0_39

503

ness for motion level tele operation as well as higher level intelligent autonomous functions" [3]. Nowadays autonomous robotic systems have at their disposal a new generation of 3D sensors, which provide 3D data of unprecedented quality [16]. In robotic systems, 3D data is used to compute some form of internal representation of the environment. In this paper, we refer to this as 3D scene representation or simply 3D representation. The improvement of 3D data available to robotic systems should pave the road for more comprehensive 3D representations. In turn, advanced 3D representations of the scenes are expected to play a major role in future robotic applications since they support a wide variety of tasks, including navigation, localization, and perception [4].

In summary, the improvement in the quality of 3D data clearly opens the possibility of building more complex scene representations. In turn, more advanced scene representations will surely have a positive impact on the overall performance of robotic systems. Despite this, complex scene representations have not yet been substantiated into robotic applications. The problem is how to process the large amounts of 3D data. In this context, classical computer graphics algorithms are not optimized to operate in real time, which is an non-negotiable requirement of the majority of robotic applications. Unless novel, efficient methodologies are introduced, which produce compact yet elaborate scene representations, robotic systems are limited to mapping the scenes in classical 2D or 2.5D representations or are restricted to off-line applications.

Very frequently, the scenarios where autonomous systems operate are urban locations or buildings. Such scenes are often characterized for having a large number of well defined geometric structures. In outdoor scenarios, these geometric structures could be road surfaces or buildings, while in indoor scenarios they may be furniture, walls, stairs, etc. We refer to the scale of these structures as a macro scale, meaning that 3D sensor may collect thousands of measurements of those structures in a single scan. A scene representations is defined by the surface primitive that is employed. For example, triangulation approaches make use of triangle primitives, while other approaches such as Poisson resort to implicit surfaces. Triangulation approaches generate surface primitives that are considered to have a micro scale, since a geometric structure of the scene could contain hundreds or thousands of triangles. Micro scale primitives are inadequate to model large scale environments because they are not compact enough.

In this paper, we present a novel methodology to compute a 3D scene representation. The algorithm uses macro scale polygonal primitives to model the scene. This means that the representation of the scene is given as a list of large scale polygons that describe the geometric structure of the environment. The proposed representation addresses the problems that were raised in previous lines: the representation is compact and can be computed much faster than most others, while at the same time providing a sufficiently accurate geometric representation of the scene from the point of view of the tasks required by an autonomous system.

Scene reconstruction is defined as the computation of a geometric 3D model from multiple measurements. These measurements could be obtained from stereo systems, range sensors, etc. It could also include the texture mapping of the generated model.

Scene reconstruction methodologies are grouped into two different approaches: surface based representations or volumetric occupancy representations. In the first, the underlying surfaces of the scene that generated the range measurements are estimated, while in the second, the range measurements are grouped into grid cells which are then as labelled free or occupied. Traditional surface based representations include several 3D triangulations methodologies, such as 3D Delaunay triangulation [10], or Ball Pivoting Algorithm (BPA) [2]. The Greedy triangulation (GT) is an approach designed for fast surface reconstruction from large noisy data sets [12]. Given an unorganized 3D point cloud, the algorithm recreates the underlying surface's geometrical properties using data re-sampling and a robust triangulation algorithm, the authors claim to achieve near real time. There are also some alternative higher order surface representation methods such as Poisson surface reconstruction [11], Orientation Inference Framework [6] or learning approaches [13]. However, most of these methods do not tackle well noisy range measurements and, above all, since these methods involve a large number of nearest neighbor queries, they are very slow to compute. One attempt to accelerate the triangulation of point clouds was done in [12], but authors report they have only achieved near real time. Volumetric occupancy representations include occupancy grids [17], elevation maps [14], multi-level surface maps [15] or octrees [18]. While these representations are easier to compute, they do not provide accurate information about the geometry of the scene. The remainder of this paper is organized as follows: section 2 presents the proposed approach, results are given in section 3 and conclusions in section 4.

2   Proposed Approach

This work proposes to explore the usage of geometric polygonal primitives to perform scene reconstruction. In other words, the idea is to describe a scene by a list of polygons. The detection of polygonal geometric primitives is simple when compared to the detection of other more complex primitives. Furthermore, given that road environments are often geometrically structured, it seems feasible to represent the 3D structure with a set of planar polygons. In addition to that, polygons are compact representations, which require only the support plane and a list of points to be described.

Geometric polygonal primitives are described by a support plane and a bounding polygon. Let $\mathcal{G}^i$ represent the *ith* polygonal geometric primitive of a given scene, with the support plane Hessian form coefficients denoted by $\mathcal{G}^i_p = \begin{bmatrix} a^i & b^i & c^i & d^i \end{bmatrix}$. The search for the support plane is done on a given input point cloud $\mathcal{P}$ using a RANSAC procedure [7]. RANSAC is an iterative method to estimate parameters of a mathematical model from a set of observed data points. The assumption is that data consists of *inliers*, i.e., data whose distribution can be explained by some set of model parameters, and *outliers*, data that does not fit the model. The input to the RANSAC algorithm is a set of observed data values, a parametrized model which is

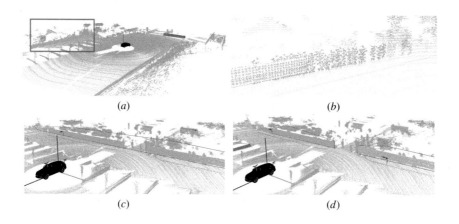

Fig. 1 Plane detection examples using RANSAC: (*a*) five best RANSAC candidates for the input point cloud in grey; (*b*) a detail of (*a*); (*c*) without using clustering; (*d*) using clustering;

fitted to the observations, and the output are the model parameters, i.e., in the case of detecting the support plane of polygonal primitives the Hessian coefficients.

Figure 1 (*a*) shows in different colors the inlier points of the five best candidates of a RANSAC search. Wall like structures are correctly detected. Figure 1 (*c*) shows the inliers (signalled in green) of a RANSAC plane detection. In this case, range measurements from two separate walls have been signalled as inliers of a single support plane. To address this issue, the set of inliers of each support plane hypothesis is used as input to a clustering procedure. Using the proposed clustering algorithm it is possible to separate the two walls into separate polygons as shown in Fig. 1 (*d*).

The computation of the bounding polygon is done after the detection of the support plane. The bounding polygon P^i is defined by a list of 2D points p:

$$P^i = [p_0, p_1 ... p_n] = \begin{bmatrix} x_0 \ x_1 \ ... \ x_n \\ y_0 \ y_1 \ ... \ y_n \end{bmatrix} \tag{1}$$

where n is the number of points in the polygon. In order to define the polygon points in $\mathbb{R}^2$ (which saves memory), a local reference system for each polygon primitive is defined, with Z axis normal to the support plane, and the orientation of the remaining axes defined arbitrarily. Polygons are computed using a 2D convex hull operation. In this work the implementation provided in [8] is used to compute the 2D convex hull, based on a non recursive version of [5], presented in [1].

The proposed detection of polygonal primitives is designed in a cascade like processing architecture, which is efficient and fast to process. The input point cloud contains a large amount of 3D points. We assume 3D points can only belong to a single polygonal primitive, which makes sense since polygonal primitive represent objects

---

Algorithm 1. Cascade configuration for the detection of geometric polygonal primitives

---

Require: $\mathcal{P}^{it=0}$, the input point cloud at iteration 0
Ensure: A list of geometric polygonal primitives $\mathbf{G} = \{\mathcal{G}^0, \mathcal{G}^1, ..., \mathcal{G}^n\}$
  Initialize number of primitives, $k \leftarrow 0$
  Initialize number of iterations, $it \leftarrow 0$
  Initialize primitives list, $\mathbf{G} \leftarrow \{\}$
  Initialize cycle break flag, $cycle\_break \leftarrow$ `false`
  while $cycle\_break =$ `false` **do**
      RANSAC search over $\mathcal{P}^k$, returns estimated plane $\hat{\mathcal{G}}_p^k$ (first guess) and inliers $\mathcal{I}^k$
      if RANSAC found a candidate **then**
         Cluster inliers point cloud $\mathcal{I}^k$ to cluster list $\mathbf{C}=\{C^0, C^1, ..., C^n\}$
         Find largest cluster, $max\_cluster = argmax_i(\mathbf{size}(C^i))$
         Set the primitive support points $\mathcal{S}^k$ to the largest cluster, $\mathcal{S}^k = C^{max\_cluster}$
         Compute accurate plane coefficients from support points, $\mathcal{G}_p^k \leftarrow$ PCA over $\mathcal{S}^k$
         Compute bounding polygon P^k, its area $\mathbf{A}(\mathrm{P}^k)$ and solidity $\mathbf{S}(\mathrm{P}^k)$
         if $\mathbf{A}(\mathrm{P}^k) > \mathbf{A}_t$ and $\mathbf{S}(\mathrm{P}^k) > \mathbf{S}_t$ **then**
            Add to primitive list, $\mathbf{G} \leftarrow \{\mathbf{G}, \mathcal{G}^k\}$
            increment number of primitives, $k \leftarrow k + 1$
         end if
         Remove support points $\mathcal{S}^k$ from $\mathcal{P}^{it}$, compute $\mathcal{P}^{it+1}$
      else
         Finish search for primitives, $cycle\_break =$ `true`
      end if
      increment number of iterations, $it \leftarrow it + 1$
  end while

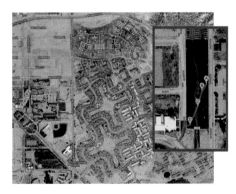

Fig. 2 Sequence collected from the Massachusetts Institute of Technology (MIT) dataset.

in the scene. Let $\mathcal{S}^k$ be the point cloud containing the support points of primitive k, and $\mathcal{P}^k$ be the input point cloud in which the primitive was searched. The input point cloud for the search of the next primitive, $\mathcal{P}^{k+1}$, is obtained by removing the support points of primitive k:

$$\mathcal{P}^{k+1} = \left\{ p \in \mathcal{P}^k \mid p \notin \mathcal{I}^k \right\}. \tag{2}$$

Since every iteration of primitive detection will conduct a search on a smaller point cloud, it is expected that the cascade configuration is capable of reducing the processing time. Algorithm 1 details the complete procedure for the detection of a set of polygonal primitives given a point cloud.

3 Results

In order to evaluate the proposed 3D processing techniques a complete dataset both with 3D laser data, cameras and accurate egomotion is required. The MIT autonomous vehicle *Talos* competed in the Darpa Urban Challenge and achieved fourth overall place. The data logged by the robot is publicly available [9]. In total, the MIT logs sum up to 315GB of data. We have collected a small sequence of 40 seconds (200 meters of vehicle movement) at the start of the race (see Fig. 2). The sequence contains a continuous stream of sensor data, but in addition we have marked five locations (A through E) which are used to facilitate the analysis of the results. Additional information on each location is given in Table 1. Figure 3 shows images from all cameras, isometric and top views of the 3D data, and a satellite photograph of location C. The proposed approach is evaluated by analysing how the scenario contained in the sequence is reconstructed.

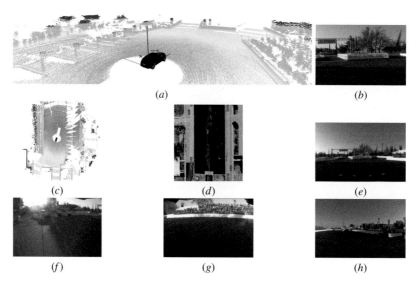

Fig. 3 Location C of the sequence: (a) isometric view; (c) top view; (b) front 6mm camera; (e) front (h) rear; (f) left (g) right; (d) satellite view of the location.

(a)   (b)

Fig. 4 Detection of geometric polygonal primitives in the data sets of sequence 1: (*a*) location *C*; (*b*) location *D*.

Table 1 Information on each of the locations in this sequence. Columns description: pt, number of points; size, memory size in mega bytes; t, mission time in seconds; d, traveled distance in meters.

Location	Location Snapshot		Sequence accumulated			
Name	pt ($\times 10^6$)	size (MB) [1]	pt ($\times 10^6$)	size (MB) [1]	t (s)	d (m)
A	1.3	15.6	1.3	15.6	1	0
B	1.3	15.6	13.0	156.0	11	75
C	1.3	15.6	26.0	312.0	21	125
D	1.3	15.6	39.0	468.0	31	140
E	1.3	15.6	52.0	624.0	41	190

[1] Computed from the number of points times the three xyz dimensions times the four bytes for each dimension (type *float32*). It is an approximate value since there are other informations on the message, such as the time stamp, the coordinate frame identification, etc.

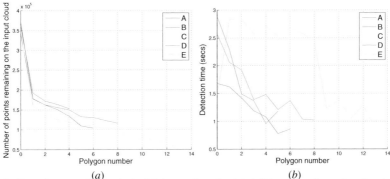

(a)   (b)

Fig. 5 Cascade processing analysis: (*a*) the number of points left to process for a given input point cloud, as a function of the index of the detected primitive; (*b*) the time it takes to perform the detection of each of the geometric polygonal primitives as a function of the primitives index.

Figure 4 shows the polygonal primitives detected at locations *C* and *D*. It is possible to observe that the majority of the relevant planes are picked up by the algorithm.

3.1 Computational Efficiency

The detection of polygonal primitives is operated in a cascade-like configuration. In other words, the algorithm will search for polygonal primitives on a given input point cloud. After the first primitive is found, all the range measurements that are explained by that primitive are removed from the input point cloud. The second primitive is then searched in a smaller point cloud and so on. Since the search for a primitive is done over a decreasing size point cloud, it is expected that the search becomes faster with the number of detected primitives. In Fig. 5 an analysis of the computation time of each primitive is displayed. Primitives with higher numbers are detected in posterior phases.

Figure 5 (a) shows the number of points remaining in the input point cloud as a function of the primitive number. Results are shown for all locations in sequence 1. The number of detected primitives varies from location to location. It is also possible to observe that, as expected, the number of remaining points decreases with the increase in the number of detected primitives. Also, the reduction in the number of points is higher for early detected primitives. Hence, since the algorithm tends to remove the largest portion of points at the early stages of the cascade processing, this means that the latter stages will also be more efficient to compute. The reason for this behaviour is the RANSAC algorithm. Because RANSAC will search for the larger consensus, it will most likely select planes that are supported by a greater number of points. In this way, RANSAC tends to select first polygons with the largest amount of primitive support points. As a consequence, the largest decreases in the input point cloud occur early in the cascade, which in turn fastens subsequent detection stages of the cascade. The detection time per primitive is shown in Fig. 5 (b). The detection time tends to decrease with the increase in polygon number, for the reasons that where previously reported.

3.2 Comparison With Other Approaches

In this section we will compare the proposed approach with three surface reconstruction methodologies: Ball Pivoting Algorithm (BPA) [2], Greedy triangulation (GT) [12] and Poisson Surface Reconstruction (POIS) [11]. Two different parameter configurations the proposed approach are used. In the first Geometric Polygonal Primitives (GPP)1, parameters are set so that only very large polygons are detected. Processing time is faster, since a lot of polygons are discarded but, on the other hand, accuracy or completeness of the scene representation should be degraded. The second alternative, GPP 2, is configured so that even small polygons are detected, which should provides a more accurate scene description at the cost of a higher computation time.

Figure 6 (a) shows that the BPA method Figure 6 (d) shows results from the GPP. Since our approach uses primitives to define macro size structures, the number of polygons used to represent the scene is small. Even though, it can be said that the most relevant polygons are part of the representation

Table 2 shows the computation times each algorithm took to reconstruct each of the locations in the sequence. The GPP methodology is the fastest one. This efficiency is related to the simplicity of the computed representation, and to the fact that RANSAC analyses only a small sample of points in the input point cloud, which means that not all input points are visited in order to reconstruct the scene, as is the case with the slower triangulation approaches.

To measure the accuracy of each reconstruction approach, the results obtained by BPA (the most accurate algorithm) are used as reference. Let X and Y be two meshes. The Hausdorff distance between those meshes $d_H(X, Y)$ is computed as:

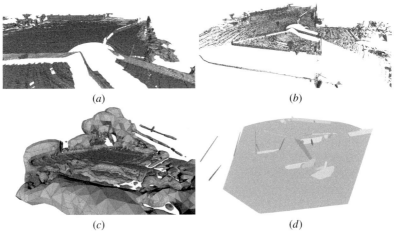

Fig. 6 Reconstruction of location E of MIT sequence: (a) BPA; (b) GT; (c) POIS; (d) GPP2.

Table 2 Comparison of the computation time of several approaches for surface reconstruction on the MIT data sets.

Sequence/	Processing time (secs)				
Location	BPA	GT	POIS	GPP 1	GPP 2
A	659.0	154.0	63.2	16.3	27.3
B	752.9	157.5	61.6	25.3	17.4
C	488.2	156.3	56.3	13.5	49.4
D	480.4	142.4	52.6	25.2	25.2
E	558.8	149.0	57.9	47.4	58.1
μ	585.9	151.8	58.3	25.5	35.5

$$d_{\mathrm{H}}(X, Y) = \max\{ \sup_{x \in X} \inf_{y \in Y} d(x, y),\ \sup_{y \in Y} \inf_{x \in X} d(x, y) \}, \tag{3}$$

where sup and inf are the supremum and infinus, respectively. Since the Hausdorff distance is computed over a set of points (sampled using a Montecarlo strategy). In this particular case a variation of the Hausdorff distance, called the one sided Hausdorff distance is used where only the $\sup_{x \in X} \inf_{y \in Y} d(x, y)$ part is computed, because we only wish to measure how distant is each approach to the ground truth and not the other way around. In this case, the X meshes are given by each of the algorithms and the Y mesh is supplied by the ground truth mesh BPA. Table 3 shows the Hausdorff distance values obtained by GT, POIS, GPP 1 and GPP 2 using BPA meshes as ground truth.

The algorithm that obtains the best results is GT. The accuracy presented by the GPP 1 and GPP 2 approaches are about 0.8 and 0.65 meters, respectively. Figure 7 shows a graphical representation of the error for all of the approaches. For each approach, the output mesh has been sampled and the points are shown with color associated to the computed one sided Hausdorff distance of each point. A Red-Green-Blue colormap is used to code the distance. Red represents zero distance and blue maximum distance. In Fig. 7 (a), corresponding to the GT approach, almost all points have red color, resulting in low mean error. The POIS approach, represented in Fig. 7 (b), shows a lot of points in blue and green color, e.g., points whose minimum distance to the ground truth sampled points was very large. This is why POIS shows low accuracy values. In the case of the GPP approaches, 7 (c) and (d), some regions of the sampled points are more prone to have large error distances, while those in red seem to perfectly fit the ground truth mesh. The reason for this is that the BPA methodology, that was selected to serve as ground truth, does not perform interpolation over occluded areas, as the GPP approaches do. Most of the errors appear in the polygonal primitive that represents the ground plane, since this is the one that suffers more from occlusion from other planes. Errors may also result from the usage of convex hulls to compute the boundary polygons. We have investigated this by using alternatives to the proposed approach where the ground plane polygon is suppressed, and where concave hulls are used. Results are shown in Table 3. We

Table 3 Comparison of the accuracy of the several approaches using BPA results as ground truth and Hausdorff distance as metric.

Location	Hausdorff distance (meters)											
	GT			POIS			GPP 1			GPP 2		
	max	mean	RMS	max	mean	RMS	max	mean	RMS	max	mean	RMS
A	11.7	0.15	0.41	14.0	1.39	2.98	7.6	1.02	1.71	7.6	0.87	1.55
B	11.8	0.12	0.37	14.1	1.39	2.99	12.7	0.94	1.77	12.6	0.81	1.62
C	12.7	0.18	0.44	13.9	1.06	2.59	8.9	0.87	1.54	8.9	0.69	1.32
D	13.8	0.10	0.40	13.9	1.90	4.00	7.6	0.86	1.47	7.6	0.69	1.28
E	12.5	0.14	0.49	14.0	1.42	3.03	14.0	1.25	2.56	14.0	1.11	2.39
μ	12.5	0.14	0.42	13.9	1.43	3.12	10.2	0.99	1.81	10.1	0.83	1.63

can observe that, with the option of ground plane suppression and concave hull, the mean accuracy of GPP 2 increases to 0.1 meters, when compared to the previous 1.63.

Figure 8 shows a visual analysis of the Hausdorff distance errors for these variations of GPP 2. It is possible to observe that regions with error, e.g., in blue and green, decrease considerably when the concave hull is used, but in particular when the ground plane polygon is discarded.

Table 4 Comparison of the Hausdorff distance accuracy of the GPP 2 approach using: the standard approach, convex hull and ground plane included (also in Table 3); the convex hull with no ground plane included; the concave hull with ground plane; and the concave hull without ground plane.

Hull	Convex Included			Convex Not included			Concave Included			Concave Not included		
Ground plane												
Location	max	mean	RMS	max	mean	RMS	max	mean	RMS	max	mean	RMS
A	7.6	0.87	1.55	1.8	0.15	0.26	6.8	0.71	1.25	1.2	0.13	0.19
B	12.6	0.81	1.62	1.5	0.11	0.19	12.6	0.53	1.09	1.1	0.08	0.14
C	8.9	0.69	1.32	1.9	0.16	0.29	6.6	0.52	0.99	1.9	0.12	0.22
D	7.6	0.69	1.28	2.2	0.14	0.26	7.3	0.59	1.13	2.1	0.11	0.21
E	14.0	1.11	2.39	1.7	0.10	0.19	8.8	0.32	0.81	1.4	0.08	0.14
μ	10.1	0.83	1.63	1.8	0.13	0.24	8.4	0.53	1.05	1.5	0.10	0.18

GPP 2 Hausdorff distance (meters)

(a)   (b)

(c)   (d)

Fig. 7 Qualitative analysis of the one sided Hausdorff distance in location C sequence 1: (a) GT; (b) POIS; (c) GPP 1; (d) GPP 2; A Red-Green-Blue color map is used to code the distance. Red represents zero distance and blue maximum distance.

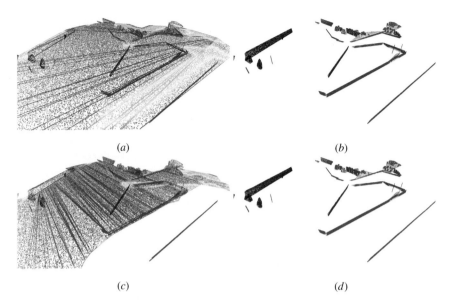

(a)   (b)

(c)   (d)

Fig. 8 Results from the Hausdorff distance obtained when using alternatives for the GPP 2 method for location *E*, sequence 1: (*a*) the standard GPP 2, with ground plane and convex hull; (*b*) discarded ground plane, convex hull; (*c*) with ground plane, concave hull; (*d*) discarded ground plane, concave hull. A Red-Green-Blue color map is used to code the distance. Red represents zero distance and blue maximum distance.

4 Conclusions

This paper proposes a novel approach to produce scene representations using the array of sensors on-board autonomous vehicles. Since roads are semi structured environments with a great deal of macro size geometric structures, we argue that the use of polygonal primitives is well suited to describe these scenes. Results have shown that the proposed approach is capable of producing accurate descriptions of the scene, and that it is considerably faster than all the approaches used in this evaluation. Future work will include the addition of texture on the polygons generated by the proposed algorithm. In this way, we expect to have the means to produce scene representations that can be used not only for standard task such as obstacle detection and motion planning, but also for more complex endeavours such as recognizing patterns in the scene.

Acknowledgments This work has been supported by the "Fundação para a Ciência e Tecnologia" under grant agreement SFRH/BD/43203/2008. A. Sappa has been partially supported by: the Spanish Government under Project TIN2014-56919-C3-2-R and the PROMETEO Project of the "Secretaría Nacional de Educación Superior, Ciencia, Tecnología e Innovación de la República del Ecuador.

References

1. Barber, C.B., Dobkin, D.P., Huhdanpaa, H.: The quickhull algorithm for convex hulls. ACM Transactions on Mathematical Software **22**(4), 469–483 (1996)
2. Bernardini, F., Mittleman, J., Rushmeier, H., Silva, C., Taubin, G.: The ball-pivoting algorithm for surface reconstruction. IEEE Transactions on Visualization and Computer Graphics **5**(4), 349–359 (1999)
3. Birk, A., Vaskevicius, N., Pathak, K., Schwertfeger, S., Poppinga, J., Buelow, H.: 3-d perception and modeling. IEEE Robotics Automation Magazine **16**(4), 53–60 (2009)
4. Burgard, W., Pfaff, P.: Editorial: Three-dimensional mapping, part 1. Journal of Field Robotics **26**(10), 757–758 (2009)
5. Bykat, A.: Convex hull of a finite set of points in two dimensions. Information Processing Letters **7**, 296–298 (1978)
6. Chen, Y.L., Lai, S.H.: An orientation inference framework for surface reconstruction from unorganized point clouds. IEEE Transactions on Image Processing **20**(3), 762–775 (2011)
7. Fischler, M.A., Bolles, R.C.: Random sample consensus: a paradigm for model fitting with applications to image analysis and automated cartography. In: ACM. Los Angeles, California, June 1981
8. Hert, S., Schirra, S.: 2D convex hulls and extreme points. In: CGAL User and Reference Manual. CGAL Editorial Board, 4th edn. (2012)
9. Huang, A.S., Antone, M., Olson, E., Fletcher, L., Moore, D., Teller, S., Leonard, J.: A High-rate, Heterogeneous Data Set from the DARPA Urban Challenge. International Journal of Robotics Research **29**(13), 1595–1601 (2011)
10. Jovanovic, R., Lorentz, R.: Compression of volumetric data using 3D delaunay triangulation. In: 2011 4th International Conference on Modeling, Simulation and Applied Optimization (ICMSAO), pp. 1–5, April 2011
11. Kazhdan, M., Bolitho, M., Hoppe, H.: Poisson surface reconstruction. In: Proceedings of the fourth Eurographics Symposium on Geometry Processing SGP 2006, pp. 61–70. Eurographics Association (2006)
12. Marton, Z.C., Rusu, R.B., Beetz, M.: On fast surface reconstruction methods for large and noisy datasets. In: Proceedings of the IEEE International Conference on Robotics and Automation (ICRA), Kobe, Japan, May 2009
13. de Medeiros Brito, A., Doria Neto, A., Dantas de Melo, J., Garcia Goncalves, L.: An adaptive learning approach for 3-d surface reconstruction from point clouds. IEEE Transactions on Neural Networks **19**(6), 1130–1140 (2008)
14. Oniga, F., Nedevschi, S.: Processing dense stereo data using elevation maps: Road surface, traffic isle, and obstacle detection. IEEE Transactions on Vehicular Technology **59**(3), 1172–1182 (2010)
15. Rivadeneyra, C., Miller, I., Schoenberg, J., Campbell, M.: Probabilistic estimation of multi-level terrain maps. In: IEEE International Conference on Robotics and Automation, ICRA 2009, pp. 1643–1648, May 2009
16. Rusu, R.B., Cousins, S.: 3D is here: point cloud library (PCL). In: IEEE International Conference on Robotics and Automation (ICRA), Shanghai, China, May 2011
17. Weiss, T., Schiele, B., Dietmayer, K.: Robust driving path detection in urban and highway scenarios using a laser scanner and online occupancy grids. In: 2007 IEEE Intelligent Vehicles Symposium, pp. 184–189, June 2007
18. Zhou, K., Gong, M., Huang, X., Guo, B.: Data-parallel octrees for surface reconstruction. IEEE Transactions on Visualization and Computer Graphics **17**(5), 669–681 (2011)

A Visible-Thermal Fusion Based Monocular Visual Odometry

Julien Poujol, Cristhian A. Aguilera, Etienne Danos, Boris X. Vintimilla, Ricardo Toledo and Angel D. Sappa

Abstract The manuscript evaluates the performance of a monocular visual odometry approach when images from different spectra are considered, both independently and fused. The objective behind this evaluation is to analyze if classical approaches can be improved when the given images, which are from different spectra, are fused and represented in new domains. The images in these new domains should have some of the following properties: i) more robust to noisy data; ii) less sensitive to changes (e.g., lighting); iii) more rich in descriptive information, among other. In particular in the current work two different image fusion strategies are considered. Firstly, images from the visible and thermal spectrum are fused using a Discrete Wavelet Transform (DWT) approach. Secondly, a monochrome threshold strategy is considered. The obtained representations are evaluated under a visual odometry framework, highlighting their advantages and disadvantages, using different urban and semi-urban scenarios. Comparisons with both monocular-visible spectrum and monocular-infrared spectrum, are also provided showing the validity of the proposed approach.

Keywords Monocular visual odometry · LWIR-RGB cross-spectral imaging · Image fusion

J. Poujol · C.A. Aguilera · E. Danos · R. Toledo · A.D. Sappa(✉)
Computer Vision Center, Edifici O, Campus UAB, Bellaterra, 08193 Barcelona, Spain
e-mail: angel.sappa@cvc.uab.es

C.A. Aguilera · R. Toledo
Computer Science Department, Universitat Autònoma de Barcelona,
Campus UAB, Bellaterra, Barcelona, Spain

B.X. Vintimilla · A.D. Sappa
Facultad de Ingeniería en Electricidad y Computación,
Escuela Superior Politécnica del Litoral, ESPOL, Campus Gustavo Galindo Km 30.5 Vía
Perimetral, P.O. Box 09-01-5863, Guayaquil, Ecuador

© Springer International Publishing Switzerland 2016
L.P. Reis et al. (eds.), *Robot 2015: Second Iberian Robotics Conference*,
Advances in Intelligent Systems and Computing 417,
DOI: 10.1007/978-3-319-27146-0_40

517

1 Introduction

Recent advances in imaging sensors allow the usage of cameras at different spectral bands to tackle classical computer vision problems. As an example of such emerging field we can mention the pedestrian detection systems for driving assistance. Although classically they have relied only in the visible spectrum [1], recently some multispectral approaches have been proposed in the literature [2] showing advantages. The same trend can be appreciated in other computer vision applications such as 3D modeling (e.g., [3], [4]), video-surveillance (e.g., [5], [6]) or visual odometry, which is the focus of the current work.

Visual Odometry (VO) is the process of estimating the egomotion of an agent (e.g., vehicle, human or a robot) using only the input of a single or multiple cameras attached to it. This term has been proposed by Nister [7] in 2004; it has been chosen for its similarity to wheel odometry, which incrementally estimates the motion of a vehicle by integrating the number of turns of its wheels over time. Similarly, VO operates by incrementally estimating the pose of the vehicle by analyzing the changes induced by the motion to the images of the onboard vision system.

State of the art VO approaches are based on monocular or stereo vision systems; most of them working with cameras in the visible spectrum (e.g., [8], [9], [10], [11]). The approaches proposed in the literature can be coarsely classified into: *feature based* methods, *image based* methods and *hybrid* methods. The feature based methods rely on visual features extracted from the given images (e.g., corners, edges) that are matched between consecutive frames to estimate the egomotion. On the contrary to feature based methods, the image based approaches directly estimate the motion by minimizing the intensity error between consecutive images. Generalizations to the 3D domain has been also proposed in the literature [12]. Finally, hybrid methods are based on a combination of the approaches mentioned before to reach a more robust solution. All the VO approaches based on visible spectrum imaging, in addition to their own limitation, have those related with the nature of the images (i.e., photometry). Having in mind these limitations (i.e., noise, sensitivity to lighting changes, etc.) monocular and stereo vision based VO approaches, using cameras in the infrared spectrum, have been proposed (e.g., [13], [14]) and more recently cross-spectral stereo based approaches have been also introduced [15]. The current work proposes a step further by tackling the monocular vision odometry with an image resulting from the fusion of a cross-spectral imaging device. In this way the strengths of each band are considered and the objective is to evaluate whether classical approaches can be improved by using images from this new domain.

The manuscript is organized as follow. Section 2 introduces the image fusion techniques evaluated in the current work together with the monocular visual odometry algorithm used as a refernce. Experimental results and comparisons are provided in Section 3. Finally, conclusions are given in Section 4.

2 Proposed Approach

This section presents the image fusion algorithms evaluated in the monocular visual odometry context. Let I_v be a visible spectrum (VS) image and I_{ir} the corresponding one from the Long Wavelength Infrared (LWIR) spectrum. In the current work we assume the given pair of images are already registered. The image resulting from the fusion will be referred to as F.

2.1 *Discrete Wavelet Transform based Image Fusion*

The image fusion based on discrete wavelet transform (DWT) consists on merging the wavelet decompositions of the given images (I_v, I_{ir}) using fusion methods applied to approximations coefficients and details coefficients. A scheme of the DWT fusion process is presented in Fig. 1. Initially, the process starts by decomposing the given images into frequency bands. They are analyzed by a fusion rule to determine which component ($D_i = \{d_1, ..., d_n\}$) is removed and which one is preserved. Finally, the inverse transform is applied to get the fused image into the spacial domain. There are different fusion rules (e.g., [16], [17]) to decide which coefficient should be fused into the final result. In the current work high order bands are preserved, while low frequency regions (i.e., smooth regions) are neglected. Figure 2 presents a couple of fused images obtained with the DWT process. Figure 2(*left*) depicts the visible spectrum images (I_v) and the corresponding LWIR images (I_{ir}) are presented in Fig. 2(*middle*). The resulting fused images (F) are shown in Fig. 2(*right*).

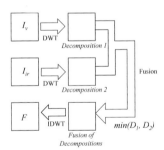

Fig. 1 Scheme of the Discrete Wavelet Transform fusion process.

2.2 *Monochrome Threshold based Image Fusion*

The monochrome threshold image fusion technique [18] just highlights in the visible image hot objects found in the infrared image. It works as follows. Firstly, an overlay image $O(x, y)$ is created using the thermal image $I_{ir}(x, y)$ and an user defined temperature threshold value τ (see Eq. 1). For each pixel value greater than the

Fig. 2 Illustrations of DWT based image fusion. (*left*) VS image. (*middle*) LWIR image. (*right*) Fused image.

threshold value τ a new customized HSV value is obtained, using a predefined H value and the raw thermal intensity for the S and V channels. In the current work the H value is set to 300—this value should be tuned according with the scenario in order to easily identify the objects associated with the target temperature:

$$O(x, y) = \begin{cases} HSV(H, I_{ir}(x, y), I_{ir}(x, y)) & \text{if } I_{ir}(x, y) > \tau \\ HSV(0, 0, 0) & \text{otherwise} \end{cases} \quad (1)$$

Secondly, after the overlay has been computed, the fused image $F(x, y)$ is computed using the visible image $I_v(x, y)$ and the overlay image $O(x, y)$ (see Eq. 2). The α value is an user defined opacity value that determines how much we want to preserve of the visible image in the fused image:

$$F(x, y) = \begin{cases} I_v(x, y)(1 - \alpha) + O(x, y)\alpha & \text{if } I_{ir}(x, y) > \tau \\ I_v(x, y) & \text{otherwise} \end{cases} \quad (2)$$

As a result we obtain an image that is similar to the visible image but with thermal clues. Figure 3 presents a couple of illustrations of the monochrome threshold image fusion process. Figure 3(*left*) depicts the visible spectrum images (I_v); the infrared images (I_{ir}) of the same scenarios are shown in Fig. 3(*middle*) and the resulting fused images (F) are presented in Fig. 3(*right*). To obtain these results the alpha was tuned to 0.3. That leads, if IR pixel intensity is higher than the temperature threshold, to a resulting pixel intensity blend by 30 percent from infrared and 70 percent from visible image.

2.3 Monocular Visual Odometry

The fused images computed above are evaluated using the monocular version of the well-known algorithm proposed by Geiger et al. in [19], which is referred to as LibVISO2.

Generally, results from monocular systems are up to a scale factor; in other words they lack of a real 3D measure. This problem affects most of monocular odometry approaches. In order to overcome this limitation, LibVISO2 assumes a fixed transformation from the ground plane to the camera (parameters given by the camera height and the camera pitch). These values are updated at each iteration by estimating the ground plane. Hence, features on the ground as well as features above the ground plane are needed for a good odometry estimation. Roughly speaking, the algorithm consists of the following steps:

Fig. 3 Illustration of monochrome threshold based image fusion. (*left*) VS image. (*middle*) LWIR image. (*right*) Fused image.

- Compute the fundamental matrix ($\mathbf{F}$) from point correspondences using the 8-point algorithm.
- Compute the essential matrix ($\mathbf{E}$) using the camera calibration parameters.
- Estimate the 3D coordinates and [$\mathbf{R}|\mathbf{t}$]
- Estimate the ground plane from the 3D points.
- Scale the [$\mathbf{R}|\mathbf{t}$] using the values of camera height and pitch obtained in previous step.

3 Experimental Results

This section presents experimental results and comparisons obtained with different cross-spectral video sequences. In all the cases GPS information is used as ground truth data to evaluate the performance of evaluated approaches. The GPS ground truth must be considered as a weak ground truth, since it was acquired using a low-cost GPS receiver. Initially, the system setup is introduced and then the experimental result are detailed.

3.1 System Setup

This section details the cross-spectral stereo head used in the experiments together with the calibration and rectification steps. Figure 4 shows an illustration of the whole platform (from the stereo head to the electric car used for obtaining the images).

The stereo head used in the current work consists of a pair of cameras set up in a non verged geometry. One of the camera works in the infrared spectrum, more precisely Long Wavelength Infrared (LWIR), detecting radiations in the range of $8 - 14\ \mu m$. The other camera, which is referred to as (VS) responds to wavelengths from about 390 to 750 nm (visible spectrum). The images provided by the cross-spectral stereo head are calibrated and rectified using [20]; a process similar to the one presented in [3] is followed. It consists of a reflective metal plate with an overlain chessboard pattern. This chessboard can be visualized in both spectrums making possible the cameras' calibration and image rectification.

The LWIR camera (Gobi-640-GigE from Xenics) provides images up to 50 fps with a resolution of 640×480 pixels. The visible spectrum camera is an ACE from Basler with a resolution of 658×492 pixels. Both cameras are synchronized using an external trigger. Camera focal lengths were set so that pixels in both images contain similar amount of information from the given scene. The whole platform is placed on the roof of a vehicle for driving assistance applications.

Once the LWIR and VS cameras have been calibrated, their intrinsic and extrinsic parameters are known, being possible the image rectification. With the above system setup different video sequences have been obtained in urban and semi-urban scenarios. Figure 5 shows the map trajectories of three video sequences. Additional information is provided in Table 1.

Fig. 4 Acquisition system (cross-spectral stereo rig on the top left) and electric vehicle used as mobile platform.

Fig. 5 Trajectories used during the evaluations: (*left*) Vid00 path; (*middle*) Vid01 path; (*right*) Vid02 path.

Table 1 Detailed characteristics of the three datasets used for the evaluation.

Name	Type	Duration (sec)	Road length (m)	Average speed (km/h)
Vid00	Urban	49.9	235	17.03
Vid01	Urban	53.6	365	24.51
Vid02	Semi-urban	44.3	370	30.06

3.2 Visual Odometry Results

In this section experimental results and comparisons, with the three video sequences introduced above (see Fig. 5 and Table 1), are presented. In order to have a fair comparison the user defined parameters for the VO algorithm (LibVISO2) have been tuned accordingly to the image nature (visible, infrared, fused) and characteristics of the video sequence. These parameters were empirically obtained looking for the best performance in every image domain. In all the cases ground truth data from GPS are used for comparisons.

Vid00 Video Sequence: it consists of a large curve in a urban scenario. The car travels more than 200 meters at an average speed of about 17 Km/h. The VO algorithm (LibVISO2) has been tuned as follow for the different video sequences (see [19] for details on the parameters meaning). In the **visible** spectrum case the bucket size has been set to 16×16 and the maximum number of features per bucket has been set to 4. The τ and match radius parameters were tuned to 50 and 200 respectively. In the **infrared** video sequence the bucket size has been also set to 16×16 but the maximum number of features per bucket has been increased to 6. Regarding τ and match radius parameters, they were set to 25 and 200 respectively. Regarding the VO with fused images the parameters were set as follow. In the video sequences obtained by the

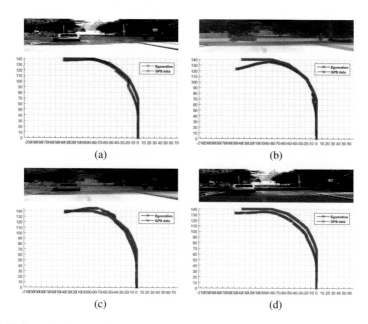

Fig. 6 Estimated trajectories for the Vid00 sequence: (*a*) Visible spectrum; (*b*) Infrared spectrum; (*c*) DWT fused images; and (*d*) Monochrome threshold fused images.

DWT fusion based approach the bucket size was set to 16×16 and the maximum number of features per bucket to 6; τ and the match radius parameters were set to 25 and 200 respectively. Finally, in the **Monochrome Threshold fusion based approach** the bucket size has been also set to 16×16 but the maximum number of features has been increased to 6. The τ and match radius parameters were tuned to 50 and 100 respectively. The refining at half resolution is disabled, since the image resolution of the cameras is small. Figure 6 depicts the plots corresponding to the different cases (visible, infrared and fused images) when they are compared with ground truth data (GPS information). Quantitative results corresponding to these trajectories are presented in Table 2. In this particular sequence, the VO computed with the visible spectrum video sequence get the best result just followed by the one obtained with the DWT video sequence. Quantitatively, both have a similar final error, on average the DWT relay on more matched points, which somehow would result in a more robust solution. The visual odometry computed with the infrared spectrum video sequence get the worst results; this is mainly due to the lack of texture in the images.

Vid01 Video Sequence: it is a simple straight line trajectory on a urban scenario consisting of about 350 meters; the car travels at an average speed of about 25 Km/h. The (LibVISO2) algorithm has been tuned as follow. In the **visible** spectrum case the bucket size was set to 16×16 and the maximum number of features per bucket has

Table 2 VO results in the **Vid00 video sequence** using images from: visible spectrum (VS); Long Wavelength Infrared spectrum (LWIR); fusion using Discrete Wavelet Transform (DWT); and fusion using Monochrome Threshold (MT).

Results	VS	LWIR	DWT	MT
Total traveled distance (m)	234.88	241.27	245	240.3
Final position error (m)	2.9	18	5.4	14.4
Average number of matches	2053	3588	4513	4210
Percentage of inliers	71.5	61.94	60	67.9

been set to 4. The variables τ and match radius parameters are respectively tuned to 25 and 200. The user defined parameters in the **Infrared** case have been set as follow. The bucket size was defined as 16×16 and the maximum number of features per bucket has been set to 50 and 200 respectively. The half resolution was set to zero. The LibVISO2 algorithm has been tuned as follow when the fused images were considered. In the **DWT fusion based approach** the bucket size was set to 16×16 and the maximum number of features per bucket set to 4. The τ and match radius parameters are respectively tuned to 25 and 200. Finally, in the **Monochrome Threshold fusion based approach** the bucket size was set to 16×16 and the maximum number of features per bucket was set to 4. The τ and match radius parameters are respectively tuned to 25 and 200. Figure 7 depicts the plots of the visual odometry computed over each of the four representations (VS, LWIR, DWT fused and Monochrome threshold fused) together with the corresponding GPS data. The visual odometry computed with the infrared video sequence gets the worst result, as can be easily appreciated in Fig. 7 and confirmed by the final position error value presented in Table 3. The results obtained with the other three representations (visible spectrum, DWT based image fusion and Monochrome Threshold based image fusion) are similar both qualitatively and quantitatively.

Vid02 Video Sequence: it is a "L" shape trajectory on a sub-urban scenario. It is the longest trajectory (370 meters) and the car has traveled faster than in the previous cases (about 30 Km/h). The (LibVISO2) algorithm has been tuned as follow. In the **visible** spectrum case the bucket size was set to 16×16 and the maximum number of features per bucket set to 4. Regarding τ and match radius parameters, they were tuned as 25 and 200 respectively. In the **infrared** case the bucket size has been set to 16×16 and the maximum number of features per bucket set to 4. τ and match radius parameters were respectively tuned to 50 and 100. In the fused image scenario the LibVISO2 algorithm has been tuned as follows. First, in the **DWT fusion based approach** the bucket size has been set to 16×16 and the maximum number of features per bucket set to 4. Like in the visible case, the τ and match radius parameters were tuned to 25 and 200 respectively. Finally, in the **Monochrome Threshold fusion based approach** the bucket size has been defined as 16×16 and the maximum number of features per bucket set to 4. The τ and match radius parameters were respectively tuned to 50 and 200. In this challenging video sequence the fused based

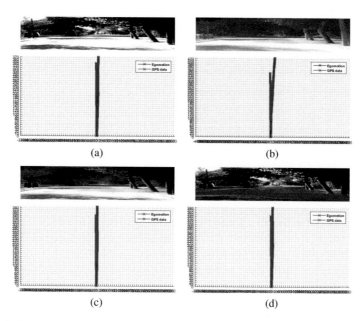

Fig. 7 Estimated trajectories for Vid01 sequence: (*a*) Visible spectrum; (*b*) Infrared spectrum; (*c*) DWT based fused image; and (*d*) Monochrome threshold based fused image.

Table 3 VO results in the **Vid01 video sequence** using images from: visible spectrum (VS); Long Wavelength Infrared spectrum (LWIR); fusion using Discrete Wavelet Transform (DWT); and fusion using Monochrome Threshold (MT).

Results	VS	LWIR	DWT	MT
Total traveled distance (m)	371.8	424	386	384
Final position error (m)	32.6	84.7	44	42.7
Average number of matches	1965	1974	2137	2060
Percentage of inliers	72.6	67.8	61.5	65.4

Table 4 VO results in the **Vid02 video sequence** using images from different spectrum and fusion approaches (VS: visible spectrum; LWIR: Long Wavelength Infrared spectrum, DWT: fusion using Discrete Wavelet Transform, MT: fusion using Monochrome Threshold).

Results	VS	LWIR	DWT	MT
Total traveled distance (m)	325.6	336.9	354.4	371.5
Final position error (m)	37.7	48.7	37.2	14.3
Average number of matches	1890	1028	1952	1374
Percentage of inliers	70	65.8	61	66

approaches get the best results (see Fig. 8). It should be highlighted that in the Monochrome Threshold fusion the error is less than half the one obtained in the visible spectrum (see values in Table 4).

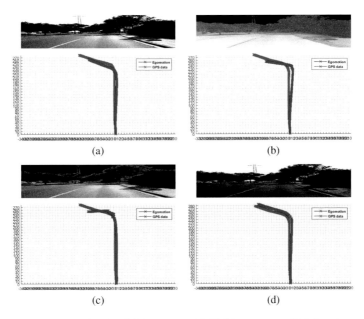

Fig. 8 Estimated trajectories for Vid02 sequence: (*a*) Visible spectrum; (*b*) Infrared spectrum; (*c*) DWT fused image; and (*d*) Monochrome threshold based fused image.

In the general, the usage of fused images results in quite stable solutions; supporting somehow the initial idea that classical approaches can be improved when the given cross-spectral images are fused and represented in new domains.

4 Conclusion

The manuscript evaluates the performance of a classical monocular visual odometry by using images from different spectra represented in different domains. The obtained results show that the usage of fused images could help to obtain more robust solutions. This evaluation study is just a first step to validate the pipeline in the emerging field of image fusion. As future work other fusion strategies will be evaluated and a more rigorous framework set up.

Acknowledgments This work has been supported by: the Spanish Government under Project TIN2014-56919-C3-2-R; the PROMETEO Project of the "Secretaría Nacional de Educación Superior, Ciencia, Tecnología e Innovación de la República del Ecuador"; and the "Secretaria d'Universitats i Recerca del Departament d'Economia i Coneixement de la Generalitat de Catalunya" (2014-SGR-1506). C. Aguilera was supported by Universitat Autónoma de Barcelona.

References

1. Geronimo, D., Lopez, A.M., Sappa, A.D., Graf, T.: Survey of pedestrian detection for advanced driver assistance systems. IEEE Transactions on Pattern Analysis and Machine Intelligence **32**(7), 1239–1258 (2010)
2. Hwang, S., Park, J., Kim, N., Choi, Y., Kweon, I.S.: Multispectral pedestrian detection: benchmark dataset and baseline. In: IEEE International Conference on Computer Vision and Pattern Recognition (2015)
3. Barrera, F., Lumbreras, F., Sappa, A.D.: Multimodal stereo vision system: 3D data extraction and algorithm evaluation. IEEE Journal of Selected Topics in Signal Processing **6**(5), 437–446 (2012)
4. Barrera, F., Lumbreras, F., Sappa, A.D.: Multispectral piecewise planar stereo using manhattan-world assumption. Pat. Recognition Letters **34**(1), 52–61 (2013)
5. Conaire, C.O., O'Connor, N.E., Cooke, E., Smeaton, A.: Multispectral object segmentation and retrieval in surveillance video. In: IEEE International Conference on Image Processing, pp. 2381–2384 (2006)
6. Denman, S., Lamb, T., Fookes, C., Chandran, V., Sridharan, S.: Multi-spectral fusion for surveillance systems. Comp. & Electrical Engineering **36**(4), 643–663 (2010)
7. Nistér, D., Naroditsky, O., Bergen, J.: Visual odometry. In: IEEE Intgernational Conference on Computer Vision and Pattern Recognition, vol. 1, pp. I–652 (2004)
8. Scaramuzza, D., Fraundorfer, F., Siegwart, R.: Real-time monocular visual odometry for on-road vehicles with 1-point RANSAC. In: IEEE International Conference on Robotics and Automation, pp. 4293–4299 (2009)
9. Tardif, J.P., Pavlidis, Y., Daniilidis, K.: Monocular visual odometry in urban environments using an omnidirectional camera. In: IEEE International Conference on Intelligent Robots and Systems IROS, pp. 2531–2538 (2008)
10. Howard, A.: Real-time stereo visual odometry for autonomous ground vehicles. In: International Conference on Intelligent Robots and Systems, pp. 3946–3952 (2008)
11. Scaramuzza, D., Fraundorfer, F.: Visual odometry. IEEE Robotics & Automation Magazine **18**(4), 80–92 (2011)
12. Comport, A.I., Malis, E., Rives, P.: Accurate quadrifocal tracking for robust 3D visual odometry. In: IEEE International Conference on Robotics and Automation, ICRA, Roma, Italy, April 10–14, 2007, pp. 40–45 (2007)
13. Chilian, A., Hirschmüller, H.: Stereo camera based navigation of mobile robots on rough terrain. In: IEEE International Conference on Intelligent Robots and Systems IROS, pp. 4571–4576. IEEE (2009)
14. Nilsson, E., Lundquist, C., Schön, T., Forslund, D., Roll, J.: Vehicle motion estimation using an infrared camera. In: 18th IFAC World Congress, Milano, Italy, 28 August–2 September 2011, pp. 12952–12957. Elsevier (2011)
15. Mouats, T., Aouf, N., Sappa, A.D., Aguilera-Carrasco, C.A., Toledo, R.: Multispectral stereo odometry. IEEE Transactions on Intelligent Transportation Systems **16**(3), 1210–1224 (2015)
16. Amolins, K., Zhang, Y., Dare, P.: Wavelet based image fusion techniques – an introduction, review and comparison. ISPRS Journal of Photogrammetry and Remote Sensing **62**(4), 249–263 (2007)
17. Suraj, A., Francis, M., Kavya, T., Nirmal, T.: Discrete wavelet transform based image fusion and de-nosing in FPGA. Journal of Electriacal Systems and Information Technology **1**, 72–81 (2014)
18. Rasmussen, N.D., Morse, B.S., Goodrich, M., Eggett, D., et al.: Fused visible and infrared video for use in wilderness search and rescue. In: Workshop on Applications of Computer Vision (WACV), pp. 1–8. IEEE, December 2009
19. Geiger, A., Ziegler, J., Stiller, C.: Stereoscan: dense 3D reconstruction in real-time. In: Intelligent Vehicles Symposium (IV) (2011)
20. Bouguet, J.Y.: Camera calibration toolbox for matlab, July 2010

Part IV
Control and Planning in Aerial Robotics

Indoor SLAM for Micro Aerial Vehicles Using Visual and Laser Sensor Fusion

Elena López, Rafael Barea, Alejandro Gómez, Álvaro Saltos, Luis M. Bergasa, Eduardo J. Molinos and Abdelkrim Nemra

Abstract This paper represents research in progress in Simultaneous Localization and Mapping (SLAM) for Micro Aerial Vehicles (MAVs) in the context of rescue and/or recognition navigation tasks in indoor environments. In this kind of applications, the MAV must rely on its own onboard sensors to autonomously navigate in unknown, hostile and GPS denied environments, such as ruined or semi-demolished buildings. This article aims to investigate a new SLAM technique that fuses laser and visual information, besides measurements from the inertial unit, to robustly obtain the 6DOF pose estimation of a MAV within a local map of the environment. Laser is used to obtain a local 2D map and a footprint estimation of the MAV position, while a monocular visual SLAM algorithm enlarges the pose estimation through an Extended Kalman Filter (EKF). The system consists of a commercial drone and a remote control unit to computationally afford the SLAM algorithms using a distributed node system based on ROS (Robot Operating System). Some experimental results show how sensor fusion improves the position estimation and the obtained map under different test conditions.

Keywords Micro Aerial Vehicles · Indoor navigation · Sensor fusion · Simultaneous Localization and Mapping · Robot Operating System

1 Introduction

The growing research on MAVs and the consequent improvement of technologies like microcomputers and onboard sensor devices, have increased the performance

E. López(✉) · R. Barea · A. Gómez · Á. Saltos · L.M. Bergasa · E.J. Molinos
Department of Electronics, Alcalá University, Alcalá, Spain
e-mail: {elena,barea,bergasa}@depeca.uah.es

A. Nemra
Ecole Militaire Polytechnique, Algiers, Algeria
e-mail: karim_nemra@yahoo.fr

© Springer International Publishing Switzerland 2016
L. Paulo Reis et al. (eds.), *Robot 2015: Second Iberian Robotics Conference*,
Advances in Intelligent Systems and Computing 417,
DOI: 10.1007/978-3-319-27146-0_41

requirements of such kind of systems. Enabled by GPS and MEMS inertial sensors, MAVs that can fly in outdoor environments without human intervention have been developed [1,2,3]. Unfortunately, most indoor environments remain without access to external positioning systems, and autonomous MAVs are very limited in their ability to operate in these areas.

Traditionally, unmanned ground vehicles operating in GPS-denied environments can rely on dead reckoning and onboard environmental sensors for localization and mapping using SLAM techniques. However, attempts to achieve the same results with MAVs have not been as successful due to several reasons: the inaccuracy and high drift of Inertial Navigation Systems (INS) compared to encoder-based dead reckoning, the limited payload for sensing and computation, and the fast and unstable dynamics of air vehicles, are the main challenges which must be tackled.

Especially, pose estimation is essential for many navigation tasks, including localization, mapping and control. The technique used depends mainly on the available on board sensors, which in aerial navigation must be carefully chosen due to payload limitations. Through their low weight and consumption, most commercial MAVs incorporate at least one monocular camera, so VSLAM (Visual SLAM) techniques have been widely used. However, most of these works have been limited to small workspaces which have definite image features and sufficient sun light. Furthermore, computational time is too high for the fast dynamics of aerial vehicles, making difficult to control them. On the other hand, despite their greater weight and consumption, range sensors such as RGB-D cameras or laser range sensors have also been used on MAVs due to their fast distance detection.

This paper focuses on fusion of laser, monocular vision and IMU (Inertial Measurement Unit) to robustly track the position of a MAV using SLAM. To face the computational requirements, the system is composed of a flight and a ground unit, so that code can be distributed in different nodes using ROS (Robot Operating System). In order to calculate pose which is relatively insensitive to errors, 2D map is generated based on laser. The estimated footprint position of the MAV is then filtered with IMU and VSLAM information using an EKF (Extended Kalman Filter) to obtain a full 6DOF pose estimation, that is demonstrated to be robust under different illumination and environmental test conditions.

The remaining part of this paper is organized as follows. Section 2 discusses related work. Section 3 describes the overall system. The SLAM approach is explained in section 4. The experimental results are presented in Section 5. Finally, it is followed by the conclusion and future work in Section 6.

2   Related Work

The most challenging part of SLAM for MAVs is to obtain the 6DOF pose of the vehicle without odometry information. To do this, different sensor sources have been suggested, such as laser range sensors [4], monocular cameras [5], stereo cameras [6] or RGB-D sensors [7,8].

Due to weight limitations, most of the works only use the onboard camera and IMU to apply VSLAM (Visual SLAM) techniques [9,10,11,12,13,14]. These systems demonstrate autonomous flight in limited indoor environments, but their approaches have been constrained to environments with specific features and lighting conditions, and thus may not work as well for general navigation in GPS-denied environments. On the other hand, the high working rate of actual laser scanners, along with their direct and accurate range detection, make them a very advantageous sensor for indoor navigation. Several works, such as [4,15,16], fuse laser and IMU measurements to obtain 2D maps and to estimate the 6DOF pose of the MAV.

However, there are very few works in which both laser and vision are used to solve the SLAM problem in MAVs. The main challenge is the computational charge, that can't be afford by the onboard processors. For example, in [17] laser and IMU are fused to estimate the 6DOF position of the robot, whilst vision is only used to loop closure in the obtained map. In [18] laser and vision are used, but to solve separately the outdoors and indoors odometry problem.

In this work, laser, vision and IMU measurements are fused to solve the SLAM problem in complex indoor environments and robustly estimate the 6DOF pose of the MAV, using a distributed system with a flight unit and a ground station.

3   System Overview

We address the problem of autonomous indoor MAV localization as a software challenge, focusing on high level algorithms integration rather than specific hardware. For this reason, we use a commercial platform with minor modifications, and an open-source development platform (ROS), so that drivers of sensors and some algorithms can be used without development.

3.1   Hardware Architecture

Our quadrotor MAV, shown in figure 1, is based on the ARDrone from Parrot [19]. This MAV can carry up to 120g of payload for about 5min. It's equipped with two cameras (one directed forward and one directed downward), an ultrasonic altimeter, a 3-axis accelerometer and 2 gyroscopes. It incorporates an onboard controller based on ARM 468 MHz processor with 128 Mb DDR Ram, with a Linux distribution. It also provides a USB port and is controlled via Wireless LAN.

To the commercial platform, we have added a Hokuyo URG-04LX laser for direct range measuring, an additional upward facing sonar and a Raspberry-PI board for reading and transmitting these sensor measurements. The Raspberry and the remote computer are both connected to the ARDrone network to control the robot.

Fig. 1 The experimental platform with onboard computation and sensing.

Although the ARDrone comes with some software for basic functionality [20], it's not open-source nor easy to modify, and so we treat the drone as a black box, using only the available W-LAN communication channels to access and control it. Specifically, these are the inputs/outputs we use in our SLAM system:

- Video channel, to receive the video stream of the forwards facing camera, with maximal supported resolution of 320x240 and frame rate of 18fps.
- Navigation channel, to read onboard sensor measurements every 5ms. The data used by our system are:

   - Drone orientation as roll, pitch and yaw angles $\left(\overline{\Phi},\overline{\Theta},\overline{\Psi}\right)$.
   - Horizontal velocity in drone's coordinate system $\left(\overline{vdx},\overline{vdy}\right)$, calculated onboard by an optical-flow based motion estimation algorithm [20].
   - Drone height $\overline{h}$, obtained from the ultrasound altimeter measurements.

- Command channel, to send the drone control packages, with the desired roll and pitch angles, yaw rotational velocity and vertical speed:

$$\mathbf{u} = \left(\hat{\Phi},\hat{\Theta},\hat{\Omega},\hat{v}z\right) \tag{1}$$

Besides this, a ROS node in the Raspberry-PI reads the Hokuyo laser scan and broadcast it through the drone's network.

3.2 Software Architecture

As it's shown in figure 2, the onboard controller and processor perform sensor readings and basic control of the MAV. The ground station executes our SLAM system and also the planning and control strategies, the last ones being out of the scope of this paper.

GROUND STATION

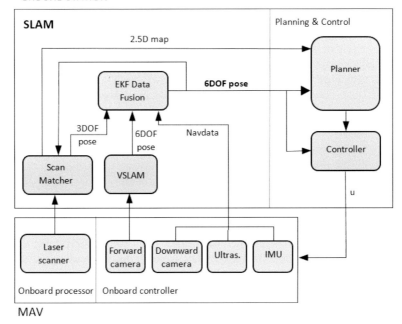

Fig. 2 The experimental platform with onboard computation and sensing.

The SLAM module consist of three major components: (1) a scan matching al-gorithm that uses laser readings to obtain a 2,5D map of the environment and a 3DOF pose estimation of the footprint of the MAV on the map; (2) a monocular visual SLAM system that obtains a 6DOF pose estimation and (3) an Extended Kalman Filter that fuses the last estimations with the navigation data provided by the onboard sensors of the MAV to obtain a robust 6DOF estimation of the posi-tion of the robot in the 2,5D map. This estimation is used, at the same time, as a priori assumption for the next scan matching step, closing the SLAM algorithm.

4   SLAM Approach

In the following subsections, we describe the modules of the SLAM system shown in figure 2.

4.1   Scan Matcher

This module aligns consecutive scans from the laser rangefinder to estimate the vehicle's motion. Although lots of scan matching techniques have been developed and applied to SLAM for ground robots moving on flat surfaces [21], most of them require odometric information that is not available in MAVs. However, the

HectorSLAM project [22], developed by the Team Hector of the Technische Universität Darmstadt, presents a system for fast online generation of 2D maps that uses only laser readings and a 3D attitude estimation system based on inertial sensing. This method requires low computational resources and it's widely used in different research groups because it is available as open source and based on ROS.

In this work, we adapt the HectorSLAM system to our scan matching module, in order to obtain a 2,5D map and a 2D estimation of the footprint pose of the MAV within the map, consisting in the (x,y) coordinates and yaw angle, that we call:

$$z_{LASER} = (x_L, y_L, \Psi_L) \tag{2}$$

The 2D pose estimation is based on optimization of the alignment of beam endpoints with the map obtained so far. The endpoints are projected into the actual map and the occupancy probabilities are estimated. The scan matching is solved using a Gaussian-Newton equation, which finds the rigid transformation that best fits the laser beams with the map. A multi-resolution map representation is used to avoid getting stuck in local minima. In addition, a 3D attitude estimation system based on the IMU measurements transforms the laser readings into a base-stabilized frame (horizontal to the ground) in order to compensate the roll and pitch movements when obtaining the 2.5D map.

Fig. 3 shows the map and footprint pose estimation obtained by the scan matcher in one of our experiments. Although HectorSLAM provides good results in confined environments, the lack of odometry information to detect horizontal movements (only attitude is obtained from the IMU and used to stabilize laser measurements) results in undesirable effects, such as shortening of corridors and not loops detection.

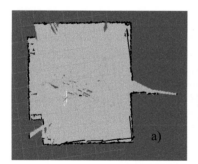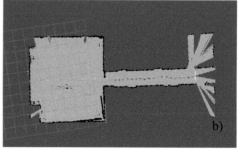

Fig. 3 Partial results of the scan matcher while moving around a room and along a corridor. The real length of the corridor is 17.6m, while it is shortened to 13.8m by the algorithm.

4.2 *Monocular SLAM*

For monocular SLAM, our solution is based on LSD-SLAM (Large-Scale Direct Monocular SLAM) [23], available as a ROS package. This is a direct (feature-less) monocular SLAM algorithm which, along with highly accurate pose estimation based on direct image alignment, reconstructs the 3D environment in real-time as pose-graph of keyframes with associated semi-dense depth maps. By the moment, and due to the high computational charge of 3D mapping, we are only using the 6DOF pose estimation of this algorithm as an input to the data fusion filter, and we use the 2,5D laser map for navigation.

Fig. 4 shows the 3D map and pose estimation obtained by the V-SLAM system in the same room of the experiment shown in fig.3. While results are good in this case, the system needs a high amount of visual characteristics that are not available in corridor areas or in dark zones, where it needs to be fused with other sensors.

Fig. 4 Partial results of the VSLAM system while moving around the room of fig. 3.a).

4.3   Data Fusion with EKF

In order to fuse all available data, we employ an Extended Kalman Filter (EKF). This EKF is also used to compensate for the different time delays in the system, as detailed described in [24], arising from wireless LAN communication and compu-tationally complex visual tracking.

The EKF uses the following state vector:

$$\chi_t := (x_t, y_t, z_t, vx_t, vy_t, vz_t, \Phi_t, \Theta_t, \Psi_t, \Omega_t)^T \quad \in \quad \mathfrak{R}^{10} \qquad (3)$$

Where (x_t, y_t, z_t) is the position of the MAV in m, (vx_t, vy_t, vz_t) the velocity in m/s, $(\Phi_t, \Theta_t, \Psi_t)$ the roll, pitch and yaw angles in deg, and Ω_t the yaw-rotational speed in deg/s, all of them in world coordinates. In the following, we define the prediction and observation models.

a) Prediction Model

The prediction model is based on the full motion model of the quadcopter's flight dynamics and reaction to control commands derived in [24]. A new calibration of the model parameters has been done because the Hokuyo laser sensor weight considerably modifies the dynamics of the system.

The model stablishes that the horizontal acceleration of the MAV is proportional to the horizontal force acting upon the quadcopter, that is, the accelerating force minus the drag force. The drag is proportional to the horizontal velocity of the quadcopter, while the accelerating force is proportional to a projection of the z-axis onto the horizontal plane, which leads to:

$$\dot{vx}_t = K_1\left(K_2\left(\cos\Psi_t \sin\Phi_t \cos\Theta_t - \sin\Psi_t \sin\Theta_t\right) - vx_t\right) \quad (4)$$

$$\dot{vy}_t = K_1\left(K_2\left(-\sin\Psi_t \sin\Phi_t \cos\Theta_t - \cos\Psi_t \sin\Theta_t\right) - vy_t\right) \quad (5)$$

Furthermore, the influence of the sent control command $\mathbf{u} = \left(\hat{\Phi}, \hat{\Theta}, \hat{\Omega}, \hat{vz}\right)$ is described by the following linear model:

$$\dot{\Phi}_t = K_3\left(K_4\hat{\Phi}_t - \Phi_t\right) \quad (6)$$

$$\dot{\Theta}_t = K_3\left(K_4\hat{\Theta}_t - \Theta_t\right) \quad (7)$$

$$\dot{\Omega}_t = K_5\left(K_6\hat{\Omega}_t - \Omega_t\right) \quad (8)$$

$$\dot{vz}_t = K_7\left(K_8\hat{vz}_t - vz_t\right) \quad (9)$$

We estimated the proportional coefficients K_1 to K_8 from data collected in a series of test flights. From eqs (4) to (9) we obtain the overall state transition function:

$$\begin{pmatrix} x_{t+1} \\ y_{t+1} \\ z_{t+1} \\ vx_{t+1} \\ vy_{t+1} \\ vz_{t+1} \\ \Phi_{t+1} \\ \Theta_{t+1} \\ \Psi_{t+1} \\ \Omega_{t+1} \end{pmatrix} \leftarrow \begin{pmatrix} x_t \\ y_t \\ z_t \\ vx_t \\ vy_t \\ vz_t \\ \Phi_t \\ \Theta_t \\ \Psi_t \\ \Omega_t \end{pmatrix} + \Delta_t \begin{pmatrix} vx_t \\ vy_t \\ vz_t \\ K_1\left(K_2\left(\cos\Psi_t \sin\Phi_t \cos\Theta_t - \sin\Psi_t \sin\Theta_t\right) - vx_t\right) \\ K_1\left(K_2\left(-\sin\Psi_t \sin\Phi_t \cos\Theta_t - \cos\Psi_t \sin\Theta_t\right) - vy_t\right) \\ K_7\left(K_8\hat{vz}_t - vz_t\right) \\ K_3\left(K_4\hat{\Phi}_t - \Phi_t\right) \\ K_3\left(K_4\hat{\Theta}_t - \Theta_t\right) \\ \Omega_t \\ K_5\left(K_6\hat{\Omega}_t - \Omega_t\right) \end{pmatrix} \quad (10)$$

b) Inertial Navigation Observation Model

This model relates the onboard measurements obtained through the navigation channel of the quadcopter described in section 3.1 (that we called "navdata" in figure 2) and the state vector. The quadcopter measures its horizontal speed $(\overline{vdx}, \overline{vdy})$ in its local coordinate system, which we transform into the world frame (vx_t, vy_t). The roll and pitch angles measured by the accelerometer are direct observations of the corresponding state variables. On the other hand, we differentiate the height measurement and the yaw measurement as observations of the respective velocities. The resulting measurement vector $\mathbf{z}_{NAVDATA}$ and observation function $h_{NAVDATA}(\mathbf{x}_t)$ are:

$$\mathbf{z}_{NAVDATA,t} = \left(\overline{vdx}, \overline{vdy}, (\overline{h}_t - \overline{h}_{t-1}), \overline{\Phi}, \overline{\Theta}, (\overline{\Psi}_t - \overline{\Psi}_{t-1}) \right)^T \tag{11}$$

$$h_{NAVDATA}(\boldsymbol{\chi}_t) := \begin{pmatrix} vx_t \cos \Psi_t - vy_t \sin \Psi_t \\ vx_t \sin \Psi_t + vy_t \cos \Psi_t \\ vz_t \\ \Phi_t \\ \Theta_t \\ \Omega_t \end{pmatrix} \tag{12}$$

c) VSLAM Observation Model

When LSD-SLAM successfully tracks a video frame, its 6DOF pose estimation is transformed from the coordinate system of the front camera to the coordinate system of the quadcopter, leading to a direct observation of the quadcopter's pose given by:

$$\mathbf{z}_{VSLAM,t} = f\left(\mathbf{E}_{DC} \mathbf{E}_{C,t} \right) \tag{13}$$

$$h_{VSLAM}(\boldsymbol{\chi}_t) := \left(x_t, y_t, z_t, \Phi_t, \Theta_t, \Psi_t \right)^T \tag{14}$$

where $\mathbf{E}_{C,t} \in SE(3)$ is the estimated camera pose, $\mathbf{E}_{DC} \in SE(3)$ the constant transformation from the camera to the quadcopter coordinate system and $f : SE(3) \rightarrow \mathfrak{R}^6$ the transformation from an element of SE(3) to the roll-pitch-yaw representation.

d) Scan Matcher Observation Model

The scan matcher maintains a 3DOF estimation of the footprint pose of the MAV, that is considered as a direct observation of the corresponding state variables, as it's shown in the following linear observation model:

$$z_{LASER,t} = (x_{L,t}, y_{L,t}, \Psi_{L,t})^T \tag{15}$$

$$H_{LASER} = \begin{bmatrix} 1 & 0 & 0 & 0 & 0 & 0 & 0 & 0 & 0 & 0 \\ 0 & 1 & 0 & 0 & 0 & 0 & 0 & 0 & 0 & 0 \\ 0 & 0 & 0 & 0 & 0 & 0 & 0 & 0 & 1 & 0 \end{bmatrix} \tag{16}$$

4.4 SLAM Integration

For best performance, information between the scan matcher and the EKF estimate is exchanged in both directions. Thus, the 6DOF pose estimate of the EKF is projected on the xy-plane and is used as start estimate for the optimization process of the scan matcher, as it's shown in figure 2.

5   Results

In order to compare the performance of the SLAM system with the different sensors separately and fused in the proposed EKF, we teleoperated the MAV along one of the rooms and corridors of our work environment while registering data from laser, camera and ARDrone sensors. Figure 5 shows the footprint projections of the estimated 6D trajectories in three different executions of the SLAM system. In red it is shown the estimated trajectory when only the prediction model and the ARDrone navdata measurements are used; in this case, due to the inaccuracy of commands and the high drift of IMU measurements, the estimation is very poor. In green colour we show the estimation when using the prediction model and the laser scan matcher observation. Although the results are better than using only the scan matcher (see fig. 3), corridor is still shortened from 17.6m to 15.2m. In blue we show the results of the entire SLAM system, using commands for prediction and laser, visual and navdata observations for correction. The map shown in fig 5 corresponds to this execution of the EKF. In this case, we obtain the best map and trajectory estimation, in which loop closure is better in the room, and the estimated corridor length is 17.4m, very close to the real one.

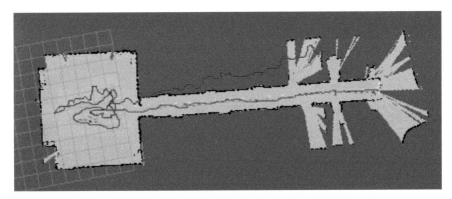

Fig. 5 Experimental results in three different executions of the SLAM system: (red) using prediction an navdata observation models; (green) using prediction and scan matcher observation model; (blue) using prediction and visual, scan matcher and navdata observation models.

6   Conclusions and Future Work

This paper shows work in progress and initial results of a indoor SLAM system for MAVs that fuses visual, laser and on board sensors to obtain a better estimation of the 6D pose of the MAV and a 2D map of the local environment. Fusing laser and vision is possible by using light algorithms running on a remote control station. In future work, we will add a large scale 3D mapping system with loop closure detection by using advanced VSLAM algorithms.

Acknowledgements   This work is supported in part by the IndexMAV Project (CCG2014/EXP-078) which is funded by the University of Alcalá and in part by the RoboCity2030 III-CM project (S2013/MIT-2748) funded by Programas de I+D en la Comunidad de Madrid and cofunded by Structured Funds of the UE.

References

1. Mellinger, D., Kumar, V.: Minimum snap trajectory generation and control for quadrotors. In: Proc. IEEE Intelligent Conference on Robotics and Automation (ICRA), pp. 2520–2525 (2011)
2. Lindsey, Q., Mellinger, D., Kumar, V.: Construction of cubic structures with quadrotor teams. In: Proceedings on Robotics: Science and Systems (RSS) (2011)
3. Kushleyev, A., Mellinger, D., Kumar, V.: Towards a swarm of agile micro quadrotors. In: Proceedings of Robotics: Science and Systems (RRS) (2012)
4. Grzonka, S., Grisetti, G., Burgard, W.: Towards a navigation system for autonomous indoor flying. In: Proceedings IEEE Intelligent Conference on Robotics and Automation (ICRA) (2009)
5. Achtelik, M., Lynen, S., Weiss, S., Kneip, L., Chli, M., Siegwart, R.: Visual-inertial SLAM for a small helicopter in large outdoor environments. In: Proceedings IEEE Conference on Intelligent Robots and Systems (IROS), pp. 2651–2652 (2012)

6. Fraundorfer, F., Heng, L., Honegger, D., Lee, L., Meier, L., Tanskanen, P., Pollefeys, M.: Vision-based autonomous mapping and exploration using a quadrotor MAV. In: Proceedings IEEE Inter. Conference on Intelligent Robots and Systems (IROS) (2012)

7. Huang, A.S., Bachrach, A., Henry, P., Krainin, M., Maturana, D., Fox, D., Roy, N.: Visual odometry and mapping for autonomous flight using and RGB-D camera. In: Proceedings IEEE Intelligent Symposium of Robotics Research (ISRR), pp. 1–16 (2011)

8. Bylow, E., Sturm, J., Kerl, C., Kahl, F., Cremers, D.: Real-time camera tracking and 3D reconstruction using signed distance functions. In: Proceedings of Robotics: Science and Systems (RSS) (2013)

9. Tournier, G.P., Valenti, M., How, J.P., Feron, E.: Estimation and control of a quadrotor vehicle using monocular vision and moiré patterns. In: Proceedings of AIAA GNC, Keystone, Colorado (2006)

10. Johnson, N.G.: Vision-assisted control of a hovering air vehicle in an indoor setting. Master's thesis, BYU (2008)

11. Kemp, C.: Visual Control of a Miniature Quad-Rotor Helicopter. Ph.D. Thesis, Churchill College, University of Cambridge (2006)

12. Ahrens, S., Levine, D., Andrews, G., How, J.P.: Vision-based guidance and control of a hovering vehicle in unknown, GPS-denied environments. In: IEEE International Conference on Robotics and Automation (ICRA), pp. 2643–2648 (2009)

13. Teulière, C., Marchand, E., Eck, L.: 3-D Model-Based tracking for UAV Indoor Localization. IEEE Transactions on Cybernetics **45**(5) (2015)

14. Brockers, R., Humenberger, M., Weiss, S., Matthies, L.: Towards autonomous navigation of miniature UAV. In: 2014 IEEE Conference on Computer Vision and Pattern Recognition Workshops (2014)

15. Moon, S., Eom, W., Gong, H.: Development of a large-scale 3D map generation system for indoor autonomous navigation flight. In: Proceedings of Asia-Pacific International Symposium on Aerospace Technology, APISAT 2014 (2014)

16. Li, R., Liu, J., Zhang, L., Hang, Y.: LIDAR/MEMS IMU integrated navigation (SLAM) method for a small UAV in indoor envir. In: Inertial Sensors and Systems (2014)

17. Shen, S., Michael, N., Kumar, V.: Autonomous Multi-Floor Indoor Navigation

18. Tomic, T., Schmid, K., Lutz, P., Domel, A.: Toward a fully autonomous UAV. In: IEEE Robotics & Automation Magazine (2012)

19. ardrone2.parrot.com

20. Bristeau, P.-J., Callou, F., Vissiere, D., Peit, N.: The navigation and control technology inside the AR. Drone micro UAV. In: 18th IFAC World Congress, pp. 1477–1484 (2011)

21. Lu, F., Milios, E.: Globally consistent range scan alignment for environment mapping. Autonomous Robots **4**, 333–349 (1997)

22. Kohlbrecher, S., Stryk, O., Meyer, J., Klingauf, U.: A flexible and scalable SLAM system with full 3D motion estimation. In: Int. Conference Safety Security and Rescue Robotics (SSRR), pp. 155–160 (2011)

23. Engel, J., Schöps, T., Cremers, D.: LSD-SLAM: large-scale direct monocular SLAM. In: Computer Vision (ECCV 2014), pp. 834–849 (2014)

24. Engel, J., Sturm, J., Cremers, D.: Accurate figure flying with a quadrocopter using onboard visual and inertial sensing. In: Proceedings of the Workshop on Visual Control of Mobile Robots (ViCoMoR) at the IEEE/RJS International Conference on Intelligent Robot Systems (IROS) (2012)

Compliant and Lightweight Anthropomorphic Finger Module for Aerial Manipulation and Grasping

Alejandro Suarez, Guillermo Heredia and Anibal Ollero

Abstract In this paper, a single DOF anthropomorphic finger module specifically designed for aerial manipulation and grasping is presented, where low weight and compliance are required features for safer performance of the aerial platform. The under-actuated mechanism consists of a high torque micro motor with a small reel for driving a tendon that moves the three joints of the finger. An elastic element maintains the finger bones tied together and extended by default, providing a compliant response against collisions with walls, the floor or any object in the environment during the grasping operation. Compliance provided by the elastic joints will also be exploited for stable object grasping and for collision detection. The development of a specific electronics for finger control instead of employing conventional servos make possible the design and application of a wider range of control strategies, including position, velocity or open-loop force control. The proposed design is validated through different grasping and control experiments.

Keywords Finger module · Compliance · Aerial manipulation and grasping

1 Introduction

The current trend in aerial robotics is to provide unmanned aerial vehicles (UAV's) like multi-rotors or autonomous helicopters with grasping and manipulation capabilities in order to extend the range of applications and scenarios where such platforms may be useful. Consider for example a group of ground robots deployed in an area of difficult access performing a certain task like inspection or maintenance. In such scenario, an UAV endowed with a robot manipulator could

A. Suarez · G. Heredia(✉) · A. Ollero
Robotics, Vision and Control Group, University of Seville, 41092 Seville, Spain
e-mail: {asuarezfm,guiller,aollero}@us.es

© Springer International Publishing Switzerland 2016
L. Paulo Reis et al. (eds.), *Robot 2015: Second Iberian Robotics Conference*,
Advances in Intelligent Systems and Computing 417,
DOI: 10.1007/978-3-319-27146-0_42

543

be used for the battery swap operation, transporting the batteries from the recharge station to the place where the ground robots are located. These activities demand a certain level of dexterity, which can be achieved designing specific grippers for such application, or, alternatively, considering a general purpose manipulator as the human hand is. In this sense, the human hand can be considered as the best manipulator in terms of dexterity, speed, force, weight and compliance.

Two approaches can be identified in the design of anthropomorphic manipulators depending on the actuations system: direct gear drive of finger joints or tendon driven actuation. The high speed multi-fingered robot hand presented in [1] makes use of harmonic drive gears and high power mini actuators for achieving accurate and power grasping of objects with different shapes. The hand, whose weight is 0.8 Kg with its three fingers and eight DOFs, is capable of catching a ball visually tracked by a pair of cameras providing a 3D estimation at 1 KHz rate. Prosthetics applications have motivated the analysis and design of biomimetic fingers driven by tendons as in [2], [3] and [4]. These works pay special attention to the identification and replication of the structure and features of the human finger, like the range of motion of the finger joints, the size and shape of the finger bones, the tendon configuration or the compliance provided by the joints. A tendon-based flexion drive mechanism consisting in a motor, a slider and a feed screw is used in [5] and [6] for building two low weight anthropomorphic hands with five and three fingers respectively, weighting 328 g and 250 g. In both cases, torsion springs are employed for finger extension, choosing the elastic constants conveniently taking into account the order when flexing the three finger joints.

Force measurement has been also treated in some works. A tactile sensor with 624 points has been incorporated to the Gifu hand II described in [7]. The hand counts with four 3-DOF fingers and a 4-DOF thumb driven by gears, with the servomotors built into the fingers and the palm. The tactile sensor is used for obtaining the contact patterns when the robot hand is grasping a spherical and a cylindrical object, comparing the results with the output obtained with the human hand. Tactile and force sensors have been also considered for the MAC-HAND shown in [8]. A tiny 6-DOF force-torque sensor is developed and integrated at each finger tip of the DLR Hand II [9]. The design principle followed here is to integrate all mechanics and electronics in the fingers and in the palm, using flexible PCBs for the connection between joints.

Aerial manipulation has received much attention in recent years. The integration between robotics manipulators and UAVs extend the scope of possible applications [10][11][12]. The use of aerial mobile manipulators opens a range of applications such as object manipulation, inspection and maintenance of industrial settings, structure construction, taking samples of material from areas difficult to access or other outdoor applications [13]. However, the limited payload of these vehicles imposes severe weight constraints to the design of the manipulator arms. Furthermore, the movement of the arm and the objects that are grasped with its end effector affect the UAV dynamics, and even may destabilize the aerial robot. The design of grasping and manipulation mechanisms for aerial applications is strongly constrained by payload, inertial and dynamic coupling limitations.

It is essential that the mass of the mechanism is low, but it is also convenient that it is distributed as close to the base of the vehicle as possible so the reaction forces and torques generated during its motion do not significantly affect the stability. On the other hand, compliance is a very desirable feature for the aerial manipulator in order to minimize the influence of contact forces propagated from the end effector to the UAV through the arm.

Grasping and manipulation capabilities with multi-rotors or helicopters have been already documented in several works. Quadrotor grasping and perching using two types of grippers, impactive and ingressive, is presented in [14]. The paper is focused on the estimation of the inertial parameters of the grasped object in two cases: hovering and in presence of aerodynamic disturbances. Dynamic equations of the vehicle relating the accelerations and inertias with the forces generated by the propellers are considered for this purpose. Helicopter stability is analyzed in [15] when contact forces are introduced in the aircraft through a compliant end-effector attached to the base employed for object grasping. The stability of a PID flight control system during object grasping and release operations is demonstrated in simulation and experimentally. Finally, reference [16] presents a control architecture for compliant interaction between a quadrotor equipped with a n-DOF manipulator and the environment.

In this paper, a 40 grams weight single DOF compliant and anthropomorphic finger module is presented. The finger module has been specifically designed for aerial grasping and manipulation and is capable of grasping different objects without requiring any other fingers. It consists of an aluminium frame where a DC micro motor is placed, and a human-size finger whose bones are tied together through a heat shrink tube that maintains the finger extended by default. A small section reel attached to motor shaft drives the nylon tendon for finger flexion, measuring the rotation of the first joint with a potentiometer. The finger module provides power and compliant grasping, so a wide variety of objects can be grasped even with a single finger, being also tolerant to collisions or impacts with the environment, as it will be shown in the paper. The mechanism has been designed considering low cost components (around 50 $ per finger module) and a simple mechanization process. The development of the control electronics removes the limitations imposed by conventional RC servos which only offer open-loop position control without any flexibility, allowing the implementation of a wider range of control methods more suited for the finger module.

The rest of the paper is organized as follows. Section 2 discusses the motivations and alternatives in the design of the presented finger module, explaining in more detail the implemented mechanism and the required electronics for its control. Section 3 covers the modeling and control of the under-actuated finger with a reel attached to the micro motor shaft that drives the tendon. Several control strategies adapted to the features of this mechanism will be presented. Section 4 validates the proposed design through a number of grasping and control experiments, while the conclusions of this work will be commented in Section 5.

2 Finger Module Design

2.1 *Motivation*

In the human hand, each finger provides four DOFs in a configuration like the one shown in Fig. 1. The thumb, where up to seven DOFs can be identified, is a special case which is out of the scope of this work. The names of the links and joints correspond to the bones of the finger. The metacarpo-phalangeal joint (MCP) connects the finger with the hand palm, allowing the adduction/abduction and the flexion/extension of the finger, with a range of motion of 40° and 90°, respectively, according to [2]. The proximal inter-phalangeal (PIP) and the distal inter-phalangeal (DIP) joints have a rotation range around 80° and 110°, although DIP joint is typically considered as under-actuated by the PIP joint. In that case, each finger would require three actuated DOFs driven by a pair of tendons; that is, six actuators.

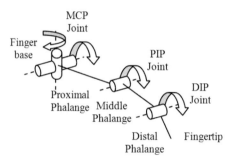

Fig. 1 Kinematic model of the four degrees of freedom anthropomorphic finger.

As this work is focused on the design of a very low weight anthropomorphic finger, it was found necessary to impose the following simplifications and constraints:

- Only flexion/extension is allowed, removing adduction/abduction at MCP joint.
- The three joints will be under-actuated by a single motor for finger flexion, driving a nylon tendon.
- The finger will stay extended by default.
- An elastic element disposed between the phalanges will cause the flexion of the finger in the following order: 1) MCP, 2) PIP, and 3) DIP.

The modular design of the fingers has a number of advantages and drawbacks with respect the design of a full hand considered as a whole. First, it simplifies and reduces the time in the design and construction process, as it avoids to waste time in the development and validation of the full hand. It also leads to highly efficient solutions as the work is focused on a more specific domain. It facilitates maintenance and replacement operations in case there is any fault, just removing the finger module from the palm frame. Finally, a number of finger modules can be considered for building hands with different configurations just changing the base frame where they are attached. On the other side, modularity usually implies higher weight, size and redundancy which could be reduced considering the hand as a whole, exploiting its geometrical structure for this purpose.

2.2   Mechanism Description

A picture of the finger module and the assembly diagram of the mechanism in the 3D model can be seen in Fig. 2 and Fig. 3. The anthropomorphic finger consists in three 8 mm U-shape aluminium profile sections of 45 mm (proximal phalange), 20 mm (middle phalange) and 15 mm (distal phalange) length. A 50 mm section of heat shrink tube maintains the three bones tied together and at the same time it acts as elastic element for PIP and DIP joints, maintaining them extended by default. The elastic constant of both joints will depend on the heating process of the heat shrink tube and in the separation between the profile sections. The attachment between the case and the aluminium sections has been reinforced with three 2 mm Ø screws. The screw at the middle phalange and a fourth screw near the midpoint of the proximal phalange correspond to the pass points of the nylon tendon that drives the finger. Two 70x15x2 mm aluminium frames and three 11x6 mm cylinders support the finger and the Pololu 298:1 micro-motor, which is fixed to the surface of the frames through double sided adhesive tape. The Murata SV01 potentiometer is placed in the external side of one of the frames and aligned with the MCP joint shaft, while an extension spring that connects the proximal phalange with the frame is used to maintain this joint extended by default. Finally, an 8 mm Ø reel with 6 mm Ø internal section is adjusted to motor shaft for tendon rolling. At this point, it is important that the nylon tendon has the minimal length in order to avoid that the cable is unrolled if the tendon becomes slack.

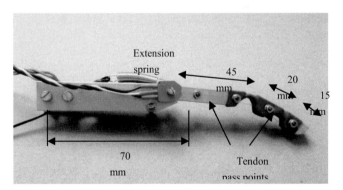

Fig. 2 Anthropomorphic finger module with compliant joints and 40 grams weight.

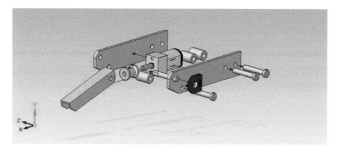

Fig. 3 View of the components of the finger module in the 3D model.

2.3 Electronics

The electronics required for the control of the finger consists in an H-bridge implemented with the LM293B integrated circuit, a STM32 VL Discovery board and a USB-to-UART device for the communications with the computer. The H-bridge is employed for the torque/speed and direction control of the Pololu 298:1 micro metal gear motor. The microcontroller board generates the digital output and the PWM signal for motor control, gets the voltage given by the potentiometer attached to MCP joint from an analog input channel, and communicates with a computer through the USB-to-UART interface for data acquisition and parameters configuration.

There are two important considerations related to the H-bridge that should be taken into account when designing the control system. First, the PWM signal taken as input represents the mean current consumed by the motor, from zero to the stall current. For example, a duty cycle of 50% would correspond to a mean current consumption of half the maximum. According to the model of the motor described in Section 3, the speed and torque of the motor will depend on the mean current provided by the H-bridge, so acting over the PWM signal it will be possible to control these signals. Secondly, changes in the direction of the motor cause high amplitude peaks on the current, which are not desirable. This happens when joint position is close to the reference and oscillates around it. To avoid this, a dead PWM zone should be defined for increasing lifespan of motor and H-bridge.

3   Modeling and Control

The intention of this section is showing how to control MCP finger joint position, velocity and torque from the PWM signal that is taken as input by the H-bridge. Two case studies should be distinguished depending on if motor is stalled because the finger is grasping an object (force-torque control) or in free running (position-velocity control). In the following, θ, $\dot{\theta}$ and τ will represent joint position, speed and torque, respectively, F and l will be tendon force and length, and r the radius of the reel that drives the tendon. Motor torque, speed and current will be denoted by T, ω and i, respectively. PIP and DIP joints are under-actuated and their positions are not measured, so they are not considered in this study. Inertia and friction terms will not be taken into account for simplicity and because they have not a significant influence in joint position control performance. The analysis presented here only affects to the flexion motion. Finger extension is performed by the elastic elements disposed at finger joints, providing a faster response than the tendon drive mechanism.

3.1   Force-Torque Relationships for Object Grasping

When the finger is grasping an object or it is completely flexed, the motor is stalled and the nylon tendon tensed. It is well known in brushed DC motor modeling that the generated torque is proportional to the supplied current. The manufacturer usually provides the stall torque and current parameters, which correspond to a situation in which motor shaft is blocked and current consumption is maximum. The H-bridge controls the mean current injected to motor coil in terms of current pulses of variable duty cycle. If the PWM signal is in the range [0, 1], from no current to stall current, then motor torque can be controlled in the following way:

$$0 \leq T = k_t \cdot i_{stall} \cdot pwm \leq T_{stall} \tag{1}$$

where $k_t = T_{stall}/i_{stall}$ is torque constant. Tendon force is directly obtained dividing motor torque between reel radius:

$$F = T/r = k_t \cdot i_{stall} \cdot pwm/r \tag{2}$$

This expression shows that tendon tension can be directly controlled with the PWM signal in open loop. It also results especially useful in the experimental identification of the torque constant, as it is easier measuring tendon tension with extension springs rather than measuring motor torque with torsion springs. On the other hand, the computation of the torque at the MCP joint depends on P_1 and P_2 points in Fig. 4. These are the contact points were the nylon tendon passes through, with a separation angle α between them. Note that P_1 will be a function of MCP joint angle θ, while P_2 is fixed:

$$\begin{aligned} \boldsymbol{P_1} &= [d \cdot \cos\theta \quad -d \cdot \sin\theta]^T \\ \boldsymbol{P_2} &= [-b \quad -c]^T \end{aligned} \tag{3}$$

The joint torque is then:

$$\tau = F \cdot d \cdot \sin\alpha \quad , \quad \alpha = \cos^{-1}\left(\frac{\boldsymbol{P_1} \cdot \boldsymbol{P_2^T}}{\|\boldsymbol{P_1}\| \cdot \|\boldsymbol{P_2}\|}\right) \tag{4}$$

The numerical values of these constants are: $b = 9 \cdot 10^{-3}$ [m], $c = 1.1 \cdot 10^{-2}$ [m], $d = 1.4 \cdot 10^{-2}$ [m], $T_{stall} = 0.49$ [N $\cdot$ m] and $i_{stall} = 1.6$ [A].

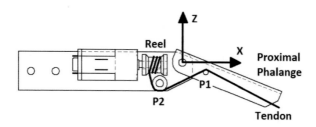

Fig. 4 Model used in the computation of the MCP joint torque. Nylon tendon passes through P_1 and P_2 before it is rolled up on the reel.

3.2 Position-Speed Relationships for Finger Positioning

As described in Section 2, in the proposed design of the finger module a potentiometer has been placed directly on MCP joint shaft for measuring proximal phalange rotation in a reliable, accurate and simple way. The other alternative would be measuring motor shaft rotation for obtaining tendon length, although this has two main drawbacks when more than one reel turn is required for finger flexion: the lack of an absolute position reference, and the necessity of counting turns. The idea now is relating MCP joint angle θ with tendon length l. Operating with (3) it results that:

$$l(\theta) = \|P_1 - P_2\| = \sqrt{b^2 + c^2 + d^2 + 2 \cdot d \cdot [b \cdot cos(\theta) - c \cdot sin(\theta)]} \quad (5)$$

As seen, it is not possible to express analytically joint angle θ as a function of tendon length l. The same happens with joint speed, which is proportional with motor speed but depends on joint angle:

$$\dot{\theta} = -\frac{l(\theta)r\omega}{d[b \sin\theta + c \cos\theta]} \quad (6)$$

As occurs with torque, motor speed ω can be controlled in open loop with the PWM signal, and thus joint speed $\dot{\theta}$, but with a low accuracy due to the nonlinearity in (6).

3.3 Control Scheme

The control architecture for the finger module has been represented in Fig. 5. The controller takes as input the joint reference θ_{ref} in rad which is scaled for obtaining the desired voltage at the potentiometer. The measurement provided by this device is subtracted for computing MCP joint error, whose sign determines the direction of motion and its amplitude the duty cycle of the PWM signal applied to the H-bridge. A simple proportional controller with constant K_p generates the required duty cycle signal from the absolute value of the joint error. Maximum tendon force and motor speed are regulated saturating the PWM signal in the range [0, 1], that is, from no current to the stall current. Finally, a dead zone is considered for preventing unnecessary changes in motor direction when joint error is low, which significantly reduces current consumption.

Note that the range of motion of the MCP joint is between zero and 90 degrees approximately. In the case θ_{ref} is above this value, the motor will continue exerting a force on the tendon, so the PIP and DIP joints will be flexed.

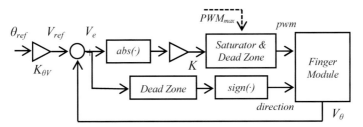

Fig. 5 Control architecture for MCP joint position and force control. Maximum joint speed and tendon force are controlled tuning the maximum PWM value in the saturator.

4 Experimental Validation

This section shows the application of the presented finger module for joint position and force control (in closed and open loop, respectively), stable object grasping, and for collision detection. A video demonstrating compliance against hammer impact, joint position control and its ability for grasping different objects can be found in [17].

4.1 Position and Force Control Results

The control scheme described in Section 3.3 is evaluated here. A stair reference for MCP joint flexion and extension motion from 0 to 90 deg is considered in Fig. 6. The upper part of the figure shows the position reference and the measured value, while the lower part represents the duty cycle applied to the H-bridge. In $t = 17$ s the reference is set to 200 deg, which is an unreachable position, as the proximal

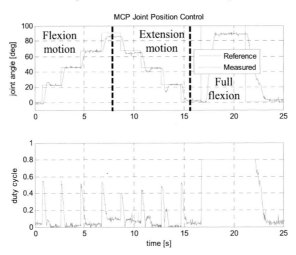

Fig. 6 MCP joint position control and PWM signal. Stair reference for flexion and extension. Proportional gain is set to $K_p = 2.5$, and PWM is saturated to 80%.

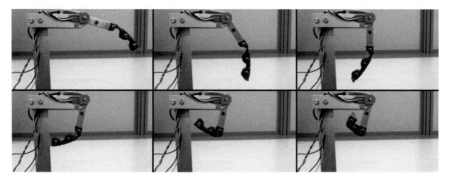

Fig. 7 Finger flexion motion performed by the three joints driven by the single DOF.

phalange mechanically blocks the rotation at 90 deg. PWM is saturated to 80%, with a dead zone of 1%, and the proportional gain set to $K_p = 2.5$. The motion sequence in full finger flexion can be seen in Fig. 7. Note that the first flexed joint is the MCP, then the PIP and finally the DIP, corresponding to lower to higher elastic constants.

Section 3.1 showed that tendon force is proportional to the duty cycle applied in the H-bridge, so MCP joint torque can be controlled in this way. For maintaining the compatibility with the position control, what it is done is saturating the PWM signal after the position controller. Fig. 8 shows joint position and torque when the finger is grasping a flexible rubber cup for an unreachable goal position of 200 deg. Saturation thresholds from 30% up to 80% at 10% steps of the stall current/torque are considered. Fig. 9 represents how the rubber cup is deformed as the threshold increases.

4.2   Object Grasping

The finger module is capable of grasping objects of different shapes and forms in a stable way by itself, without requiring any other fingers. Fig. 10 shows four examples

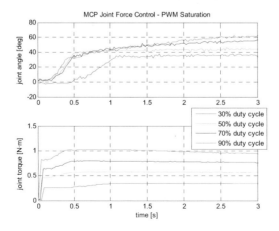

Fig. 8 MCP joint position and torque in force control mode for different saturation values.

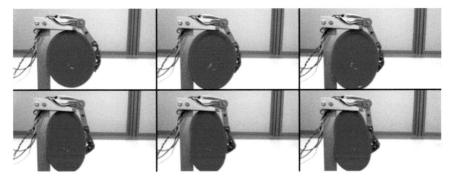

Fig. 9 Open-loop force control grasping a flexible rubber cup. PWM is saturated from 30% (left upper picture) up to 80% (right lower picture) at 10% steps.

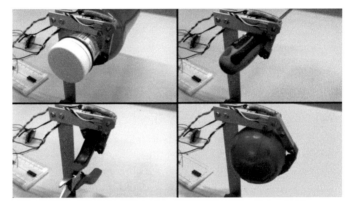

Fig. 10 Stable grasping of different objects: bottle, screwdriver, pliers and ball.

in which the finger module is grasping a bottle, a screw driver, a plier and a ball. This is so thanks to its three compliant and underactuated joints that adapt finger to object contour, exerting a constant torque on the three joints when the motor is stalled.

4.3 Impact and Collision Detection

If the finger is not grasping any object, it can be used for detecting and reacting against collisions of the manipulator with the environment in a safe way as joint compliance will cause the passive contraction of the finger. This may be especially useful in the navigation of the aerial platform in narrow areas. For this purpose, the finger would remain extended and the MCP joint position controller disabled. Any impact of the finger will cause a deviation in joint position that could be easily detected defining a constant threshold. Fig. 11 shows the response of the MCP joint when the finger is hit with a hammer (between $t = 0$ and $t = 5$ s), and when it is flexed due to a frontal collision (from $t = 6$ s). Two thresholds of ± 20 deg are considered in such a way that when the measured joint angle exceeds any of them, a collision is detected.

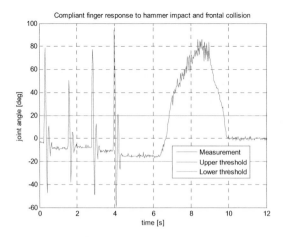

Fig. 11 MCP joint response to hammer impact and to frontal collision, detected when the measured joint angle exceeds the upper or lower threshold.

5   Conclusion

This work has presented the design and experimental validation of a compliant and low weight (40 grams) anthropomorphic finger module specifically designed for aerial manipulation and grasping. The three joints of the finger are underactuated and driven by a nylon tendon that is rolled up on a small section reel attached to a micro motor shaft. The finger module has been built using low cost components and materials, with a simple and robust aluminium frame. The development of the low level DC motor electronics has allowed position and force/torque control in closed and open loop, respectively. Experiments in position and force control, object grasping and in collision detection have validated the proposed mechanism.

Acknowledgments This work has been supported by the ARCAS Project, funded by the European Comission under the FP7 ICT Programme (ICT-2011-287617), and by the Ministerio de Educación, Cultura y Deporte FPU Program.

References

1. Namiki, A., Imai, Y., Ishikawa, M., Kaneko, M.: Development of a high-speed multi-fingered hand system and its application to catching. In: Proc. of the IROS 2003, Las Vegas, USA, October 27–31, 2003, pp. 2666–2671 (2003)
2. Bundhoo, V., Haslam, E., Birch, B., Park, E.J.: A shape memory alloy-based tendon-driven actuation system for biomimetic artificial fingers, part I: design and evaluation. Robotica **27**(1), 131–146 (2009)

3. Bundhoo, V., Park, E.J.: Design of an artificial muscle actuated finger towards biomimetic prosthetic hands. In: Proc. of the 12th Int. Conf. on Advanced Robotics (ICAR 2005), Seattle, USA, July 2005, pp. 368–375 (2005)

4. Xu, Z., Kumar, V., Matsuoka, Y., Todorov, E.: Design of an anthropomorphic robot finger system with biomimetic artificial joints. In: Proc. of the 4th Int. Conf. on Biomedical Robotics and Biomechatronics, Roma, Italy, June 24–27, 2012, pp. 568–574 (2012)

5. Takaki, T., Omata, T.: High-performance anthropomorphic robot hand with grasping-force-magnification mechanism. In: Proc. of the ICRA 2009, Kobe, Japan, pp. 2339–2345 (2009)

6. Zollo, L., Roccella, S., Guglielmelli, E., Carrozza, M., Dario, P.: Biomechatronic design and control of an anthropomorphic artificial hand for prosthetic and robotic applications. IEEE/ASME Trans. on Mechatronics 12(4), 418–429 (2007)

7. Kawasaki, H., Komatsu, T., Uchiyama, K.: Dexterous anthropomorphic robot hand with distributed tactile sensor: Gifu hand II. IEEE/ASME Trans. on Mechatronics 7(3), 296–303 (2002)

8. Cannata, G., Maggiali, M.: An embedded tactile and force sensor for robotic manipulation and grasping. In: Proc. of the 5th Int. Conf. on Humanoid Robots, Tsukuba, Japan, pp. 80–85 (2005)

9. Butterfass, J., Grebenstein, M., Liu, H., Hirzinger, G.: DLR-Hand II: next generation of a dexterous robot hand. In: Proc. of the ICRA 2001, Seoul, Korea, May 2001, pp. 101–114 (2001)

10. Kondak, K., Huber, F., Schwarzbach, M., Laiacker, M., Sommer, D., Bejar, M., Ollero, A.: Aerial manipulation robot composed of an autonomous helicopter and a 7 degrees of freedom industrial manipulator. In: Proc. of the ICRA 2014, pp. 2107–2112 (2014)

11. Jimenez-Cano, A., Martin, J., Heredia, G., Ollero, A., Cano, R.: Control of an aerial robot with multi-link arm for assembly tasks. In: Proc. of the ICRA 2013, pp. 4916–4921 (2013)

12. Orsag, M., Korpela, C., Oh, P.: Modelling and Control of MM-UAV: Mobile Manipulating Unmanned Aerial Vehicle. Journal of Intelligent and Robotic Systems 69, 227–240 (2013)

13. Heredia, G., Jimenez-Cano, A.E., Sanchez, I., Llorente, D., Vega, V., Braga, J., Acosta, J.A., Ollero, A.: Control of a multirotor outdoor aerial manipulator. In: Proc. of the IROS 2014, pp. 3417–3422 (2014)

14. Mellinger, D., Lindsey, Q., Shomin, M., Kumar, V.: Design, modelling, estimation and control for aerial grasping and manipulation. In: Proc. of the IROS 2011, pp. 2668–2673, September 2011

15. Pounds, P.E., Dollar, A.M.: Stability of Helicopters in Compliant Contact Under PD-PID Control. IEEE Transactions on Robotics 30(6), 1472–1486 (2014)

16. Giglio, G., Pierri, F.: Selective compliance control for an Unmanned Aerial Vehicle with a robotic arm. In: 22nd Mediterranean Conference of Control and Automation (MED 2014), pp. 1190–1195, June 2014

17. http://grvc.us.es/staff/alejandro/robot2015/CompliantFinger.mp4

Robust Visual Simultaneous Localization and Mapping for MAV Using Smooth Variable Structure Filter

Abdelkrim Nemra, Luis M. Bergasa, Elena López, Rafael Barea, Alejandro Gómez and Álvaro Saltos

Abstract The work presented in this paper is a part of research work on autonomous navigation for Micro Aerial Vehicles (MAVs). Simultaneous Localization and Mapping (SLAM) is crucial for any task of MAV navigation. The limited payload of the MAV makes the single camera as best solution for SLAM problem. In this paper the Large Scale Dense SLAM (LSD-SLAM) pose is fused with inertial data using Smooth Variable Structure Filter which is a robust filter. Our MAV-SVSF-SLAM application is developed under Linux using Robotic Operating System (ROS) so that the code can be distributed in different nodes, to be used in other applications of guidance, control and navigation. The proposed approach is validated first in simulation, then experimentally using the Bebop Quadrotor in indoor and outdoor environment and good results have been obtained.

Keywords Micro Aerial Vehicles · Simultaneous Localization and Mapping · Autonomous navigation · Data fusion · Robust filters · Sensor fusion · Robotic Operating System

1 Introduction

Self-localization of unmanned vehicles (UV) is still considered as a fundamental problem in autonomous vehicle navigation. The problem has been tackled in the recent past for different platforms and a number of efficient techniques were proposed as presented in Cox and Wilfong [1]. Many research works about optimal

A. Nemra
Ecole Militaire Polytechnique, Algiers, Algeria
e-mail: karim_nemra@yahoo.fr

L.M. Bergasa · E. López(✉) · R. Barea · A. Gómez · Á. Saltos
Department of Electronics, Alcalá University, Alcalá, Spain
e-mail: {bergasa,elena,barea}@depeca.uah.es

© Springer International Publishing Switzerland 2016   557
L.P. Reis et al. (eds.), *Robot 2015: Second Iberian Robotics Conference*,
Advances in Intelligent Systems and Computing 417,
DOI: 10.1007/978-3-319-27146-0_43

and robust data fusion have been presented in literature, Toledo-Moreo et al [2] Leonard and Durrant-Whyte [3], Toth [4]. More recently, driven by the interest of military applications and planetary explorations, researchers have turned their attention towards localization of vehicles moving in hostile and unknown environments [5, 6]. In these cases, a map of the environment is not available and hence a more complex simultaneous localization and mapping (SLAM) problem must be faced and solved.

Various techniques exist for robot pose estimation, including mechanical odometer, GPS, LASER, INS and cameras[7]. However in recent years the use of computer vision has always attracted researchers because of the analogy that we can do with the human tracking system.

SLAM has often been performed using a single camera, stereo cameras and recently, with the apparition of depth (RGBD) camera, RGBD-SLAM became an interesting solution.

Solve the Visual SLAM for Mini Aerial Vehicle (MAV) is a real challenge especially in GPS denied regions. Many alternatives of Aerial SLAM are implemented using LASER, Vision, and RGBD camera. However majority of these solutions cannot be used for MAV which can carry a limited payload (between 60-100 g). Thus, the best solution of the MAV- SLAM is to use a single camera.

The main contribution of this paper is to propose a robust solution for 6DoF-MAV-SLAM by fusing the pose given by the Large Scale Dense SLAM, and that given by the inertial navigation system using the Smooth Variable Structure Filter (SVSF). The algorithm is developed under Linux using ROS (Robotic Operating System) so that the code can be distributed in different nodes, to be used in other applications of control and navigation. The proposed solution should be able to: localize the MAV in indoor environment, should be robust face any photometric and geometric change and can maintain suitable pose accuracy even when the visual information is lost.

This paper is organized as follows, in section 2, previous work on single camera SLAM for MAV are presented. In section 3 process and observation model for IMU/LSD-SLAM algorithm are detailed. The Smooth Variable Structure Filter (SVSF) is presented in section 4. After that, the proposed algorithms are implemented and validated in simulation and experimentally in section 5. Finally, in section 6, conclusion and perspectives are given.

2   Single Camera SLAM for MAV

Real-time monocular Simultaneous Localization and Mapping (SLAM) have become very popular research topics. Two major reasons are (1) their use in robotics, in particular to navigate unmanned aerial vehicles (UAVs) [8, 9, 10], and (2) augmented and virtual reality applications slowly making their way into the mass-market.

2.1   Direct Methods SLAM

Direct Visual Odometry (VO) methods are very interesting for real time application by optimizing the geometry directly on the image intensities, which enables using all information in the image.

2.2 Large Scale Direct SLAM

Large-Scale Direct SLAM (LSD-SLAM) method, not only locally tracks the motion of the camera, but allows building consistent, large-scale maps of the environment. The method uses direct image alignment coupled with filtering-based estimation of semi-dense depth maps as originally proposed in [11]. The global map is represented as a pose graph consisting of keyframes as vertices with 3D similarity transforms as edges, elegantly incorporating changing scale of the environment and allowing detection and correction of the accumulated drift. The method runs in real-time on a CPU and even on a modern Smartphone [11]. The algorithm LSD SLAM consists of three major components: tracking, depth map estimation and map optimization as visualized in Figure. 1:

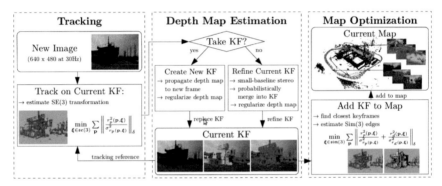

Fig. 1 Overview over the complete LSD-SLAM algorithm [11].

2.3 LSD-SLAM Limitations

Large-Scale Direct monocular SLAM (LSD-SLAM), have many advantages, firstly, it requires only a single camera, secondly, it constructs high quality map at different scales and finally and more importantly it can be run in real time on a CPU and even on a modern Smartphone. However, the LSD-SLAM algorithm suffers from several limits. Then, solving the MAV-SLAM using LSD-SLAM presents many drawbacks: - Pose and Map constructed up to scale, - very sensitive to photometric changes, - the observed scene must contain different textures and depths, - pure rotation motion cannot be estimated, - sensitive to dynamic scene. As solution, and in order to improve the performances (robustness, accuracy) of the LSD-SLAM we propose to fuse the LSD-SLAM position with inertial data using an Inertial Measurement Unit (IMU).

3 IMU/LSD-SLAM Fusion

In our work, we are using the Quadrotor Bebop which has a camera, IMU and GPS (Figure. 4). For Indoor applications the GPS signal is not available.

Therefore, fuse IMU with camera data for localization and mapping is very interesting especially if the application run in real time. Furthermore, the presence of
IMU data makes our MAV-SLAM able to maintain right 6DOF pose even when
the visual information is not available. The proposed SLAM solution is developed
under Linux using ROS (Robotic Operating System) which gives it many advantages (Parallelism, Real time, 3D map view).

3.1   Process Model (Inertial Measurement Unit)

The sate vector to be used in our SLMA problem is given by:

$$\hat{X}_k = [x \quad y \quad z \quad u \quad v \quad w \quad q_1 \quad q_2 \quad q_3 \quad q_4]^t \tag{1}$$

Where: (x y z) the position of the UAV in the navigation frame, (u v w) :
the velocities of the UAV in the body frame, (q_1 q_2 q_3 q_4) : are the quaternion to represents the MAV orientation.

The cinematic model of the MAV is given by:

$$\hat{X}_k = f(\hat{X}_{k-1}) + w_k \tag{2}$$

$$\hat{X}_k = \begin{bmatrix} p^n(k) \\ v^n(k) \\ \psi^n(k) \end{bmatrix} = \begin{bmatrix} p^n(k-1) + v^n(k)\Delta t \\ v^n(k-1) + [C_b^n(k-1)f^b(k) + g^n]\Delta t \\ \psi^n(k-1) + E_b^n(k-1)\omega^b(k)\Delta t \end{bmatrix} + w_k \tag{3}$$

w_k is the process noise, p^n, v^n and ψ^n are position velocity and orientation in
the navigation frame. Where f^b and ω^b are the body-frame referenced vehicle
accelerations and rotation rates which are provided by inertial sensors on the vehicle and g^n is the acceleration due to gravity. C_b^n and E_b^n are respectively the direction cosine matrix and rotation rate transformation matrix between the body
and navigation frames.

3.2   Observation Model (LSD-SLAM Pose)

The position given by the LSD-SLAM will be used as observation, as follows:

$$Z_k = h(\hat{X}_k)$$

$$Z_k = \begin{bmatrix} x \\ y \\ z \\ q_1 \\ q_2 \\ q_3 \\ q_4 \end{bmatrix} = \begin{bmatrix} 1 & 0 & 0 & 0 & 0 & 0 & 0 & 0 & 0 & 0 \\ 0 & 1 & 0 & 0 & 0 & 0 & 0 & 0 & 0 & 0 \\ 0 & 0 & 1 & 0 & 0 & 0 & 0 & 0 & 0 & 0 \\ 0 & 0 & 0 & 0 & 0 & 0 & 1 & 0 & 0 & 0 \\ 0 & 0 & 0 & 0 & 0 & 0 & 0 & 1 & 0 & 0 \\ 0 & 0 & 0 & 0 & 0 & 0 & 0 & 0 & 1 & 0 \\ 0 & 0 & 0 & 0 & 0 & 0 & 0 & 0 & 0 & 1 \end{bmatrix} \cdot [\hat{X}_k] \tag{4}$$

As can be seen from Eq.4, the observation model is linear which is suitable for the
Smooth Variable Structure Filter (SVSF) which is the subject of the next section.

4 Smooth Variable Structure Filter

While EKF-SLAM and FastSLAM are the two most popular solution methods for SLAM problem, newer alternatives, which offer much potential, have been proposed, including the use of the unscented Kalman filter (UKF) proposed by Julier and Uhlmann in SLAM [12]. Unlike the (EKF), the (UKF) uses a set of chosen samples to represent the state distribution. The UKF-SLAM avoids the calculation of the Jacobean and Hessian matrices but also obtain higher approximation accuracy with the unscented transformation (UT) [13]. However, for high-dimensional systems, the computation time is still heavy; thus, the filter converges slowly.

However, the above filters are all based on the framework of the Kalman filter (KF); it can only achieve a good performance under the assumption that the complete and exact information of the process model, observation model and noise distribution have to be a priori known.

To overcome some of these limitations, we propose to use the SVSF filter to solve the SLAM problem. Based on sliding mode concepts the first version of the variable structure filter (VSF) was introduced in 2003 [15]. On 2007, the smoothing variable structure filter (SVSF) which is relatively a new filter is introduced [14]. It is a predictor/corrector estimator based on sliding mode control theory and variable structure estimation concepts. In its old form, the SVSF is not a classical filter in the sense that it does not have a covariance matrix.

4.1 SVSF Principal

Essentially this method makes use of the variable structure theory and sliding mode concepts. It uses a switching gain to converge the estimates to within a boundary of the true state values (i.e., existence subspace shown in Figure. 2).

The SVSF has been shown to be stable and robust to modeling uncertainties and noise, when given an upper bound on the level of un-modeled dynamics and noise. The SVSF method is model based and may be applied to differentiable linear or nonlinear dynamic equations. An augmented form of the SVSF was presented in [16], which includes a full derivation for the filter using covariance matrices.

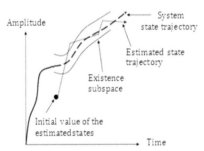

Fig. 2 The SVSF estimation concepts [15].

The basic estimation concept of the SVSF is shown in Figure 2. Some initial values of the estimated states are made based on probability distributions or designer knowledge. An area around the true system state trajectory is defined as the existence subspace. Through the use of the SVSF gain, the estimated state will be forced to within this region. Once the value enters the existence subspace, the estimated state is forced into switching along the system state trajectory.

The estimation process is iterative and may be summarized by the following set of equations (for control or estimation problem) [16]. Like the KF, the system model is used to calculate a priori state as follows:

$$X_{0/0} = X_0$$
$$E_{z,0/0} = E_{z,0}$$
$$\hat{X}_{k+1/k} = f\left(\hat{X}_{k/k}, U_k\right) \tag{5}$$
$$\hat{Z}_{k+1/k} = h(\hat{X}_{k+1/k}) \tag{6}$$
$$E_{z,k+1/k} = Z_{k+1} - \hat{Z}_{k+1/k} \tag{7}$$
$$K_{k+1} = diag\left[\left(\left|E_{z,k+1/k}\right|_{Abs} + \gamma\left|E_{z,k/k}\right|_{Abs}\right) \circ sat(\bar{\psi}^{-1}E_{z,k+1/k})\right]\left[diag(E_{z,k+1/k})\right]^{-1} \tag{8}$$

Where $\circ$ signifies Schur (or element-by-element) multiplication, the superscript $+$ refers to the pseudo inverse of a matrix and $\bar{\psi}^{-1}$ is a diagonal matrix constructed from the smoothing boundary layer vector ψ, such that:

$$\bar{\psi}^{-1} = [Diag(\psi)]^{-1} = \begin{bmatrix} \frac{1}{\psi_1} & 0 & 0 \\ 0 & \ddots & 0 \\ 0 & 0 & \frac{1}{\psi_m} \end{bmatrix} \tag{9}$$

$$sat\left(\bar{\psi}^{-1}E_{z_{k+1/k}}\right) = \begin{cases} 1, & E_{zi,k+1/k}/\psi_i \geq 1 \\ E_{zi,k+1/k}/\psi_i, & -1 < E_{zi,k+1/k}/\psi_i < 1 \\ -1, & E_{zi,k+1/k}/\psi_i \leq -1 \end{cases} \tag{10}$$

$$\hat{X}_{k+1/k+1} = \hat{X}_{k+1/k} + K_{k+1}E_{z,k+1/k} \tag{11}$$
$$\hat{Z}_{k+1/k+1} = h(\hat{X}_{k+1/k+1}) \tag{12}$$
$$E_{z,k+1/k+1} = Z_{k+1} - \hat{Z}_{k+1/k+1} \tag{13}$$

Two critical variables in this process are the a priori and a posteriori measurements error estimates, defined by (Eq.6 and 7). Note that (Eq.7) is the a posteriori measurement error estimates from the previous time step, and is used only to calculate the SVSF gain. The value of the smoothing boundary layer width vector ψ reflects the level of uncertainties and errors modelling affecting our observation model.

5   Results and Discussion

To validate the performances of the proposed approaches, we first evaluate the performances of the SVSF filter in simulation. After that, the SVSF-SLAM will be validated using the new Bebop (Figure 4).

5.1   Simulation Results

In this simulation we consider a MAV navigating using an inertial measurement unit and LSD-SLAM pose. We assume that (between t = 600s and t = 800s) the MAV passes by a uniform region (or dark region) and no feature is detected. In other word, during this region the LSD-SLAM gives completely a wrong position (red triangle in Fig.3) with large uncertainty. As can be seen from Figure 3 when the LSD-SLAM position is lost the MAV keep follows the IMU position (blue trajectory). When t >800s the visual information return back and the LSD-SLAM pose will be fused with the IMU position again using SVSF filter.

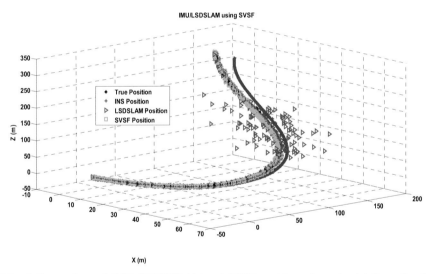

Fig. 3 Comparison of IMU, LSDVSLAM and SVSF position when visual information is not available

5.2   Experimental Results

The proposed solution of SVSF-SLAM is validated using the new MAV of parrot called Bebop (Figure 4). This latter is commercialized in December 2014 and have many advantages compared to the AR Drone2.

- **MAV Bebop Main Features**
  - Capture 1080p Video and 14MP Photos, 3-Axis Sensor (accelerometer, gyroscope and magnetometer), Linux-Based Flight Computer, Dual-core CPU, ARM 9, quad-core GPU, and 8GB of flash memory.

Fig. 4 The experimental platform the new Parrot Bebop

Experiment 1 (Indoor):
In this experiment the Quadrotor Bebop navigates in indoor environment (Figure 5). Figure 6, shows the acquired image (a), the tracked features (b) and the corresponding covariance for these features (c) using color representation. As can be seen, in this algorithm, corners and edges are selected as features, which is very important for robustness. Moreover, in the first image (c) where depth is initialized randomly then, its covariance was large (Red color), and during navigation using stereovision (Multiple view) this covariance decreases (green color).

Fig. 5 Picture of Laboratory Alcàla University

(a)   (b)   (c)

Fig. 6 Experimental validation of the IMU/LSD-SLAM algorithm (a) level gray image, (b) Features extraction and depth estimation, (c) Colour representation of the estimated depth variance.

Figure 7 shows the pose of the Bebop and the constructed map during navigation. The red camera represents the current position where the blue one represents the keyframe position.

Experiment 2 (Outdoor):
In this experiment the Quadrotor Bebop navigates in outdoor environment, the UAH University (Figure 8).

Figure 9 shows the pose of the Bebop and the constructed map during navigation. From Figure 9 we can observe that a good map is constructed with the MAV-SLAM as well as the followed trajectory.

Fig. 7 Bebop localization and 3D map building using IMU/LSD-SLAM

Fig. 8 Picture of the UAH University (Alcàla)

Fig. 9 Localization and 3D map building using MAV-SLAM algorithm

Experiment 3 (LSD-SLAM Pose Lost):

In this experiment which is done indoor (no GPS signal) we want to validate the robustness of our algorithm SVSF-IMU/LSD-SLAM. For that, during Bebop navigation we will hide the Bebop camera for a while (20 s) to see how the IMU/LSD-SLAM algorithm can maintain a suitable pose estimation.

Fig. 10 Acquired image by the Bebop camera (missing of visual information for 20 s)

Figure 10 shows few acquired images by the Bebop, as can be seen some of them are completely ambiguous. Figure 11 shows the Bebop pose and the constructed map using only the LSD-SLAM algorithm, it is clear that the algorithm cannot estimate any camera (Bebop) pose when the image information is not available. Let's see what SVSF-IMU/LSD-SLAM filter can do to solve this problem. To improve the quality of the pose estimation inertial data will be used.

Fig. 11 Pose estimation and 3D map construction using LSD-SLAM (dashed ellipse images missing, LSD-SLAM cannot estimate any Bebop position).

Figure 12 shows the angular rates and the accelerations -following the three axis- given by the IMU during the Quadrotor navigation, when, Figure 13 shows the Bebop position given only by integrating IMU data, this latter is good for short term but it suffers from drift for long term navigation. Figure 14 shows the pose and the map of the IMU/LSD-SLAM algorithm. However because the LSD-SLAM pose is given up to unknown scale, this latter is estimated by calculating the rate between IMU-pose and LSD-SLAM-pose for few iterations in the beginning, in our case the scale factor ≈ 6. As can be seen from Figure 14, even when visual information is not available (dark region) our algorithm maintains a suitable position of the Bebop Quadrotor.

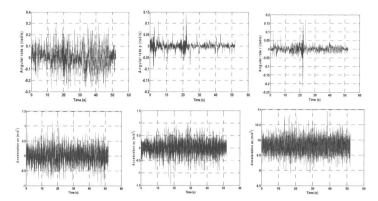

Fig. 12 Angular rates and accelerations given by the IMU

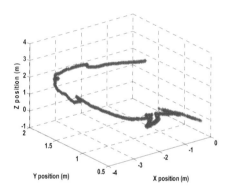

Fig. 13 Bebop position given by the IMU

Fig. 14 Pose estimation and 3D map construction using IMU/LSD-SLAM (dashed ellipse visual information is not available, IMU/LSD-SLAM can estimate the Bebop position)

6   Conclusion

The work presented in this paper is a part of autonomous navigation for MAV. The objective was to propose an online SLAM solution for an MAV (Parrot Bebop). For that purpose, the LSD-SLAM algorithm is fused with the IMU data using SVSF filter to improve the robustness of the proposed SLAM solution, and make it able to estimate accurate pose even if the visual information is not available (images without textures, dark area...). Finally, the proposed approach is validated first in simulation, then with experiment data in indoor and outdoor environment and good results have been obtained. The next objective of our work is to implement a new version of the MAV-SLAM which should be able to estimate accurate position even in dynamic environment.

Acknowledgements This work is supported by the Index MAV Project (CCG2014/EXP-078), which is funded by the University of Alcalá, and by the RoboCity2030 III-CM project (S2013/MIT-2748) funded by Programas de I+D en la Comunidad de Madrid and co-funded by Structured Funds of the UE.

References

1. Cox, I.J., Wilfong, G.T.: Autonomous robot vehicles. Springer Verlag, New York (1990)
2. Toledo-Moreo, R., Zamora-Izquierdo, M.A., Ubeda-Minarro, B., et al.: High-integrity IMM-EKF-based road vehicle navigation with low-cost GPS/SBAS/INS. IEEE Trans. Intell. Trans. Syst. **8**(3) (2007)
3. Leonard, J.J., Durrant-Whyte, H.F.: Directed sonar sensing for mobile navigation. Kluwer Academic Publisher, Boston (1992)
4. Toth, C.: Sensor integration in airborne mapping. IEEE Trans. Instrum. Meas. **51**(6), 1367–1373 (2002)
5. Lacroix, S., Mallat, A., Chatilla, R., et al.: Rover self localization in planetry-like environments. In: 5th International Symposium on Artificial Intelligence, Robotics and Automation in Space, Noordwijk, The Netherlands, June 1999
6. Olson, C.F.: Probabilstic self-localization for mobile robots. IEEE Trans. Robot. Automat. **16**(1), 55–66 (2000)
7. Mark Theodore Draelos: The Kinect Up Close: Modifications for Short-Range Depth Imaging, Article, North Carolina State University (2012)
8. Forster, C., Pizzoli, M., Scaramuzza, D.: SVO: fast semi-direct monocular visualn odometry. In: Intl. Conf. on Robotics and Automation (ICRA) (2014)
9. Engel, J., Sturm, J., Cremers, D.: Camera-based navigation of a low-cost quadrocopter. In: Intl. Conf. on Intelligent Robot Systems (IROS) (2012)
10. Achtelik, M., Weiss, S., Siegwart, R.: Onboard IMU and monocular vision based control for MAVs in unknown in- and outdoor environments. In: Intl. Conf. on Robotics and Automation (ICRA) (2011)
11. Engel, J., Schops, T., Cremers, D.: LSD-SLAM: Large-Scale Direct Monocular SLAM. ECCV (2014)

12. Yan, X., Zhao, C., Xiao, J.: A novel fast-SLAM algorithm based on iterated unscented Kalman filter. In: Proceedings of the IEEE International Conference on Robotics and Biomimetics (ROBIO 2011), pp. 1906–1911 (2011)
13. Nemra, A., Aouf, N.: Robust INS/GPS Sensor Fusion for UAV Localization Using SDRE Nonlinear Filtering. IEEE Sensors Journal **10**(4), April 2010
14. Habibi, S.R.: The Smooth Variable Structure Filter. Proceedings of the IEEE **95**(5), 1026–1059 (2007)
15. Habibi, S.R., Burton, R.: The Variable Structure Filter. Journal of Dynamic Systems, Measurement, and Control (ASME) **125**, 287–293 (2003)
16. Gadsden, S.A., Habibi, S.R.: A new form of the smooth variable structure filter with a covariance derivation. In: IEEE Conference on Decision and Control, Atlanta, Georgia (2010)

Tanker UAV for Autonomous Aerial Refueling

Jesús Martín, Hania Angelina, Guillermo Heredia and Aníbal Ollero

Abstract Increasing flight endurance of Unmanned Aerial Vehicles (UAVs) is a main issue for many applications of these aircrafts. This paper deals with air to air refueling between UAVs. Relative estimation using only INS/GPS system is not sufficiently accurate to accomplish an autonomous dock for aerial refueling using a boom system in the tanker.

In this paper we propose a quaternion based relative state estimator to fuse GPS and INS sensor data of each UAV with vision pose estimation of the receiver obtained from the tanker. Simulated results validate the approach and are the starting point for ground and flight tests in the next months.

Keywords UAV · Refueling · Formation flight · Boom

1 Introduction

In aerial refueling, there are two commonly used methods: the probe and drogue refueling system and the boom and receptacle refueling system [1]. In each case, the operation procedure is different. Firstly, in the probe and drogue method, the tanker aircraft acts as a passive element releasing a long flexible hose that trails behind and below the plane through which the fuel flows. At the end of the hose there is a cone-shaped component, where the receiver is hooked, known as a drogue or basket. The receiver aircraft extends a device called a probe, which is a

J. Martín(✉) · G. Heredia · A. Ollero
Robotics, Vision and Control Group, University of Seville, Seville, Spain
e-mail: {jesus.martin,guiller,aollero}@us.es

H. Angelina
Airbus Defence and Space, Madrid, Spain
e-mail: hania.angelina@airbus.com

A. Ollero—This work has been supported by Airbus Defence & Space under the project Open Innovation SAVIER and the MarineUAS PROJECT (H2020-ITN-642153).

© Springer International Publishing Switzerland 2016   571
L.P. Reis et al. (eds.), *Robot 2015: Second Iberian Robotics Conference*,
Advances in Intelligent Systems and Computing 417,
DOI: 10.1007/978-3-319-27146-0_44

rigid arm placed usually on one side of the airplane. As the tanker flies in a race-track, with no control on the drogue, the receiver aircraft flies behind and below the tanker aircraft and so that the probe of the receiver aircraft docks with the drogue from the tanker. Once the docking is accomplished, fuel is pumped through the hose, and the two aircrafts fly in formation until the fuel transfer is complete. Once the refueling is completed, the receiver aircraft then decelerates to undock the probe out of the drogue.

In the boom and receptacle refueling method, the approach change. Here the tanker is an active part in the refueling maneuver due to the possibility of boom position control. The boom can be assimilated to a long, rigid tube, fitted to the rear of the tanker aircraft. It normally has three degrees of freedom: pitch, yaw and a telescoping extension. Small wings enable the boom to be controlled into a receptacle of the receiver aircraft. The receiver aircraft is equipped with a receiver socket fitted onto the top of the aircraft, on its center line and usually either behind or close to the front of the cockpit. The receiver socket connects to the aircraft fuel system to redistribute the fuel into the receiver tanks. The boom has a nozzle which docks into the socket. During refueling operations, the tanker aircraft flies in a racetrack and level attitude at constant speed, while the receiver takes different standards positions behind and below the tanker before reaching the contact one. As the receiver pilot flies in formation with the tanker, the boom operator in the tanker's tail uses a joystick to move the boom and extend the telescoping component to connect the boom's nozzle to the receiver. Once docked, an electrical signal is passed between the boom and receiver, the valves in both the boom and the receiver are opened. Pumps on the tanker drive fuel through the boom fuel pipeline and into the receiver. When refueling is complete, the valves are closed and the boom is retracted before the receiver break the formation.

Compared to boom based refueling system, drogue based is simpler and more flexible, can be implemented in multiple types of aircrafts, and allows multiple aircraft being refueled. However, the drogue provides a lower fuel transfer rate. In the drogue based system the docking maneuver are made by the receiver aircraft, which is a demanding task for the receiver pilot. In boom based system, the receiver pilot's workload is slightly lower. The boom based system was designed for USAF heavy bomber refueling due to the higher fuel transfer rate. However, the tanker can only service one receiving aircraft at a time. Furthermore, the space and weight associated with the boom assembly limit the types of aircraft that can be equipped with this system.

The aerial refueling maneuver is divided into three phases: the pre-refueling or approach phase, the refueling phase, and the separation phase. In the approach phase, the receiver aircraft approaches the tanker aircraft from below and behind and gets connected with the tanker. This phase is also divided in three steps: wing position, pre-contact and contact. During the refueling phase, fuel is pumped from the tanker aircraft into the receiver aircraft. The receiver aircraft tries to hold a stationary position relative to the tanker aircraft to maintain the connection between drogue and probe, or boom and receptacle. This phase can also be called "station keeping". The separation phase begins as soon as fuel transfer ends. The receiver aircraft decelerates and becomes detached from the tanker aircraft.

Tanker wake turbulence makes flying the receiver aircraft during aerial refueling, especially during the first two phases, more difficult than under normal flight. Furthermore, as the receiver aircraft approaches the tanker aircraft in drogue based system, the relative position of the hose and drogue fluctuates due to wind gusts and turbulence. Then, it is not a trivial task to make the connection between the drogue and the probe. For a manned aircraft, such difficulties can be overcome by a pilot's agility. For UAVs, these difficulties impose challenges that should resolved through appropriated automatic flight control system design.

The developing of aerial refueling requires advances in boom based system for many reasons [2] [3] including higher flow rates, lighter work load for the receiver pilot, simpler equipment for the receiver aircraft, insensitivity to perturbations, easier docking operation and possibility to automatize the tanker operation.

The system proposed in this paper is composed of an UAV with an autonomous boom, equipped with a computer vision system and a refueling computer.

The boom is designed with CATIA and XFLR5 CFD. Two control surfaces are installed in a "H" configuration on the back of the boom and are used to control the boom attitude in pitch and roll [3] [4].

Aerial refueling is a sustained close formation flight. Close formation flight is a problem that need an accurate relative state estimate and robust formation guidance that utilizes the estimated relative state of the tanker and receiver aircraft. The simplest method to obtain the relative positioning is to compare the GPS coordinates from the tanker to the receiver. This method is useful in high separation formation flight due to accuracy is in the order of meters, but it is not enough for aerial refueling. In addition to this error, a very important problem is the time synchronization error between the measurements of both UAVs. High horizontal vehicle dynamics and data losses increase this error.

Aerial refueling needs an accuracy of around 0,4 meters. In order to achieve this, directly observed measurements should be used. Computer vision techniques previously applied in formation flight include active visual contours [5] [6], silhouette based techniques [7] and feature extraction [8] [9]. On the other hand, the drawback of vision systems is the high influence of occlusion, incorrect matching, cluttering and dynamic lighting conditions. To compensate these drawbacks and develop a robustness and accurate aerial refueling estimator, it is necessary to use information from inertial sensor, magnetometers, GPS and barometer.

An unscented Kalman filter (UKF) is proposed for the on board AHRS (Attitude and heading reference system) of each UAV and the relative state estimation [10] [11]. Each UAV has a UKF based AHRS to fuse on board GPS, inertial sensors, magnetometer and barometer. Furthermore, on board of the tanker, refueling estimator use an UKF to fuse vision data, receiver state vector and measurements, and tanker state vector and measurements [12]. The unscented Kalman filter (UKF) has several advantages over the extended Kalman filter (EKF). The derivation of Jacobians is not necessary, provides at least second-order nonlinear approximation instead of the first-order EKF, is more robust to initial errors and computation requirements are almost the same. In the aerial refueling problem, initial errors are particularly important due to the large difference in accuracy between the GPS and vision-based measurements, hence the resilience of the filter to initial error is very important. A downside of the UKF is that a quaternion parametrization of the attitude results in a non-unit quaternion estimate when the mean is computed. In our case we use generalized Rodrigues parameters (GRPs) to represent the attitude error. In our system, receiver and tanker are flying in formation. Visual markers are in the receiver and the tanker has mounted a camera pointing backwards with 5° of negative pitch. The boom root is 10 centimeters down of the tanker center of gravity to minimize the effect of the boom movement in the tanker.

2 Problem Formulation of Relative Estimation in Tanker Frame

This section describes the complete relative estimation algorithm First, a review of state estimation of each individual UAV be done through a filter UKF, and subsequently extended this algorithm to estimate the relative between the tanker and receiver. Subsequently adding vision measures will be explained.

2.1 Problem Formulation of Relative Estimation in Tanker Frame

As in[11] an UKF based state estimator is used, the individual state estimation x is obtained by measurements fusion of the differents sensor of each UAV. This process is based in two steps, prediction and observation. During the prediction stage the UAV dynamic model is used to predict the next time step state vector. The inputs, u, are the angular rates provided by the three axis gyros and the acceleration provided by the three axis accelerometer in body axis. In the observation phase measurements from the GPS, three axis magnetometer and barometer are used to correct the estimate

$$x = [\, P \; V \; \boldsymbol{q} \; a \; a_b \; \omega_b \,]^T \qquad (1)$$

$$u = \left[\tilde{a}_b \; \tilde{\omega}_{ib}^b \right]^T \qquad (2)$$

where the position $P = [\, X \; Y \; Z \,]^T$ and velocity $V = [\, v_x \; v_y \; v_z \,]^T$ are expressed in the NED (north-east-down) navigation frame, using as origin the ground station position. The quaternion $\boldsymbol{q} = [\, q_0 \; q_1 \; q_2 \; q_3 \,]^T$ describes the aircraft attitude used to form the rotation matrix C_b^n, which is used to transform from body to navigation frame. The definition of C_b^n and all other inertial mechanisation equations, and a full derivation of the INS mechanisation equations can be found in [13]. In the state propagation, the following inertial navigation mechanisation equations are used

$$\dot{x} = \begin{bmatrix} \dot{P} \\ \dot{V} \\ \dot{q} \\ \dot{\omega}_B \\ \dot{a}_b \end{bmatrix} = \begin{bmatrix} V \\ C_b^n \tilde{a} - (2\omega_{ie}^n + \omega_{en}^n) \times V + g^E \\ \frac{1}{2}\tilde{\Omega}_{\tilde{\omega}}q \\ n_{\omega b} \\ n_{a b} \end{bmatrix} \quad (3)$$

where $\tilde{a}_b$ and $\tilde{\omega}_{ib}^b$ are the IMU accelerometer and gyro varying bias, g^e is the Earth gravity vector in the navigation frame and ω_{ie}^n is the rotation of Earth in the navigation frame [13]. It is assumed that the navigation frame is fixed, and therefore ω_{en}^n is a zero vector, a_b and ω_b are defined as a random walk where $n_{\omega b}$ and n_{a_b} are zero-mean Gaussian random variables, $\tilde{a}$ and $\tilde{\omega}$ are the bias and noise corrected imu measurements, and n_a and n_ω are the accelerometer and gyro measurement noise terms. The gyro measurement are corrected by the Earth´s rotation.

In the state observation phase, the different sensor sampling rates have been taken into account. Thus, $h^{gps}[x, k]$ is used when GPS measurement is available whereas $h^{no\ gps}[x, k]$ is taken into account when the magnetometer and pressure observations are available.

Then,

$$h^{gps}[x, k] = \begin{bmatrix} \bar{P}^{gps} \\ \bar{V}^{gps} \end{bmatrix} = \begin{bmatrix} P^- + C_b^n r_{gps} + n_{P_{gps}} \\ V^- + C_b^n \tilde{\omega} \times r_{gps} + n_{V_{gps}} \end{bmatrix} \quad (4)$$

where r_{gps} is the position of the GPS antenna relative to the centre of gravity, $n_{P_{gps}}$ and $n_{V_{gps}}$ are the measurement noise terms for the GPS position and velocity.

Moreover,

$$h^{no\ gps}[x, k] = \begin{bmatrix} \bar{h} \\ \bar{\psi} \end{bmatrix} = \begin{bmatrix} h_0 - Z^- + n_h \\ \tan^{-1}\left(\frac{2(q_0 q_3 + q_1 q_2)}{1 - 2(q_2^2 + q_3^2)}\right) + n_\psi \end{bmatrix} \quad (5)$$

where n_h and n_ψ are the measurement noise terms of the pressure altitude and the heading.

Raw data sensor have to be preprocessed before use in the observation models presented previously. GPS measurements are in geodetic frame, and are converted in to the navigation frame using the transformation in [13]. Then, heading $\tilde{\psi}$ is calculated from the observed magnetic vector, $\tilde{H}$ by first de-rotating through roll, ϕ, and pitch, θ using (6) and (7) and then calculating $\tilde{\psi}$ using

$$\tilde{M}_x = \tilde{H}_x \cos\theta + \tilde{H}_y \sin\theta \cos\phi + \tilde{H}_z \cos\phi \sin\theta \quad (6)$$

$$\tilde{M}_y = \tilde{H}_y \cos\phi + \tilde{H}_z \sin\phi \quad (7)$$

$$\tilde{\psi} = atan2\left(-\tilde{M}_y, \tilde{M}_x\right) + \psi_{dec} \quad (8)$$

Altitude above mean sea level (MSL), $\tilde{h}$, is now calculated using the atmospheric pressure. First, the pressure at MSL, p_{MSL}, is estimated using

$$p_{MSL} = p_0 \left(1 - \frac{L\,h_0}{T_0}\right)^{\frac{-gM}{R\,L}} \tag{9}$$

where p_0 is the initial pressure and h_0 is the initial MSL height, as observed by the GPS. Finally, $\tilde{h}$ is determined using

$$\tilde{h} = \frac{T_0}{L}\left(1 - \left(\frac{\tilde{p}}{p_{MSL}}\right)^{\frac{R\,L}{g\,M}}\right) \tag{10}$$

The equations of the prediction and update stage are omitted, but can be found in [10], [14], [15].

2.2   Estimation

This section describes the relative state estimation framework to be used in the tanker to estimate the receiver position, in a turbulent outdoor environment, with change in wind direction and intensity and with different and dynamic lighting conditions. Our approach employs computer vision to provide an additional measurement. In this way it is possible to provide an accurate and reliable state estimate which is capable to maintain the accuracy during transient visual outages, but degrades during extended outages. The relative state, $x_{r|t}$ contains the position $P_{r|t}$, the relative velocity $V_{r|t}$ and a pressure bias h_b. The state are expressed as the tanker with respect to the leader, in the navigation frame and centred on the tanker. h_b is estimated to account for bias in the measurement of the barometric altitude.

$$x_{r|t} = \begin{bmatrix} P_{r|t} & V_{r|t} & h_b \end{bmatrix}^T \tag{11}$$

$$u_{l|f} = \begin{bmatrix} \widetilde{a_r} & q_r & \tilde{a}_t & q_t & 0 \end{bmatrix}^T \tag{12}$$

$$Q_{l|f} = diag\begin{bmatrix} \sigma_{\tilde{a}_r}^2 & \sigma_{qr}^2 & \sigma_{\tilde{a}_t}^2 & \sigma_{qt}^2 & \sigma_{h_B}^2 \end{bmatrix} \tag{13}$$

To avoid convergence problems to incorrect and ambiguous states, during vision updates, q_r and q_t have been excluded from $x_{r|t}$, q_r, and q_t is estimated in each aircraft and transmitted between both of them

$$\dot{x}_{r|t} = \begin{bmatrix} \dot{P}_{r|t} \\ \dot{V}_{r|t} \\ \dot{h}_B \end{bmatrix} = \begin{bmatrix} V_{r|t} \\ C_r^n \tilde{a}_r - C_t^n \tilde{a}_t \\ 0 \end{bmatrix} \tag{14}$$

As in the individual on board estimator described in the previous section, the relative UKF based refueling estimator uses of GPS measurements, positioning and velocities, and barometric altitude. Using individual absolute measurement increase the time synchronization between data samples, especially in the horizontal movements where the dynamics are faster. Receiver data is transmitted to tanker wirelessly, so latency should be taken into account. Milliseconds delays in data transmission involve meters of error in positioning.

In this work the problem of storing the tanker data with time stamp, is approached using TOW (time of week) of the GPS and then wait to the receiver data with an adequate TOW. All the measurements are delayed by communications latency, and therefore the error is a function of the latency and relative position dynamics.

In case of matching of the receiver and tanker GPS measurements the following relative estimator correction is applied

$$h^{gps}\left(x_{r|t}^-, k\right) = \begin{bmatrix} P_{r|t}^- \\ V_{r|t}^- \end{bmatrix} \tag{15}$$

Taking into account that the vertical relative velocity is low, the relative altitude corrections are applied each time step,

$$h\left(x_{r|t}^-, k\right) = -Z_{r|t}^- - h_b \tag{16}$$

2.3   Vision System

As in [11] our system uses a features based method where visual markers are in the receiver, in a known position. Receiver is observed by the tanker camera. It is critical that most of the markers can be seen during all the refueling maneuver. Pose can be calculated directly using the n correspondences between the markers position of the receiver ς_j^r, the 2D observations $\bar{\delta}_j$ and the camera parameters. This requires $n \geq 3$ for a solution and $n \geq 4$ for a unique solution. To solve this PnP problem we used POSIT [16]. The drawback of the of this vision system are the effects in the pose estimation   of matching fails, occlusions or a partial part of the target outside of the FOV. These fails are very problematic during the approximation maneuver or the docking   phase.

As in [11] we propose a tightly-coupled design with use the n raw 2D marker observations of the receiver, where $\bar{\delta}_j = \begin{bmatrix} u_j v_j \end{bmatrix}^T, j = 1, \dots, n$. To guarantee the observability in the UKF $n = 4$ and $n \geq 3$ is required [8] [17]. The expected observations $\bar{\delta}_j, j = 1, \dots, m$ are calculated firstly transforming ς_j^r from receiver´s body frame to the world frame, using 17.

$$\varsigma_j^t = C_n^t \left(C_r^n \varsigma_j^r + P_{r|t} \right) \tag{17}$$

Then, using (18), vision sensor extrinsic parameters transform ς_j^t to the camera frame

$$\varsigma_j^c = \begin{bmatrix} C_t^c & P_{t|c} \end{bmatrix} \begin{bmatrix} \varsigma_j^f \\ 1 \end{bmatrix} \tag{18}$$

Where $P_{t|c}$ and C_t^c are the translation and rotation from the followers body frame to the camera frame. C_t^c include the camera mounting Euler angles and the axes transformation. $\bar{\delta}_j$ is calculated using K, a matrix which encapsulates the camera intrinsic parameters, focal length, aspect ratio, principal point and distortion. The final vision measurement model is provided in (20).

$$\begin{bmatrix} \bar{\delta}_j \\ 1 \end{bmatrix} = K \begin{bmatrix} \frac{\varsigma_{x_j}^c}{\varsigma_{z_j}^c} \\ \frac{\varsigma_{y}^c}{\varsigma_{z_j}^c} \\ 1 \end{bmatrix} \tag{19}$$

$$h^{vis}\left(x_{\overline{r|t}}, k\right) = \begin{bmatrix} \bar{\delta}_1 & \bar{\delta}_2 & \dots & \bar{\delta}_n \end{bmatrix}^T \tag{20}$$

3   Refueling Boom Modelling

The refueling boon has three degrees of freedom, pitch, yaw and extension. In our case, extension can be neglected, so pitch angle (θ_{boom}) about the y axis and yaw angle (β_{boom}) have been considered for boom modeling. The boom is modeled in Simmechanics to create a coupled model of the tanker and the boom.

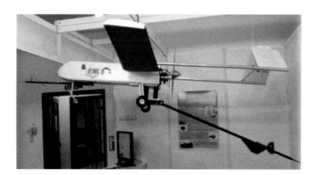

Fig. 1 GRVC Tanker & Boom prototype

In our boom two independent control surfaces are installed symmetrically to control in pitch and yaw angle independently, the wingspan of elevator are 650mm and the rudder are 110mm. During the refueling maneuver it is necessary an automatic control system to manipulate the boom via rudder and elevator. The objective of the automatic control is to maintain the adequate attitude to dock with the receiver UAV.

During the modelling of the system, the following assumptions are taken into account: boom mass is constant, fuel flow is neglected, moment of inertia are constant, the boom is stiff, the aerodynamics forces of boom are equally distributed, the application point is constant in all the flight envelope and the joints are friction free.

According to the two degree-of-freedom and the stiffener of the boom, the equation of motion are written in the form shown in (21) and (22).

$$M_\theta = |L_{og}|(G_z - D \sin \theta_{boom}) - |L_{of}|L \cos \theta_{boom} = J_\theta \ddot{\theta} \tag{21}$$

$$M_\psi = |L_{og}|G_y \sin \theta_{boom} + Y|L_{of}| \sin \theta_{boom} = -J_\phi \ddot{\psi} \tag{22}$$

where L_{og} are the distance between joint and centre of gravity and L_{of} is the distance between the joint and the force application point. G_y and G_z are the gravity force, that can be decomposed into three directions in the boom body coordinate system, two of which are

$$G_y = mg \sin \psi \tag{23}$$

$$G_z = mg \cos \theta \cos \psi \tag{24}$$

Lift force consists of fixed-boom contribution, elevator contribution and extension-boom contribution, and are represented with L. Lateral force, Y, comes from the rudders, and drag force, D, from the whole boom.

$$L = \frac{1}{2} \rho S_e V_\infty^2 C_l \tag{25}$$

$$D = \frac{1}{2} \rho S_e V_\infty^2 C_{d0_e} + \frac{1}{2} \rho S_r V_\infty^2 C_{d0_r} \tag{26}$$

$$Y = \frac{1}{2} \rho S_r V_\infty^2 C_Y \tag{27}$$

When boom and receiver are docked, the boom movement obeys to the tanker and receiver relative movement. Boom is designed in CATIA V5 and the aerodynamics coefficients, such as lift, drag, and moment coefficients are calculated using XFLR5.

Our boom has an onboard controller with a 6 DOF IMU and analog angle sensor in the joint. Two methods of attitude estimation are developed, a direct read from the analog sensor in the joint, and with the on board IMU with a linear Kalman filter for the attitude estimation. Pitch and yaw command are calculated with the information obtained from the relative state estimator.

$$\theta_{cmd} = \tan^{-1}(X_{rel}/Z_{rel}) - \theta_{tanker} \tag{28}$$

$$\psi_{cmd} = \tan^{-1}(X_{rel}/y_{rel}) \tag{29}$$

Finally, a PID controller is used for attitude control of the boom, depending the angles references. This angles, depends of the boom flying phase, that are divided in deploy, operation and recovery of the boom.

4   Simulations

This section present a precise simulation environment to test different estimators and algorithms focused in aerial refueling. The Simulator provides a controllable environment to compare initial results prior to flight test. Tanker and receiver are modeled as a 6DOF non-linear system. The actuators are modeled with first order time lag, saturation limits and rate limits.   The environment is modeled using earth gravity models, magnetic field models, winds and atmospherics models. UAV and boom sensor suite are modeled with bias, Gaussian noise and cross coupling. The GPS model also incorporates Gauss-Markov noise.   Telemetry between tanker and receiver includes latency and the possibility of data packets loss.

Vision sensor was modeled with a resolution of 1920x1080 pixels and a FOV 70°x42° at 30fps. FOV is an important parameter in camera selection, because the need of tracking all the markers during the refueling maneuver. To calculate relative pose, firstly requires simulate pixel coordinates. The coordinates were calculated using (17) to (19) and the position and attitudes of the tanker and the receiver. Then white noise and a random order in the pixel stream were added to simulate the simulated measurements.

Figure 2 and 3 represent the relative attitude and position between tanker and receiver respectively. The green one represents the true data obtained from the simulator, the red represents the raw data of each UAV, the intense blue represents the vision measurements and the light blue the estimated data.

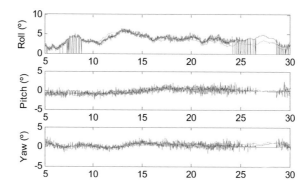

Fig. 2 Relative attitude

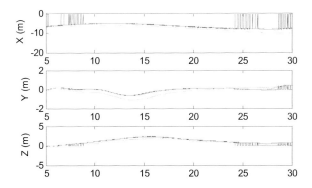

Fig. 3 Relative position

During simulations vision dropouts occurred. These can be seen when POSIT measurement goes to zero. The vertical lines indicate the beginning and end of the dropout. These vision failures can be attributed to the markers outside the tanker FOV. As can be seen, despite failures, the accuracy of the estimator remains virtually intact.

Figure 4 shows relative velocities between UAVs. The intense blue is the true data obtained from the simulator, the red one is the raw data and the green one is the estimator data.

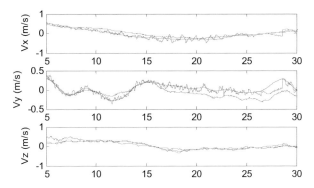

Fig. 4 Relative velocities

Figure 5 shows the boom control commands obtained from the relative state estimator, and the boom controller response. The boom controller works properly during all the refueling maneuver with an almost negligible error.

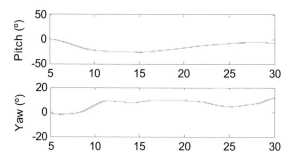

Fig. 5 Boom attitude

5   Conclusions

This paper presented the first step for an autonomous tanker UAV with a vision based estimator based in the unscented Kalman filter. Some modifications were done to enable quaternions to be used in the unscented Kalman filter. The attitude error, parametrized by GRPs was estimated for this application. A complete refueling environment was modeled, with 6dof non-linear UAV models and the boom model attached to the tanker. Simulations validated the estimator approach and demonstrated an adequate relative state estimation for the aerial refueling. The UKF based AHRS of each UAV, also demonstrated improvements. The boom controller compensates the oscillation between aircrafts. Future work will focus on real implementation, ground and flight testing,   and vision system improvements to provide a robustn relative state estimation.

References

1. Mao, W., Eke, F.O.: A survey of the dynamics and control of Aircraft During Aerial Refueling. Nonlineas Dynamics and Systems Theory **8**(4), 375–388 (2008)
2. Nalepka, J.P., Hinchman, J.L.: Automated aerial refueling: extending the effectiveness of unmanned air vehicles. In: AIAA Modeling and Simulation Technologies Conference, San Francisco, CA, pp. 240–247 (2005)
3. van't Riet, R., Thomas, F.R.: KC-10A refueling boom control system. In: Proceedings of the IEEE National Aerospace and Electronics Conference, Dayton, OH, pp. 354–361 (1980)
4. Smith, J.J., Kunz, D.L.: Simulation of the dynamically coupled KC-135 tanker and refueling boom. In: 2007 AIAA Modeling and Simulation Technologies Conference, SC (2007)
5. Doebbler, J., Valasek, J., Monda, M., Schaub, H.: Boom and receptacle autonomous air refueling using a visual pressure snake optical sensor. In: AIAA Atmospheric Flight Mechanics Conference and Exhibit. American Institute of Aeronautics and Astronautics

6. Johnson, E.N., Watanabe, Y., Ha, J., Calise, A.J., Tannenbaum, A.R.: Image processing, estimation, guidance, and flight test of vision-based formation flight. In: Proceedings of the 3rd International Symposium on Innovative Aerial/Space Flyer (2006)
7. Khansari-Zadeh, S.M., Saghafi, F.: Vision-based navigation in autonomous close proximity operations using neural networks. IEEE Transactions on Aerospace and Electronic Systems **47**(2), 864–883 (2011)
8. Williamson, W.R., Glenn, G.J., Dang, V.T., Speyer, J.L., Stecko, S.M., Takacs, J.M.: Sensor fusion applied to autonomous aerial refueling. Journal of Guidance, Control, and Dynamics **32**(1), 262–275 (2009)
9. Mahboubi, Z., Kolter, Z., Wang, T., Bower, G., Ng, A.Y.: Camera based localization for autonomous UAV formation flight. In: Proceedings of the AIAA Conference (2011)
10. Julier, S.J., Uhlmann, J.K., Durrant-Whyte, H.F.: A new approach for filtering nonlinear systems. In: American Control Conference, vol. 3, pp. 1628–1632 (2012)
11. Wilson, D.B., Goktogan, A.H., Sukkarieh, S.: A vision based relative navigation framework for formation flight. In: IEEE International Conference on Robotics & Automation of the 2014, pp. 4988–4995
12. Lee, D.R., Pernicka, H.: Vision-based relative state estimation using the unscented kalman filter. International Journal of Aeronautical and Space Sciences, 24–36 (2011)
13. Rogers, R.M.: Applied Mathematics in Integrated Navigation Systems, 3rd ed., ser. AIAA Education Series. AIAA (2007)
14. Oh, S., Johnson, E.N.: Relative motion estimation for vision-based formation flight using unscented kalman filter. In: AIAA Guidance, Navigation and Control Conference and Exhibit. AIAA (2007)
15. Crassidis, J., Markley, F.: Unscented filtering for spacecraft attitude estimation. In: AIAA Guidance, Navigation, and Control Conference. AIAA
16. Dementhon, D.F., Davis, L.S.: Model-based object pose in 25 lines of code. International Journal of Computer Vision **15**(1–2), 123–141 (1995)
17. Fosbury, A., Crassidis, J.: Relative navigation of air vehicles. Journal of Guidance, Control, and Dynamics **31**(4), 824–834 (2008)

Task Allocation for Teams of Aerial Robots Equipped with Manipulators in Assembly Operations

Jorge Muñoz-Morera, Ivan Maza, Carmelo J. Fernandez-Agüera and Anibal Ollero

Abstract The work presented in this paper is part of the autonomous planning architecture of a team of aerial robots equipped with on-board robotic arms. The mission of the team is the construction of structures in places where the access is difficult by conventional means, which is the scenario considered in the framework of the ARCAS European Project. This paper presents a planning engine for this context. From the 3D CAD model of the structure an assembly planner computes the required assembly tasks, which are the inputs for the system. These tasks are assigned to the available aerial robots by a task allocation planner, which computes an assignment and improves it step by step by communicating with a symbolic planner. The symbolic planner estimates the cost of the sequence of actions needed in the mission execution for the given assignment. This paper presents simulation results that show the feasibility of the approach and a comparison between different solvers.

Keywords Assembly planning · Symbolic planning · Task planning · Task allocation · Multiple aerial robots

1 Introduction and Related Work

The work described in this paper is part of the ARCAS European Project [1] funded by the European Commission. One of the goals of this project is to build a structure by using a team of aerial robots equipped with on-board manipulators. The practical

J. Muñoz-Morera · I. Maza(✉) · C.J. Fernandez-Agüera · A. Ollero
Grupo de Robótica, Visión y Control, Universidad de Sevilla,
Camino de Los Descubrimientos s/n, 41092 Sevilla, Spain
e-mail: {jorgemunoz,imaza,aollero}@us.es, cjtech90@gmail.com

A. Ollero—This work has been partially supported by the ARCAS Project (FP7-ICT-287617) funded by the EU FP7 and the RANCOM (P11-TIC-7066) and AEROMAIN (DPI2014-59383-C2-1-R) national projects.

© Springer International Publishing Switzerland 2016
L.P. Reis et al. (eds.), *Robot 2015: Second Iberian Robotics Conference*,
Advances in Intelligent Systems and Computing 417,
DOI: 10.1007/978-3-319-27146-0_45

interest of this system can be found in situations where it is required to build a structure in places with difficult access by conventional means.

Although Unmanned Aerial Systems (UAS) have been designed and developed for performing military missions, currently the trend is to extend their applicability to civil mission such as firefighting, critical infrastructure protection or remote surveillance, among many others. The multiple variety of platforms, control systems and ground-based equipment [2, 3], and the heterogeneity of the communication devices have made difficult the interoperability among different systems.

Several works [4, 5, 6] have addressed cooperation in teams of aerial robots for multi-purpose missions. However, in those papers the dexterous manipulation with aerial robots was not present. This manipulation is needed to change the orientation of the parts to avoid obstacle collision during transportation or to assemble the parts correctly in the structure. The use of aerial robots allows to perform assembly tasks in any point of the 3D space, which can represent a relevant advantage compared to ground robots in areas of difficult access. Assembly planning is the process of creating a detailed assembly plan to craft a whole product from separate parts by taking into account the final product geometry, available resources to manufacture that product, fixture design, feeder and tool descriptions, etc. Efficient assembly plans can reduce time and costs significantly. The assembly planning problem has been shown to be an NP-complete problem [7] and covers three main assembly subproblems: sequence planning, line balancing, and path planning. Reference [8] presents a survey on assembly sequencing from a combinatorial and geometrical perspective.

Most existing classical planners can be classified into two categories [9]: planners which use domain-dependent knowledge and planners which use domain-independent knowledge. The former can exploit their specific knowledge to guide the planning process and solve larger problems faster than other planners, with the disadvantage of needing a person who gives the knowledge on how to solve the problems. Such knowledge can be designed using temporal logic (TLPlan [10] and TALplanner [11]) or task decomposition (SHOP2 [12], SIPE-2 [13], O-PLAN [14]). On the other hand, a planner that uses domain-independent knowledge (SGPlan [15], FastDownward [16], LPG [17]) does not need specific knowledge so the domain formalization is simpler, with the disadvantage that the performance of the planner may be lower than that of domain-dependent planners. The integration of both types of planners, which let use their advantages and avoid their disadvantages, is a matter of study, and some works in that direction can be read in [18] and [19].

This paper presents a planning engine to solve a general structure assembly problem with a team of aerial robots. This problem is formally presented in Sect. 2. From the 3D CAD model of the structure the assembly planner presented in Sect. 3 computes the required assembly tasks, taking into account the manipulation needed by the robotic arms to insert the parts in the structure. These tasks are assigned to the available aerial robots by the task allocation planner described in Sect. 4. This planner optimizes the computed assignment by calling the symbolic planner explained in Sect. 3, which estimates the cost of the sequence of actions needed in the mission execution for the given assignment and gives feedback to the task allocation planner in the search of a better assignment. Section 5 explains with more detail the connection between the task allocation planner and the symbolic planner. Section 6 includes simulation results that show the feasibility of the approach and a comparison between different solver configurations. Section 7 closes the paper with the conclusions and future work.

2 Problem Statement

Let us consider a mission $\mathcal{M}$ consisting on the assembly of a structure composed by several parts initially located around the environment. The parts have to be assembled on specific locations by a team of n aerial robots starting the mission from their home locations. Then the mission is composed by a set of assembly tasks $\mathcal{T}$. Each of the parts has a weight and a dependency list consisting on the tasks that must be done prior to its assembly. Let us define $\mathcal{L}$ as the set of stock parts locations, $\mathcal{L}'$ as the set of locations where the parts have to be assembled and $\mathcal{H}$ as the home locations of the aerial robots. The objective is to assemble all the parts on their locations minimizing the travel flight times of the vehicles and exploiting the potential parallelism that can be achieved using multiple aerial robots.

The implicit combinatorial problem can be expressed by the edges of a graph $G(V, E)$ considering the following notation:

- $\mathcal{T} = \{t_1, t_2, ..., t_m\}$ is a set of m assembly tasks.
- $P_i \subseteq \mathcal{T}$ is the set of preconditions for the i-th task, i.e. the subset of tasks that must have been done prior to the execution of that task.
- n is the number of aerial robots.
- $\mathcal{L} = \{l_1, l_2, ..., l_m\}$ is the set of stock parts locations and $\mathcal{L}' = \{l'_1, l'_2, ..., l'_m\}$ is the set of final assembly locations.
- $\mathcal{H} = \{h_1, h_2, ..., h_n\}$ is the set of aerial robots home locations.
- $V = \{\mathcal{L} \cup \mathcal{L}' \cup \mathcal{H}\} = \{v_1, v_2, ..., v_{2m+n}\}$ is the set of vertices of the G graph.
- $E = \{(v_i, v_j)|v_i, v_j \in V; i < j\}$ is the edge set.
- $R_k = \{r_1, r_2, ..., r_s\} \subseteq V$ is the route for the k-th aerial robot, composed by a subset of s_k vertices from V.
- Cost c_{r_i, r_j} is a non-negative travel time between vertex r_i and r_j, where $c_{r_i, r_j} = c_{r_j, r_i}$.
- $p = \{p_1, p_2, ..., p_n\}$ is a vector with the maximum payloads of the aerial robots.
- $w = \{w_1, w_2, ..., w_m\}$ is a vector containing the weights of the parts.
- $q = \{q_1, q_2, ..., q_m\}$ is a vector containing the times on which the parts are finally assembled.

The problem consists in determining a set $\mathcal{R}$ of routes with minimal cost and a vector q of minimal task assembly times, with each route starting at the home locations of the vehicles, such that every vertex in $\mathcal{L}$ is visited at least by one vehicle and followed by its subsequent vertex in $\mathcal{L}'$, without exceeding the payload of each vehicle and respecting the preconditions for each of the parts. The possibility of visiting the same location with multiple aerial robots is given by the fact that some parts should be carried cooperatively by more than one aerial robot if they are too heavy. The problem described is similar to the well-known Vehicle Routing Problem (VRP) [20]. For the k-th aerial robot, the cost of a route is given by

$$C(R_k) = \sum_{i=1}^{s_k-1} c_{r_i, r_{i+1}}, \tag{1}$$

where $r_1 \in \mathcal{H}$, $r_i \in V$ and $r_j \in \mathcal{L} \implies r_{j+1} \in \mathcal{L}'$. Considering that up to two aerial robots can cooperatively transport a single heavy part, this route R_k is feasible if $(p_k \geq w_j) \vee (\exists R_z | r_j \in R_z \wedge (p_k + p_z) \geq w_j)$, i.e. the weight of each part does not exceed the maximum payload of the aerial robot transporting it or there is another available aerial robots so that both can transport it cooperatively. For each part w_j in this route it must be met in addition that $\forall t_s \in P_j : q_s < q_j$, i.e. all the parts from its set of preconditions must have been assembled before that part.

The goal of the planner is to minimize the total travel time $\sum_{i=1}^{n} C(R_i)$ of the feasible routes executing all the assembly tasks of the mission and balance the workload of the different aerial robots.

3 Assembly and Symbolic Planners

The input for the planning engine is the 3D CAD model of the structure that has to be build. This model is loaded by the assembly planner, whose purpose is to obtain an assembly plan with all the assembly tasks needed to build the structure. The assembly plan is computed by using the *assembly-by-disassembly* technique, consisting on finding a plan to disassemble the whole structure and reversing the order of its operations to get the final assembly plan. In that plan, each task represents the assembly of one specific part of the structure and contains the preconditions that must be met prior to its execution, namely the assembly tasks that must be done before the assembly of that specific part. The only requirement to execute an assembly task from the plan is to have its preconditions met. That feature makes the computed assembly plan totally independent on the number of vehicles available for its execution.

Some of the tasks in the computed assembly plan could be executed in parallel to reduce the whole mission assembly time. This information is implicitly encoded in the plan as the tasks preconditions. At any given moment, all the assembly tasks that have their preconditions met could be executed in parallel. In addition to the minimization of the assembly time, a low level plan must be computed for each aerial robot. Feasible paths must be computed for the navigation of the robots, both for the part picking and the part assembling. Also, those assembly tasks that must be executed cooperatively between robots require synchronization among the involved aerial robots.

In this paper, the JSHOP2 symbolic planner [12] has been applied to order the assembly tasks and compute a low level plan for each aerial robot. JSHOP2 is a domain-independent planning system based on Hierarchical Task Networks (HTN) that decomposes the tasks into smaller subtasks and so on, until obtaining low-level tasks that can not be further decomposed and thus can be directly performed. An assembly domain has been designed for the JSHOP2 symbolic planner, which receives the assembly plan and computes an ordering of the assembly tasks, as well as low-level plans for each of the involved aerial robots. For difficulties on transporting a section of the structure composed of several parts among multiple aerial robots, the domain restricts the assembly tasks to be only single parts that are assembled on the main structure. To compute the low-level plans, the symbolic planner needs an assignment of assembly tasks to aerial robots which is in turn computed by the task allocation planner presented in the next section. After computing the low-level plans the symbolic planner makes a cost estimation for the execution, whose value will be used by the task allocation planner to try to find a better assignment.

A more detailed description for the assembly and the symbolic planners can be found in [21].

4   Task Allocation Planner

The planner chosen to solve the assignment problem presented in Sect. 2 was OptaPlanner [22], an open source, multi-platform planning engine written in Java and released under Apache Software License. OptaPlanner is aimed to solve planning problems with resource usage optimization. It is capable of generating near-optimal plans by applying optimization heuristics and meta-heuristics combined with score calculation. One of his main advantages is that the solver's algorithm is highly configurable. In OptaPlanner it is possible to use different heuristics and meta-heuristics algorithms applied in sequence, so that the user can select the most suitable for the problem in question. The optimization is done in base of a score calculation that is computed after all the planning entities have been assigned. This score determines the suitability of the last computed solution: if after searching for a new solution the new score is worse than the score calculated in the previous solution, then the last solution is discarded and the process continues trying to generate a solution with a better score.

4.1   Problem Domain

In OptaPlanner the entities of the real world that must be assigned are called *planning entities*, and are represented as Java classes. These planning entities have one or more *planning variables*, represented as Java attributes that take different *planning values* over the planning process. To solve an assignment problem, each of the planning entities must have been given a valid planning value for each of its planning variables.

For the problem defined in this section the domain has been defined as follow: the planning entities are the assembly tasks. Each assembly task represents a part that is stored on a specific location and that must be assembled on a certain location by an aerial robot. Each assembly task has defined a chained planning variable. This chained planning variable takes as value the assembly task that has been assigned before the actual. All chains start from the positions where the aerial robots are landed, so the first assigned task will allways point to an aerial robot, the second assigned task will point to the first assigned task or to another aerial robot, and so on. By this way when all planning entities have been assigned, those that are present in the same chain are considered to be assigned to the same aerial robot.

4.2 Solver Phases

As it was mentioned before, the OptaPlanner solver can use multiple optimization algorithms in sequence. Each of the optimization algorithms used is called a solver phase. During the execution of the solver there is never more than one solver phase executing at the same time, so a solver phase only starts when the previous phase has finished. There are three types of solver phases that can be used in the Opta-Planner solver: Construction Heuristics (CH), Metaheuristics (MH) and Exhaustive Search (ES).

The CH solver phase builds an initial solution in a short time. The solution computed is not always feasible, but it tries to find it fast so that the following solver phases can finish the search of a feasible one by starting from that initial solution. There are different algorithms that can be used as CH. One common characteristic of them is that when a CH assigns a planning entity, that assignment remains unchanged until the end of the algorithm. This is the main reason that makes the CH algorithms find solutions that may be unfeasible: no re-planning is done at this phase.

The MH solver phase is based on different types of local search algorithms. Local search starts from the initial solution computed by the CH phase and evolves it into a mostly better and better solution. At each solution, it evaluates a number of moves between the planning entities and applies the most suitable to step to the next solution, whose score may be better, equal or worse than the previous. Allowing as solution a new one which has a worse score than the previous is important because it avoids getting stuck in local minima. The local search does not use a search tree, but a search path. When finding a new solution all possible moves are evaluated but unless it is the chosen move, it does not investigate further the rest of possible solutions. That makes the local search very scalable, but it may not find never the optimal solution.

Finally, the ES solver phase does not depend on previous phases and it is applied alone. Brute force or the branch and bound algorithms are available. These methods guarantee the find of the optimal solution for a problem, but are poorly scalable so are not usually chosen to solve real problems.

4.3 Solution Moves

To create a new solution the solver generates all moves that is possible to do from the last solution and selects a subset of them to be evaluated. A move is a change in the value of a planning variable for one or more planing entities, which results in a new assignment. From the selected subset of moves the solver chooses the one which produces the better score and applies it to the current solution to generate a new one. The chosen move is called step.

The solver can be configured to use multiple classes that allow to do different types of moves among the planning entities, called selectors. It is possible to use any of the available selectors on any problem domain, although the suitability of the selectors depends on the problem type. The most common selectors used in different problem types are the following:

– Change move: selects one planning entity and modify the value of its planning variable.
– Swap move: selects two planning entities and swap the value of its planning variables.
– Subchain Change move: selects a subchain of planning entities and moves them to a different chain or to a different position inside the same chain where they were selected.
– Subchain Swap move: selects two subchains of planning entities from different chains or from the same chain and swaps its positions.

Over the selectors is possible to apply filters. A filter is a condition that is checked over each of the moves generated by the selector. If the move does not meet the condition it is then discarded and another different move is generated again, until obtaining a valid one. With this mechanism is possible to discard early those moves that are known to produce a bad score before being evaluated by the solver, saving computation time and better assisting the search of the solution.

5 Connecting the Task Allocation Planner and Symbolic Planner

To compare the suitability of the different solutions computed along the task allocation process, a score-based calculation is done after each new solution is found. This score is based on the definition of three types of constraints with different levels of relevance. Given a new solution, its score consists on the sum of the broken constraints for each of the constraint types defined for the problem. The constraint types are:

– Hard-constraints: these are constraints that must not be broken in any case. A broken hard-constraint will lead to an unfeasible plan, so its sum must be zero.

The hard-constraints defined for the assembly problem say that the weight of a part must not exceed the payload capability of the aerial robot it has been assigned to.
– Medium-constraints: these are constraints that are desirable to be broken the less as possible. Its importance is under the importance of the hard-constraints but above the importance of the soft-constraints. The one defined measures the potential parallelism that can be achieved with the current assignment, desired to be as high as possible.
– Soft-constraints: these are the constraints with lower priority. They have the lowest impact when broken, but still they must be minimized. The soft-constraint defined tries to minimize the distance travelled by all the aerial robots.

After a new solution is found, the score related to the new assignment is calculated. The sums that come from the hard and medium constraints can be computed from inside the task allocation planner. However that is not the case for the soft-constraints, as they measure the distance travelled by the aerial robots during the assembly. That distance can only be estimated if the order of the assembly operations is previously known, but as commented in Sect. 3 this order is not specified explicitly in the assembly plan. To solve this, the assignment computed by the task allocation planner is given as input to the symbolic planner. The symbolic planner computes an order for the assembly tasks and gives an estimation of the distance travelled by the aerial robots during the assembly for the given assignment. This estimation is returned to the task allocation planner, which uses it as the soft-constraint value closing the score calculation loop.

By this way, the optimization of the assignment is done cooperatively between the task allocation planner and the symbolic planner while preserving each of them its own domain.

6 Simulation Results

In this section a representative mission will be used to measure the plan quality and computation times of the whole planning engine comparing different solver configurations. In this mission, a fleet composed of four aerial robots equipped with robotic arms is used. The environment modelled for the mission can be seen on Fig. 1. The mission is composed by eleven assembly tasks corresponding to the eleven parts that conform the final structure.

The workflow of the planning engine is as follows: the entry point to the system is the 3D CAD environment model that contains the initial state (stock parts and home locations of the aerial robots) and the 3D CAD model of the structure to be built. The assembly planner reads the 3D CAD model of the structure and generates a valid assembly plan. Given the assembly plan, the task allocation planner solves the assignment problem and tries to optimize it by communicating with the symbolic planner, which in turn finds a low-level plan for each of the aerial robots and returns a cost that is used by the task allocation planner in the optimization process.

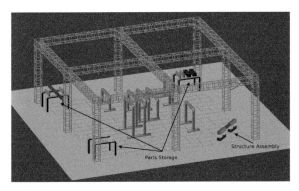

Fig. 1 Environment modelled for the mission. The parts are initially stored over tables in three different stock areas and finally assembled in the lower right corner of this figure.

The baseline solver configuration has the following features:

- One Construction Heuristic solver phase, configured with a First Fit Decreasing algorithm. This algorithm sorts the planning entities by decreasing difficulty and assigns the planning entity which produces the best score at each time, until no left are unassigned.
- One Local Search solver phase, configured with a Late Acceptance algorithm. This algorithm tries to optimize the assignment by doing one move and accepting it if does not decrease the score or leads to a score that is at least the score of a fixed number of steps ago.
- Four different move selectors, included inside a fifth union selector that selects with the same probability at each step one of the others to generate the next move.

From the baseline solver configuration template, a total of seven new solver configurations were created to test them against the former. All of them have been created with exactly the same Construction Heuristic and Local Search algorithm as the baseline, as well as the same move selectors, but differ in two main characteristics. The first is that all the new configurations have the sub-chain move selectors to discard as moves the reverse of a given selected sub-chain, reducing thus the number of possible moves to do between steps. The second is that a filter has been specifically implemented for each of the new configurations to try to discard moves that are known to produce bad scores and that lead to a waste of computation time. So, the main difference among the baseline configuration and the test configuration are the use of filters.

The descriptions of the solver configurations used in the simulation besides the baseline are the following:

- Change Filter: a filter is defined to avoid moving a planning entity from a position in a chain to another position in the same chain. This is because the order of the planning entities inside the same chain is finally computed by the symbolic planner

Table 1 Simulation results for each of the solvers, obtained for the example mission. The Hard, Medium and Soft columns show the sum of the broken constraints for each constraint type, given as negative values. The score computation rate is the average score calculation count per second for each solver configuration. The simulation was run on a machine with an Intel i7 CPU at 2 GHz and 8GB RAM.

Solver	Hard	Medium	Soft	Score computation rate
ChangeSwap Filter	0	0	-320	6/s
Swap Filter	0	0	-329	2/s
SubChainChange Filter	0	0	-330	2/s
Change Filter	0	0	-338	5/s
SubChainSwap Filter	0	0	-338	2/s
Default	0	0	-339	3/s
Nearby Filter	0	0	-339	1/s
Nearby CS Filter	0	-1	-358	1/s

Fig. 2 Simulation screenshot of an assembly task done by an aerial robot. All parts have a handle from which the aerial robots can take them and a cavity beneath that let them be stacked. For the simulation, a developed middleware with grasp, motion and path-planning capabilities was used to avoid obstacle collisions and perform the pick and place operations. A video of the mission can be downloaded from http://grvc.us.es/robot2015plan.

and not by the task allocation planner, and thus these moves are redundant because will all produce the same score.

- Swap Filter: a filter is defined to avoid swapping planning entities from the same chain.
- ChangeSwap Filter: a combination of the two previous filters.
- SubChainChange Filter: a filter is defined to avoid moving a sub-chain of planning entities from a position in a chain to another position in the same chain.
- SubChainSwap Filter: a filter is defined to avoid swapping sub-chains of planning entities from the same chain.
- Nearby Filter: this configuration tries to make moves among those planning entities that are near each other in distance terms, to try to minimize the travelled distance.
- NearbyCS Filter: a combination of the previous filter and the ChangeSwap filter.

The score and performance results obtained after running the simulation for each of the eight solver configurations can be seen on Table 1. The solver configuration that achieved the best score was the ChangeSwap Filter configuration, improving the score obtained by the default configuration in 19 units for the Soft constraints.

With respect the average calculate count per second, which measures how fast the solver is, the ChangeSwap Filter configuration achieved again the highest ratio, doing two times the score calculate count per second achieved by the default configuration. The two configurations that used the nearby technique achieved the worst values, obtaining only one per second.

Figure 2 show a screenshot of the mission execution during the simulation.

7   Conclusions and Future Work

The system developed performs task assignment and scheduling given the number of aerial robots in order to increase the parallelism and cooperation in the mission. The main contribution of the paper is the successful connection of the task allocation planner with the symbolic planner in the context of structure assembly missions.

However the approach has been tested in missions involving multiple simulated aerial robots. In future work the goal is to execute the mission with real aerial robots in order to find more realistic and complex contexts to test the architecture developed.

References

1. Kondak, K., Krieger, K., Albu-Schaeffer, A., Schwarzbach, M., Laiacker, M., Maza, I., Rodriguez-Castano, A., Ollero, A.: Closed-loop behavior of an autonomous helicopter equipped with a robotic arm for aerial manipulation tasks. International Journal of Advanced Robotic Systems **10**(145), 1–9 (2013). http://dx.doi.org/10.5772/53754
2. Perez, D., Maza, I., Caballero, F., Scarlatti, D., Casado, E., Ollero, A.: A ground control station for a multi-UAV surveillance system. Journal of Intelligent and Robotic Systems **69**(1–4), 119–130 (2013). http://dx.doi.org/10.1007/s10846-012-9759-5
3. Maza, I., Caballero, F., Molina, R., Peña, N., Ollero, A.: Multimodal interface technologies for UAV ground control stations. a comparative analysis. Journal of Intelligent and Robotic Systems **57**(1–4), 371–391 (2010). http://dx.doi.org/10.1007/s10846-009-9351-9
4. Lindsey, Q., Mellinger, D., Kumar, V.: Construction of cubic structures with quadrotor teams. In: Proceedings of Robotics: Science and Systems, Los Angeles, CA, USA, June 2011
5. Maza, I., Munoz-Morera, J., Caballero, F., Casado, E., Perez-Villar, V., Ollero, A.: Architecture and tools for the generation of flight intent from mission intent for a fleet of unmanned aerial systems. In: 2014 International Conference on Unmanned Aircraft Systems (ICUAS), May 2014, pp. 9–19. IEEE (2014). http://dx.doi.org/10.1002/rob.20383
6. Maza, I., Caballero, F., Capitan, J., de Dios, J.M., Ollero, A.: A distributed architecture for a robotic platform with aerial sensor transportation and self-deployment capabilities. Journal of Field Robotics **28**(3), 303–328 (2011). http://dx.doi.org/10.1002/rob.20383
7. L. Kavraki, J.-C. Latombe, R. H. Wilson, "On the complexity of assembly partitioning," Information Processing Letters, 48(5), pp. 229–235, 1993. http://www.sciencedirect.com/science/article/pii/002001909390085N

8. Jimenez, P.: Survey on assembly sequencing: a combinatorial and geometrical perspective. Journal of Intelligent Manufacturing, 1–16 (2011). http://www.scopus.com/inward/record.url? eid=2-s2.0-79961144486&partnerID=40&md5=33af318d197c04cd5079192aab49a916

9. Ingrand, F., Ghallab, M.: Robotics and Artificial Intelligence: A perspective on deliberation functions. AI Communications 27(1), 63–80 (2014)

10. Bacchus, F., Kabanza, F., Sherbrooke, U.D.: Using temporal logics to express search control knowledge for planning. Artificial Intelligence **116**, 2000 (2000)

11. Kvarnström, J., Doherty, P.: TALplanner: A temporal logic based forward chaining planner. Annals of mathematics and Artificial Intelligence **30**, 2001 (2001)

12. Nau, D., Ilghami, O., Kuter, U., Murdock, J.W., Wu, D., Yaman, F.: Shop2: An HTN planning system. Journal of Artificial Intelligence Research **20**, 379–404 (2003)

13. Wilkins, D.E.: Practical Planning: Extending the Classical AI Planning Paradigm. Morgan Kaufmann Publishers Inc., San Francisco (1988)

14. Currie, K., Tate, A., Bridge, S.: O-Plan: the open planning architecture (1990)

15. Chen, Y., Wah, B.W., Hsu, C.W.: Temporal planning using subgoal partitioning and resolution in SGPlan. J. of Artificial Intelligence Research **26**, 369 (2006)

16. Helmert, M.: The fast downward planning system. Journal of Artifical Intelligence Research, 191–246 (2006)

17. Gerevini, A., Saetti, A., Serina, I.: Planning through stochastic local search and temporal action graphs. Journal of Artifical Intelligence Research **20**, 239–290 (2003)

18. Gerevini, A., Kuter, U., Nau, D., Saetti, A., Waisbrot, N.: Combining domain-independent planning and HTN planning: The Duet Planner (2008)

19. Shivashankar, V., Alford, R., Kuter, U., Nau, D.: The GoDel Planning System: A more perfect union of domain-independent and hierarchical planning

20. Dantzig, G.B., Ramser, J.H.: The truck dispatching problem. Management Science **6**(1), 80–91 (1959). http://dx.doi.org/10.1287/mnsc.6.1.80

21. Munoz-Morera, J., Maza, I., Fernandez-Aguera, C.J., Caballero, F., Ollero, A.: Assembly planning for the construction of structures with multiple UAS equipped with robotic arms. In: 2015 International Conference on Unmanned Aircraft Systems (ICUAS), June 2015, pp. 1049–1058. IEEE (2015)

22. Red Hat open source community: OptaPlanner (2014). http://www.optaplanner.org/ (accessed December 12, 2014)

A Proposal of Multi-UAV Mission Coordination and Control Architecture

Juan Jesús Roldán, Bruno Lansac, Jaime del Cerro and Antonio Barrientos

Abstract Multi-UAV missions are complex systems that may include a fleet of UAVs, a crew of operators and different computers and interfaces. Currently, an important challenge is the reduction of the number of operators that is required for performing a multi-UAV mission. This challenge can be addressed by increasing the autonomy of fleets and providing capabilities of operators to the interfaces. This paper presents a proposal of control architecture for multi-UAV missions. This architecture shares some elements with centralized and distributed approaches and it has three layers: mission, task and action. The mission layer is implemented in the GCS and performs the mission planning and operator interfacing. Meanwhile, the task and action layers are located in the UAVs and perform respectively the task planning and executing. This architecture is applied to a simulation environment that reproduce a competitive scenario with two fleets of UAVs.

Keywords Multi-UAV · Multi-robot · Coordination · Control · Architecture · Mission · Task · Action · Layer · Simulator · Game · Competitive · Scenario

1 Introduction

Nowadays, the application of Unmanned Aerial Vehicles (UAVs) in multiple areas is a reality. These areas include agriculture [18], environmental monitoring [4], surveillance and reconnaissance [2], mapping [6], intervention in disaster areas [1], etc.

Nevertheless, the rise of UAVs also poses a series of challenges both short and long term. These challenges can be summarized in three points: The pass from

J.J. Roldán(✉) · B. Lansac · J. del Cerro · A. Barrientos
Centre for Automation and Robotics (UPM-CSIC), Madrid, Spain
e-mail: {jj.roldan,j.cerro,antonio.barrientos}@upm.es, b.lansac@alumnos.upm.es
http://www.car.upm-csic.es/

© Springer International Publishing Switzerland 2016   597
L.P. Reis et al. (eds.), *Robot 2015: Second Iberian Robotics Conference*,
Advances in Intelligent Systems and Computing 417,
DOI: 10.1007/978-3-319-27146-0_46

UAV to fleet control, the increase of UAV autonomy and the reduction of operator workload [3] [9].

The architecture of control of the fleet is fundamental to address these challenges. Literature contains multiple proposals of control architectures for multi-robot systems (a diverse set is contained in references [14], [11], [19], [7], [13] and [17]). According to their nature, these proposals can be classified in a gradual scale from centralized to distributed architectures.

- Centralized architectures: The Ground Control Station (GCS) performs all the functions of data reception and processing, decision-making, mission planning, task generation and allocation, path and payload planning and command sending. Meanwhile, the UAVs only follow the paths, execute the payload actions and collect data. These architectures have some advantages (ease of design, implement and use), but also some disadvantages (scalability to multiple UAVs and potential overload of the GCS).
- Distributed architectures: The fleet or the UAVs perform the functions of data collection and processing, decision-making, task planning and assignment (e.g. competitive [12] or cooperative [10] algorithms), path and payload planning, path following and payload executing. Meanwhile, the GCS is only an interface among the fleet and the operator, which provides the operator with information of the mission and sends the commands of the operator to the fleet. These architectures also present some advantages (mainly the modularity and adaptability to complex scenarios) and some disadvantages (need of coordination scheme and dependency on communications).

In practice, most of proposals of multi-robot control architectures are hybrid, i.e. a compromise solution with elements of both centralized and distributed models. These proposals consider centralized and distributed functions and assign the first ones to GCS and the second ones to UAVs. These architectures can be adapted to different scenarios, missions, systems and fleets.

This paper presents a proposal of coordination and control architecture for multi-UAV missions. This approach follows a hybrid scheme and consists of three layers (mission, task and action) located in GCS (mission) or UAVs (task and action). Table 1 shows a comparison among it and other proposals that are included in recent publications. The coordination and control architecture is described with detail in the next sections and applied to a simulation environment.

2   Control and Coordination Architecture

The analysis performed in previous works [16] defined a hierarchy with three levels for multi-UAV missions: mission, task and action. The mission is a set of objectives that require a series of resources, the task is a sequence of actions performed by a resource to achieve an objective, and the action is any UAV maneuver or payload

Table 1 State of art of coordination and control architectures for multi-robot systems.

Proposal	Class	Organization	Application
Matellán et al. [14]	Hybrid	2 levels: Task agenda (C) / reactive skills (D)	Simulated and real teams of robots (game)
Khaleghi et al. [11] (A)	Centralized	3 modules: Detection, tracking and motion planning (C)	Simulated and real teams with UAVs and UGVs (surveillance)
Khaleghi et al. [11] (B)	Distributed	3 modules: Detection, tracking and motion planning (D)	Simulated and real teams with UAVs and UGVs (surveillance)
Zheng et al. [19]	Hybrid	5 layers: Interaction, organization and coordination (C) / execution (D)	Real team of robots (game)
Lang et al. [7]	Distributed	4 concepts: sensor, actuator and controller (D) / server (C)	Simulated and real teams of heterogeneous robots (game)
Marino et al. [13]	Hybrid	3 levels: Supervisor (C) / action-behavior and robot (D)	Simulated swarm and real team with 3 UGVs (patrol)
Teichteil et al. [17]	Hybrid	2 layers: Deliberative (C) / reactive (D)	Rotorcraft (search and rescue)
Roldán et al. (Here)	Hybrid	3 layers: mission (C) / task and action (D)	Simulated team of UAVs (game)

utilization. The proposal of this paper follows this scheme and defines three control layers: mission, task and action (as shown in figure 1).

2.1 Mission Layer

The mission layer is the most complex component of the architecture and supposes the main contribution of this paper. This layer requires a certain intelligence in order to establish a strategy for the mission and to assign the tasks to the robots. Two approaches have been studied for the mission layer: one that starts from robots and another one that starts from tasks.

Robot Centered Approach: This approach is focused on robots and works asynchronously: it generates and assigns a task when a robot is free. The strategies are planned by the combination of heuristic rules and random factor. The process of mission planning is shown in figure 2 and explained below.

At first, the system analyzes the situation of the mission and classifies it into a set of possible situations. This set of situations must be defined by the operators according to mission analysis and previous experience.

Within each situation, there are different tasks that can be launched for achieving different objectives. The system computes a probability threshold for each task

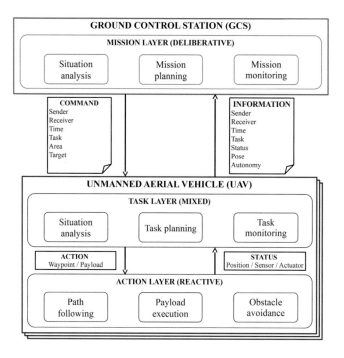

Fig. 1 Proposal of coordination and control architecture for multi-UAV missions.

$(p(T_i))$ taking into account information about mission (M) and fleet (F). For instance, if a UAV is near a target, the probability of launching a task with this UAV and this target will be high. On the other hand, if the autonomy of a UAV is low to reach a target, the probability of launching a task with this UAV and this target will be low. Once the thresholds are computed, the limits for each task $(f_{i-1}$ and f_i for $T_i)$ are defined through the different probabilities. Later, a random number $(r \in [0, 1])$ is generated and a task (T_S) is selected.

$$p(T_i) = f(M, F) \tag{1}$$

$$f_i = \frac{\sum_{j=0}^{i} p(T_j)}{\sum_{j=0}^{n} p(T_j)} \tag{2}$$

$$f_{i-1} < r \le f_i \Rightarrow T_S = T_i \tag{3}$$

This proposal for mission layer is flexible due to the need of adaptation to different scenarios, missions and fleets. On the one hand, it allows the introduction of two

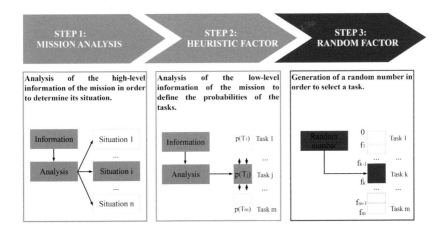

Fig. 2 Mission layer: robot centered approach.

sets of rules related to situations and tasks. On the other hand, it also allows the implementation of machine learning algorithms to adjust the thresholds of tasks, such as gradient descent and genetic algorithms.

Task Centered Approach: This approach is focused on tasks and works synchronously: it generates the tasks that are required and assigns them to the robots that are free with a certain frequency. The work of this layer is shown in figure 3 and its nodes and structures are defined below.

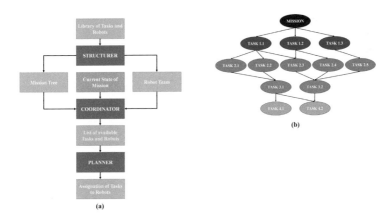

Fig. 3 (a) Mission layer: task centered approach, (b) Example of mission tree.

The structurer manages a library with the different tasks and robots and generates a mission tree and a robot team according to the mission. The mission tree (figure 3) is a structure that contains the tasks and their precedence relations. The task of the tree named $T_{i,j}$ is the j task of the i level.

The coordinator combines the information of mission tree and robot team with the real-time data of the mission (completed or pending tasks and busy or free robots). It decides the tasks of mission tree that must be performed (T) and the robots that are available to perform them (R).

Finally, the planner operates with these T and R datasets and performs the allocation of tasks to robots. The assignation process has three stages:

- Evaluation of each pair (task, robot) to determine their compatibility. This operation uses information as the domain of robots and tasks (ground or air) and the features of robots (e.g. speed and autonomy). This information is collected in a score matrix, which can be converted to a preference matrix, where each task has a preference order for robots and vice versa.
- Transformation of T and R datasets in order to obtain the same dimensionality. When there are more tasks than robots, the algorithm should choose the most relevant tasks. On the other hand, when there are more robots than tasks, the algorithm can employ multiple robots for single tasks.
- Application of Gale-Shapley algorithm [5] considering the analogy among this situation and the classic stable marriage problem [8]. This method is described in algorithm 1.

Algorithm 1. Gale-Shapley algorithm applied to task assignment.

```
function GALE- SHAPLEY(T, R)
    while tᵢ unassigned do
        for each tᵢ do
            for each rⱼ (preference order) do
                if rⱼ unassigned then
                    Pair(tᵢ, rⱼ)
                else
                    if rⱼ assigned to tₖ (less preference) then
                        Break(tₖ, rⱼ)
                        Pair(tᵢ, rⱼ)
                    end if
                end if
            end for
        end for
    end while
    return Pairs(T,R)
end function
```

2.2 Task Layer

The task layer receives the commands from the mission layer. Once a message is received, the task layer extracts the task and splits it into basic tasks, as shown in table 2. For instance, transport complex task consists of a sequence of *Capture*, *Go to waypoint* and *Release* basic tasks. Then the layer performs the planning of these tasks with all the required actions: path and payload planning.

During task execution, the task layer collects information about the task status (e.g. achieved, pending and failed targets) and robot status (e.g. location and autonomy). This information is used to the control and supervision of the task and allows the replanning when it is necessary. All this information is sent to the mission layer in order to define the strategy for the mission and performs the coordination of the UAVs.

2.3 Action Layer

The action layer performs the task execution (mainly, the path following and the payload execution) and has reactive capabilities (such as avoidance of obstacles and threats). This layer is connected to task layer, receives the commands for their execution (waypoints and payload commands) and sends the information about the results (UAV position, sensor measures and actuator states).

3 Multiple Mini-Robot Simulator

Multiple Mini-Robot Simulator (MMRS) is an environment developed to reproduce multi-UAV missions for testing different components, such as mission models, control systems and interfaces. MMRS has been developed taking into account the following issues:

– Mission complexity: The complexity of mission and scenario should cover not only current but also expected multi-UAV missions and scenarios.
– Competitive scenario: A competitive scenario with equality of forces is chosen in order to generate adverse conditions for the fleet. This kind of scenarios require to consider not only the planning and execution of a plan, but also the reaction to the changes in the scenario and the movements of the other fleet.
– Modular simulator: The modularity allows to adapt the simulator to different application by adding and removing their components.

A multi-UAV game based on the Quidditch of Harry Potter novels is posed in order to meet the requirements of complexity and competition (figure 4). In the game, two fleets of three UAVs and a coach compete for the maximum score. There are three

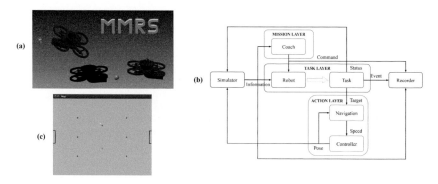

Fig. 4 Multiple Mini-Robot Simulator (MMRS): a) Concept, b) Scheme, c) Simulation.

balls with different functions: the first one is launched to goal to score 10 points, the second one is launched to opponents to stop them and the third one is captured to score 50 points. The duration of the match is 200 seconds.

On the other hand, the simulator has been developed by using Robot Operating System (ROS) [15] in order to meet the requirement of modularity (figure 4). ROS allows to build a modular structure with nodes that process the information and topics that perform the communication. The simulator can reproduce different games by launching more or less nodes and it can adapt to different scenarios, missions and fleet by developing new nodes.

As seen in figure 4, the architecture and the concepts of this work have been applied to the simulator. As the main contribution is related to mission layer, the implementation of this layer in the simulator is described with more details.

The coach node performs the role of mission layer with the approach focused on robots. This approach allows generate and assigning new tasks when the robots finish their tasks, in contrast to the other one that can leave some robots unused in this kind of missions.

It analyzes the situation of the game and classifies it into six categories: *Begin*, *Attack*, *Defense*, *Transition*, *Emergency* or *Finish*. Each category presents a set of possible tasks that can be launched according to the context of the game. For instance, when the team is attacking, the UAV that controls the Quaffle can advance with it, pass to a teammate or shoot to goal. On the other hand, the UAVs that does not control the Quaffle can be unmarked to receive a pass, support their teammate, use the Bludger against an opponent or directly search the Snitch. The table 2 contains all the situations of the game and their possible tasks.

Meanwhile, robot and task nodes make up the task layer. They decompose the complex tasks into basic tasks (as shows the table 2) and perform their planning and monitoring. Last, navigation and controller nodes performs the role of action layer. They control the movement of robots and the use of sensors and actuators.

Table 2 Coordination and control architecture applied to simulator.

Situation	Complex task	Basic tasks
Begin	Begin	Begin
Attack	Drive	Go to Waypoint
	Pass	Release
	Shoot	Release
	Unmark	Go to Waypoint
	Support	Support
	Bludger	Go to Waypoint-Capture-Release
	Snitch	Surveillance-Reconnaissance-Tracking-Capture
Defense	Keep	Go to Waypoint
	Mark	Tracking
	Press	Tracking-Capture
	Block	Go to Waypoint-Capture
	Bludger	Go to Waypoint-Capture-Release
Transition	Quaffle	Go to Waypoint-Capture
	Bludger	Go to Waypoint-Capture-Release
	Snitch	Surveillance-Reconnaissance-Tracking-Capture
Emergency	Maintenance	Maintenance
Finish	Finish	Finish

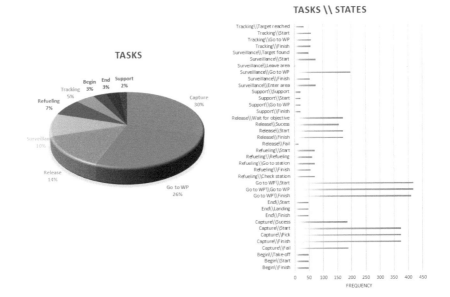

Fig. 5 Tasks and states executed in matches.

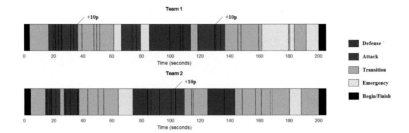

Fig. 6 Situations for both teams over a match.

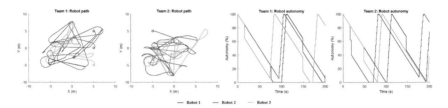

Fig. 7 Path and autonomy of robots in a match.

4 Results

The first tests with the simulator were made to analyze and model the game and optimize the performance of the architecture. Sixteen matches were performed to obtain significant information for making conclusions. Some results that are relevant for understanding the work of simulator are shown in figures 5, 6 and 7.

Figure 5 shows the distribution of tasks and states over the matches. As shown, the most used task is *Capture*, followed by *Go to Waypoint* and *Release*. Meanwhile, figure 6 shows the evolution of a match and the situations for the teams. Each block is the result of the analysis of mission situation performed by the coach. As can be seen, the situations of attack and defense are alternative, what indicates that the strategies are working correctly. Finally, figure 7 depicts the paths of robots and their autonomy. As shown, most of the game is developed in center positions and the robots must perform maintenance operations to keep their autonomy during the match.

5 Conclusions

In this work, a coordination and control architecture is developed for multi-UAV missions. It represents a hybrid approach between centralized and distributed theoretical schemes and has three layers: mission layer (which is located in GCS and follows a deliberative model), task layer (which is located in the UAVs and has a mixed behavior) and action layer (which is also in the UAVs and follows a reactive model).

Two approaches are presented for mission layer: one that combines heuristic rules with random factor and another that establish an analogy between task allocation and stable marriage problem.

In contrast to other proposals of literature, this architecture is especially designed for managing multiple UAVs from a GCS. It allows increasing the autonomy of fleets and transferring functions from operators to interfaces, what can be decisive to reduce the workload of the operators and improve the feasibility of multi-UAV missions.

The architecture is applied to a multi-UAV simulator with a competitive scenario. The simulations show that the architecture is suitable for multi-UAV missions even with adverse conditions.

Acknowledgments This work is framed on the SAVIER (Situational Awareness VIrtual Environment) Open Innovation Project, which is both supported and funded by Airbus Defence & Space.

The research leading to these results has received funding from the RoboCity-2030-III-CM project (Robótica aplicada a la mejora de la calidad de vida de los ciudadanos. fase III; S2013/MIT-2748), funded by Programas de Actividades I+D en la Comunidad de Madrid and cofunded by Structural Funds of the EU, and from the DPI2014-56985-R project (Protección robotizada de infraestructuras críticas) funded by the Ministerio de Economía y Competitividad of Gobierno de España.

References

1. Adams, S.M., Friedland, C.J.: A survey of unmanned aerial vehicle (UAV) usage for imagerycollection in disaster research and management. In: 9th International Workshop on Remote Sensing for Disaster Response (2011)
2. Avellar, G.S.C., Thums, G.D., Lima, R.R., Iscold, P., Torres, L.A.B., Pereira, G.A.S.: On the development of a small hand-held multi-UAV platform for surveillance and monitoring. In: 2013 International Conference on Unmanned Aircraft Systems (ICUAS), pp. 405–412. IEEE (2013)
3. Cummings, M.L., Bruni, S., Mercier, S., Mitchell, P.J.: Automation architecture for single operator, multiple UAV command and control. Technical report, DTIC Document (2007)
4. Dunbabin, M., Marques, L.: Robots for environmental monitoring: Significant advancements and applications. IEEE Robotics & Automation Magazine **19**(1), 24–39 (2012)
5. Gale, D., Shapley, L.S.: College admissions and the stability of marriage. The American Mathematical Monthly **69**(1), 9–15 (1962)
6. Garzón, M., Valente, J., Zapata, D., Barrientos, A.: An aerial-ground robotic system for navigation and obstacle mapping in large outdoor areas. Sensors **13**(1), 1247–1267 (2013)
7. Guedes Lang, R., Da Silva, I.N., Romero, R., et al.: Development of distributed control architecture for multi-robot systems. In: 2014 Joint Conference on Robotics: SBR-LARS Robotics Symposium and Robocontrol (SBR LARS Robocontrol), pp. 163–168. IEEE (2014)
8. Iwama, K., Miyazaki, S.: A survey of the stable marriage problem and its variants. ICKS International Conference on Informatics Education and Research for Knowledge-Circulating Society **1**, 131–136 (2008)
9. Jacobs, B., De Visser, E., Freedy, A., Scerri, P.: Application of intelligent aiding to enable single operator multiple UAV supervisory control (2010)

10. Jin, Y., Minai, A.A., Polycarpou, M.M.: Cooperative real-time search and task allocation in UAV teams. In: Proceedings of the 42nd IEEE Conference on Decision and Control, vol. 1, pp. 7–12. IEEE (2003)
11. Khaleghi, A.M., Xu, D., Minaeian, S., Li, M., Yuan, Y., Liu, J., Son, Y.J., Vo, C., Mousavian, A., Lien, J.M.: A comparative study of control architectures in UAV/UGV-based surveillance system. In: Proceedings of the 2014 Industrial and Systems Engineering Research Conference (2014)
12. Lemaire, T., Alami, R., Lacroix, S.: A distributed tasks allocation scheme in multi-UAV context. In: 2004 IEEE International Conference on Robotics and Automation, ICRA 2004, vol. 4, pp. 3622–3627. IEEE (2004)
13. Marino, A., Parker, L.E., Antonelli, G., Caccavale, F.: A decentralized architecture for multi-robot systems based on the null-space-behavioral control with application to multi-robot border patrolling. Journal of Intelligent & Robotic Systems $71(3–4)$, 423–444 (2013)
14. Matellán, V., Borrajo, D.: ABC2 an agenda based multi-agent model for robots control and cooperation. Journal of Intelligent and Robotic Systems $32(1)$, 93–114 (2001)
15. Quigley, M., Conley, K., Gerkey, B., Faust, J., Foote, T., Leibs, J., Wheeler, R., Ng, A.Y.: Ros: an open-source robot operating system. In: ICRA Workshop on Open Source Software, vol. 3, p. 5 (2009)
16. Roldán, J.J., del Cerro, J., Barrientos, A.: A proposal of methodology for multi-UAV mission modeling. In: 23rd IEEE Mediterranean Conference on Automation and Robotics (MED), pp. 1–7 (2015)
17. Teichteil-Königsbuch, F., Fabiani, P.: A multi-thread decisional architecture for real-time planning under uncertainty. In: 3rd ICAPS 2007 Workshop on Planning and Plan Execution for Real-World Systems (2007)
18. Valente, J., Sanz, D., Barrientos, A., del Cerro, J., Ribeiro, A., Rossi, C.: An air-ground wireless sensor network for crop monitoring. Sensors $11(6)$, 6088–6108 (2011)
19. Zheng, W., Wang, T., Xi, X.: Development of an internet-based multi-robot teleoperation system. In: Proceedings of the 2006 IEEE International Conference on Mechatronics and Automation, pp. 329–333. IEEE (2006)

LiDAR-Based Control of Autonomous Rotorcraft for Inspection of Pole-Shaped Structures

Bruno J. Guerreiro, Carlos Silvestre and Rita Cunha

Abstract This paper addresses the problem of trajectory tracking control of autonomous rotorcraft relative to pole-shaped structures using LiDAR sensors. The proposed approach defines an alternative kinematic model, directly based on LiDAR measurements, and uses a trajectory-dependent error space to express the dynamic model of the vehicle. An LPV representation with piecewise affine dependence on the parameters is adopted to describe the error dynamics over a set of predefined operating regions. The synthesis problem is stated as a continuous-time $\mathcal{H}_2$ control problem, solved using LMIs and implemented within the scope of gain-scheduling control theory. The performance of the proposed control method is validated with comprehensive simulation results.

Keywords Trajectory tracking · Sensor-based control · LiDAR · UAV

1 Introduction

The current procedures for infrastructure inspection still require the on-site presence of highly specialized workers, which mainly use visual inspection aided by dedicated hydraulic or rope access equipment. However, the increasingly availability

B.J. Guerreiro(✉) · C. Silvestre
Department of Electrical and Computer Engineering, Faculty of Science and Technology, University of Macau, Av. da Universidade, Taipa, Macau, China
e-mail: bguerreiro@isr.ist.utl.pt, csilvestre@umac.mo

C. Silvestre
Instituto Superior Técnico, Universidade de Lisboa, Lisboa, Portugal

B.J. Guerreiro · R. Cunha
Institute for Systems and Robotics, Instituto Superior Técnico, Universidade de Lisboa, Av. Rovisco Pais 1, 1049-001 Lisboa, Portugal
e-mail: rita@isr.ist.utl.pt

© Springer International Publishing Switzerland 2016
L.P. Reis et al. (eds.), *Robot 2015: Second Iberian Robotics Conference*,
Advances in Intelligent Systems and Computing 417,
DOI: 10.1007/978-3-319-27146-0_47

609

of non-invasive inspection techniques are enabling the automatic detection of non-visible hollows, cracks, and other relevant infrastructure defects. Having rotary-wing unmanned aerial vehicles (RUAV), such as those depicted in Fig. 1, equipped with these technologies would increase the quality and frequency of the inspections, as well as decrease the inspection and operational costs. Nonetheless, as a consequence of the unreliability of GPS near large infrastructures and buildings, even the state-of-the-art RUAV technologies are not able to fly autonomously, demanding highly specialized operators on-site maintaining line-of-sight with the vehicle.

Fig. 1 Aerial robotic vehicles developed at ISR/IST for infrastructure inspection, featuring vibration isolated custom avionics, horizontal and vertical LiDARs, remotely controlled SLR Camera, and high precision GPS recievers.

To address these drawbacks, several authors have focused on using nonlinear techniques, such as backstepping [3], or nonlinear model predictive control [5]. An alternative to tackle the design of controllers for vehicles with complex and highly nonlinear dynamic models, is to resort to gain-scheduling control theory [9], as adopted in this paper. The proposed approach relies on the formulation of a nonlinear LiDAR-based kinematics for the position of the vehicle relative to the structure to be inspected, assuming for that purpose that a 2-D LiDAR provides horizontal profiles of the environment and a range sensor provides the vertical distance to the ground. A trajectory-dependent error space is then defined to express the dynamic model of the vehicle, which should be driven to zero by the trajectory tracking controller to be designed.

In combination with gain-scheduling techniques, linear parameter varying (LPV) models and linear matrix inequalities (LMIs) are frequently used for the design of each controller, which together constitute powerful tools for tackling complex nonlinear control problems. Several examples in the literature attest its level of success, such as [2, 11], whereas further theoretical details can be found in [4] and [10]. The flight envelope of the vehicle is partitioned into a set of overlapping regions of operation, and for each of these regions an LPV representation with piecewise affine dependence on the parameters is considered to accurately model the error dynamics. By imposing an affine parameter dependent structure, a continuous-time state feedback controller can be derived in order to guarantee stability and $\mathcal{H}_2$-norm performance bound over a polytopic set of parameters using a finite number of LMIs.

Based on this result, a controller is synthesized for each of the operating regions and the overall controller is implemented within the framework of gain-scheduling control theory using the D-methodology to switch between controllers [7].

The paper is organized as follows. The model of a general rotorcraft is presented in Section 2, while in Section 3 the LiDAR kinematics and a sensor-based trajectory-dependent error space are introduced. The control synthesis results and implementation details are provided in Section 4, preceding the simulation results that are presented in Section 5. Finally, the conclusions and directions of further work are given in Section 6.

2   Dynamic Model

This section summarizes the rotorcraft dynamic model, in particular, for quadrotor helicopters. A comprehensive coverage of rotorcraft flight dynamics, with main focus on traditional helicopters, can be found in [8], while particular details can be found in [1, 6] and references therein.

The vehicle is modeled as a rigid-body driven by forces and moments applied to its center of mass that include the contribution of the rotors, fuselage, and gravity. The usual frames are also considered: an Earth-fixed reference frame $\{E\}$, using the north-east-down (NED) convention, and a body-fixed frame $\{B\}$, defined with origin at the vehicle's center of mass, with the x-axis pointing forward, the y-axis pointing right, and the z-axis pointing down.

2.1   Rotorcraft Dynamics

Considering $\mathcal{SE}(3) := \mathbb{R}^3 \times \mathcal{SO}(3)$ as the special Euclidean group in 3-D space, then, the pair $\left({}^E\mathbf{p}_B, {}^E_B\mathbf{R} \right) \in \mathcal{SE}(3)$ denotes the configuration of $\{B\}$ relative to $\{E\}$, also denoted by $(\mathbf{p}, \mathbf{R})$, for the sake of notation simplicity. The rotation matrix $\mathbf{R}$ can also be parameterized by the ZYX Euler angles $\boldsymbol{\lambda} = \begin{bmatrix} \phi\ \theta\ \psi \end{bmatrix}^T, \theta \in (-\frac{\pi}{2}, \frac{\pi}{2}), \phi, \psi \in \mathbb{R}$. These parameters can be readily defined by using $\mathbf{R} = \mathbf{R}_z(\psi)\, \mathbf{R}_y(\theta)\, \mathbf{R}_x(\phi)$, where, for instance, $\mathbf{R}_x(.)$ yields the basic rotation of a given angle about the x-axis. In addition, let the linear and angular velocities of $\{B\}$ relative to $\{E\}$, expressed in $\{B\}$, be given by $\mathbf{v} = \begin{bmatrix} u\ v\ w \end{bmatrix}^T$ and $\boldsymbol{\omega} = \begin{bmatrix} p\ q\ r \end{bmatrix}^T$, respectively. Then, the kinematic equations of motion of a generic rigid-body can be written as

$$\begin{cases} \dot{\mathbf{p}} = \mathbf{R}\,\mathbf{v} & \text{(1a)} \\ \dot{\boldsymbol{\lambda}} = \mathbf{Q}(\phi, \theta)\,\boldsymbol{\omega} & \text{(1b)} \end{cases}$$

where matrix $\mathbf{Q}$ relates the vehicle angular velocity with the time derivative of the Euler angles.

The rotorcraft nonlinear dynamics can be obtained by defining the external forces and moments as functions of the control vector, $\mathbf{u}$, the linear and angular velocity of the vehicle, and the wind velocity vector, $\mathbf{v}_w$, described in $\{B\}$. Most rotorcraft vehicles, such as traditional helicopters and most multirotor vehicles, can only generate a thrust force along the vehicle's vertical axis in addition to three independent torques, one for each body rotation axis. In this model, the roll and pitch proprietary closed-loop dynamics are modeled as a first-order linear system with time constant $1/a_2$, and such that the actuation signals u_ϕ and u_θ can be seen as angular references. On the other hand, the yaw actuation u_r is regarded as an angular velocity reference, as the yaw dynamics of the vehicle is also considered to be a first-order linear system with time constant $1/a_1$. This simplified inner-loop system can be expressed as

$$\begin{cases} \dot{r} = a_1(-r + u_r) & \text{(2a)} \\ \dot{\lambda} = \mathbf{A}_\lambda^\lambda \lambda + \mathbf{A}_\lambda^r r + \mathbf{B}_\lambda \mathbf{u} & \text{(2b)} \end{cases}$$

where $\mathbf{A}_\lambda^\lambda = -a_2 \mathbf{I}_{xy}$, $\mathbf{A}_\lambda^r = \mathbf{e}_z$, $\mathbf{e}_z := \begin{bmatrix} 0 & 0 & 1 \end{bmatrix}^T$, $\mathbf{B}_\lambda = a_2 \begin{bmatrix} \mathbf{I}_{xy} & \mathbf{0}_{3\times 1} \end{bmatrix}$, $\mathbf{I}_{xy} = \text{diag}(1, 1, 0)$, and the constants a_1 and a_2 can be identified for each vehicle and inner-loop pair. The actuation vector is defined as $\mathbf{u} = \begin{bmatrix} u_\phi & u_\theta & u_r & u_T \end{bmatrix}^T \in \mathbb{R}^4$, where u_T denotes the actuation signal that controls the generation of the vertical thrust force.

Finally, considering the vehicle mass m and the tensor of inertia $\mathbf{I}_B$, the velocity dynamics state-space equation can be written as

$$\dot{\mathbf{v}} = -\mathbf{S}(\boldsymbol{\omega})\mathbf{v} + \mathbf{g}(\phi, \theta) - m^{-1}(u_T \mathbf{e}_z + \mathbf{f}_D(\mathbf{v}, \mathbf{v}_w)) \qquad (3)$$

where the earth's gravity acting on the body is denoted by $\mathbf{g}(\phi, \theta)$, $\mathbf{f}_D(\mathbf{v}, \mathbf{v}_w)$ accounts for the drag effects acting on the vehicle, $\boldsymbol{\omega}$ is related $\dot{\phi}$ and $\dot{\theta}$, according to (1b), and the operator $\mathbf{S}(.)$ denotes the skew-symmetric matrix such that $\mathbf{S}(\mathbf{a})\mathbf{b}$ represents the cross product $\mathbf{a} \times \mathbf{b}$, for some $\mathbf{a}, \mathbf{b} \in \mathbb{R}^3$.

It should be noted that, for simplicity of presentation, the yaw angle kinematic equation in (2b) is an approximated version of (1b), for which the values of ϕ and θ are considered to be sufficiently small. Also, the disturbance input $\mathbf{v}_w$ can be modeled as a random process with a predefined power spectral density (PSD) such as the von Karman turbulence model, or using the statistical discrete gust approach (SDG), essentially employed to describe more structured disturbances (see [8] and references therein for further details).

2.2 Equilibrium Trajectories

The dynamic equilibrium or trimming trajectories of a rigid-body requires that the sum of the external forces and the inertial forces acting on the body are zero, implying that $\dot{\mathbf{v}}_c = \mathbf{0}$, $\dot{\boldsymbol{\omega}}_c = \mathbf{0}$, and also that $\dot{\mathbf{u}}_c = \mathbf{0}$, $\dot{\phi}_c = 0$, and $\dot{\theta}_c = 0$ (where the subscript

$(.)_c$ denotes the respective variable at trimming). It can be seen that the trimming yaw angle, $\psi_c(.)$, can change without violating the equilibrium condition, nonetheless satisfying $\dot{\psi}_c = r_c$, yielding a constant yaw rate, $\dot{\psi}_c = u_{r_c}$. As the roll and pitch trimming values must also be constant, Eq. (2b) imply that $\phi_c = u_{\phi_c}$ and $\theta_c = u_{\theta_c}$.

Further consider an additional frame, denoted as the horizontal body-fixed frame, $\{H\}$, such that $\mathbf{p}_H = \mathbf{p}$ denotes the position of frame $\{H\}$ described in frame $\{E\}$, $^E_H\mathbf{R} = \mathbf{R}_z(\psi)$ is the rotation from frame $\{H\}$ to frame $\{E\}$, and $^H_B\mathbf{R} = \mathbf{R}_y(\theta)\,\mathbf{R}_x(\phi)$ the rotation from frame $\{B\}$ to frame $\{H\}$. A versatile parametrization can be introduced by using the complete velocity vector, described in the horizontal body-fixed frame $\{H\}$, $^H\mathbf{v}_c = \mathbf{R}_y(\theta_c)\,\mathbf{R}_x(\phi_c)\,\mathbf{v}_c$, combined with the yaw angular velocity, $\dot{\psi}_c$, yielding

$$\boldsymbol{\xi} = \begin{bmatrix} ^H\mathbf{v}_c^T & \dot{\psi}_c \end{bmatrix}^T = \begin{bmatrix} ^Hu_c & ^Hv_c & ^Hw_c & \dot{\psi}_c \end{bmatrix}^T.$$

3 Sensor-Based Error Dynamics

In the case of the close inspection of an infrastructure, the visibility of the GPS satellite constellation is compromised, whereas the accuracy of LiDAR sensors is increased with the proximity to the infrastructure. Therefore, it is assumed that there is no access to absolute position measurements, but accurate estimates of the body velocities and attitude are available. Without loss of generality, the LiDAR is assumed to be at the center of mass and aligned with the body-fixed frame, as the transformation between the LiDAR frame and $\{B\}$ is known.

3.1 Pier Geometry and Position

Consider that at each sampling instant, the horizontally mounted LiDAR provides a set of n_L data points in frame $\{B\}$, from which a subset is selected as the region of interest for control purposes, $(^B\rho_i, {}^B\alpha_i)$, for $i = 1, \ldots, n$, where $^B\rho_i$ stands for the range distance measurement and $^B\alpha_i$ is the respective bearing. As the detection of invariants is more challenging in frame $\{B\}$, the original measurements are projected into frame $\{H\}$ and denoted as (ρ_i, α_i), which are functions of the original measurements as well as the roll and pitch angles.

The shape of the inspection target is assumed to be similar to a vertical cylindrical volume, which for the sake of simplicity, will be referred to as the pier for the remainder of this work. It is straightforward to see that the intersection of any vertical cylindrical pier with the xy-plane of frame $\{H\}$ yields a circle. In this way, the laser measurements (ρ_i, α_i) that represent the pier also represent an arc of a circle. Using basic notions of geometry, it can be seen that, given any point outside the circle and in the same plane, there are only two lines that pass through this point and are tangent to the circle, as shown in Fig. 2. These lines are approximated by the first and last laser measurement that intersects the pier, provided that the angular resolution of the laser is sufficiently high, which is typically below 0.25 degrees. Therefore, the center of the pier can be estimated using the first and last LiDAR measurements that intersect the pier, respectively (ρ_1, α_1) and (ρ_n, α_n), where n is the number of laser measurements that intersect the pier. The expressions for the equivalent range and

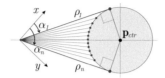

Fig. 2 Laser measurements intersecting the pier.

angle of the center of the pier, respectively denoted as ρ and α, are given by

$$\rho = \frac{\rho_1 + \rho_n}{2 \cos(\frac{\alpha_n - \alpha_1}{2})} \quad , \quad \alpha = \frac{\alpha_1 + \alpha_n}{2} . \tag{4}$$

The fact that only the first and last measurements of the laser are used to estimate the center of the pier indicates that if its shape is not exactly a cylinder, the algorithm will still be valid as the resulting detection errors can be seen as a bounded disturbance signal under mild conditions.

In order to have a complete 3-D position measurement, an additional altitude measurement is necessary. For this purpose, an additional range measurement laser can be used to acquire the distance to the ground along the z-axis of the sensor frame $\{B\}$. For simplicity, it is assumed that this sensor is located at the origin of $\{B\}$, and that the ground surrounding the inspection area is planar and horizontal. As such, the original distance-to-ground measurement in frame $\{B\}$, denoted as $^B z_g$, can be easily projected into the z-axis of frame $\{H\}$ using the expression $z_g = {}^B z_g \cos \phi \cos \theta$. Note that the horizontal ground assumption can be lifted with the use of barometric sensors or a scanning LiDAR, such that a representation of the terrain and an altitude can be measured.

3.2 LiDAR Kinematics

The equations of motion for the LiDAR kinematics are naturally divided into the horizontal plane and the vertical axis kinematics, corresponding to the two different laser sensors previously introduced. Nonetheless, they can be considered as a single sensor that provides the position of the center of the pier at ground level, $\mathbf{p}_p$, which can be expressed as $\mathbf{p}_p = \mathbf{R}_z(\alpha) \begin{bmatrix} \rho & 0 & z_g \end{bmatrix}^T$ or, alternatively, as $\mathbf{p}_p = {}^H_E\mathbf{R}({}^E\mathbf{p}_{ctr} - {}^E\mathbf{p}_B)$. Defining the LiDAR-based position vector as $\boldsymbol{\eta} = \begin{bmatrix} \rho & \alpha & z_g \end{bmatrix}^T$, after some algebraic computations with the time derivatives of the above expressions, the new kinematics equation can be written as

$$\dot{\eta} = \mathbf{r}_\eta(r) - \mathbf{I}_\eta(\rho)\,\mathbf{R}_z^T(\alpha)\,\mathbf{R}_y(\theta)\,\mathbf{R}_x(\phi)\,\mathbf{v}\;, \tag{5}$$

where $\mathbf{r}_\eta(r) = \begin{bmatrix} 0 & -r & 0 \end{bmatrix}^T$, ${}_B^H\mathbf{R} = \mathbf{R}_y(\theta)\,\mathbf{R}_x(\phi)$, and $\mathbf{I}_\eta(a) = \mathrm{diag}\!\left(\begin{bmatrix} 1 & 1/a & 1 \end{bmatrix}\right)$, for all $a \neq 0$. Thus, this new position kinematics equation can be used to replace (1a) and formulate a trajectory dependent error space to be used in a sensor-based control synthesis.

3.3 Error Dynamics

Consider the vehicle equations of motion presented in (3), (2), and the LiDAR-based position kinematics (5), denoting the trimming value LiDAR-based position vector as η_c. For the envisioned applications, the trimming trajectories of interest are z-aligned helices and hover, which are characterized by constant values of range and bearing of the pier center, ρ_c and α_c. This subset of the possible helicopter trimming trajectories, $\mathcal{E}_L \subset \mathcal{E}$, can be described by the parameterization $\boldsymbol{\xi}_L = \begin{bmatrix} \rho_c & \alpha_c & {}^Hw_c & \dot{\psi}_c \end{bmatrix}^T$, where ρ_c and α_c are, respectively, the trimming distance and azimuth of the pier in frame $\{H\}$, while ${}^Hw_c$ is the desired z-component of the velocity vector described in $\{H\}$. This parameterization can be mapped onto the general parameterization introduced in Section 2.2.

Defining the state vector as $\mathbf{x} = \begin{bmatrix} \mathbf{v}^T, r, \boldsymbol{\eta}^T, \boldsymbol{\lambda}^T \end{bmatrix}^T \in \mathbb{R}^{10}$, the sensor-based error vector can be simply defined as $\mathbf{x}_e = \mathbf{x} - \mathbf{x}_c$, where $\mathbf{x}_c$ is the desired trajectory. Let also $\mathbf{u}_e = \mathbf{u} - \mathbf{u}_c$ and $\mathbf{v}_{w_e} = \mathbf{v}_w$, which considers that there is no disturbance at trimming, $\mathbf{v}_{w_c} = 0$. Then, in the new error coordinate system, the linearization of the rigid-body dynamics and sensor-based kinematics given by (3), (2), and (5) along a trimming trajectory is time invariant, and that the nonlinear error dynamics can then be expressed as

$$\dot{\mathbf{x}}_e = \mathbf{f}_e(\mathbf{x}_e, \mathbf{u}_e, \mathbf{v}_w, \mathbf{x}_c, \mathbf{u}_c) = \begin{bmatrix} \dot{\mathbf{v}} \\ \dot{r} \\ \dot{\boldsymbol{\eta}} - \dot{\boldsymbol{\eta}}_c \\ \dot{\boldsymbol{\lambda}} - \dot{\boldsymbol{\lambda}}_c \end{bmatrix} \tag{6}$$

noting that $\dot{\boldsymbol{\eta}}_c = \begin{bmatrix} 0 & 0 & {}^Hw_c \end{bmatrix}^T$ and $\dot{\boldsymbol{\lambda}}_c = \begin{bmatrix} 0 & 0 & \dot{\psi}_c \end{bmatrix}^T$.

The linearization of (6) about the origin, or equivalently, the linearization of (3), (2), and (5) about the trimming trajectory can be expressed in the generalized error space as

$$\delta\dot{\mathbf{x}}_e = \mathbf{A}_e^{\xi}\,\delta\mathbf{x}_e + \mathbf{B}_e^{\xi}\,\delta\mathbf{u}_e + \mathbf{B}_{w_e}^{\xi}\,\delta\mathbf{v}_w \tag{7}$$

where $\mathbf{A}_e^{\xi} = \frac{\partial}{\partial\mathbf{x}}\mathbf{f}_e(.)\big|_{\xi}$ is a constant matrix for each trimming trajectory defined by $\boldsymbol{\xi}$, and similar definitions are considered for $\mathbf{B}_e^{\xi}$ and $\mathbf{B}_{w_e}^{\xi}$. Considering a small enough region of operation, the nonlinear system can be accurately approximated by an LPV, that continuously depend on the parameter vector $\boldsymbol{\xi}$. Introducing the output vector to be controlled $\mathbf{z}$, the general exogenous input vector $\mathbf{w}$, and redefining the state and input vectors as $\mathbf{x} = \delta\mathbf{x}_e$ and $\mathbf{u} = \delta\mathbf{u}_e$, an LPV system can be defined as

$$
\begin{cases}
\dot{\mathbf{x}} = \mathbf{A}(\boldsymbol{\xi})\mathbf{x} + \mathbf{B}_w(\boldsymbol{\xi})\mathbf{w} + \mathbf{B}(\boldsymbol{\xi})\mathbf{u} & \text{(8a)} \\
\mathbf{z} = \mathbf{C}(\boldsymbol{\xi})\mathbf{x} + \mathbf{D}(\boldsymbol{\xi})\mathbf{w} + \mathbf{E}(\boldsymbol{\xi})\mathbf{u} & \text{(8b)}
\end{cases}
$$

It follows that there is a linear time invariant plant (8), associated with each trimming trajectory $\boldsymbol{\xi} \in \mathcal{E}_L$, for which a linear controller can be designed.

4 Controller Design

In this section an LMI approach is used to tackle the continuous-time state feedback $\mathcal{H}_2$ synthesis problem for a polytopic LPV system such as (8). This system is parameterized by $\boldsymbol{\xi}$, which is a possibly time-varying parameter vector and belongs to the convex set $\mathcal{E}^j = \mathrm{co}(\mathcal{E}_0^j)$. Here, the operator co(.) denotes the convex hull of the elements of the argument set, $\mathcal{E}_0^j = \{\boldsymbol{\xi}_1, \ldots, \boldsymbol{\xi}_{n_j}\}$, where $\boldsymbol{\xi}_1$ to $\boldsymbol{\xi}_{n_j}$ are the vertices of a polytope.

It can be seen that testing for stability or solving the synthesis problem without any further result, involves an infinite number of LMIs. As such, several different structures for LPV systems have been proposed which reduce the problem to that of solving a finite number of LMIs. In this section, an affine polytopic description is adopted, which can also be used to model a wide spectrum of systems and, as shown in the results presented in Section 5, is an adequate choice for the system at hand.

Definition 1 (Affine Polytopic LPV System). *The system (8) is said to be a polytopic LPV system if the system matrix* $\mathbf{P}(\boldsymbol{\xi}) = \begin{bmatrix} \mathbf{A}(\boldsymbol{\xi}) & \mathbf{B}_w(\boldsymbol{\xi}) & \mathbf{B}(\boldsymbol{\xi}) \\ \mathbf{C}(\boldsymbol{\xi}) & \mathbf{D}(\boldsymbol{\xi}) & \mathbf{E}(\boldsymbol{\xi}) \end{bmatrix}$ *verifies* $\mathbf{P}(\boldsymbol{\xi}) \in$ $\mathrm{co}\left(\mathbf{P}_1, \ldots, \mathbf{P}_{n_r}\right)$ *for all* $\boldsymbol{\xi} \in \mathcal{E}^j$, *where* $\mathbf{P}_i = \begin{bmatrix} \mathbf{A}_i & \mathbf{B}_{w_i} & \mathbf{B}_i \\ \mathbf{C}_i & \mathbf{D}_i & \mathbf{E}_i \end{bmatrix}$, *for all* $i = 1, \ldots, n_j$. *Moreover, if* $\mathcal{E}^j$ *is a polytopic set, such as* $\mathcal{E}^j = \mathrm{co}\left(\mathcal{E}_0^j\right)$, $\mathcal{E}_0^j = \{\boldsymbol{\xi}_1, \ldots, \boldsymbol{\xi}_{n_j}\}$, *and* $\mathbf{P}(\boldsymbol{\xi})$ *depends affinely on* $\boldsymbol{\xi}$, *then* $\mathbf{P}_i = \mathbf{P}(\boldsymbol{\xi}_i)$ *for all* $i = 1, \ldots, n_j$, *i.e., the vertices of the parameter set can be uniquely identified with the vertices of the system.*

With this polytopic structure and the results proposed in [10, Proposition 1.19], the quadratic stability of an affine polytopic LPV system can be established if a Lyapunov equation of the closed-loop system can be satisfied for all $\boldsymbol{\xi} \in \mathcal{E}_0^j$.

The $\mathcal{H}_2$ synthesis problem can be described as that of finding a control matrix $\mathbf{K}$ that stabilizes the closed-loop system and minimizes the $\mathcal{H}_2$-norm of $\mathbf{T}_{zw}(\boldsymbol{\xi})$, denoted by $\|\mathbf{T}_{zw}(\boldsymbol{\xi})\|_{\mathcal{H}_2}$. It is assumed that matrix $\mathbf{D}(\boldsymbol{\xi}) = 0$ in order to guarantee that $\|\mathbf{T}_{zw}(\boldsymbol{\xi})\|_{\mathcal{H}_2}$ is finite for every internally stabilizing and strictly proper controller.

The following theorem is used for controller design and relies on results available in [4] and [10], after being rewritten for the case of polytopic LPV systems. In the following, tr (.) denotes the trace of the argument matrix.

Theorem 1 (Polytopic stability). *If there are real matrices* $\mathbf{X} = \mathbf{X}^T \succ 0$, $\mathbf{Y} \succ 0$, *and* $\mathbf{W}$ *such that*

$$\begin{bmatrix} \mathbf{A}(\boldsymbol{\xi})\mathbf{X} + \mathbf{X}\mathbf{A}^T(\boldsymbol{\xi}) + \mathbf{B}(\boldsymbol{\xi})\mathbf{W} + \mathbf{W}^T\mathbf{B}^T(\boldsymbol{\xi}) & \mathbf{B}_w(\boldsymbol{\xi}) \\ \mathbf{B}_w^T(\boldsymbol{\xi}) & -\mathbf{I} \end{bmatrix} \prec 0 \tag{9a}$$

$$\begin{bmatrix} \mathbf{Y} & \mathbf{C}(\boldsymbol{\xi})\mathbf{X} + \mathbf{E}(\boldsymbol{\xi})\mathbf{W} \\ \mathbf{X}\mathbf{C}^T(\boldsymbol{\xi}) + \mathbf{W}^T\mathbf{E}^T(\boldsymbol{\xi}) & \mathbf{X} \end{bmatrix} \succ 0 \tag{9b}$$

$$\text{tr}(\mathbf{Y}) < \alpha^2 \tag{9c}$$

for all $\boldsymbol{\xi} \in \mathcal{E}_0^j$, *where the static feedback controller is defined as* $\mathbf{K} = \mathbf{W}\mathbf{X}^{-1}$, *then, the closed-loop system is quadratically stable and there exists an upper-bound* α *for the continuous-time* $\mathcal{H}_2$*-norm of the closed-loop operator* $\mathbf{T}_{zw}(\boldsymbol{\xi})$ *for all* $\boldsymbol{\xi} \in \mathcal{E}^j$, *i.e.,* $\|\mathbf{T}_{zw}(\boldsymbol{\xi})\|_{\mathcal{H}_2} < \alpha$, $\forall \boldsymbol{\xi} \in \mathcal{E}^j$.

With this result, for which the proof omitted for lack of space, the optimal solution for the continuous-time $\mathcal{H}_2$ control problem is approximated through the minimization of α subject to the LMIs of Theorem 1.

To guarantee that the closed-loop system has zero steady-state error in position and yaw angle, the integral of the output vector $\mathbf{y}_e = \begin{bmatrix} \boldsymbol{\eta}_e^T & \psi_e \end{bmatrix}^T$ is included in the design as a performance output, augmenting the state dynamics with the equation $\dot{\mathbf{x}}_i = \mathbf{y}_e$. To meet the design requirements, the weighting function associated with the integral state is chosen as $\mathbf{W}_1 = \text{diag}(7, 4, 1, 1)$ and the state weight is given by $\mathbf{W}_3 = \text{diag}(0.01\mathbf{I}_4, 3, 1, 5, \mathbf{0}_2, 10)$. Dynamic weights are used for the actuation vector $\mathbf{u}$ as well as for the disturbance process $\mathbf{w}$ to represent the Von Karman disturbance model transfer functions.

As the control synthesis presented above is only valid for a specific operating region, here represented by $\mathcal{E}^j \subset \mathcal{E}$, the set of all possible trimming trajectories, $\mathcal{E}$, is partitioned into overlapping regions of operation, $\mathcal{E}^j$ for $j = 1, \ldots, n_r$, such that the union of all operating regions completely covers the set of all possible trimming trajectories, i.e., $\mathcal{E}^1 \cup \mathcal{E}^2 \cup \cdots \cup \mathcal{E}^{n_r} = \mathcal{E}$. The controller implementation scheme is based on the D-methodology, comprehensively described in [7], which moves all integrators to the plant input, and adds differentiators where they are needed to preserve the transfer functions and the stability characteristics of the closed-loop system. Thus, using the D-methodology implementation with the synthesized controllers for each of the considered operating regions allows the controller to switch between regions without having to interpolate between the individual controllers.

5 Results

This section provides realistic simulation results for the sensor-based trajectory tracking controller proposed in this paper, obtained in the Matlab®/ Simulink simulation
environment using the nonlinear dynamic quadrotor model, as detailed in Section 2.
Three simulations of the same scenario were carried out to enable the comparison of
data considering: (a) both wind disturbance and LiDAR measurement noise, (b) only
wind disturbance, and (c) without any disturbance or noise. In addition to the wind
disturbance noise generated using the Von Karman disturbance model, the simulation
scenario presented below also includes a discrete wind gust with amplitude 2.5 m/s,
rising time of 1 s, constant direction in the earth-fixed frame, which is applied at time
$t = 25$ s. Furthermore, the LiDAR measurement noise is simulated as a band-limited
white noise with standard deviation of 0.1 cm, 0.07 deg, and 0.1 cm, respectively, for
the measurements of ρ, α, and z_{hg}.

The gain-switching controller design considers $n_r = 128$ overlapping polytopic
regions, $\mathcal{E}^j$ for all $j = 1, \ldots, n_r$, with $2^4 = 16$ vertices each, and defined partitioning
$\rho_c \in [0.4, 2]$ m into 2 intervals; $\alpha_c \in [-130, 130]$ deg into 8 intervals; $\dot{\psi}_c \in$
$[-30, 30]$ deg/s into 4 intervals; and $w_c \in [-0.5, 0.5]$ m/s into 2 intervals. The
state-space matrices of the continuous-time system (8) are approximated by affine
functions on the parameters $\boldsymbol{\xi} \in \mathcal{E}^j$ using least squares fitting. The analysis of the
errors introduced by this approximation shows that the average maximum errors
throughout the regions is 5.7% in the case of matrix $\mathbf{A}$, and 8.8% for matrix $\mathbf{B}_w$,
while matrix $\mathbf{B}$ has no approximation errors as it does not depend on the trimming
parameters.

Considering the envisioned applications for automatic infrastructure inspection,
in the results provided in this paper the vehicle is required to track a trajectory
composed of: i) take-off and stationary hover at the initial position relative to the pier:
$\boldsymbol{\xi}_L = \begin{bmatrix} 0.7 & 0 & 0 & 0 \end{bmatrix}^T$ in the respective units m, deg, deg/s, and m/s; ii) an ascending helix
with three loops facing the pier: $\boldsymbol{\xi}_L = \begin{bmatrix} 0.7 & 0 & 24 & -0.03 \end{bmatrix}^T$; and iii) a stationary hover at
final position: $\boldsymbol{\xi}_L = \begin{bmatrix} 0.7 & 0 & 0 & 0 \end{bmatrix}^T$. The results of the sensor-based trajectory tracking
simulation scenario are represented in Fig. 3, featuring the laser-based position and
attitude tracking errors, the control action, the wind disturbances, and the transitions
between regions of operation. The moments when the vehicle takes off from the
ground, transitions from the initial hover to the ascending helix trajectory, when the
wind gust is applied, as well as when the vehicle transitions from the ascending helix
to a stationary hover at the final position are also represented in these figures with
vertical lines, respectively, at $t = 0.0$ s, $t = 12.5$ s, $t = 30$ s, and $t = 57.5$ s.

It can be seen that the vehicle has no problem in the transitions from hover to
the ascending helix and vice-versa, converging to the desired trajectory, as well
as attenuating the effects of the wind disturbance and LiDAR measurement noise,
when they exist. This can be observed in Fig. 3(c), noting that the tracking errors
without (with) disturbance and measurement noise are lower than 1.3 cm (6 cm) in
ρ, 5.5 deg (8.5 deg) in α, and 1.2 cm (7.6 cm) in z_{hg}. After the wind gust depicted
in Fig. 3(e) is applied, the actuation has to adapt to the constant direction of the

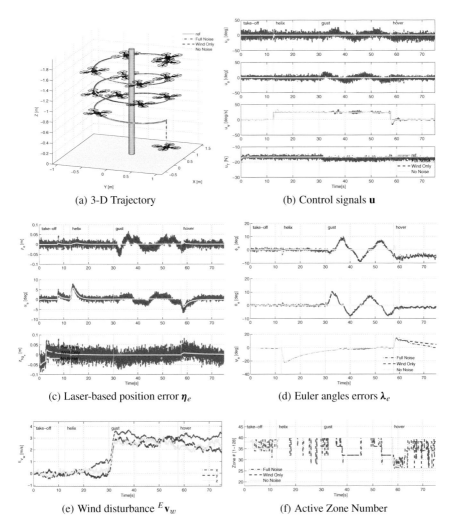

(a) 3-D Trajectory

(b) Control signals **u**

(c) Laser-based position error $\boldsymbol{\eta}_e$

(d) Euler angles errors $\boldsymbol{\lambda}_e$

(e) Wind disturbance $^E\mathbf{v}_w$

(f) Active Zone Number

Fig. 3 LiDAR-based trajectory tracking control simulation with different noise configurations (with both wind disturbance and LiDAR measurement noise, only with wind disturbance, and without measurement noise nor wind disturbance).

wind gust (in frame $\{E\}$) while changing the vehicle direction to track the reference trajectory, as can be seen in Fig. 3(b). Nonetheless, the tracking errors remain within the same order of magnitude as those of the transitions from and to hover, even under the influence of wind disturbance.

Regarding the operating region transitions, presented in Fig. 3(f), it can be seen that the higher number of transitions is concentrated on the final part of the trajectory, where it must maintain a stationary hover. When no noise or disturbance is applied, the transitions between operating regions (and their respective controller gains) only occur when there is a transition in the desired trajectory. As the LiDAR measurement noise and the wind disturbances are applied, the number of transitions increase, consistently showing that the overall gain-switching controller is effective in stabilizing the vehicle and converging to the desired trajectory, even under demanding perturbations.

6 Concluding Remarks

This paper presented the design and validation of a LiDAR-based trajectory tracking control methodology for autonomous rotorcraft, considering that no absolute position solution, such as GPS, is available for control. The major contributions of this work are the introduction of a LiDAR-based nonlinear kinematics, formulated in 3-D space, and the definition of a trajectory-dependent error space to express the dynamic model of the vehicle and the sensor-based kinematics. The effectiveness of the proposed control laws was assessed in a realistic simulation environment with a nonlinear model of the vehicle and realistic mission scenarios. The obtained results clearly indicate that the proposed methodologies are well suited to be used for automatic inspection of large infrastructures using autonomous rotorcraft.

Future work includes the experimental validation and performance evaluation of the proposed controllers during automatic inspection operation near real world infrastructures. Further research on the transitions between these two types of controllers and the stability conditions for the transitions is also considered to be necessary.

Acknowledgments This work was supported by the Macao Science and Technology Development Fund under Grant FDCT/048/2014/A1, by the University of Macau Project MYRG2015-00126-FST, and by Fundação para a Ciência e Tecnologia (FCT) through ISR, under LARSyS UID/EEA/50009/2013 project. The work of Rita Cunha was supported by the FCT Investigator Programme under Contract IF/00921/2013.

References

1. Cunha, R.: Advanced Motion Control for Autonomous Air Vehicles. Ph.D. thesis, Instituto Superior Técnico, Universidade Técnica de Lisboa, Lisbon, Portugal, June 2007
2. Cunha, R., Antunes, D., Gomes, P., Silvestre, C.: A path-following preview controller for autonomous air vehicles. In: AIAA Guidance, Navigation and Control Conference. AIAA, Keystone, CO, August 2006
3. Frazzoli, E., Dahleh, M., Feron, E.: Trajectory tracking control design for autonomous helicopters using a backstepping algorithm. In: Proceedings of the 2000 American Control Conference, vol. 6, pp. 4102–4107 (2000)
4. Ghaoui, L., Niculescu, S.I.: Advances in Linear Matrix Inequality Methods in Control. Society for Industrial and Applied Mathematics, SIAM, Philadelphia, PA (1999)
5. Guerreiro, B.J., Silvestre, C., Cunha, R.: Terrain avoidance nonlinear model predictive control for autonomous rotorcraft. Journal of Intelligent & Robotic Systems **68**(1), 69–85 (2012)

6. Guerreiro, B.J.: Sensor-based Control and Localization of Autonomous Systems in Un-known Enviornments. Ph.D. thesis, Instituto Superior Técnico, University of Lisbon (2013) (unpublished)
7. Kaminer, I., Pascoal, A., Khargonekar, P.P., Coleman, E.E.: A velocity algorithm for the im-plementation of gain-scheduled controllers. Automatica **31**(8), 1185–1191 (1995)
8. Padfield, G.D.: Helicopter Flight Dynamics: The Theory and Application of Flying Qualities and Simulation Modeling. AIAA Education Series, AIAA, Washington DC (1996)
9. Rugh, W.J., Shamma, J.S.: Research on gain scheduling. Automatica **36**(10), 1401–1425 (2000). Survey Paper
10. Scherer, C., Weiland, S.: Lecture notes on linear matrix inequality methods in control. Dutch Institute of Systems and Control (2000)
11. Silvestre, C., Pascoal, A., Kaminer, I.: On the design of gain-scheduled trajectory tracking contollers. International Journal of Robust and Nonlinear Control **12**(9), 797–839 (2002)

A Predictive Path-Following Approach for Fixed-Wing Unmanned Aerial Vehicles in Presence of Wind Disturbances

Alessandro Rucco, A. Pedro Aguiar, Fernando Lobo Pereira
and João Borges de Sousa

Abstract In this paper we address the path-following problem for fixed-wing Unmanned Aerial Vehicles (UAVs) in presence of wind disturbances. Given a desired path with a specified airspeed profile assigned on it, the goal is to follow the desired maneuver while minimizing the control effort. We propose a predictive path following control scheme based on trajectory optimization techniques, to compute feasible UAV trajectories, combined with a sample-data Model Predictive Control (MPC) approach, to handle the wind field. By explicitly addressing the wind field, the UAV exploits the surrounding environment thus extending its capabilities in executing the desired maneuver. We provide numerical computations based on straight line and circular paths under various wind conditions. The computations allow us to highlight the benefits of the proposed control scheme.

Keywords Path-following · Trajectory optimization · Model Predictive Control · Virtual Target Vehicle · Unmanned Aerial Vehicle

1 Introduction

In the past few decades, with advances in embedded processors, sensors, actuators, and control systems, Unmanned Aerial Vehicles (UAVs) have been successfully deployed in various mission scenarios. Of particular interest is the execution of long endurance missions as, e.g., in applications for surveillance and security. Such missions can be addressed by exploiting the energy from the surrounding environment, see, e.g., [1], [2], and [3]. This motivates the design of trajectory optimization tools

---

A. Rucco(✉) · A.P. Aguiar · F.L. Pereira · J.B. de Sousa
Research Center for Systems and Technologies (SYSTEC),
Faculty of Engineering of the University of Porto (FEUP), and
Institute for Systems and Robotic (ISR), 4200-465 Porto, Portugal
e-mail: {alessandrorucco,pedro.aguiar,flp,jtasso}@fe.up.pt

© Springer International Publishing Switzerland 2016
L.P. Reis et al. (eds.), *Robot 2015: Second Iberian Robotics Conference*,
Advances in Intelligent Systems and Computing 417,
DOI: 10.1007/978-3-319-27146-0_48

for the online computation of feasible trajectories that explicitly take into account (and, therefore, exploit) atmospheric flow field.

Many control schemes have been introduced in the literature to tackle the wind field and improve the UAVs performance. In [4], the authors proposed a path following method based on a vector field approach. Using Lyapunov stability criteria, the authors provided asymptotic decay of path following errors in the presence of constant wind disturbances. In [5], a path following strategy for fixed wing UAVs based on the theory of nested saturation is proposed. The wind field is assumed to be constant. The effectiveness of the proposed strategy has been demonstrated through numerical computations and flight tests. A path planning strategy combined with a sliding surface controller for the surveillance of multiple waypoints is proposed in [6]. The wind is composed by a known constant plus a (small) time-varying component. In [7], a sliding mode controller and a receding-horizon controller are proposed and compared. Both controllers are evaluated with respect to wind disturbances. In these works, however, there are no considerations of optimality in the paths (maneuvers) to be followed by the UAVs. Recently, in [3] a control architecture for dynamic soaring is described. It is based on wind field prediction, trajectory planning and a low-level path-following controller. Polynomial splines are used to model the complex wind field. On the other hand, the wind model is based on a parameter estimation approach. In [8], based on nonlinear optimal control techniques, an offline trajectory optimization strategy for UAVs is proposed. In particular, the strategy is based on a Virtual Target Vehicle (VTV) perspective where a virtual target is introduced. Moreover, the strategy allows one to avoid singularities that occur in some path following methods described in the literature (for instance, when the vehicle is approaching the desired path with a perpendicular direction); however, the strategy does not take into account the wind. In Figure 1, we compare the optimal paths for a desired straight line maneuver in a scenario without wind, Figure 1a, and with constant wind field, Figure 1b. We use the problem formulation and strategy proposed in [8]. In order to highlight the wind influence on the (local) optimal trajectory, we only change the initial condition: the UAV is traveling along the desired path in the wrong direction. By comparing the optimal paths (see solid green lines), we can observe that the wind strongly affects the optimal trajectory. Moreover, the UAV exploits the wind in order to follow the desired path and, at the same time, minimize the control effort (thus increasing its endurance). These numerical computations provide evidence that the wind must be taken into account when trying to compute optimal UAV trajectories for, e.g., long endurance missions.

In this paper, we extend and adapt the trajectory optimization strategy presented in [8] for path-following in presence of time-varying wind. We provide an optimal control problem formulation of the path-following problem in presence of wind by using a VTV approach. In order to design a controller capable of executing the desired maneuver in presence of time-variant wind, we exploit the potential of the Model Predictive Control (MPC) approach. We do not model the wind field (thus keeping a reasonable level of complexity), yet we use a sampled-data MPC architecture. At every sampling time, a measurement of the wind is assumed to be available and a trajectory optimization technique, [9] and [10], is run for the current prediction

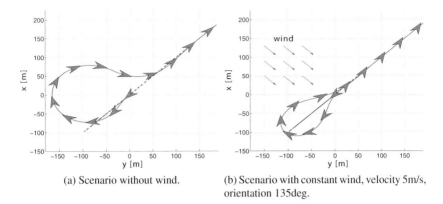

(a) Scenario without wind.

(b) Scenario with constant wind, velocity 5m/s, orientation 135deg.

Fig. 1 Optimal paths for a straight line maneuver (optimization strategy proposed in [8]). The UAV is displayed at every 3 seconds (in green triangles).

horizon. Only the first values of the resulting optimal control solution is then applied to the system, and the rest of the solution is discarded. This procedure is then repeated at each sampling instant, thus resulting in a sample-data MPC approach. We provide numerical computations to show the effectiveness of the proposed predictive path-following strategy.

The rest of the paper is organized as follows. In Section 2, we introduce the UAV model, introduce the VTV approach and formulate the path-following problem in terms of the new set of coordinates. In Section 3 we describe the MPC path-following and the optimal control based strategy to solve the constrained optimal control problem. In Section 4 we provide a set of numerical computations highlighting the effectiveness of the proposed predictive path-following approach.

2 Problem Formulation

In this section we address the path following problem. We briefly introduce the Virtual Target Vehicle approach and rewrite the UAV model with respect to a new set of coordinates. We formulate the path following problem with respect to the new dynamical system.

2.1 Constrained UAV Model

The vehicle dynamics is based on a simplified flight dynamics model called Co-ordinated Flight Vehicle (CFV) widely used in the literature, see, e.g., [8], [11] and [12]. The CFV model is described by ordinary differential equations of reasonable complexity and, at the same time, captures the nonlinear dynamics of maneuvering flight, [11]. The planar UAV motion can be described by

$$
\begin{aligned}
\dot{x} &= v_a \cos \psi + v_w \cos \psi_w \\
\dot{y} &= v_a \sin \psi + v_w \sin \psi_w, \\
\dot{\psi} &= \frac{g \tan \phi}{v_g}, \\
\dot{v}_a &= u_1, \\
\dot{\phi} &= u_2,
\end{aligned}
\tag{1}
$$

where (x, y) is the longitudinal and lateral position (with respect to the global spatial frame), ψ is the heading angle, v_a is the air-relative velocity (airspeed), v_g is the ground-relative velocity (ground speed), ϕ is the roll angle, g denote the gravity acceleration, and v_w and ψ_w are the velocity and orientation of the wind, respectively. The airspeed and the ground speed are related by

$$
\begin{aligned}
v_g \cos \chi &= v_a \cos \psi + v_w \cos \psi_w, \\
v_g \sin \chi &= v_a \sin \psi + v_w \sin \psi_w,
\end{aligned}
\tag{2}
$$

where χ is the course angle. Using the law of sines (see Figure 2), the heading and the course angles are related by

$$
\psi = \chi - \arcsin \left(\frac{v_w}{v_a} \sin (\psi_w - \chi) \right).
\tag{3}
$$

This equation will be used for the development of the optimization-based path-following strategy.

We take the control inputs of the planar UAV to be the longitudinal acceleration, $u_1 = \dot{v}_a$, and the roll rate, $u_2 = \dot{\phi}$. Due to physical limitations or security specifications, state and input constraints are imposed on the UAV model. The following state-input constraints are taken into account:

$$
\begin{aligned}
v_{a\,min} &\leq v_a \leq v_{a\,max}, \\
|\phi| &\leq \phi_{max}, \\
|u_1| &\leq u_{1max}, \\
|u_2| &\leq u_{2max}.
\end{aligned}
\tag{4}
$$

The constraints parameters used in the paper are based on the ones given in [8, 12], i.e., $v_{min} = 18\text{m/s}$, $v_{max} = 25\text{m/s}$, $\phi_{max} = 24\text{deg}$, $u_{1max} = 0.2\text{m/s}^2$, $u_{2max} = 28\text{deg/s}$.

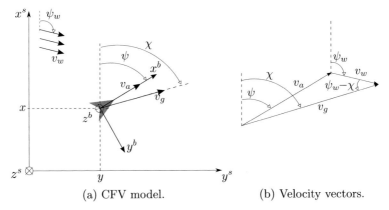

(a) CFV model.   (b) Velocity vectors.

Fig. 2 Coordinated Flight Vehicle model. The figure shows the quantities used to describe the model.

2.2 Error Dynamics: Virtual Target Vehicle Approach

Given a desired geometric path $(\bar{x}_d(s), \bar{y}_d(s))$ parameterized by the variable $s \in \mathbb{R}$ and the initial condition of the CFV, our goal is to compute a feasible trajectory (i.e., it must satisfy the dynamics of the nonlinear system (1) and the constraints (4)) that accurately follows the desired path with a specified desired aispeed along the path $\bar{v}_{ad}(s)$. Let us consider a Virtual Target Vehicle (VTV) that is constrained to travel along the desired path, see Figure 3. Given the arc-length parametrization of the VTV path, $s_{vtv} \mapsto (\bar{x}_{vtv}(s_{vtv}), \bar{y}_{vtv}(s_{vtv}))$, we choose a new set of coordinates. Note that the bar symbol indicates that a quantity is expressed as a function of the arc-length rather than time. The coordinates of the CFV (x, y) can be defined with respect to the position of the VTV, see Figure 3,

$$\begin{bmatrix} x \\ y \end{bmatrix} = \begin{bmatrix} \bar{x}_{vtv} \\ \bar{y}_{vtv} \end{bmatrix} + R_z(\bar{\chi}_{vtv}) \begin{bmatrix} e_x \\ e_y \end{bmatrix}, \tag{5}$$

where e_x and e_y are the longitudinal and lateral error coordinates, respectively, and $R_z(\bar{\chi}_{vtv})$ is the rotation matrix transforming vectors from the error frame into the global spatial frame. Following the derivation in [8], we get the expression of $\dot{e}_x$ and $\dot{e}_y$ as

$$\begin{aligned} \dot{e}_x &= v_a \cos(\psi - \bar{\chi}_{vtv}) + w_x \cos \bar{\chi}_{vtv} + w_y \sin \bar{\chi}_{vtv} - (1 - e_y \bar{\sigma}_{vtv}) \dot{s}_{vtv}, \\ \dot{e}_y &= v_a \sin(\psi - \bar{\chi}_{vtv}) - w_x \sin \bar{\chi}_{vtv} + w_y \cos \bar{\chi}_{vtv} - e_x \bar{\sigma}_{vtv} \dot{s}_{vtv}. \end{aligned} \tag{6}$$

Defining the local heading angle as $\mu = \psi - \bar{\chi}_{vtv}$ and the VTV's velocity as an additional control input $u_3 = \dot{s}_{vtv}$, the nonlinear system (1) can be written with respect to the new set of coordinates $(\mathbf{x}, \mathbf{u}) = (e_x, e_y, \mu, v_a, \phi, s_{vtv}, u_1, u_2, u_3)$:

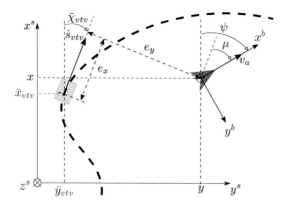

Fig. 3 Planar tracking error coordinates. The bold triangle and the light rectangle indicate the real UAV and the VTV, respectively. The bold dash line indicates the desired path.

$$\dot{e}_x = v_a \cos \mu + w_x \cos \bar{\chi}_{vtv} + w_y \sin \bar{\chi}_{vtv} - (1 - e_y \bar{\sigma}_{vtv}) u_3 \,,$$
$$\dot{e}_y = v_a \sin \mu - w_x \sin \bar{\chi}_{vtv} + w_y \cos \bar{\chi}_{vtv} - e_x \bar{\sigma}_{vtv} u_3 \,,$$
$$\dot{\mu} = \frac{g \tan \phi}{v_g} - \bar{\sigma}_{vtv} u_3 \,,$$
$$\dot{v}_a = u_1 \,, \tag{7}$$
$$\dot{\phi} = u_2 \,,$$
$$\dot{s}_{vtv} = u_3 \,.$$

We now formulate the path-following problem with respect to the new dynamical system.

Problem 1 (Path-Following). Consider the UAV model described by (7), and let $(\bar{x}_d(s), \bar{y}_d(s))$ be a given desired spatial path and $\bar{v}_{ad}(s)$ a given desired airspeed assignment, both parameterized by $s \in \mathbb{R}$. Design feedback control laws for the longitudinal acceleration u_1, the roll rate u_2, and the rate of progression u_3 of the VTV along the desired path such that the error coordinates, e_x and e_y, and the course angle error, $(\chi - \chi_d)$, converge to zero as $t \to \infty$, and the airspeed, v_a, converges to the desired one as $t \to \infty$.

3   MPC Path Following

In this section we describe the predictive path following control scheme used to solve the path following problem. The control technique used here is a sampled-data MPC architecture combined with an optimization-based path-following strategy. At every sampling time, new information on the system (i.e., wind) becomes available and a trajectory optimization technique computes the solution of a finite horizon optimal control problem. Nonlinear least square method for the optimization of trajectory functionals with constraints is well suited for addressing the Problem 1. We consider the following optimization problem

$$\min_{\mathbf{x}(\cdot),\mathbf{u}(\cdot)} \frac{1}{2} \int_{t_k}^{t_k+T} \left(\|\mathbf{x}(t) - \mathbf{x}_d(t)\|_Q^2 + \|\mathbf{u}(t) - \mathbf{u}_d(t)\|_R^2 \right) dt$$

$$+ \frac{1}{2} \|\mathbf{x}(t_k + T) - \mathbf{x}_d(t_k + T)\|_{P_1}^2 \qquad (8)$$

$$\text{subj. to (7),} \quad \textit{dynamics constraints}, \forall t \in [t_k, t_k + T],$$

$$\text{(4),} \quad \textit{state/input constraints}, \forall t \in [t_k, t_k + T],$$

where $(\mathbf{x}_d(\cdot), \mathbf{u}_d(\cdot))$ is a suitable desired curve, $T > 0$ is the prediction horizon, and Q, R and P_1 are positive definite weighting matrices. Terminal cost in (8) is used to achieve stability of the proposed MPC approach. In particular, for a suitable choice of the matrices Q, R, P_1, and the prediction horizon T, we can guarantee stability of the proposed MPC approach. We refer the reader to [13] for a more detailed discussion.

Here, we propose to solve the optimal control problem (8) using the PRojection Operator based Newton method for Trajectory Optimization (PRONTO), [9], combined with the barrier function relaxation developed in [10]. This is a direct method based for solving continuous time optimal control problems. A key feature is that it explicitly handles constraints in the state-input, thus exploring complex nonlinear behavior of the system. PRONTO, being an iterative descent method, guarantees the convergence to a local minimum of the optimal control problem (8). The solution is a trajectory (i.e., it satisfies the dynamics constraints (7) and the constraints (4)) that is close in the L_2 norm to the desired curve $(\mathbf{x}_d(\cdot), \mathbf{u}_d(\cdot))$. We set the desired curve as follows. The desired longitudinal and lateral error coordinates, e_{xd} and e_{yd}, are set to zero. The desired airspeed, v_{ad}, is assigned. We use the relation between the heading and the course angles given in (3) for setting the desired local heading angle, that is $\mu_d = -\arcsin\left(\frac{v_w}{v_{ad}} \sin(\psi_w - \chi_d)\right)$, where the desired course angle, χ_d, is assigned (it can be obtained from the desired path). It is worth noting that, at each sampling time, the wind components, v_w and ψ_w, are assumed to be known. The desired roll angle is obtained by setting $\dot{\mu} = 0$ in (7), that is $\phi_d = \arctan\left(\frac{v_{gd}^2 \sigma_d}{g}\right)$, where the desired ground speed is given by (2), i.e., $v_{gd}^2 = (v_{ad} \cos \psi_d + v_w \cos \psi_w)^2 + (v_{ad} \sin \psi_d + v_w \sin \psi_w)^2$, the desired heading angle is $\psi_d = \mu_d + \chi_d$, and the desired curvature, σ_d, is assigned (once again, the desired curvature can be obtained from the desired path). The desired velocity of the VTV, u_{3des}, is set to be equal to the ground sped v_{gd}. Finally, in order to take into account a penalty on the control effort, the desired longitudinal acceleration and roll rate, u_{1des} and u_{2des}, are set to zero. This aspect allows the UAV to reduce the power consumption, which, as highlighted in the introduction, is an important aspect in view of long endurance missions.

4 Numerical Computations

In this section we provide numerical computations showing the effectiveness of the proposed predictive path following control scheme. We start with the straight line maneuver with constant wind. This computation allows us to highlight some features of the optimization-based path-following strategy. Then, we consider the loiter maneuver in presence of time-variant wind. This scenario allows us to highlight the main advantage of the MPC approach.

4.1 Straight Line Maneuver

Similarly to the computations shown in [8] and in [14], we set as desired curve the straight line shown in Figure 4a, see dash-dot line. Given the initial position $(x, y) = (0, 100)$ and orientation $\psi = \frac{\pi}{2}$ of the actual vehicle, the goal is to follow the desired straight line path with a desired airspeed, $v_{ad} = 19$m/s, along it. We run the MPC strategy by setting the following weighting matrices $Q = diag([0.01, 0.01, 1.0, 0.1, 0.001, 1e-09])$, $R = diag([1.0, 0.1, 0.0001])$. We make use of a relatively short update time $(t_{k+1} - t_k) = 1$ second and a fixed prediction horizon time $T = 15$ seconds. The wind is assumed constant with $v_w = 5$m/s and $\psi_w = 135$deg.

Figure 4 show the trajectory of the UAV. It is worth noting that such a maneuver allows us to highlight an important aspect of the optimization-based path-following strategy and some features of the CFV model. First, we highlight that Problem 1 is solved. The error coordinates, Figure 4a, the course angle error, Figure 4b, converge to zero, and the airspeed converges to the desired one, see Figures 4c. Second, we observe that the UAV is "drifting" laterally, see the UAV orientation at the end of the maneuver in Figure 4a. This behavior is due to the presence of the (high) wind and it is a well-known phenomenon in aviation, the so-called crosswind landing. This (drifting) maneuver is performed when the aircraft is approaching the runway for landing and there is a significant component of the wind that is perpendicular to the runway.

4.2 Loiter Maneuver

As an application for surveillance/monitoring missions (main motivation of the paper), we provide numerical computations based on the loiter maneuver (i.e., loitering around a desired waypoint). We set as desired curve the circular path centered at the origin with a radius of 100m. Again, the desired airspeed is $v_{ad} = 19$m/s. The initial UAV position is $(x, y) = (0, 125)$, with course angle $\chi = \frac{\pi}{2}$, and airspeed $v_a = 19$m/s.

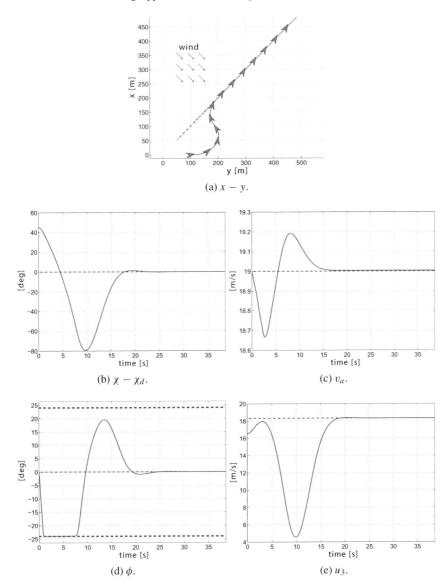

(a) $x - y$.

(b) $\chi - \chi_d$.

(c) v_a.

(d) ϕ.

(e) u_3.

Fig. 4 Straight line maneuver with constant wind field. The desired (dash-dot blue line) and the CFV (solid green line) trajectories are shown. Constraints in dash red line.

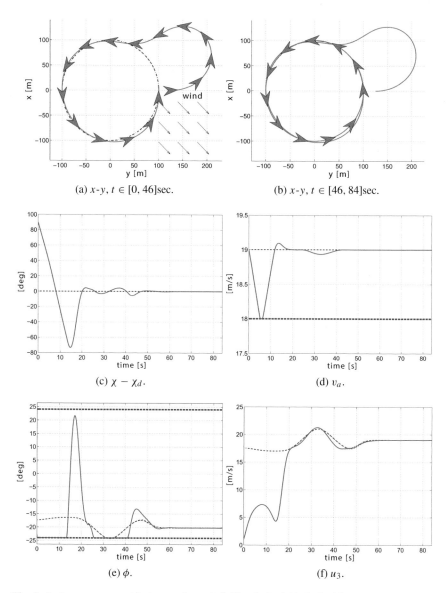

(a) x-y, $t \in [0, 46]$sec.

(b) x-y, $t \in [46, 84]$sec.

(c) $\chi - \chi_d$.

(d) v_a.

(e) ϕ.

(f) u_3.

Fig. 5 Loiter maneuver with time-variant wind. The desired (dash-dot blue line) and the CFV (solid green line) trajectories are shown. Constraints in dash red line.

The trajectory is shown in Figure 5. We assume time-variant wind. At the beginning of the maneuver, we consider $v_w = 2$m/s and $\psi_w = 135$deg. At time $t = 40$sec, the wind gradually decrease and becomes zero at $t = 60$sec. Next, we highlight two features of the proposed control approach. First, due to the initial condition of the actual vehicle, the VTV starts the maneuver with a low velocity, i.e., $u_3 = 1$m/s, see Figure 5f. Roughly speaking, the VTV is "waiting" the actual vehicle while it is performing an aggressive maneuver, i.e., it is performing a manevuer at the maximum of its capabilities. Thus, the roll angle and the airspeed are constrained, see Figures 5d and 5e. This is an important feature of the VTV approach that allows us to improve the convergence of the actual path to the desired one. Second, due to the wind, i.e., for $t \in [0, 40]$sec, the actual vehicle is not able to follow (exactly) the desired path, see Figure 5a. After a transient behavior, the error coordinates e_x and e_y do not converge to zero, yet oscillate around zero. Also, the airspeed is not constant as the desired one, Figure 5d. This behavior is due to the dynamics constraints and the wind. In fact, we highlight that the constraint on the roll angle is active at about $t \in [26, 41]$sec, see Figure 5e. On the other hand, thanks to the optimization-based strategy, the CFV is performing an ellipse path instead of the desired circular path (see Figure 5a), thus exploiting the wind. For $t \in [60, 84]$ the wind is zero, the roll angle is not constrained and the CFV is able to follow the desired path (see Figure 5b) with the desired airspeed along it, thus solving the path-following problem.

Finally, it is interesting to show the effect of the wind velocity on the path. In Figure 6, we show the path performed by the CFV for different magnitude of the wind. Here, for the clarity of the presentation, we consider the CFV initially following the maneuver with no error. For $v_w = 1$m/s, the desired maneuver is feasible and the CFV executes it accurately, see Figure 6a. For $v_w = 3$m/s and $v_w = 5$m/s, the desired maneuver is unfeasible (due to the dynamic constraints and the wind) and the CFV exploits the wind in order to both minimize the error coordinates and the control effort, see Figures 6b and 6c, respectively.

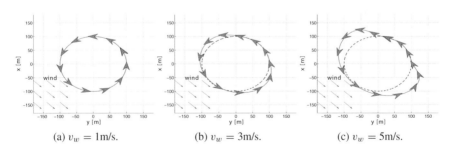

(a) $v_w = 1$m/s.　　　　(b) $v_w = 3$m/s.　　　　(c) $v_w = 5$m/s.

Fig. 6 Loiter maneuver for different magnitude of the wind.

5 Conclusions

In this paper, we proposed a predictive path following control scheme based on trajectory optimization techniques combined with a sample-data Model Predictive Control (MPC) approach for fixed-wing Unmanned Aerial Vehicles (UAVs). The wind is taken into account by exploiting the MPC approach. The control scheme is a versatile controller that responds to changes in the surrounding environment, thus extending the UAV capabilities in executing the desired maneuver. This can be seen as a preliminary step in the direction of the design of advanced control tools for improving UAV endurance. We provided numerical computations thus showing the effectiveness of the proposed control scheme. The numerical computations allowed us to capture important dynamic aspects of the vehicle as well as highlight the main features of the proposed control scheme.

References

1. Langelaan, J.W.: Tree-based trajectory planning to exploit atmospheric energy. In: American Control Conference, pp. 2328–2333 (2008)
2. Bencatel, R., Kabamba, P., Girard, A.: Perpetual dynamic soaring in linear wind shear. AIAA Journal of Guidance, Control, and Dynamics 37(5), 1712–1716 (2014)
3. Bird, J., Langelaan, J., Montella, C., Spletzer, J., Grenestedt, J.: Closing the loop in dynamic soaring. In: AIAA Guidance, Navigation, and Control Conference, pp. 2014–0263 (2014)
4. Nelson, D.R., Barber, D.B., McLain, T.W., Beard, R.W.: Vector field path following for miniature air vehicles. IEEE Transactions on Robotics 23(3), 519–529 (2007)
5. Beard, R., Ferrin, J., Humpherys, J.: Fixed wing UAV path following in wind with input constraints. IEEE Transactions on Control Systems Technology 22(6), 2103–2117 (2014)
6. McGee, T.G., Hedrick, J.K.: Path planning and control for multiple point surveillance by an unmanned aircraft in wind. In: American Control Conference, pp. 4261–4266 (2006)
7. Jackson, S., Tisdale, J., Kamgarpour, M., Basso, B., Hedrick, J.K.: Tracking controllers for small UAVs with wind disturbances: theory and flight results. In: IEEE Conf. on Decision and Control, pp. 564–569 (2008)
8. Rucco, A., Aguiar, A., Hauser, J.: Trajectory optimization for constrained UAVs: a virtual target vehicle approach. In: 2015 International Conference on Unmanned Aircraft Systems (ICUAS), pp. 236–245, June 2015
9. Hauser, J.: A projection operator approach to the optimization of trajectory functionals. In: IFAC World Congress, Barcelona (2002)
10. Hauser, J., Saccon, A.: A barrier function method for the optimization of trajectory. In: IEEE Conf. on Decision and Control, pp. 864–869 (2006)
11. Hauser, J., Hindman, R.: Aggressive flight maneuvers. In: IEEE Conf. on Decision and Control, pp. 4186–4191 (1997)
12. Prodan, I., Olaru, S., Bencatel, R., De Sousa, J.B., Stoica, C., Niculescu, S.-I.: Receding horizon flight control for trajectory tracking of autonomous aerial vehicles. Control Engineering Practice 21(10), 1334–1349 (2013)
13. Fontes, F.A.C.C., Vinter, R.B.: Nonlinear model predictive control: specifying rates of exponential stability without terminal state constraints. In: IFAC Meeting on Advanced Control of Chemical Processes, pp. 821–826 (2000)
14. Sujit, P., Saripalli, S., Sousa, J.: Unmanned aerial vehicle path following: A survey and analysis of algorithms for fixed-wing unmanned aerial vehicless. IEEE Control Systems Magazine 34(1), 42–59 (2014)

An Efficient Method for Multi-UAV Conflict Detection and Resolution Under Uncertainties

David Alejo, José Antonio Cobano, G. Heredia and A. Ollero

Abstract This paper presents a efficent conflict detection and resolution (CDR) method for an aerial vehicle sharing airspace with other aerial vehicles. It is based on a conflict detection (CD) algorithm (axis-aligned minimum bounding box) and conflict resolution (CR) algorithm (genetic algorithms) to find safe trajectories. Monte-Carlo estimation is used to evaluate the best predicted trajectories considering different sources of uncertainty such as the wind, the inaccuracies in the vehicle model and limitations of on-board sensors and control system. Simulations are performed in different scenarios and conditions of wind to test the method.

Keywords Path planning · Unmanned Aerial Vehicle · Uncertainty · Conflict Detection · Conflict Resolution

1 Introduction

Unmanned Aerial Vehicles (UAVs) have attracted a significant interest in a wide range of applications [1][2], like search and rescue, surveillance, fire monitoring, structure assembly, etc. Morevoer, the safety plays an important role in missions with multiple aerial vehicles. In these missions, the problem is to maintain as much as possible the trajectories planned to perform the missions, but also maintaining a minimum separation with other UAVs or obstacles in spite of uncertainties and unexpected perturbations. Therefore, a collision detection and resolution system (CDR) is needed in these systems.

D. Alejo(✉) · J.A. Cobano · G. Heredia · A. Ollero
Robotics, Vision and Control Group (University of Seville),
Camino de Los Descubrimientos s/n, 41092 Seville, Spain
e-mail: {dalejo,jcobano,guiller,aollero}@us.es
http://grvc.us.es

© Springer International Publishing Switzerland 2016   635
L.P. Reis et al. (eds.), *Robot 2015: Second Iberian Robotics Conference*,
Advances in Intelligent Systems and Computing 417,
DOI: 10.1007/978-3-319-27146-0_49

General concepts and methods on safe path planning can be applied in order to solve this problem. UAV trajectory planning algorithms and CDR methods have been studied extensively. A detailed survey on the former is presented in [3], and [4] reviews papers on the latter.

Several techniques have been proposed in the literature, including potential fields [5], evolutionary algorithms (EAs) [6], particle swarm optimization [7], graph search like A* and D* [8] and Rapidly-exploring Random Trees (RRT) [9]. The general problem of path planning considering moving obstacles is NP-hard [10]. Some differential constraints given by the model of the aircraft should be considered to make the generated paths feasible. Sampling-based techniques, as opposed to combinatorial planning, are usually preferred in these NP-hard problems. These planning schemes match particularly well when the solution space is hard to model or unknown a priori because of its dynamic nature. The application of EAs is an efficient and effective alternative for this problem [9].

However, most of the papers do not consider sources of uncertainty to compute the trajectories, which could be very important in small UAVs. A path planning method for mobile robots taking into account the uncertainties in the estimation of the robot position by means of Partially Observable Markov Decision Process (POMDPs) is presented in [11]. However, the existence of other robots in the environment is not studied. In this paper, an algorithm based on the Monte-Carlo method is used to predict UAV trajectories and compute the uncertainty of the trajectories of each UAV, considering the atmospheric conditions, the UAV model used for prediction, and the limitations of the sensors.

This paper proposes an efficient method to plan safe paths. The planning algorithm is based on a genetic algorithm. Furthermore, the proposed method considers the uncertainty of each trajectory computed using Monte-Carlo methods. The information obtained from this method is used to evaluate which one is the best collision-free trajectory. Note that when an UAV detects a collision it changes its trajectory maintaining its cruise speed to solve the collision without cooperating with the rest of UAVs.

The method is based on the one proposed in [12]. The main contributions of the paper are related to the following improvements: change of the conflict detection algorithm to reduce the computational load, an anytime approach is considered, the execution time is bounded and the number of Monte-Carlo simulations is reduced. All these improvements influence on the execution time of the method by considerably reducing it with respect to [12].

The paper is organized in five sections. Section 2 states the problem definition. Sections 3 and 4 describe the proposed methods. Section 5 shows the simulations carried out and Section 6 details the conclusions.

2 Problem Formulation

The problem considered is the collision-free path planning for an UAV entering an airspace with other UAVs and static obstacles or forbidden regions. Each UAV has contracted a collision-free trajectory and when another UAV comes in this airspace with an initial flight plan, a potential collision may be detected. The rest of UAVs and static obstacles can be a threat to the UAV so a technique that ensures a minimum separation between the UAV and them is necessary.

Some assumptions are introduced in order to simplify the problem. First, the solution will be obtained by adding intermediate waypoints to the initial flight plan. The path will be defined by a sequence of waypoints. Second, it is also assumed that velocity changes are not allowed, so UAVs move at cruise speed in the CR maneuver. Last, altitude level changes are neither allowed, resulting in planar motion. Constant altitude is often determined by mission constraints, such as sensor resolution or radar visibility.

The information that the UAV needs in order to solve the problem is the following: location of the static obstacles and radii, flight plan of the rest of UAVs, parameters of the model of each UAV, wind model parameters, and initial path of the UAV considered. The initial and solution path should have the same starting and goal points. Therefore the proposed method can be considered as a centralized algorithm. In order to be implemented in a decentralized fashion, two considerations have to be taken into account. First, at least the position of other UAVs should be known by each nearby UAV in order to execute the algorithm, if the flight plans are not available, the nearby UAV could be assumed as performing a stationary flight in the near future. Also, few indications about the flight intent of the UAVs could be interchanged. Second, due to the stochastic nature of the proposed algorithm, the solutions obtained by each UAV should be interchanged in order to execute the same CR maneuver. A usual approach is to execute the CR maneuver with lowest cost.

The uncertainty is taken into account by means of Monte-Carlo simulations that lead to a statistical representation of the state of each UAV. This statistical representation is used to predict the probability of collision.

3 Safe Path Planning Method

The proposed planning method for CDR is split into two blocks: the detection algorithm (CD) and the conflict resolution (CR) method based on GA.

3.1 Conflict Detection Algorithm

The CD algorithm is based on axis-aligned minimum bounding box. This technique has a low execution time and needs few parameters to describe the system.

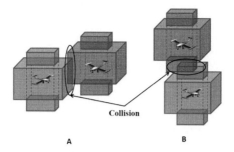

Fig. 1 Detection algorithm.

A security envelope is defined around each UAV in order to avoid collisions between them. Each UAV security envelope is approximated by two boxes joined together: a horizontal box and another vertical box that covers the aerial robot (see Figure 1). Each box is defined by the intersection of three intervals, one by axis. The minimum separation, S, can be calculated as the half sum of the length of the boxes. A collision is detected when two boxes overlap, which is produced when the intersection between each interval is not the empty set. Thus, the 3D problem is reduced to three problems of interval overlapping, one for each coordinate axis. Let us consider the intervals in one coordinate $A = [A_i, A_e]$ and $B = [B_i, B_e]$. The condition of overlapping for this coordinate is given by:

$$(A_e > B_i) \wedge (A_i < B_e) \tag{1}$$

3.2 Evolutionary Algorithm

EAs include a huge set of general purpose optimization algorithms. This set includes the Genetic Algorithm (GA), Genetic Programming (GP), Evolutionary Programming (EP) and many more. All of these algorithms share in common that are biologically inspired; i.e. they use a strategy that resembles one approach found in the nature such as evolution and natural selection.

GA is a global optimization technique that, under the proper conditions, is able to converge to the global optimal of a criteria function under arbitrary constraints [13]. The proposed approach in this paper is to generate a sequence of waypoints that define a path to solve a conflict in such a way that the generated plans are safe, i.e. no collisions can be produced during the flight.

Figure 2 represents a basic GA flow diagram. The individuals represent a waypoint sequence, i. e. a possible path for the UAV to follow. Firstly, an initial population is generated. This is done by sampling the search space. In particular, a uniform distribution has been selected in order to sample the search space.

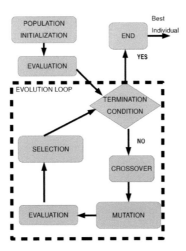

Fig. 2 Basic flow diagram of the Genetic Algorithm (GA).

Once the initial population has been generated, the following step is to evaluate the fitness of each candidate with the cost function proposed in Eq. 2.

$$J = \omega_1 L_i + \omega_2 S^j + \omega_3 \sum_{i=1}^{M} \theta_i + \omega_4 C \tag{2}$$

where L_i is the length of the ith path, S^j is the separation between the ith path and the jth static obstacle, and C is a variable that is set to 1 if a collision occurs with another UAV or the static obstacles; or to 0 if no collision takes place. $\omega_i \forall 1 \le i \le 4$ are the weights. A collision with a static obstacle or another UAV occurs when d_j is less than the required safety distance, d_{req}, where d_j is the separation between an UAV and the jth static obstacle. θ_i is the turn radius associated to waypoint i^{th}.

The crossover and mutation operators are considered in the GA. By iteration of the selection and reproduction processes, the GA ends up computing a near-optimal trajectory path that ensures a collision-free trajectory path.

In the simulations, the open source galib toolbox[1] has been used for performing GA optimization.

4 Uncertainty Considerations

The initial path of each UAV is assumed as known. In particular, this paper considers that each UAV have to fly to different Points of Interest (PoIs) for monitoring

---

[1] GAlib: A C++ Library of Genetic Algorithm Components. http://lancet.mit.edu/ga/

purposes [14]. Then, an algorithm based on the Monte-Carlo method is used to compute the uncertainty of the trajectories of each UAV, considering the atmospheric conditions, the UAV model used for prediction, and the limitations of the sensors and on board control system. In case of small UAVs, the most important source of uncertainty during the flight is the change in the atmospheric conditions, mainly the wind.

Figure 3 illustrates the proposed safe CDR method with uncertainty considerations. It starts with an evaluation of the current trajectories simulated by Monte-Carlo checking for possible collisions. If a collision is detected from the predicted trajectories, an iterative CDR planning loop starts. The planning loop involves the generation of a new path using GA and a later evaluation of the path by using the Monte-Carlo method. Note that the safety distance σ_{req} will depend on the stage of the algorithm. First, a worst case is solved in order to achieve a fast solution (Eq. 3). If a solution is found, a narrower distance is proposed in Eq. 4. If the solution proposed by GA is found to be conflictive by the CD method, σ_{req} goes back to Eq. 3.

$$\sigma_{req} = \sigma_{max,1}\sigma_{max,coll} + \sigma_{safe} \tag{3}$$

$$\sigma_{req} = \sigma_{max,1} + \sigma_{safe} \tag{4}$$

where $\sigma_{max,1}$ is the maximum obtained deviation of the controlled UAV estimated by the CD block; $\sigma_{max,coll}$ is the max. deviation of the UAV that is in conflict with the controlled UAV; and σ_{safe} is the minimum safety distance between UAVs.

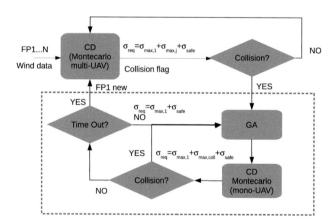

Fig. 3 Flow diagram of the proposed CDR method with uncertainty considerations.

4.1 Monte-Carlo Method

The Monte-Carlo method that has been implemented is detailed in Algorithm 1. It starts by generating an initial set of possible states of the UAV. Then, a simulation is started for each possible state using the stochastic model detailed in Section 4.2. This procedure is then repeated for each UAV.

Algorithm 1. Monte-Carlo method

Generate a set of possible initial states $S = x_i(0) \forall i \leq n$ of size n according to the estimated state and its uncertainty

$d_{max} \leftarrow 0$

for $j = 1$ to prediction horizon **do**

  for $i = 1$ to n **do**

    $x_i(j) \leftarrow model(x_i(j-1))$

  end for

  Evaluate the mean state $\overline{x(j)}$

  Evaluate the maximum deviation to the mean state $d_{max}(j)$

  if $d_{max}(j) > d_{max}$ **then**

    $d_{max} \leftarrow d_{max}(j)$

  end if

end for

Thus, the sources of uncertainty are considered to evaluate the trajectory. In the next iterations, the maximum deviation obtained in this stage for all UAVs is used in other stages for collision detection.

4.2 Model

The application of the Monte-Carlo method requires a stochastic model of both the UAV and the atmosphere. In order to decrease the computation load for real time implementation and to carry out studies, a simple kinematic UAV model based on the unicycle vehicle has been used in this paper:

$$\dot{x} = v_i cos(\psi) + \omega_\rho cos(\omega_\phi) \tag{5}$$

$$\dot{y} = v_i sin(\psi) + \omega_\rho sin(\omega_\phi) \tag{6}$$

$$\dot{\psi} = \alpha_\psi (\psi^c - \psi) \tag{7}$$

where (x, y) represent the 2D coordinates, v_i is the cruise speed of the vehicle and ψ is the heading of the UAV. α_ψ is a parameter that depends on the characteristics of the UAV. ψ^c is the heading reference to the control system. The following constraint are used:

$$-\dot{\psi}_{max} \leq \dot{\psi} \leq \dot{\psi}_{max} \tag{8}$$

where ψ_{max} is a positive constant that depends on both the dynamics of the UAV and the behavior of the path controller.

The atmospheric model includes the wind vector speed modulus, ω_ρ, and direction, ω_ψ. The wind can be modeled without loss of generality as a Normal distribution of mean and standard deviation $\overline{\omega_\phi}$ and ω_ψ^σ, respectively. On the other hand, wind speed distribution is known to fit well with a Weibull distribution at low altitudes [15]:

$$f(x; k, c) = \frac{k}{c} \left(\frac{x}{c}\right)^{k-1} e^{-(x/c)^k} \forall x \geq 0 \qquad (9)$$

Even though the Weibull Distribution is usually determined by the shape factor, k, and the scale factor, c. Nevertheless, the wind speed mean $\overline{\omega_\rho}$ and standard deviation ω_ρ^σ will be shown for clarity in Section 5, as they are more understandable parameters than k and c. An approximate relationship between the mean and standard deviation of the velocity of the wind can be obtained by using the empirical method detailed in [16]:

$$k = \left(\frac{\omega_\rho^\sigma}{\overline{\omega_\rho}}\right)^{-1.086} \qquad (10)$$

$$c = \frac{\overline{\omega_\rho}}{\Gamma\left(1 + \frac{1}{k}\right)} \qquad (11)$$

where $\Gamma(x)$ is the Gamma function.

Due to the stochastic nature of this model, each specific simulation is affected by different wind disturbances, i.e. different samples of the direction and speed distributions above will replace ω_ρ and ω_ϕ in Eq. 5 and 6.

Eqs. 5-7 model the behavior of the UAV taking into account the heading references and the uncertainty of the wind. The parameters of the model can be estimated using real flight data. The full model takes into account not only the above parameters and equations but also high level control considerations about the waypoint tracking control that will generate the reference ψ^c.

5   Simulations

The algorithms have been run in a PC with a CPU Intel Core i7-3770@3.4GHz equipped with 16 GB of RAM. Kubuntu OS 14.04. The code has been written in the C++ language and compiled with gcc-4.8.2.

Different scenarios with up to ten UAVs (UAV1-UAV10) and up to eight obstacles have been considered. The same scenarios as in [12] have been considered to better analyze the improvements obtained in this work. The scenarios are defined from the UAVs and obstacles shown in Figure 4: Scenario 1 (S1) with UAV1-UAV5 and OBS1;and, Scenario 2 (S2): All UAVs and all obstacles.

The number of simulations in each analysis is fifteen. The default data associated to the wind is: $\overline{\omega_\rho} = 6m/s$, $\omega_\rho^\sigma = 1m/s$, $\overline{\omega_\phi} = 0rad$, $\omega_\phi^\sigma = 0.5rad$.

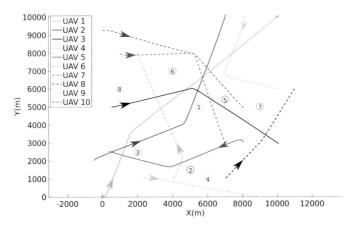

Fig. 4 Considered scenario: UAV 1 is commanded to travel from (0,0) to (10000,10000). Static obstacles are represented in circles. An example of solution generated by the proposed method is shown in green continuous line.

5.1 Dependency of the Execution Time with the Number of UAVs and Obstacles

Figure 5 shows the dependency of the mean execution time with the number of UAVs and obstacles. The considered scenario is S2.

Note that the mean execution time increases almost linearly with the number of UAVs. On the other hand, the number of obstacles practically has not effect on the execution time. It is important to highlight that the times are reduced in approximately 130s in the case of obstacles over the results obtained in [12], and thus the execution time of the proposed method spends only 10% in comparison. In spite of executing the algorithm in a more powerful computer, most of the improvements regarding to execution time are achieved by the modifications in the proposed method. In addition, the execution time increases at a lower rate with the number of UAVs, when compared with the method proposed in [12].

5.2 Different Wind Conditions

The execution time and goodness of the obtained solution when the wind conditions vary ($\overline{\omega_\rho}$ and $\overline{\omega_\phi}$) in S1 are also analyzed. First, Figure 6 shows the mean execution time of the algorithm with different values of $\overline{\omega_\rho}$. Note that this time increases as $\overline{\omega_\rho}$ increases. However, differences are barely noticeable as the execution time is inside the range [16.1, 16.9]s. The times are reduced on 120s.

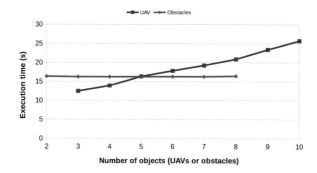

Fig. 5 Mean execution time of the algorithm when varying the number of UAVs and number of obstacles in S2.

Second, Figure 7 represents the mean execution time of the algorithm with different values of $\overline{\omega_\phi}$. Note that, in general, the execution time is almost independent of the wind direction. In fact, all the results are confined into the $[15.1, 16.9]s$ range. As a remark, the execution time is slightly higher in two angles: $\overline{\omega_\rho} = -45, 135deg$). It might be produced because of the difficulty to perform collision avoidance in these particular cases. In any case, the differences are not noticeable and a reduction of $180s$ is found in most cases.

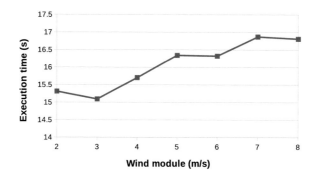

Fig. 6 Mean execution time of the algorithm when varying the wind module in S1.

5.3 Execution Time Distribution

The results regarding the execution time presented so far do not distinguish the stage of the algorithm in which the time is being spent. Therefore, the efforts of both CD and CR blocks have been taking into consideration.

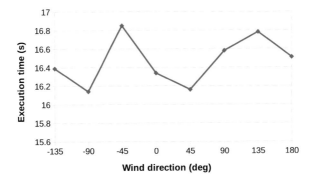

Fig. 7 Mean execution time of the algorithm when varying wind direction in S1.

In the next study, the execution times obtained in S2 with default wind conditions at each stage of the algorithm are separated in order to provide an detailed analysis of the complexity of each stage of the method.

Figure 8 represents the mean timeline which specifies the sequence in which the algorithms are executed and the duration of each stage. These results have been obtained by taking the mean execution time obtained in ten executions of S2 with obstacles OBS1 and OBS2 for each given number of UAVs in the system. The wind conditions are the default ones.

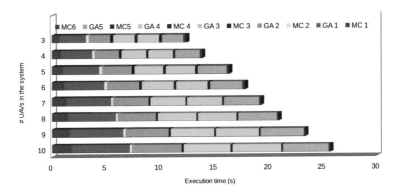

Fig. 8 Sequence of execution of the proposed method with varying number of UAVs.

Note that the first CD call is much longer than the subsequent calls and its execution time depends on the number of UAVs in the system in a great deal. In contrast, the next calls to the Monte-Carlo algorithm are much shorter and their execution time do not depend on the number of UAVs. This makes sense taking into account that not all trajectories are modified. Therefore, the whole system has only to be simulated in the first call. In the rest of CD calls only the trajectory of UAV 1 has to be calculated.

Regarding to the CR block, it is executed 5 times, performing 6 GA iterations each time to achieve an acceptable solution (please refer to [12]). The execution time of this block in the first call is noticeable higher than in the next calls. This may be produced by the amount of time performing the GA initialization with the data associated to the problem, memory management and the population initialization. The GA execution time remains constant in the subsequent calls.

6 Conclusions

In this paper, a novel and efficient method for safe CDR under uncertainties have been proposed. The method is a non-collaborative conflict detection and resolution method that takes into account the uncertainties associated with the estimation of the trajectories of UAVs in outdoors. Several improvements have been performed with respect to [12]. Extensive simulation results have been obtained that demonstrate the reduction of the execution time.

Future work will include real-time CDR experiments in outdoor environments and the paralelization of the algorithms for real-time applications.

Acknowledgements This work was partially supported by the European Commission Horizon 2020 Programme under the MARINEUAS (SI-1378/2015) project and FP7 ICT Programme under the EC-SAFEMOBIL (288082) project, and the AEROMAIN project funded by the Ministerio de Economa y Competitividad of the Spanish Government (DPI2014-59383-C2-1-R). Autors would like to thank Roberto Conde for his contribution in the original method.

References

1. Ollero, A.: Aerial robotics cooperative assembly system (ARCAS): first results. In: Aerial Physically Acting Robots (AIRPHARO) Workshop, IROS 2012, Vilamoura, Portugal, October 7–12, 2012
2. Beard, R.W., McLain, T., Nelson, D., Kingston, D., Johanson, D.: Decentralized cooperative aerial surveillance using fixed-wing miniature UAVs. Proceedings of the IEEE **94**(7), 1306–1324 (2006)
3. Goerzen, C., Kong, Z., Mettler, B.: A Survey of Motion Planning Algorithms from the Perspective of Autonomous UAV Guidance. Journal of Intelligent Robot Systems **57**, 65–100 (2010)
4. Kuchar, J.K., Yang, L.C.: A review of conflict detection and resolution modeling methods. IEEE Transactions on Intelligent Transportation Systems **1**, 179–189 (2000)
5. Khatib, O.: Real-Time Obstacle Avoidance for Manipulator and Mobile Robots. International Journal of Robotics Research **5**(1), 90–98 (1986)
6. Conde, R., Alejo, D., Cobano, J.A., Viguria, A., Ollero, A.: Conflict detection and resolution method for cooperating unmanned aerial vehicles. Journal of Intelligent and Robotic Systems **65**, 495–505 (2012)
7. Alejo, D., Cobano, J., Heredia, G., Ollero, A.: Conflict-free 4D trajectory planning in unmanned aerial vehicles for assembly and structure construction. Journal of Intelligent and Robotic Systems **73**, 783–795 (2014)

8. Stentz, A.: Optimal and efficient path planning for partially-known environments. In: Proceeding of the International Conference on Robotics and Automation, Los Alamitos, CA, vol. 4, pp. 172–179 (2002)
9. Kuffner, J., LaValle, S.: RRT-connect: an efficient approach to single-query path planning. In: IEEE International Conference on Proceedings of the Robotics and Automation, ICRA 2000, vol. 2, pp. 995–1001 (2000)
10. Szczerba, R.J.: Threat neeting for real-time, intelligent route planners. In: Proc. of the IEEE Symposium Information, Decision and Control, Adelaide, Australia, pp. 377–382 (1999)
11. Roy, N., Gordon, G., Thrun, S.: Planning under uncertainty for reliable health care robotics. In: The 4th International Conference on Field and Service Robotics, July 14–16, 2003
12. Cobano, J.A., Conde, R., Alejo, D., Ollero, A.: Path planning based on genetic algorithms and the monte-carlo method to avoid aerial vehicle collisions under uncertainties. In: Proceedings in the IEEE International Conference on Robotics and Automation, Shanghai (China), pp. 4429–4434, May 9–13, 2011
13. Goldberg, D.E.: Genetic Algorithms in Search, Optimization and Machine Learning, 1st edn. Addison-Wesley Longman Publishing Co., Inc., Boston (1989)
14. Alejo, D., Cobano, J.A., Heredia, G., Martnez de Dios, J.R., Ollero, A.: Efficient trajectory planning for WSN data collection with multiple UAVs in cooperative robots and sensor networks. Studies in Computational Intelligence. Springer International Publishing Switzerland, pp. 53–75 (2015)
15. Lun, I.Y., Lam, J.C.: A study of weibull parameters using long-term wind observations. Renewable Energy **20**(2), 145–153 (2000)
16. Indhumathy, D., Seshaiah, C., Sukkiramathi, K.: Estimation of weibull parameters for wind speed calculation at Kanyakumari in India. International Journal of Innovative Research in Science, Engineering and Technology **3**, 8340–8345 (2014)

Part V
Communication Aware Robotics

Network Interference on Cooperative Mobile Robots Consensus

Daniel Ramos, Luis Oliveira, Luis Almeida and Ubirajara Moreno

Abstract In this work we present the integration between a robot cooperative control strategy and a wireless network simulated with OMNeT++. We use a consensus control strategy to carry out a rendez-vous task where information is shared among a group of robots. These robots are then simulated in a MANET environment with a TDMA-based protocol to minimize message collisions. We consider two cases in this work: a fixed pre-determined topology, which does not accept new links, and a dynamic topology that creates new links as robots get within communication range. We show the impact of the network on the control strategy performing a rendez-vous task, considering both topologies. In particular, there is a considerable degradation of the rendez-vous task if care is not taken when deploying the cooperative control strategy, e.g. the initial message collisions due to desynchronized slot start. Finally, we compare these simulation results with those from a Matlab implementation of the control strategy using a typical simplified network model. The difference reveals the importance of using more accurate network models such as those of OMNeT++.

Keywords Mobile robots · Decentralized control · Cooperative strategy · Network protocols · Topology control · MANET

1 Introduction

Robotic tasks such as coverage, exploration, and cooperative transport, can take advantage of a group of cooperative mobile robots to achieve a better overall performance. In these scenarios, a fully distributed solution is frequently desired,

L. Oliveira · L. Almeida
Instituto de Telecomunicações/FEUP, University of Porto, Porto, Portugal
email: luisfnqoliveira@gmail.com, lda@fe.up.pt

D. Ramos(✉) · U. Moreno
DAS, Universidade Federal de Santa Catarina, Florianopolis, Brazil
email: daniel8484@gmail.com, ubirajara.f.moreno@ufsc.br

© Springer International Publishing Switzerland 2016
L.P. Reis et al. (eds.), *Robot 2015: Second Iberian Robotics Conference*,
Advances in Intelligent Systems and Computing 417,
DOI: 10.1007/978-3-319-27146-0_50

651

consequently a decentralized decision architecture is needed and inter-robot communications play a vital role [1]. Ideally, the communication among the robots is bandwidth efficient, has low latency and has no obstructions [2]. However, this does not apply to real networks, especially in the case of Mobile Ad-hoc NETworks (MANET), which have limited bandwidth, propagation delays, access collisions, and limited range [3].

There are several approaches to deal with non-reliable communication in cooperative robotics: avoiding explicit communication completely (i.e. using local sensors, only), implementing a fault tolerant control strategy on each robot, implementing a reliable network protocol to avoid message collision or yet, simply ignoring all communication faults by considering an ideal communication medium. In this work we consider real explicit communication thus discarding the first and last approaches.

Nevertheless, there are solutions for network faults in the remaining approaches. The control approach makes the robot navigation robust enough to tolerate transient lack of information due to errors. In [9], the author developed a consensus strategy integrated with a predictive algorithm, which were capable of recovering from occasional network faults. In a similar way, the work in [11] uses a predictive adaptive control to minimize the effects of channel faults for wireless sensors and actuators.

In the network protocol approach, the solutions seek to solve the problem in the network layer through transmission control to avoid message collisions. In [12], the authors analyze the information flow and the impact of message exchange over a network for a group of robots. In [2] the authors propose a time division multiple access (TDMA) protocol with loose synchronization for MANET that does not need clock synchronization. Other recent work concludes that not only the network and its protocols have significant impact on a robot task, but also that the network cannot be simply modelled as percentage of faults [6].

In Robotics, many works still consider an ideal communication medium between the robots [6] or a few occasional faults, only. However, some initiatives already oppose to this trend, studying the network in a robotic context integrating robot models in network simulators. For example, the integration between NS2 and Arena Robotic Simulator [13], NS2 and Matlab [14], NS2 and ARGOS [6]. Despite showing the impact of the network in simple robots and tasks, these works do not evaluate the impact in complex control models.

Recent surveys on network simulators [20][21] show little performance difference among them and that most can be used for MANET simulations with good accuracy. NS3 has its merits as consuming less CPU and less memory, but OMNET++ has a better graphical user interface and thus, allowing a better debugging than the others. However, OMNET++ is not a network simulator, but a framework that gives the tools to make a network simulation. This framework also allows implementing non-network related algorithms as the robots movement and cooperative control strategy.

OMNET++ has received special attention of researchers and has been extensively used for several applications. It was integrated with Matlab [15] to simulate large wireless sensor networks (WSN) [16], to evaluate energy consumption in WSN [17] and to test new network synchronization protocols for robots and wireless sensors [18]. These works show that OMNET++ is a resourceful tool for different applications and through add-ons such as the INET framework, it has a suitable representation of real networks.

One aspect that deserves particular attention for our proposes is the network topology. In fact, the robots are constantly moving and creating a dynamic topology that adds or removes links according to the communication range. Moreover, from a control point of view, it may be interesting to use just a subset of the available links, for example, to limit propagation of stale information or to free bandwidth. In this case, OMNeT++ is also an adequate tool since it can integrate a simulation of the actual robots navigation, thus generating dynamic topologies, making possible to evaluate complex behaviors, an evaluation that would not be possible otherwise.

In this work, we evaluate a consensus control strategy combined with predictive control (RHC - Receding Horizon Control) for a group of cooperative mobile robots communicating with a MANET protocol based on time division multiple access (TDMA) with loose synchronization [8]. This MANET protocol remains stable even with dynamic topologies and automatically reconfigures in case of network separation in sub-networks. The objective of our work is to show the impact of a network on the control strategy performing a rendez-vous task, comparing with a typical simplistic communication model. We also show that it is possible to improve the results and reduce the impact of the network if a correct set of techniques and protocols are chosen together. In the next section, we detail the network and protocols used in this work, followed by a brief explanation of the RHC consensus control strategy. The simulations are detailed in section 4 followed by the results in section 5 and the conclusion in section 6.

2 Mobile Ad-Hoc Network

A MANET is a wireless mobile local area network that does not rely on a base station to coordinate messages exchanges in the network, as the nodes forward packets to and from each other on their own.

In this work, we are using the IEEE 802.11b standard, which is part of the 802.11 series of WLAN standards from IEEE. This standard uses CSMA/CA (carrier sense multiple access with collision avoidance) for medium access control. Although the "b" series was replaced for faster 802.11 models like the "g", we considered it as worst-case scenario in a wireless network, with maximum bandwidth of 11Mb/s. Devices using 802.11b experience interference from other products operating in the 2.4 GHz band, which includes microwave ovens, Bluetooth and ZigBee devices and cordless telephones. For the higher network layers we use the UDP protocol on transport layer and IP protocol model for network layer.

In our work, we consider that there is an overlay transmission control layer that synchronizes the robots so that they access the network essentially in exclusive mode within the team, thus avoiding collisions among team members. We call this protocol RATDMA for its clock-free reconfigurable and adaptive synchronization.

2.1 RATDMA

Reconfigurable and Adaptive TDMA [2] is a technique that achieves synchronization using message exchange. When a robot receives a message from its neighbors, it estimates the delay of that message and uses that information to delay its next transmission slot. This corresponds to a phase shift of the TDMA round that tends to reduce the impact of persisting periodic interference while maximizing the time between transmissions of the robots in the team.

With this synchronization scheme, we avoid the need for clock synchronization but the TDMA round effective duration will vary. Thus, we call it a loose synchronization. For example, consider a team of three robots (Fig. 1) and a delay that affects robot one. In this situation, robot two will correct this delay in the next round and the robot three in second round after. Thus, two rounds after the initial delay that affected robot 1, the team will be synchronized again with a correspondingly delayed round, which corresponds to a phase shift of the cycle phase.

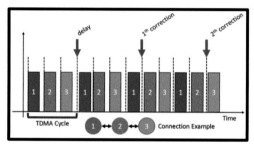

Fig. 1 RATDMA example for three robots and one delay affecting robot one. The delay is corrected by the neighbor robot on the subsequent cycle and by the third on the second cycle.

3   RHC Consensus for Rendez-vous

To ensure the cooperation between the robots, a decentralized coordination method is necessary which in this work consists in a consensus technique combined with a Receding Horizon Control (RHC), developed in [9].

The consensus problem assumes an explicit communication between the robots, in which each robot disseminates some of its local information to its neighbors. Each robot makes decisions individually, using the neighbors information and its decisions must be coherent with the rest of the group.

The communication structure can be modelled by an oriented graph $G = (V, A, w)$, where V is a set of nodes, A is the set of links between the nodes and w the weight of each link. The nodes represent the robots and the links represent the physical communication links between the robots. The set A can be presented in a matricial form denoted Adjacency Matrix, with elements $a_{ij} = w_{ij}$ when there is a communication link between the nodes i and j, and $a_{ij} = 0$ when there is no communication.

To accomplish the rendez-vous, all the robots must agree on a meeting point. Consequently, the robots must exchange some information, namely their current position.

3.1 RHC Consensus Formulation

Considering x_i the pose vector of the current robot, x_j the received pose from its neighbor j and x_r the received reference, a function cost (Eq. 1) can be formulated using the adjacency matrix elements a_{ij} and an additional element indicating if the robot receives the reference directly, a_{ir}.

$$\dot{x}_i[k] = -\sum_{j=1}^{N} a_{ij}(x_i - x_j) - a_{ir}(x_i - x_r) \tag{1}$$

Equation 1 can be modified to a quadratic form, obtaining an initial cost function J with three terms (Eq. 2), which uses the notation $\|x\|_Q^2 \equiv x^T \cdot Q \cdot x$, the element γ_i is a ponderation factor for the robot velocity variation Δv_i and p the prediction steps.

$$J_i = \sum_{j=1}^{N}\sum_{k=1}^{p} \|x_i[k] - x_j[k]\|_{a_{ij}}^2 + \sum_{k=1}^{p} \|x_i[k] - x_r[k]\|_{a_{ir}}^2 + \sum_{k=1}^{p} \|\Delta v_i[k]\|_{\gamma_i}^2 \tag{2}$$

In this formulation, the first term represents the rendez-vous consensus, which makes the robots agree about a meeting point. The second term is the reference tracking -when present- and the third term penalizes the velocity variation, making trajectories smoother.

The results of the optimization are the control signals (velocity in x and y coordinates), which are utilized in a first order dynamic equation to determine the movement of the robot (Eq. 3), where h is the temporal step.

$$x_i[k] = x_i[k-1] + v_i[k]h \tag{3}$$

As we use a decentralized approach, the robots do not know the control signals of their neighbors during the prediction part and thus, an approximation was made by using $v_j[k] \approx v_j[0]$ and $x_j[k] \approx x_j[0]$, $\forall k = 1, ..., p$, which basically corresponds to considering the position and velocity of the neighboring robots constant during the prediction steps.

3.2 Matricial Form

As shown in [9], Eq. 3 can be rewritten in matricial form and by transforming it in terms of a quadratic function of the form $\frac{1}{2}x^T Q x + c^T x$, we obtain Eqs. 4, 5 and 6.

$$J_i = \frac{1}{2}V_i^T H_i V_i + f_i V_i \tag{4}$$

$$H_i = \sum_{\substack{j=1 \\ j \neq i}}^{n} T^T a_{ij} T + T^T a_{ir} T + T'^T \gamma_i T' \tag{5}$$

$$f_i = \sum_{\substack{j=1 \\ j \neq i}}^{n} (X_{i,0}^T - X_{j,0}^T) a_{ij} T + (X_{i,0}^T - X_{r,0}^T) a_{ir} T - V_{i,0}^T \gamma_i T' \tag{6}$$

Note that T is an $p \times p$ lower triangular matrix composed by h. Using these formulations, the optimization problem can be stated as Eq. (7) to find the optimum value in each iteration in each robot, where v_i^x and v_i^y are the velocities in x and y coordinates and $v_{\max f}, v_{\max b}$ are the maximum saturation velocities (forward and backward) of the robots. The problem is solved by using specific libraries, described in section 4.

$$\min \frac{1}{2}[V_i^x \quad V_i^y] \begin{bmatrix} H_i & 0 \\ 0 & H_i \end{bmatrix} \begin{bmatrix} V_i^x \\ V_i^y \end{bmatrix} + [f_i^x \quad f_i^y] \begin{bmatrix} V_i^x \\ V_i^y \end{bmatrix}$$
$$s.t. \quad v_{\max b}^x \leq v_i^x \leq v_{\max f}^x$$
$$v_{\max b}^y \leq v_i^y \leq v_{\max f}^y \tag{7}$$

3.3 Distance Constraints

In this work, we go beyond the work in [9] summarized in the previous section, by also considering that the distance between robots is constrained. This is needed to enforce a minimum distance between neighbors thus avoiding collisions and a maximum distance within the communication range, thus avoiding breaking communication links during the rendez-vous task. This is simply described in Eq. 8, where $r_{ij,min}$ is the minimum distance, $||d_{ij}||$ is the distance between two robots and $r_{ij,max}$ the maximum distance.

$$r_{ij,min} \leq ||d_{ij}|| \leq r_{ij,max} \tag{8}$$

Similarly to the matricial formulation in section 3.2, there is a need to approximate $x_{j,k}$ during the k prediction steps and we also opted to consider it constant and equal to $\tilde{x}_{j,k} = x_{j,0}$ and thus resulting in $\tilde{d}_{ij} = \tilde{x}_{j,k} - x_{i,0}$. The distance predicted by robot i must respect these limits and use a simple scalar projection displacement $\Delta x_{i,k} = h \cdot v_{i,1}$, thus obtaining the final formulation as Eq. 9 for minimum distance and Eq. 10 for maximum distance.

$$-\frac{h}{||\tilde{d}_{ij}||}[\tilde{d}_{ij}^x \quad \tilde{d}_{ij}^y] \begin{bmatrix} v_{i,1}^x \\ v_{i,1}^y \end{bmatrix} + ||\tilde{d}_{ij}|| - r_{i,j\,min} \geq 0 \tag{9}$$

$$\frac{h}{||\tilde{d}_{ij}||}[\tilde{d}_{ij}^x \quad \tilde{d}_{ij}^y] \begin{bmatrix} v_{i,1}^x \\ v_{i,1}^y \end{bmatrix} - ||\tilde{d}_{ij}|| + r_{i,j\,max} \geq 0 \tag{10}$$

If the temporal step h is small enough, the approximation is acceptable [19].

4   Simulations

In this work, we carry out two different simulations: one with OMNET++ using the algorithms described so far and one as a Matlab implementation of the cooperative control strategy with a simplistic network model which considers a perfect channel with a clear range and a percentage of losses applied randomly. The latter simulation is found in several works on Control and Robotics and we use it to assess its accuracy.

The choice for OMNeT++ was due to its GUI, its modularity and because it is a general-purpose discrete-event based simulation framework. Strictly, it is not a network simulator, but it can perform network simulations correctly and accurately with the INET and Mobility Framework add-ons [20].

For both simulations we use the same initial network topology (Fig. 2), and the same initial conditions and parameters (Table 1). We use no complex topology control but the distance constraints allow us to enforce a topology that is increasingly dense during the execution of the rendez-vous task. In particular, there should be no link disruptions, just new links being established as the robots get closer. Therefore, we consider two cases, the normal dynamic topology that results from the robots motion, and a static topology equal to the initial topology, in which new links are rejected by the control approach.

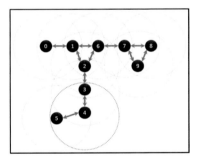 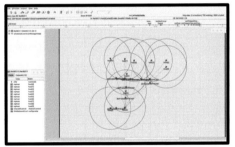

Fig. 2 Snapshot of the simulation initial topology on the left and the   OMNeT++ simulation case with messages being exchanged, on the right.

Table 1. Simulation parameters (* - some cases only).

Number of robots (N)	10
Initial positions X	[600; 800; 800; 800; 800; 600; 900; 1100; 1300; 1300]
Initial positions Y	[430; 430; 630; 830; 1030; 1030; 450; 450; 450; 550]
Velocities saturation	10cm/s and -10cm/s
Communication range	250cm
Prediction steps (p)	10
Link weight (a)	1
Velocity variation weight (λ)	1
Control time step	100ms
*TDMA cycle	100ms
*Jitter	Random (0...0.1ms)

4.1   *Matlab Simulation*

This simulation follows the implementation in [9] and executes the same cooperative control strategy with the referred constraints. To solve the quadratic optimization problem, it uses the standard *quadprog* function from Matlab 2014a.

We execute this simulation with two network models. In the first case, robots send messages at the same time to any neighbor robot within their communication range, and these messages do not have any delay or faults. The second case considers a simple TDMA for messages, suffering influence of robots' position and cycle slot. Both cases are executed with the two topology cases resulting in four cases. These cases reproduce the most common approaches followed in many of the simulation based work in Control and Robotics.

4.2   *OMNeT++ Simulation*

The OMNeT++ simulation environment is shown in Fig. 2. The simulation is organized in a set of independent modules. There is a module to move the robot (as a low-level controller), another to make the cooperative consensus, and another for topology and communication network control. OMNeT++ modularity allows changing any of these modules without the need to rebuild the entire code. Moreover, we solve the optimization problem using a library called QuadProg++ [22] for C++, which we integrated easily in simulation.

To evaluate the impact of each network feature, we selectively disabled each of those features in order to create several study cases, which we describe in detail in the next section. For all cases we considered both fixed and dynamic topologies.

5   Results

The simulation cases consist in a combination of different configurations for seven features as described below and shown in Table 2.

- Matlab: means it was a Matlab simulation;
- OMNeT++: means it was an OMNeT++ simulation;
- Network: means we are using the INET network protocol stack module (OSI layers, delays, IEEE 802.11 and radio signal models);
- TDMA: means we use a common TDMA, with clock synchronization;
- RATDMA: means we use the RATDMA protocol, with loose synchronization;
- Jitter: means we insert random clock errors in each slot starting time;
- Starting Time: describes different transmission modes:
  - Ideal: robots transmit exactly at the right slot time;
  - Real: robots transmit with a random delay between 0 and 3 ms in relation with their correct slot time;
  - Close: robots start transmitting almost at same time in a random order that differs from initial established slot order;
  - Together: the robots start transmitting exactly at the same time.

Table 1 Simulation cases parameters.

Cases	Matlab	OMNeT++	Network	TDMA	RATDMA	Jitter	Starting Time
1	X			X			Ideal
2	X						Together
3		X		X			Ideal
4		X	X	X			Ideal
5		X	X		X		Ideal
6		X	X		X	X	Ideal
7		X	X		X	X	Real
8		X	X		X	X	Close
9		X	X		X	X	Together Hold
10		X	X	X		X	Real
11		X	X	X		X	Close

In these experiments we removed the minimum distance constraint since the robots were reduced to dimensionless points in space, and analyzed the final convergence point and the time it took to reach this point. The stop criterion was when the standard deviation of the robots current position to the consensus point was less than one.

For the cases with fixed topology (Fig. 3 and Fig. 4), the convergence point and the time for convergence was approximately similar for both Matlab and OMNeT++ simulations, as long as a message collision avoidance mechanism, TDMA or RATDMA in this case, was in place and the starting conditions were similar (cases 1, 3-6 plus 7 and 10). The latter two cases, 7 and 10, considered delayed transmissions in the beginning of the slots, e.g. caused by late transmissions in progress at that moment, but without any visible impact given the relatively small but realistic delays of 0 to 3ms. This is somehow expected because the presence of a TDMA mechanism approximates the network from an ideal network in what concerns collisions.

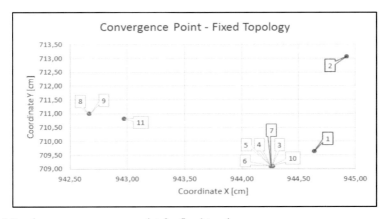

Fig. 3 Rendez-vous convergence point for fixed topology cases.

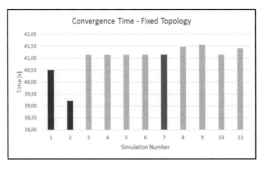

Fig. 4 Time to reach rendez-vous for fixed topology cases.

When the Matlab simulation is run with a simplistic network model (all transmitting at the same time but ignoring collisions – case 2), the results are erroneously better in convergence time.

Case 2 can also be compared with case 9 in which the automatic round reconfiguration of RATDMA is used to handle the situation in which all robots start transmitting together. This situation, initially, causes many collisions but as the protocol creates slots and allocates them to the robots the collisions disappear. This initial period of poor network performance causes a visible degradation in the convergence point and time. This is, however, a realistic case in which several robots are triggered simultaneously and the TDMA round is automatically formed by the RATDMA protocol. Note that there is no equivalent case with common (static) TDMA because when all robots are triggered together, they will be always transmitting outside their slots, thus colliding and mutually destroying their messages, leading to rendez-vous task failure.

Finally, two realistic but bad situations are presented in cases 8 and 11. As the information order is changed but there is not any initial collision, the TDMA will not cause any collision until the jitter dislocate the slots long enough to cause it. However, RATDMA will reallocate the slots and correct the transmitting order, but with some collisions as the slots overlay each other.

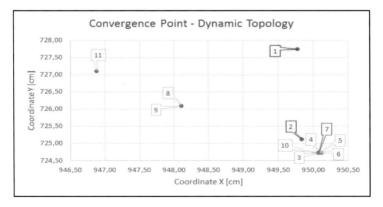

Fig. 5 Rendez-vous convergence point for dynamic topology cases.

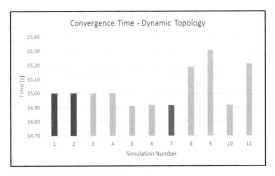

Fig. 6 Time to reach rendez-vous for dynamic topology cases.

The results of the simulations with dynamic topology are presented in Fig. 5 and Fig. 6. One clear difference to the previous set of simulations is that there is more variation in the convergence time. This is expected since in favorable cases robots move quickly to a point where the number of inter-robot links is rapidly increased, leading to a faster information dissemination. Conversely, unfavorable cases require more time before robots can reach a point with large link density, thus, taking longer to reach the point of the fast information dissemination.

Nonetheless, in both cases the benefits of this link densification arises from topology control implemented through the maximum distance constraint that guarantees a non-decreasing number of links as the team progresses to the convergence point. Consequently, the convergence time is visibly lower in the dynamic topology cases.

The main difference between Matlab and OMNeT++ simulations is that Matlab is more optimistic about the robots connections, i.e. Matlab automatically creates new links as soon as robots enter in the defined communication range, even if a robot did not send any message. On the other side, OMNeT++ is more realistic and only adds a new communication link when a robot receives a message from the new neighbor. As the robots are constantly moving, the link creation happens in different moments and thus, having different results in convergence point. This also slightly affects the convergence time, having approximately a round difference between the results, mainly because of the stop condition not being achieved in the same round.

Still with respect to the unfavorable cases (8, 9 and 11), it is important to highlight that, despite the link densification effect of the dynamic topology, the performance is still strongly affected by the concrete network protocol and its initial conditions. Once again, using simplistic network models such as cases 1 and 2 leads to misleading better results in the cooperative task.

6   Conclusion

In this work we performed the integration of network protocols with a complex cooperative strategy, and we assessed the impact of the network protocols on the cooperative task. Results show that using simplistic network models, such as those typically used in Matlab simulations by the Control and Robotics communities, may lead to performance levels in cooperative tasks that are unrealistically better than in real scenarios with real networks. Our results also show that properly configuring the robots before starting a cooperative task, mainly concerning their synchronization, can lead to improved global performance.

The main difficulty in this work was to calibrate both simulations to have similar results without the network aspects, mainly because of the difference in optimization libraries and programming languages. Moreover, several aspects of the OMNeT++ simulation had to be completely implemented, as the low level control for the robots and the movement display.

For future work, we will extend this study to evaluate the impact of alien traffic in the network, i.e. produced by sources outside the team, and consider different collaborative tasks, such as reference consensus and formation control.

References

1. Rocha, R.: Building Volumetric Maps with Cooperative Mobile Robots and Useful Information Sharing: a Distributed Control Approach based on Entropy, Ph.D. thesis, Faculty of Engineering of University of Porto, Portugal (2006)
2. Oliveira, L., Almeida, L., Santos, F.: A loose synchronisation protocol for managing RF ranging in mobile ad-hoc networks. In: RoboCup 2011: Robot Soccer World Cup XV. Lecture Notes in Computer Science, vol. 7416, pp. 574–585 (2011)
3. Baer, P.A.: Platform-Independent Development of Robot Communication Software, Ph.D. Thesis, University of Kassel, Germany (2008)
4. Zhang, Z., et al.: COSMO: CO-simulation with MATLAB and OMNeT++ for indoor wireless networks. In: IEEE Global Telecommunications Conference (2010)
5. Khan, A., Othmana, S.: A Performance Comparison of Network Simulators for Wireless Networks. ArXiv:1307.4129, Cornell University (2013)
6. Kudelski, L.M., et al.: RoboNetSim: An integrated framework for multi-robot and network simulation. Robotics and Autonomous Systems **61**, 483–496 (2013)
7. Pekkarinen, E.: Wireless Sensor Network Simulation With OMNeT++. Master dissertation, Tampere University of Technology, Finland (2014)
8. Oliveira, L., et al.: A loose synchronization protocol for managing RF ranging in mobile ad-hoc networks. In: RoboCup 2011. Lecture Notes in Computer Science, vol. 7416, pp. 574–585 (2012)
9. Ordoñez, B., Moreno, U.F., Cerqueira, J., Almeida, L.: Generation of trajectories using predictive control for tracking consensus with sensing and connectivity constraint. In: Koubaa, A., Khelil, A. (eds.) Cooperative Robots and Sensor Networks. Springer Studies in Computational Intelligence, vol. 507, pp. 19–37 (2014)
10. Baer, P.A.: Platform-Independent Development Of Robot Communication Software, Ph.D. Thesis, faculty of Electrical Engineering and Computer Science of the University of Kassel, Germany (2008)
11. Gil, P., Nunes, G., Santos, A., Cardoso, A.: Adaptive model based predictive networked control over WSAN with tolerance to transmission faults on the forward channel. In: 5th International Conference on Sensing Technology, pp. 341–346 (2011). ISSN 2156-8065

12. Wei, C., Hindriks, K., Jonker, C.M.: The role of communication in coordination protocols for cooperative robot teams. In: Proceedings of the 6th International Conference on Agents and Artificial Intelligence, pp. 28–39 (2014). doi:10.5220/0004758700280039
13. Ye, W., et al.: Evaluating control strategies for wireless-networked robots using an integrated robot and network simulation. In: Proc. of the IEEE International Conference on Robotics and Automation (ICRA), pp. 2941–2947 (2001)
14. Pohjola, M., et al.: Wireless control of mobile robot squad with link failure. In: Proceedings of the 6th International Symposium on Modeling and Optimization in Mobile, Ad Hoc, and Wireless Networks (WiOpt), pp. 648–656 (2008)
15. Zhang, Z., et al.: COSMO: CO-Simulation with MATLAB and OMNeT++ for Indoor Wireless Networks. In: Proc. of the IEEE Globecom. IEEE Communications Society (2010)
16. Guan, Z.Y.: A Reliability Evaluation of Wireless Sensor Network Simulator: Simulation vs. Testbed. Master thesis, UNITEC New Zealand (2011)
17. Pekkarinen, E.: Wireless Sensor Network Simulation With OMNeT++. Master thesis, Tampere University Of Technology, Finland (2014)
18. Stratulat, B., et al.: Wireless synchronization protocols for collaborative robotic and sensor environments. In: 19th International Workshop on Robotics in Alpe-Adria-Danube Region – RAAD, Budapest, Hungary (2010)
19. Correia, F.L., De, B., Ordoñez, B., Cerqueira, J., Almeida, L., Moreno, U.F.: Controle de Veículos Autônomos em Formação com Seguimento de Referência Utilizando Consenso e RHC. Congresso Brasileiro de Automática - CBA, Brazil (2014)
20. Wiengärtner, E., et al.: A performance comparison of recent network simulators, Distributed Systems Group RWTH, Aachen University, Aachen, Germany (2009)
21. Pujeri, U.R., Palanisamy, V.: Survey of Various Open Source Network Simulators. International Journal of Science and Research (IJSR), December 2014. ISSN: 2319-7064
22. QuadProg++ Project. http://quadprog.sourceforge.net/ (accessed June 2015)

A FIPA-Based Communication Infrastructure for a Reconfigurable Multi-robot System

Thomas M. Roehr and Satia Herfert

Abstract This paper presents a high-level communication infrastructure to deal with dynamically changing reconfigurable multi-robot systems. The infrastructure builds upon official standards of the Foundation for Intelligent Physical Agents (FIPA). FIPA standards have been successfully applied in a variety of multi-agent frameworks, but they have found little application in the domain of robotics. This paper introduces an implementation that can complement existing robotic communication frameworks and allows the robotics community to take better advantage of multi-agent research efforts. We present the essential components of the infrastructure and show its interoperability using the widely known multi-agent framework JADE.

Keywords Multi-robot system · Distributed communication · FIPA

1 Introduction

Robustness and reliability are key aspects for the design of robots that have to operate autonomously in remote places. Single robotic systems are widely applied, but reliability can often be achieved only by increasing redundancy. In contrast, multi-robot systems (MRSs) inherently offer a higher level of redundancy and are thus naturally suited for application scenarios where systems are exposed to unforeseen risks of outage. This holds even more when looking at reconfigurable multi-robot systems (RMRSs), e.g., [20] presents a RMRS which consists of multiple robotic systems which can be dynamically extended using modular so-called payload-items (cf. Figure 1). The modularity of the system is achieved by introducing a standardized

T.M. Roehr(✉)
DFKI GmbH Robotics Innovation Center, Bremen, Germany
e-mail: thomas.roehr@dfki.de

S. Herfert—No current affiliation. Contribution provided during his time as research assistent at DFKI GmbH Robotics Innovation Center.

© Springer International Publishing Switzerland 2016
L.P. Reis et al. (eds.), *Robot 2015: Second Iberian Robotics Conference*,
Advances in Intelligent Systems and Computing 417,
DOI: 10.1007/978-3-319-27146-0_51

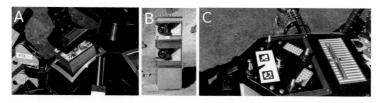

Fig. 1 A modular, reconfigurable multi-robot system [20], A: Sherpa robot's manipulator approaching a payload-item, B: stack of two camera and one power payload-item, C: electro-mechanical interface on the back of the legged robot CREX

electro-mechanical interface (EMI) that allows to connect two systems. Payload-items serve as generic containers which can host sensors; they comprise two EMIs.

Due to its modularity an RMRS offers more flexibility to change its morphology and to cope with dynamically arising challenges. In order to exploit the redundancy of a RMRS an effective coordination mechanism has to be put in place, e.g., to dynamically build a payload-stack from a number of payload-items. Additionally, individual systems can appear or disappear in the communication infrastructure either as part of a nominal operation, e.g., as result of a merge of two systems, or as part of a non-nominal outage. At the foundation of organizing the RMRS presented in [20] lies a communication infrastructure that accounts for dynamically changing robotic systems which establishes a peer-2-peer communication network. Compared to [20] we have significantly revised the infrastructure's components to improve the applicability. This paper details the revised infrastructure and its benefits for robotic cooperation. The following Section 2 presents the state of the art and background information to motivate the approach. Section 3 provides a detailed presentation of the essential components. Subsequently, we present an evaluation in Section 4 and illustrate a typical application and interoperability. We conclude with a discussion of the benefits and limitations of the presented approach.

2   Background

A core activity of robotics is the integration of hardware and software, and managing communication within a robotic system takes a significant share of this work. Nowadays, robotic middlewares or frameworks such as Robot Operating System (ROS) [17] or Robot Construction Kit (Rock) [11] are key to effectively implementing robots. These frameworks typically rely on specialized and modular components that are inter-connected to form an information processing network designed to solve a given problem. The mentioned frameworks allow the construction of distributed systems, but focus on setting up single robotic systems. ROS and Rock, for example, depend on the availability of a centralized naming service to access and connect components; while the former relies on an instance of the so-called ROS Master, the later requires CORBA [13] to provide the naming service. Several strategies have been applied by these frameworks to mitigate these effects, but support for fully distributed systems with no single point of failure has not been achieved. MRSs use communication patterns such as broadcasting [14], blackboards [1] or cloud-based [8] communication, mixed with publish-subscribe mechanisms [12] or peer-to-peer solutions. Though these communication patterns have their individual benefits, a main challenge for completely decoupled robots resides in setting up and maintaining the inter-communication channel. Additionally, distributed agent-based systems use auctions as a common interaction pattern to solve the resource allocation problem [21]. However, communication in all the robotic frameworks mentioned above is based on stateless message exchange and does not naturally extend to auctions, which have a state-based interaction pattern. In order to facilitate the development of physical agents the Foundation for Intelligent Physical Agents (FIPA) [6] has devised a set of standards, defining – among other things – an abstract agent architecture including message formats and interaction protocols. The abstract architecture consists of a set of infrastructure services such as an agent directory with *white pages* and *yellow pages* functionality, i.e., a component that manages the information about all known agents. Though FIPA is now a standards committee of IEEE, the application of these standards in the domain of MRSs is rarely seen. In contrast, a broad application is found in the area of multi-agent systems (MASs), e.g., Java Agent DEvelopment Framework (JADE) [3] offers a reference implementation of FIPA standards and it is widely used in the multi-agent community. The best known implementations of FIPA standards, such as FIPA-OS [16] and JADE [3] are Java-based – a factor that might hinder a broader application in robotics where C/C++ implementations are often predominant. Mobile-C [4] offers a C-based implementation, but sets its focus on a different use case: the transition of software-agents between multiple machines by relying on a C/C++ interpreter. Our communication architecture builds upon FIPA standards as an outcome of the MAS community. At the same time we close the aforementioned implementation gap.

3 A FIPA-Based Communication Infrastructure

Reconfiguration in the context of MRS refers to a change of morphology, i.e., two or more previously separated systems are physically merged to form a so-called *coalition*. More formally, a RMRS consists of a set of physical agents $A = \{a_1, a_2, \ldots, a_n\}$, which can form coalitions $C \subseteq A$. Each coalition again represents a single physical actor. A distinct set of coalitions is typically denoted a *coalition structure CS*, and each agent can be interpreted as a coalition C of size $|C| = 1$ [24]. The overall flexibility of the RMRS directly depends on the modularity of the system and the fact that these systems can be (almost) arbitrarily combined.

This paper (cf. Figure 2) introduces a high-level communication layer that can complement existing frameworks such as ROS and Rock to build fully distributed systems and support the operations of an RMRS and its dynamics. The main requirements are: (i) transparent handling of coalitions that join or leave the communication

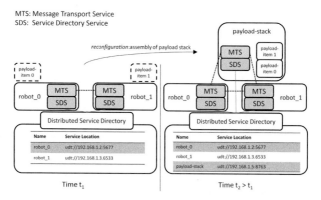

Fig. 2 Service infrastructure of a reconfigurable system, before and after reconfiguration, i.e., assembling two initially inactive payload-items to one active payload-stack

infrastructure, (ii) interoperability and standardization of high-level communication, (iii) communication protocols for coordination of reconfiguration, and (iv) applicability to ARM-based devices such as payload-items [20].

In a typical MRS the coalition structure is fixed and a single robot leaving the infrastructure can be considered a non-nominal event. In contrast, a RMRS takes advantage of a flexible coalition structure so that agents joining and leaving the system are part of nominal operation. Furthermore, one of the key features of the RMRS is extensibility and due to a standardized interface new functionality for the overall system can be independently designed. However, to guarantee interoperability communication has to be standardized as well; this does account for the inter-robot communication and does not extend to the internal communication of a single robot. We assume that the high-level communication layer acts as a control channel and allows for coordination of multiple systems, e.g., to perform reconfiguration.

We take FIPA [6] as guideline since it applies to a fully distributed scenario and is therefore well suited as a guideline to our design inter-robot communication. In particular, we reuse the concept of a message-transport service (MTS) including the related Agent Communication Language (ACL), and the service-directory service (SDS) in order to establish a robust agent communication channel. FIPA-based communication accounts for interaction protocols as patterns of message flows between multiple robots based on so-called 'performatives' such as *request* or *query-when*. This facilitates the design of coordination protocols involving operators or autonomous agents, and in the context of RMRS serves as basis for automated negotiation for reconfiguration. The essential backbone of this communication infrastructure comprises the SDS and the MTS, which are described in the following and are implemented in our software library *fipa_services* [19]. Compared to [20] the following changes have been made to increase applicability of the infrastructure: (i) encapsulation of main services into a C++ library, (ii) support for message envelopes (including bit-efficient and XML encoding), (iii) support for additional representation types for messages (support was limited to *bit-efficient*), and (iv) support for selecting transports.

3.1 Service Directory Service

An SDS is a mandatory element of the FIPA Abstract Architecture [6] and all robots (including payload-items) in our infrastructure (cf. Figure 2) run an SDS, which overall forms a decentralized Distributed Service Directory (DSD). The DSD decouples the modular systems in the RMRS, since robots can register and deregister to this DSD dynamically by attaching or detaching to an MTS. The usage of a DSD eliminates the single-point of failure that comes with most existing robotic middlewares, but instead the DSD becomes a critical service overall for an RMRS. The current implementation builds upon Avahi [15] – a component for zeroconf [10] and service discovery that is standard to most Linux-based systems.

3.2 Message Transport Service

Along with the SDS the FIPA specification requires an MTS. The single purpose of the MTS is to deliver messages (wrapped in envelopes) to attached clients or to another MTS which can forward this message to one of its connected clients. Figure 2 shows the overall communication infrastructure. The MTS retrieves a receiver's transport address from the DSD and a receiver's name is interpreted as a regular expression for this search. This allows for a natural support of broadcasting and multicasting using wildcards in the receiver name. Each MTS stamps a handled messages to prevent looping of messages. Connections are established in a lazy fashion; the MTS tries to connect to the remote MTS only if a message is directed to a remote client. Each MTS can draw from a list of supported transports to transfer messages in-between two MTS – currently, *fipa_services* supports TCP and UDT [7]. If multiple transports can be used, they are tried one after another: here UDT first, TCP second.

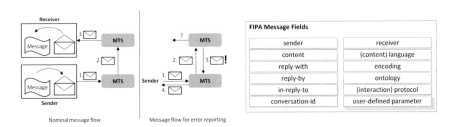

Fig. 3 (left) Message handling for a nominal and an error scenario, (right) FIPA Message structure

Notification about failed delivery is part of the MTS functionality and all available transports are tried in order to deliver a message. If a receiver cannot be found in the DSD or the delivery fails at whatever MTS, an error response is triggered. This error message is propagated back to the sender (if it is still available), using the reverse

Table 1 Supported FIPA representations (evaluated in Section 4)

Element	Representation types
message	bit-efficient, XML, string
envelope	bit-efficient, XML

communication path. Relay is not part of this infrastructure, but achieved by applying dedicated meshing protocols.

Figure 3 outlines the message flow for the nominal message delivery and a failed delivery. The error scenario considers that delivery fails at the MTS of the (expected) receiving client. Messages comprise a set of standard fields (cf. Figure 3) and can be serialized using different representation types. Table 1 lists the currently supported representation types.

A message is put into an envelope which is then forwarded to an MTS; the envelope contains the serialized message as its payload. This allows the MTS to operate in a content-agnostic way, and decoding of the messages and its content only needs to be done by the final receiver. The results in Section 4 will illustrate this characteristic.

3.3 Conversation Monitor

FIPA also describes the application of interaction protocols, which make use of performatives defined in a message. An interaction protocol describes a message flow using a state transition system $\sum = (S, P, R, \gamma)$, where $S = \{s_0, s_1, \dots\}$ is the set of conversation states, $P = \{accept\text{-}proposal, agree, \dots\}$ is the set of performatives, and $R = \{r_1, r_2, \dots\}$ is the set of roles. The set of roles typically consists only of a sender and receiver label. The state transition function is $\gamma : S \times P \times R \rightarrow 2^S$.

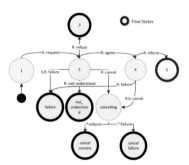

Fig. 4 State machine for the request protocol [6] including default states and transitions as well as roles: receiver R, sender S and wildcard .* representing any other client. Transitions are labeled with roles (disjunctive) and performatives required to match.

   To take advantage of interaction protocols, we implemented a conversation monitor to validate state transitions that are triggered by incoming and outgoing messages.
Validation is done with respect to a single agent which is involved in the conversation. To model interaction protocols we use a custom extension of State Chart XML
(SCXML) [23] and embed default transitions for error handling and cancellation of
interaction protocols to reduce the modeling effort for users. Figure 4 illustrates how
the accounting for default states affects a simple request protocol. This functionality
allows to outline and validate inter-robot coordination, especially for reconfiguration
and cooperation between multiple robotic systems.

4   Evaluation of Fipa_Services

In this section we evaluate the performance of our *fipa_services* library regarding
potential effects on bandwidth and computation.. The evaluation shows the impact
of the selection of the representation type (cf. Table 1) and the applicability to embedded systems. The representation-dependent overhead for messages wrapped into
an envelope is illustrated in Figure 5, and shows significant differences between the
different representation types.

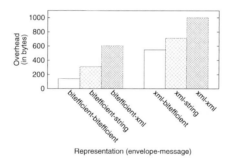

Fig. 5 Net overhead for a message with content size of 1 byte and the header containing 99 byte
information filling all message fields (cf. Figure 3)

   Figure 6 illustrates the message encoding and decoding performance on a PC with
an Intel CORE i7 2.1 GHz, 12 GB RAM and a Gumstix with ARM Cortex-A7 CPU
720 MHz, 256 MB RAM *; the evaluation is based on 1000 encoding and decoding
cycles that are measured using software timers.
   This evaluation shows that bit-efficient message encoding is best suited for environments with limited bandwidth. It comes with a slight performance drawback
for small messages compared to the XML representation, but performs better for

---

* All modular payload-items host a Gumstix Overo Fire.

messages with large content. The string representation shows the worst performance for large messages on the Gumstix, while it remains ahead on XML on the PC.

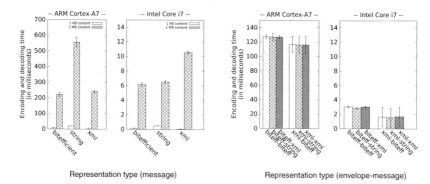

Fig. 6 (left) Message handling performance for 1000 cycles of encoding and decoding and 1 KB and 1 MB sized contents, (right) Evaluation of envelope handling for 1000 cycles of encoding and decoding and each envelope containing a message with 1 MB content

Figure 6 presents the performance for different combinations of envelope and message representations. The performance on the ARM-based system illustrates the achieved effect that an envelope's payload does not need to be decoded for transport on an MTS. Bit-efficient encoding for envelopes comes with a small performance drawback, but is also exhibits a significantly smaller standard deviation.

This implementation [19] fully supports bit-efficient encoding and is to our best knowledge the only publicly available implementation of FIPA's final standard; an implementation for the experimental standard exists in the form of a JADE plugin, but is limited to the message representation [9]. A fair comparison against the reference framework JADE could not be made due to the warm-up behavior of the Java Virtual Machine (JVM), i.e., code optimization is done at runtime. The influence of this feature in an application scenario will have to be investigated. However, initial findings indicate that after warm-up JADE will show a superior performance compared to *fipa_services*.

The benefits of using FIPA-based communication can be attributed to the partial standardization of coordination. All components in *fipa_services* have been developed to allow for simple integration into various robotic frameworks. We have embedded *fipa_services* into Rock and illustrate the application within this robotic framework for coordination of robotic reconfiguration and its interoperability in the following.

4.1   Application

To illustrate the application of the infrastructure we take two examples of robotic coordination: a multi-robot reconfiguration process and distributed mutual exclusion.

Multi-robot Reconfiguration. High-level coordination is required to perform physical reconfiguration, i.e. merging, of two or more robots. To apply coordination the definition of a content language can introduce decoupling which allow for a more generic application. Here, we complemented our communication architecture using a content language to control a robot's actions; the content language is text-based and human-readable to permit interpretation on any system and facilitate debugging. The content language also allows querying available robot actions, along with the status of running actions. The content language enables robots to control each other and has been successfully applied as part of the unrevised infrastructure for the docking process in [20] in order to merge two systems: one master robot guiding another client robot to connect to one of the master's electro-mechanical interface (EMI) (cf. Figure 1). We successfully applied our revised infrastructure to perform the visually guided docking of the Sherpa robot's manipulator to the CREX robot [20]. Sherpa's manipulator was moved to a fixed position, and the CREX robot was remotely guided by the Sherpa robot.

Distributed Mutual Exclusion. A main feature of an RMRS is the exchange of resources. But to perform automated reconfiguration as mentioned in the previous application scenario, the participating resources should be exclusively reserved beforehand. We implemented Ricart-Agrawala's [18] non-token-based and Suzuki-Kasami's [22] token-based algorithm using *fipa_services*. Both algorithms cannot deal with a dynamically changing set of agents or agent-failure. To overcome this limitation the detection of resource owners and agent-failure have been added to the implementation. Detection of the resource owner relies on a query message and the owner's response that is sent as broadcast. The detection of agent-failure is based upon sending heartbeat messages at low-frequency (0.2 Hz) in-between participants that depend upon each other, i.e., the (physical) owner of a resource, the lock holder and system waiting to lock the resource. This allows to mark a given resource to be unreachable and trigger further error handling. This approach leads to three overlapping interactions between multiple agents: (1) probing, (2) discovery of the owner of a resource, and (3) performing the actual locking. The three types of interactions are modeled with corresponding protocols. Figure 7 illustrates the protocol for the (extended) Ricart-Agrawla's algorithm. When an agent acquires or releases the lock it communicates this information to the resource owner using the performatives *confirm* and *disconfirm*. In addition to using interaction protocols, the use of a content language, e.g., for Suzuki-Kasami's algorithm to exchange the token between different agents, provides a further mean to separate and validate the flow of messages.

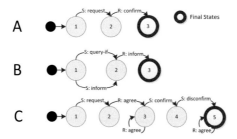

Fig. 7 Three separate interaction protocols that are part of the (extended) Ricart-Agrawala's algorithm; A: probing of inter-dependent agents, B: discovery of the resource owner (owner information is sent as broadcast), C: the message flow when performing the actual locking algorithm. Failure handling is embedded into default transitions (which are not depicted) and *S* denotes the conversation initiator's role and *R* the recipient's role.

Modeling interaction protocols allows verification of the message flow at runtime and facilitates debugging. For our distributed algorithms protocol compliance of all participating agents is enforced by using the conversation monitor (cf. Section 3.3).

4.2 *Interoperability*

Our motivation to use standards is to achieve extensibility and interoperability, e.g., due to the standardization of messages and infrastructure we can connect Rock to JADE using the existing extension mechanisms [2] and connecting JADE to the DSD. Since the DSD is based on Avahi it uses multi-cast DNS for communication. Registering services within JADE's infrastructure can thus be achieved by using JmDNS [5]. A problem was encountered upon registration of JADE agents since dots in service names cannot be handled by JmDNS; dots in agent names had to be replaced by the question mark wildcard for registration (cf. Figure 8). However, this change is transparent to users and agents, which can refer to other agents only by their real name. Agents from both ecosystems communicate via XML encoded envelopes and messages. The bit-efficient encoding implementations of the two systems turned out to be incompatible – as mentioned, the only corresponding JADE plugin implements the experimental standard, while *fipa_services* implements the latest standard. To validate the integration we placed an EchoAgent in the JADE infrastructure and attached a client *rock_agent* to an MTS in the Rock infrastructure; all components resided on the same local machine (Intel CORE i7 2.1 GHz). Figure 8 illustrates the resulting services registered in the DSD. The average round-trip time measured for a sequence of 10000 conversations and a message content size of 100 bytes was: 24 ± 5.6 ms. For future applications, this integration allows to use JADE's infrastructure and tooling for high-level coordination in parallel to Rock's infrastructure and tools.

Fig. 8 Reuse of existing tooling: JADE's management GUI and the default avahi-discovery GUI to browse visible services, i.e., the set of registered services

5 Conclusion

An RMRS requires coordinating reconfiguration at various levels: (i) finding available resources, (ii) allocating resources, (iii) performing required reconfiguration maneuvers, and (iv) handling errors. This demands a structured approach and tool support in order to reach practical and scalable solutions. The presented infrastructure provides a basis using standardized components and allows verification of robot communication by relying on FIPA standards. The implemented libraries are open-source [19] and hosted along with Rock.

The criticism of [24] points out the semiformality of FIPA, but still acknowledges its practicality. Our experience confirms this standpoint and with *fipa_services* we provide the robotics community with easier access to these benefits with the intention to facilitate the application of multi-robot systems. In addition, while FIPA provides an architectural template, the individual services such as the DSD can remain completely out of the FIPA context.

Obviously, specific interaction scenarios can be easily solved by creating special message types and the presented infrastructure adds overhead for abstraction and standardization. However, in order to maintain a truly extensible and fully distributed RMRS, we advocate the presented approach.

Acknowledgements This work was supported by the German Space Agency (DLR) and conducted under grant agreement 50RA1301 (BMWi).

References

1. BBN Technologies: Cougaar Architecture Document (2004). http://cougaar.org/doc/11_4/online/CAD_11_4.pdf
2. Bellifemine, F., Caire, G., Greenwood, D.: Developing Multi-Agent Systems with JADE. Wiley, Chichester (2007). doi:10.1002/9780470058411
3. Bellifemine, F., Poggi, A., Rimassa, G.: JADE-A FIPA-compliant agent framework. In: Proceedings of the 4th International Conference on Practical Application of Agents and Multi-Agent Technology (PAAM), pp. 97–108 (1999)

4. Chen, B., Cheng, H.H., Palen, J.: Mobile-C: a mobile agent platform for mobile C-C++ agents. Software: Practice and Experience (Software Pract Ex) **36**(15), 1711–1733 (2006). doi:10.1002/spe.v36:15

5. Cheshire, S.: JmDNS (2011). http://jmdns.sourceforge.net/

6. Foundation of Intelligent Physical Agents: FIPA Abstract Architecture Specification (2011). http://www.fipa.org/specs/fipa00001/SC00001L.pdf

7. Gu, Y., Grossman, R.L.: UDT: UDP-based data transfer for high-speed wide area networks. Computer Networks **51**, 1777–1799 (2007). doi:10.1016/j.comnet.2006.11.009

8. Hartanto, R., Eich, M.: Reliable, cloud-based communication for multi-robot systems. In: The 6th Annual IEEE International Conference on Technologies for Practical Robot Applications (TePRA-2014). IEEE (2014)

9. Heikki, H., Laukkanen, M.: How to use the Bit-efficient ACL (BE-ACL) encoding with JADE (2015). http://jade.tilab.com/doc/tutorials/BEFipaMessage.html

10. IETF Zeroconf Working Group: Zero Configuration Networking (Zeroconf) (2011). http://zeroconf.org/

11. Joyeux, S., Schwendner, J., Roehr, T.M.: Modular Software for an Autonomous Space Rover. In: The 12th International Symposium on Artificial Intelligence, Robotics and Automation in Space (i-SAIRAS 2014) (2014)

12. Object Management Group: Distributed Data Service (DDS) (2014). http://www.omg.org/spec/DDS/

13. Object Management Group: Common Object Request Broker Architecture (2015). http://www.corba.org/

14. Parker, L.E.: ALLIANCE: An architecture for fault tolerant multirobot cooperation. IEEE Transactions on Robotics and Automation **14**, 220–240 (1998). doi:10.1109/70.681242

15. Poettering, L., Lloyd, T., Estienne, S.: Avahi (2012). http://www.avahi.org

16. Poslad, S., Buckle, P., Hadingham, R.: The FIPA-OS agent platform: Open source for open standards. In: 5th International Conference and Exhibition on the Practical Application of Intelligent Agents and Multi-Agents (2000). http://fipa-os.sourceforge.net

17. Quigley, M., Conley, K., Gerkey, B., Faust, J., Foote, T.B., Leibs, J., Wheeler, R., Ng, A.Y.: ROS: an open-source Robot Operating System. ICRA Workshop on Open Source Software (2009)

18. Ricart, G., Agrawala, A.K.: An optimal algorithm for mutual exclusion in computer networks. Communications of the ACM **24**, 9–17 (1981). doi:10.1145/358527.358537

19. Roehr, T.M.: Rock Multiagent (2015). https://github.com/rock-multiagent

20. Roehr, T.M., Cordes, F., Kirchner, F.: Reconfigurable Integrated Multirobot Exploration System (RIMRES): Heterogeneous Modular Reconfigurable Robots for Space Exploration. Journal of Field Robotics **31**(1), 3–34 (2014). doi:10.1002/rob.21477

21. Shoham, Y., Leyton-Brown, K.: Multiagent Systems: Algorithmic, Game-Theoretic, and Logical Foundations. Cambridge University Press, New York (2009)

22. Suzuki, I., Kasami, T.: A distributed mutual exclusion algorithm. ACM Transactions on Computer Systems **3**(4), 344–349 (1985). doi:10.1145/6110.214406

23. W3C: State Chart XML (SCXML): State Machine Notation for Control Abstraction (2015). http://www.w3.org/TR/scxml

24. Weiss, G. (ed.): Multiagent Systems, 2nd edn. MIT Press (2009)

Multi-robot Optimal Deployment Planning Under Communication Constraints

Yaroslav Marchukov and Luis Montano

Abstract In this paper, we address the problem of optimal multi-robot team deployment while maintaining communication for all the robots. The objective is to execute the mission of reaching several goals with minimal number of robots, as well as reducing the total distance travelled to reach the goals. Therefore, we develop an algorithm that computes some secondary or virtual goals to move robots enhancing the coverage over the map. Due to the presence of obstacles, we study the use of different criteria in order to add more flexibility to the optimization in terms of travelled distance or relay nodes saving.

Keywords Communication constraints · Graph connectivity · Multi-robot · Optimal deployment

1 Introduction

In last decades many techniques were developed in order to control multi-robot teams executing some deployment tasks. These techniques pursue different purposes, as control of formations [7][9], when a team of robots must perform a task without losing communication and the goals are close one to each other. These methods are common in *master-slave* configurations, where a leader plan its trajectory and the rest of the team uses some reactive method to follow it [11].

Other type of task is exploration, which consists on covering some terrain. In order to cover maximum area, the multi-robot team is separated [10]. These tasks may be executed in different ways, planning the path of each robot to make it concur sometimes with other robots and share information [4] or assuring to maintain

Y. Marchukov(✉) · L. Montano
Aragon Institute for Engineering Research (I3A), University of Zaragoza, Zaragoza, Spain
e-mail: {yamar,montano}@unizar.es

© Springer International Publishing Switzerland 2016
L.P. Reis et al. (eds.), *Robot 2015: Second Iberian Robotics Conference*,
Advances in Intelligent Systems and Computing 417,
DOI: 10.1007/978-3-319-27146-0_52

connectivity all the time [2][5]. However, it is necessary to be cautious planning the paths of every robot, as the error might lead to the loss of some member of the team.

Recently this kind of problems became the solution to many deployment problems and many new algorithms were developed. A recent line of research in this topic is an efficient channel modelling to have a better knowledge of the signal strength at every point of the map [1][13]. Others make an effort developing robust algorithms to keep communication, based on the minimal binary rate [5], bit error ratio [13] or RSSI to assure packet transmission [2]. However, some of these methods perform robot trajectories where a leader robot reaches the goals, while the only task of the rest of the team is providing communication to the leader. This leads to suboptimal task execution, this is, employing more robots than requires the mission and compute paths that visit the same points several times.

Many algorithms were developed in order to avoid these situations. Works like [12] propose a method to seek optimal positions unifying disconnected groups of robots to recover the communication. This method uses a weighted graph to approach clusters of mobile agents. The weight is computed depending on different metrics as movement cost, number of agents and distance. Others as [3], compute trajectories for a robot that uses the signal of a team of independent agents to walk through the map. This technique resolves scenarios which require intermittent communication with some members of the team. It uses a planner to compute the best waypoints of the grid, so that obtains the best positions for the robot to maintain communication at each time step. The work in [4] is similar in spirit, but consists in some area exploration of a multi-robot team having periodic connectivity between the agents. Thus, it searches positions where the robots could have *line-of-sight* and share information.

The scenario to be solved is a set of distributed goals, which must be visited. Therefore, a multi-robot team, with a static base station, deploys and visits every goal as fast as possible and ensuring the connectivity of each agent with the rest of the team. Due to the short range of the base station, some agents are used as links supplying signal to the distant robots. We focus in some works, which solve similar scenarios, as in [3][4][12], but using another approach: the use of a realistic propagation model instead of the *line-of-sight*, consideration of both, connected and disconnected nodes, connection of every robot to the base station, through direct link or using other robots, the control of each relay agent to ease the trajectory to other robots. We develop a dynamic model, inspired by the optimal router positioning problem [6], where the router locations are optimized in order to cover all the map using the minimal number of routers.

Thus, we propose a method, to be executed before the trajectory planning algorithm, which computes some secondary or virtual goals. These new goals have the purpose of increasing the coverage over the map placing some robots in these positions. Thus, the contributions of our work are: (i) assure that all the robots reach primary goals with communication, (ii) minimizing the number of agents performing the task of link, in other words, the team reserve more robots for primary goals, and (iii) reducing the distance to primary goals and avoiding the undesirable trajectories to reach the best coverage positions.

The rest of the paper is organized as follows. In Section II we describe our problem. Section III describes a graph solution to deployment problem and show some simulated graph configurations. In Section IV we consider the optimization of the formed graph and its dynamic control, in order to maintain the connectivity between the team visiting all the nodes and the links, followed by the simulated results.

2 Problem Statement

Consider N nodes in a static m-dimensional space $\mathbb{R}^m$ and let $\mathbf{X} \in \mathbb{R}^m$ denote the position of each goal. Let denote M as the available robots, with $M \leq N$, that have to achieve the mission of reaching the goals and serve as communication relays to improve the connectivity. The lower number of robots executing relay tasks enable higher number of robots to accomplish primary tasks and, consequently, finish the mission with lower time. Let define $\mathbf{x} \in \mathbf{X}$ as all the goal positions, which could be reached with connectivity only positioning a robot at each goal. Thus, we define $\mathcal{G}(\mathbf{x}) = (\mathcal{V}, \mathcal{E}(\mathbf{x}))$ as a dynamic graph where:

- $\mathcal{V} = \{1, ..., n\}$ is a set of vertices representing all connected goals $n \in N$, with n=1 denoting the BS and $\mathbf{x}(0)$ its position.
- $\mathcal{E}(\mathbf{x}) = \{(i, j)|p_{ij}(\mathbf{x}) \geq \gamma\}$ is an edges set containing the locations of the n vertices $\mathbf{x} \in \mathbf{X}$, where p_{ij} is the signal strength between node i and node j and γ is a constant threshold.

The transmission channel is considered symmetric, thus $p_{ij} = p_{ji}$. We consider all the goals $\mathbf{x}$ connected because there exists a path from BS to every node n, formed by adjacent vertices in the graph and expressed as $P_n = \{v_1, \dots, v_n\}$. This way, the distance to some node n is expressed as the sum of Euclidean distances between every pair of vertices that form a path to n:

$$d_n = \sum_k ||\mathbf{x}_{k+1} - \mathbf{x}_k|| \tag{1}$$

where k represents every node of a unique path P_n. On the other hand, the set of those $\mathbf{X}$ which do not have a path from the BS are considered disconnected $\mathbf{x}_{disc}$.

Problem 1 (Goal Connection): Given the initial connected graph $\mathcal{G}(\mathbf{x})$, compute the new distribution of $\mathbf{x}$ so that $\mathbf{x}_{disc} \in \mathbf{x}$. In other words, we want to find a new $\mathcal{G}'(\mathbf{x})$ configuration that connects all N nodes.

As the mission is to reach goals which are not in the coverage area or are achievable using suboptimal configuration, the idle robots change their goals and supply signal to the next set of reachable goals. Therefore, it is necessary to know which nodes perform the link task and which ones are the linked nodes at each moment of the mission.

Problem 2 (Optimal Goal Linking): Given an initial graph, compute a tree $\mathcal{T}(\mathbf{x}) \in \mathcal{G}(\mathbf{x})$ so that the path to every node contains minimal number of nodes. At the same time, the costs of the edges, denoted as w, are taken into consideration for minimizing the distance or the number of nodes.

$$\mathcal{T}^* = \underset{\mathcal{T}}{\operatorname{argmin}} \sum_{u \in \mathbf{U}} \left(|\mathbf{P}_u| + \sum_{(i,j) \in P_u} w(i, j) \right) \tag{2}$$

where $\mathbf{U} : \{\mathbf{x} \in \mathcal{T} | \mathbf{dp}(\mathbf{x}) > 2\}$ represents all the unreachable goals directly from the base station, $\mathbf{dp}$ is the depth of each node in the tree, $\mathbf{P}_x$ is a set of paths from the root of the tree to the node x, $|\mathbf{P}_x|$ is the number of nodes in each path, (i, j) is each consecutive pair of nodes conforming the path P_x and costs w are chosen according to different criteria, described in the next section.

Let us define: $\mathbf{x}_{visible}$ as a set of nodes which are in the coverage area and may be visited $\mathbf{x}_{visible} : \{\mathbf{x} | \mathbf{dp}(\mathbf{x}_{visit}) + 1\}$, $\mathbf{x}_{visit}$ as a set of visited primary goals, $\mathbf{x}_v$ as virtual goals that provide connectivity to other primary goals. Thus, we say the node is visited when some robot has reached this goal and only the base station is considered as visited goal at $t = 0$. The visible nodes are descendants of the visited goals, that is, $\mathbf{dp}(\mathbf{x}_{visible}) = \mathbf{dp}(\mathbf{x}_{visit}) + 1$.

Thus, visited primary goals are extracted from the goal list and new virtual goals $\mathbf{x}_v$ are inserted as secondary tasks. This can be formulated as:

$$\mathbf{x}_k^+ \leftarrow (\mathbf{x}_k \setminus \mathbf{x}_{visit}) \cup \mathbf{x}_v \tag{3}$$

where $\mathbf{x}_k$ and $\mathbf{x}_k^+$ are the nodes before and after optimization respectively.

Problem 3 (Optimal positions): Let $\mathbf{x}_{visible}^+$ be all desirable visible goals after visiting the computed virtual goals. Formulated as $\mathbf{x}_{visible}^+ : \{\mathbf{x} | \mathbf{dp}(\mathbf{x}_v) + 1\} \cup \{\mathbf{x}_{disc} | p(\mathbf{x}_v, \mathbf{x}_{disc}) \geq \gamma\}$. We determine all the virtual goals that conduct us to maximize the number of visible nodes and guarantee their connectivity:

$$\mathbf{x}_v^* = \underset{\mathbf{x}_v}{\operatorname{argmax}} (|\mathbf{x}_{visible}^+| \ni p(\mathbf{x}_v, \mathbf{x}(\mathbf{dp}(\mathbf{x}_v) - 1) \geq \gamma) \tag{4}$$

All the process of optimization is depicted in a simple example in Fig. 1. The crosses represent all the primary goals, that is, all the locations that should be visited. The discontinuous line depicts the possible links between nodes and the continuous line depicts communication links. In 1(a) the initial nodes distribution is represented, the minimal depth tree is obtained with equation (2). Some nodes are initially disconnected, so can not be visited with communication. Therefore, using (3) and (4) the optimal position of the virtual goals $\mathbf{x}_v$ is computed, so that they maximize the number of visible nodes and connect disconnected ones; Fig. 1(b).

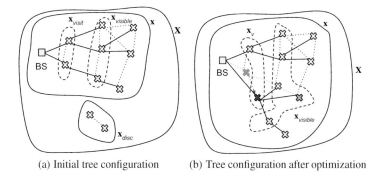

(a) Initial tree configuration   (b) Tree configuration after optimization

Fig. 1 Optimization process. In 1(a) the visited nodes $\mathbf{x}_{visit}$ provide signal to the nodes of next level of the tree, $\mathbf{x}_{visible}$, and some nodes are disconnected $\mathbf{x}_{disc}$ from the rest of the team. Fig. 1(b) shows the location for the optimized virtual nodes $\mathbf{x}_v$ providing signal to the next level of the tree. As the upper node didn't find better location thus it keeps at the same position, and the lower node, highlighted as a red cross, changed its location in order to connect the disconnected nodes ($\mathbf{x}_{disc}$) and reduce the depth of some nodes, depicted with green cross.

3 Weighted Graphs

In this section, we describe our method for obtain a first graph configuration. First, in section 3.1 we define the path-loss model, used to compute the possible links between the nodes. This contributes with binary information to each edge, 1 when exists a link and 0 when is no connection between the nodes. In order to add more flexibility to the tree configuration, the different ways to weight the graph and the tree computation are defined in section 3.2 and, finally, we show some simulated results in section 3.3.

3.1 Path-Loss Model

As defined in Section 2, the existence of the edges is dependent on the signal strength. There are many models to obtain this parameter, all of them depend on the attenuation because of the distance between transmitter and receiver, shadowing due to traverse obstacles, and the multipath effect due to the reflections produced in the environment. Many of these models are empirical and only use distance information. In order to provide more realistic approach to the problem, we take into consideration a geometric information of the scenario for the shadowing effect. So, the received power including obstacles is calculated as in [8]. This work determine a Multi-Wall-and-Floor model (MWF) considering the attenuation due to the number of traversed walls and floors.

$$L_{MWF} = L_o + 10n \log_{10}(d) + \sum_{i=1}^{I} \sum_{k=1}^{K_{wi}} L_{wik} + \sum_{j=1}^{J} \sum_{k=1}^{K_{fj}} L_{fjk} \qquad (5)$$

where L_o is the path loss at a distance of 1m, n is a path-loss exponent, d is representing the distance between transmitter and receiver, L_{wik} is the attenuation due to the material type i and $k - th$ traversed wall, I is the number of traversed walls, K_{wi} the number of traversed walls of category i, L_{fjk} is the attenuation due to the material type j and $k - th$ traversed floor, J is the number of traversed floors, K_{fj} the number of traversed floors of category j.

Therefore, eq. 5 is used to compute the received signal power and then, using a threshold, the edges of the graph are obtained.

3.2 Weight Function

Algorithm 1 summarizes all the procedure of the graph configuration and the computation of the links of every node. For notation, $\mathbf{p}$ represents all the primary goal positions and their link $\mathbf{l}$, so $\mathbf{p} = [\mathbf{X}^T \ \mathbf{l}]^T$, γ is a constant threshold for communication defined by user, and n is the path-loss exponent. Initially the value of each link l is fixed to zero, once the tree is constructed, the values are assigned to every node. As $\mathbf{X}$ contains both the connected and disconnected goal locations, the link values of the disconnected are fixed to infinite. The signal strength is computed in order to obtain the connectivity graph and the obstacles are included in the computation using (5), line 3 of the algorithm. Nevertheless, this results in a binary information, 1 value when exists a link and 0 when there is no connection between the nodes. As described in Section 2, our purpose is to obtain the shortest paths in the graph, i.e. minimize the number of nodes in each branch of the tree. Furthermore, we are interested in using some criterion for local minimization of distance or links. Therefore, a weight function with these restrictions is computed to each edge of the same depth of the tree using (2). Thus, the costs of the edges are more flexible and vary in the interval [0,1]. The higher values are interpreted as the worst option and the lower ones are the potential candidates to pertain to a trees branch. This procedure is depicted in lines 6-9 of the algorithm, where l represents a depth of the tree and $\mathbf{G}_l$ are all the possible connections or edges between link level l and linked level $l + 1$.

The *cost_edges* function in line 8 calculates the costs following three different criteria:

- Connectivity: number of descendants of a node. Using this criterion the linked nodes are connected to nodes which have the highest number of possible connections.
- Exclusiveness: degree of a node, i.e. number of edges incident to the vertex or number of possible links providing communication to the node. With this criterion we force the nodes to connect to links which are the unique able to provide connectivity to some nodes.

Algorithm 1. Graph configuration

Data: Goal positions **p**, threshold γ, path-loss exponent n
Result: Minimum Depth Tree **T**
1 **Function** $graph\_conf$ (**p**)
2   **foreach** $p \in$ **p do**
3 | **G** $\leftarrow pathloss(p, \mathbf{p}, \gamma, n)$
4   **end**
5   **dp** $\leftarrow compute\_depth(\mathbf{G})$
6   **for** $l = 1, .., max(\mathbf{dp})$ **do**
7 | **w** $\leftarrow cost\_edges(\mathbf{G}_l,)$
8   **end**
9   **T** $\leftarrow graph\_lower\_cost(\mathbf{w}, \mathbf{p}(0))$
10 **end**

– Distance: the distance between relay and linked nodes. This criterion is used for distance reduction.

Hence, the defined criteria is used in the *cost_edges* function, formulated as:

$$\mathbf{w}_t = \alpha_c \mathbf{w}_c + \alpha_e \mathbf{w}_e + \alpha_d \mathbf{w}_d \tag{6}$$

where $\alpha_c, \alpha_e, \alpha_d$ represent the weighting coefficient of the different criteria.

Finally, in the line 9 is executed a shortest path algorithm. The root of the tree is the base station, which is the unique member of the team able to communicate with all the members. As all the obtained costs are positive in presence of link and null in its absence, Dijkstra algorithm is used for the shortest path calculation.

3.3 *Configuring Connection Tree*

We configure the graph finding the shortest path to each node in terms of graph depth as in (2), that is, minimizing the number of components participating in every path. This assures to save up the maximum number of robots for primary tasks. In order to provide a realistic approach to the simulations some real features are extracted from [1] and [8]. As our problem is deployment of multi-robot teams in urban or building scenarios, only the walls are considered for the MWF propagation model in (5). The attenuation due to the walls is simplified and we consider 10dB of attenuation due to each traversed wall. The path-loss exponent is considered as in free space, $n = 2$ [8]. As proved in [1], the RSSI should remain at least at -70dBm in order to assure 100% of PDR (Packet Delivery Ratio).

An example for a tree configuration, using different criteria in equation 6 described in 3.2, is represented in Fig. 2. The obtained configurations depict different tree configurations due to the weight criterion. The first configuration in 2(a) was obtained applying only the path-loss function explained in 3.1, thus Dijkstra algorithm

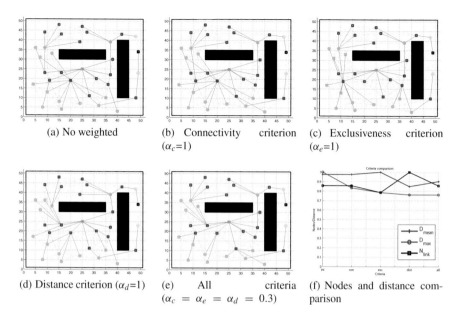

(a) No weighted

(b) Connectivity criterion (α_c=1)

(c) Exclusiveness criterion (α_e=1)

(d) Distance criterion (α_d=1)

(e) All criteria ($\alpha_c = \alpha_e = \alpha_d = 0.3$)

(f) Nodes and distance comparison

Fig. 2 Tree configuration using different criteria: Fig. 2(a)-2(e) highlight configurations adopting different criteria, each color of the nodes represents different tree depth, cyan dotted lines represent possible links and blue lines represent chosen links. 2(f) depicts the normalized values of link nodes (N_l) and normalized mean and maximum distance (D_{mean} and D_{max}) of each variable for different criteria.

computed a random connection between available nodes, with edges values equal to 0 or 1. The next two criteria, in 2(b)-2(c), have node saver performance, that is, they use the minimal number of link nodes. The distance performance is exposed in 2(d). And finally, the equally weighted criterion is performed in 2(e). We can observe, that every criteria perform better than no weighted scenario, which compute configurations with large paths and wasting nodes. Connectivity and exclusiveness save up relay nodes in exchange of larger paths. Otherwise, the distance criterion computes shortest paths, but uses more nodes as relay. The results of the equally distributed weights reach a compromise between the distance and links saving, thus, this combination can be adopted in scenarios that present difficulties due to the obstacles.

4 Optimal Deployment

The previous algorithm generates a first plan of node distribution and connections to optimize the tree using different criteria. This section describes the technique to enhance the coverage over the map, so that apart of the primary goals some members of the team executes secondary tasks, that is, giving support to other robots. Therefore,

---

Algorithm 2. Main algorithm

Data: Initial positions **p**
Result: Optimized points **p**
1 $\mathbf{T} \leftarrow graph\_conf(\mathbf{p})$
2 $\mathbf{dp} \leftarrow compute\_depth(\mathbf{T})$
3 $\mathbf{p} \leftarrow \mathbf{T}$
4 **for** $l = 2, .., max(\mathbf{dp})$ **do**
5 $\quad \mathbf{p} \leftarrow sort(\mathbf{p}_{visit})$
6 $\quad \mathbf{p}_v \leftarrow compute\_vp(\mathbf{p}, l)$
7 $\quad \mathbf{p} = \mathbf{p} \backslash \mathbf{p}_{visit}$
8 $\quad \mathbf{p} = \mathbf{p} \cup \mathbf{p}_v$
9 **end**

---

the virtual goals which accomplish this purpose are computed in order to find the optimal locations for the coverage tasks. As a result of the algorithm, a minimum number of robots used as relays once they have finished its primary task and the maximum number of robots are free to achieve the next primary task, so minimizing the total time of the mission to reach all the goals. In Fig 1 a solution can be seen for the example. Thus, in this section we present the algorithm of the virtual goals search in 4.1, and show some results of different scenarios for different criteria in 4.2.

4.1 Algorithm

All the iterative process of optimization is described in Algorithm 2 and it consists in finding the solution to equation (3). As described in Section 3, the first step is to configure the graph, i.e. obtain the links or edges between every node. Hence, the depth of each node is computed and the visible nodes are selected in order to plan the trajectory to these points. The algorithm is executed for all the nodes belonging to the same depth of the tree. In section 2 we established that only the base station is considered as visited at the beginning. Thus, when some primary nodes are considered visited, the secondary goals are computed. The visited nodes are sorted according to its occupation, that is, a node is idle if it doesn't have descendants in the tree and is occupied or busy if some node is its descendant, depicted in line 5. Then, the first ones looking for its optimal position are the idle robots. Once the virtual goals are achieved, some robots are positioned in this locations and provide signal to other members, established as visible. This procedure is repeated until optimize all the depths of the tree.

The computation of virtual goals is described in the Algorithm 3. The first step is to delimit the area which contains all the possible optimal positions. This area must be reachable from the lower and higher levels of the tree, that is, the intersection area between the signals of all the nodes of inferior and superior depths to the optimized depth, denoted as $\mathbb{S}_l$. As optimization consists in maximizing both, the visible connected and disconnected nodes as in (4), the intersection area includes the signal of all the reachable disconnected goals. Line 2 represents the intersection of all the described areas. This area represents potential locations to move robots for coverage of the following primary goals. Furthermore, the function includes the obstacles information using (5) in order to discard shadow zones, where the signal suffers fadings due to traversing of obstacles.

Once the area is computed, each of these points is a candidate to be a virtual goal, including the initial visited goal, expressed as $\mathbf{p}_l$. There are cases where it doesn't exist any virtual goal $\mathbf{x}_v$ which is able to improve the present configuration with $\mathbf{x}_{visit}$. Therefore every point of $\mathbb{S}_l$ is analyzed and inserted into the points vector $\mathbf{p}$. Then the Minimum Depth Tree $\mathbf{T}$ is computed using Algorithm 1 and the depth of every node is obtained. The process is depicted in lines 3-7 of the algorithm. Greedy criterion is adopted in order to maximize the number of linked nodes, so that, the amount of the linked nodes is stored, $\mathbf{ld}$ in line 7. The nodes with highest number of linked nodes are selected to be candidates for virtual goals $\mathbf{x}_v$. This procedure leads us to compute the solution of (4), thus the number of visible nodes is maximized and, at the same time, communication is preserved. These maximized visible nodes include disconnected nodes and nodes which pertain to higher tree depths, now achievable employing minimal number of link nodes.

From the list all the possible link candidates $\mathbf{pl}$, in line 9, the algorithm selects the nodes which are closer to the original visited nodes, in order to reduce the distance of displacement of each robot.

Algorithm 3. Virtual points search

    Data: Goal points $\mathbf{p}$, level to optimize l
    Result: Virtual points $\mathbf{vp}$
1 **Function** $compute\_vp(\mathbf{p}, l)$
2     $\mathbb{S}_l = (\mathbb{S}_{level<l} \cap \mathbb{S}_{level>l} \cap \mathbb{S}_{disc}) \notin (\mathbb{S}_{obst} \cup \mathbb{S}_{shadow})$
3   **foreach** $p \in \mathbb{S}_l$ **do**
4         $\mathbf{p} \leftarrow insert(p)$
5         $\mathbf{T} \leftarrow graph\_conf(\mathbf{p})$
6         $\mathbf{dp} \leftarrow compute\_depth(\mathbf{T})$
7         $\mathbf{ld} \leftarrow |\mathbf{dp}(l+1)|$
8   **end**
9     $\mathbf{pl} \leftarrow \{\mathbb{S}_l | max(\mathbf{ld})\}$
10    $\mathbf{d} \leftarrow dist(\mathbf{p}_l, \mathbf{pl})$
11    $\mathbf{vp} \leftarrow \{\mathbf{pl} | min(\mathbf{d})\}$
12 **end**

4.2 Optimal Tree Configuration and Discussion

In this section some simulated results of virtual goals search are presented in order to validate the algorithm. As in section 3.3, the parameters are fixed in $\gamma = -70$dBm, $n = 2$ and the strength of the transmitted signal is $P_{tx} = -42$dBm. The algorithm is executed in the scenario depicted in Fig. 3(a). The simulations were executed for several number of randomly distributed goals over the map. The disconnected nodes should be reached with communication and the depth of the tree, should be reduced, so that reducing the number of links of every branch. The search of virtual goals is executed using different weight coefficients, in order to compare the performance of the algorithm for different points distribution and the obstacles geometry. Thus, the values of the coefficients are depicted in the Table 1. A video with several examples is attached for better understanding of the algorithm[1].

The evaluation of every point of the map in the process of optimization makes the problem intractable in terms of time. Therefore, the intersection of the coverage area, line 2 of Algorithm 3 reduces substantially the number of evaluated points.

The algorithm computes virtual goals that reduce the number of link nodes or distance to some distant nodes, that is, reducing the path distance to these nodes, defined in equation 1. Independently of the optimization criterion, the optimized goals perform better deployment than initial distributions in both terms, distance and number of nodes. This is depicted in Fig. 3. In Fig. 3(b) the node metric is depicted and Fig. 3(c) represents the distance of each level of the tree. The minimal number of the required nodes (N_{min}) up to a given level is computed as the sum of the nodes used as relay (N_{link}) and the number of branches up to this level, which correspond to the minimal number of robots executing primary task at this level. Notice, that the maximum value of the required nodes corresponds to all the necessary nodes for the mission. The distances to every goal are computed as the path to this goal, using eq.(1), thus the mean(D_{mean}) and maximum (D_{max}) distances for each level are obtained. The Tree depth minimization for optimized results is observed. Due to the maximization of visible and minimization of disconnected nodes, eq. (4), dynamic tree is reconfigured in every tree depth, reducing the depth of the nodes in the tree, thus reaching the goals earlier (N_{visit}). This optimization is performed at the expense of using more link nodes at the same tree level, represented with green in Fig. 3(b). In the Fig. 3(c) the distance of the same depth is higher, because the dynamic tree has more visible nodes at each iteration of the algorithm, so that these nodes are achievable earlier.

The connectivity and exclusiveness criteria often perform similar results, although compute different weights. This is because of the dense distribution of the nodes. That is, when a link node is close to a group of nodes to be linked, it has many possible links with the next depth of the tree, so it has high connectivity. At the same

Table 1 Weight coefficients

	ini	con	exc	dist	con+dist	exc+dist	con+exc	all
α_c	0	1	0	0	0.5	0	0.5	0.3
α_e	0	0	1	0	0	0.5	0.5	0.3
α_d	0	0	0	1	0.5	0.5	0	0.3

---

[1] The video can be found in http://robots.unizar.es/data/videos/robot15yamar.avi

time, it is probably that this node is the unique able to provide signal to some distant node of the group, so it has higher exclusiveness than the other nodes of the same depth.

Figure 4 shows the average results obtained from 50 randomly distributed scenarios with 10, 25 and 50 goal nodes. The results highlight the performance of the three criteria and different weight coefficients of each one. Both, the connectivity and exclusiveness, reach the purpose of link nodes saving. And the distance considers the shortest path in exchange of employing more nodes as relay. We observe a trend that in scenarios with higher node density, the differences between different criteria are more remarkable. This is due to amount of the nodes, that is, in these scenarios there are many choices for the possible locations of virtual goals.

Maps with many obstacles perform better combining distance and any of node saving criterion. The connectivity approach saves more relay nodes in high goal density scenarios. On the contrary, in these scenarios, the combination of exclusiveness and distance criteria shows to accomplish the task with better balance of link nodes and distance.

Due to the geometric distribution of the obstacles the different criteria demonstrate perform suboptimal solutions, specially the distance criterion. The paths formed by the links between the nodes may traverse some wall and do not take into account the cost of surrounding this wall. Thus, one limitation of our method, the algorithm may compute points which will not be reachable with the shortest path or might conduct to scenarios where all the disconnected nodes can not be connected, due to the obstacles distribution. As future work it may be profitable to compute a penalization factor dependant of the obstacle density. This factor will be included in every traversing wall edges, so that, avoid to build a path through walls which are hard to surround.

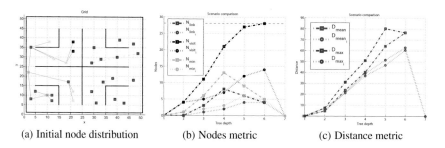

(a) Initial node distribution   (b) Nodes metric   (c) Distance metric

Fig. 3 Comparison between initial tree in 3(a) performance and dynamic tree at each tree level with virtual goals for 28 random goals. The base station is depicted with green square, located at the position (10,10), and the disconnected nodes are represented with blue squares. Circles in Fig. 3(b) and 3(c) depict no optimized tree results and squares depict the dynamic tree performance in which all the nodes are connected. Fig. 3(b) highlights the nodes usage, in blue the number of link nodes (N_{link}), in black the number of visited nodes with communication (N_{visit}) and in green, the minimal number of required nodes for every tree depth (N_{min}). In Fig. 3(c) the results of distance metric are represented, mean (D_{mean}) and maximum (D_{max}) distance, in meters, for each tree depth are represented in blue and red respectively.

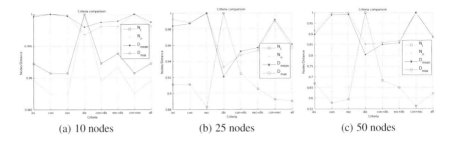

(a) 10 nodes　　　　　　(b) 25 nodes　　　　　　(c) 50 nodes

Fig. 4 Average results for different number of randomly distributed nodes using the weights of Table 1. Every figure depicts normalized values with respect to the value of that variable for all the criteria: the total number of nodes used as links (N_l), the total number of nodes required for the mission (N_n), the mean and maximum distance travelled by the team (D_{mean} and D_{max}, respectively).

Our algorithm carry out an optimal deployment with greedy nature. That is, it attempts to connect the higher number of nodes and, at the same time, it minimize the distance to the optimal position. So that, if a better position for a node is not found, this node remain in the same location. For future work some informative positions may be computed, instead of minimal distance, in order to direct the node to positions that will be beneficial for future steps of optimization process.

Another parameter to study is the task allocation when starts the algorithm. We tested our method using the criterion of the "idles first", described in section 4.1. With this approach the nodes that do not perform the task of link are the first to be optimized. It is practical for initial scenarios with few nodes per tree level. But dense distributions of nodes at the same levels may lead to configurations where virtual goals found for a robot, which could be covered with other, considered busier, but closer one.

5 Conclusion

We propose an algorithm for optimal deployment of a multi-robot team, consisting in the computation of virtual goals for covering the next goals. Therefore, the purpose of these goals is to guarantee that every member of the team executes its task with communication, positioning a relay in these points. At the same time, these locations reduce the number of link nodes maximizing the coverage over the map. The simulated results show that the algorithm allows flexibility in terms of relay nodes or distance depending on the scenario. The Multi-Wall-and-Floor model is used for the graph edges computation, including obstacles information. Dijkstra algorithm solves the shortest path problem, in order to build the minimal depth tree. A weight function is applied to the edges, allowing to adapt the best criteria depending on the scenario. The robots visit the primary goals with communication if some relay is located at the virtual goals. Thus as future work, a planning trajectory algorithm must be included in order to control and maintain, if it is possible, the communication during the trajectories execution.

References

1. Rizzo, C., Tardioli, D., Sicignano, D., Riazuelo, L., Villaroel, J.L., Montano, L.: Signal-based deployment planning for robot teams in tunnel-like fading environments. International Journal of Robotics Research (2013)
2. Tardioli, D., Mosteo, A.R., Riazuelo, L., Villaroel, J.L., Montano, L.: Enforcing network connectivity in robot team missions. The International Journal of Robotics Research **29**(4), 460–480
3. Flushing, E.F., Kudelski, M., Gambardella, L.M., Caro, G.: Spatial prediction of wireless links and its application to the path control of mobile robots. In: 9th IEEE International Symposium on Industrial Embedded Systems pp. 218–227 (2014)
4. Hollinger, G., Singh, S.: Multi-robot coordination with periodic connectivity. IEEE Transactions on Robotics **28**(4)
5. Fink, J., Ribeiro, A., Kumar, V.: Robust control of mobility and communications in autonomous robot teams. IEEE Access **1** (2013)
6. Ficco, M., Esposito, C., Napolitano, A.: Calibrating indoor positioning systems with low efforts. IEEE Transactions on Mobile Computing **13**, 737–751 (2014)
7. Ji, M., Egenrstedt, M.: Distributed coordination control of multiagent systems while preserving conectedness. IEEE Transactions on Robotics **23**(4) (2007)
8. M.Lott, Forkel, I.: A multi-wall-and-floor model for indoor radio propagation. In: Proc. 53rd Vehicular Technology Conference IEEE VTS, pp. 464–468 (2001)
9. Zavlanos, M.M., Egerstedt, M.B., Pappas, G.J.: Graph theoretic connectivity control of mobile robot networks. Proceedings of the IEEE (2011)
10. Rooker, M.N., Birk, A.: Multi-robot exploration under the constraints of wireless networking. Control Engineering Practice **15**(4), 435–445
11. Urcola, P., Riazuelo, L., Lázaro, M.T., Montano, L.: Cooperative navigation using environment compliant robot formations. In: IEEE/RSJ International Conference on Intelligent Robots and Systems (2008)
12. Anderson, S.O., Simmons, R., Goldberg, D.: Maintaining line-of-sight communications networks between planetary rovers. In: IEEE/RSJ International Conference on Intelligent Robots and Systems, pp. 2266–2272 (2003)
13. Yan, Y., Mostofi, Y.: Robotic router formation in realistic communication environments. IEEE Transactions on Robotics **28**(4) (2012)

Guaranteeing Communication for Robotic Intervention in Long Tunnel Scenarios

Carlos Rizzo, Domenico Sicignano, L. Riazuelo, D. Tardioli, Francisco Lera, José Luis Villarroel and L. Montano

Abstract In tunnel-like environments such as road tunnels or mines, a team of networked mobile robots can provide surveillance, search and rescue, or monitoring services. However, these scenarios pose multiple challenges from the robotics and from the communication points of view. Structurally, tunnels are much more longer than they are wide, and in the communication context, the multipath propagation causes strong fading phenomena. While the former can be addressed implementing routing techniques that allow multi-hop communication, fadings are unavoidable and represent a serious issue. However, under certain transmitter-receiver configurations, these fadings are predictable and periodic. On this basis, in this work we present a set of solutions to improve the communications between a base station and a robot taking advantage of spatial diversity and link-quality-aware navigation. These proposals are tested by means of simulations and real-world experiments carried out in a long railway tunnel, involving mobile and fixed nodes towards an application for monitoring purposes.

Keywords Relay deployment · Tunnel-like environments · Wireless propagation · Communications · Spatial diversity · Navigation

1 Introduction

In multi-robot operations, including surveillance, monitoring and rescue, the mission success depends on different tasks, such as motion control, navigation, localization

C. Rizzo · D. Sicignano(✉) · L. Riazuelo · D. Tardioli · F. Lera · J.L. Villarroel · L. Montano
Robotics, Perception and Real-Time Group (RoPeRT),
Aragón Institute for Engineering Research (I3A), Universidad de Zaragoza, Zaragoza, Spain
e-mail: {crizzo,doxsi,riazuelo,dantard,lera,jlvilla,montano}@unizar.es

D. Tardioli
Centro Universitario de la Defensa, Zaragoza, Spain

© Springer International Publishing Switzerland 2016
L.P. Reis et al. (eds.), *Robot 2015: Second Iberian Robotics Conference*,
Advances in Intelligent Systems and Computing 417,
DOI: 10.1007/978-3-319-27146-0_53

691

and communication. In the last years, the latter has attracted increasing interest in the robotics community, specially in real-world missions, due to the challenge of maintaining network connectivity among the robots.

In [10], the authors develop a system to deploy robots which simultaneously achieve mission tasks and maintain connectivity in urban scenarios. The system reactively constrains the motions of the robots depending on the measured signal to maintain the connectivity at all moment. In [1], the authors propose and implement an adaptive solution that positions a team of robot routers to provide communication coverage to an independent set of client robots, without previous knowledge of the clients' positions. The system is able to position the robotic routers to satisfy the robotic client demands, while adapting to changes in the environment and fluctuations in the wireless channels.

However, electromagnetic waves do not propagate in tunnels as in regular indoor scenarios nor in free space, even if Line-of-Sight (LoS) is maintained between emitter and receiver. In [6, 7], the authors introduce a technique to deploy a team of robots to carry out remote inspections and radiation surveys in different areas of the underground facilities at the European Organization for Nuclear Research (CERN). Different radio signal propagation models and stochastic estimation techniques are used to enhance the wireless communication qualities. However, the experiments in the cited works were performed in segments shorter than 100 m in length. In long tunnels, the multipath propagation affects the radio signal. Free space solutions or regular indoor propagation models do not work well, and therefore special techniques must be adopted in order to ensure reliable communication between the robots. This aspect must be taken into account when planning a communications-aware robotic application to avoid jeopardizing the success of the mission.

The contribution of this work lies in a set of techniques that allow an efficient and effective way of guaranteeing the network connectivity between a robot or a team of robots and a base station, deployed inside long tunnel scenarios.

The remaining of the paper is organized as follows: in the next Section, the fundamentals of electromagnetic propagation in tunnels are analyzed. Then, in Section 3, the test scenario and experimental setup are described. In Section 4, different techniques to guarantee the network connectivity are presented and tested by means of real-world experiments and simulations. Finally, Section 5 summarizes the main conclusions and future work.

2   Radio Wave Propagation in Tunnels

Propagation in tunnels differs from regular indoor and outdoor scenarios. For operating frequencies high enough with free space wavelength much smaller than the tunnel cross-section dimensions, tunnels behave as hollow dielectric waveguides. If an emitting antenna is placed inside a long tunnel, the spherical radiated wavefronts will be multiply scattered by the surrounding walls. The superposition of all these scattered waves is itself a wave that propagates in one dimension —along the tunnel

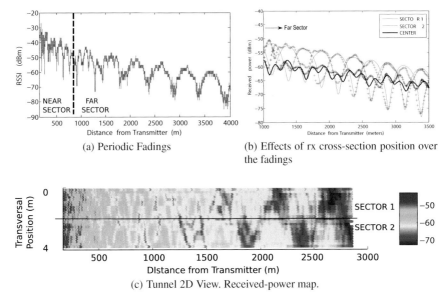

(a) Periodic Fadings

(b) Effects of rx cross-section position over the fadings

(c) Tunnel 2D View. Received-power map.

Fig. 1 Measured Received Power at 2.4 GHz. The transmitter was kept fixed and the receiver was displaced along 4 km from the transmitter. In (b), the same experiment was repeated for three different receiver's cross-section positions: left half, center and right half. The solid lines represent the modal theory simulations, and the dotted lines the experimental results. Finally, in (c), the whole cross-section is shown. The signal was sampled at 0.125 m in the transverse dimension and 0.1 m in the longitudinal dimension.

length— with a quasi-standing wave pattern in the transversal dimensions. This allows extending the communication range, but affects the signal with the appearance of strong fadings.

There are many different possible transversal standing wave patterns for a given tunnel shape. Each one is called a *mode* and has its own wavelength, close to — but different from— the free space one, and with its own decay rate. For a detailed explanation, a good online source can be found in [5].

The electromagnetic field radiated from the antenna starts propagating along the tunnel distributed among many of the possible propagating modes supported by this waveguide. After a long-enough travel distance, only the lowest order modes survive (i.e. those with the lowest attenuation rate), giving rise to the slow fadings in the so-called *far sector*. These fadings are caused by the pairwise interaction between the propagating modes, and therefore, the higher the number of modes, the more complex the fading structure. On the transmitter side, the position of the antenna allows to maximize or minimize the power coupled to a given mode, favoring the interaction between certain modes, and allowing to produce a specific fading structure.

In this specific work, given the tunnel dimensions and transmitter-receiver setup, the dominant modes are the first three modes (the ones that survive in the far sector). By placing the transmitter antenna close to a tunnel wall, we maximize the power coupled to the first and second modes while minimizing the excitation of the third one. In the receiving side, this produces a strictly periodic fading structure. The superposition of the first and second propagation modes (called EH_{11}^y and EH_{21}^y respectively) creates a 512 m periodic fading structure (Fig. 1(a)). In the very center of the tunnel there is no contribution from the second mode, and the third mode (EH_{31}^y), with lower energy, becomes observable, creating another fading structure with a different period. However, the received power associated to the fadings maxima is lower compared to the previous fading structure. The situation is illustrated in Fig. 1(b), which shows the data collected by having one antenna in each half of the tunnel, and another located exactly in the center. It can be seen that there is a spatial phase difference of 180 degrees between both halves of the tunnel (i.e. a maximum of one fading matches a minimum of the other) caused by the transversal structure of the second mode. Moreover, Fig. 1(c) shows an upper 2D view of the tunnel, collecting all the data in both the longitudinal and transversal directions, where the signal power is represented by a color according to its value (from blue to red, with blue being a higher received power). It can be noticed the same spatial phase among the fadings belonging to the same half of the tunnel, and the spatial phase difference of 180 degrees between both halves. For a more detailed explanation, see [8].

In this work, we take advantage of this spatial phase difference between the longitudinal fadings in terms of spatial diversity, as well as with navigation techniques, with the final goal of guaranteeing a link-quality above a threshold.

3   Experimental Setup

All the experiments were carried out inside a tunnel using a robotic vehicle to collect the information, and an actual communication was established among the nodes of the network. The environment, the robotic vehicle used and the communication protocol in charge of connecting all the nodes belonging to the network are described below.

The Environment. The Somport tunnel was selected as the location to carry out the experiments. This is an old out-of-service railway tunnel representative of long straight tunnels common in transport or mine applications. The 7.7 km straight tunnel connects Spain and France through the central Pyrenees. It has a horseshoe-shape cross section, approximately 5 m high and 4.65 m wide (Fig. 2(a)). It also has small emergency shelters every 25 m (which are 1 m wide, 1.5 m high, and 0.6 m in depth), and 17 lateral galleries, each of them more than 100 m long and of the same height as the tunnel.

The Robotic Vehicle and Communication Nodes. To test the concepts, a robotic vehicle and communication nodes (establishing an ad-hoc network) were used. The communication nodes (base station - r_0, transmitter - r_1, and repeaters - r_2 and r_3)

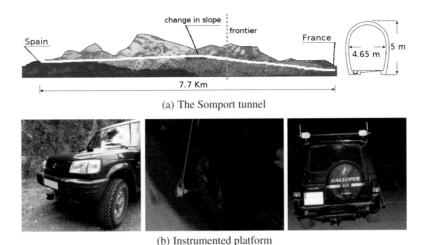

(a) The Somport tunnel

(b) Instrumented platform

Fig. 2 Experimental Setup

are composed by a Laptop with a WiFi network card (Ralink rt2770 chipset), and their position will be specified in each experiment.

An all terrain vehicle was used as the mobile platform. It was equipped, besides with one or two communication nodes (depending on the experiment), with two 0.5 degree resolution odometers and a SICK LMS 200 Laser Range Finder (for localization and mapping purposes). The equipped platform can be seen in Fig. 2(b).

The Communication Protocol. To carry out the experiments and verify that the communication was being effective, we used the RT-WMP [11], in charge of routing the packets among the nodes belonging to the network. This is a protocol for MANETs that supports time-sensitive traffic, guaranteeing that the end-to-end message delay has a bounded and known duration. The protocol also provides global static message priorities and supports multi-hop communications. Its routing is based on the link quality among the nodes: the protocol chooses the safer path to route the messages using the information contained in the Link Quality Matrix (LQM) —a weighted adjacency matrix that contains information about the topology of the network in terms of link quality— which is shared among all the nodes. Each link is assigned a weight that depends on the Received Signal Strength Indicator (RSSI) perceived by the receiver nodes. Higher values are assigned to lower qualities and lower values to higher ones, and the Dijkstra algorithm is then used to compute the path that offers the higher probability for a message to reach its destination.

4 Proof of Concepts: Guaranteeing a Good Quality Link

Due to the known correlation between the RSSI and the Packet Delivery Ratio (PDR), the received power must be maintained above certain threshold to avoid packet loss [13]. To deal with the periodic fadings in the Somport tunnel and maintain the connectivity between a team of robots and a base station (located at the entrance of the tunnel), we presented a fading-crossing navigation technique in [9], where the robots used each others as relays to go across the fadings.

However, in this work, we propose other solutions that depends on specific resources availability. Particularly, we consider approaches that rely, on one hand, on large-scale spatial-diversity schemes and, on the other hand, on propagation-aware navigation strategies.

4.1 Taking Advantage of Large-Scale Spatial Diversity

Lienard et al. [3] propose to replace the concept of spatial diversity with modal diversity on the basis that an increase in the channel capacity is related to the presence of several propagating modes. As the number of modes rapidly decreases with the distance, the authors claim that a full benefit from the spatial diversity can not be obtained at distances far from the transmitter [4]. However, our approach differs in the fact that we aim to take advantage of the transversal structure of the fadings (i.e. the spatial phase delay between both halves of the tunnel) to establish a secure zone for communication. That is, instead of increasing the channel capacity, we ensure a link quality above a certain threshold to maintain the communication with a base station.

4.1.1 Spatial Diversity on the Receiver

In commercial devices equipped with spatial diversity capabilities, the separation between two antennas is in the order of few centimeters (0.125 m at most in the case of 2.4 GHz devices) to overcome the temporal fadings (i.e. temporal variations of the channel). However, by inspecting Fig. 1(c), it can be noticed that in order to take full advantage of the transverse structure of the fadings, larger scale spatial diversity schemes must be implemented (in the order of decimeters or even meters). Moreover, what is relevant is the antenna position in the tunnel cross section.

In the specific setup for this work (transmitter close to the tunnel wall), the tunnel is divided in halves according to the spatial phase difference of the fadings (Fig. 1(c)). By using two antennas and placing them in each half of the tunnel, we can provide a better quality link with the base station in comparison to the case of placing both antennas in the same half of the tunnel.

To test the concepts in a real situation, we placed two communication nodes in the moving platform 1.5 m apart from each other.

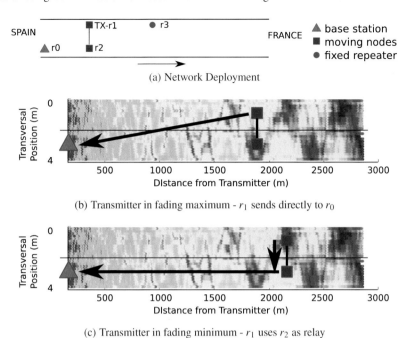

(a) Network Deployment

(b) Transmitter in fading maximum - r_1 sends directly to r_0

(c) Transmitter in fading minimum - r_1 uses r_2 as relay

Fig. 3 Experiment exploiting spatial diversity on the robot.

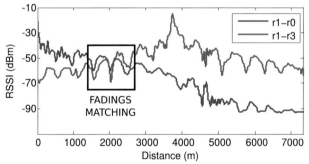

Fig. 4 Measured left-antenna RSSI for the base station and mid-tunnel repeater signals in a one-way travel of the mobile platform along the full tunnel lenght. Notice the good matching of the fading maxima of the two sources.

Consider Fig. 3: with the two nodes on the mobile platform active, a straight line travel path along the tunnel center guarantees that when one of the transmitting antennas is located in a fading minimum, the other will be in a maximum at the other side of the tunnel center and viceversa. The expected behavior using the multihop protocol cited before is that when r_1 (the transmitter node) is on a fading maximum, it sends the data directly to r_0 (base station), as shown in Fig. 3(b). On the other hand, when the transmitter is on a minimum, r_2, the spatial diversity node, is on a maximum, and r_1 uses this node as relay to send the data to r_0 (Fig. 3(c)).

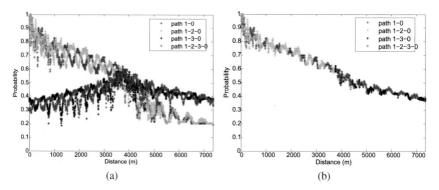

Fig. 5 Probability of the different links (a) and Probability of packets reaching destination and the path followed (b).

Using only a base station at the entrance of the tunnel would be enough to cover around one half of the tunnel length. As shown in Fig. 3(a), to extend the coverage along all the tunnel length the right strategy is to place repeater(s) in position(s) in which the base station emitter produces a fading maximum, so that the max/min pattern of the repeaters overlaps with that of the base station. In fact, in our scenario a single repeater, r_3, was placed 4 km away from the Spain gate —as stated before, in a spot where the base station creates a fading maximum— and it has proven to be sufficient to guarantee the link quality between the base station and the mobile platform. Fig. 4 shows the signals measured along the tunnel full length. The repeater position was estimated a priori from the known periodic fading structure of the base station Radio Frequency (RF) field.

Fig. 5 shows the result of the experiments. Fig. 5(a) presents the probability —from the protocol point of view— of the different links during the trip of the mobile platform through the whole tunnel. The protocol calculates the probability as a function of the number of hops and the quality of each link, and selects the path with highest probability. As expected, the probability of the paths that include node r_2 (the on-board relay) are always shifted about 180 degrees with respect to those that do not. Fig. 5(b) shows the routes followed by the messages during the experiment which corresponds with the envelop of the four curves just cited. At the beginning the protocol uses alternatively the paths $r_1 \rightarrow r_0$ and $r_1 \rightarrow r_2 \rightarrow r_0$ while when the mobile platform passes the relay r_3 the protocol switches automatically to $r_1 \rightarrow r_3 \rightarrow r_0$ and $r_1 \rightarrow r_2 \rightarrow r_3 \rightarrow r_0$.

As a practical application, the system is able to build a map of the tunnel in the base station in real time, using the odometry and laser data provided by the robotic vehicle. The laser sensor information (1.51KB) is sent to the base station at 4.6 Hz and contains 360 measurements of a 180? of field of view. The odometry data (0.72KB) is sent to the base station at 10 Hz and contains the position and velocity of the vehicle. Using techniques based on SLAM algorithms [2], the computer placed on the base station (a Getac B3000 Rugged laptop with a core i7 processor) builds the map of the environment. Fig. 6 shows three different areas of the environment

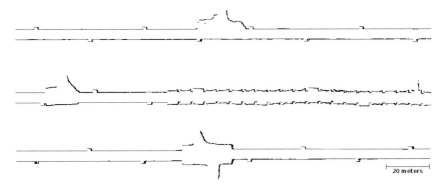

20 meters

Fig. 6 Three different areas of the map built.

mapped, highlighting the most relevant features of the map built. The whole map is around 7.7 km and the areas showed are around 200 m.

4.1.2 Spatial Diversity on the Infrastructure

If otherwise the mobile platform carries only one node, the strategy to produce better coverage is to deploy a repeater shifted half fading period (256 m in this case) from the base station, as shown in Fig. 7. With this arrangement, on the same side of the tunnel the fading maxima of the base station signal coincide with the minima of the repeater signal and viceversa (see Fig. 8(a)). If only one transmitter is used (with no shifted repeater), the robot would has to traverse the fading minima and its link with the base station could fall below an acceptable threshold (fixed at -60 dBm in the figure to obtain a PDR close to 100%) causing data loss and even a disconnection of the network. However, by taking advantage of two 180 degrees shift between the fadings, the robot can benefit from the maxima of each one of them and maintain a link quality above the threshold.

The expected behavior is that when r_1 is in a fading maximum with respect to the base station, it sends the data directly to r_0. On the other hand, when r_1 is in a fading minimum with respect to the base station, it means it is in a maximum with respect to r_2 (the repeater), and uses this node as relay to send the data. As r_0 and r_2 are fixed, and separated 256 m, there is always a constant and good quality link between them. Figure 8(b) shows the result of the experiment. As expected the protocol chooses alternatively the paths $r_1 \rightarrow r_0$ and $r_1 \rightarrow r_2 \rightarrow r_0$ as a function of the RSSI (and thus of the probability) guaranteeing the optimal routing at all moment.

4.2 Through Navigation Strategies

On the other hand, if only a base station with no repeaters is available, and if the robot is unable to carry two receivers separated a distance in the order of decimeters,

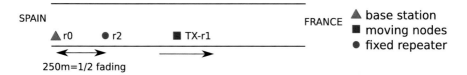

Fig. 7 Experiment exploiting spatial diversity on tunnel.

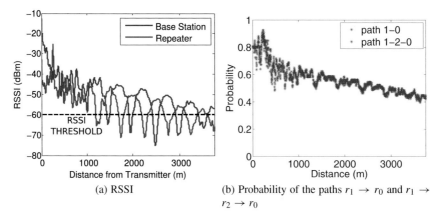

(a) RSSI

(b) Probability of the paths $r_1 \rightarrow r_0$ and $r_1 \rightarrow r_2 \rightarrow r_0$

Fig. 8 Measured single antenna RSSI for the base station and mid-tunnel repeater signals in a one-way travel of the mobile platform along half tunnel lenght. Notice the good matching of the fading maxima of the base station signal with the minima of the repeater one.

it is evident that traveling in a straight line and traversing the fadings will cause the robot to fall in hazardous areas in terms of communications (fadings minima), where a RSSI below a threshold causes data loss and may even result in a completely loss of connectivity between the robot and the base station. In this situation, the transversal structure of the fadings can be taken into account to design a navigation strategy capable of improving the communication with the base station.

Intuitively, it can be seen in Fig. 9(a) that the robot should follow an *s-like* or *zig-zag* trajectory (depicted as a blue dashed line) , traveling from the fadings maxima in one half of the tunnel (*sector 1*) to maxima in the other half (*sector 2*), avoiding the valley zones.

In order to locate the fadings maxima, the proposed navigation can benefit from RF-source localization approaches such as those presented in [14, 15], taking into account some considerations. Particularly in [12], the authors sample the RSS and estimate the 2D gradient through gently oscillatory navigation paths to guide a ground robot to a RF source.

The case of 2D navigation inside a tunnel is slightly different, though. In the near sector, the robot must actually move away from the base station, in the direction in which the RSSI decreases. Once in the far sector, the fadings maxima can be considered as the RF sources in the previous approaches. In order to avoid local maxima, a global navigation goal is set.

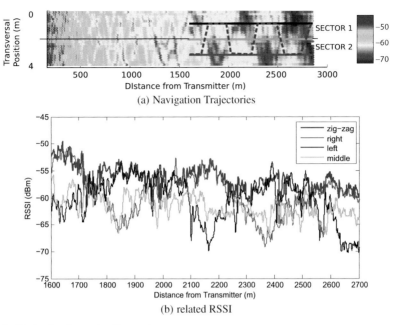

(a) Navigation Trajectories

(b) related RSSI

Fig. 9 Navigation Strategies Comparison.

To validate this proposal, some simulations have been performed over the real RF coverage map. A pioneer robot navigates 1 km in the presence of slow fadings (far sector), using different navigation strategies.

Fig. 9(b) shows the RSSI comparing the zig-zag navigation path with others, such as the most intuitive navigation path for tunnels: travelling in a straight line. It can be seen that when travelling in straight lines along each half of the tunnel, the 512 m fading are perceived. When the robot travels through the tunnel center, no deep fadings appear, but the received power is lower compared to the fadings maxima. At last, by travelling in zig-zag the robot takes advantage of the maxima related to the fadings in each half of the tunnel, ensuring a better link quality in comparison to the previous cases.

5   Conclusions

In this work, we have presented a set of solutions to improve and guarantee the communication between a robot and a base station in long tunnel scenarios, where the signal is affected by strong fading phenomena that can jeopardize the reliability of the communication with data loss or even disconnection of the network. We have shown that in tunnel scenarios the transversal structure of the large-scale fadings cannot be ignored and that commercial small-scale spatial diversity is not effective.

Also, we have proposed a set of methods that take into account the predictable behavior of the fadings and take advantage of their characteristic with the end of offering a better communication quality. We also proposed a method to select the more convenient positions where to place repeaters and, on the other hand, a zig-zag navigation strategy that allows to extend the communication range without the need for additional hardware.

The deployment methods were tested in a real-world scenario using communication nodes and a multi-hop protocol. As a result, we demonstrated that the routing algorithm chooses the best path taking advantage of the fadings maxima, guaranteeing at every moment the communication with the base station. Also, the simulation results of the navigation strategy show that the link quality between the mobile node and the base station is maintained at a higher level compared to other cases, such as traveling in straight lines either at each side of the tunnel, or at the center. Some topics remain open for future work, such as performing autonomous navigation using signal quality maps, and extend these solutions to more complex underground scenarios.

Acknowledgments This work has been funded by the projects TELOMAN DPI2012-32100 (CI-CYT, Spanish Gov.), SIRENA (CUD 2013-05) and DGA-FSE (T04).

References

1. Gil, S., Kumar, S., Katabi, D., Rus, D.: Adaptive communication in multi-robot systems using directionality of signal strength. In: Proceedings of the International Symposium on Robotics Research (2013)
2. Grisetti, G., Stachniss, C., Burgard, W.: Improved techniques for grid mapping with rao-blackwellized particle filters. IEEE Transactions on Robotics **23** (2007)
3. Lienard, M., Degauque, P., Molina-Garcia-Pardo, J.M.: Wave propagation in tunnels in a MIMO context-a theoretical and experimental study. Comptes Rendus Physique **7**(7), 726–734 (2006)
4. Molina-Garcia-Pardo, J., Lienard, M., Degauque, P., Dudley, D., Juan-Llacer, L.: Interpretation of MIMO channel characteristics in rectangular tunnels from modal theory. IEEE Transactions on Vehicular Technology **57**(3), 1974–1979 (2008)
5. Orfanidis, S.J.: Electromagnetic waves and antennas. Rutgers University (2014). http://www.ece.rutgers.edu/~orfanidi/ewa/
6. Parasuraman, R., Fabry, T., Molinari, L., Kershaw, K., Castro, M.D., Masi, A., Ferre, M.: A multi-sensor rss spatial sensing-based robust stochastic optimization algorithm for enhanced wireless tethering. Sensors **14**(12), 23970–24003 (2014)
7. Parasuraman, R., Masi, A., Ferre, M.: Wireless Communication Enhancement Methods for Mobile Robots in Radiation Environments. Radio communication for robotic application at CERN. Ph.D. thesis, Madrid, Polytechnic U. (september 2014), presented October 17, 2014
8. Rizzo, C., Lera, F., Villarroel, J.: Transversal fading analysis in straight tunnels at 2.4 GHz. In: 2013 13th International Conference on ITS Telecommunications (ITST), pp. 313–318, November 2013
9. Rizzo, C., Tardioli, D., Sicignano, D., Riazuelo, L., Villarroel, J.L., Montano, L.: Signal-based deployment planning for robot teams in tunnel-like fading environments. The International Journal of Robotics Research **32**(12), 1381–1397 (2013)
10. Tardioli, D., Mosteo, A., Riazuelo, L., Villarroel, J., Montano, L.: Enforcing Network Connectivity in Robot Team Missions. The International Journal of Robotics Research **29**(4), 460–480 (2010)

11. Tardioli, D., Sicignano, D., Villarroel, J.L.: A wireless multi-hop protocol for real-time applications. Computer Communications **55**, 4–21 (2015)
12. Twigg, J., Fink, J., Yu, P., Sadler, B.: RSS gradient-assisted frontier exploration and radio source localization. In: 2012 IEEE International Conference on Robotics and Automation (ICRA), pp. 889–895, May 2012
13. Vlavianos, A., Law, L., Broustis, I., Krishnamurthy, S., Faloutsos, M.: Assessing link quality in IEEE 802.11 wireless networks: which is the right metric? In: IEEE 19th International Symposium on Personal, Indoor and Mobile Radio Communications, PIMRC 2008, pp. 1–6, september 2008
14. Wadhwa, A., Madhow, U., Hespanha, J., Sadler, B.: Following an RF trail to its source. In: 2011 49th Annual Allerton Conference on Communication, Control, and Computing (Allerton), pp. 580–587, September 2011
15. Yu, P.L., Twigg, J.N., Sadler, B.M.: Radio signal strength tracking and control for robotic networks. In: Proc. SPIE, vol. 8031, pp. 803116–803116-12 (2011)

Visual Surveillance System with Multi-UAVs Under Communication Constrains

P. Ramon, Begoña C. Arrue, J.J. Acevedo and A. Ollero

Abstract In this paper, it is proposed a visual surveillance system for multiple UAVs under communication constrains. In previous works, a dynamic task allocation algorithm was designed for assigning patrolling and tracking tasks between multiple robots. The idea was to assign the intruders dynamically among the robots using one-to-one coordination technique. However due to communication constrains, every UAVs could store different information. In this paper, local information about targets is obtained by a visual algorithm that detects moving objects during surveillance tasks using fixed low-cost monocular RGB-camera connected to an on-board computer. This system was tested in a urban surveillance scenario, implemented in an indoor test-bed, under EC-SAFEMOBIL project.

Keywords UAVs · Multi-agent · Surveillance · Computer vision · Motion detection · Tracking

1 Introduction and Related Work

Unmanned Aerial Vehicles (UAVs) have recently aroused great interests in civil and commercial applications for monitoring, surveillance and disaster response. UAV is a type of very complex system which integrates different hardware components, such as camera, Global Positioning System(GPS), Inertial Measurement Unit (IMU), controller, and different software components, such as image processing, path plan-

P. Ramon(✉) · B.C. Arrue · J.J. Acevedo · A. Ollero
GRVC, Escuela Superior de Ingeniera, Av. de Los Descubrimientos S/n, Sevilla, Spain
e-mail: {pabramsor,barrue,aollero}@us.es, jacevedo@isr.ist.utl.pt
http://www.grvc.us.es

B.C. Arrue and A. Ollero are with Grupo de Robotica, Vision y Control, Universidad de Sevilla, Spain.
J.J. Acevedo is with Instituto de Sistemas e Robotica, Instituto Superior Tecnico, Lisboa, Portugal.

© Springer International Publishing Switzerland 2016
L.P. Reis et al. (eds.), *Robot 2015: Second Iberian Robotics Conference*,
Advances in Intelligent Systems and Computing 417,
DOI: 10.1007/978-3-319-27146-0_54

ning and inner loop control. Due to repetitive tasks in remote and/or hazardous environments, UAV is very promising to play more important roles in many applications and recent developments have proven the benefits in different ways. Equipped with cameras and other sensors, these autonomously flying robots can quickly sense the environment from a birds eye view or transport goods from one place to another.

For some applications, it is beneficial if a team of coordinated UAVs rather than a single UAV is employed. This emerging technology is still at an early stage and, consequently, significant research and development efforts are needed.

In large scenarios, an opened communication between all the robots could be no possible. Therefore, a distributed approach would be more appropriate in these cases. Also, decentralized and distributed approaches offer increased robustness, adaptability and scalability (as proposed in [1]). These methods usually rely on the interchange of a reduced amount of variables (so called *coordination variables*) required to obtain a solution in a cooperative manner. In [2], it is proposed a dynamic and decentralized task allocation algorithm based on the one-to-one coordination to solve the multi-target tracking task allocation even under communications constraints. The idea is to perform an allocation process between each pair of contacting robots. They join their presently assigned tasks, such as they both share the same information about the targets to be tracked by them. Then, based on the information of both aerial robots, each one in a independent manner can obtain the same solution and update its list of assigned targets to track.

Communication infrastructure consists on hardware components (robots and servers/ground-stations) and their software. Internally, each robot has an module that is responsible communications. First communication network is local communication between robots. This is a narrow-band and short-range communication. The algorithms in [2] take account these constrains. First network or short-range network (red dotted line **a** in Fig. 1), is used by aerial robots to exchange *coordination variables* [8] (the minimum information that the robots require to share to solve the problem from a distributed manner), which will be explained in section 2.2. Also a second network (solid blue line **b** in Fig. 1) links aerial robots through the *cloud* with servers or ground stations. Thanks to this connection the robot can update the global plan. In our experiments, each task is completely independent and refers to a set of robots patrolling a perimeter and tracking multiple targets. In above, the explained system refers to the behavior and procedure of aerial robots in a single task.

This paper proposes a low-cost on-board real time vision system [14, 15] for target/intruder detection. This information is stored locally and shared while communication is available. The vision system provides local information about intruders to UAVs. Afterwards, due to one-to-one communication, these data are shared to others robots. This approach was tested during experiments for cooperative (dynamic task allocation) perimeter surveillance between aerial robots [1, 2] under the European Project EC-SAFEMOBIL, which addresses an urban surveillance scenario.

The paper is organized as follows. The system description is in section 2, which describes the Vision System and dynamic task allocation. Validation results are presented in section 3.

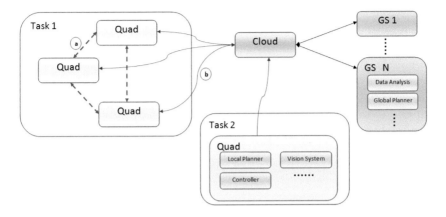

Fig. 1 Communication architecture

2 System Description

Let a set of n aerial robots be $Q := Q_1, Q_2, ..., Q_n$ (Fig. 4). Robots' mission is patrolling a perimeter S in order to detect and track m possible intruders $I := I_1, I_2, ..., I_m$ tacking account communication constrains between aerial robots. Every i aerial robots and j intruder can have different capabilities as described in [2]. Also, in order to provide to aerial robots information about the environment, they were supplied with a single fixed camera.

Therefore, this paper consider two problems that have to be approached in a distributed manner tacking into account that robots can only communicate asynchronously:

– Visual perception problem. It is responsible of detecting and tracking visually targets using on-board system to ensure performance during denied communication.
– Distributed task Allocation. It ensures task assignation of targets between aerial robots in order tu maximize the number of targets being tracked.

2.1 Vision System

In this section a tracking system to provide to the aerial robots information about the intruders is presented. This information, is used locally until the robot meet another one. Then they share the information and automatically re-assign tasks in order to achieve the global plan. However, these communications is not continuous, so it's essential to have a fast algorithm for acquiring information to have the data updated as much as possible.

Feature Detection and Optical Flow. Since Horn-Schunck formulation [7] of his optical flow algorithm, many of optical flow algorithms [3, 6, 12, 13] appeared. We aimed to achieve as much information as possible but under the premise of a Real Time application. In order to reach that, we use a pyramidal implementation of Lucas Kanade feature tracker algorithm (PYR-LK) [4]. This algorithm is supplied with a set of "key features" and calculate its motion. PYR-LK is not a dense optical flow estimator. So, on the one hand, partial information about the movement on the scene can be lost. But on the other hand, is faster than dense algorithms.

The set of features is grouped by a *cluster algorithm* using position and velocity. These clusters filter noise features in order to obtain better results in following steps.

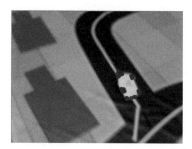

Fig. 2 Results of optical flow detection

Feature Transformation and Particle Filtering. We have implemented a particle filter that uses data obtained by the camera combined with aerial robot IMU information.Features obtained by previous step are transformed from 2D to 3D by using a simple pinhole model. An example of pixels' transformation is shown in picture Fig. 3a.

Every particle (of particle filter) simulates a possible target. These particles use a simple car motion model. Fig. 3b shows results in one frame of a tracking sequence.

Collecting Particle Filter Results. Particle filter provides a disjointed set of particles distributed around the targets. In order to use this information, we use the probabilistic clustering method called Gaussian Mixture Model Expectation-Maximization algorithm or GMMEM [5, 11]. GMM is a representation model for a complex probabilistic functions as a mixture of simple models (Gaussians in this case). Let be $p(x|\theta)$ the probability distribution of a random variable x with parameters θ.

A key advantage of using Particle filter and afterwards GMMEM rather than other filters as Kalman Filters is that they allow multiple targets intrinsically. It also provides probabilistic information about the possible detected objects. Fig. 3c and 3d show results of single and multiple target tracking using particle filter and clustered with GMMEM in order to get a belief of target's position.

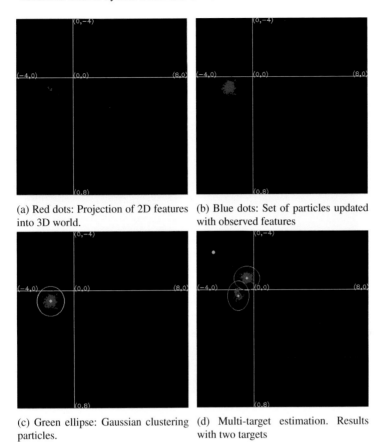

(a) Red dots: Projection of 2D features into 3D world.

(b) Blue dots: Set of particles updated with observed features

(c) Green ellipse: Gaussian clustering particles.

(d) Multi-target estimation. Results with two targets

Fig. 3 Representation of results

2.2 Dynamic Task Allocation

In the one-to-one coordination technique [2] when two robots meet, they set out a reduced version of the whole problem only with the information they have. This technique represents a good approach in dynamic scenarios and under communication constraints. The proposed solution requires that each robot Q_i stores locally information, so called *coordination variables* (a vector containing information about its own status and capabilities, segment of perimeter to patrol, known number of aerial robots, known targets, an ordered plan composed by the tasks that have been allocated to the robot and the next way-point to visit). Communication between robots is ensured along the perimeter of surveillance. If a robot detects a target, it leaves the surveillance task and start tracking the intruder. Convergence of this algorithm was proved previous work [1, 2].

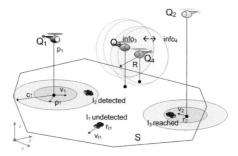

Fig. 4 Aerial robots patrolling to detect and follow potential intruder [2]. Q_i Aerial robot i; P_i Position of aerial robot i; I_j Intruder j; r_j Position of robot j; R Communication range; S Perimeter of surveillance

When two robots meet, they share their information about the known targets, their own states and capabilities. As both robots have the same information about targets and the same allocation functions to solve the problem, both compute the same solution in an independent manner. Therefore, each one obtains the plan for both and can execute its own plan. This assignment function should consider the targets to be allocated and the features of both robots. When a robot looses sight of the target which it was tracking, it sets the target as unassigned in its target list, updates the assigning time, takes this target out from its plan and recompute its own plan, according to its new target list. The robots without assigned targets perform surveillance task on the perimeter to detect new targets. This process is repeated continuously.

3 Validation Results

The algorithm was tested using datasets recorded in a testbed with line-follower cars and quads (HummingBird) with onboard computers (odroids U3) and Logitech cameras C920 (Fig. 5). The testbed has limited size, so in order to test communication constraint data transmission between aerial robots were limited by range assigned manually. As mentioned, convergence of dynamic task allocation system was proved in previous work [2]. Also scalability issues were studied under EC-SAFEMOBIL European project using up to 5 aerial robots on real experiments and up to 20 in simulations.

Regarding feature detection, two algorithms were tested in order to achieve better performance and computing time. Particularly Shi-Tomasi feature detector [10] and Features from accelerated segment test (or FAST) [9] were tested. The table 1 shows relevant results of test cases:

First conclusion is that FAST algorithm is faster than Shi-Tomasi. So in order to ensure a real time application is a better option. A second conclusion is that FAST is more sensitive that Shi-Tomasi because it doesn't filter the number of particles.

(a) Drone + CPU   (b) Testbed Scenario

Fig. 5 Set up

Then, it needs a feature filter afterwards. At last, we tested feature detectors with an additional gaussian filter. However, it increases computing time and does not improve the performance of the feature detectors. In conclusion, best result is obtained with FAST combined with feature filter.

The following paragraphs describes results and conclusions of the test in the CATEC's testbed. First and second experiments were recorded using the testbed VICON system and only one possible target was in the scene. In order to test higher vibrations and possible looseness of targets, a third experiment was performed, with an UAV controlled manually by a pilot. Furthermore a second moving intruder was added in the scene to test multi-target feature.

Table 2 shows a summary of results. Both, percentage of cpu and memory usage remain stable. Only computational time seems to have variations mainly during feature detection and optical flow computation due to the increased number of features.

Fig. 6 illustrates position estimation resulted from whole process. Red line is the ground truth of the vehicle position. Blue crosses are the estimated position at every frame. Fig 6b, shows the results of a second experiment where the vehicle cross under a bridge (bridge is shown in fig. 5b). Also results of experiments 3 (fig. 6c) show the behavior of the algorithm with bridge occlusion and when the vehicle comes out of the camera.

Table 1 Feature detection comparison

Algorithm	Parameters	Avg. N. features	Avg. Time
Shi-Tomasi	25^1	24.3	0.0366
Shi-Tomasi	50^1	36.75	0.0367
FAST	-	$40|360^2$	$0.015|0.03^2$
FAST+filter	50^1	35.64	0.0152
blur+FAST+filter	50^1	22.83	0.0183

[1] Max. allowed no. features by the filter.
[2] Left: High vibrations. Right: No vibrations.

Table 2 Time summary

Experiment	AverageTime	FPS	%CPU	%memory
1	0.0415	24.1253	~ 83.2	~ 0.7
2	0.0433	23.1051	~ 82.7	~ 0.7
3	0.0466	21.4667	~ 83.3	~ 0.7

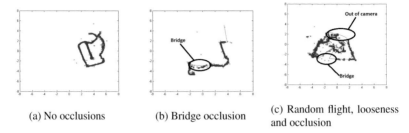

(a) No occlusions   (b) Bridge occlusion   (c) Random flight, looseness and occlusion

Fig. 6 Experimental Results

4   Conclusions and Future Works

In this paper, we have presented an algorithm to generate information about moving targets tracked by UAVs. This local information ensures achieving local tasks. In addition, global convergence of plans is ensured thank to the dynamic task allocation of [2]. The proposed system is robust to changes in the initial conditions. Experimental and simulation results validate the approach and shows some relevant features.

Furthermore, the vision system is executed on-board at real time providing updated and useful information to robots. In future work, we aim to improve motion feature detection to allow better and dense results without importantly increasing computation time. Parallel implementation of this algorithm and dynamic parameter programming seems to be the best option. Finally, all the results of the vision system were obtained from in-door experiments. In order to prepare the system for the real life, out-door experiments need to be done.

We also plan to implement visual perception as long-time vision memory and machine learning algorithm to apply better filtering on results and making possible to differentiate tracked objects on the environment.

Acknowledgement We would like to thank Miguel Angel Trujillo, Yamnia Rodriguez and Irene Alejo for their support with the experiments. This work has been developed in the framework of the project EC-SAFEMOBIL (FP7-ICT-288082) EU-funded project, the AEROMAIN(DPI2014-59383-C2-1-R) Spanish National Research project and the project MarineUAS (H2020-MSCA-ITN-2014). Also, J.J. Acevedo, was also partially supported by strategic funding LARSyS (FCT [UID/EEA/ 50009/ 2013]).

References

1. Acevedo, J., Arrue, B., Maza, I., Ollero, A.: Cooperative large area surveillance with a team of aerial mobile robots for long endurance missions. Journal of Intelligent and Robotic Systems, 1–17. doi:10.1007/s10846-012-9716-3
2. Acevedo, J.J.: Cooperation of multiple heterogeneous aerial robots in surveillance missions. PhD thesis, Escuela Tecnica Superior de Ingenieria, december 2014
3. Barron, J., Fleet, D., Beauchemin, S.: Performance of optical flow techniques. International Journal of Computer Vision 12(1), 43–77 (1994)
4. Bouguet, J.-Y.: Pyramidal implementation of the lucas kanade feature tracker description of the algorithm (2000)
5. Fern, X.Z., Lin, W.: Cluster ensemble selection. Statistical Analysis and Data Mining 1(3), 128–141 (2008)
6. Fleet, D.J., Langley, K.: Recursive filters for optical flow. IEEE Transactions on Pattern Analysis and Machine Intelligence 17(1), 61–67 (1995)
7. Horn, B., Schunk, B.: Determining optical flow. Artificial Intelligence 20 (1981)
8. McLain, T.W., Beard, R.W.: Coordination variables, coordination functions, and cooperative timing missions. Journal of Guidance, Control, and Dynamics 28(1), 150–161 (2005)
9. Rosten, E., Drummond, T.W.: Machine learning for high-speed corner detection.In: ECCV, pp. I:430–I:443 (2006)
10. Shi, J., Tomasi, C.: Good features to track. In: IEEE Conference on Computer Vision and Pattern Recognition (CVPR 1994), Seattle, June 1994
11. Smyth, P.: Model selection for probabilistic clustering using cross-validated likelihood. Statistics and Computing 10(1), 63–72 (2000)
12. Sun, D., Roth, S., Black, M.J.: Secrets of optical flow estimation and their principles. In: CVPR, pp. 2432–2439. IEEE Computer Society (2010)
13. Sun, D., Roth, S., Lewis, J.P., Black, M.J.: Learning opticalflow. In: Forsyth, D.A., Torr, P.H.S., Zisserman, A. (eds.) ECCV (3). Lecture Notes in Computer Science, vol. 5304, pp. 83–97. Springer (2008)
14. Lian, F.-L., Yu, T.-H., Fuh, C.-S.: 3d indoor environment construction from monocular camera on quadricopter. Conference on Computer Vision, Graphics and Image Processing (2014)
15. Duan, H., Bi, Y.: Implementation of autonomous visual tracking and landing for a low-cost quadrotor. International Journal for Light and Electron Optics (2013)

Part VI
Educational Robotics

Simulation of a System Architecture for Cooperative Robotic Cleaning

Hugo Costa, Pedro Tavares, Joana Santos, Vasco Rio and Armando Sousa

Abstract The increase of the use of Autonomous Vehicles in different types of environments leads to an improvement of the Localization and Navigation algorithms. The goal is to increase the levels of efficiency, security and robustness of the system, minimizing the tasks completion time.

The application of cleaning robots in domestic environments have several advantages however some improvements should be performed in order to develop a robust system. Also in large spaces one robot doesn't achieve the desired performance in terms of robustness to faults and efficiency in the cleaning process. Considering a fleet of autonomous robots, this process could be improved. The purpose of our paper is the presentation of an architecture for management a fleet of cleaning robots, considering a complete coverage path planning for large and structured environments. Compartments are found in a grid-like decomposition and an area coverage strategy are evolved (optimized) by using Genetic Algorithms. The Task allocation module is based on Auctions strategy, thus obtaining cooperation under dynamic constraints in complex environments. The case study optimizes the number of robots involved in the cooperative cleaning of a full building in the campus, based on its real architectural plans.

Keywords Multi-robot cooperation · Cleaning · Path planning · Area-covering · Genetic algorithms

H. Costa(✉) · P. Tavares · V. Rio · A. Sousa
FEUP - Faculty of Engineering, University of Porto, Porto, Portugal
e-mail: {ee10112,ee10131,ee04196}@fe.up.pt

J. Santos · A. Sousa
INESC TEC - INESC Technology and Science (formerly INESC Porto), Porto, Portugal
e-mail: {ee09133,asousa}@fe.up.pt

© Springer International Publishing Switzerland 2016
L.P. Reis et al. (eds.), *Robot 2015: Second Iberian Robotics Conference,*
Advances in Intelligent Systems and Computing 417,
DOI: 10.1007/978-3-319-27146-0_55

717

1 Introduction

Cleaning Robots belongs to the large group of service robots, which have the ability to perform basic housework tasks but also take care of people like a vigilant. Vacuum Cleaning Robots have gained relevance among household users. However, commercial applications typically are designed to be applied in small spaces and offices. In large environments like airports and train stations with a lot of free space, the cleaning results are better than in environments with many obstacles and compartments. Considering the structure and the dimensions of the space, a fleet of cleaning robots guarantees a more efficient cleaning in terms of tasks execution time and complete area coverage than a single robot. Some works using low-cost robotic vacuum cleaners are presented in the Literature Review. [4] presents a robust localization system considering the low-precision of the sensors used. [12] was presented a prototype multi-robot system of vacuum cleaning robots. The robot platform used was the iRobot Roomba vacuum cleaners which already have a simple algorithm for cleaning an area.

In this paper we propose a centralized architecture to manage a fleet of cleaning robots. The validation of the proposed methodology was made, using real architectural plants of a full building with many compartments. The scheduling of the tasks in a multi-robot system is a typically problem that have been studied in the State-of-Art. Several works uses a market-based mechanism [8] [7], where robots exchange their tasks over time, in order to minimize the total cost function. This framework is advantageous mainly for static environments. Similarly, in [9] each robot as the capacity of sharing part of its task with another robot, in order to an efficiently tasks allocation. Other methods to solve the Multi-Robot Task Allocation is based on Bid-Auctions, like in [6] and [10]. In [6] a dynamic bid auction finds a task allocation that reduces two parameters, the task completion time as well as the cost incurred by individual robots. The strategy used in the CleanSim is also based in auction Strategy ensuring cooperation in complex and structured environments. Cleaning Tasks require a robust area coverage method which builds a trajectory capable to cover every unoccupied area. In [5] is proposed a method based on genetic algorithms that guarantees the complete area coverage path planning. Genetic Algorithms solves many optimization problems and are based on an iterative approach inspired in the evolution via natural selection. Basically, initial solutions are randomly generated, called chromosome or genome, and then these populations are evaluated according with fitness functions. These functions includes the parameters which the system wants to minimize. Individuals with higher fitness values have higher probability of passing to following generations. Another interesting approach to solve optimization problems, are based in neural networks, like in Article [3]. The dynamics of each neuron is given by a shunting neural equation. Each robot treats the other robots as moving obstacles [11]. The application of cleaning methods based in genetic algorithms or neural networks requires that the area to be cleaning has been divided into elementary sub-areas, called Cellular Decomposition, [2] and [4].

2 System Architecture

The first step was to develop a modular architecture that allow us to increment or decrement the number of robots available and the map information. Three main systems comprises the developed framework:

- Navigation
- CTPS: Central Task Planning System
- Cleaning Method

Figure 1 presents the interactions between these modules for a multi-robot approach, considering N robots. CleanSim denotes the simulator implemented in python that will be exposed in Section 3.

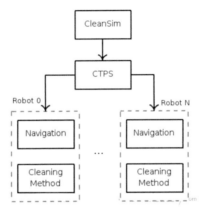

Fig. 1 General Architecture of CleanSim.

In order to implement a cooperative cleaning system was selected a centralized architecture, which is composed by a team of robots and a CTPS (Central Task Planning System). All the robots report to the CTPS and this one decides what each robot should do and in which way. These reports are not only made by the robot position and state of tasks, but also by its internal condition and the avoidance of non-predicted obstacles. This centralized system presents many advantages:

- Event log;
- Easy to change the number of robots;
- World state consultation by a robot;
- Keep the processing power away from the robots allowing their simplicity;
- Prevent the network channel from over overuse because the robots cannot communicate between them;
- Let you easily create a parallel between the simulation and a practical situation

To implement the proposed solution a data structure were carefully designed, with two major goals: modularity and expansion, as we can see in the Figure 2. The data structure is composed by two main groups, Robot and Building. The Robot is a simple model of a differential one with the capability of receiving and performing simple orders. The Building is a group of Floors, each one with Doors, Rooms and Dirtiness. All this elements are disposed in graph allowing global navigation between Rooms and Local navigation inside the Room. Each individual room has its own area and cleaning model. The robots are assigned to the rooms which should be cleaned by an action according with several parameters that will be explained in Section 2.3.

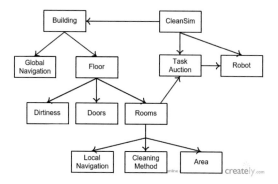

Fig. 2 Data Structure of the developed architecture.

2.1 Navigation

The navigation was divided into two sub-modules, a Local Navigation and a Global Navigation. Local navigation allows the robot navigates between two points inside the same room, being impossible to cross the door. Global navigation allows the robot to travel between rooms (door to door). This implementation simplifies a large and complex problem into two small and easier ones reducing drastically the processing time. This gain more importance when the rooms are smaller because the time to calculate the path could be significant or even overcome the time needed to clean the room.

Both approaches calculates the minimum path between two points under a graph which contains all robot's accessible points. It is only considered that the robot can travel horizontally, vertically and in the diagonal direction (45°). Dijkstra Algorithm was applied to find the shortest path, if it exists.

The higher room simulated in the Local Navigation Module have $25m \times 16m$, which generates a graph with 2520 nodes. In this case, the algorithm calculates the minimum path in $40ms$.

Global Navigation uses the same method that the Local, however in order to minimize the time complexity, in the graph building the doors isolate large group

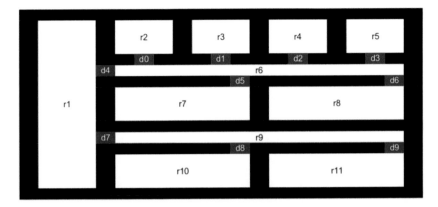

Fig. 3 Graph for the Global navigation. Notes that each room is represented as only one node.

of cells. Each room is represented with only one node in the graph of the Global Navigation, reducing the problem complexity to $n_{doors} + n_{rooms}$.

Throw the conjugation of this two navigation methods, the robot can travel between every two map points efficiently.

2.2 Central Task Planning System

At this point, the robots are already capable of traveling between rooms and cleaning them with a good strategy.

The tasks' allocation to each robot is handled by the CTPS with a dynamic auction. This approach is based on the resources' allocation according with the system needs. In our case, robots are resources and compartments are the needs. Available robots are auctioned for cleaning a compartment. A robot choose the room that placed the highest best, which is defined by Equation 1. This function is based on the following parameters:

- d_R - Distance from the robot to the bid compartment.
- d_{Rs} - Distance of the other robots to the compartment.
- t_{clean} - Cleaning Time for a compartment.

$$n_{robot} = -d_R + \frac{d_{Rs}}{N_{robots} - 1} + \frac{t_{clean}}{k} \qquad (1)$$

Equation 1 guarantees the better solution considering:

- the nearest robot;
- the minimization of the bidding value even with many available robots;
- and giving higher priority to the larger rooms reducing the global time;

2.3 *Cleaning Strategy*

Cleaning robots should be capable of sharing spaces with humans, which mean that its trajectories should be predictable. In order to achieve this feature the cleaning area is divided into rectangular areas, called elementary areas, creating restricted spaces where the robot can navigate increasing its predictability. A Cleaning area comprises the set of nodes belonging to a compartment that should be considered by the robot during the cleaning process. An area is considered clean when the robot go through all these points. These areas are defined applying a Table Based Decomposition algorithm, exposed in Algorithm 4.

- $l[\ ]$, $c[\ ]$: arrays with the lines and rows transitions between wall and passage.
- A_{start}: initial starting area with l rows and c columns.

1: $[l[\ ], c[\ ]] = FindTransition();$
2: $A_{start} = AreaDefinition(l[\ ], c[\ ]);$
    $\{a_{start}$ have $n_l + 1 \times n_c + 1$ areas. Each resultant areas are represented by one pixel called cell.$\}$

3: **loop**
4:    $IncreaseNumberCell();$
     $\{$Increase the number of cells$\}$
5:    **if** $GroupCells()$ **then**
6:      **return**
      $\{$Try to group in every way each cell until it is not possible to group any more.$\}$
7:    **end if**
8:    $GenerateNewAreas();$
     $\{$Obtain new areas of each cell.$\}$
9: **end loop**

Fig. 4 Table Based Decomposition Algorithm

Algorithm 4 divides the map as a set of elementary areas to be cleaned. The next step was to find the best order which these areas should be cleaned as well as the optimal/sub-optimal path to navigates and clean each area. Genetic Algorithms were used in order to find a near optimal solution that minimizes the cost function. In our framework the implementation of Genetic Algorithms was advantageous mainly if we consider the total number of possible combinations.

Considering a compartment with n areas, the number of cleaning order possibilities is given by:

$$C_{order} = !n; \tag{2}$$

Each area can be cleaned using 16 different methods (see Table 1), so the number of possibilities to clean an area is:

$$C_{path} = 16^n; \tag{3}$$

Table 1 Combinations of cleaning methods for each area

		gene[0]	gene[1]	gene[2]	gene[3]
0000	0	circular	anti clockwise	top	right
0001	1	circular	anti clockwise	top	left
0010	2	circular	anti clockwise	bot	right
0011	3	circular	anti clockwise	bot	left
0100	4	circular	clockwise	top	right
0101	5	circular	clockwise	top	left
0110	6	circular	clockwise	bot	right
0111	7	circular	clockwise	bot	left
1000	8	S	y Axis	top	right
1001	9	S	y Axis	top	left
1010	10	S	y Axis	bot	right
1011	11	S	y Axis	bot	left
1100	12	S	x Axis	top	right
1101	13	S	x Axis	top	left
1110	14	S	x Axis	bot	right
1111	15	S	x Axis	bot	left

The combination of Equations 2 and 3 gives the total number of possibilities:

$$C = C_{order} \times C_{path} = n! \times 16^n \qquad (4)$$

A genetic algorithm comprises 4 layers which are genes, chromosomes, individuals and population. In our approach, each compartment is an individual to which should be assigned a cleaning method. Always that a robot navigates to a compartment, need to know how the better way to clean that. This information is sent by the CTPS module with a chromosome. A chromosome is constituted by n sets of 4 genes (*cleanMethod()*), according with the Table 1. Each gene represents the order which the areas should be cleaned (*sortGene()*).

- **CleanMethod():** Cleaning Method for each area. This is composed of 4 boolean genes where a mutation corresponds to a value inversion.
- **SortGene():** Sort a gene with the order which the areas of a compartment should be cleaned. It's an array between $[0, n - 1]$, where n represents the number of areas of a compartment. Each gene mutation has a probability of occurrence.

An initial solution was considered in the *SortGene()* function in order to accelerate the convergence of the algorithm. This initial solution was generated using a greedy solution. Besides the solution found wasn't the optimal solution, in our approach a sub-optimal solution is enough.

3 Simulator

In order to a better validation and evaluation of our framework, it was created a simulator, called CleanSim.

3.1 Environment Acquisition

In order to keep the modularity and adaptability of the system, the map is obtained by processing CAD files from the building. The processing algorithm occurs in three steps, first, if needed, the user edits the file to add or to change necessary information. Then, all the unnecessary layers are removed and the image is converted to bitmap. In this format the doors are identified and painted as blues, all the cleaning areas should be white and the rest in a different color. Also a dirtiness layer is created, where the pixel intensity represents the amount of dirt. Secondly, the image is imported to the simulator and scanned to identify all the objects, door and rooms, through a labeling technique with connectivity-8. Then, among all the identified objects, the system searches for the objects that are huddled together creating a graph that will be used to create paths. The results from this process are saved into hard-drive allowing the program to import these results instead of calculate them every time it is switched on.

3.2 Dirtiness Simulation

In order to simulates the dirtiness, two images were overlaid to the map, as can be seen in the Figure 5.

The first image (shown in red) represents the variation of the dirtiness, according with Equation 5. $S_{x,y}$ denotes the dirtiness in each pixel. $P_{x,y}$ gives the derivative of the function $S_{x,y}$ in order to the time. The second layer is the dirtiness of each pixel on the initial time instant.

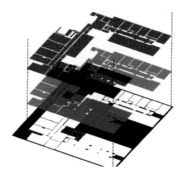

Fig. 5 Representation of the layers used to the dirtiness simulation.

$$P_{x,y} = k \times \frac{dS_{x,y}}{dt} \qquad (5)$$

4 Results

Some experiments were performed at different architecture levels. Firstly, will be presented the navigation results and the advantages of the proposed modular architecture. Then, genetic algorithms for area covering and the task allocation system will be detailed.

4.1 Navigation

The path planning was separated in global navigation and local navigation, both of them based on a navigation grid. This approach splits a complex problem in to small ones, allowing the system to be scaled to more complex facilities including elevators and several buildings. Also, with this implementation it is possible to use different path planning algorithms for different robots and aggregate several parts of the building if necessary. In this way, the navigation can be adapted to the global and local needs.

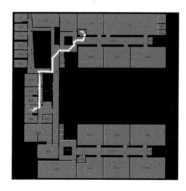

Fig. 6 Navigation between two rooms.

In Figure 6 is represented the path produced by the Navigation system. Notes that the grid resolution and the fact that the robot can only moves at multiple angles of $45°$ limits the solution. Comparing the presented result with the optimal one, if we have an infinitesimal grid, the total length of the trajectory would pass from $32.2m$ to $30.7m$ representing a gain of 5%.

4.2 Area Covering

Genetic Algorithms were used to the optimization of the path made by the robot in order to completely clean a room. Genetic evolution is a method that after a few iterations can present a good solution to the problem. It is necessary to find an advantageous number of iterations to stop the algorithm evolution. The strategy is based on an evolution of 4^n times a population of 20 individuals, measuring the time variations. Each population of 20 individuals represent current solutions, and the combinations of them allows to find more solutions which will be future generations.

Table 2 Genetic Evolution on a Case Study Generic Floor

Generations	1	4	16	64	256	1024
Individuals	20	80	360	1280	5120	20480
Time(s)	10294	10213	10103	9978	9874	9860
$\frac{\Delta T}{n_{generations}}(s)$	0	27.167	9.167	2.609	0.542	0.018

As presented in Table 2, the first generation gets a time variation of $27s$ approximately and in the last generation this time is reduced to $0.018s$. Predicting the same behavior, it is expectable that in the next 1024 generations the total time variation would be around $0.47s$. In the first 1024 generations we obtain a reduction of time in $434s$ and by duplicating them the gain is only 0.11%. It's possible to conclude that for this practical case, 1024 generations are enough.

4.3 Task Allocation

To allocate a task to a robot we propose an auction method, where the rooms bid an available robot. The robot is attributed to the winner. Biding value is a combination

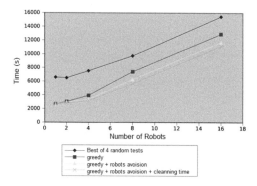

Fig. 7 Comparative performance of different task allocation methodologies.

of three factors: proximity to the robot (greedy), distance of other robots (robots avoidance) to the room and cleaning time. Several simulations were performed with different number of robots and biding strategies.

Figure 7 represents the total unproductive time taken by the robots cleaning a building. The active cleaning time is constantly $9857s$. As we can observe the non-productive time is always above $2000s$ increasing with the number of robots. This time is given by the traveling time between the starting positions until the room that should be cleaned. Also the firsts robots to finish, have to wait for the last.

5   Conclusion

The purpose of this work is the presentation of a modular architecture which manages a fleet of cleaning robots. Three main systems were developed: Navigation Module (which was divided into local and global), the Central Task Planning and a Cleaning Technique that uses genetic algorithms to improve the efficiency of the cleaning method. All modules were implemented in a centralized processing unit that is aware of everything and capable of bidirectional communication with each individual robot. In order to make the system suitable for different building, the map is not static, being directly imported from a CAD file with minor modifications. The main contributions of this project are the addition of genetic approaches to solve the cleaning problem and the usage of an auction task allocation and division in a dynamic way. The simulation experiments were performed using a developed simulator, called *CleanSim*, which includes the simulation of the map dirtiness. As future work, the CleanSim environment should be improved, adding random factors like peoples, obstacles and doors. These elements will allow to create more realistic simulations.

Acknowledgments This work is partially financed by the ERDF – European Regional Development Fund through the COMPETE Programme (operational programme for competitiveness) and by National Funds through the FCT – Fundação para a Ciência e a Tecnologia (Portuguese Foundation for Science and Technology) within project ≪FCOMP-01-0124-FEDER-037281≫.

References

1. Luo, C., Yang, S.X.: A real-time cooperative sweeping strategy for multiple cleaning robots. In: IEEE International Symposium on Intelligent Control (2002)
2. Oh, J.S., Choi, Y.H., Park, J.B., Zheng, Y.F.: Complete coverage navigation of cleaning robots using triangular-cell-based map. IEEE Transactions on Industrial Electronics (2004)
3. Luo, C., Yang, S.X., Stacey, D.A.: Real-time path planning with deadlock avoidance of multiple cleaning robots. In: IEEE International Conference on Robotics and Automation (2003)
4. Pinheiro, P., Cardozo, E., Wainer, J., Rohmer, E.: Cleaning Task Planning for an Autonomous Robot in Indoor Places with Multiples Rooms. International Journal of Machine Learning and Computing (2015)

5. Jimenez, P.A., Shirinzadeh, B., Nicholson, A., Alici, G.: Optimal area covering using genetic algorithms. In: IEEE/ASME International Conference on Advanced Intelligent Mechatronics (2007)
6. Dasgupta, R.: A Dynamic-bid Auction Algorithm for Cooperative, Distributed Multi-Robot Task. Technical Report, Department of Computer Science University of Nebraska at Omaha (2009)
7. Dias, M., Anthony, S.: A free market architecture for distributed control of a multirobot system. In: 6th International Conference on Intelligent Autonomous Systems (2000)
8. Dias, M., Anthony, S.: Opportunistic optimization for market-based multirobot control. In: IEEE/RSJ International Conference on Intelligent Robots and Systems (2002)
9. Robert, Z., Stentz, A.: Complex task allocation for multiple robots. In: IEEE International Conference on Robotics and Automation (2005)
10. Robert, Z.: An Auction-Based Approach to Complex Task Allocation for Multirobot Teams. PhD Thesis, The Robotics Institute Carnegie Mellon University Pittsburgh, Pennsylvania
11. Santos, J., Costa, P., Rocha, L.F., Moreira, A.P., Veiga, G.: Time enhanced A*: towards to the development of a new approach for multi-robot coordination. In: IEEE International Conference on Industrial Technology (ICIT) (2015)
12. Nikitenko, A., Grundspenkis, J., Liekna, A., Ekmanis, M., Kulikovskis, G., Andersone, I.: Multi-robot system for vacuum cleaning domain. In: 12th International Conference, PAAMS, Salamanca, Spain (2014)

Learning Robotics for Youngsters - The RoboParty Experience

A. Fernando Ribeiro, Gil Lopes, Nino Pereira and José Cruz

Abstract The involvement of children and adolescents in robotics is on demand by the many robotics events and competitions all over the world. This non-deterministic world is more attractive, fun, hands-on and with real results than computer virtual simulations and 3D worlds. It is important, by different reasons, to involve people of all ages in an area that some consider the future of mankind and an opportunity to increase the low rate of engineers globally. Robotics competitions at this level are essentially based on teaching motion and programming skills by using Lego™ based robots and a set of challenges to overcome. This paper presents a different approach that is being used by Minho University in order to attract STEM candidates into these fields, with visible success and excellent results. The event is called RoboParty® and teaches children, adolescents and adults, from any area, how to build a robot from scratch, using electronics, mechanics and programming during three non-stop days.

Keywords Learning robotics · STEM · RoboParty · Educational robotics

1 Introduction

Robotics is becoming youngsters desire because it involves the future, technology, and is fun. They are everywhere and doing all sort of tasks from industry to services, in health, sports, space travelling, house keeping, etc. Human beings rely more and more on robotic machines as most things are now taken for granted. This creates the interest of young people, to explore and to be engaged sooner on the creation, development and deployment of robotics.

There are many robotics initiatives like First Lego League, RoboCup, Eurobot amongst others, and these are being taken globally in the form of competitions.

A.F. Ribeiro(✉) · Gil Lopes · Nino Pereira · José Cruz
Departamento de Electrónica Industrial, Universidade do Minho, Guimarães, Portugal
e-mail: {fernando,gil}@dei.uminho.pt, {martins,jcruz}@sarobotica.pt

© Springer International Publishing Switzerland 2016
L. Paulo Reis et al. (eds.), *Robot 2015: Second Iberian Robotics Conference*,
Advances in Intelligent Systems and Computing 417,
DOI: 10.1007/978-3-319-27146-0_56

When young people is involved or a competition refers the word "Junior", it is commonly a Lego™ based competition where high level programming and Lego assembly bricks, shafts, wheels and chains are used to build a robot to accomplish the objectives.

Even though there are some companies developing robotic kits these normally involve only the mechanics and not the electronics, or they come mostly assembled. In some competitions it is recognised that robots end up being assembled and programmed by the teachers or parents, and children are only left to start/stop the robot in the competition field. The competition and children's participation is not focused on the pedagogical side but on competitiveness. Also, the information about the electronics is in most cases not known due to be a proprietary technology.

Teaching children general principles of electronics is a way to make them understand the important aspects of the different electronic components, how they operate, where to touch to avoid electric shocks or to avoid damaging the components. It makes their life easier to deal and to operate with electronic equipment and it entices some enthusiasts on following an engineering career on electronics or robotics.

"Lack of engineers" is a buzzword filling newspapers headlines in the past few years [1, 2], and robotics industry is no exception. The reasons are many but one possible solution is to foster youngster's curiosity and interest for robotics engineering related areas, motivating them to learn in a structured way and forcing them to a hands-on-science experience [3-5]. A pedagogical approach is essential in order to allow their interest on robotics and engineering in general to naturally grow. Postponing this "lack of engineers" problem to future generations will compromise our own future.

The Laboratory of Automation and Robotics (LAR) [6] from University of Minho (Guimaraes – Portugal) [7], has developed mobile autonomous robots and participated on many national and international robotics competitions for the past 18 years, with special attention dedicated to the worldwide robotics challenge RoboCup [8], both on the Middle Size League and RoboCup@Home League.

Also, a large demand for talks and demonstrations in primary and secondary schools from all over the country about robotics is been requested to this group along the years. The robotics subject is on demand by the amount of Hollywood films on this theme, the robotics events announced and shown by the media, the console game industry with appealing games on robotics, the evolution on the industry of toys and their capabilities and the affordable robotic kits available in the shops nowadays.

The Robotics group at University of Minho has decided to take a step further and motivate youngsters to the robotics field, and instead of helping them competing with robots they invite them to come to the University of Minho in groups of 4 people to learn how to build a small robot. Teams of students with a tutor (teacher, parent or any adult responsible for the team) would get together for three non-stop days in a single event, where they would be taught how to build and program a robot with their own hands, with lectures specially created for their young ages, by experts on robotics. The experience was a complete success and this paper describes this experience.

2   RoboParty – Edutainment Robotics

The event's main objective is to teach robotics to people who have no knowledge on robotics, in an entertaining and fun way, highly practical hands-on approach, in a friendly and helping environment with balanced and decompressing breaks for ludic, entertaining and sports activities. Starting from ages of 11, the event is being populated by people of all ages but mainly focused at youngsters with a peak at around 16-18 years old.

Participants are guided and supervised by experts on robotics by fostering their enthusiasm in science, technology, engineering and mathematics (STEM) studies, as they only need to bring a sleeping bag, a laptop computer, the desire to learn robotics and a good state of mind. In the end, they take home a mobile robot built by themselves, beginners level knowledge on electronics, programming, mechanics, and a will to improve their knowledge (as well as new friendships, souvenirs from RoboParty, lots of pictures and good memories for a possible new future).

Each team is made of four people, being one of them the team leader, adult and responsible for the whole team. This leader can be a teacher, a parent or any other person responsible for the youngsters, willing to also learn and participate in the event fully. The participants are encouraged to share knowledge and ideas with other teams in order to establish good links of friendship, exchanging contacts and creating future relations for common projects and robotics events participation.

2.1   The Facilities

In order to accommodate around about five hundred people including participants, volunteers and staff, a large space is necessary and it is held in the University sports hall since its first edition (Figure 1). Tables are arranged to allow four teams work in the same workspace. This promotes information sharing encouraging their relationship. Each table has its mains sockets underneath for laptops and soldering irons.

Fig. 1 RoboParty working area (left) and sleeping area (right) at the University sports hall

The working area takes about 2/3 of the overall pavilion area and the other 1/3 is left for multipurpose activities such as entertainment and ludic activities during day and as a sleeping area at night, as shown in Figure 1.

The University sports hall provides good conditions for the type of event such as large space, central heating/air conditioning, toilet facilities for a large amount of people, dressing and shower rooms, spa and sauna area and a fitness lounge. It also provides for security with surveillance cameras, electronic entrance control, file cabinets with lockers and a reception. Outside the building there is a large car parking space. The University canteen provides lunch and dinner meals for the participants, and breakfast is provided by the coffee shop next to the sports hall. The spacious University gardens are a good place for relaxation and walking.

For an event like RoboParty a great deal of staff is necessary and a group of over 60 volunteers participates providing a precious help during the three days. They are mainly students of the Industrial Electronics degree.

Lectures on how to build the robot, mechanics, soldering the electronic components, programming, some history of robotics and national and international robotics competitions, servicing robots and other subjects, are taken in the multipurpose area.

Fig. 2 Lecture on how to build the robot

Robot demonstrations and display of new high-tech gadgets is reserved to a special area next to the working space.

The entertaining/sports activities are very popular; such as the indoor Aircraft Modelling, Basketball, Football, Tennis table, Badminton, Wood Ball, Taekwondo, Yoga, Judo, Karate, Capoeira, Stretches, Cardio Session, Triathlon indoor, Golf, Quick Chess, circus activities, Ballroom dance, Archery, Horse Riding and Scuba Diving. Each participant decides on which activities to participate. Professionals on each activity (indoor or outdoor) are present to follow, guide and teach participants.

2.2   The RoboParty Image

The RoboParty image was created by a professional designer who produced Ruminho (Robotics at University of MINHO), the event's mascot. It consists of a friendly two wheels robot, with two robotic arms and large eyes. Based on the Ruminho mascot, several products were created (Fig. 3) such as the event's T-shirt offered to all the participants, the RFID badge used for check-in at the working area entrance, some trophies given to the winners on some of the robot trials or sports competitions, the participation certificate, the Ruminho teddy bear and advertising posters sent to schools months before.

A web site is available to provide all the information required, to advertise the event and to allow team registration. Parents and tutors can use the web site to find all the information in order to understand all the procedures involved, security aspects and the event rules. The web site also provides pictures and movies of previous editions so people can have an idea of what to expect. During the event the gallery is populated with pictures and movies taken during the day, along with a webcam streaming of the working area so parents at home can enjoy the event live.

Fig. 3 RoboParty image used in different products

3   The RoboParty Event Step-by-Step

3.1   The Teaching

Lectures are given to participants in different stages according to the milestone achieved by the teams. The initial lectures are based on soldering advices and tips, robot mechanical assembly and mainly, how to recognize the different electronic components, their polarity (if exists) and how and where they should fit in the PCB. All consideratins regarding safety, organization and optimization are also referred. This set of lectures is given on the first day, right after the teams receive their robotic kit and these are specially suitable for the participants' young age. Lectures are provided in an informal environment in the multipurpose area as shown in Fig. 2.

In order to increase the participants' knowledge on robotics and other areas of science, two speakers are invited every year to talk about their expertise in their research area. This way, youngsters can become aware of some state of the art robots, ideas and backgrounds. The venue is at the University main auditorium of Azurem Campus.

The most relevant part of RoboParty is the hands-on work of soldering and mechanical assembly of the robot. According to many youngsters this is the part they most like to do. It takes almost one third of the whole event and teams are invited to enrol all team members in this task. This leads that everyone experiments soldering, handling electronic components, using screwdrivers, pliers and other tools. Tutors can help on these tasks by guiding the team members accordingly to avoid errors, damage of components and hurting themselves by wrongly handling a tool or the soldering iron. Fig. 4 shows two young boys sharing the soldering and dividing the task in two: while one is soldering the components, the other holds the cutting pliers to cut the excess of the component pins. This is common practice in many teams as the way they share the workload of each task. Once in a while they swap positions.

Fig. 4 Participants building electronics (left) and programming (right)

Team collaboration and tutoring is very important as referred previously. In order to stimulate team members, tutors should put questions forward to create dialog, sharing of information and subject discussion in the group.

Another important aspect is the female participation. It has been observed that the female gender participation is increasing, and some teams are majorly female based. Fig. 5 shows an example of a team made of young girls.

Fig. 5 Team of young girls (being helped by a volunteer)

Robot programming is the next task, after finishing the soldering and the mechanical parts of the robot. The programming lectures are given after the teams have finished building their robot and it is commonly on the second day. This task consists of installing a program on the team's computer that will provide the necessary environment to develop, compile and communicate with the robot to upload the binary code. The language used was BASIC for PICAXE until 2012, but from 2013 onwards the PCB uses an Arduino UNO as the brain (programmed in C language).

The first steps to sensor feedback are performed at this stage. Fine-tuning the infrared sensor distance is made first using the robot's test program for this sensor. It is at this point that the interaction with an operating robot begins and typically, a sheet of white paper is used to test it. They look very excited on their first time they interact with the robot and watching its reaction. Their proud relies on the fact that it was a machine built by themselves and it performs well. The test program can sense if the frontal infrared sensor detects an obstacle on the left or on the right side, by rotating the robot wheels in opposition. The eyes of the proud builders shine when the sheet of paper is moved from left to right and they observe a reaction to their action meaning that after all, they have accomplished something.

As a complementary part of the programming lecture, the example code supplied with the robot's CD has demonstrated to be a good guide for the user's first steps. Most teams tend to use this example code and proliferate from there as a starting point, only changing small parts of the code. Their mind for programming is being structured at this time and code reusing is something they embrace easily.

Three tasks are then defined on each team, depending on the extra sensors they acquire with the robotic kit. The first task is to program the robot to perform the obstacle avoidance challenge, i.e., with the frontal infrared sensors, the robot can be programmed to avoid hitting walls. The obstacle trial is a small circuit, based on walls where the robot must accomplish the circuit without touching the walls at the minimum time possible (Fig. 6). Programming creativity and a bit of luck are the ingredients for this task and youngsters react quite well to the challenge.

3.2   The Robotic Kit "Bot'n Roll"

A special purpose robotic kit was developed for this event, by SAR – Solutions for Automation and Robotics [10] and University of Minho. This kit was named Bot'n Roll ONE [11]. The robot assembly has three major steps: mechanics build up, electronics soldering and robot programming. All necessary parts are in the box. The idea behind the project was to create a small, affordable and expandable robotic kit that even a child could assemble it. The electronic components to be used are discrete and normal size components, such as resistors, capacitor, diodes, voltage regulators and microchips. The printed-circuit board (PCB) was designed to be easily handled, where the components can be positioned with plenty of space around them. Space was not a constraint but a necessity. This would allow youngsters to learn how to fit the components, soldering and cutting the component's pin extension free of hassles. Some more complex components were supplied as a finished part, namely the communication board. This board contains surface mount components and converts the RS232 signals from the robot's processor to USB signal in order to enable the robot to be connected to a computer. In that case, the board is supplied assembled and soldered, with the pins ready to be fit into the robot's PCB and soldered afterwards.

The mechanical parts are also easy to assemble. A set of motors and wheels has to be screwed to each other and then screwed to the robot's base. A third caster wheel has also to be screwed to the robot's base in the back of it to make it horizontal.

All the wiring and connectors were thought also to be easy to handle and colourful to avoid any doubts. Some care should be taken and participants are warned on the danger of mixing the cables, but no harmful situation occurs if they are mixed but a damaged component. The power switch is positioned in an accessible place and the wires have to be soldered to it. The robot's battery is positioned underneath and supported with Velcro to allow a fast replacement for charging.

Fig. 6 Obstacle avoidance (left) and race of champions (right) challenges

The second task is on the constructive and creative side where the team decorates their robot. Manual skills are very important to attain the intended objectives.

Sometimes teams dedicate the time of a single member to the task during the whole event, one that is more skilled in craftsman work. The result can be overwhelming and great achievements have been made. Fig. 7 shows some examples robots built. Mechanics can be adapted in order to adapt a toy on top in order to move an arm or leg synchronised with the wheels, for example. Teams are very creative.

The third task is the dance challenge. Programming for this task involves creativity for choreography. The team is allowed to pick a song and to program their robot to the song rhythm. Some teams use the mixture of their robot dressing with a favourite song. All together, with some funny robot movements, the 90 seconds performance on the stage can make the crowd laugh out loud. This last challenge makes teams to join the previous tasks into a single one: craftsman work, choreography, robot programming and loads of creativity. Since the working area of the event is open space, and four teams share the same workbench, it is interesting to observe how youngsters learn with each other, sharing the best ideas, having their own ideas, in a friendly and peaceful environment.

Fig. 7 Some decorated robots ready for the Dance challenge

The SAR staff were aware of the organisation's intention and helped on the launch and support of this event. Nowadays, this robotic kit is commercially available and the company is still involved in the event. At each edition, a workshop is reserved inside the working area where they can provide close support to the needs of the participants. Having such young people handling for the first time of their lives small electronic components, one could expect to have damaged components, bad soldering or a component soldered in the wrong place. SAR company's technicians are there during the whole event as they provide the last resort help to ensure every single robot will work in the end.

The kit contains all mechanical and electronic parts required to build the robot. The Fig. below shows the components and the robot fully assembled. After RoboParty, the robot belongs to the team and they can take it home/school.

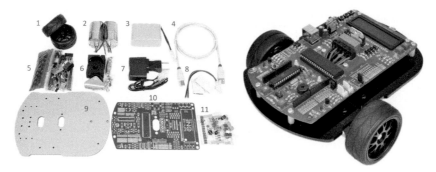

Fig. 8 Bot'n roll box contents (left) and Bot'n Roll fully assembled (right)

3.3   Extra Sensors

The robotic kit comes with the basic frontal infrared sensors allowing the robot to avoid obstacles. Some extra sensors designed for the kit can be acquired before or during the event. One very required sensor is the line following. With this sensor, attached underneath, the robot can be programmed to follow a dark line in a contrasting background. Together with obstacle avoidance, the robot becomes a complete autonomous traveller. A special track and challenge was created in order to allow teams to program and experiment their robot to achieve the best times (Fig. 6).

3.4   Entertaining/Sports Activities

A non-stop event for three days is exhausting and relaxing moments are necessary. Having that in mind, RoboParty has a set of activities that allows participants to have a break and to enjoy themselves with some ludic activities, and be part of them. Participants can register (for free). these activities provides mental relaxation so that when they go back to work, all the problems seem to have gone. There are indoor and outdoor sports activities. Fig. 9 shows a participant learning to play Croquet.

Fig. 9 RoboParty participant learning Croquet as an outdoor activity

The RoboParty Organisation is always trying to bring new and unknown entertainment or sport activities to the event, at least new to the majority of the participants. Scuba diving and horse riding were two activities, for example, that a great majority of the youngsters tried at RoboParty for the first time in their lives. It is an opportunity to practice engineering and robotics together with these activities, in a single event.

3.5 Checking in at the Event

Participants are instructed to bring a sleeping bag and a computer. On their arrival, a bag is provided to the team leader at the check-in desk containing RFID badges, each with the participant's passport style photograph to allow them access the facilities, a RoboParty T-Shirt to each participant, a city map, the event's program, the robotic Kit and a CD containing the instructions (with a video) on how to build the robot. All under aged participants bring a permit form signed by the parents, giving them permission to attend the event. Some basic tools are also necessary and apart from the ones supplied with the kit, extra tools such as a soldering iron and pliers must be brought from home, or acquired at the event in the form of a tool kit available as an extra. The web site contains a description of all necessary tools to build up the robot.

3.6 Help of Volunteers

An event like RoboParty involving over 500 participants, during 3 non-stop days, needs a good amount of staff. Being a tight budget event, volunteers were the solution found to use as staff members that could guide, help, reply to any calls or necessities the participants may have. This precious help is what makes RoboParty event a success. This task is performed by the Industrial Electronics students from University of Minho, organised by the IEEE student branch of the University. Their help starts before the event, preparing the working area layout, and finishes after the event dismantling and tiding up everything. During the event, they are the perfect help to the teams providing a close guidance at the workbench. Since their T-Shirts have different colours they are very easy to spot and teams frequently call them on any question they might have. These volunteers are the first filter or triage for any hardware problem, before taking the robot to the skilled technicians from SAR (who developed the robot), on their workshop in the working area. These volunteers support the entire party program as they guide people between activities, at lunch and dinner times to the canteen, taking care of the participants as if they were their family. The participants rapidly learn that they have someone in the party to whom they can trust. On average around about 60 volunteers are registered to support the event.

4   Results

4.1   Overall Results

Some results can be drawn after nine editions. RoboParty Organisation has defined a maximum of 100 teams of four people due to lack of space considerations and that number is reached every edition. There is a waiting list managedin the last month. This way, the event has always full house creating a fantastic atmosphere. Participants come from all over the country including the Portuguese islands.

As a matter of interest, an increase on the number of teams from the same school has been observed every year. In other words, a school that has participated for the first time with a single team tends to register more teams in the following year. Another interesting aspect is the family participation, not linked to any school. Some families have embraced the RoboParty spirit and participated in the event as a family, to learn about robotics, highlighting a family of four (parents and two adolescent sons) that live around 400 kms from the venue, who participated on four consecutive editions of the event. RoboParty has been subject to some studies to assess its results. One of them is described on [12].

5   Conclusions

Every year RoboParty has full house, which makes it very rewarding. Over 3500 different persons already participate and assembled a robot at RoboParty, and learned some concept of electronics, tried out some mechanical assembling and learned how to program a robot (both in BASIC PICAXE and C for Arduino). Around about 1000 robots have been built in the last 9 editions, by participants from all over the country   and some foreign countries like USA, Denmark, Angola, Ireland, etc.

Participants ranging from 9 years old up to 65 years old, participated with success. And their teachers (adult to accompany the team) come from all knowledge areas like Informatics, History, Gymnastics, English teachers, Philosophy, etc.

Most of the adults responsible from teams come year after year, with different students.

The three days are very tiring but participants leave very happy, carrying on their hands a robot build by themselves, which will permit them to continue their learning process.

The challenges are also very motivating, and make them work hard to have their robots ready and properly working.

Most participants use their robot to participate on National and International robotics events, re-programming their robots, and some teams even achieved very good results nationally and internationally.

All robots leave RoboParty working, since there is a staff team that fixes any major mistakes teams could have made.

This event is proving that it is possible to motivate youngster to STEM areas (Science, Technology, Engineering and Mathematics).

Video overview available on https://www.youtube.com/watch?v=ke4N7EJ McVY

Acknowledgment The RoboParty Organisation wishes to thank the company SAR – Solutions for Automation and Robotics, who have made all efforts in developing this project, the University of Minho and especially all the staff from the sports hall, which whom the organisation of the event would be impossible and to the IEEE student branch at Minho University and their volunteers our deep appreciation for their time and efforts in all the support for this event.

References

1. Still lack of 36,000 engineers in Germany in German Economy section (2009). http://just4business.eu/2009/06/still-lack-of-36000-engineers-in-germany/
2. Wilson, R.: Lack of engineers threatens UK economy. Electronics Weekly.com (2007). http://www.electronicsweekly.com/Articles/2007/07/12/41796/Lack-of-engineers-threatens-UK-economy.htm
3. Ribeiro, F.: Building a robot to use in school: teachers and students learning together. In: Proceedings of the International Conference on Hands-on Science: Formal and Informal Science Education (HSCI 2008), Olinda-Recife, Brazil (2008)
4. Ribeiro, F.: New ways to learn science with enjoyment: robotics as a challenge. In: Proceedings of the International Conference on Hands-on Science: Science for All, Quest for Excellence (HSCI 2009), Ahmedabad, Índia (2009)
5. Ribeiro, F., Lopes, G.: Summer on campus - learning robotics with fun. In: 7th International Conference on Hands-on Science - Bridging the Science and Society Gap (HSCI 2010), Rethymno, Greece (2010)
6. Laboratory of Automation and Robotics (2009). http://www.robotica.dei.uminho.pt
7. University of Minho (2009). http://www.uminho.pt
8. RoboCup Federation (2010). http://www.robocup.org
9. RoboParty (2010). www.roboparty.org
10. SAR - Solutions for Automation and Robotics. http://www.sarobotica.pt/
11. Bot'n roll One - The RoboParty robot. http://www.botnroll.com/
12. Soares, F., Leão, C., Santos, S., Ribeiro, F., Lopes, G.: An Early Start in Robotics – K-12 Case-Study. International Journal of Engineering Pedagogy (iJEP) **1**(1), 50–56 (2011)

Creating a Multi-robot Stage Production

Junyun Tay, Somchaya Liemhetcharat and Manuela Veloso

Abstract A multi-robot stage production is novel and challenging as different robots have to communicate and coordinate to produce a smooth performance. We made a multi-robot stage production possible using the NAO humanoid robots and the Lego Mindstorms NXT robots with a group of undergraduate women who had programming experience, but little experience with robots. The undergraduates from around the world were participating in a three day workshop – Opportunities for Undergraduate Research in Computer Science (OurCS), organized by the School of Computer Science from Carnegie Mellon University that provide opportunities for these undergraduates to work on computing-related research problems. They were given twelve and a half hours over a span of three days to familiarize themselves with the robots, plan the storyboard of the performance, program the robots, generate a multi-robot performance and create a presentation on what they learned and did. In this paper, we describe the tools and infrastructure we created to support the creation of a multi-robot stage production within the allocated time and explain how the time in the workshop was allocated to enable the undergraduates to complete the multi-robot stage production.

1 Introduction

A successful multi-robot stage production is extremely difficult, especially when different robot platforms are used and multiple robots have to communicate and coordinate with little human intervention to produce a coherent performance. Using

J. Tay(✉) · M. Veloso
Carnegie Mellon University, Pittsburgh, USA
e-mail: {junyun,veloso}@cmu.edu

J. Tay
Nanyang Technological University, Singapore, Singapore

S. Liemhetcharat
Institute for Infocomm Research, Singapore, Singapore
e-mail: liemhet-s@i2r.a-star.edu.sg

© Springer International Publishing Switzerland 2016
L.P. Reis et al. (eds.), *Robot 2015: Second Iberian Robotics Conference*,
Advances in Intelligent Systems and Computing 417,
DOI: 10.1007/978-3-319-27146-0_57

743

multiple robot platforms requires familiarity of the robots' capabilities and a good understanding of how to program the robots. Manually controlling each robot individually requires a lot of practice between humans and is especially tedious when the robots have multiple degrees of freedom (DOF), e.g., the NAO humanoid robot has 21 DOF. Automating the behaviors of the robots, multi-robot communication and coordination are crucial to the synchronization of the robots' performance. In this paper, we explain the tools and infrastructure we created and describe our plan to support such an endeavor.

OurCS is organized to provide opportunities for undergraduate women to have a chance to work on computing-related problems and attend talks to learn about life in graduate school. We wanted to enable the participants to have a chance to work with multiple robots and different robot platforms within the short period of time that they were given. We conceived the idea of a multi-robot stage production, but no script was provided to the participants. The participants were allowed to exercise their creativity, yet at the same time, were limited by the capabilities of the robots. Hence, without a strong understanding of the robot's hardware and software constraints, they would be unable to produce an entertaining multi-robot stage production. The participants were asked to sign up for the project beforehand and assigned to the research project based on available slots. At the end of the workshop, the participants had to produce a 5-8 minutes stage production and presentation to share what they learned.

We chose two robot platforms – the Lego Mindstorms NXTs and the NAO humanoid robots (Fig. 1). The Lego Mindstorms NXTs are commercially available in stores, but not the NAO humanoid robots as they are much more expensive. The NAO humanoid robots are not configurable, unlike the Lego Mindstorm NXTs, but are more complex in the number of DOF. To enable the Lego Mindstorms NXTs to communicate one another, we used the XBee radios, which is not packaged with the Lego Mindstorms NXT. Having these two robot platforms allow the participants to have a unique experience that is difficult to get elsewhere. We also aimed to ensure that every participant has a chance to work on each robot platform and learn about the individual hardware and software capabilities. The Lego Mindstorms NXTs were already constructed with a arm that can rotate, wheels to move around and XBee radios to communicate with one another. Extra software functions were written in RobotC to enable the NXTs to send robot to robot messages. For the NAO humanoid robot, external software functions were written to allow motions to be easily synchronized with speech, music and LEDs in the eyes of the NAO humanoid robot. We planned our tools and infrastructure such that the participants were able to use these software functions to create a performance that enable the robots to communicate, synchronize the motions of the robots with text-to-speech or music and change the LEDs of the NAO robots.

All the participants had programming experience, but only some had experience with the Lego Mindstorms NXTs. None of them had worked with the NAO humanoid robot. At the end of the workshop, all of them had a chance to work on all the robots and put together an entertaining and informative eight-minute presentation to explain what they learned and did. We successfully enabled a group of undergraduates with little experiences with robots to create a multi-robot stage production in twelve and a

(a) NAO humanoid robot: 21 rotational joints; LEDs in the eyes; speakers in the ears; text-to-speech.

(b) Lego Mindstorms NXT robot: Two motors for the wheels; one motor for the arm; XBee communication module; internal speaker.

Fig. 1 Robots and their features. Images are not to scale.eps

half hours using two NAO humanoid robots and three Lego Mindstorms NXT robots. It is rare to see opportunities for participants to experience working on multiple robots, let alone different robotic platforms, specifically Lego Mindstorms NXTs and NAO humanoid robots in our case. Careful planning is needed to allow the participants to complete a multi-robot stage production within a short period of time and to gain useful insights about multi-robot coordination and communication, lessons that can never be learned with a single robot.

In Section 2, we review what others have done in the area of educational robotics. Next, in Section 3, we describe the preparations we made to facilitate learning of multiple robot platforms, multi-robot communication and coordination within a short period of time. We explain the rationale and the inner workings behind the tools and infrastructure created so that the participants can focus on creating the motions and behaviors of the robots based on the storyboard they planned and use abstractions of robot functions without in-depth knowledge of the code for multi-robot communication and coordination. Following that, in Section 4, we list the learning objectives and the activities that we planned out for the participants to enable them to effectively learn about the hardware and software of the robots. In Section 5, we recount the actual day-to-day activities of the participants. Lastly, we outline what the participants learned and discuss our insights from the planning to the implementation and execution of the multi-robot stage production in Section 6.

2 Related Work

Carnegie Mellon Robotics Academy developed materials to teach multi-robot communications [3], using robot platforms such as LEGO Mindstorms NXT, the VEX robotics kit and the Arduino family of processors and ROBOTC. ROBOTC was used as the programming language as ROBOTC can be used for these robotic platforms with little to no changes in the code. Avanzato discussed the use of low-cost educational robot platforms and high-level software support for existing educational robot platforms based on the materials from Carnegie Mellon Robotics Academy [1]. Avanzato also described a "multi-robot design challenge for a regional robot contest, multi-robot classroom and laboratory activities, and a programmable controller for multi-robot communication are presented" [1]. Specifically, four autonomous mobile robots, comprising one NXT and three VEX robots were designed and implemented to cooperatively explore a maze with the goal of extinguishing multiple candles that were randomly located [1].

Lego Mindstorms has always been used as a educational tool to teach robotics due to its low cost and simplicity. Casini et al. have developed a remote laboratory for multi-robot systems using LEGO Mindstorms NXT technology [4]. Remote users can design control laws in Matlab and test them by performing experiments remotely using the team of robots available in the experimental setup. There is a global vision system, which "simulates different types of sensors and communication architectures" [4]. Users will only be able to remotely control the NXTs robots but not interact with the robots directly. Others used software such as LabView [6]. Franklin and Parker proposed "Overwatch as an inexpensive educational tool for teaching and experimenting in multi-robot systems" [5] and used Scribblers as the robot platform to experiment with a multi-robot system [5]. McLurkin et al. proposed using Rice r-one mobile robots for multi-robot curriculum and described the courses that they implemented using their platform [8] to teach multi-robot concepts. Others used e-pucks which are designed for education in engineering [9]. Many robotic platforms and software have been developed for multi-robot education. All these robot platforms are mostly low cost and less complex than humanoid robots. By using the Lego Mindstorms NXT and the NAO humanoid robots, we allow the participants to have a unique experience with two very different robot platforms in terms of hardware and software.

3   Preparations

In this section, we describe the tools and infrastructure we prepared before the workshop to support the multi-robot stage production. The tools and infrastructure were built upon previous work and described in Section 3.1 for the NAO humanoid robots and in Section 3.2 for the Lego Mindstorms NXT robots.

3.1   NAO Humanoid Robots

Our RoboCup team, CMurfs (Carnegie Mellon United Robots For Soccer) [7], participated in the RoboCup Standard Platform League competition using the NAO humanoid robots to play soccer. We show the Cognitive Agent, a part of our RoboCup code architecture, which processes the Robot State and generates a Robot Command.

A Robot Command comprises a Motion Command (e.g., walk, perform a motion), a Speech Command (to be used by the text-to-speech engine), and an LED command (to display colors on the robot). More details of our code architecture can be found in [7]. The Cognitive Agent consists of various components, namely:

- Agent Manager is a component that passes the information between components.
- Game State Manager is used to update the game state based on the messages received from the Game Controller, as well as button presses on the NAO robot. The Game Controller is used in a game in the Standard Platform League of RoboCup to synchronize the game state across all NAO robots in the game. The game state consists of states such as the start of the game, penalty shot etc.
- Vision receives the camera images and analyzes the images to determine features such as ball, lines, goal posts, robots etc. that are important in the game.
- Localization determines the global position and orientation of the robot in the game using features from Vision and odometry based on motion commands.
- World Model keeps track of the robot's hypotheses of the ball and the information shared by the other robots (teammates).
- Behaviors generate robot commands based on the features and information available and also determine the messages to send to teammates. In order to make decisions, Behaviors is built on a Finite State Machine (FSM) model, which is described in detail in Section 4.2. A FSM is made up of several states and the transitions between states are based on the conditions programmed.
- Log allows messages to be sent across the network to a Remote Display Client. This component is useful for debugging the states and information available on all the robots.
- Communications is in charge of sending and receiving messages between robots and a Remote Display Client.

Each component has their own input and output interfaces and allows us to easily configure the use of each component. We can easily enable and disable components based on the needs. This flexible and configurable architecture enables us to use the code from our RoboCup team as the functionalities we require to support the a multi-robot stage production exist. We disable the use of the Game Stage Manager, Vision, Localization, World Model and Log in the workshop. The rest of the components enable us to achieve the following: (1) communication between the NAO robots; (2) ability to execute motions; (3) set the LEDs in the eyes; (4) text-to-speech capabilities.

We replaced the Game Manager with a new component, called the Puppet Master. We created a "puppet master" program, *OurCS Command*, shown in Fig. 2, where the participants could use the computer to send messages to the robots to start or stop behaviors of the robots via a user interface shown in Fig. 2. The user interface shown in Fig. 2 reads in a list of commands from a text file before it starts up. These commands can be selected and sent to all the robots as messages. The robots that can receive the messages are listed in the list of robots shown in Fig. 2. The list of robots and the IP addresses of the robots can also be configured via another text file. The participants can also debug the states of the FSM running in the robots. We used UDP communication between *OurCS Command* and the NAOs, but the

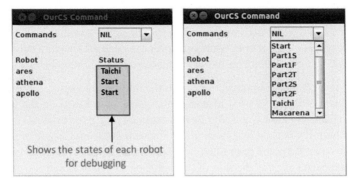

Fig. 2 Snapshot of the Puppet Master UI (left) and the list of commands sent (right).

Algorithm 1. Snippet of sample code showing a FSM and MacroAction in action

```
1:  if fsm.inState(Part1F) then
2:    if fsm.isNewState() then
3:      // assumes that the actionListStatus = UnLoaded
4:      actionList1F.clear()
5:      MacroAction tempAction(StaticAction::actShakeHead, "Terrible moves, N X T", "",
        0xFF0000, 0xFF0000, Action::BothEyes)
6:      if MY_ROBOT_NAME == ROBOT1 then
7:        actionList1F.push_back(tempAction)
8:      end if// Thse two lines must be included to load MacroActions
9:      currentActionList = &actionList1F
10:     actionListStatus = Loaded
11:   else
12:     if actionListStatus==UnLoaded then
13:       // loaded the action list and executing
14:       if commandToExecute=="Part2T" then
15:         // commandToExecute is the command sent by Puppet Master
16:         fsm.trans(Part2T, "Done with Part1F")
17:         continue
18:       end if
19:     end if
20:   end if
21: end if
```

---

message protocol was abstracted from the students. In this way, the students focused on creating messages and how the messages influenced the FSM transitions between states, without worrying about how the messages were transmitted from the computer to the robot. We also added specific robot-robot and computer-robot messages so that the NAOs and the Puppet Master can communicate the states in the FSM.

We also created a component that automatically exports motions from Chore-graphe into a format compatible with our RoboCup code. Choregraphe is a piece of software created by Aldebaran Robotics, the manufacturer of the NAO humanoid robot, that allows users to easily generate motions on the NAO. We will explain in detail how Choregraphe is used in Section 4.1.

To enable the participants to program the NAO to execute motions, play a wave file, use the text-to-speech capabilities, and change the LED colors in the eyes simultaneously, we created a function wrapper: MacroAction(Motion, TextToSay, WaveFileName, LeftEyeColor, RightEyeColor). In this way, the students could focus on creating behaviors (sequences of macro actions) for the performance, without having to handle synchronizing different types of actions or multiple threads and processes on the NAO. The students can use the function wrapper instead. of sending 4 separate commands.

We prepared two laptops to be used with two NAO robots respectively. We installed NaoQi, Choregraphe and our RoboCup code on the laptops. We also prepared sample code to illustrate how the FSM and MacroAction classes can be used. The participants can refer to these code or edit the code to suit their needs. The sample code is shown in Algorithm 1.

3.2 Lego Mindstorms NXT Robots

We previously created curriculum for students at the K-12 level to learn about multi-robot concepts using a variety of robot platforms such as the Lego Mindstorms NXT. We used ROBOTC, a programming language that can be compiled and downloaded to different platforms with little or no changes. Therefore, for the multi-robot stage performance, the participants could code in RobotC to actuate the robot and to pass messages from one robot to another. We built upon their experiences and curriculum developed for multi-robot communication and synchronization of motions. The lessons that we will use are described in detail later. We built two Lego Mindstorms NXT with three wheels where two are actuated and an actuated arm which can rotate.

4 Planning of Day-to-Day Activities

In this section, we describe the day-to-day activities that we planned for the participants. We formed three groups of students where two groups work on the NAO robots and one group works on the NXT robots. The groups are formed based on the participants' interest. Each group is also led by a mentor, who is proficient in the robot platform and software. We plan to spend about four hours per day with the participants. On Day 2, we swap the groups so that everyone has a chance to work on the NAO and NXT robots.

4.1   Day 1

At the start of the day, we gave an introduction of the task – a multi-robot stage production. We gave a general overview of the hardware available on the NAOs and NXTs. We explained that the robots are limited in terms of perception because we disabled the use of sensors such as cameras, ultrasound sensors etc on the NAOs, except touch sensors on the NXTs. The use of cameras and vision are too complex and do not help the participants to learn about multi-robot communication and coordination.

Next, we explained the use of actuators and DOF of a robot. We highlight that the NAO robot is more complex in terms of the number of DOF where the NAO has 21 DOF whereas the NXT has 3 DOF. The NAO robot also has more sensors such as cameras, ultrasound, gyroscopes and accelerometer, touch sensors such as chest buttons and foot bumpers, LEDs in the eyes and text-to-speech capabilities. Though the Mindstorm NXT set includes many sensors such as light sensors, ultrasound sensors, we only added touch sensors to the NXTs that we built.

Following that, we break the participants into three groups, so that two groups will work on creating motions on the NAO robot and one group will work on creating motions on the NXT. At the end of the day, the participants should create a basic outline of the stage production. We described what we teach the participants in detail.

Choregraphe - Creating Motions on a Single NAO Robot. To generate a motion on the NAO, we introduced the concept of a keyframe (static pose) and interpolation time shown as a timeline to create animated motions on the NAO (Figure 3). A motion is made up of several keyframes and timings to interpolate between keyframes are also defined. We taught the students how to use Choregraphe, a software to create motions easily provided by Aldebaran Robotics, the manufacturer of the NAO. We explained that robots have angular joint limits and maximum joint angular velocities that cannot be exceeded. The angular joint limits are shown in Choregraphe with sliders for each joint and the end of each slide of the slider shows the minimum and maximum joint angle. We also taught the participants how to record each static pose as a keyframe and create a sequence of motions.

Stability of a humanoid robot is extremely important as compared to a wheeled robot, especially if the NAO humanoid falls, since the cost of repairing the NAO humanoid robot is high. Therefore, we emphasized the balance of the NAO humanoid robot. We attached a harness to the NAO humanoid robot when the participants create motions so that they can hold on to the robot and ensure that it doesn't fall.

Commands to Actuate a Lego Mindstorm NXT. We first taught how motions on the NXTs are generated by actuating each motor independently at different speeds, and provided sample code of basic motions. The students then created new motions such as moving in an arc, and spinning the NXT's "arm".

We then introduced how the NXTs could communicate using the XBee radios. We explained the concept of using a common language, so that the robots understood

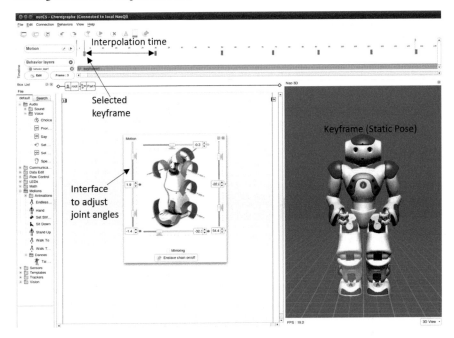

Fig. 3 Choregraphe

messages that were sent and received. The students implemented a sequence of actions on two NXTs, where they sent messages to each other and took turns executing actions. Next, the students created a follow-the-leader sequence, where one robot would select a random action, inform the other robot, and both robots would execute the same action together. Thus, the students successfully understood the basics of multi-robot communication and coordination.

To synchronize the motions on the NXTs with the storyboard of the multi-robot stage production, the students also used a "puppet master" to switch action sequences within the NXTs. Within each sequence of actions, the NXTs communicated to take turns performing actions, and synchronize to perform identical actions together.

4.2   Day 2

For Day 2, the participants learn how to convert the motions of the NAO robot to code. They will also learn about Finite State Machine and to use the function wrapper, MacroAction. The participants also learn about multi-robot communication, specifically how to pass messages between the NAOs and the Puppet Master and how to pass messages between the NXTs. We did not prepare code to send messages between NAOs and NXTs as we believe that the participants will be able to pick up multi-robot communication concepts from what we already prepared.

Finite State Machine. A finite state machine (FSM) enables them to implement their storyboard: (1) A FSM has a finite number of states and the story can be broken down into sequential steps; (2) The transition from one state to another is initiated by a triggering event or condition, e.g., a message received by the NAO from the "puppet master" (explained below), or the end of a sequence of motions. We also created code templates and sample code to provide examples on how to use our code, so that they could easily create behaviors for the NAOs without worrying about the code syntax. Although we provided the functionality of enabling message-passing between robots, the undergraduates did not use them as they decided to use the "puppet master" program to coordinate the robots. The participants had to code in C++, but were provided code templates as examples to follow.

Based on the curriculum and code developed for multi-robot communication, we list the lessons from [2] and summarized the multi-robot communication lessons:

1. Message passing: To communicate, the robot must be able to send and receive a message that comprise a string of limited number of characters. Relaying messages from one robot to another is not as simple as it looks. It requires the robot to send a message repeatedly to ensure that the message is received using the function void SendStringRepeated. When a message is sent as a string, the robot must be able to check if a message exists and read the message.
2. Guaranteed message delivery: To ensure that a message is sent and received, we have a function called SendStringConsistently where a message is guaranteed to be received by the recipients. We have another function SendStringConsistentlyTo where the message is guaranteed to be received by a particular robot.
3. Condition to initiate a motion: In multi-robot coordination, it is common for one robot to wait for a message to be received from another before a motion is initiated.
4. Map a message string to an action: After the participants learn how to actuate the robot, we teach the participants about mapping a string to an action. The string can be sent from one robot to another as a message and once the string is received, the robot can be actuated to perform an action (motion). This lesson is built on the previous three lessons.
5. Parameterized message string to action: Instead of defining fixed actions (motions) to be performed, we can parameterize the action. For example, for the robot to rotate by a certain number of degrees, we can pass the parameter 45 in the message so that the robot will rotate by 45 degrees.

4.3   Day 3

For Day 3, we plan to allow time for the participants to finish working on their stage production and do a full rehearsal before their presentation. We emphasize the importance of practice at the venue as behaviors of the robots may differ at the venue due to different conditions such as texture of carpet, causing variations in the motions.

Also, the participants will have to practice their presentations and to synchronize the actions between the NAOs and the NXTs since there was no message passing between the NAOs and the NXTs. The NAOs can be synchronized via commands sent from the Puppet Master and the NXTs can be synchronized via commands sent from a computer or messages sent from another NXT.

5 Actual Day-to-Day Activities

We describe the day-to-day activities that actually happened during the workshop. The participants only spent a few hours each day working on the robots as they had other talks and activities scheduled. We divided them into three groups and each group of participants is led by a mentor, who is familiar with the robot platform and software. Each group of participants only has 2-3 undergraduates. The participants had no script to follow except that the task was to produce a multi-robot stage performance. Hence, the participants had to learn, plan and execute a multi-robot performance using the NXTs and the NAOs and summarize what they learned in a 5-8 minutes presentation.

5.1 Day 1

On the first day, the participants spent four hours and forty-five minutes working on the robots. They familiarized themselves with the NAOs and NXTs and created motions on the robots. Lastly, the participants discussed and designed an outline of the multi-robot stage production. They found the music for the production and created motions for the stage production. They came up with the idea of a NAO tai-chi master teaching another NAO and two NXTs tai-chi. The robots then ended with the Macarena dance.

5.2 Day 2

On the second day, the participants spent four hours working on the robots. The participants created a detailed outline and wrote them on the whiteboard (Fig. 4). Later, the participants refined the timeline to determine the timings of each event in the stage production and is listed here:

1. Introduction (30 seconds)
2. Tai chi master / student practice (90 seconds)
3. Tai chi demonstration by the two NAOs and the Lego NXTs (60 seconds)
4. Explanation of the NAOs (60 seconds)
5. Explanation of the NXTs (60 seconds)
6. Macarena (60 seconds)

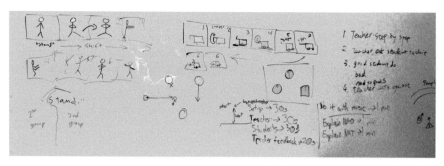

Fig. 4 Snapshot of Script

After the participants planned the storyboard, they assigned responsibilities to each team member, such as the sections to present in the PowerPoint presentation. Following that, we taught them the concept of a finite state machine and multi-robot communication concepts such as message passing from one robot to another. The participants created motions for the Macarena dance and learned to incorporate the keyframe motions into the code. They learned how to use the function wrapper MacroAction and how to check for messages from the Puppet Master. The participants also swapped groups to learn about another robot platform that they did not learn on Day 1.

At the end of the day, the participants showed a demonstration of the motions created for the NAOs and the NXTs. However, the NAO-to-NXT communications had not been implemented, hence they sent commands to each robot type independently.

5.3   Day 3

On the third day, the participants spent three hours and forty-five minutes working on the robots. The participants finished the Tai Chi demonstration and added the Macarena dance into code. They also produced a PowerPoint presentation explaining what they learned. They also had to practice synchronizing the commands sent from the NAO and NXT Puppet Masters so that the robots appear synchronized in their performance. A summary of their work during the three day workshop, multi-robot stage production and presentation can be viewed at https://youtu.be/0y9cG9lnKOk.

6   Conclusion

The participants learned a lot from the workshop, which included all the concepts we wanted to impart in terms of multi-robot communication and coordination. They also learned a new model – Finite State Machine – and found them easy to apply in the workshop to initiate different states based on the messages sent. They also learned that creating stable motions for a humanoid robot is extremely difficult. Walking and moving legs apart such as sliding the legs across the floor is intuitive and easy for humans, but difficult to achieve using keyframe motions. They also learned that sliding legs apart for a humanoid robot on different carpet textures is difficult, hence they created a motion that lifts one leg up slightly before putting the leg down to create a sliding leg motion. They also learned that rehearsals are important as the behaviors of the robot may differ in different environments. They had to edit the motions of the robot due to the differences in carpet textures at the presentation venue. The feedback from the participants were that they learned a lot from these experiences with multiple robots and some wanted to continue pursuing graduates studies in robotics. One of the participants even spent a summer with us working on RoboCup.

In this paper, we addressed the challenges of enabling undergraduates with little robotic experience to create a multi-robot stage production in twelve and a half hours. We described how the time was structured and how concepts and the software infrastructure were abstracted so that the students focused on the stage production. We divided the concepts we wanted to teach into manageable sizes, and students can apply what they learned immediately. The concepts were easy to understand based on their prior programming experiences. The students could also be creative in generating a script for the performance which keeps them motivated as they were involved from the planning to the execution of their plan. Hence, by sharing our experiences and how we prepare for this workshop to support a multi-robot stage production in a short amount of time, we hope that others can learn from our experiences.

Acknowledgments We thank Brian Coltin for his guidance of the students who participated in the workshop. Junyun Tay is part of the NTU-CMU Dual PhD Programme in Engineering (Robotics). The views and conclusions contained herein are those of the authors only.

References

1. Avanzato, R.L.: Multi-robot communication for education and research. In: 2013 ASEE Annual Conference. ASEE Conferences (2013). https://peer.asee.org/22304
2. Carnegie Mellon Robotics Academy: FIRE Wiki. http://www.robotc.net/firewiki/index.php/Main_Page
3. Carnegie Mellon Robotics Academy: Multi-Robot Communications. http://www.cs2n.org/activities/multi-robot-communications
4. Casini, M., Garulli, A., Giannitrapani, A., Vicino, A.: A lego mindstorms multi-robot setup in the automatic control telelab. In: Proceedings of 18th IFAC World Congress, p. 2 (2011)
5. Franklin, D.M., Parker, L.E.: Overwatch: an educational testbed for multi-robot experimentation. In: 26th International Florida Artificial Intelligence Research Society Conference (FLAIRS) (2013)
6. de Gabriel, J.M.G., Mandow, A., Fernández-Lozano, J., García-Cerezo, A.: Using lego nxt mobile robots with labview for undergraduate courses on mechatronics. IEEE Trans. Education **54**(1), 41–47 (2011)
7. Liemhetcharat, S., Coltin, B., Tay, J., Veloso, M.: CMurfs 2011 team description paper. In: Proc. RoboCup 2011 CD (2011)

8. McLurkin, J., Rykowski, J., John, M., Kaseman, Q., Lynch, A.: Using multi-robot systems for engineering education: Teaching and outreach with large numbers of an advanced, low-cost robot. IEEE Transactions on Education **56**(1), 24–33 (2013)
9. Mondada, F., Bonani, M., Raemy, X., Pugh, J., Cianci, C., Klaptocz, A., Magnenat, S., Zufferey, J.-C., Floreano, D., Martinoli, A.: The e-puck, a robot designed for education in engineering. In: Proceedings of the 9th Conference on Autonomous Robot Systems and Competitions, pp. 59–65 (2009)

Robotics: Using a Competition Mindset as a Tool for Learning ROS

Valter Costa, Tiago Cunha, Miguel Oliveira, Heber Sobreira
and Armando Sousa

Abstract In this article, a course that explores the potential of learning ROS using a collaborative game world is presented. The competitive mindset and its origins are explored, and an analysis of a collaborative game is presented in detail, showing how some key design features lead participants to overcome the challenges proposed through cooperation and collaboration. The data analysis is supported through observation of two different game simulations: the first, where all competitors were playing solo, and the second, where the players were divided in groups of three. Lastly, the authors reflect on the potentials that this course provides as a tool for learning ROS.

Keywords ROS · Teaching · Course · Education

1 Introduction

Many university level programs are obligated to prepare students for professional employment while simultaneously providing academic rigor. Therefore, it is critical that the courses students are enrolled in develop the required skills to succeed in a team-oriented environment within the competitive market that is the professional industry.

Human competition is a contest where two or more people strive for a goal that cannot be shared, usually resulting in a winner and a loser. Individuals and/or groups are then in a position where they must achieve a certain outcome. For example, in

V. Costa(✉) · T. Cunha · A. Sousa
FEUP - Faculty of Engineering, University of Porto, Porto, Portugal
e-mail: {ee09115,ee10203,asousa}@fe.up.pt

M. Oliveira · H. Sobreira · A. Sousa
INESC TEC - INESC Technology and Science (formerly INESC Porto), Porto, Portugal
e-mail: m.riem.oliveira@gmail.com, heber.m.sobreira@inescporto.pt

© Springer International Publishing Switzerland 2016
L.P. Reis et al. (eds.), *Robot 2015: Second Iberian Robotics Conference*,
Advances in Intelligent Systems and Computing 417,
DOI: 10.1007/978-3-319-27146-0_58

757

V. Costa et al.

most team sport competitions, teams engage for the purpose of winning matches
to take first place in a tournament [1]. While some people claim that competition
can enrich the person's learning experience (e.g. Verhoeff 1997 [2], Lawrence 2004
[3], and Fulu 2007 [4]), other authors (Lamet al. 2001 [5], Vockell 2004 [6]) beg to
differ. This course proposes an approach to teaching that offers several key points
such as giving the students a better understanding of their work expectations and a
way to evaluate their performance, opening the lines of communication between the
students and the teacher to boost feedback, actively engage the students in their own
learning and providing the teacher with increased information about the students
problems which allows the teacher to adjust the teaching as the course progresses.

2 Robotics Operative System

As the scale of robots and robotic challenges continues to grow it becomes a daunting
task to program. The code is extensive, the varying hardware parts makes the code reusability close to zero, and the projects require large-scale software integration efforts
[7]. To meet this demand many frameworks were developed by the academia [8]. It's
in this context that ROS (Robotics Operative System) appears. ROS was designed
to face specific challenges encountered when developing large-scale service robots.
Various efforts at Stanford University in the mid-2000s and by Willow Garage in
2007 provided significant resources to extend these concepts [9]. The design goals
ROS presents in [7, 10]

- **Multi-lingual;**
- **Tools-based;**
- **Peer-to-peer;**
- **Thin;**
- **Free and Open-Source.**

A very important aspect on project continuation is the amount of time developers
spend programming. Usually, due to ease of use, programming time, or other, pro-
grammers code in various languages. ROS allows four different languages in each
process/node: C++, LISP, Octave and Python. To manage the complexity of ROS,
a *micro kernel* design In an effort to manage the complexity of ROS, we have was
adopted, where a large number of small tools were used to build and run the various
ROS components, opposite to a single one. Another design goal important to refer is
the Peer-to-peer. "A system built using ROS consists of a number of processes, potentially on a number of different hosts, connected at runtime in a peer-to-peer topology"
[7]. This means that ROS has a distributed topology. Lastly, it is well known that
most existing robotics software projects contain drivers or algorithms which could
be reusable outside of the scope of the project they were created. However, it is not
always easy to separate that code from the rest of the software so that it can be used
in a different context seamlessly. To combat this tendency, ROS encourages all driver
and algorithm development to occur in standalone libraries.

There are four fundamental concepts of the ROS implementation that must be detailed to minimally understand ROS. Those are:

- **Nodes:** are processes that run a given application. As ROS was designed to be modular, a regular architecture uses several nodes;
- **Messages:** are the way by which nodes communicate. Messages are published to topics that can be read by nodes. Standard types like integer, floating point, boolean, etc. are supported, as are arrays of those same types. Messages can include nested structures and arrays, like C structs. Message types use standard ROS naming conventions [11].
- **Topics:** are buses over which nodes exchange messages. Nodes are not aware of which topic they are communicating with. There can be multiple publishers and subscribers to a topic. Topics are intended for unidirectional, streaming communication [12]. Each topic is appended by the ROS message type used to publish to it, and nodes can only receive messages with a matching type. ROS currently supports TCP/IP-based and UDP-based message transport [12].
- **Services:** Request / reply communications are done via services, which are defined by a pair of messages: one for the request and one for the reply. A providing ROS node offers a service under a string name, and a client calls the service by sending the request message and awaiting the reply. Client libraries usually present this interaction to the programmer as if it were a remote procedure call [13].

All the concepts above are based on the publisher-subscriber pattern. In software architecture, this is a messaging pattern where senders, called the publishers, do not program the messages to be sent to the receivers, called subscribers. Instead, those are sent to another topic without knowledge of what, if any, subscribers there may be. Likewise, subscribers express interest in one or more topics, and only receive messages that are of interest (matching types), without knowledge of what, if any, publishers exist.

There are two main advantages to this approach. The first, *Loose coupling*, which is a direct consequence of the unawareness of the publishers when subscribing to a topic and vice-versa. This allows the topics to be ignorant of the system topology, because their normal operations is not affected regardless of the other part. Comparing this paradigm to a normal client-server, we know that a client cannot proceed with its functions if the server stops working. The second advantage is that this model provides the ability to scale the system (*scalability*) more easily than normal client-server paradigms.

The largest disadvantage ROS presents is the fact that it has a significant learning curve [11]. This article promotes a competitive approach to teach ROS using a game. The next section discusses some other resources for learning ROS.

ROS organizes its software through packages. Packages are collection of files, generally including executables and other files for a given application.

3 Other Learning Resources for ROS

Apart from the official documentation provided in ROS website [11], other efforts have been made to try to diminish ROS learning curve. Several books were published teaching how to install and work around with ROS, with some examples [14, 15]. In the academic environment, several online courses are accessible for learning ROS [16, 17, 18]. ROS grants one fully documented example in the wiki. That simulator is called TurtleSim [19]. Doing this example ensures the ROS installation was successful. To start turtlesim, open 3 terminals and in each type:

1. roscore
2. rosrun turtlesim turtlesim_node
3. rosrun turtlesim turtle_teleop_key

The separate terminals allow separate processes to run simultaneously. If all goes well the user can control the turtle on the screen.

4 Designing the Competitive Approach

While the turtlesim example is simple, it explains ROS principles really well. To detail how to use the competitive mindset when teaching ROS, an explanation of how to design this competitive approach must be given. Human competition is a contest where two or more people strive for a goal that cannot be shared, usually resulting in a winner and a loser [20].

Game-based learning is a useful foundation for incorporating a competitive mindset. Games have been developed for coursework in several areas, most notably computer science [21, 22, 23], with an overall positive response from both students and instructors. Games are effective in stimulating a positive attitude towards learning from the students, increasing the levels of interest, motivation and problem-solving mindset [24]. Games promote active learning, thus enabling students to remember fundamental knowledge, applying it to new domains and leading to the integration of knowledge and new, creative connections between previously acquired ideas [25]. Additionally, they can be a platform for particularly effective learning techniques, such as distributed repetition, while preventing the students from becoming bored [24, 26]. Additionally, games can be included in the coursework without excessive overload to the instructor [23], making them a sustainable educational tool.

Competition is a factor of most games, and can lead to increased motivation in the students. Competition promotes a competitive drive to be better than their peers, but it also provides an effective form of evaluation against peers. These reactions foster a stronger commitment to the learning objectives, added interest and motivation to study more advanced material [3, 27]. Competitions in teams are also an effective way to further collaborative learning [28].

While the work of Lawrence et al. advocates direct competition between students [3], such methods gain less consensus from the students than the use of gaming as a learning method [27] and some students feel uncomfortable [29]. As an alternative, a parallel offer of an indirect competition, in which the student is judged against a single standard rather than interacting with other opponents, may benefit students that feel less confident [29]. Another option is to hold the competition at the end of the course and frame rewards of positive results as "extra points". This approach allows all students to share the learning results and to avoid the disinterest of unsuccessful players, and has been applied with success to education in Game Theory [28].

5 ROS Course Tutorial

The course objective is for each participant to develop code for a robot participating in a chasing tail game variant. In the game three or more teams each with 3 or more players choose an animal class (dog, cat, cheetah or turtle) to represent their player and have to implement a strategy to either chase or run away from other players. A simulated referee must verify if a player has managed to catch another, as well as, generate values for the movement of each class (e.g. Turtle moves slow, the Cheetah moves fast). All code is developed using Qt Creator and the simulation can be seen using RVIZ.

The presented problem can be detailed in three fases:

- Establish player, referee and world communication.
- Produce the movement of the player, applying the strategy and the class mechanics.
- Represent the player in the game map.

The first step to start the course is to install and validate the ROS installation.

The ROS installation that was used was ROS Indigo using Catkin. The distribution can be installed by following the tutorial on the ROS Wiki [30].

5.1 IDE Configuration

Qt Creator must be launched from terminal to enable environment variables. As mentioned before, the software is organized by packages, which are self-sufficient units of compilable code. Module programming eases the process of code production and garantees task independence. In this case, a unique package suffices and will store all the code. After launching qt Creator, open a new terminal and write: catkin create pkg player1 roscpp std_msgs This code generates automatically a folder containing a CMakefile as well as two folders, devel and src. In order to use Qt Creator in conjunction with ROS, an edition as to be made to the CMakeLists.txt file.

Contents of the CMakeLists.txt:

```
cmake_minimum_required(VERSION 2.8.3)
project(catkin_workspace)
include(/opt/ros/indigo/share/catkin/cmake/toplevel.cmake)
```

5.2 Configuration Steps

The first step to engage the students is the creation of a subscriber node. This node will subscribe to the referee messages, such as, the velocity each player possesses. A tutorial to create the subscriber is available at the ROS WIKI [31]. Next, in order to visualize the movement of each player, a marker is needed. A marker is an object in the world (game map). This can be done following the example available in [32] and creating spherical players. To simulate the player's movement a TF was used that is updated when a new message arrives. a TF (short for transform) is a package that lets the user keep track of multiple coordinate frames over time. tf maintains the relationship between coordinate frames in a tree structure buffered in time, and lets the user transform points, vectors, etc between any two coordinate frames at any desired point in time [33]. To visualize the player's movement and their tf RVIZ is used. In this simulator, the players must configure what they want to see (in this case the other players). RVIZ is a tool that comes with the ROS distribution and can be launched by opening a terminal and typing:

```
rosrun rviz rviz
```

The referee is responsible for maintaining the game state, defining the time of the round. It publishes the messages of each animal node, such as the speed of each animal, score and their life, as seen on figure 1. It also checks if any player was hunted, updating the scoreboard if a animal is killed, then re-spawning them after a specified time. The code for the referee is available in [34].

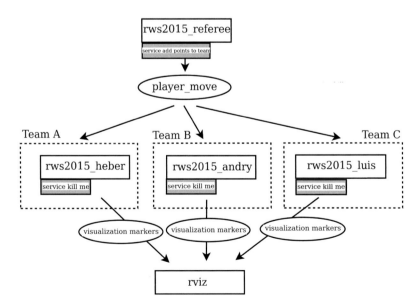

Fig. 1 Block diagram of the game architecture.

After the game configuration, it was given to the players liberty to program their own player using whatever strategy they felt was best. Some players chose to hunt other participants, others focused on running away, and some applied mixed strategies. Figure 2 shows the game window and the teams playing.

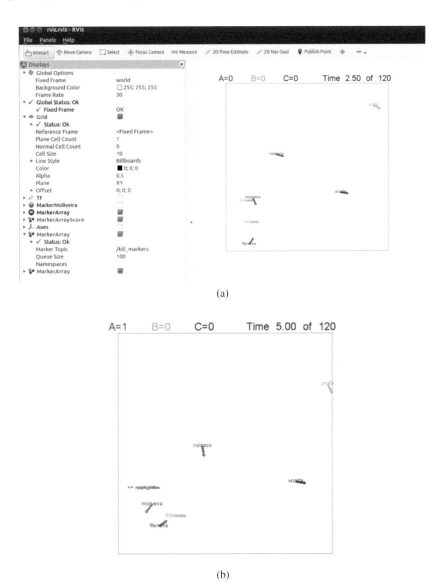

Fig. 2 Game Example

6   User Survey

To ascertain the prior knowledge of ROS and the accessibility and acceptance of the course, a survey was made to inquire the participants of the ROS course. The list of questions asked is shown next:

1. Have you had contact with ROS before?
2. Did the procedure seemed adequate? (Is the sequence of steps logical?)
3. Would it be necessary to have more written documentation?
4. Do you agree with the learning process being based on a competitive mindset?
5. Did you finished all the proposed assignments?
6. Identify strengths and weaknesses of this course. Suggest improvements, if any.

Questions 1 to 5 were answered in a five-point rating scale composed of Strongly Agree(5) - Agree(4) - Neutral(3) - Disagree(2) - Strongly Disagree(1), whereas the last question is an open-ended one. Figure 3 shows the answers of questions 1 to 5 of the survey, composed of a population of 9 people.

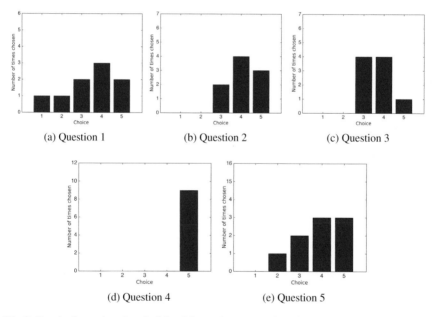

(a) Question 1   (b) Question 2   (c) Question 3

(d) Question 4   (e) Question 5

Fig. 3 Result of questions 1 to 5 of the elaborated survey on the ROS course

More than half had some prior contact with ROS, although in a beginner level. Almost everyone agrees that the course was well structured, overcoming the learning curve. Even so, more detailed documentation on the course is requested, finding the existing one is not enough. Everyone unanimously believes that structuring through a competitive exercise is effective to learn ROS. Nevertheless, not everyone finished all the designated tasks. From the answers of question 6, some key points were established, such as the use of the "Chasing Tail" game as the objective is well accepted, encouraging the learning of ROS and the course could be longer, with a more extensive game to develop.

7   Conclusions and Lessons Learned

This course proposes a novel teaching approach to ROS. There's a lack of initiatives that focus on teaching ROS, because it's a highly specific robotic framework and because the learning curve is steep. However, the course propelled the use of competition to overcome that barrier. The participants inquired felt however that the course should've been longer and more documentation should exist (other than the ROS official one).

Acknowledgments This work is partially financed by the ERDF – European Regional Development Fund through the COMPETE Programme (operational programme for competitiveness) and by National Funds through the FCT – Fundação para a Ciência e a Tecnologia (Portuguese Foundation for Science and Technology) within project FCOMP-01-0124-FEDER-037281.

References

1. Burton-Chellew, M.N., Ross-Gillespie, A., West, S.A.: Cooperation in humans: competition between groups and proximate emotions. Evolution and Human Behavior **31**(2), 104–108 (2010). http://dx.doi.org/10.1016/j.evolhumbehav.2009.07.005
2. Verhoeff, T.: The Role of Competitions in Education. Future world: Educating for the 21st century, 1–10 (1997). http://olympiads.win.tue.nl/ioi/ioi97/ffutwrld/competit.html
3. Lawrence, R.: Teaching data structures using competitive games. IEEE Transactions on Education **47**(4), 459–466 (2004)
4. Fulu, I.: Enhancing: Learning through Competitions. School of InfoComm Technology, Ngee Ann Polytechnic (2007)
5. Lam, S.-F., Yim, P.-S., Law, J.S.F., Cheung, R.W.Y.: The effects of competition on achievement motivation in Chinese classrooms. The British Journal of Educational Psychology **74**(Pt 2), 281–296 (2004)
6. Vockell, E.: Educational Psychology: A Pratical Approach. Purdue University Calumet, on-line book (2004)
7. Quigley, A.N.M., Gerkey, B., Conley, K., Faust, J., Foote, T., Leibs, J., Berger, E., Wheeler, R., Quigley, B.G.M.: ROS: an open-source Robot Operating System (2009)
8. Garousi, V.: Experience in Developing a Robot Control Software **4**(1), 3–13 (2011)
9. ROS History (2015). http://www.ros.org/history/
10. ROS (2015). http://www.ros.org/
11. ROS Messages. http://wiki.ros.org/Messages
12. ROS Topics (2015). http://wiki.ros.org/Topics
13. ROS Services (2015). http://wiki.ros.org/Services
14. O'Kane, J.M.: A Gentle Introduction to ROS (2013)

15. Martinez, A., Fernández, E.: Learning ROS for Robotics Programming (2013). http://books.
    google.com/books?hl=en&lr=&id=2ZL9AAAAQBAJ&oi=fnd&pg=PT12&dq=Learning+
    ROS+for+Robotics+Programming&ots=VJMhUZ_xwN&sig=N0nBv1htLn3BuBwYb0cP1p
    ZmBJ8
16. ClearPathRoboticsROS. http://www.clearpathrobotics.com/blog/how-to-guide-ros-101/
17. Learning ROS I (2015). http://u.cs.biu.ac.il/~yehoshr1/89-685/
18. Learning ROS II (2015). http://www.cs.cornell.edu/Courses/cs4758/2013sp/courseinfo.html
19. Learning ROS III (2015). http://wiki.ros.org/turtlesim
20. Cantador, I., Conde, J.M.: Effects of competition in education: a case study in an e-learning
    environment. In: ADIS International Conference e-Learning 2010 (2010)
21. Adams, J.: Chance-It: an object-oriented capstone project for CS-1. ACM SIGCSE Bulletin
    (1998). http://dl.acm.org/citation.cfm?id=273140
22. Katrin Becker, A.C.: Teaching with Games: The Minesweeper and Asteroids Experience. De-
    partment of Computer Science, University of Calgary, 2400 University Drive NW, Calgary
23. Hill, J.M.D., Ray, C.K., Blair, J.R.S., Carver, C.A.: Puzzles and games. ACM SIGCSE Bulletin
    35(1), 182 (2003)
24. Prensky, M.: Digital game-based learning. Computers in Entertainment **1**(1), 21 (2003)
25. Mcgovern, A., Tidwell, Z., Rushing, D.: Teaching introductory artificial intelligence through
    java-based games. Artificial Intelligence, 1729–1736 (2011)
26. Dunlosky, J., Rawson, K.A., Marsh, E.J., Nathan, M.J., Willingham, D.T.: Improving Stu-
    dents' Learning With Effective Learning Techniques: Promising Directions From Cognitive
    and Educational Psychology. Psychological Science in the Public Interest **14**(1), 4–58 (2013).
    http://psi.sagepub.com/lookup/doi/10.1177/1529100612453266
27. Ribeiro, P., Simoes, H., Ferreira, M.: Teaching Artificial Intelligence and Logic Programming
    in a Competitive Environment. Informatics in Education **8**(1), 85–100 (2009)
28. Burguillo, J.C.: Using game theory and Competition-based Learning to stimulate student mo-
    tivation and performance. Computers and Education **55**(2), 566–575 (2010)
29. Mcgovern, A., Trytten, D.: Making In-Class Competitions Desirable For Marginalized Groups,
    pp. 31–33 (2013)
30. indigo/Installation/Ubuntu - ROS Wiki. http://wiki.ros.org/indigo/Installation/Ubuntu
31. ROS/Tutorials/WritingPublisherSubscriber(c++) - ROS Wiki (2015). http://wiki.ros.org/ROS/
    Tutorials/WritingPublisherSubscriber(c%2B%2B)
32. rviz/DisplayTypes/Marker - ROS Wiki. http://wiki.ros.org/rviz/DisplayTypes/Marker
33. tf/Tutorials/Introduction to tf - ROS Wiki. http://wiki.ros.org/tf/Tutorials/Introductiontotf
34. mikemriem / rws2015_moliveira – Bitbucket. https://bitbucket.org/mikemriem/rws2015_
    moliveira

The Khepera IV Mobile Robot: Performance Evaluation, Sensory Data and Software Toolbox

Jorge M. Soares, Iñaki Navarro and Alcherio Martinoli

Abstract Taking distributed robotic system research from simulation to the real world often requires the use of small robots that can be deployed and managed in large numbers. This has led to the development of a multitude of these devices, deployed in the thousands by researchers worldwide. This paper looks at the Khepera IV mobile robot, the latest iteration of the Khepera series. This full-featured differential wheeled robot provides a broad set of sensors in a small, extensible body, making it easy to test new algorithms in compact indoor arenas. We describe the robot and conduct an independent performance evaluation, providing results for all sensors. We also introduce the Khepera IV Toolbox, an open source framework meant to ease application development. In doing so, we hope to help potential users assess the suitability of the Khepera IV for their envisioned applications and reduce the overhead in getting started using the robot.

1 Introduction

A wide range of mobile robotic platforms can be found nowadays in the market and across robotic laboratories worldwide. Some of them have been used by many researchers, reaching a critical mass that elevated them to *de facto* standards in their domain of use [5, 10, 11]. Convergence to these shared platforms has been argued to improve collaboration and repeatability, allowing for easy validation or refutation of algorithms under study [1, 18].

J.M. Soares(✉) · I. Navarro · A. Martinoli
Distributed Intelligent Systems and Algorithms Laboratory,
École Polytechnique Fédérale de Lausanne (EPFL), 1015 Lausanne, Switzerland
e-mail: jorge.soares@epfl.ch

J.M. Soares
Laboratory of Robotics and Systems in Engineering and Science, Instituto Superior Técnico,
University of Lisbon, Av. Rovisco Pais, 1049-001 Lisbon, Portugal

© Springer International Publishing Switzerland 2016
L.P. Reis et al. (eds.), *Robot 2015: Second Iberian Robotics Conference*,
Advances in Intelligent Systems and Computing 417,
DOI: 10.1007/978-3-319-27146-0_59

Fig. 1 The Khepera IV robot (image courtesy of K-Team).

The success of a robotic platform does not depend solely on its technical qualities, but also on set of accompanying tools, such as libraries, management scripts, and suitable simulators. Among small indoor robots, one platform that achieved widespread acceptance is the Khepera III [13, 15]. Released in 2006, the Khepera III has seen over 600 sales to 150 universities worldwide and has been used in hundreds of publications. In our lab, it has been successfully employed across diverse research topics, including odor sensing [17], navigation and localization [13], formation control [6], flocking [12], and learning [3], as well as numerous student projects.

In this paper we present and test the new Khepera IV robot designed by K-Team[1], the successor to the Khepera III. Released in January 2015, the Khepera IV is a differential wheeled mobile robot with a diameter of 14 cm (see Fig. 1). It is equipped with 12 infrared sensors, five ultrasound sensors, two microphones, and a camera. Proprioceptive sensors include two wheel encoders and an inertial measurement unit (IMU). Wireless communication can be accomplished using Bluetooth or 802.11b/g, and processing takes place in an Gumstix embedded computer running GNU/Linux.

We perform an exhaustive test of the sensors and actuators in order to understand their performance and create an accurate model of the robot. The data collected in the process is made freely available to other researchers, who will be able to use it for deriving and validating their own models. In addition, we present an open source toolbox designed in our lab, composed of a collection of scripts, programs and code modules for the Khepera IV robot, enabling fast application development and making it easier to run multi-robot experiments. Both the datasets and the Khepera IV Toolbox are available for download on our website[2].

The remainder of this article is organized as follows. In Section 2 we describe in detail the Khepera IV. Section 3 focuses on an exhaustive performance test of the different sensors and actuators of the robot. In Section 4 we introduce two software packages that complement the Khepera IV. Finally, Section 5 draws the conclusions about the Khepera IV robot.

---

[1] http://www.k-team.com

[2] http://disal.epfl.ch/robots/khepera4

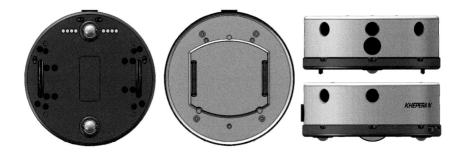

Fig. 2 Bottom, top, front, and left views of the robot (image courtesy of K-Team).

2 Technical Description

The Khepera IV is a small differential wheeled robot designed for indoor use. It is shaped like a cylinder, with a diameter of 14.08 cm and a ground-to-top height of 5.77 cm (wheels included). Its outer shell is composed of two hard plastic parts with slots for the sensors and actuators. Inside, it follows a stacked PCB design. The complete robot weighs 566 g. Figure 2 shows the technical drawings for the robot.

The two actuated wheels are 42 mm in diameter (including the O-rings that act as tires) and are centered on each side of the robot, spaced 10.54 cm apart. Two caster ball transfer units, at the front and at the back, provide the remaining contact points. This solution results in 0.5-1 mm of ground clearance, making the robot very stable but preventing its use on any surface that is not effectively flat and smooth.

2.1 Electronics

The brain of the robot is a Gumstix Overo FireSTORM COM, an off-the-shelf embedded computer that carries a Texas Instruments DM3730 800MHz ARM Cortex-A8 Processor with a TMS320C64x Fixed Point DSP core, 512 MB of DDR LPDRAM, and 512 MB of NAND Flash memory. The robot ships with a pre-installed 4 GB microSD card for user programs and data. A Wi2Wi W2CBW003C transceiver provides both 802.11b/g (WiFi) and Bluetooth 2.0+EDR capabilities using internal antennas.

Low-level aspects of the robot are managed by a dsPIC33FJ64 GS608 microcontroller that builds a bridge between the embedded computer and the built-in hardware. Additional devices can be connected via an extension bus on top of the robot, as well as an external USB port.

Energy is provided by a 3400 mAh 7.4 V lithium-ion polymer battery. The battery is not swappable and can be charged in approximately 5 hours using the charging jack. Support is also provided for charging from the extension bus (allowing for the use of external, stackable battery packs) and from a set of contacts under the body of the robot (designed for automatic charging stations).

2.2 Sensors and Actuators

The Khepera IV robot is equipped with a rich set of sensing devices:

- Twelve Vishay Telefunken TCRT5000 reflective optical sensors. Eight of these sensors are equally spaced in a ring around the robot body, while four of them are downward-facing. When in proximity mode, the sensors emit a wavelength of 950 mm and their published range is 2-250 mm. They may also operate in passive mode and measure ambient light. The sampling frequency for the infrared sensors is 200 Hz, regardless of the mode of operation.
- Five Prowave 400PT12B 40 kHz ultrasonic transceivers. The sensors' published range is of 25-200 cm with a beam angle of 85° at -6 dBm, and a sensor can be sampled every 20 ms. The effective sampling rate depends on the number of sensors enabled, ranging from 50 Hz for a single sensor to 10 Hz if the whole set is in use.
- A center-mounted single-package ST LSM330DLC iNEMO inertial measurement unit (IMU), featuring a 3D accelerometer and a 3D gyroscope. The accelerometer is configured to a ± 2 g range and a 100 Hz data rate, and the gyroscope is configured to a ± 2000 dps range and a 95 Hz data rate. Data are read by the micro-controller in groups of 10, therefore a new set of accelerometer readings is available every 100 ms and a new set of gyroscope readings is available every 105 ms.
- Two Knowles SPU0414HR5H-SB amplified MEMS microphones, one on each side. The omnidirectional microphones have a gain of 20 dB and a frequency range of 100-10000 Hz. The rated SNR is 59 and the sensitivity is -22 dBV at 1 kHz.
- One front-mounted Aptina MT9V034C12ST color camera with a 1/3" WVGA CMOS sensor, yielding a resolution of 752x480 px. The robot comes with a fixed-focus 2.1 mm lens with IR cut filter, mounted on a M12x0.5 thread. The specified fields of view are 150° diagonal, 131° horizontal and 101° vertical.

Motion capabilities are provided by two Faulhauber 1717 DC motors, one driving each wheel. Each motor has 1.96 W nominal power, transferred through two gearboxes with 38:1 overall gear ratio and 66.3 % overall efficiency, yielding 1.3 W usable power per wheel. The motors are paired with Faulhaber IE2-128 high-resolution encoders, with a full wheel revolution corresponding to 19456 pulses. This yields approximately 147.4 pulses per millimeter of wheel displacement. The motor speed is regulated through pulse width modulation (PWM), and the motors can be set to different modes: closed-loop speed control, speed-profile control and position control, as well as open loop.

The robots are equipped with three RGB LEDs, mounted on the top of the robot in an isosceles triangle, with light guides to the top shell. The LED color can be controlled with 6-bit resolution on each channel, making them useful for tracking and identification. Finally, a PUI Audio SMS-1308MS-2-R loudspeaker, with nominal power 0.7 W, SPL 88 dBA and frequency range 400-20000 Hz can be used for communication or interaction.

2.3 Extension Boards

The native functionality of the robot can be extended through the use of generic USB or Bluetooth devices, or by designing custom boards plugging into the KB-250 bus. This 100-pin connection provides power, I^{2}C, SPI, serial and USB buses, as well as more specific lines for, e.g., LCD or dsPIC interfacing. K-Team commercializes several boards, including a gripper, a laser range finder, and a Stargazer indoor localization module.

The interface is compatible with that of the Khepera III, and existing boards should work with no alterations. Our lab has, in the past, developed several boards that we are now using with the new robot, including an infrared relative localization board [14], an odor sensing board [8] and a 2D hot-wire anemometer [8], as well as a power conversion board.

Mechanically, however, the different shape of the robot shell may require changes to existing hardware. Boards with large components on the underside can be paired with an additional spacer, while boards inducing significant stress on the connectors should be attached either magnetically or using the screw-in mounting points. Depending on their size and construction, boards may obstruct the view to the tracking LEDs.

3 Performance Assessment

A core part of this paper is the evaluation conducted for the Khepera IV robot and its sensors. This work serves two purposes: informing potential users of the expected behavior and performance of each component, and allowing for the development of robot models. While data is presented here in summarized form, the datasets for each experiment are available on our website.

To this effect, we have undertaken a diligent effort to test all relevant sensors and actuators. We benchmarked the on-board computer, providing an idea of how much algorithmic complexity the robot can handle. We determined the response of the infrared and ultrasound sensors, to determine operational ranges and make it possible to define sensor models. We looked into the camera distortion and the microphone directionality. We assessed the accuracy of the odometry, and provided a superficial analysis of the IMU signals. We have also tested the motor response and the robot's energy use.

Table 1 nbench results for Khepera IV, Khepera III and workstation

	Khepera IV	Khepera III	Workstation
Numeric sort	212.92	168.76	2218.40
Fourier	856.75	184.46	43428.00
Assignment	3.65	1.10	52.65

3.1 Computation

The computational performance of the embedded Overo Firestorm COM computer
was assessed using nbench, a GNU/Linux implementation of the BYTEmark benchmark suite. The same tests were run in the Khepera IV, the Khepera III and a typical
mid-range desktop computer, equipped with an Intel Core i7 870 CPU. For both
Khepera robots, we used the precompiled binaries available on the Gumstix ipkg
repositories, while for our reference computer the program was compiled from source
using gcc 4.8.2. The results are presented in Table 1.

The three tests are respectively representative of integer arithmetic, floating point
arithmetic and memory performance. The benchmark evaluates single-core performance, and therefore does not benefit from the additional cores on the desktop machine. Furthermore, it is not optimized for the DSP extensions on the FireSTORM.

In comparison to the Khepera III, there is a very significant increase in floating
point and memory performance, enabling the implementation of more complex algorithms on-board. However, performance is still limited when compared to a desktop
computer, which may justify offloading computations if, for instance, undertaking
extensive video processing.

3.2 Infrared Sensors

To determine the response of the infrared sensors, the robot was placed in a dimly lit
room next to a smooth, light-colored wall, with the front sensor directly facing the
wall. The robot was moved along a perpendicular line to the wall, in steps of 1 cm
from 0 cm (i.e. touching the wall) up to 10 cm, and in steps of 2 cm up to a maximum
distance of 30 cm. For each distance, 5000 sensor readings were collected. The data
is presented in Fig. 3a in box plot form.

Due to the very low variation in readings, most boxes in the plot degenerate to
line segments. Below 4 cm, the sensor saturates at the maximum reading (1000). In
the range of 4-12 cm the response follows a smooth curve and, for longer distances,
the measured value is indistinguishable from background noise in the absence of
obstacles.

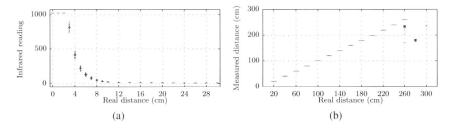

Fig. 3 (a) Box plot of the infrared sensor response. (b) Box plot of the ultrasound sensor response. For distances greater than 260 cm, the sensor consistently return the code for no-hit (1000) with only outliers present in the plot.

3.3 Ultrasound Sensors

A similar protocol was followed for the ultrasound sensors, measured against the same target. All sensors were disabled except for the front-facing one, in order to maximize the sampling rate. The robot was placed at distances from 20 cm up to 300 cm, in steps of 20 cm. For each distance, 5000 measurements were obtained. The data is presented in Fig. 3b in box plot form.

The sensor is accurate and precise, with typical sub-cm standard error across the entire published range. Above the 250 cm published range, sensor performance degrades rapidly, with large numbers of ghost detections, generally the product of ground reflections. From 280 cm, the sensor mostly reports no hits, as expected. During the initial experiments, we observed some problems with multiples of the actual distance being returned when the obstacle was positioned at 60 cm distance. These appear to be due to multi-path additive effects involving the floor, and disappeared when tested with a different floor material.

We experimentally determined the ultrasound sensor beam angle, which is of approximately 92° at 1 m, matching the specifications.

3.4 Camera

An example image capture using the robot camera in a well-lit room is presented in Fig. 4a. 3DF Lapyx was used to process a set of 33 full-resolution (752 x 480 px) images of a checkerboard pattern taken with the robot camera and extract its intrinsic calibration parameters. The calibration results using Brown's distortion model are included in Table 2. Over the entire set of images, this calibration results in a mean square reprojection error of 0.251978 pixels.

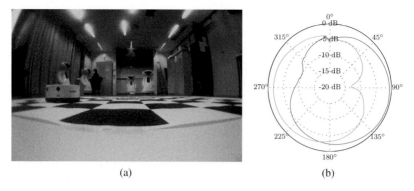

(a)   (b)

Fig. 4 (a) Example image capture by the robot camera. (b) Directivity pattern for the left and right microphones using a 1 kHz source. The maximum recorded amplitude was taken as the reference.

Table 2 Camera calibration parameters, using Brown's distortion model.

Focal length	F_x	380.046		Radial distortion	K_1	-0.332931
	F_y	379.116			K_2	0.113039
Principal point	C_x	393.38			K_3	0
	C_y	273.695			P_1	-0.000124994
Skew		0			P_2	0.000195704

3.5 Microphones

The microphones were tested using an omni-directional sound source emitting a 1 kHz tone. The robot was placed one meter away, and slowly rotated in place while capturing both microphones. The resulting wave files were bandpass filtered to remove motor noise and extract the reference tone. Figure 4b shows the directionality diagrams for each microphone, with a clear lateral main lobe for each microphone.

3.6 Odometry

The odometry was tested by having the robot move a significant distance while calculating its position by integrating the wheel encoder data. Two paths were tested: a square with one-meter sides, and a circle with one-meter diameter. Multiple experiments were run for each path, with robots moving at approximately 20% of the top speed. The surface, wheels and rollers were cleaned, and the odometry was calibrated before the experiments. The calibration was performed as described in the odometry calibration example from [7], which consists of a simplified version of the Borenstein method [2].

An overhead camera was used with SwisTrack [9] to capture the ground truth at 20 Hz, while the robot odometry was polled at 100 Hz. The camera was calibrated using Tsai's method, reporting a mean distance error of 2 mm over the 16 calibration

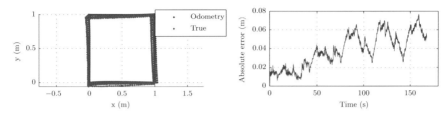

Fig. 5 Odometry-derived and ground truth tracks of the robot while describing four laps around a one-meter square, and associated absolute error over time.

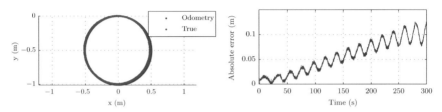

Fig. 6 Odometry-derived and ground truth tracks of the robot while going in a one-meter diameter circle for five minutes (approximately 14.5 laps), and associated absolute error over time.

points. A realistic estimate of the maximum error across the entire arena is in the order of 1 cm.

The origin of the trajectory was matched by subtracting the initially measured position from the ground truth tracks, and the initial heading was matched by minimizing the cumulative absolute error over the first 5 % of position measurements. The error metric is the Euclidean distance between the position estimated using the odometry and the actual position of the robot.

The square experiment consisted of describing five laps around a square with side length 1 m, totaling 20 m per experiment (discounting in-place corner turns). The trajectory was programmed as a set of waypoints, with the position being estimated using the odometry alone. As such, in this test, the odometry information is used in the high-level control loop. An example run and resulting absolute error is presented in Fig. 5.

The circle experiment consisted of five minutes spent describing a circle of diameter 1 m, totaling approximately 14.5 laps and 46 m. The robot was set to a constant speed at the beginning of the experiment, and the odometry was logged but not used for the high-level control. The encoder information is, however, still used by the motor PID controller to regulate the speed. An example run and resulting absolute error is presented in Fig. 6.

Each experiment was repeated five times, in both the clockwise and counterclockwise directions. The results are summarized in Table 3. For every set of experiments, we take both the average absolute error over the trajectory and the maximum recorded error.

Table 3 Mean and maximum absolute error for the odometry experiments. Each row is the result of five runs.

		Average error		Maximum error	
		μ (m)	σ (m)	μ (m)	σ (m)
Square	Clockwise	0.033	0.006	0.105	0.063
	Counterclockwise	0.033	0.006	0.104	0.038
Circle	Clockwise	0.056	0.031	0.127	0.067
	Counterclockwise	0.093	0.027	0.206	0.054

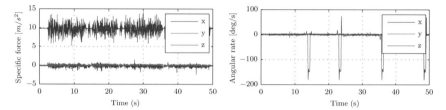

Fig. 7 Accelerometer and gyroscope signals for a single square path. The robot starts moving at $t = 2$ s and briefly pauses at the end of each segment. The peaks in z angular rates correspond to the corners of the square.

Position estimation using odometry is, by nature, subject to error accumulation over time. However, and while the error does show an upward tendency, it is obvious from figures 5-6 that it is not monotonically increasing. As such, the average and maximum error provide more useful information than the final error.

The error is larger for the circle experiments, as is the error variation. This is partially expected, due to the longer experiment length and the larger fraction of circular motion.

The mechanical set-up of the robots appears to be very sensitive to minor imperfections or dirt on the floor. Namely, the spheres easily get clogged after some hours of use, and even in seemingly flat surfaces the robot sometimes loses traction, significantly reducing odometry performance. Nevertheless, the odometry is very accurate, typically achieving maximum errors in the order of 0.5 % of the distance traveled.

3.7   IMU

Inertial sensors were tested by having the robot describe a square trajectory similar to the one used in the previous section, with added pauses before and after each in-place corner turn. The sensors were logged at their maximum frequency, and the captured signals are plotted in Fig. 7. Separately, the scale and bias of the accelerometers were calibrated using the procedure in [4], while for the gyroscopes the initial bias averaged over 200 samples was subtracted from the measurements.

There is no visually apparent structure in the accelerometer data, while in the gyroscope data the corner turns are easily observable. The fact that the robot is

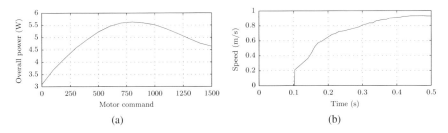

Fig. 8 (a) Self-reported overall power as a function of motor speed commands. Propulsion power is the component over the 2.95 W baseline. (b) Motor speed over time, following a maximum speed request at $t = 0.1$ s. The speeds were obtained by capturing the encoder values differences at 100 Hz.

inherently unstable in pitch, due to the 4-point support design, creates significant noise in the $z - x$ accelerometer measurements due to the changing orientation of the gravity vector.

Superficial analysis of the data suggests that the IMU can be used for pitch and roll estimation and, while the robot is not equipped with an absolute heading sensor, the yaw delta can be estimated over short time frames. Position and velocity estimation, on the other hand, was not accurate enough to be of use even when using Kalman-based techniques with zero-velocity updates [16].

Given the high quality odometry, the IMU seems of little use for single-sensor dead-reckoning but may complement the odometry in a sensor fusion approach. Perhaps more realistically, it can be useful for attitude estimation, alternative interaction modes, and vibration monitoring.

3.8 Motors

Figure 8a shows the propulsion power curve, measured for a robot rotating in place at different speed commands. Starting at a baseline consumption of 2.95 W with the robot idle, power increases to a maximum of 5.63 W for speed 800, then decreasing to 4.64 W at full speed.

The maximum speed achievable in speed control mode was determined by having the robots move in a straight line while tracking their position with SwisTrack. For a speed command of 1400, the robot achieved a speed of 0.93 m/s. At higher speed commands, approaching the saturation limit of 1500, the robot is unable to move in a straight line and the trajectory begins to curve, with no appreciable increase in linear speed.

The high torque-to-weight ratio allows the Khepera IV to quickly accelerate to the desired speed. Figure 8b shows the results of requesting the maximum speed at

Table 4 Self-reported overall power for different activities.

Activity	Power (W)
Idle	2.95
CPU load	3.12
Motors (50%)	5.58
Motors (100%)	4.63

Activity	Power (W)
Camera	3.12
Ultrasound	3.00
Infrared	2.95
IMU	2.95

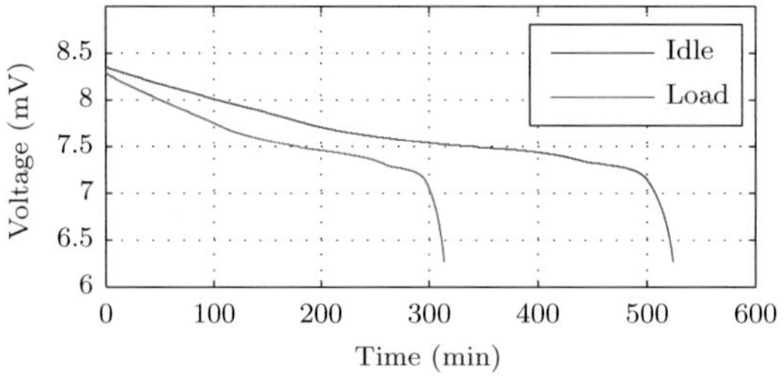

Fig. 9 Battery voltage over time, for idle and full load situations.

time $t = 0.1$ s, obtained by sampling the wheel encoders at 100 Hz period. The robot accelerates to 90 % of the top speed in 0.24 s, and achieves top speed in 0.339 s. This results in an average acceleration $a = 2.74$ m/s^2 to 100 %.

3.9 Power

The number and type of devices activated and in use can significantly influence energy use and limit autonomy. To estimate the potential impact, we used the built-in power management features to measure the energy use when performing different activities. Note that the numbers in Table 4 do not necessarily reflect the power used by individual components; for instance, when using the camera, the increase in power is mostly justified by the heavy load placed on the CPU.

Some sensors, such as the infrared and IMU, use negligible energy when actively queried. The largest chunk of power corresponds to idle consumption and is independent of the devices in use. This is, in large part, due to the number of devices that are, by default, initialized on boot, including the 802.11 and Bluetooth radios. Motor power is strongly dependent on the load, and will vary depending on the robot weight, terrain, changes in speed, and obstacles in contact with the robot.

We have also performed a long-term test intended to assess the maximum autonomy of the robots, both in an idle situation (no devices active or programs running) and in a full load situation (camera video streaming to computer, motors moving at full speed, ultrasound sensors active). The voltage decay over time is shown in Fig. 9.

An idle robot starting from a full charge fails after approximately 8.7 h of continuous use, while a robot using maximum power last approximately 5.2 h. This is in accordance with previous results showing high idle consumption.

4 Software

There are, at the moment, two open source libraries for Khepera IV application development: K-Team has developed `libkhepera`, which ships with the robots, and our lab has developed the Khepera IV Toolbox, which we make available as an open source package. Both provide similar base functionality, and are improvements on older libraries for the Khepera III robot. The two libraries are independent, and programs using them can coexist in the same robot, although it is not recommended that they run in parallel.

4.1 *libkhepera*

`libkhepera` is distributed with the Khepera IV. It allows complete control over the robot hardware, generally at a fairly low level. It allows the user to configure devices, read sensor data and send commands. These operations can be accomplished using simple wrapper functions in the main library. Most functions return primitive data types, although some still output data in unstructured buffers.

Conceptually, the library should provide an easy upgrade path for those using the Khepera III equivalent. However, improvements and simplifications in the new library will require changes to existing applications. Full documentation is provided with the library. The outstanding limitation of `libkhepera` is the lack of higher-level constructs, often forcing the user to write verbose code and re-implement frequently used functionality.

4.2 *Khepera IV Toolbox*

The Khepera IV Toolbox is an evolution of the Khepera III Toolbox [7], with which it shares a significant portion of the code. The initial motivation for its development was providing a straightforward API that could be used with little concern for the underlying details, while also fixing some usability and technical constraints of the robot-provided library.

At a basic level, it provides the same functionality as `libkhepera`, albeit in a different shell. We have developed the API in a way that minimizes the number of lines of code, trying to provide simpler functions that yield, with a single call, the desired result. Most querying functions fill C structures that neatly package complex data.

In addition to the core functionality, the toolbox provides higher level modules that implement support for frequent tasks. These include:

– NMEA, which allows easy processing of NMEA-formatted messages
– Measurement, which handles periodic measurement taking and supports arbitrary data sources
– OdometryTrack, which integrates wheel odometry information to provide a position estimate
– OdometryGoto, which supports waypoint navigation using this position estimate

There are also modules for facilitating I^2C communication and for each of our custom boards. It is easy to extend the toolbox with additional reusable modules, and the build system makes it trivial to include them in applications.

The toolbox also provides an extensive set of scripts that expedite building, deploying and running applications. These scripts take multiple robot IDs as argument and perform actions such as uploading programs, executing them, and getting the resulting logs, allowing a user to coordinate experiments using relatively large fleets of robots from a single command line.

5 Conclusion

In this paper, we have presented the Khepera IV mobile robot and assessed the performance of each of its parts. The robot clearly improves upon its predecessor, packing powerful motors, more complete sensing capabilities, a capable computer, and a long lasting battery, all inside a smaller and more stable shell.

The odometry is very accurate in non-slippery floors, making the less precise IMU not very useful for navigation applications. The ultrasound sensors were found to be precise along their entire operating range, while the infrared sensors have somewhat limited range, creating a blind area between 12-20 cm in our experiments; these values depend, of course, on the materials and environmental conditions, and longer ranges can be obtained using specialized infrared-reflective material. The camera and microphones provide good quality information, and are valuable additions to the robot. Among the limitations, the restricted ground clearance has the greatest impact, making the robot unfit for anything but flat surfaces.

We have also presented two software libraries for the robot, including our own, open source, Khepera IV Toolbox. This library makes it easy to develop applications, and enables the user to easily control multiple robots. It provides a clear upgrade path for users working with the Khepera III robot and the corresponding Toolbox.

We have made the Khepera IV Toolbox code publicly available, together with the raw datasets for all our experiments. In this way, we intend to help our colleagues develop their own robot models and jump-start development on the Khepera IV, reducing the platform overhead for future research and educational use.

Acknowledgments We thank Claude-Alain Nessi and K-Team for their cooperation in the work leading up to this paper and the material provided, and Thomas Lochmatter for his past work on the Khepera III Toolbox and advice.

References

1. Bonsignorio, F.P., Hallam, J., del Pobil, A.P., Madhavan, R.: The role of experiments in robotics research. In: ICRA Workshop on the Role of Experiments in Robotics (2010)
2. Borenstein, J., Feng, L.: Measurement and correction of systematic odometry errors in mobile robots. IEEE Transactions on Robotics and Automation **12**(6), 869–880 (1996)
3. Di Mario, E., Martinoli, A.: Distributed particle swarm optimization for limited time adaptation with real robots. Robotica **32**(02), 193–208 (2014)
4. Frosio, I., Pedersini, F., Borghese, N.A.: Autocalibration of MEMS accelerometers. IEEE Transactions on Instrumentation and Measurement **58**(6), 2034–2041 (2009)
5. Gouaillier, D., Hugel, V., Blazevic, P., Kilner, C., Monceaux, J., Lafourcade, P., Marnier, B., Serre, J., Maisonnier, B.: Mechatronic design of NAO humanoid. In: IEEE International Conference on Robotics and Automation, pp. 769–774 (2009)
6. Gowal, S., Martinoli, A.: Real-time optimized rendezvous on nonholonomic resource-constrained robots. In: International Symposium on Experimental Robotics. Springer Tracts in Advanced Robotics, vol. 88, pp. 353–368 (2013)
7. Lochmatter, T.: Khepera III toolbox (wikibook). http://en.wikibooks.org/wiki/Khepera_III_Toolbox (accessed September 16, 2015)
8. Lochmatter, T.: Bio-inspired and probabilistic algorithms for distributed odor source localization using mobile robots. PhD thesis 4628, École Polytechnique Fédérale de Lausanne (2010)
9. Lochmatter, T., Roduit, P., Cianci, C., Correll, N., Jacot, J., Martinoli, A.: SwisTrack - a flexible open source tracking software for multi-agent systems. In: IEEE/RSJ International Conference on Intelligent Robots and Systems, pp. 4004–4010 (2008)
10. Mondada, F., Franzi, E., Guignard, A.: The development of Khepera. In: Proceedings of the First International Khepera Workshop, pp. 7–14 (1999)
11. Mondada, F., Bonani, M., Raemy, X., Pugh, J., Cianci, C., Klaptocz, A., Magnenat, S., Zufferey, J.C., Floreano, D., Martinoli, A.: The e-puck, a robot designed for education in engineering. In: Proceedings of the 9th Conference on Autonomous Robot Systems and Competitions, Portugal, pp. 59–65 (2009)
12. Navarro, I., Matía, F.: A framework for collective movement of mobile robots based on distributed decisions. Robotics and Autonomous Systems **59**(10), 685–697 (2011)
13. Prorok, A., Arfire, A., Bahr, A., Farserotu, J., Martinoli, A.: Indoor navigation research with the Khepera III mobile robot: an experimental baseline with a case-study on ultra-wideband positioning. In: Proceedings of the IEEE International Conference on Indoor Positioning and Indoor Navigation (2010)
14. Pugh, J., Raemy, X., Favre, C., Falconi, R., Martinoli, A.: A fast onboard relative positioning module for multirobot systems. IEEE/ASME Transactions on Mechatronics **14**(2), 151–162 (2009)
15. Schmidt, L., Buch, B., Burger, B., Chang, S.H., Otto, J.A., Seifert, U.: Khepera III mobile robot practical aspects. Tech. rep., University of Cologne, Cologne (2008). http://www.uni-koeln.de/phil-fak/muwi/sm/research/k3/download/k3_description.pdf
16. Colomar, D.S., Nilsson, J.O., Händel, P.: Smoothing for ZUPT-aided INSs. In: Proceedings of the International Conference on Indoor Positioning and Indoor Navigation (2012). doi:10.1109/IPIN.2012.6418869
17. Soares, J.M., Aguiar, A.P., Pascoal, A.M., Martinoli, A.: A distributed formation-based odor source localization algorithm: design, implementation, and wind tunnel evaluation. In: IEEE International Conference on Robotics and Automation, Seattle, WA, USA, pp. 1830–1836 (2015)
18. Takayama, L.: Toward a science of robotics: goals and standards for experimental research. In: Robotics: Science and Systems, Workshop on Good Experimental Methodology in Robotics (2009)

Author Index